AWS D1.1/D1.1M:2020
An American National Standard

Approved by the
American National Standards Institute
December 9, 2019

Structural Welding Code—Steel

24th Edition

Supersedes AWS D1.1/D1.1M:2015

Prepared by the
American Welding Society (AWS) D1 Committee on Structural Welding

Under the Direction of the
AWS Technical Activities Committee

Approved by the
AWS Board of Directors

Abstract

This code covers the welding requirements for any type of welded structure made from the commonly used carbon and low-alloy constructional steels. Clauses 1 through 11 constitute a body of rules for the regulation of welding in steel construction. There are eight normative and eleven informative annexes in this code. A Commentary of the code is included with the document.

ISBN Print: 978-1-64322-087-1
ISBN PDF: 978-1-64322-088-8
© 2020 by American Welding Society
All rights reserved
Printed in the United States of America

Photocopy Rights. No portion of this standard may be reproduced, stored in a retrieval system, or transmitted in any form, including mechanical, photocopying, recording, or otherwise, without the prior written permission of the copyright owner.

Authorization to photocopy items for internal, personal, or educational classroom use only or the internal, personal, or educational classroom use only of specific clients is granted by the American Welding Society provided that the appropriate fee is paid to the Copyright Clearance Center, 222 Rosewood Drive, Danvers, MA 01923, tel: (978) 750-8400; Internet: <www.copyright.com>.

Statement on the Use of American Welding Society Standards

All standards (codes, specifications, recommended practices, methods, classifications, and guides) of the American Welding Society (AWS) are voluntary consensus standards that have been developed in accordance with the rules of the American National Standards Institute (ANSI). When AWS American National Standards are either incorporated in, or made part of, documents that are included in federal or state laws and regulations, or the regulations of other governmental bodies, their provisions carry the full legal authority of the statute. In such cases, any changes in those AWS standards must be approved by the governmental body having statutory jurisdiction before they can become a part of those laws and regulations. In all cases, these standards carry the full legal authority of the contract or other document that invokes the AWS standards. Where this contractual relationship exists, changes in or deviations from requirements of an AWS standard must be by agreement between the contracting parties.

AWS American National Standards are developed through a consensus standards development process that brings together volunteers representing varied viewpoints and interests to achieve consensus. While AWS administers the process and establishes rules to promote fairness in the development of consensus, it does not independently test, evaluate, or verify the accuracy of any information or the soundness of any judgments contained in its standards.

AWS disclaims liability for any injury to persons or to property, or other damages of any nature whatsoever, whether special, indirect, consequential, or compensatory, directly or indirectly resulting from the publication, use of, or reliance on this standard. AWS also makes no guarantee or warranty as to the accuracy or completeness of any information published herein.

In issuing and making this standard available, AWS is neither undertaking to render professional or other services for or on behalf of any person or entity, nor is AWS undertaking to perform any duty owed by any person or entity to someone else. Anyone using these documents should rely on his or her own independent judgment or, as appropriate, seek the advice of a competent professional in determining the exercise of reasonable care in any given circumstances. It is assumed that the use of this standard and its provisions is entrusted to appropriately qualified and competent personnel.

This standard may be superseded by new editions. This standard may also be corrected through publication of amendments or errata, or supplemented by publication of addenda. Information on the latest editions of AWS standards including amendments, errata, and addenda is posted on the AWS web page (www.aws.org). Users should ensure that they have the latest edition, amendments, errata, and addenda.

Publication of this standard does not authorize infringement of any patent or trade name. Users of this standard accept any and all liabilities for infringement of any patent or trade name items. AWS disclaims liability for the infringement of any patent or product trade name resulting from the use of this standard.

AWS does not monitor, police, or enforce compliance with this standard, nor does it have the power to do so.

Official interpretations of any of the technical requirements of this standard may only be obtained by sending a request, in writing, to the appropriate technical committee. Such requests should be addressed to the American Welding Society, Attention: Director, Standards Development, 8669 NW 36 St, # 130, Miami, FL 33166 (see Annex T). With regard to technical inquiries made concerning AWS standards, oral opinions on AWS standards may be rendered. These opinions are offered solely as a convenience to users of this standard, and they do not constitute professional advice. Such opinions represent only the personal opinions of the particular individuals giving them. These individuals do not speak on behalf of AWS, nor do these oral opinions constitute official or unofficial opinions or interpretations of AWS. In addition, oral opinions are informal and should not be used as a substitute for an official interpretation.

This standard is subject to revision at any time by the AWS D1 Committee on Structural Welding. It must be reviewed every five years, and if not revised, it must be either reaffirmed or withdrawn. Comments (recommendations, additions, or deletions) and any pertinent data that may be of use in improving this standard are requested and should be addressed to AWS Headquarters. Such comments will receive careful consideration by the AWS D1 Committee on Structural Welding and the author of the comments will be informed of the Committee's response to the comments. Guests are invited to attend all meetings of the AWS D1 Committee on Structural Welding to express their comments verbally. Procedures for appeal of an adverse decision concerning all such comments are provided in the Rules of Operation of the Technical Activities Committee. A copy of these Rules can be obtained from the American Welding Society, 8669 NW 36 St, # 130, Miami, FL 33166.

This page is intentionally blank.

Personnel

AWS D1 Committee on Structural Welding

A. W. Sindel, Chair	TRC Solutions
T. L. Niemann, Vice Chair	Fickett Structural Solutions, LLC
R. D. Medlock, 2nd Vice Chair	High Steel Structures, LLC
J. A. Molin, Secretary	American Welding Society
U. W. Aschemeier	Subsea Global Solutions
E. L. Bickford	IISI
T. M. Burns	Thom Burns Consulting, LLC
H. H. Campbell, III	Pazuzu Engineering
R. D. Campbell	Bechtel
B. Connelly	Eustis Engineering, LLC
R. B. Corbit	APTIM (Retired)
M. E. Gase	Midwest Steel, Incorporated
M. A. Grieco	MA Department of Transportation
J. J. Kenney	Shell International E&P
J. H. Kiefer	ConocoPhillips (Retired)
J. R. Kissell	Trinity Consultants
B. Krueger	Los Alamos National Laboratory
V. Kuruvilla	Lexicon, Incorporated
J. Lawmon	American Engineering and Manufacturing, Incorporated
N. S. Lindell	Project & Quality Solutions
D. R. Luciani	Canadian Welding Bureau
P. W. Marshall	Moonshine Hill Proprietary Systems Engineering
M. J. Mayes	Terracon Consultants
D. L. McQuaid	D. L. McQuaid & Associates, Incorporated
J. Merrill	TRC Solutions
D. K. Miller	The Lincoln Electric Company
J. B. Pearson, Jr.	ALRV Consultant, LLC
D. D. Rager	Rager Consulting, Incorporated
T. J. Schlafly	American Institute of Steel Construction
R. E. Shaw, Jr.	Steel Structures Technology Center, Incorporated
M. M. Tayarani	Pennoni Associates, Incorporated
P. Torchio, III	Williams Enterprises of GA, Incorporated (Retired)

Advisors to the D1 Committee on Structural Welding

N. J. Altebrando	STV Incorporated
F. G. Armao	The Lincoln Electric Company
G. L. Fox	Consultant
H. E. Gilmer	HRV Conformance Verification Associates, Incorporated
G. J. Hill	G.J. Hill & Associates
M. L. Hoitomt	Consultant
C. W. Holmes	Modjeski & Masters, Inc. (Retired)
G. S. Martin	Retired

Advisors to the D1 Committee on Structural Welding (Continued)

D. C. Phillips	*Retired*
P. G. Kinney	*Sandia National Laboratories*
J. W. Post	*J. W. Post & Associates, Incorporated*
R. W. Stieve	*Parsons Corporation*
K. K. Verma	*Consultant*

Active Past AWS D1 Main Committee Chairs

D. L. McQuaid	*D.L. McQuaid & Assoc. Incorporated*
D. K. Miller	*The Lincoln Electric Company*
D. D. Rager	*Rager Consulting, Incorporated*

AWS D1Q Subcommittee on Steel

T. J. Schlafly, Chair	*American Institute of Steel Construction*
P. Torchio, III, Vice Chair	*Williams Enterprises of GA, Incorporated (Retired)*
J. A. Molin, Secretary	*American Welding Society*
U. W. Aschemeier	*Subsea Global Solutions*
M. Bernasek	*C-Spec*
E. L. Bickford	*IISI*
H. H. Campbell, III	*Pazuzu Engineering*
R. V. Clarke	*Retired*
M. E. Gase	*Midwest Steel, Incorporated*
H. E. Gilmer	*HRV Conformance Verification Associates, Incorporated*
R. L. Holdren	*Arc Specialties*
W. S. Houston	*Stanley Black & Decker–Nelson Stud Welding*
J. J. Kenney	*Shell International E&P*
J. H. Kiefer	*Conoco Philips (Retired)*
P. G. Kinney	*Sandia National Laboratories*
L. Kloiber	*LeJeune Steel Company*
V. Kuruvilla	*Lexicon, Incorporated*
D. R. Luciani	*Canadian Welding Bureau*
P. W. Marshall	*Moonshine Hill Proprietary Systems Engineering*
R. P. Marslender	*Kiewit Offshore Services, LTD.*
G. S. Martin	*Retired*
M. J. Mayes	*Terracon Consultants*
J. Merrill	*TRC Solutions*
J. I. Miller	*Chevron*
S. P. Moran	*American Hydro Corporation*
T. C. Myers	*ExxonMobil*
J. C. Nordby	*Entergy*
D. D. Rager	*Rager Consulting, Incorporated*
R. E. Shaw, Jr.	*Steel Structures Technology Center, Incorporated*
A. W. Sindel	*TRC Solutions*
J. L. Warren	*McDermott*

Advisors to the D1Q Committee on Steel

N. J. Altebrando	*STV Incorporated*
B. M. Butler	*Walt Disney World Company*
J. W. Cagle	*Schuff Steel*
B. Capers	*Walt Disney World Company*
H. A. Chambers	*SNH Market Consultants*
M. A. Grieco	*MA Department of Transportation*

Advisors to the D1Q Committee on Steel (Continued)

J. Guili	*Tru-Weld Equipment Company*
C. W. Hayes	*The Lincoln Electric Company*
C. W. Holmes	*Modjeski & Masters, Inc. (Retired)*
M. J. Jordan	*Johnson Plate and Tower Fabrication*
J. E. Koski	*Stud Welding Products Incorporated*
N. S. Lindell	*Project & Quality Systems*
D. L. McQuaid	*D. L. McQuaid & Associates, Incorporated*
R. D. Medlock	*High Steel Structures*
D. K. Miller	*The Lincoln Electric Company*
J. A. Packer	*University of Toronto*
J. B. Pearson, Jr.	*ALRV Consultant, LLC*
D. C. Phillips	*Retired*
J. W. Post	*J. W. Post & Associates, Incorporated*
R. W. Stieve	*Parsons Corporation*
M. M. Tayarani	*MA Department of Transportation*
S. J. Thomas	*Consultant*
K. K. Verma	*Consultant*
P. Workman	*Tru-Weld*
D. A. Wright	*Wright Welding Technologies*

D1Q Subcommittee Task Group on Design

T. J. Schlafly, Co-Chair	*American Institute of Steel Construction*
D. K. Miller, Co-Chair	*The Lincoln Electric Company*
T. Green, Vice Chair	*Wiss, Janney, Elstner Associates*
D. B. Ferrell	*Ferrell Engineering, Incorporated*
M. J. Jordan	*Johnson Plate and Tower Fabrication*
J. J. Kenney	*Shell International E & P*
L. A. Kloiber	*LeJeune Steel Company*
P. W. Marshall	*Moonshine Hill Proprietary Systems Engineering*
J. M. Ocel	*Federal Highway Administration*
J. A. Packer	*University of Toronto*
J. B. Pearson, Jr.	*ALRV Consultant, LLC*
R. E. Shaw, Jr.	*Steel Structures Technology Center, Incorporated*
R. H. R. Tide	*Wiss, Janney, Elstner Associates*

Advisors to the D1Q Subcommittee Task Group on Design

B. Capers	*Walt Disney World Company*
J. Desjardins	*Bombardier Transportation*

D1Q Subcommittee Task Group on Prequalification

C. Zanfir, Chair	*Canadian Welding Bureau*
L. M. Bower, Vice Chair	*NCI Building Systems*
W. J. Bell	*Atlantic Testing Laboratories*
H. H. Campbell, III	*Pazuzu Engineering*
M. D. Florczykowski	*Precision Custom Components*
D. R. Luciani	*Canadian Welding Bureau*
P. W. Marshall	*MHP Systems Engineering*
D. K. Miller	*The Lincoln Electric Company*
J. I. Miller	*Chevron*

D1Q Subcommittee Task Group on Prequalification (Continued)

S. P. Moran	*American Hydro Corporation*
J. C. Norby	*Entergy*
R. E. Shaw, Jr.	*Steel Structures Technology Center, Incorporated*
A. W. Sindel	*TRC Solutions*
P. Torchio, III	*Williams Enterprises of Georgia, Incorporated (Retired)*

D1Q Subcommittee Task Group on Qualification

T. C. Myers, Chair	*ExxonMobil*
S. J. Findlan, Vice Chair	*CB&I Power*
M. Bernasek	*C-Spec*
E. L. Bickford	*IISI*
T. R. Blissett	*Accurate Weldment Testing, LLC*
M. J. Harker	*Idaho National Laboratory*
R. L. Holdren	*Arc Specialties*
J. J. Kenney	*Shell International E & P*
J. H. Kiefer	*ConocoPhillips Company (Retired)*
R. P. Marslender	*Kiewit Offshore Services, Ltd.*
J. R. McGranaghan	*Arcosa Meyers Utility Structures*
D. W. Meyer	*ESAB Welding & Cutting Products*
J. D. Niemann	*Kawasaki Motors Manufacturing Corporation USA*
D. D. Rager	*Rager Consulting, Incorporated*
A. W. Sindel	*TRC Solutions*
D. A. Stickel	*Caterpillar, Incorporated*
B. M. Toth	*CB&I*
K. K. Welch	*Miller Electric Manufacturing Company*

Advisors to the D1Q Subcommittee Task Group on Qualification

G. S. Martin	*Retired*
D. C. Phillips	*Retired*
K. K. Verma	*Consultant*

D1Q Subcommittee Task Group on Fabrication

J. I. Miller, Chair	*Chevron*
M. E. Gase, Vice Chair	*Midwest Steel, Incorporated*
S. E. Anderson	*Anderson Inspections*
L. N. Bower	*NCI Building Systems*
H. H. Campbell, III	*Pazuzu Engineering*
R. V. Clarke	*Retired*
H. E. Gilmer	*HRV Conformance Verification Associates, Incorporated*
M. A. Grieco	*Massachusetts Department of Transportation*
C. Carbonneau	*Tishman AECOM*
J. H. Kiefer	*ConocoPhillips Company (Retired)*
P. G. Kinney	*Sandia National Laboratories*
V. Kuruvilla	*Lexicon, Incorporated*
G. S. Martin	*Retired*
E. S. Mattfield	*New York City Department of Buildings*
R. D. Medlock	*High Steel Structures, LLC*
J. E. Mellinger	*Pennoni Associates, Incorporated*
R. L. Mertz	*Alta Vista Solutions*
R. E. Monson	*Pennoni*

Advisors to the D1Q Subcommittee Task Group on Fabrication

W. G. Alexander	*WGAPE*
B. Anderson	*Molex Incorporated*
J. W. Cagle	*C. P. Buckner Steel Erection, Incorporated*
G. L. Fox	*Consultant*
G. J. Hill	*G. J. Hill & Associates*
R. L. Holdren	*ARC Specialties*
C. W. Holmes	*Modjeski & Masters, Incorporated (Retired)*
D. L. McQuaid	*D. L. McQuaid & Associates, Incorporated*
J. E. Myers	*Consultant*
J. W. Post	*J. W. Post and Associates, Incorporated*
T. J. Schlafly	*American Institute of Steel Construction*
J. F. Sokolewicz	*Trinity Industries, Incorporated*
K. K. Verma	*Consultant*

D1Q Subcommittee Task Group on Inspection

P. G. Kinney, Chair	*Sandia National Laboratories*
J. J. Kinsey, Vice Chair	*Caltrop Corporation*
M. E. Gase, 2 Vice Chair	*Midwest Steel, Incorporated*
S. E. Anderson	*Anderson Inspections*
U. W. Aschemeier	*Subsea Global Solutions*
R. V. Clarke	*Retired*
J. A. Cochran	*Kiewit Corporation*
J. M. Davis	*NDE-Olympus NDT-University Ultrasonics*
P. A. Furr	*Consultant*
H. E. Gilmer	*HRV Conformance Verification Associates, Incorporated*
C. W. Hayes	*The Lincoln Electric Company*
P. T. Hayes	*GE Inspection Technologies, LP*
J. K. Hilton	*KTA-Tator, Incorporated*
N. S. Lindell	*Product of Quality Solutions*
G. S. Martin	*Retired*
E. S. Mattfield	*New York City Department of Buildings*
J. E. Mellinger	*Pennoni Associates, Incorporated*
J. Merrill	*TRC Solutions*
R. L. Mertz	*Alta Vista Solutions*
R. E. Monson	*Pennoni Associates, Incorporated*
J. B. Pearson, Jr.	*ALRV Consultant, LLC*
C. E. Pennington	*Nova*
R. E. Stachel	*HRV Conformance Verification Associates, Incorporated*
K. J. Steinhagen	*Consultant*

Advisors to the D1Q Subcommittee Task Group on Inspection

D. A. Dunn	*PSI (Retired)*
J. J. Edwards	*DOT Quality Services*
G. J. Hill	*G. J. Hill & Associates*
R. K. Holbert	*Alstom Power*
J. H. Kiefer	*ConocoPhillips Company (Retired)*
C. A. Mankenberg	*Shell International Exploration & Production*
D. L. McQuaid	*D.L. McQuaid & Associates, Incorporated*
D. G. Yantz	*Canadian Welding Bureau*

D1Q Subcommittee Task Group on Stud Welding

W. S. Houston, Chair	*Stanley Black & Decker–Nelson Stud Welding*
U. W. Aschemeier, Vice Chair	*Subsea Global Solutions*
R. D. Campbell	*Bechtel*
A. D. D'Amico	*Consultant*
B. C. Hobson	*Image Industries*
I. W. Houston	*Stanley Black & Decker–Nelson Stud Welding*
J. E. Koski	*Stud Welding Products, Incorporated*
D. R. Luciani	*Canadian Welding Bureau*
C. W. Makar	*Cox Industries*
S. P. Moran	*American Hydro Corporation*
P. Torchio, III	*Williams Enterprises of Georgia, Incorporated (Retired)*
J. S. Wirtz	*Stone & Webster, Incorporated*
P. Workman	*Tru-Weld Equipment Company*

Advisors to the D1Q Subcommittee Task Group on Stud Welding

C. B. Champney	*Stanley Black & Decker–Nelson Stud Welding*
J. Guili	*Tru-Weld Equipment Company*
R. Schraff	*Stanley Black & Decker–Nelson Stud Welding*
M. M. Tayarani	*Pennoni Associates, Incorporated*

D1Q Standing Task Group on Tubulars

J. J. Kenney, Chair	*Shell International E & P*
M. A. Grieco, Vice Chair	*Massachusetts Department of Transportation*
E. L. Bickford	*IISI*
N. M. Choy	*California Department of Transportation*
R. V. Clarke	*Retired*
D. B. Ferrell	*Ferrell Engineering, Incorporated*
R. B. Fletcher	*Atlas Tube*
P. A. Huckabee	*Gill Engineering Associates, Incorporated*
L. A. Kloiber	*LeJeune Steel Consultant*
C. Long	*Consultant*
P. W. Marshall	*Moonshine Hill Proprietary Systems Engineering*
R. P. Marslender	*Kiewit Offshore Services, Ltd.*
J. P. McCormick	*University of Michigan*
R. L. Mertz	*Alta Vista Solutions*
R. E. Monson	*Pennoni Associates, Incorporated*
K. T. Olson	*FORSE Consulting-Steel Tube Institute*
J. A. Packer	*University of Toronto*
R. Sougata	*Sougata Roy, LLC*
R. Sause	*ATLSS Center Lehigh University*

Advisors to the D1Q Standing Task Group on Tubulars

J. J. Edwards	*DOT Quality Services*
V. Kuruvilla	*Lexicon, Incorporated*
M. J. Mayes	*Terracon Consultants*
R. D. Medlock	*High Steel Structures, LLC*
T. L. Niemann	*Fickett Structural Solutions, LLC*
D. D. Rager	*Rager Consulting, Incorporated*
T. J. Schlafly	*American Institute of Steel Construction*
A. W. Sindel	*TRC Solutions*

D1M Standing Task Group on New Materials

M. D. Kerr, Chair	*McDermott*
T. J. Schlafly, Vice Chair	*American Institute of Steel Construction*
R. S. Caroti	*Arcelor Mittal*
C. Haven	*Hobart Brothers Company*
D. A. Koch	*Washington State Universtiy*
V. Kuruvilla	*Lexicon, Incorporated*
R. D. Medlock	*High Steel Structures, LLC*
D. W. Meyer	*ESAB Welding & Cutting Products*
T. M. Nelson	*LTK Engineering Services*
P. R. Niewiarowski	*Sargent & Lundy, LLC*
D. D. Rager	*Rager Consulting, Incorporated*
J. L. Schoen	*Nucor-Yamato Steel*
J. L. Warren	*McDermott*

Advisors to the D1M Standing Task Group on New Materials

B. Capers	*Walt Disney World Company*
S. C. Finnigan	*Arcelor Mittal*
C. W. Hayes	*The Lincoln Electric Company*
M. L. Hoitomt	*Consultant*
J. B. Pearson, Jr.	*ALRV Consultant, LLC*
D. C. Phillips	*Retired*
J. W. Post	*J. W. Post & Associates, Incorporated*
D. Rees-Evans	*Steel Dynamics*
A. W. Sindel	*TRC Solutions*

This page is intentionally blank.

Foreword

This foreword is not part of AWS D1.1/D1.1M:2020, *Structural Welding Code—Steel*,
but is included for informational purposes only.

The first edition of the *Code for Fusion Welding and Gas Cutting in Building Construction* was published by the American Welding Society in 1928 and called Code 1 Part A. It was revised in 1930 and 1937 under the same title. It was revised again in 1941 and given the designation D1.0. D1.0 was revised again in 1946, 1963, 1966, and 1969. The 1963 edition published an amended version in 1965, and the 1966 edition published an amended version in 1967. The code was combined with D2.0, *Specifications for Welding Highway and Railway Bridges*, in 1972, given the designation D1.1, and retitled AWS *Structural Welding Code*. D1.1 was revised again in 1975, 1979, 1980, 1981, 1982, 1983, 1984, 1985, 1986, 1988, 1990, 1992, 1994, 1996, 1998, 2000, 2002, 2004, 2006, 2008 and 2010. A second printing of D1.1:2010 was published in 2011. From 1972 to 1988, the D1.1 code covered the welding of both buildings and bridges.

In 1988, AWS published its first edition of AASHTO/AWS D1.5, *Bridge Welding Code*; coincident with this, the D1.1 code changed references of buildings and bridges to statically loaded and dynamically loaded structures, respectively, in order to make the document applicable to a broader range of structural applications. After the publishing of the 2010 edition, it was decided that the AWS *Structural Welding Code—Steel* would be published on a five year revision cycle instead of a two year revision cycle. This was done in order to sync the publication cycle of AWS Structural Welding Code-Steel with the publication cycles of the AISC Steel Building Specification and the International Building Code. This 2020 edition is the 24th edition of D1.1.

Changes in Code Requirements, underlined text in the clauses, subclauses, tables, figures, or forms indicates a change from the 2015 edition. A vertical line in the margin of a table or figure also indicates a change from the 2015 edition.

The following is a summary of the most significant technical changes contained in D1.1/D1.1M:2020:

Summary of Changes

Clause/Table/Figure/Annex	Modification
Clause 1	
Clause 2	This is a new clause listing normative references. It replaces subclause 1.9 and Annex S from the previous edition.
Clause 3	This is a new clause that provides terms and definitions specific to this standard. It replaces subclause 1.3 and Annex J from the previous edition.
Clause 4	Clause 4 was presented as Clause 2 in the previous edition. Annex A Figures in the previous edition were incorporated into Clause 4.
Clause 5	Clause 5 was presented as Clause 3 in the previous edition. The Clause has also been restructured to follow the normal progression of writing a prequalified WPS. Table 5.2 has been editorially renamed and reorganized to list WPS essential variables. Additional requirements have been added when using shielding gases and a new Table 5.7 was added on shielding gases. New materials have been added to Tables 5.3 and 5.8.
Clause 6	Clause 6 was presented as Clause 4 in the previous edition. Revisions include the requirements for the qualification of WPSs using waveform technology. All the CVN testing requirements have been added to Table 6.7, so they now are all contained in a single place. The WPS retest requirements have been clarified. The PJP Groove weld clause has been reorganized to clarify the qualification of PJP Groove welds using the Joint Details in Figure 5.2. Part D of the Clause has been reorganized to better align the testing procedures and qualification of CVNs with the order that they would be accomplished.

(Continued)

Summary of Changes (Continued)

Clause/Table/Figure/Annex	Modification
Clause 7	Clause 7 was presented as Clause 5 in the previous edition. Revisions were made to the weld restoration of base metal with mislocated holes.
Clause 8	Clause 8 was presented as Clause 6 in the previous edition. Revisions were made to the qualification requirements for inspection personnel to ensure that all welding inspectors are qualified. The Engineer's responsibilities as it relates to Inspection were also clarified. Digital radiography has been added to Radiographic Testing. The limitations for geometric unsharpness have been added to the Code and the equation has been revised to match the equation in ASME *Boiler and Pressure Vessel Code*, Section V, Article 2. The methodology to determine the attenuation factor has been updated to reflect that UT instruments are now capable of reporting a fractional value for dB.
Clause 9	Clause 9 was presented as Clause 7 in the previous edition. The code was updated to require the manufacturer's permanent identification on headed studs and deformed anchor bars. Revisions were made to provide weld procedure requirements for fillet welding of studs.
Clause 10	Clause 10 was presented as Clause 9 in the previous edition. The calculations for static strength of welded tubular connections were removed in deference to AISC design provisions.
Clause 11	Clause 11 was presented as Clause 8 in the previous edition.
Annex A	Annex A was presented as Annex B in the previous edition.
Annex B	Annex B was presented as Annex H in the previous edition.
Annex D	Annex D was presented as Annex F in the previous edition.
Annex E	Annex E was presented as Annex D in the previous edition.
Annex F	Annex F was presented as Annex E in the previous edition.
Annex H	New Annex that addresses phased array ultrasonic testing (PAUT)
Annex J	Annex J was presented as Annex M in the previous edition.
Annex K	Annex K was presented as Annex P in the previous edition.
Annex L	Annex L was presented as Annex T in the previous edition.
Annex M	Annex M was presented as Annex U in the previous edition.
Annex N	Annex N was presented as Annex K in the previous edition.
Annex O	Annex O was presented as Annex Q in the previous edition.
Annex P	Annex P was presented as Annex L in the previous edition.
Annex Q	Annex Q was presented as Annex O in the previous edition.
Annex R	Annex R has been modified to contain preliminary design of circular tube connections previously contained in the Tubular Structures clause as well as ovalizing parameter alpha.
Annex T	Annex T was presented as Annex N in the previous edition.
C-Annex H	Commentary was added for Annex H

Commentary. The Commentary is nonmandatory and is intended only to provide insightful information into provision rationale.

Normative Annexes. These annexes address specific subjects in the code and their requirements are mandatory requirements that supplement the code provisions.

Informative Annexes. These annexes are not code requirements but are provided to clarify code provisions by showing examples, providing information, or suggesting alternative good practices.

Index. As in previous codes, the entries in the Index are referred to by subclause number rather than by page number. This should enable the user of the Index to locate a particular item of interest in minimum time.

(Continued)

Errata. It is the Structural Welding Committee's Policy that all errata should be made available to users of the code. Therefore, any significant errata will be published in the Society News Section of the *Welding Journal* and posted on the AWS web site at: http://www.aws.org/standards/page/errata.

Suggestions. Your comments for improving AWS D1.1/D1.1M:2015, *Structural Welding Code—Steel* are welcome. Submit comments to the Secretary of the D1Q Subcommittee, American Welding Society, 8669 NW 36 St, # 130, Miami, FL 33166.

This page is intentionally blank.

Table of Contents

Page No.

Personnel v
Foreword xiii
List of Tables xxii
List of Figures xxiv

1. **General Requirements** 1
   1.1 Scope 1
   1.2 Standard Units of Measurement 1
   1.3 Safety Precautions 2
   1.4 Limitations 2
   1.5 Responsibilities 2
   1.6 Approval 3
   1.7 Mandatory and Nonmandatory Provisions 3
   1.8 Welding Symbols 3

2. **Normative References** 4

3. **Terms and Definitions** 7

4. **Design of Welded Connections** 17
   4.1 Scope 17

   Part A—Common Requirements for Design of Welded Connections (Nontubular and Tubular Members) 17
   4.2 General 17
   4.3 Contract Plans and Specifications 17
   4.4 Effective Areas 18

   Part B—Specific Requirements for Design of Nontubular Connections (Statically or Cyclically Loaded) 21
   4.5 General 21
   4.6 Stresses 21
   4.7 Joint Configuration and Details 23
   4.8 Joint Configuration and Details—Groove Welds 23
   4.9 Joint Configuration and Details—Fillet Welded Joints 24
   4.10 Joint Configuration and Details—Plug and Slot Welds 25
   4.11 Filler Plates 25
   4.12 Built-Up Members 25

   Part C—Specific Requirements for Design of Nontubular Connections (Cyclically Loaded) 26
   4.13 General 26
   4.14 Limitations 26
   4.15 Calculation of Stresses 26
   4.16 Allowable Stresses and Stress Ranges 26
   4.17 Detailing, Fabrication, and Erection 28
   4.18 Prohibited Joints and Welds 29
   4.19 Inspection 30

5. **Prequalification of WPSs** 62
   5.1 Scope 62

   Part A—WPS Development 62
   5.2 General WPS Requirements 62

Part B—Base Metal 63
   5.3   Base Metal 63

Part C—Weld Joints 63
   5.4   Weld Joints 63

Part D—Welding Processes 65
   5.5   Welding Processes 65

Part E—Filler Metals and Shielding Gases 65
   5.6   Filler Metal and Shielding Gas 65

Part F—Preheat and Interpass Temperature Requirements 66
   5.7   Preheat and Interpass Temperature Requirements 66

Part G—WPS Requirements 67
   5.8   WPS requirements 67

Part H—Postweld Heat Treatment 67
   5.9   Postweld Heat Treatment 67

6. Qualification 124
   6.1   Scope 124

Part A—General Requirements 124
   6.2   General 124
   6.3   Common Requirements for WPS and Welding Personnel Performance Qualification 125

Part B—Welding Procedure Specification (WPS) Qualification 125
   6.4   Production Welding Positions Qualified 125
   6.5   Type of Qualification Tests 125
   6.6   Weld Types for WPS Qualification 126
   6.7   Preparation of WPS 126
   6.8   Essential Variables 126
   6.9   WPS Requirements for Production Welding Using Existing Non-Waveform or Waveform WPSs 127
   6.10 Methods of Testing and Acceptance Criteria for WPS Qualification 128
   6.11 CJP Groove Welds 130
   6.12 PJP Groove Welds 130
   6.13 Fillet Welds 131
   6.14 Plug and Slot Welds 131
   6.15 Welding Processes Requiring Qualification 132

Part C—Performance Qualification 132
   6.16 General 132
   6.17 Type of Qualification Tests Required 133
   6.18 Weld Types for Welder and Welding Operator Performance Qualification 133
   6.19 Preparation of Performance Qualification Forms 133
   6.20 Essential Variables 133
   6.21 CJP Groove Welds for Nontubular Connections 134
   6.22 Extent of Qualification 134
   6.23 Methods of Testing and Acceptance Criteria for Welder and Welding Operator Qualification 134
   6.24 Method of Testing and Acceptance Criteria for Tack Welder Qualification 135
   6.25 Retest 135

Part D—Requirements for CVN Toughness Testing 136
   6.26 General: CVN Testing 136
   6.27 CVN Tests 137
   6.28 Combining FCAW-S with Other Welding Processes in a Single Joint 138
   6.29 Reporting 138

7.	**Fabrication**	188
	7.1 Scope	188
	7.2 Base Metal	188
	7.3 Welding Consumables and Electrode Requirements	188
	7.4 ESW and EGW Processes	190
	7.5 WPS Variables	191
	7.6 Preheat and Interpass Temperatures	191
	7.7 Heat Input Control for Quenched and Tempered Steels	191
	7.8 Stress-Relief Heat Treatment	191
	7.9 Backing	192
	7.10 Welding and Cutting Equipment	193
	7.11 Welding Environment	193
	7.12 Conformance with Design	193
	7.13 Minimum Fillet Weld Sizes	193
	7.14 Preparation of Base Metal	193
	7.15 Reentrant Corners	195
	7.16 Weld Access Holes, Beam Copes, and Connection Material	196
	7.17 Tack Welds and Construction Aid Welds	196
	7.18 Camber in Built-Up Members	197
	7.19 Splices	197
	7.20 Control of Distortion and Shrinkage	197
	7.21 Tolerance of Joint Dimensions	198
	7.22 Dimensional Tolerance of Welded Structural Members	199
	7.23 Weld Profiles	201
	7.24 Technique for Plug and Slot Welds	202
	7.25 Repairs	202
	7.26 Peening	203
	7.27 Caulking	204
	7.28 Arc Strikes	204
	7.29 Weld Cleaning	204
	7.30 Weld Tabs	204
8.	**Inspection**	216
	Part A—General Requirements	216
	8.1 Scope	216
	8.2 Inspection of Materials and Equipment	218
	8.3 Inspection of WPSs	218
	8.4 Inspection of Welder, Welding Operator, and Tack Welder Qualifications	218
	8.5 Inspection of Work and Records	218
	Part B—Contractor Responsibilities	219
	8.6 Obligations of the Contractor	219
	Part C—Acceptance Criteria	219
	8.7 Scope	219
	8.8 Engineer's Approval for Alternate Acceptance Criteria	219
	8.9 Visual Inspection	219
	8.10 Penetrant Testing (PT) and Magnetic Particle Testing (MT)	219
	8.11 Nondestructive Testing (NDT)	220
	8.12 Radiographic Testing (RT)	220
	8.13 Ultrasonic Testing (UT)	221
	Part D—NDT Procedures	222
	8.14 Procedures	222
	8.15 Extent of Testing	223

    Part E—Radiographic Testing (RT) 223
    8.16 RT of Groove Welds in Butt Joints..223
    8.17 RT Procedures 224
    8.18 Examination, Report, and Disposition of Radiographs 226

    Part F—Ultrasonic Testing (UT) of Groove Welds...227
    8.19 General 227
    8.20 Qualification Requirements 227
    8.21 UT Equipment 228
    8.22 Reference Standards..228
    8.23 Equipment Qualification...229
    8.24 Calibration for Testing 229
    8.25 Testing Procedures 230
    8.26 Preparation and Disposition of Reports 231
    8.27 Calibration of the UT Unit with IIW Type or Other Approved Reference Blocks (Annex G) 232
    8.28 Equipment Qualification Procedures 233
    8.29 Discontinuity Size Evaluation Procedures 235
    8.30 Scanning Patterns..235
    8.31 Examples of dB Accuracy Certification..236

    Part G—Other Examination Methods...236
    8.32 General Requirements...236
    8.33 Radiation Imaging Systems 236
    8.34 Advanced Ultrasonic Systems 237
    8.35 Additional Requirements 237

9. Stud Welding 270
    9.1 Scope 270
    9.2 General Requirements..270
    9.3 Mechanical Requirements..271
    9.4 Workmanship/Fabrication 271
    9.5 Technique 272
    9.6 Stud Application Qualification Requirements 273
    9.7 Production Control...274
    9.8 Fabrication and Verification Inspection Requirements 274
    9.9 Manufacturers' Stud Base Qualification Requirements..275

10. Tubular Structures 283
    10.1 Scope 283

    Part A—Design of Tubular Connections..283
    10.2 Design Criteria 283
    10.3 Identification and Parts of Tubular Connections 285
    10.4 Symbols 285
    10.5 Weld Design...285
    10.6 Thickness Transition...286
    10.7 Material Limitations...287

    Part B—Prequalification of Welding Procedure Specifications (WPSs)...287
    10.8 Fillet Weld Requirements...287
    10.9 PJP Requirements...287
    10.10 CJP Groove Weld Requirements 288

    Part C—Welding Procedure Specification (WPS) Qualification 288
    10.11 Common Requirements for WPS and Welding Personnel Performance Qualification 288
    10.12 Production Welding Positions Qualified 288
    10.13 Type of Qualification Tests, Methods of Testing, and Acceptance Criteria
            for WPS Qualification 289

10.14	CJP Groove Welds for Tubular Connections	289
10.15	PJP and Fillet Welds Tubular T-, Y-, or K-Connections and Butt Joints	290

Part D—Performance Qualification ... 290

10.16	Production Welding Positions, Thicknesses, and Diameters Qualified	290
10.17	Weld Types for Welder and Welding Operator Performance Qualification	291
10.18	CJP Groove Welds for Tubular Connections	291
10.19	PJP Groove Welds for Tubular Connections	291
10.20	Fillet Welds for Tubular Connections	291
10.21	Methods of Testing and Acceptance Criteria for Welder and Welding Operator Qualification	292

Part E—Fabrication ... 292

10.22	Backing	292
10.23	Tolerance of Joint Dimensions	293

Part F—Inspection ... 293

10.24	Visual Inspection	293
10.25	NDT	293
10.26	UT	293
10.27	RT Procedures	294
10.28	Supplementary RT Requirements for Tubular Connections	294
10.29	UT of Tubular T-, Y-, and K-Connections	294

11. Strengthening and Repair of Existing Structures ... 347

11.1	Scope	347
11.2	General	347
11.3	Base Metal	347
11.4	Design for Strengthening and Repair	347
11.5	Fatigue Life Enhancement	348
11.6	Workmanship and Technique	348
11.7	Quality	348

Annexes ... 349
Annex A (Normative)—Effective Throats of Fillet Welds in Skewed T-Joints ... 351
Annex B (Normative)—Guideline on Alternative Methods for Determining Preheat ... 353
Annex D (Normative)—Temperature-Moisture Content Charts ... 365
Annex E (Normative)—Flatness of Girder Webs—Statically Loaded Structures ... 369
Annex F (Normative)—Flatness of Girder Webs—Cyclically Loaded Structures ... 373
Annex G (Normative)—Qualification and Calibration of UT Units with Other Approved Reference Blocks ... 379
Annex H (Normative)—Phased Array Ultrasonic Testing (PAUT) ... 383
Annex I (Normative)—Symbols for Tubular Connection Weld Design ... 399
Annex J (Informative)—Sample Welding Forms ... 403
Annex K (Informative)—Contents of Prequalified WPS ... 425
Annex L (Informative)—Filler Metal Strength Properties ... 427
Annex M (Informative—AWS A5.36 Filler Metal Classifications and Properties ... 439
Annex N (Informative)—Guide for Specification Writers ... 455
Annex O (Informative)—UT Examination of Welds by Alternative Techniques ... 457
Annex P (Informative)—UT Equipment Qualification and Inspection Forms ... 475
Annex Q (Informative)—Local Dihedral Angle ... 485
Annex R (Informative)—Ovalizing Parameter Alpha ... 491
Annex S (Informative)—List of Reference Documents ... 497
Annex T (Informative)—Guidelines for the Preparation of Technical Inquiries for the Structural Welding Committee ... 499

Commentary ... 501
List of AWS Documents on Structural Welding ... 619
Index ... 621

List of Tables

Table		Page No.
4.1	Effective Size of Flare-Groove Welds Filled Flush.	31
4.2	Z Loss Dimension (Nontubular)	31
4.3	Allowable Stresses	32
4.4	Equivalent Strength Coefficients for Obliquely Loaded Fillet Welds	33
4.5	Fatigue Stress Design Parameters	34
5.1	Prequalified WPS Requirements	68
5.2	Essential Variables for Prequalified WPSs	69
5.3	Approved Base Metals for Prequalified WPSs	70
5.4	Filler Metals for Matching Strength for Table 5.3, Groups I, II, III, and IV Materials	73
5.5	Minimum Prequalified PJP Weld Size (S)	77
5.6	Filler Metal Requirements for Exposed Bare Applications of Weathering Steels	77
5.7	Prequalified WPS Shielding Gas Options for GMAW electrodes conforming to AWS A5.18/A5.18M	77
5.8	Prequalified Minimum Preheat and Interpass Temperature	78
6.1	WPS Qualification—Production Welding Positions Qualified by Plate, Pipe, and Box Tube Tests	139
6.2	WPS Qualification—CJP Groove Welds: Number and Type of Test Specimens and Range of Thickness Qualified	140
6.3	WPS Qualification—PJP Groove Welds: Number and Type of Test Specimens and Range of Thickness Qualified	140
6.4	WPS Qualification—Fillet Welds: Number and Type of Test Specimens and Range of Thickness Qualified	141
6.5	PQR Essential Variable Changes Requiring WPS Requalification for SMAW, SAW, GMAW, FCAW, and GTAW	142
6.6	PQR Essential Variable Changes Requiring WPS Requalification for ESW or EGW	145
6.7	PQR Supplementary Essential Variable Changes for CVN Testing Applications Requiring WPS Requalification for SMAW, SAW, GMAW, FCAW, GTAW, and ESW/ EGW	147
6.8	Table 5.1, Table 6.9, and Unlisted Steels Qualified by PQR	148
6.9	Code-Approved Base Metals and Filler Metals Requiring Qualification per Clause 6	149
6.10	Welder and Welding Operator Qualification—Production Welding Positions Qualified by Plate Tests	154
6.11	Welder and Welding Operator Qualification—Number and Type of Specimens and Range of Thickness and Diameter Qualified	155
6.12	Welding Personnel Performance Essential Variable Changes Requiring Requalification	157
6.13	Electrode Classification Groups	157
6.14	CVN Test Temperature Reduction	158
6.15	Charpy V-Notch Test Acceptance Criteria for Various Sub-Size Specimens	158
6.16	Filler Metal Essential Variables—FCAW-S Substrate/Root	159
7.1	Allowable Atmospheric Exposure of Low-Hydrogen Electrodes	205
7.2	Minimum Holding Time	205
7.3	Alternate Stress-Relief Heat Treatment	205
7.4	Limits on Acceptability and Repair of Mill Induced Laminar Discontinuities in Cut Surfaces	206
7.5	Camber Tolerance for Typical Girder	206
7.6	Camber Tolerance for Girders without a Designed Concrete Haunch	207
7.7	Minimum Fillet Weld Sizes	207
7.8	Weld Profiles	208
7.9	Weld Profile Schedules	208

8.1	Visual Inspection Acceptance Criteria	239
8.2	UT Acceptance-Rejection Criteria (Statically Loaded Nontubular Connections and Cyclically Loaded Nontubular Connections in Compression)	240
8.3	UT Acceptance-Rejection Criteria (Cyclically Loaded Nontubular Connections in Tension)	241
8.4	Hole-Type IQI Requirements	242
8.5	Wire IQI Requirements	242
8.6	IQI Selection and Placement	242
8.7	Testing Angle	243
8.8	UT Equipment Qualification and Calibration Requirements	245
9.1	Mechanical Property Requirements for Studs	277
9.2	Minimum Fillet Weld Size for Small Diameter Studs	277
10.1	Fatigue Stress Design Parameters	297
10.2	Available Stresses in Tubular Connection Welds	298
10.3	Stress Categories for Type and Location of Material for Circular Sections	300
10.4	Fatigue Category Limitations on Weld Size or Thickness and Weld Profile (Tubular Connections)	302
10.5	Z Loss Dimensions for Calculating Prequalified PJP T-, Y-, and K-Tubular Connection Minimum Weld Sizes	302
10.6	Joint Detail Applications for Prequalified CJP T-, Y-, and K-Tubular Connections	303
10.7	Prequalified Joint Dimensions and Groove Angles for CJP Groove Welds in Tubular T-, Y, and K-Connections Made by SMAW, GMAW-S, and FCAW	304
10.8	WPS Qualification—Production Welding Positions Qualified by Plate, Pipe, and Box Tube Tests	305
10.9	WPS Qualification—CJP Groove Welds: Number and Type of Test Specimens and Range of Thickness and Diameter Qualified	306
10.10	WPS Qualification—PJP Groove Welds: Number and Type of Test Specimens and Range of Thickness Qualified	308
10.11	WPS Qualification—Fillet Welds: Number and Type of Test Specimens and Range of Thickness Qualified	308
10.12	Welder and Welding Operator Qualification—Production Welding Positions Qualified by Pipe and Box Tube Tests	309
10.13	Welder and Welding Operator Qualification—Number and Type of Specimens and Range of Thickness and Diameter Qualified	310
10.14	Tubular Root Opening Tolerances, Butt Joints Welded Without Backing	312
10.15	Visual Inspection Acceptance Criteria	313
10.16	Hole-Type IQI Requirements	314
10.17	Wire IQI Requirements	314
10.18	IQI Selection and Placement	314
A.1	Equivalent Fillet Weld Leg Size Factors for Skewed T-Joints	352
B.1	Susceptibility Index Grouping as Function of Hydrogen Level "H" and Composition Parameter P_{cm}	357
B.2	Minimum Preheat and Interpass Temperatures for Three Levels of Restraint	357
E.1	Intermediate Stiffeners on Both Sides of Web	370
E.2	No Intermediate Stiffeners	370
E.3	Intermediate Stiffeners on One Side Only of Web	371
F.1	Intermediate Stiffness on Both Sides of Web, Interior Girders	374
F.2	Intermediate Stiffness on One Side Only of Web, Fascia Girders	375
F.3	Intermediate Stiffness on One Side Only of Web, Interior Girders	376
F.4	Intermediate Stiffness on Both Sides of Web, Fascia Girders	377
F.5	No Intermediate Stiffeners, Interior or Fascia Girders	378
H.1	Essential Variables for PAUT	393
H.2	PAUT Acceptance Criteria	393
H.3	Discontinuity Classification	393
M.1	AWS A5.36/A5.36M Carbon Steel Electrode Classifications with Fixed Requirements	441
M.2	AWS A5.36/A5.36M Tension Test Requirements	442
M.3	AWS A5.36/A5.36M Charpy Impact Test Requirements	442
M.4	Electrode Usability Characteristics	443

M.5	AWS A5.36/A5.36M Composition Requirements for Shielding Gases	445
M.6	Weld Metal Chemical Composition Requirements	446
M.7	AWS A5.20/A5.20M Procedure Requirements for "D" Optional Supplemental Designator	448
M.8	AWS A5.36/A5.36M Procedure Requirements for "D" Optional Supplemental Designator	448
M.9	Comparison of Classifications of AWS A5.18, A5.20, A5.28, and A5.29 Specifications to AWS A5.36 Fixed and Open Classifications for Multiple-Pass FCAW and GMAW—Metal Cored Electrodes	449
O.1	Acceptance-Rejection Criteria	464
R.1	Terms for Strength of Connections (Circular Sections)	494
	Commentary	505
C-5.1	Typical Current Ranges for GMAW-S on Steel	534
C-10.1	Structural Steel Plates	596
C-10.2	Structural Steel Pipe and Tubular Shapes	597
C-10.3	Structural Steel Shapes	597
C-10.4	Classification Matrix for Applications	598
C-10.5	CVN Testing Conditions	598
C-10.6	CVN Test Values	599
C-10.7	HAZ CVN Test Values	599
C-11.1	Guide to Welding Suitability	608
C-11.2	Relationship Between Plate Thickness and Burr Radius	608

List of Figures

Figure		Page No.
4.1	Fillet Weld	51
4.2	Unreinforced Bevel Groove Weld	52
4.3	Bevel Groove Weld with Reinforcing Fillet Weld	52
4.4	Bevel Groove Weld with Reinforcing Fillet Weld	53
4.5	Unreinforced Flare Bevel Groove Weld	53
4.6	Flare Bevel Groove Weld with Reinforcing Fillet Weld	54
4.7	Maximum Fillet Weld Size Along Edges in Lap Joints	54
4.8	Transition of Thicknesses (Statically Loaded Nontubular)	55
4.9	Transversely Loaded Fillet Welds	55
4.10	Minimum Length of Longitudinal Fillet Welds at End of Plate or Flat Bar Members	56
4.11	Termination of Welds Near Edges Subject to Tension	56
4.12	End Return at Flexible Connections	57
4.13	Fillet Welds on Opposite Sides of a Common Plane	57
4.14	Thin Filler Plates in Splice Joint	58
4.15	Thick Filler Plates in Splice Joint	58
4.16	Allowable Stress Range for Cyclically Applied Load (Fatigue) in Nontubular Connections (Graphical Plot of Table 4.5)	59
4.17	Transition of Butt Joints in Parts of Unequal Thickness (Cyclically Loaded Nontubular)	60
4.18	Transition of Width (Cyclically Loaded Nontubular)	61
5.1	Prequalified CJP Groove Welded Joint Details (Dimensions in Inches)	82
5.1	Prequalified CJP Groove Welded Joint Details (Dimensions in Millimeters)	93
5.2	Prequalified PJP Groove Welded Joint Details (Dimensions in Inches)	104
5.2	Prequalified PJP Groove Welded Joint Details (Dimensions in Millimeters)	112
5.3	Prequalified Fillet Weld Joint Details (Dimensions in Inches)	120
5.3	Prequalified Fillet Weld Joint Details (Dimensions in Millimeters)	121
5.4	Prequalified Skewed T-Joint Details (Nontubular)	122

5.5	Prequalified CJP Groove, T-, and Corner Joint	123
5.6	Weld Bead in which Depth and Width Exceed the Width of the Weld Face	123
6.1	Positions of Groove Welds	160
6.2	Positions of Fillet Welds	161
6.3	Positions of Test Plates for Groove Welds	162
6.4	Positions of Test Plate for Fillet Welds	163
6.5	Location of Test Specimens on Welded Test Plates—ESW and EGW—WPS Qualification	164
6.6	Location of Test Specimens on Welded Test Plate Over 3/8 in [10 mm] Thick—WPS Qualification	165
6.7	Location of Test Specimens on Welded Test Plate 3/8 in [10 mm] Thick and Under—WPS Qualification	166
6.8	Face and Root Bend Specimens	167
6.9	Side Bend Specimens	168
6.10	Reduced-Section Tension Specimens	169
6.11	Guided Bend Test Jig	170
6.12	Alternative Wraparound Guided Bend Test Jig	171
6.13	Alternative Roller-Equipped Guided Bend Test Jig for Bottom Ejection of Test Specimen	171
6.14	All-Weld-Metal Tension Specimen	172
6.15	Fillet Weld Soundness Tests for WPS Qualification	173
6.16	Test Plate for Unlimited Thickness—Welder Qualification and Fillet Weld Consumable Verification Tests	174
6.17	Test Plate for Unlimited Thickness—Welding Operator Qualification and Fillet Weld Consumable Verification Tests	174
6.18	Location of Test Specimen on Welded Test Plate 1 in [25 mm] Thick—Consumables Verification for Fillet Weld WPS Qualification	175
6.19	Optional Test Plate for Unlimited Thickness—Horizontal Position—Welder Qualification	176
6.20	Test Plate for Limited Thickness—All Positions—Welder Qualification	177
6.21	Optional Test Plate for Limited Thickness—Horizontal Position—Welder Qualification	178
6.22	Fillet Weld Root Bend Test Plate—Welder or Welding Operator Qualification—Option 2	179
6.23	Method of Rupturing Specimen—Tack Welder Qualification	180
6.24	Butt Joint for Welding Operator Qualification—ESW and EGW	180
6.25	Fillet Weld Break and Macroetch Test Plate—Welder or Welding Operator Qualification Option 1	181
6.26	Plug Weld Macroetch Test Plate—Welder or Welding Operator Qualification and WPS Qualification	182
6.27	Fillet Weld Break Specimen—Tack Welder Qualification	183
6.28	CVN Test Specimen Locations	184
6.29	Macroetch Test Assemblies for Determination of PJP Weld Size	185
6.30	Intermix Test Plate	186
6.31	Interface Scribe Line Location	187
6.32	Intermix CVN Test Specimen Location	187
7.1	Edge Discontinuities in Cut Material	209
7.2	Weld Access Hole Geometry	210
7.3	Workmanship Tolerances in Assembly of Groove Welded Joints	211
7.4	Requirements for Weld Profiles	212
8.1	Discontinuity Acceptance Criteria for Statically Loaded Nontubular and Statically or Cyclically Loaded Tubular Connections	247
8.2	Discontinuity Acceptance Criteria for Cyclically Loaded Nontubular Connections in Tension (Limitations of Porosity and Fusion Discontinuities)	251
8.3	Discontinuity Acceptance Criteria for Cyclically Loaded Nontubular Connections in Compression (Limitations of Porosity or Fusion-Type Discontinuities)	255
8.4	Hole-Type IQI	259
8.5	Wire IQI	260
8.6	RT Identification and Hole-Type or Wire IQI Locations on Approximately Equal Thickness Joints 10 in [250 mm] and Greater in Length	261
8.7	RT Identification and Hole-Type or Wire IQI Locations on Approximately Equal Thickness Joints Less than 10 in [250 mm] in Length	261

8.8	RT Identification and Hole-Type or Wire IQI Locations on Transition Joints 10 in [250 mm] and Greater in Length.	262
8.9	RT Identification and Hole-Type or Wire IQI Locations on Transition Joints Less than 10 in [250 mm] in Length	263
8.10	RT Edge Blocks	263
8.11	Transducer Crystal	264
8.12	Qualification Procedure of Search Unit Using IIW Reference Block	264
8.13	Typical IIW Type Block	265
8.14	Qualification Blocks	266
8.15	Plan View of UT Scanning Patterns	268
8.16	Transducer Positions (Typical)	269
9.1	Dimension and Tolerances of Standard-Type Headed Studs	278
9.2	Typical Tension Test Fixture	279
9.3	Torque Testing Arrangement and Table of Testing Torques	280
9.4	Bend Testing Device	281
9.5	Suggested Type of Device for Qualification Testing of Small Studs	282
10.1	Allowable Fatigue Stress and Strain Ranges for Stress Categories, Tubular Structures for Atmospheric Service	315
10.2	Parts of a Tubular Connection	316
10.3	Fillet Welded Lap Joint (Tubular)	319
10.4	Transition of Thickness of Butt Joints in Parts of Unequal Thickness (Tubular)	320
10.5	Fillet Welded Prequalified Tubular Joints Made by SMAW, GMAW, and FCAW	321
10.6	Prequalified Joint Details for PJP T-, Y-, and K-Tubular Connections	322
10.7	Prequalified Joint Details for CJP T-, Y-, and K-Tubular Connections	325
10.8	Definitions and Detailed Selections for Prequalified CJP T-, Y-, and K-Tubular Connections	326
10.9	Prequalified Joint Details for CJP Groove Welds in Tubular T-, Y-, and K-Connections—Standard Flat Profiles for Limited Thickness	327
10.10	Prequalified Joint Details for CJP Groove Welds in Tubular T-, Y-, and K-Connections—Profile with Toe Fillet for Intermediate Thickness	328
10.11	Prequalified Joint Details for CJP Groove Welds in Tubular T-, Y-, and K-Connections—Concave Improved Profile for Heavy Sections or Fatigue	329
10.12	Positions of Test Pipe or Tubing for Groove Welds	330
10.13	Positions of Test Pipes or Tubing for Fillet Welds	331
10.14	Location of Test Specimens on Welded Test Pipe—WPS Qualification	332
10.15	Location of Test Specimens for Welded Box Tubing—WPS Qualification	333
10.16	Pipe Fillet Weld Soundness Test—WPS Qualification	334
10.17	Tubular Butt Joint—Welder Qualification with and without Backing	335
10.18	Tubular Butt Joint—WPS Qualification with and without Backing	335
10.19	Acute Angle Heel Test (Restraints not Shown)	336
10.20	Test Joint for T-, Y-, and K-Connections without Backing on Pipe or Box Tubing (≥ 6 in [150 mm] O.D.)—Welder and WPS Qualification	337
10.21	Test Joint for T-, Y-, and K-Connections without Backing on Pipe or Box Tubing (< 4 in [100 mm] O.D.)—Welder and WPS Qualification	338
10.22	Corner Macroetch Test Joint for T-, Y-, and K-Connections without Backing on Box Tubing for CJP Groove Welds—Welder and WPS Qualification	339
10.23	Location of Test Specimens on Welded Test Pipe and Box Tubing—Welder Qualification	340
10.24	Class R Indications	341
10.25	Class X Indications	343
10.26	Single-Wall Exposure—Single-Wall View	344
10.27	Double-Wall Exposure—Single-Wall View	344
10.28	Double-Wall Exposure—Double-Wall (Elliptical) View, Minimum Two Exposures	345
10.29	Double-Wall Exposure—Double-Wall View, Minimum Three Exposures	345
10.30	Scanning Techniques	346
B.1	Zone Classification of Steels	359

B.2	Critical Cooling Rate for 350 HV and 400 HV	359
B.3	Graphs to Determine Cooling Rates for Single-Pass SAW Fillet Welds	360
B.4	Relation Between Fillet Weld Size and Energy Input	363
D.1	Temperature-Moisture Content Chart to be Used in Conjunction with Testing Program to Determine Extended Atmospheric Exposure Time of Low-Hydrogen SMAW Electrodes	366
D.2	Application of Temperature-Moisture Content Chart in Determining Atmospheric Exposure Time of Low-Hydrogen SMAW Electrodes	367
G.1	Other Approved Blocks and Typical Transducer Position	381
H.1	Phased Array Imaging Views	394
H.2	Example of a Supplemental Reference Block	394
H.3	Example Standard Reflector Locations in Weld Mockup	395
H.4	Sensitivity Levels	395
H.5	Example of Time Based Linearity Verification	396
H.6	Linearity Verification Report Form	397
M.1	AWS A5.36/A5.36M Open Classification System	454
O.1	Standard Reference Reflector	465
O.2	Recommended Calibration Block	465
O.3	Typical Standard Reflector (Located in Weld Mock-Ups and Production Welds)	466
O.4	Transfer Correction	467
O.5	Compression Wave Depth (Horizontal Sweep Calibration)	467
O.6	Compression Wave Sensitivity Calibration	468
O.7	Shear Wave Distance and Sensitivity Calibration	468
O.8	Scanning Methods	469
O.9	Spherical Discontinuity Characteristics	470
O.10	Cylindrical Discontinuity Characteristics	470
O.11	Planar Discontinuity Characteristics	471
O.12	Discontinuity Height Dimension	471
O.13	Discontinuity Length Dimension	472
O.14	Display Screen Marking	472
O.15	Report of UT (Alternative Procedure)	473
R.1	Simplified Concept of Punching Shear	495
R.2	Reliability of Punching Shear Criteria Using Computed Alpha	495
R.3	Definition of Terms for Computed Alpha	496
	Commentary	501
C-4.1	Balancing of Fillet Welds About a Neutral Axis	521
C-4.2	Shear Planes for Fillet and Groove Welds	521
C-4.3	Eccentric Loading	522
C-4.4	Load Deformation Relationship for Welds	522
C-4.5	Example of an Obliquely Loaded Weld Group	523
C-4.6	Graphical Solution of the Capacity of an Obliquely Loaded Weld Group	524
C-4.7	Single Fillet Welded Lap Joints	525
C-5.1	Examples of Centerline Cracking	534
C-5.2	Details of Alternative Groove Preparations for Prequalified Corner Joints	535
C-5.3	Oscillograms and Sketches of GMAW-S Metal Transfer	535
C-6.1	Type of Welding on Pipe That Does Not Require Pipe Qualification	540
C-7.1	Examples of Unacceptable Reentrant Corners	554
C-7.2	Examples of Good Practice for Cutting Copes	554
C-7.3	Permissible Offset in Abutting Members	554
C-7.4	Correction of Misaligned Members	555
C-7.5	Typical Method to Determine Variations in Girder Web Flatness	555
C-7.6	Illustration Showing Camber Measurement Methods	556
C-7.7	Measurement of Flange Warpage and Tilt	557
C-7.8	Tolerances at Bearing Points	558
C-8.1	90° T- or Corner Joints with Steel Backing	572

C-8.2	Skewed T- or Corner Joints	572
C-8.3	Butt Joints with Separation Between Backing and Joint	573
C-8.4	Effect of Root Opening on Butt Joints with Steel Backing	573
C-8.5	Resolutions for Scanning with Seal Welded Steel Backing	574
C-8.6	Scanning with Seal Welded Steel Backing	574
C-8.7	Illustration of Discontinuity Acceptance Criteria for Statically Loaded Nontubular and Statically or Cyclically Loaded Tubular Connections	575
C-8.8	Illustration of Discontinuity Acceptance Criteria for Statically Loaded Nontubular and Statically or Cyclically Loaded Tubular Connections 1-1/8 in [30 mm] and Greater, Typical of Random Acceptable Discontinuities	576
C-8.9	Illustration of Discontinuity Acceptance Criteria for Cyclically Loaded Nontubular Connections in Tension	577
C-9.1	Allowable Defects in the Heads of Headed Studs	581
C-10.1	Illustrations of Branch Member Stresses Corresponding to Mode of Loading	599
C-10.2	Improved Weld Profile Requirements	600
C-10.3	Upper Bound Theorem	600
C-10.4	Yield Line Patterns	601
C-11.1	Microscopic Intrusions	609
C-11.2	Fatigue Life	609
C-11.3	Toe Dressing with Burr Grinder	610
C-11.4	Toe Dressing Normal to Stress	610
C-11.5	Effective Toe Grinding	611
C-11.6	End Grinding	611
C-11.7	Hammer Peening	612
C-11.8	Toe Remelting	612

Structural Welding Code—Steel

1. General Requirements

1.1 Scope

This code contains the requirements for fabricating and erecting welded steel structures. When this code is stipulated in contract documents, conformance with all provisions of the code shall be required, except for those provisions that the Engineer (see 1.5.1) or contract documents specifically modifies or exempts.

The following is a summary of the code clauses:

1. General Requirements. This clause contains basic information on the scope and limitations of the code, key definitions, and the major responsibilities of the parties involved with steel fabrication.

2. Normative References. This clause contains a list of reference documents that assist the user in implementation of this code or are required for implementation.

3. Terms and Definitions. This clause contains terms and definitions as they relate to this code.

4. Design of Welded Connections. This clause contains requirements for the design of welded connections composed of tubular, or nontubular, product form members.

5. Prequalification of WPSs. This clause contains the requirements for exempting a Welding Procedure Specification (WPS) from the WPS qualification requirements of this code.

6. Qualification. This clause contains the requirements for WPS qualification and the performance qualification tests required to be passed by all welding personnel (welders, welding operators, and tack welders) to perform welding in accordance with this code.

7. Fabrication. This clause contains general fabrication and erection requirements applicable to welded steel structures governed by this code, including the requirements for base metals, welding consumables, welding technique, welded details, material preparation and assembly, workmanship, weld repair, and other requirements.

8. Inspection. This clause contains criteria for the qualifications and responsibilities of inspectors, acceptance criteria for production welds, and standard procedures for performing visual inspection and nondestructive testing (NDT).

9. Stud Welding. This clause contains the requirements for the welding of studs to structural steel.

10. Tubular Structures. This clause contains exclusive tubular requirements. Additionally, the requirements of all other clauses apply to tubulars, unless specifically noted otherwise.

11. Strengthening and Repair of Existing Structures. This clause contains basic information pertinent to the welded modification or repair of existing steel structures.

1.2 Standard Units of Measurement

This standard makes use of both U.S. Customary Units and the International System of Units (SI). The latter are shown within brackets ([ ]) or in appropriate columns in tables and figures. The measurements may not be exact equivalents; therefore, each system must be used independently.

1.3 Safety Precautions

Safety and health issues and concerns are beyond the scope of this standard and therefore are not fully addressed herein. It is the responsibility of the user to establish appropriate safety and health practices. Safety and health information is available from the following sources:

American Welding Society:

(1) ANSI Z49.1, *Safety in Welding, Cutting, and Allied Processes*

(2) AWS Safety and Health Fact Sheets

(3) Other safety and health information on the AWS website

Material or Equipment Manufacturers:

(1) Safety Data Sheets supplied by materials manufacturers

(2) Operating Manuals supplied by equipment manufacturers

Applicable Regulatory Agencies

Work performed in accordance with this standard may involve the use of materials that have been deemed hazardous, and may involve operations or equipment that may cause injury or death. This standard does not purport to address all safety and health risks that may be encountered. The user of this standard should establish an appropriate safety program to address such risks as well as to meet applicable regulatory requirements. ANSI Z49.1 should be considered when developing the safety program.

1.4 Limitations

The code was specifically developed for welded steel structures that utilize carbon or low alloy steels that are 1/8 in [3 mm] or thicker with a minimum specified yield strength of 100 ksi [690 MPa] or less. The code may be suitable to govern structural fabrications outside the scope of the intended purpose. However, the Engineer should evaluate such suitability, and based upon such evaluations, incorporate into contract documents any necessary changes to code requirements to address the specific requirements of the application that is outside the scope of the code. The Structural Welding Committee encourages the Engineer to consider the applicability of other AWS D1 codes for applications involving aluminum (AWS D1.2), sheet steel equal to or less than 3/16 in [5 mm] thick (AWS D1.3), reinforcing steel (AWS D1.4), stainless steel (AWS D1.6), strengthening and repair of existing structures (AWS D1.7), seismic supplement (AWS D1.8), and titanium (AWS D1.9). The AASHTO/AWS D1.5 *Bridge Welding Code* was specifically developed for welding highway bridge components and is recommended for those applications.

1.5 Responsibilities

1.5.1 Engineer's Responsibilities. The Engineer shall be responsible for the development of the contract documents that govern products or structural assemblies produced under this code. The Engineer may add to, delete from, or otherwise modify, the requirements of this code to meet the particular requirements of a specific structure. All requirements that modify this code shall be incorporated into contract documents. The Engineer shall determine the suitability of all joint details to be used in a welded assembly.

The Engineer shall specify in contract documents, as necessary, and as applicable, the following:

(1) Code requirements that are applicable only when specified by the Engineer.

(2) All additional NDT that is not specifically addressed in the code.

(3) Extent of verification inspection, when required.

(4) Weld acceptance criteria other than that specified in Clause 8.

(5) CVN toughness criteria for weld metal, base metal, and/or HAZ when required.

(6) For nontubular applications, whether the structure is statically or cyclically loaded.

(7) Which welded joints are loaded in tension.

(8) All additional requirements that are not specifically addressed in the code.

(9) For OEM applications, the responsibilities of the parties involved.

1.5.2 Contractor's Responsibilities. The Contractor shall be responsible for WPSs, qualification of welding personnel, the Contractor's inspection, and performing work in conformance with the requirements of this code and contract documents.

1.5.3 Inspector's Responsibilities

1.5.3.1 Contractor Inspection. Contractor inspection shall be supplied by the Contractor and shall be performed as necessary to ensure that materials and workmanship meet the requirements of the contract documents.

1.5.3.2 Verification Inspection. The Engineer shall determine if Verification Inspection shall be performed. Responsibilities for Verification Inspection shall be established between the Engineer and the Verification Inspector.

1.6 Approval

All references to the need for approval shall be interpreted to mean approval by the Authority Having Jurisdiction or the Engineer.

1.7 Mandatory and Nonmandatory Provisions

1.7.1 Code Terms "Shall," "Should," and "May." "Shall," "should," and "may" have the following significance:

1.7.1.1 Shall. Code provisions that use "shall" are mandatory unless specifically modified in contract documents by the Engineer.

1.7.1.2 Should. The word "should" is used to recommend practices that are considered beneficial, but are not requirements.

1.7.1.3 May. The word "may" in a provision allows the use of optional procedures or practices that can be used as an alternative or supplement to code requirements. Those optional procedures that require the Engineer's approval shall either be specified in the contract documents, or require the Engineer's approval. The Contractor may use any option without the Engineer's approval when the code does not specify that the Engineer's approval shall be required.

1.8 Welding Symbols

Welding symbols shall be those shown in AWS A2.4, *Standard Symbols for Welding, Brazing, and Nondestructive Examination*. Special conditions shall be fully explained by added notes or details.

2. Normative References

The documents listed below are referenced within this publication and are mandatory to the extent specified herein. For undated references, the latest edition of the referenced standard shall apply. For dated references, subsequent amendments to, or revisions of, any of these publications do not apply.

American Welding Society (AWS) Standards:

AWS A2.4, *Standard Symbols for Welding, Brazing, and Nondestructive Examination*

AWS A3.0M/A3.0, *Standard Welding Terms and Definitions, Including Terms for Adhesive Bonding, Brazing, Soldering, Thermal Cutting, and Thermal Spraying*

AWS A4.3, *Standard Methods for Determination of the Diffusible Hydrogen Content of Martensitic, Bainitic, and Ferritic Steel Weld Metal Produced by Arc Welding*

AWS A5.01M/A5.01:2013 (ISO 14344:2010 MOD), *Procurement Guidelines for Consumables—Welding and Allied Processes—Flux and Gas Shielded Electrical Welding Processes*

AWS A5.1/A5.1M:2012, *Specification for Carbon Steel Electrodes for Shielded Metal Arc Welding*

AWS A5.5/A5.5M:2014, *Specification for Low-Alloy Steel Electrodes for Shielded Metal Arc Welding*

AWS A5.12M/A5.12:2009 (ISO 6848:2004 MOD), *Specification for Tungsten and Oxide Dispersed Tungsten Electrodes for Arc Welding and Cutting*

AWS A5.17/A5.17M-97 (R2007), *Specification for Carbon Steel Electrodes and Fluxes for Submerged Arc Welding*

AWS A5.18/A5.18M:2005, *Specification for Carbon Steel Electrodes and Rods for Gas Shielded Arc Welding*

AWS A5.20/A5.20M:2005, *Specification for Carbon Steel Electrodes for Flux Cored Arc Welding*

AWS A5.23/A5.23M:2011, *Specification for Low-Alloy Steel Electrodes and Fluxes for Submerged Arc Welding*

AWS A5.25/A5.25M-97 (R2009), *Specification for Carbon and Low-Alloy Steel Electrodes and Fluxes for Electroslag Welding*

AWS A5.26/A5.26M-97 (R2009), *Specification for Carbon and Low-Alloy Steel Electrodes for Electrogas Welding*

AWS A5.28/A5.28M:2005, *Specification for Low-Alloy Steel Filler Metals for Gas Shielded Arc Welding*

AWS A5.29/A5.29M:2010, *Specification for Low-Alloy Steel Electrodes for Flux Cored Arc Welding*

AWS A5.30/A5.30M:2007, *Specification for Consumable Inserts*

AWS A5.32M/A5.32:2011 (ISO 14175:2008 MOD), *Welding Consumables—Gases and Gas Mixtures for Fusion Welding and Allied Processes*

AWS A5.36/A5.36M:2012, *Specification for Carbon and Low-Alloy Steel Flux Cored Electrodes for Flux Cored Arc Welding and Metal Cored Electrodes for Gas Metal Arc Welding*

AWS B5.1, *Specification for the Qualification of Welding Inspectors*

AWS B4.0, *Standard Methods for Mechanical Testing of Welds*

AWS C4.1-77 (R2010), *Criteria for Describing Oxygen-Cut Surfaces and Oxygen Cutting Surface Roughness Gauge*

AWS D1.0, *Code for Welding in Building Construction*

AWS D1.8/D1.8M, *Structural Welding Code—Seismic Supplement*

AWS D2.0, *Specification for Welded Highway and Railway Bridges*

AWS QC1, *Standard for AWS Certification of Welding Inspectors*

ANSI Z49.1, *Safety in Welding, Cutting, and Allied Processes*

American Institute of Steel Construction (AISC) Standards:

ANSI/AISC 360, *Specification for Structural Buildings*

American Petroleum Institute (API) Standards:

API 2W, *Specification for Steel Plates for Offshore Structures, Produced by Thermo-Mechanical Control Processing*

API 2Y, *Specification for Steel Plates, Quenched and- Tempered, for Offshore Structures*

American Society of Mechanical Engineers (ASME) Standards:

ASME *Boiler and Pressure Vessel Code*, Section V, Article 2

ASME B46.1, *Surface Texture (Surface Roughness, Waviness, and Lay)*

American Society for Nondestructive Testing (ASNT) Standards:

ASNT CP–189, *ASNT Standard for Qualification and Certification of Nondestructive Personnel*

ASNT *Recommended Practice No.* SNT-TC-1A, *Personnel Qualification and Certification in Nondestructive Testing*

American Society for Testing and Materials (ASTM) Standards:

All ASTM base metals listed in Table 5.3 and Table 6.9 are found in ASTM 01.04, *Steel—Structural, Reinforcing, Pressure Vessel Railway*, ASTM 01.03, *Steel-Plate, Sheet, Strip, Wire; Stainless Steel Bar*, or ASTM 01.01, *Steel-Piping, Tubing, Fittings*

ASTM A6, *Standard Specification for General Requirements for Rolled Structural Steel Bars, Plates, Shapes, and Sheet Piling*

ASTM A109, *Standard Specification for Steel, Strip, Carbon, (0.25 Maximum Percent), Cold–Rolled*

ASTM A370, *Mechanical Testing of Steel Products*

ASTM A435, *Specification for Straight Beam Ultrasonic Examination of Steel Plates*

ASTM A673, *Specification for Sampling Procedure for Impact Testing of Structural Steel*

ASTM E23, *Standard Methods for Notched Bar Impact Testing of Metallic Materials, for Type A Charpy (Simple Beam) Impact Specimen*

ASTM E92, *Test Method for Vickers Hardness of Metallic Materials*

ASTM E94, *Standard Guide for Radiographic Examination Using Industrial Radiographic Film*

ASTM E140, *Hardness Conversion Tables for Metals*

ASTM E165, *Test Method for Liquid Penetrant Examination*

ASTM E709, *Guide for Magnetic Particle Inspection*

ASTM E747, *Controlling Quality of Radiographic Testing Using Wire Parameters*

ASTM E1032, *Radiographic Examination of Weldments Using Industrial X-Ray Film*

ASTM E1254, *Standard Guide for Storage of Radiographs and Unexposed Industrial Radiographic Films*

ASTM E2033, *Standard Practice for Radiographic Examination using Computed Radiology (Photostimulable Luminescence Method)*

ASTM E2445, *Standard Practice for Performance Evaluation and Long-Term Stability of Computed Radiology Systems*

ASTM E2698, *Standard Practice for Radiological Examination Using Digital Detector Arrays*

ASTM E2699, *Standard Practice for Digital Imaging and Communication in Nondestructive Evaluation (DICONDE) for Digital Radiographic (DR) Test Methods.*

ASTM E2737, *Standard Practice for Digital Detector Array Performance Evaluation and Long-Term Stability*

Canadian Standards Association (CSA) Standards:

CSA W178.2, *Certification of Welding Inspectors*

International Institute of Welding (IIW) Ultrasonic Reference Block

The Society for Protective Coatings (SSPC) Standards:

SSPC-SP2, *Hand Tool Cleaning*

3. Terms and Definitions

AWS A3.0M/A3.0, *Standard Welding Terms and Definitions, Including Terms for Adhesive Bonding, Brazing, Soldering, Thermal Cutting, and Thermal Spraying*, provides the basis for terms and definitions used herein. However, the following terms and definitions are included below to accomodate usage specific to this document.

The terms and definitions in this glossary are divided into three categories: (1) general welding terms compiled by the AWS Committee on Definitions and Symbols; (2) terms, defined by the AWS Structural Welding Committee, which apply only to UT, designated by (UT) following the term; and (3) other terms, preceded by asterisks, which are defined as they relate to this code.

For the purposes of this document, the following terms and definitions apply:

A

*alloy flux. A flux upon which the alloy content of the weld metal is largely dependent.

*all-weld-metal test specimen. A test specimen with the reduced section composed wholly of weld metal.

*amplitude length rejection level (UT). The maximum length of discontinuity allowed by various indication ratings associated with weld size, as indicated in Tables 8.2 and 8.3.

*angle of bevel. See **bevel angle**.

arc gouging. Thermal gouging that uses an arc cutting process variation to form a bevel or groove.

as-welded. The condition of weld metal, welded joints, and weldments after welding, but prior to any subsequent thermal, mechanical, or chemical treatments.

*attenuation (UT). The loss in acoustic energy which occurs between any two points of travel. This loss may be due to absorption, reflection, etc. (In this code, using the shear wave pulse-echo method of testing, the attenuation factor is 2 dB per inch of sound path distance after the first inch.)

automatic welding. Welding with equipment that requires only occasional or no observation of the welding, and no manual adjustment of the equipment controls. Variations of this term are **automatic brazing, automatic soldering, automatic thermal cutting,** and **automatic thermal spraying**.

*auxiliary attachments. Members or appurtenances attached to main stress-carrying members by welding. Such members may or may not carry loads.

average instantaneous power (AIP), *waveform-controlled welding*. The average of products of amperages and voltages determined at sampling frequencies sufficient to quantify waveform changes during a welding interval.

axis of a weld. See **weld axis**.

B

backgouging. The removal of weld metal and base metal from the weld root side of a welded joint to facilitate complete fusion and CJP upon subsequent welding from that side.

backing. A material or device placed against the back side of the joint, or at both sides of a weld in ESW and EGW, to support and retain molten weld metal. The material may be partially fused or remain unfused during welding and may be either metal or nonmetal.

backing pass. A weld pass made for a backing weld.

backing ring. Backing in the form of a ring, generally used in the welding of pipe.

backing weld. Backing in the form of a weld.

*backup weld (tubular structures). The initial closing pass in a CJP groove weld, made from one side only, which serves as a backing for subsequent welding, but is not considered as a part of the theoretical weld (Figures 10.9 through 10.11, Details C and D).

back weld. A weld made at the back of a single groove weld.

base metal. The metal or alloy that is welded, brazed, soldered, or cut.

bevel angle. The angle between the bevel of a joint member and a plane perpendicular to the surface of the member.

box tubing. Tubular product of square or rectangular cross section. See **tubular**.

*brace intersection angle, θ (tubular structures). The acute angle formed between brace centerlines.

*Building Code. The building law or specification or other construction regulations in conjunction with which this code is applied.

NOTE: In the absence of any locally applicable building law or specifications or other construction regulations, it is recommended that the construction be required to comply with the Specification for Structural Steel Buildings (AISC).

butt joint. A joint between two members aligned approximately in the same plane.

butt weld. A nonstandard term for a weld in a butt joint. See **butt joint**.

C

*cap pass. One or more weld passes that form the weld face (exposed surface of completed weld). Adjacent cap passes may partially cover, but not completely cover, a cap pass.

*caulking. Plastic deformation of weld and base metal surfaces by mechanical means to seal or obscure discontinuities.

complete fusion. Fusion over the entire fusion faces and between all adjoining weld beads.

complete joint penetration (CJP). A joint root condition in a groove weld in which weld metal extends through the joint thickness.

*Computed radiography (CR).

*CJP groove weld (statically and cyclically loaded structures). A groove weld which has been made from both sides or from one side on a backing having CJP and fusion of weld and base metal throughout the depth of the joint.

*CJP groove weld (tubular structures). A groove weld having CJP and fusion of weld and base metal throughout the depth of the joint or as detailed in Figures 10.4 10.7 through 10.11, and 10.19. A CJP tubular groove weld made from one side only, without backing, is allowed where the size or configuration, or both, prevent access to the root side of the weld.

complete penetration. A nonstandard term for **CJP**. See **complete joint penetration**.

construction aid weld. A weld made to attach a piece or pieces to a weldment for temporary use in handling, shipping, or working on the structure.

consumable guide ESW. See **ESW, consumable guide**.

continuous weld. A weld that extends continuously from one end of a joint to the other. Where the joint is essentially circular, it extends completely around the joint.

*contract documents. Any codes, specifications, drawings, or additional requirements that are contractually specified by the Owner.

*Contractor. Any company, or that individual representing a company, responsible for the fabrication, erection manufacturing or welding, in conformance with the provisions of this code.

*Contractor's Inspector. The duly designated person who acts for, and in behalf of, the Contractor on all inspection and quality matters within the scope of the code and of the contract documents.

corner joint. A joint between two members located approximately at right angles to each other in the form of an L.

*cover pass. See **cap pass**.

CO₂ welding. A nonstandard term for **GMAW** with carbon dioxide shielding gas.

crater. A depression in the weld face at the termination of a weld bead.

* Charpy V-notch (CVN).

D

* digital detector array (DDA).

*decibel (dB) (UT). The logarithmic expression of a ratio of two amplitudes or intensities of acoustic energy.

*decibel rating (UT). See preferred term **indication rating**.

defect. A discontinuity or discontinuities that by nature or accumulated effect (for example total crack length) render a part or product unable to meet minimum applicable acceptance standards or specifications. This term designates rejectability.

defective weld. A weld containing one or more defects.

*defect level (UT). See **indication level**.

*defect rating (UT). See **indication rating**.

depth of fusion. The distance that fusion extends into the base metal or previous bead from the surface melted during welding.

*dihedral angle. See **local dihedral angle. discontinuity.** An interruption of the typical structure of a material, such as a lack of homogeneity in its mechanical or metallurgical, or physical characteristics. A discontinuity is not necessarily a defect.

downhand. A nonstandard term for **flat welding position**.

*direct radiography (DR).

*drawings. Refers to plans design and detail drawings, and erection plans.

E

*edge angle (tubular structures). The acute angle between a bevel edge made in preparation for welding and a tangent to the member surface, measured locally in a plane perpendicular to the intersection line. All bevels open to outside of brace.

*effective length of weld. The length throughout which the correctly proportioned cross section of the weld exists. In a curved weld, it shall be measured along the weld axis.

electrogas welding (EGW). An arc welding process that uses an arc between a continuous filler metal electrode and the weld pool, employing approximately vertical welding progression with backing to confine the molten weld metal. The process is used with or without an externally supplied shielding gas and without the application of pressure.

electroslag welding (ESW). A welding process that produces coalescence of metals with molten slag that melts the filler metal and the surfaces of the workpieces. The weld pool is shielded by this slag, which moves along the full cross section of the joint as welding progresses. The process is initiated by an arc that heats the slag. The arc is then extinguished by the conductive slag, which is kept molten by its resistance to electric current passing between the electrode and the workpieces.

ESW, consumable guide. An electroslag welding process variation in which filler metal is supplied by an electrode and its guiding member.

*end return. The continuation of a fillet weld around a corner of a member as an extension of the principal weld.

*Engineer. A duly designated individual who acts for and in behalf of the Owner on all matters within the scope of the code.

F

*fatigue. Fatigue, as used herein, is defined as the damage that may result in fracture after a sufficient number of stress fluctuations. Stress range is defined as the peak-to-trough magnitude of these fluctuations. In the case of stress reversal, stress range shall be computed as the numerical sum (algebraic difference) of maximum repeated tensile and compressive stresses, or the sum of shearing stresses of opposite direction at a given point, resulting from changing conditions of load.

faying surface. The mating surface of a member that is in contact with or in close proximity to another member to which it is to be joined.

filler metal. The metal or alloy to be added in making a welded, brazed, or soldered joint.

fillet weld leg. The distance from the joint root to the toe of the fillet weld.

fin. A defect in a bar or other rolled section caused by the steel spreading into the clearance between the rolls. This produces a thick overfill which, if rolled again, usually becomes a lap.

flare-bevel-groove weld. A weld in the groove formed between a joint member with a curved surface and another with a planar surface.

*flash. The material which is expelled or squeezed out of a weld joint and which forms around the weld.

flat welding position. The welding position used to weld from the upper side of the joint at a point where the weld axis is approximately horizontal, and the weld face lies in an approximately horizontal plane.

flux cored arc welding (FCAW). An arc welding process that uses an arc between a continuous filler metal electrode and the weld pool. The process is used with shielding gas from a flux contained within the tubular electrode, with or without additional shielding from an externally supplied gas, and without the application of pressure.

*flux cored arc welding—gas shielded (FCAW-G). A flux cored arc welding process variation in which additional shielding is obtained from an externally supplied gas or gas mixture.

*flux cored arc welding—self shielded (FCAW-S). A flux cored arc welding process where shielding is exclusively provided by a flux contained within the tubular electrode.

fusion. The melting together of filler metal and base metal (substrate), or of base metal only, to produce a weld.

fusion line. The boundary between weld metal and base metal in a fusion weld.

*fusion-type discontinuity. Signifies slag inclusion, incomplete fusion, incomplete joint penetration, and similar discontinuities associated with fusion.

fusion zone. The area of base metal melted as determined on the cross section of a weld.

G

gas metal arc welding (GMAW). An arc welding process that uses an arc between a continuous filler metal electrode and the weld pool. The process is used with shielding from an externally supplied gas and without the application of pressure.

gas metal arc welding-short circuit arc (GMAW-S). A gas metal arc welding process variation in which the consumable electrode is deposited during repeated short circuits.

gas pocket. A nonstandard term for porosity.

Geometric unsharpness U_g). The fuzziness or lack of definition in a radiographic image resulting from the combination of source size, object-to-film distance, and source-to-object distance.

gouging. See thermal gouging.

groove angle. The total included angle of the groove between workpieces.

*groove angle, θ (tubular structures). The angle between opposing faces of the groove to be filled with weld metals, determined after the joint is fit-up.

groove face. The surface of a joint member included in the groove.

groove weld. A weld made in the groove between the workpieces.

H

heat-affected zone (HAZ). The portion of the base metal whose mechanical properties or microstructure have been altered by the heat of welding, brazing, soldering, or thermal cutting.

horizontal fixed position (pipe welding). The position of a pipe joint in which the axis of the pipe is approximately horizontal, and the pipe is not rotated during welding (see Figures 6.1, 6.2, and 10.12).

horizontal welding position, *fillet weld.* The welding position in which the weld is on the upper side of an approximately horizontal surface and against an approximately vertical surface (see Figures 6.1, 6.2, 6.3, and 6.4).

*horizontal reference line (UT). A horizontal line near the center of the UT instrument scope to which all echoes are adjusted for dB reading.

horizontal rotated position (pipe welding). The position of a pipe joint in which the axis of the pipe is approximately horizontal, and welding is performed in the flat position by rotating the pipe (see Figures 6.1, 6.2, and 10.12).

*hot-spot strain (tubular structures). The cyclic total range of strain which would be measured at the point of highest stress concentration in a welded connection.

NOTE: When measuring hot-spot strain, the strain gage should be sufficiently small to avoid averaging high and low strains in the regions of steep gradients.

I

*image quality indicator (IQI). A device whose image in a radiograph is used to determine RT quality level. It is not intended for use in judging the size nor for establishing acceptance limits of discontinuities.

*indication (UT). The signal displayed on the oscilloscope signifying the presence of a sound wave reflector in the part being tested.

*indication level (UT). The calibrated gain or attenuation control reading obtained for a reference line height indication from a discontinuity.

*indication rating (UT). The decibel reading in relation to the zero reference level after having been corrected for sound attenuation.

intermittent weld. A weld in which the continuity is broken by recurring unwelded spaces.

interpass temperature. In a multipass weld, the temperature of the weld area between weld passes.

J

joint. The junction of members or the edges of members that are to be joined or have been joined.

joint penetration. The distance the weld metal extends from the weld face into a joint, exclusive of weld reinforcement.

joint root. That portion of a joint to be welded where the members approach closest to each other. In cross section, the joint root may be either a point, a line, or an area.

L

lap joint. A joint between two overlapping members in parallel planes.

*layer. A stratum of weld metal or surfacing material. The layer may consist of one or more weld beads laid side by side.

*leg (UT).** The path the shear wave travels in a straight line before being reflected by the surface of material being tested. See sketch for leg identification. Note: Leg I plus leg II equals one V-path.

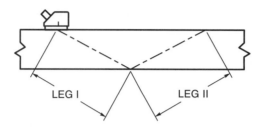

leg of a fillet weld. See **fillet weld leg.**

*****local dihedral angle, θ (tubular structures).** The angle, measured in a plane perpendicular to the line of the weld, between tangents to the outside surfaces of the tubes being joined at the weld. The exterior dihedral angle, where one looks at a localized section of the connection, such that the intersecting surfaces may be treated as planes.

M

manual welding. Welding with the torch, gun or electrode holder held and manipulated by hand. Accessory equipment, such as part motion devices and manually controlled filler material feeders may be used. See **automatic welding, mechanized welding,** and **semiautomatic welding.**

mechanized process (XXXX-ME). An operation with equipment requiring manual adjustment by an operator in response to visual observation, with the torch, gun, wire guide assembly, or electrode holder held by a mechanical device. See **mechanized welding.**

mechanized welding (W-ME). See **mechanized process.**

*****Magnetic particle testing (MT).**

N

*****nondestructive testing (NDT).** The process of determining acceptability of a material or a component in accordance with established criteria without impairing its future usefulness.

*****node (UT).** See **leg.**

*****nominal tensile strength of the weld metal.** The tensile strength of the weld metal indicated by the classification number of the filler metal (e.g., nominal tensile strength of E60XX is 60 ksi [420 MPa]).

O

*****Original Equipment Manufacturer (OEM).** A single Contractor that assumes some or all of the responsibilities assigned by this code to the Engineer.

overhead welding position. The welding position in which welding is performed from the underside of the joint (see Figures 6.1, 6.2, 6.3, and 6.4).

overlap, *fusion welding.* The protrusion of weld metal beyond the weld toe or weld root.

*****Owner.** The individual or company that exercises legal ownership of the product or structural assembly produced to this code.

oxygen cutting (OC). A group of thermal cutting processes that severs or removes metal by means of the chemical reaction between oxygen and the base metal at elevated temperature. The necessary temperature is maintained by the heat from an arc, an oxyfuel gas flame, or other source.

oxygen gouging. Thermal gouging that uses an oxygen cutting process variation to form a bevel or groove.

P

*parallel electrode. See SAW.

partial joint penetration (PJP). Joint penetration that is intentionally less than complete.

pass. See weld pass.

peening. The mechanical working of metals using impact blows.

*pipe. Hollow circular cross section produced or manufactured in accordance with a pipe product specification. See tubular.

*piping porosity (ESW and EGW). Elongated porosity whose major dimension lies in a direction approximately parallel to the weld axis.

*piping porosity (general). Elongated porosity whose major dimension lies in a direction approximately normal to the weld surface. Frequently referred to as *pin holes* when the porosity extends to the weld surface.

plug weld. A weld made in a circular hole in one member of a joint fusing that member to another member. A fillet-welded hole shall not be construed as conforming to this definition.

porosity. Cavity-type discontinuities formed by gas entrapment during solidification or in a thermal spray deposit.

positioned weld. A weld made in a joint that has been placed to facilitate making the weld.

*postweld heat treatment. Any heat treatment after welding.

preheating. The application of heat to the base metal immediately before welding, brazing, soldering, thermal spraying, or cutting.

preheat temperature, *welding*. The temperature of the base metal in the volume surrounding the point of welding immediately before welding is started. In a multiple-pass weld, it is also the temperature immediately before the second and subsequent passes are started.

*Liquid penetrant testing (PT).

*Postweld heat treatment (PWHT).

Q

qualification. See welder performance qualification and **WPS qualification.**

R

random sequence. A longitudinal sequence in which the weld bead increments are made at random.

*reference level (UT). The decibel reading obtained for a horizontal reference-line height indication from a reference reflector.

*reference reflector (UT). The reflector of known geometry contained in the IIW reference block or other approved blocks.

reinforcement of weld. See **weld reinforcement.**

*resolution (UT). The ability of UT equipment to give separate indications from closely spaced reflectors.

root face. That portion of the groove face within the joint root.

root gap. A nonstandard term for **root opening. root of joint.** See **joint root**.

root of weld. See **weld root.**

root opening. A separation at the joint root between the workpieces.

*Radiographic testing (RT).

S

*single electrode. One electrode connected exclusively to one power source which may consist of one or more power units.

*parallel electrode. Two electrodes connected electrically in parallel and exclusively to the same power source. Both electrodes are usually fed by means of a single electrode feeder. Welding current, when specified, is the total for the two.

*multiple electrodes. The combination of two or more single or parallel electrode systems. Each of the component systems has its own independent power source and its own electrode feeder.

*scanning level (UT). The dB setting used during scanning, as described in Tables 8.2 and 8.3.

semiautomatic welding. Manual welding with equipment that automatically controls one or more of the welding conditions.

*shelf bar. Steel plates, bars, or similar elements used to support the overflow of excess weld metal deposited in a horizontal groove weld joint.

Shielded metal arc welding (SMAW). An arc welding process with an arc between a covered electrode and the weld pool. The process is used with shielding from the decomposition of the electrode covering, without the application of pressure, and with filler metal from the electrode.

shielding gas. Protective gas used to prevent or reduce atmospheric contamination.

single-welded joint. A joint that is welded from one side only.

size of weld. See weld size.

slot weld. A weld made in an elongated hole in one member of a joint fusing that member to another member. The hole may be open at one end. A fillet welded slot shall not be construed as conforming to this definition.

*sound beam distance (UT). See sound path distance.

*sound path distance (UT). The distance between the search unit test material interface and the reflector as measured along the centerline of the sound beam.

*__storage phosphor imaging plate (SPIP)__.

spatter. The metal particles expelled during fusion welding that do not form a part of the weld.

stringer bead. A type of weld bead made without appreciable weaving motion.

*stud base. The stud tip at the welding end, including flux and container, and 1/8 in [3 mm] of the body of the stud adjacent to the tip.

*stud welding (SW). An arc welding process that produces coalescence of metals by heating them with an arc between a metal stud, or similar part, and the other workpiece. When the surfaces to be joined are properly heated, they are brought together under pressure. Partial shielding may be obtained by the use of a ceramic ferrule surrounding the stud. Shielding gas or flux may or may not be used.

Submerged arc welding (SAW). An arc welding process that uses an arc or arcs between a bare metal electrode or electrodes and the weld pool. The arc and molten metal are shielded by a blanket of granular flux on the workpieces. The process is used without pressure and with filler metal from the electrode and sometimes from a supplemental source (welding rod, flux, or metal granules).

T

tack weld. A weld made to hold parts of a weldment in proper alignment until the final welds are made.

*tack welder. A fitter, or someone under the direction of a fitter, who tack welds parts of a weldment to hold them in proper alignment until the final welds are made.

*tandem. Refers to a geometrical arrangement of electrodes in which a line through the arcs is parallel to the direction of welding.

thermal gouging. A thermal cutting process variation that removes metal by melting or burning the entire removed portion, to form a bevel or groove.

throat of a fillet weld.

actual throat. The shortest distance between the weld root and the face of a fillet weld.

theoretical throat. The distance from the beginning of the joint root perpendicular to the hypotenuse of the largest right triangle that can be inscribed within the cross section of a fillet weld. This dimension is based on the assumption that the root opening is equal to zero.

throat of a groove weld. A nonstandard term for **groove weld size**.

T-joint. A joint between two members located approximately at right angles to each other in the form of a T.

toe of weld. See **weld toe**.

*****transverse discontinuity.** A weld discontinuity whose major dimension is in a direction perpendicular to the weld axis "X," see Annex P, Form P-11.

total instantaneous energy (TIE), *waveform-controlled welding*. The sum of products of amperages, voltages and time intervals determined at sampling frequencies sufficient to quantify waveform changes during a welding interval.

*****tubular.** A generic term that refers to sections including pipe products (see **pipe**) and the family of square, rectangular, and round hollow-section products produced or manufactured in accordance with a tubular product specification. Also referred to as hollow structural section (HSS).

*****tubular connection.** A connection in the portion of a structure that contains two or more intersecting members, at least one of which is a tubular member.

*****tubular joint.** A joint in the interface created by a tubular member intersecting another member (which may or may not be tubular).

U

*****unacceptable discontinuity**. See **defect**.

undercut. A groove melted into the base metal adjacent to the weld toe or weld root and left unfilled by weld metal.

*****Ultrasonic testing (UT).**

V

*****Verification Inspector.** The duly designated person who acts for, and in behalf of, the Owner on all inspection and quality matters designated by the Engineer.

vertical welding position. The welding position in which the weld axis, at the point of welding, is approximately vertical, and the weld face lies in an approximately vertical plane (see Figures 6.1, 6.2, 6.3, and 6.4).

*****vertical position (pipe welding).** The position of a pipe joint in which welding is performed in the horizontal position and the pipe is not rotated during welding (see Figures 6.1, 6.2, and 10.12).

*****V-path (UT).** The distance a shear wave sound beam travels from the search unit test material interface to the other face of the test material and back to the original surface.

W

weave bead. A type of weld bead made with transverse oscillation.

weld. A localized coalescence of metals or nonmetals produced by heating the materials to the welding temperature, with or without the application of pressure or by the applications of pressure alone and with or without the use of filler material.

weldability. The capacity of a material to be welded under the imposed fabrication conditions into a specific, suitably designed structure and to perform satisfactorily in the intended service.

weld axis. A line through the length of a weld, perpendicular to and at the geometric center of its cross section.

weld bead. A weld resulting from a pass. See **stringer bead** and **weave bead**.

welder. One who performs a manual or semiautomatic welding operation.

welder certification. Written certification that a welder has produced welds meeting a prescribed standard of welder performance.

welder performance qualification. The demonstration of a welder's ability to produce welds meeting prescribed standards.

weld face. The exposed surface of a weld on the side from which welding was done.

welding. A joining process that produces coalescence of materials by heating them to the welding temperature, with or without the application of pressure or by the application of pressure alone, and with or without the use of filler metal. See also the Master Chart of Welding and Allied Processes in the latest edition of AWS A3.0.

welding machine. Equipment used to perform the welding operation. For example, spot welding machine, arc welding machine, and seam welding machine.

welding operator. One who operates adaptive control, automatic, mechanized, or robotic welding equipment.

welding sequence. The order of making the welds in a weldment.

weld pass. A single progression of welding along a joint. The result of a pass is a weld bead or layer.

weld reinforcement. Weld metal in excess of the quantity required to fill a joint.

weld root. The points, as shown in cross section, at which the root surface intersects the base metal surfaces.

weld size.

   fillet weld size. For equal leg fillet welds, the leg lengths of the largest isosceles right triangle that can be inscribed within the fillet weld cross section. For unequal leg fillet welds, the leg lengths of the largest right triangle that can be inscribed within the fillet weld cross section.

NOTE: When one member makes an angle with the other member greater than 105°, the leg length (size) is of less significance than the effective throat, which is the controlling factor for the strength of the weld.

   groove weld size. The joint penetration of a groove weld.

weld tab. Additional material that extends beyond either end of the joint, on which the weld is started or terminated.

weld toe. The junction of the weld face and the base metal.

weldment. An assembly whose component parts are joined by welding.

WPS qualification. The demonstration that welds made by a specific procedure can meet prescribed standards.

***welding procedure specification (WPS).** A document providing the detailed methods and practices involved in the production of a weldment.

4. Design of Welded Connections

4.1 Scope

This clause contains requirements for design of welded connections. It is divided into three parts as follows:

Part A—Common Requirements for Design of Welded Connections (Nontubular and Tubular Members)

Part B—Specific Requirements for Design of Nontubular Connections (Statically or Cyclically Loaded). The requirements shall apply in addition to the requirements of Part A.

Part C—Specific Requirements for Design of Nontubular Connections (Cyclically Loaded). When applicable, the requirements shall apply in addition to the requirements of Parts A and B.

Part A
Common Requirements for Design of Welded Connections (Nontubular and Tubular Members)

4.2 General

This part contains requirements applicable to the design of all welded connections of nontubular and tubular structures, independent of loading.

4.3 Contract Plans and Specifications

4.3.1 Plan and Drawing Information. Complete information regarding base metal specification designation (see 5.3 and 6.8.3) and location, type, size, and extent of all welds shall be clearly shown on the contract plans and specifications, hereinafter referred to as the contract documents. If the Engineer requires specific welds to be performed in the field, they shall be designated in the contract documents. The fabrication and erection drawings, hereinafter referred to as the shop drawings, shall clearly distinguish between shop and field welds.

4.3.2 Notch Toughness Requirements. If notch toughness of welded joints is required, the Engineer shall specify the minimum absorbed energy with the corresponding test temperature for the filler metal classification to be used, or the Engineer shall specify that the WPSs be qualified with CVN tests. If WPSs with CVN tests are required, the Engineer shall specify the minimum absorbed energy, the test temperature and whether the required CVN test performance is to be in the weld metal, or both in the weld metal and the HAZ (see 6.2.1.3 and Clause 6, Part D).

4.3.3 Specific Welding Requirements. The Engineer, in the contract documents, and the Contractor, in the shop drawings, shall indicate those joints or groups of joints in which the Engineer or Contractor require a specific assembly order, welding sequence, welding technique or other special precautions. See 7.4.1 and C-7.4.1 for limitations on the application of ESW and EGW welding.

4.3.4 Weld Size and Length. Contract design drawings shall specify the effective weld length and, for PJP groove welds, the required weld size "(S)." For fillet welds and skewed T-joints, the following shall be provided on the contract documents.

(1) For fillet welds between parts with surfaces meeting at an angle between 80° and 100°, contract documents shall specify the fillet weld leg size.

(2) For welds between parts with the surfaces meeting at an angle less than 80° or greater than 100°, the contract documents shall specify the effective throat.

End returns and hold-backs for fillet welds, if required by design, shall be indicated on the contract documents.

4.3.5 Shop Drawing Requirements. Shop drawings shall clearly indicate by welding symbols or sketches the details of groove welded joints and the preparation of base metal required to make them. Both width and thickness of steel backing shall be detailed.

4.3.5.1 PJP Groove Welds. Shop drawings shall indicate the weld groove depths "D" needed to attain weld size "(S)" required for the welding process and position of welding to be used.

4.3.5.2 Fillet Welds and Welds in Skewed T-Joints.

The following shall be provided on the shop drawings:

(1) For fillet welds between parts with surfaces meeting at an angle between 80° and 100°, shop drawings shall show the fillet weld leg size,

(2) For welds between parts with surfaces meeting at an angle less than 80° or greater than 100°, the shop drawings shall show the detailed arrangement of welds and required leg size to account for effects of joint geometry and, where appropriate, the Z-loss reduction for the process to be used and the angle,

(3) End returns and hold-backs.

4.3.5.3 Welding Symbols. The contract documents shall show CJP or PJP groove weld requirements. Contract documents do not need to show groove type or groove dimensions. The welding symbol without dimensions and with "CJP" in the tail designates a CJP weld as follows:

The welding symbol without dimension and without CJP in the tail designates a weld that will develop the adjacent base metal strength in tension and shear. A welding symbol for a PJP groove weld shall show dimensions enclosed in parentheses below "(S_1)" and/or above "(S_2)" the reference line to indicate the groove weld sizes on the arrow and other sides of the weld joint, respectively, as shown below:

4.3.5.4 Prequalified Detail Dimensions. The joint details described in 5.4.2 and 10.9 (PJP) and 5.4.1 and 10.10 (CJP) have repeatedly demonstrated their adequacy in providing the conditions and clearances necessary for depositing and fusing sound weld metal to base metal. However, the use of these details shall not be interpreted as implying consideration of the effects of welding process on base metal beyond the fusion boundary nor suitability of the joint detail for a given application.

4.3.5.5 Special Details. When special groove details are required, they shall be detailed in the contract documents.

4.3.5.6 Specific Inspection Requirements. Any specific inspection requirements shall be noted on the contract documents.

4.4 Effective Areas

4.4.1 Groove Welds

4.4.1.1 Effective Length. The maximum effective weld length of any groove weld, regardless of orientation, shall be the width of the part joined, perpendicular to the direction of tensile or compressive stress. For groove welds transmitting shear, the effective length is the length specified.

4.4.1.2 Effective Size of CJP Groove Welds. The weld size of a CJP groove weld shall be the thickness of the thinner part joined. An increase in the effective area for design calculations for weld reinforcement shall be prohibited. Groove weld sizes for T-, Y-, and K-connections in tubular construction are shown in Table 10.7.

4.4.1.3 Minimum Size of PJP Groove Welds. PJP groove welds shall be equal to or greater than the size "(S)" specified in 5.4.2.3(1) unless the WPS is qualified in conformance with Clause 6.

4.4.1.4 Effective Size of Flare-Groove Welds. The effective size of flare-groove welds when filled flush shall be as shown in Table 4.1, except as allowed by 6.12.4. For flare-groove welds not filled flush, the underfill U shall be deducted. For flare-V-groove welds to surfaces with different radii R, the smaller R shall be used. For flare-groove welds to rectangular tubular sections, R shall be taken as two times the wall thickness.

4.4.1.5 Effective Area of Groove Welds. The effective area of groove welds shall be the effective length multiplied by the effective weld size.

4.4.2 Fillet Welds

4.4.2.1 Effective Length (Straight). The effective length of a straight fillet weld shall be the overall length of the full size fillet, including end returns. No reduction in effective length shall be assumed in design calculations to allow for the start or stop crater of the weld.

4.4.2.2 Effective Length (Curved). The effective length of a curved fillet weld shall be measured along the centerline of the effective throat.

4.4.2.3 Minimum Length. The minimum length of a fillet weld shall be at least four times the nominal size, or the effective size of the weld shall be considered not to exceed 25% of its effective length.

4.4.2.4 Intermittent Fillet Welds (Minimum Length). The minimum length of segments of an intermittent fillet weld shall be 1-1/2 in [38 mm].

4.4.2.5 Maximum Effective Length. For end-loaded fillet welds with a length up to 100 times the leg dimension, it is allowed to take the effective length equal to the actual length. When the length of end-loaded fillet welds exceeds 100 but not more than 300 times the weld size, the effective length shall be determined by multiplying the actual length by the reduction coefficient β.

$$\beta = 1.2 - 0.2\left(\frac{L}{100w}\right) \le 1.0$$

where

β = reduction coefficient
L = actual length of end-loaded weld, in [mm]
w = weld leg size, in [mm]

When the length exceeds 300 times the leg size, the effective length shall be taken as 180 times the leg size.

4.4.2.6 Calculation of Effective Throat. For fillet welds between parts meeting at angles between 80° and 100° the effective throat shall be taken as the shortest distance from the joint root to the weld face of a 90° diagrammatic weld (see Figure 4.1). For welds in acute angles between 60° and 80° and for welds in obtuse angles greater than 100°, the weld leg size required to provide the specified effective throat shall be calculated to account for geometry (see Annex A). For welds in acute angles between 60° and 30°, leg size shall be increased by the Z loss dimension to account for the uncertainty of sound weld metal in the root pass of the narrow angle for the welding process to be used (see 4.4.3).

4.4.2.7 Reinforcing Fillet Welds. The effective throat of a combination PJP bevel groove weld and a fillet weld shall be the shortest distance from the joint root to the weld face of the diagrammatic weld minus 1/8 in [3 mm] for any groove detail requiring such deduction (see Figures 4.2, 4.3, 4.4 and 5.2).

The effective throat of a combination of PJP flare bevel groove weld and a fillet weld shall be the shortest distance from the joint root to the weld face of the diagrammatic weld minus the deduction for incomplete joint penetration (see Table 4.1, Figures 4.5, 4.6 and 5.2).

4.4.2.8 Minimum Size. The minimum size fillet weld shall not be smaller than the size required to transmit the applied load nor that provided in 7.13.

4.4.2.9 Maximum Weld Size in Lap Joints. The maximum fillet weld size detailed along the edges of base metal in lap joints shall be the following:

(1) the thickness of the base metal, for metal less than 1/4 in [6 mm] thick (see Figure 4.7, Detail A).

(2) 1/16 in [2 mm] less than the thickness of the base metal, for metal 1/4 in [6 mm] or more in thickness (see Figure 4.7, Detail B), unless the weld is designated on the shop drawing to be built out to obtain full throat thickness for a leg size equal to the base metal thickness. In the as-welded condition, the distance between the edge of the base metal and the toe of the weld may be less than 1/16 in [2 mm] provided the weld size is clearly verifiable.

4.4.2.10 Effective Area. The effective area shall be the effective weld length multiplied by the effective throat.

4.4.3 Skewed T-Joints

4.4.3.1 General. T-joints in which the angle between joined parts is greater than 100° or less than 80° shall be defined as skewed T-joints. Prequalified skewed T-joint details are shown in Figure 5.4. The details of joints for the obtuse and acute sides may be used together or independently depending upon service conditions and design with proper consideration for effects of eccentricity.

4.4.3.2 Welds in Acute Angles Between 80° and 60° and in Obtuse Angles Greater than 100°. When welds are deposited in angles between 80° and 60° or in angles greater than 100° the contract documents shall specify the required effective throat. The shop drawings shall clearly show the placement of welds and the required leg dimensions to satisfy the required effective throat (see Annex A).

4.4.3.3 Welds in Angles Between 60° and 30°. When welding is required in an acute angle that is less than 60° but equal to or greater than 30° [Figure 5.4(D)], the effective throat shall be increased by the Z-loss allowance (Table 4.2). The contract documents shall specify the required effective throat. The shop drawings shall show the required leg dimensions to satisfy the required effective throat, increased by the Z-loss allowance (Table 4.2) (see Annex A for calculation of effective throat).

4.4.3.4 Welds in Angles Less than 30°. Welds deposited in acute angles less than 30° shall not be considered as effective in transmitting applied forces except as modified for tubular structures in 10.14.4.2.

4.4.3.5 Effective Length. The effective length of skewed T-joints shall be the overall length of the full size weld. No reduction shall be assumed in design calculations to allow for the start or stop of the weld.

4.4.3.6 Minimum Weld Size. The requirements of 4.4.2.8 shall apply.

4.4.3.7 Effective Throat. The effective throat of a skewed T-joint in angles between 60° and 30° shall be the minimum distance from the root to the diagrammatic face, less the Z loss reduction dimension. The effective throat of a skewed T-joint in angles between 80° and 60° and in angles greater than 100° shall be taken as the shortest distance from the joint root to the weld face.

4.4.3.8 Effective Area. The effective area of skewed T-joints shall be the specified effective throat multiplied by the effective length.

4.4.4 Fillet Welds in Holes and Slots

4.4.4.1 Diameter and Width Limitations. The minimum diameter of the hole or the width of slot in which a fillet weld is to be deposited shall be no less than the thickness of the part in which it is made plus 5/16 in [8 mm].

4.4.4.2 Slot Ends. Except for those ends which extend to the edge of the part, the ends of the slot shall be semicircular or shall have the corners rounded to a radius not less than the thickness of the part in which it is made.

4.4.4.3 Effective Length. For fillet welds in holes or slots, the effective length shall be the length of the weld along the centerline of the throat.

4.4.4.4 Effective Area. The effective area shall be the effective length multiplied by the effective throat. In the case of fillet welds of such size that they overlap at the centerline when deposited in holes or slots, the effective area shall not be taken as greater than the cross-sectional area of the hole or slot in the plane of the faying surface.

4.4.5 Plug and Slot Welds

4.4.5.1 Diameter and Width Limitations. The minimum diameter of the hole or the width of slot in which a plug or slot weld is to be deposited shall be no less than the thickness of the part in which it is made plus 5/16 in [8 mm.]. The maximum diameter of the hole or width of slot shall not exceed the minimum diameter plus 1/8 in [3 mm] or 2-1/4 times the thickness of the part, whichever is greater.

4.4.5.2 Slot Length and Shape. The length of the slot in which slot welds are to be deposited shall not exceed ten times the thickness of the part in which it is made. The ends of the slot shall be semicircular or shall have the corners rounded to a radius not less than the thickness of the part in which it is made.

4.4.5.3 Effective Area. The effective area of plug and slot welds shall be the nominal area of the hole or slot in the plane of the faying surface.

4.4.5.4 Minimum Depth of Filling. The minimum depth of filling of plug and slot welds shall meet the following requirements:

(1) for slot or plug welds in material 5/8 in [16 mm] thick or less, the thickness of the material.

(2) for slot or plug welds in materials over 5/8 in [16 mm] thick, one-half the thickness of the material or 5/8 in [16 mm], whichever is greater.

In no case is the minimum depth of filling required to be greater than the thickness of the thinner part being joined.

Part B
Specific Requirements for Design of Nontubular Connections
(Statically or Cyclically Loaded)

4.5 General

The specific requirements of Part B together with the requirements of Part A shall apply to all connections of nontubular members subject to static loading. The requirements of Parts A and B, except as modified by Part C, shall also apply to cyclic loading.

4.6 Stresses

4.6.1 Calculated Stresses. The calculated stresses to be compared with the allowable stresses shall be nominal stresses determined by appropriate analysis or stresses determined from the minimum joint strength requirements that may be specified in the applicable design specifications which invoke this code for design of welded connections.

4.6.2 Calculated Stresses Due to Eccentricity. In the design of welded joints, the calculated stresses to be compared with allowable stresses, shall include those due to design eccentricity, if any, in alignment of connected parts and the position, size and type of welds, except as provided in the following:

For statically loaded structures, the location of fillet welds to balance the forces about the neutral axis or axes for end connections of single-angle, double-angle, and similar members is not required. In such members, weld arrangements at the heel and toe of angle members may be distributed to conform to the length of the various available edges.

4.6.3 Allowable Base Metal Stresses. The calculated base metal stresses shall not exceed the allowable stresses specified in the applicable design specifications.

4.6.4 Allowable Weld Metal Stresses. The calculated stresses on the effective area of welded joints shall not exceed the allowable stresses given in Table 4.3 except as allowed by 4.6.4.2, 4.6.4.3, and 4.6.4.4. The use of 4.6.4.2 shall be limited to the analysis of a single linear fillet weld or fillet weld groups consisting of parallel linear fillet welds all loaded at the same angle.

4.6.4.1 Stress in Fillet Welds. Stress in fillet welds shall be considered as shear applied to the effective area for any direction of applied load.

4.6.4.2 Alternative Allowable Fillet Weld Stress. For a single linear fillet weld or fillet weld groups consisting of parallel linear fillet welds all loaded at the same angle and loaded in plane through the centroid of the weld group, the allowable stress may be determined by Formula (1):

Formula (1)

$$F_v = 0.30\, F_{EXX}\, (1.0 + 0.50\, \sin^{1.5} \Theta)$$

where

F_v = allowable unit stress
F_{EXX} = electrode classification number, i.e., electrode strength classification
Θ = angle between the direction of force and the axis of the weld element, degrees

4.6.4.3 Instantaneous Center of Rotation. The allowable stresses in weld elements within a weld group that are loaded in-plane and analyzed using an instantaneous center of rotation method to maintain deformation compatibility and the nonlinear load-deformation behavior of variable angle loaded welds shall be the following:

$$F_{vx} = \Sigma F_{vix}$$
$$F_{vy} = \Sigma F_{viy}$$
$$F_{vi} = 0.30\, F_{EXX}\, (1.0 + 0.50\, \sin^{1.5} \Theta)\, F(\rho)$$
$$F(\rho) = [\rho\, (1.9 - 0.9\rho)]^{0.3}$$
$$M = \Sigma\, [F_{viy}\, (x) - F_{vix}\, (y)]$$

where

F_{vx} = Total internal force in x direction
F_{vy} = Total internal force in y direction
F_{vix} = x component of stress F_{vi}
F_{viy} = y component of stress F_{vi}
M = Moment of internal forces about the instantaneous center of rotation
ρ = Δ_i/Δ_m ratio of element "i" deformation to deformation in element at maximum stress
Δ_m = $0.209\, (\Theta + 6)^{-0.32}\, W$, deformation of weld element at maximum stress, in [mm]
Δ_u = $1.087\, (\Theta + 6)^{-0.65}\, W$, <0.17 W, deformation of weld element at ultimate stress (fracture), usually in element furthest from the instantaneous center of rotation, in [mm]
W = leg size of the fillet weld, in [mm]
Δ_i = deformation of weld elements at intermediate stress levels, linearly proportioned to the critical deformation based on distance from instantaneous center of rotation, in [mm] = $r_i \Delta_u / r_{crit}$
x = x_i component of r_i
y = y_i component of r_i
r_{crit} = distance from instantaneous center of rotation to weld element with minimum Δ_u/r_i ratio, in [mm]

4.6.4.4 Concentrically Loaded Weld Groups. Alternatively, for the special case of a concentrically loaded weld group, the allowable shear stress for each weld element may be determined using Formula (2) and the allowable loads of all elements calculated and added.

Formula (2) $F_v = 0.30\, C\, F_{EXX}$

where

F_v = allowable unit stress
F_{EXX} = nominal tensile strength of filler metal
C = the equivalent strength coefficient for obliquely loaded fillet weld, chosen from Table 4.4.

4.6.5 Allowable Stress Increase. Where the applicable design specifications allow the use of increased stresses in the base metal for any reason, a corresponding increase shall be applied to the allowable stresses given herein but not to the stress ranges allowed for base metal or weld metal subject to cyclic loading.

4.7 Joint Configuration and Details

4.7.1 General Considerations. Welded connections shall be designed to satisfy the strength and stiffness or flexibility requirements of the general invoking specifications.

4.7.2 Compression Member Connections and Splices

4.7.2.1 Connections and Splices Designed to Bear Other than Connections to Base Plates. Unless otherwise specified in contract documents, column splices which are finished to bear shall be connected by PJP groove welds or by fillet welds sufficient to hold the parts in place. Where compression members other than columns are finished to bear at splices or connections welds shall be designed to hold all parts in alignment and shall be proportioned for 50% of the force in the member. The requirements of Table 5.5 or Table 7.7 shall apply.

4.7.2.2 Connections and Splices Not Finished to Bear Except for Connections to Base Plates. Welds joining splices in columns and splices and connections in other compression members which are not finished to bear, shall be designed to transmit the force in the members, unless CJP welds or more restrictive requirements are specified in contract documents or governing specifications. The requirements of Table 5.5 or Table 7.7 shall apply.

4.7.2.3 Connections to Base Plates. At base plates of columns and other compression members, the connection shall be adequate to hold the members securely in place.

4.7.3 Base Metal Through-Thickness Loading. T and corner joints whose function is to transmit stress normal to the surface of a connected part, especially when the base metal thickness of the branch member or the required weld size is 3/4 in [20 mm] or greater, shall be given special attention during design, base metal selection and detailing. Joint details which minimize stress intensity on base metal subject to stress in the through-thickness direction shall be used where practical. Specifying weld sizes larger than necessary to transmit calculated stress shall be avoided.

4.7.4 Combinations of Welds. Except as provided herein, if two or more welds of different type (groove, fillet, plug, slot) are combined to share the load in a single connection, the capacity of the connection shall be calculated as the sum of the individual welds determined relative to the direction of applied load. This method of adding individual capacities of welds does not apply to fillet welds reinforcing PJP groove welds (see Figures 4.3, 4.4, and 4.6).

4.7.5 Butt, Corner, and T-Joint Surface Contouring. Fillet welds may be applied over CJP and PJP groove welds in butt joints joining parts of unequal width or thickness, corner, and T-joints for the purpose of contouring weld face or to reduce stress concentrations. When such surface contouring fillet welds are used in statically loaded applications, the size need not be more than 5/16 in [8 mm]. The fillet-like reinforcement on the surface of T and corner joint groove welds that naturally occurs shall not be cause for rejection nor need it be removed provided it does not interfere with other elements of the construction. No minimum contour radius need be provided.

4.7.6 Weld Access Holes. When weld access holes are required, they shall be sized to provide clearances necessary for deposition of sound weld metal. The shape and size requirements of 7.16.1 shall apply. The designer and detailer shall recognize that holes of the minimum required size may affect the maximum net area available in the connected base metal.

4.7.7 Welds with Rivets or Bolts. Connections that are welded to one member and bolted or riveted to the other shall be allowed. When bolts and welds share the load on a common faying surface, strain compatibility between the bolts and welds shall be considered (see commentary).

4.8 Joint Configuration and Details—Groove Welds

4.8.1 Transitions in Thicknesses and Widths. For statically loaded structures, surface contouring fillet welds need not be provided. When surface contouring fillet welds are required by the Engineer, they shall be specified in the contract documents (see Figure 4.8).

4.8.2 Partial Length CJP Groove Weld Prohibition. Intermittent or partial length CJP groove welds shall be prohibited except that members built-up of elements connected by fillet welds may have groove welds of limited length at points of localized load application to participate in the transfer of localized load. The groove weld shall extend at uniform size for at least the length required to transfer the load. Beyond this length, the groove shall be made with a transition in depth to zero over a distance not less than four times its depth. The groove shall be filled flush before application of the fillet weld.

4.8.3 Intermittent PJP Groove Welds. Intermittent PJP groove welds, flare bevel, and flare-groove welds may be used to transfer shear stress between connected parts.

4.8.4 Weld Tab Removal. For statically loaded nontubular structures, weld tabs need not be removed. When removal is required, or when finishing to surface requirements other than that described by 7.14.8, the requirements shall be specified in the contract documents.

4.9 Joint Configuration and Details—Fillet Welded Joints

4.9.1 Lap Joints

4.9.1.1 Transverse Fillet Welds. Transverse fillet welds in lap joints transferring stress between axially loaded parts shall be double-fillet welded (see Figure 4.9) except where deflection of the joint is sufficiently restrained to prevent opening under load.

4.9.1.2 Minimum Overlap. The minimum overlap of parts in stress-carrying lap joints shall be five times the thickness of the thinner part, but not less than 1 in [25 mm]. Unless out-of-plane deflection of the parts is prevented, they shall be double fillet welded (see Figure 4.9) or joined by at least two transverse lines of plug or slot welds or two or more longitudinal fillet or slot welds.

4.9.2 Longitudinal Fillet Welds. If longitudinal fillet welds are used alone in lap joints of end connections of flat bar or plate members, the length of each fillet weld shall be no less than the perpendicular distance between them (see Figure 4.10). The transverse spacing of longitudinal fillet welds used in end connections shall not exceed 16 times the thickness of the thinner connected part unless suitable provision is made (as by intermediate plug or slot welds) to prevent buckling or separation of the parts. The longitudinal fillet welds may be either at the edges of the member or in slots. The design of connections using longitudinal fillet welds for members other than flat bar cross sections shall be as provided in the general design specifications.

4.9.3 Fillet Weld Terminations

4.9.3.1 General. Fillet weld terminations may extend to the ends or sides of parts or may be stopped short or may have end returns except as limited by the following cases:

4.9.3.2 Lap Joints Subject to Tension. In lap joints in which one part extends beyond the edge or side of a part subject to calculated tensile stress, fillet welds shall terminate not less than the size of the weld from the start of the extension (see Figure 4.11).

4.9.3.3 Maximum End Return Length. Welded joints shall be arranged to allow the flexibility assumed in the connection design. If the outstanding legs of connection base metal are attached with end returned welds, the length of the end return shall not exceed four times the nominal size of the weld (see Figure 4.12 for examples of flexible connections).

4.9.3.4 Transverse Stiffener Welds. Except where the ends of stiffeners are welded to the flange, fillet welds joining transverse stiffeners to girder webs shall start or terminate not less than four times nor more than six times the thickness of the web from the web toe of the web-to-flange welds.

4.9.3.5 Opposite Sides of a Common Plane. Fillet welds on the opposite sides of a common plane shall be interrupted at the corner common to both welds (see Figure 4.13), except as follows:

When joints are required to be sealed, or when a continuous weld is needed for other reasons, the contract documents shall specify where these welds are required to be continuous.

4.9.4 Fillet Welds in Holes or Slots. Fillet welds in holes or slots in lap joints may be used to transfer shear or to prevent buckling or separation of lapped parts. Minimum spacing and dimensions of holes or slots for fillet welds shall conform to the requirements of 4.4.4.1, 4.4.4.2, 4.9.1, 4.9.2, and 4.10. These fillet welds may overlap subject to the limitation provisions of 4.4.4.4. Fillet welds in holes or slots are not considered to be plug or slot welds.

4.9.5 Intermittent Fillet Welds. Intermittent fillet welds may be used to transfer stress between connected parts.

4.10 Joint Configuration and Details—Plug and Slot Welds

4.10.1 Minimum Spacing (Plug Welds). The minimum center-to-center spacing of plug welds shall be four times the diameter of the hole.

4.10.2 Minimum Spacing (Slot Welds). The minimum center-to-center spacing of lines of slot welds in a direction transverse to their length shall be four times the width of the slot. The minimum center-to-center spacing in a longitudinal direction shall be two times the length of the slot.

4.10.3 Prequalified Dimensions. Dimensions for prequalified plug and slot welds are described in 4.4.5 and 5.4.4.

4.10.4 Prohibition in Quenched and Tempered Steels. Plug and slot welds shall be prohibited in quenched and tempered steels with specified minimum F_y greater than 70 ksi [490 MPa].

4.11 Filler Plates

Wherever it is necessary to use filler plates in joints required to transfer applied force, the filler plates and the connecting welds shall conform to the requirements of 4.11.1 or 4.11.2, as applicable.

4.11.1 Thin Filler Plates. Filler plates less than 1/4 in [6 mm] thick shall not be used to transfer stress. When the thickness of the filler plate is less than 1/4 in [6 mm], or when the thickness of the filler plate is greater than 1/4 in [6 mm] but not adequate to transfer the applied force between the connected parts, the filler plate shall be kept flush with the edge of the outside connected part, and the size of the weld shall be increased over the required size by an amount equal to the thickness of the filler plate (see Figure 4.14).

4.11.2 Thick Filler Plates. When the thickness of the filler plate is adequate to transfer the applied force between the connected parts, the filler plate shall extend beyond the edges of the outside connected base metal. The welds joining the outside connected base metal to the filler plate shall be sufficient to transmit the force to the filler plate, and the area subject to applied force in the filler plate shall be adequate to avoid overstressing the filler plate. The welds joining filler plate to the inside connected base metal shall be sufficient to transmit the applied force (see Figure 4.15).

4.11.3 Shop Drawing Requirement. Joints requiring filler plates shall be completely detailed on shop and erection drawings.

4.12 Built-Up Members

4.12.1 Minimum Required Welding. If two or more plates or rolled shapes are used to build up a member, sufficient welding (fillet, plug, or slot type) shall be provided to make the parts act in unison but not less than that which may be required to transmit the calculated stress between the parts joined.

4.12.2 Maximum Spacing of Intermittent Welds

4.12.2.1 General. Except as may be provided by 4.12.2.2 or 4.12.2.3, the maximum longitudinal spacing of intermittent welds connecting a plate component to other components shall not exceed 24 times the thickness of the thinner plate nor exceed 12 in [300 mm]. The longitudinal spacing between intermittent fillet welds connecting two or more rolled shapes shall not exceed 24 in [600 mm].

4.12.2.2 Compression Members. In built-up compression members, except as provided in 4.12.2.3, the longitudinal spacing of intermittent fillet weld segments along the edges of an outside plate component to other components shall not exceed 12 in [300 mm] nor the plate thickness times $0.730 \sqrt{E/F_y}$ (F_y = specified minimum yield strength and E is Young's modulus of elasticity for the type of steel being used.) When intermittent fillet weld segments are staggered along opposite edges of outside plate components narrower than the width provided by the next sentence, the spacing shall not exceed 18 in [460 mm] nor the plate thickness times $1.10 \sqrt{E/F_y}$. The unsupported width of web, cover plate, or diaphragm plates, between adjacent lines of welds, not exceed the plate thickness times $1.46 \sqrt{E/F_y}$. When unsupported transverse spacing exceeds this limit, but a portion of its width no greater than $1.46 \sqrt{E/F_y}$ times the thickness would satisfy the stress requirement, the member shall be considered acceptable.

4.12.2.3 Unpainted Weathering Steel. For members of unpainted weathering steel exposed to atmospheric corrosion, if intermittent fillet welds are used, the spacing shall not exceed 14 times the thickness of the thinner plate nor 7 in [180 mm].

Part C
Specific Requirements for Design of Nontubular Connections
(Cyclically Loaded)

4.13 General

4.13.1 Applicability. Part C applies only to nontubular members and connections subject to cyclic load, within the elastic range, of frequency and magnitude sufficient to initiate cracking and progressive failure (fatigue). The provisions of Part C provide a method for assessing the effects of repeated fluctuations of stress on welded nontubular structural elements which shall be applied to minimize the possibility of a fatigue failure.

4.13.2 Other Pertinent Provisions. The provisions of Parts A and B shall apply to the design of members and connections subject to the requirements of Part C.

4.13.3 Engineer's Responsibility. The Engineer shall provide either complete details, including weld sizes, or shall specify the planned cycle life and the maximum range of moments, shears, and reactions for the connections in contract documents.

4.14 Limitations

4.14.1 Stress Range Threshold. No evaluation of fatigue resistance shall be required if the live load stress range is less than the threshold stress range, FTH (see Table 4.5).

4.14.2 Low Cycle Fatigue. Provisions of Part C are not applicable to low-cycle loading cases which induce calculated stresses into the inelastic range of stress.

4.14.3 Corrosion Protection. The fatigue strengths described in Part C are applicable to structures with suitable corrosion protection, or subject only to mildly corrosive environments such as normal atmospheric conditions.

4.14.4 Redundant–Nonredundant Members. This code no longer recognizes a distinction between redundant and nonredundant members.

4.15 Calculation of Stresses

4.15.1 Elastic Analysis. Calculated stresses and stress ranges shall be nominal, based upon elastic stress analysis at the member level. Stresses need not be amplified by stress concentration factors for local geometrical discontinuities.

4.15.2 Axial Stress and Bending. In the case of axial stress combined with bending, the maximum combined stress shall be that for concurrent applied load cases.

4.15.3 Symmetrical Sections. For members having symmetrical cross sections, the connection welds shall preferably be arranged symmetrically about the axis of the member, or if symmetrical arrangement is not practical, the total stresses including those resulting from joint eccentricity shall be included in the calculation of the stress range.

4.15.4 Angle Members. For axially stressed angle members, the center of gravity of the connecting welds shall lie between the line of the center of gravity of the angle's cross section and the center of the connected leg, in which case the effects of eccentricity may be ignored. If the center of gravity of the connecting weld lies outside this zone, the total stresses, including those resulting from eccentricity of the joint from the center of gravity of the angle, shall be included in the calculation of the stress range.

4.16 Allowable Stresses and Stress Ranges

4.16.1 Allowable Stresses. The calculated unit stresses in welds shall not exceed the allowable stresses described in Table 4.3.

4.16.2 Allowable Stress Ranges. Stress range is defined as the magnitude of fluctuation in stress that results from the repeated application and removal of the live load. In the case of stress reversal, the stress range shall be computed as the numerical sum of the maximum repeated tensile and compressive stresses or the sum of maximum shearing stresses of opposite direction at a given point, resulting from differing arrangement of live load. The calculated range of stress shall not exceed the maximum computed by Formulas (3) through (6), as applicable (see Figure 4.16 for graphical plot of Formulas (3) through (6) for stress categories A, B, B', C, D, E, E', and F).

For categories A, B, B', C, D, E, and E', the stress range shall not exceed F_{SR} as determined by Formula (3).

Formula (3)

$$F_{SR} = \left(\frac{C_f}{N}\right)^{0.333} \geq F_{TH} \text{ (ksi)}$$

$$F_{SR} = \left[\left(\frac{C_f \times 329}{N}\right)^{0.333} \geq F_{TH} \text{ (MPa)}\right]$$

in which:

F_{SR} = Allowable stress range, ksi [MPa]
C_f = Constant from Table 4.5 for all categories except category F.
N = Number of cycles of stress range in design life.
  = Cycles per day × 365 × years of design life.
F_{TH} = Threshold fatigue stress range, that is the maximum stress range for infinite life, ksi [MPa]

For stress category F, the stress range shall not exceed F_{SR} as determined by Formula (4).

Formula (4)

$$F_{SR} = \left(\frac{C_f}{N}\right)^{0.167} \geq F_{TH} \text{)ksi)}$$

$$F_{SR} = \left[\left(\frac{C_f \times 11 \times 10^4}{N}\right)^{0.167} \geq F_{TH} \text{ (MPa)}\right]$$

in which:

C_f = Constant from Table 4.5 for Category F

For tension-loaded plate elements at cruciform, T and corner joint details with CJP welds, PJP welds, fillet welds or combinations of the preceding, transverse to the direction of stress, the maximum stress range on the cross section of the tension-loaded plate element shall be determined by (a), (b), or (c) as follows:

(a) **For the cross section of a tension-loaded plate element,** the maximum stress range on the base metal cross section at the toe of the weld governed by consideration of crack initiation from the toe of the weld, the stress range shall not exceed FSR as determined by Formula (3), Category C, which shall be equal to:

$$F_{SR} = \left(\frac{44 \times 10^8}{N}\right)^{0.333} \geq 10 \text{ (ksi)}$$

$$F_{SR} = \left[\left(\frac{14.4 \times 10^{11}}{N}\right)^{0.333} \geq 68.9 \text{ (MPa)}\right]$$

(b) **For end connections of tension-loaded plate elements using transverse PJP welds,** with or without reinforcing or contouring fillet welds, the maximum stress range on the base metal cross section at the toe of the weld governed by consideration of crack initiation from the root of the weld shall not exceed FSR as determined by Formula (5).

Formula (5)

$$F_{SR} = \left| R_{PJP} \left(\frac{44 \times 10^8}{N} \right)^{0.333} \right| \text{ (ksi)}$$

$$F_{SR} = \left[R_{PJP} \left(\frac{14.4 \times 10^{11}}{N} \right)^{0.333} \text{ (MPa)} \right]$$

In which:

R_{PJP} = Reduction factor for reinforced or nonreinforced PJP joints

$$R_{PJP} = \frac{0.65 - 0.59(2a/t_p) + 0.72(w/t_p)}{t_p^{0.167}} \leq 1.0 \text{ (for in)}$$

$$\frac{1.12 - 1.01(2a/t_p) + 1.24(w/t_p)}{t_p^{0.167}} \leq 1.0 \text{ (for mm)}$$

$2a$ = the length of the nonwelded root face in the direction of the thickness of the tension-loaded plate

t_p = the thickness of tension loaded plate element (in or mm)

w = the leg size of the reinforcing or contouring fillet, if any, in the direction of the thickness of the tension-loaded plate (in or mm)

(c) For end connections of tension-loaded plate elements using a pair of fillet welds, the maximum stress range on the base metal cross section at the toe of the weld governed by consideration of crack initiation from the root of the weld due to tension on the root shall not exceed F_{SR} as determined by Formula (6). Additionally, the shear stress range on the throat of the weld shall not exceed F_{SR} by Formula (4) for Category F.

Formula (6)

$$F_{SR} = R_{FIL} \left(\frac{44 \times 10^8}{N} \right)^{0.333} \text{ (ksi)}$$

$$F_{SR} = \left[R_{FIL} \left(\frac{14.4 \times 10^{11}}{N} \right)^{0.333} \text{ (MPa)} \right]$$

In which:

R_{FIL} = Reduction Factor for joints using a pair of transverse fillet welds only

$$R_{FIL} = \frac{0.06 + 0.72 \left(\frac{w}{t_p} \right)}{t_p^{0.167}} \leq 1.0 \text{ (for in)}$$

$$R_{FIL} = \frac{0.10 + 1.24 \left(\frac{w}{t_p} \right)}{t_p^{0.167}} \leq 1.0 \text{ (for mm)}$$

4.17 Detailing, Fabrication, and Erection

4.17.1 Transitions in Thickness and Width

4.17.1.1 Butt-Joint Thickness Transitions. Butt joints between parts having unequal thicknesses and subject to cyclic tensile stress shall have smooth transitions between offset surfaces at a slope of no more than 1 in 2-1/2 with the surface of either part. The transition may be accomplished by sloping the weld surfaces, by chamfering the thicker part, or by a combination of the two methods (see Figure 4.17).

4.17.1.2 Butt-Joint Width Transitions. Butt joints between parts of unequal width subject to cyclic stress into the tensile range shall have a smooth transition between the offset edges at a slope of not more than 1 in 2-1/2 with the edge of either part or shall be provided with a transition having a 24 in [600 mm] minimum radius tangent to the narrower part at the center of the butt joint (see Figure 4.18). An increased stress range may be used for steels having a yield strength greater than 90 ksi [620 MPa] with details incorporating the radius.

4.17.2 Backing

4.17.2.1 Welds for Attaching Steel Backing. Requirements for welds for attaching steel backing and whether the backing shall be removed or left in place shall be determined as described in 4.17.2.2, 4.17.2.3, 4.17.2.4, and the stress range categories of Table 4.5. The Engineer shall note the fatigue stress category on the contract drawings. The Contractor shall note on the shop drawings the required location, the weld detail to be used, whether the tack welds shall be inside the groove or shall be allowed to be outside the groove, and whether the backing shall be allowed to remain in place or are to be removed to provide for the intended stress range category.

4.17.2.2 CJP T and Corner Joints Made from One Side. Welds for attaching backing may be made inside or outside the joint groove. Backing for joints subject to cyclic transverse tension (fatigue) loading shall be removed and the back side of the joint finished consistent with face weld. Any unacceptable discontinuity discovered or caused by removal shall be repaired to the acceptance criteria of this code.

4.17.2.3 CJP Butt Joints. Welds for attaching backing may be inside or outside the groove unless restricted in the stress category description. Tack welds located outside the joint groove shall terminate not closer than 1/2 in [12 mm] from the edge of the connected part. Backing may remain in place or be removed unless restricted in the stress category used in design.

4.17.2.4 Longitudinal Groove Welds and Corner Joints. Steel backing, if used, shall be continuous for the full length of the joint. Welds for attaching backing may be inside or outside the groove (see 7.9.1.2).

4.17.3 Contouring Weld at Corner and T-Joints. In transverse corner and T-joints subject to tension or tension due to bending, a single pass contouring fillet weld, not less than 1/4 in [6 mm] in size shall be added at reentrant corners.

4.17.4 Flame Cut Edges. Flame cut edges need not be dressed provided they meet the roughness provisions of 7.14.8.3.

4.17.5 Transversely Loaded Butt Joints. For transversely loaded butt joints, weld tabs shall be used to provide for cascading the weld termination outside the finished joint. End dams shall not be used. Weld tabs shall be removed and the end of the weld finished flush with the edge of the member.

4.17.6 Fillet Weld Terminations. In addition to the requirements of 4.9.3.3 the following applies to weld terminations subject to cyclic (fatigue) loading. For connections and details with cyclic forces on outstanding elements of a frequency and magnitude that would tend to cause progressive failure initiating at a point of maximum stress at the end of the weld, fillet welds shall be returned around the side or end for a distance not less than two times the nominal weld size (see Figure 4.12).

4.18 Prohibited Joints and Welds

4.18.1 One-Sided Groove Welds. Groove welds, made from one side only without backing or made with backing, other than steel, that has not been qualified in conformance with Clause 6 shall be prohibited except that these prohibitions for groove welds made from one side shall not apply to the following:

(1) Secondary or nonstress carrying members.

(2) Corner joints parallel to the direction of calculated stress between components of built-up members

4.18.2 Flat Position Groove Welds. Bevel-groove and J-groove welds in butt joints welded in the flat position shall be prohibited where V-groove or U-groove joints are practicable.

4.18.3 Fillet Welds Less than 3/16 in [5 mm]. Fillet welds less than 3/16 in [5 mm] shall be prohibited.

4.18.4 T- and Corner CJP Welds with Backing Left in Place. T- and corner CJP welds subject to cyclic transverse tension stress with the backing bar left in place shall be prohibited.

4.19 Inspection

Fatigue categories B and C require that the Engineer ensure that CJP groove welds subject to cyclic transverse applied stress into the tensile range be inspected using radiographic testing (RT) or ultrasonic testing (UT).

Table 4.1
Effective Size of Flare-Groove Welds Filled Flush (see 4.4.1.4)

Welding Process	Flare-Bevel-Groove	Flare-V-Groove
SMAW and FCAW-S	5/16 R	5/8 R
GMAW[a] and FCAW-G	5/8 R	3/4 R
SAW	5/16 R	1/2 R

[a] Except GMAW-S

Note: R = radius of outside surface

Table 4.2
Z Loss Dimension (Nontubular) (See 4.4.3.3)

Dihedral Angle Ψ	Position of Welding—V or OH			Position of Welding—H or F		
	Process	Z(in)	Z(mm)	Process	Z(in)	Z(mm)
$60° > \Psi \geq 45°$	SMAW	1/8	3	SMAW	1/8	3
	FCAW-S	1/8	3	FCAW-S	0	0
	FCAW-G	1/8	3	FCAW-G	0	0
	GMAW	N/A	N/A	GMAW	0	0
$45° > \Psi \geq 30°$	SMAW	1/4	6	SMAW	1/4	6
	FCAW-S	1/4	6	FCAW-S	1/8	3
	FCAW-G	3/8	10	FCAW-G	1/4	6
	GMAW	N/A	N/A	GMAW	1/4	6

Table 4.3
Allowable Stresses (see 4.6.4 and 4.16.1)

Type of Applied Stress	Allowable Stress	Required Filler Metal Strength Level
CJP Groove Welds		
Tension normal to the effective area[a]	Same as base metal	Matching filler metal shall be used[b]
Compression normal to effective area	Same as base metal	Filler metal with a strength level equal to or one classification (10 ksi [70 MPa]) less than matching filler metal may be used.
Tension or compression parallel to axis of the weld[c]	Not a welded joint design consideration	Filler metal with a strength level equal to or less than matching filler metal may be used
Shear on effective area	0.30 × classification tensile strength of filler metal except shear on the base metal shall not exceed 0.40 × yield strength of the base metal	
PJP Groove Welds		
Tension normal to the effective area	0.30 × classification tensile strength of filler metal	Filler metal with a strength level equal to or less than matching filler metal may be used
Compression normal to effective area of weld in joints designed to bear	0.90 × classification tensile strength of filler metal, but not more than 0.90 × yield strength of the connected base metal	
Compression normal to effective area of weld in joints not designed to bear	0.75 × classification tensile strength of filler metal	
Tension or compression parallel to axis of the weld[c]	Not a welded joint design consideration	
Shear parallel to axis of effective area	0.30 × classification tensile strength of filler metal except shear on the base metal shall not exceed 0.40 × yield strength of the base metal	
Fillet Welds		
Shear on effective area or weld	0.30 × classification tensile strength of filler metal except that the base metal net section shear area stress shall not exceed 0.40 × yield strength of the base metal[d, e]	Filler metal with a strength level equal to or less than matching filler metal may be used
Tension or compression parallel to axis of the weld[c]	Not a welded joint design consideration	
Plug and Slot Welds		
Shear parallel to the faying surface on the effective area[f]	0.30 × classification tensile strength of filler metal	Filler metal with a strength level equal to or less than matching filler metal may be used

[a] For definitions of effective areas, see 4.4.
[b] For matching filler metal to base metal strength for code approved steels, see Table 5.3, Table 5.4, and Table 6.9.
[c] Fillet welds and groove welds joining components of built-up members are allowed to be designed without regard to the tension and compression stresses in the connected components parallel to the weld axis although the area of the weld normal to the weld axis may be included in the cross-sectional area of the member.
[d] The limitation on stress in the base metal to 0.40 × yield point of base metal does not apply to stress on the diagrammatic weld leg; however, a check shall be made to assure that the strength of the connection is not limited by the thickness of the base metal on the net area around the connection, particularly in the case of a pair of fillet welds on opposite sides of a plate element.
[e] Alternatively, see 4.6.4.2, 4.6.4.3, and 4.6.4.4. Footnote d (above) applies.
[f] The strength of the connection shall also be limited by the tear-out load capacity of the thinner base metal on the perimeter area around the connection.

Table 4.4
Equivalent Strength Coefficients for Obliquely Loaded Fillet Welds (see 4.6.4.4)

Load Angle for the Element Being Analyzed	Load Angle for Weld Element with Lowest Deformation Capability						
Θ	C (90)	C (75)	C (60)	C (45)	C (30)	C (15)	C (0)
0	0.825	0.849	0.876	0.909	0.948	0.994	1
15	1.02	1.04	1.05	1.07	1.06	0.883	
30	1.16	1.17	1.18	1.17	1.10		
45	1.29	1.30	1.29	1.26			
60	1.40	1.40	1.39				
75	1.48	1.47					
90	1.50						

Note: The weld element with the lowest deformation capability will be the element with the greatest load angle. Linear interpolation between adjacent load angles is permitted.

Table 4.5
Fatigue Stress Design Parameters (see 4.14.1)

Description	Stress Category	Constant C_f	Threshold F_{TH} ksi [MPa]	Potential Crack Initiation Point	Illustrative Examples
	**Section 1—Plain Material Away from Any Welding**
1.1 Base metal, except non-coated weathering steel, with as-rolled or cleaned surfaces. Flame cut edges with surface roughness value of 1000 μin [25 μm] or less, but without reentrant corners.	A	250×10^8	24 [165]	Away from all welds or structural connections	1.1/1.2 (A), (B)
1.2 Non-coated weathering steel base metal with as-rolled or cleaned surfaces. Flame cut edges with surface roughness value of 1000 μin [25 μm] or less, but without reentrant corners.	B	120×10^8	16 [110]	Away from all welds or structural connections	
1.3 Member with reentrant corners at copes, cuts, block-outs or other geometrical discontinuities, except weld access holes.				At any external edge or at hole perimeter	1.3 (A), (B), (C)
R ≥ 1 in [25 mm] with radius formed by predrilling, subpunching and reaming or thermally cut and ground to a bright metal surface	C	44×10^8	10 [69]		
R ≥ 3/8 in [10 mm] and the radius need not be ground to a bright metal surface	E'	3.9×10^8	2.6 [18]		

(Continued)

Table 4.5 (Continued)
Fatigue Stress Design Parameters (see 4.14.1)

Description	Stress Category	Constant C_f	Threshold F_{TH} ksi [MPa]	Potential Crack Initiation Point	Illustrative Examples
Section 1—Plain Material Away from Any Welding (Cont'd)					
1.4 Rolled cross sections with weld access holes made to the requirements of 4.17.4 and 7.16.1.				1.4	
Access hole R ≥ 1 in [25 mm] with radius formed by predrilling, subpunching and reaming or thermally cut and ground to a bright metal surface	C	44×10^8	10 [69]	At reentrant corner of weld access hole	
Access hole R ≥ 3/8 in [10 mm] and the radius need not be ground to a bright metal surface	E'	3.9×10^8	2.6 [18]		
1.5 Members with drilled or reamed holes.				1.5 In net section originating at side of the hole	
Holes containing pre-tensioned bolts	C	44×10^8	10 [69]		
Open holes without bolts	D	22×10^8	7 [48]		
Section 2—Connected Material in Mechanically Fastened Joints—Not Used[a]					

(Continued)

Table 4.5 (Continued)
Fatigue Stress Design Parameters (see 4.14.1)

Description	Stress Category	Constant C_f	Threshold F_{TH} ksi [MPa]	Potential Crack Initiation Point	Illustrative Examples
Section 3—Welded Joints Joining Components of Built-Up Members					
3.1 Base metal and weld metal in members without attachments built-up of plates or shapes connected by continuous longitudinal CJP groove welds, backgouged and welded from second side, or by continuous fillet welds.	B	120×10^8	16 [110]	From surface or internal discontinuities in weld	3.1
3.2 Base metal and weld metal in members without attachments built up of plates or shapes, connected by continuous longitudinal CJP groove welds with left-in place continuous steel backing, or by continuous PJP groove welds.	B'	61×10^8	12 [83]	From surface or internal discontinuities in weld	3.2

(Continued)

Table 4.5 (Continued)
Fatigue Stress Design Parameters (see 4.14.1)

Section 3—Welded Joints Joining Components of Built-Up Members (Cont'd)

Description	Stress Category	Constant C_f	Threshold F_{TH} ksi [MPa]	Potential Crack Initiation Point
3.3 Base metal at the ends of longitudinal welds that terminate at weld access holes in connected built-up members, as well as weld toes of fillet welds that wrap around ends of weld access holes.				3.3
Access hole R ≥ 1 in [25 mm] with radius formed by predrilling, subpunching and reaming or thermally cut and ground to a bright metal surface	D	22×10^8	7 [48]	From the weld termination into the web or flange
Access hole R ≥ 3/8 in [10 mm] and the radius need not be ground to a bright metal surface	E'	3.9×10^8	2.6 [18]	
3.4 Base metal at ends of longitudinal intermittent fillet weld segments.	E	11×10^8	4.5 [31]	In connected material at start and stop locations of any weld
3.5 Base metal at ends of partial length welded cover plates narrower than the flange having square or tapered ends, with or without welds across the ends.				3.5
Flange thickness ≤ 0.8 in [20 mm]	E	11×10^8	4.5 [31]	In flange at toe of end weld (if present) or in flange at termination of longitudinal weld
Flange thickness > 0.8 in [20 mm]	E'	3.9×10^8	2.6 [18]	

(Continued)

Table 4.5 (Continued)
Fatigue Stress Design Parameters (see 4.14.1)

Description	Stress Category	Constant C_f	Threshold F_{TH} ksi [MPa]	Potential Crack Initiation Point	Illustrative Examples
Section 3—Welded Joints Joining Components of Built-Up Members (Cont'd)					
3.6 Base metal at ends of partial length welded cover plates or other attachments wider than the flange with welds across the ends.				In flange at toe of end weld or in flange at termination of longitudinal weld or in edge of flange	3.6
Flange thickness ≤ 0.8 in [20 mm]	E	11×10^8	4.5 [31]		
Flange thickness > 0.8 in [20 mm]	E'	3.9×10^8	2.6 [18]		
3.7 Base metal at ends of partial length welded cover plates wider than the flange without welds across the ends.				In edge of flange at end of cover plate weld	3.7
Flange thickness ≤ 0.8 in [20 mm]	E'	3.9×10^8	2.6 [18]		
Flange thickness > 0.8 in [20 mm] is not permitted					
Section 4—Longitudinal Fillet Welded End Connections					
4.1 Base metal at junction of axially loaded members with longitudinally welded end connections. Welds are on each side of the axis of the member to balance weld stresses.				Initiating from end of any weld termination extending into the base metal	4.1
t ≤ 0.5 in [12 mm]	E	11×10^8	4.5 [31]		
t > 0.5 in [12 mm]	E'	3.9×10^8	2.6 [18]		

(Continued)

Table 4.5 (Continued)
Fatigue Stress Design Parameters (see 4.14.1)

Description	Stress Category	Constant C_f	Threshold F_{TH} ksi [MPa]	Potential Crack Initiation Point	Illustrative Examples
Section 5—Welded Joints Transverse to Direction of Stress					
5.1 Weld metal and base metal in or adjacent to CJP groove welded splices in plate, rolled shapes or built-up cross sections with no change in cross section with welds ground essentially parallel to the direction of stress and inspected in accordance with 4.19.	B	120×10^8	16 [110]	From internal discontinuities in weld metal or along fusion boundary	5.1
5.2 Weld metal and base metal in or adjacent to CJP groove welded splices with welds ground essentially parallel to the direction of stress at transitions in thickness or width made on a slope no greater than 1:2-1/2 and inspected in accordance with 4.19. $F_y < 90$ ksi [620 MPa] $F_y \geq 90$ ksi [620 MPa]	B B'	120×10^8 61×10^8	16 [110] 12 [83]	From internal discontinuities in weld metal or along fusion boundary or at start of transition when $F_y \geq 90$ ksi [620 MPa]	5.2
5.3 Base metal and weld metal in or adjacent to CJP groove welded splices with welds ground essentially parallel to the direction of stress at transitions in width made on a radius of not less than 24 in [600 mm] with the point of tangency at the end of the groove weld and inspected in accordance with 4.19.	B	120×10^8	16 [110]	From internal discontinuities in weld metal or along the fusion boundary	5.3

(Continued)

Table 4.5 (Continued)
Fatigue Stress Design Parameters (see 4.14.1)

Description	Stress Category	Constant C_f	Threshold F_{TH} ksi [MPa]	Potential Crack Initiation Point	Illustrative Examples
Section 5—Welded Joints Transverse to Direction of Stress (Cont'd)					
5.4 Weld metal and base metal in or adjacent to CJP groove welds in T- or corner joints or splices, without transitions in thickness or with transition in thickness having slopes no greater than 1:2-1/2, when weld reinforcement is not removed and inspected in accordance with 4.19.	C	44×10^8	10 [69]	From weld extending into base metal or along weld metal	5.4

(Continued)

Table 4.5 (Continued)
Fatigue Stress Design Parameters (see 4.14.1)

Description	Stress Category	Constant C_f	Threshold F_{TH} ksi [MPa]	Potential Crack Initiation Point	Illustrative Examples
Section 5—Welded Joints Transverse to Direction of Stress (Cont'd)					
5.5 Base metal and weld metal in or adjacent to transverse CJP groove welded butt splices with backing left in place. Tack welds inside groove Tack welds outside the groove and not closer than 1/2 in [12 mm] to edge of base metal	D E	22×10^8 11×10^8	7 [48] 4.5 [31]	From the toe of the groove weld or the toe of the weld attaching backing when applicable	5.5 (A) (B) (C) (D)

(Continued)

Table 4.5 (Continued)
Fatigue Stress Design Parameters (see 4.14.1)

Description	Stress Category	Constant C_f	Threshold F_{TH} ksi [MPa]	Potential Crack Initiation Point	Illustrative Examples
Section 5—Welded Joints Transverse to Direction of Stress (Cont'd)					5.6
5.6 Base metal and weld metal at transverse end connections of tension-loaded plate elements using PJP groove welds in butt. T- or corner joints, with reinforcing or contouring fillets. F_{SR} shall be the smaller of the toe crack or root crack stress range.					
Crack initiating from weld toe:	C	44×10^8	10 [69]	Initiating from weld toe extending into base metal	
Crack initiating from weld root:	C'	Formula (4)	None	Initiating at weld root extending into and through weld	

(Continued)

Table 4.5 (Continued)
Fatigue Stress Design Parameters (see 4.14.1)

Section 5—Welded Joints Transverse to Direction of Stress (Cont'd)

Description	Stress Category	Constant C_f	Threshold F_{TH} ksi [MPa]	Potential Crack Initiation Point	Illustrative Examples
5.7 Base metal and weld metal at transverse end connections of tension-loaded plate elements using a pair of fillet welds on opposite sides of the plate. F_{SR} shall be the smaller of the toe crack or root crack allowable stress range.				5.7	
Crack initiating from weld toe:	C	44×10^8	10 [69]	Initiating from weld toe extending into base metal	
Crack initiating from weld root:	C"	Formula (4)	None	Initiating at weld root extending into and through weld	
5.8 Base metal of tension loaded plate elements and on built-up shapes and rolled beam webs or flanges at toe of transverse fillet welds, adjacent to welded transverse stiffeners.	C	44×10^8	10 [69]	From geometrical discontinuity at toe of fillet extending into base metal	5.8

(Continued)

Table 4.5 (Continued)
Fatigue Stress Design Parameters (see 4.14.1)

Section 6—Base Metal at Welded Transverse Member Connections

Description	Stress Category	Constant C_f	Threshold F_{TH} ksi [MPa]	Potential Crack Initiation Point	Illustrative Examples
6.1 Base metal of equal or unequal thickness at details attached by CJP groove welds subject to longitudinal loading only when the detail embodies a transition radius, R, with the weld termination ground smooth and inspected in accordance with 4.19.				6.1 Near point of tangency of radius at edge of member	
R ≥ 24 in [600 mm]	B	120×10^8	16 [110]		
6 in ≤ R < 24 in [150 mm ≤ R < 600 mm]	C	44×10^8	10 [69]		
2 in ≤ R < 6 in [50 mm ≤ R < 150 mm]	D	22×10^8	7 [48]		
R < 2 in [50 mm]	E	11×10^8	4.5 [31]		

(Continued)

Table 4.5 (Continued)
Fatigue Stress Design Parameters (see 4.14.1)

Description	Stress Category	Constant C_f	Threshold F_{TH} ksi [MPa]	Potential Crack Initiation Point	Illustrative Examples
Section 6—Base Metal at Welded Transverse Member Connections (Cont'd)					
6.2 Base metal at details of equal thickness attached by CJP groove welds subject to transverse loading with or without longitudinal loading when the detail embodies a transition radius, R, with the weld termination ground smooth and inspected in accordance with 4.19.					6.2
6.2(a) When weld reinforcement is removed:					
$R \geq 24$ in [600 mm]	B	120×10^8	16 [110]	Near points of tangency of radius or in the weld or at fusion boundary of member or attachment	
6 in $\leq R < 24$ in [150 mm $\leq$ R < 600 mm]	C	44×10^8	10 [69]		
2 in $\leq R < 6$ in [50 mm $\leq$ R < 150 mm]	D	22×10^8	7 [48]		
$R < 2$ in [50 mm]	E	11×10^8	4.5 [31]		
6.2(b) When weld reinforcement not removed:					
$R \geq 6$ in [150 mm]	C	44×10^8	10 [69]	At toe of the weld either along edge of member or the attachment	
2 in $\leq R < 6$ in [50 mm $\leq$ R < 150 mm]	D	22×10^8	7 [48]		
$R < 2$ in [50 mm]	E	11×10^8	4.5 [31]		

(Continued)

Table 4.5 (Continued)
Fatigue Stress Design Parameters (see 4.14.1)

Description	Stress Category	Constant C_f	Threshold F_{TH} ksi [MPa]	Potential Crack Initiation Point	Illustrative Examples
Section 6—Base Metal at Welded Transverse Member Connections (Cont'd)					6.3
6.3 Base metal at details of unequal thickness attached by CJP groove welds, subject to transverse loading with or without longitudinal loading, when the detail embodies a transition radius, R, with the weld termination ground smooth and inspected in accordance with 4.19.					
6.3(a) When weld reinforcement is removed:					
R > 2 in [50 mm]	D	22×10^8	7 [48]	At toe of weld along edge of thinner material	
R ≤ 2 in [50 mm]	E	11×10^8	4.5 [31]	In weld termination in small radius	
6.3(b) When weld reinforcement is not removed:					
Any radius	E	11×10^8	4.5 [31]	At toe of weld along edge of thinner material	

(Continued)

Table 4.5 (Continued)
Fatigue Stress Design Parameters (see 4.14.1)

Description	Stress Category	Constant C_f	Threshold F_{TH} ksi [MPa]	Potential Crack Initiation Point	Illustrative Examples
Section 6 — Base Metal at Welded Transverse Member Connections (Cont'd)					
6.4 Base metal of equal or unequal thickness, subject to longitudinal stress at transverse members, with or without transverse stress, attached by fillet or PJP groove welds parallel to direction of stress when the detail embodies a transition radius, R, with weld termination ground smooth.				Initiating in base metal at the weld termination or at the toe of the weld, extending into the base metal	6.4
R ≥ 2 in [50 mm]	D	22×10^8	7 [48]		
R ≤ 2 in [50 mm]	E	11×10^8	4.5 [31]		

(Continued)

Table 4.5 (Continued)
Fatigue Stress Design Parameters (see 4.14.1)

Section 6—Base Metal at Welded Transverse Member Connections (Cont'd)

Description	Stress Category	Constant C_f	Threshold F_{TH} ksi [MPa]	Potential Crack Initiation Point	Illustrative Examples
7.1 Base metal subject to longitudinal loading at details with welds parallel or transverse to the direction of stress with or without transverse load on the detail, where the detail embodies no transition radius, and with detail length, a, in direction of stress and thickness of the attachment, b:				Initiating in base metal at the weld termination or at the toe of the weld, extending into the base metal	7.1
a < 2 in [50 mm] for any thickness, b	C	44×10^8	10 [69]		
2 in [50 mm] ≤ a ≤ lesser of 12b or 4 in [100 mm]	D	22×10^8	7 [48]		
a > lesser of 12b or 4 in [100 mm] when b ≤ 0.8 in [20 mm]	E	11×10^8	4.5 [31]		
a > 4 in [100 mm] when b > 0.8 in [20 mm]	E'	3.9×10^8	2.6 [18]		
7.2 Base metal subject to longitudinal stress at details attached by fillet or PJP groove welds, with or without transverse load on detail, when the detail embodies a transition radius, R, with weld termination ground smooth.				Initiating in base metal at the weld termination, extending into the base metal	7.2
R > 2 in [50 mm]	D	22×10^8	7 [48]		
R ≤ 2 in [50 mm]	E	11×10^8	4.5 [31]		

(Continued)

Table 4.5 (Continued)
Fatigue Stress Design Parameters (see 4.14.1)

Description	Stress Category	Constant C_f	Threshold F_{TH} ksi [MPa]	Potential Crack Initiation Point	Illustrative Examples
Section 6—Base Metal at Welded Transverse Member Connections (Cont'd)					
8.1 Base at steel headed stud anchors attached by fillet weld or automatic stud welding.	C	44×10^8	10 [69]	At toe of weld in base metal	8.1
8.2 Shear on throat of any fillet weld, continuous or intermittent, longitudinal or transverse.	F	150×10^{10} Formula (3)	8 [55]	Initiating at the root of the fillet weld, extending into the weld	8.2

(Continued)

Table 4.5 (Continued)
Fatigue Stress Design Parameters (see 4.14.1)

Section 6—Base Metal at Welded Transverse Member Connections (Cont'd)

Description	Stress Category	Constant C_f	Threshold F_{TH} ksi [MPa]	Potential Crack Initiation Point	Illustrative Examples
8.3 Base metal at plug or slot welds.	E	11×10^8	4.5 [31]	Initiating in the base metal at the end of the plug or slot weld, extending into the base metal	8.3
8.4 Shear on plug or slot welds.	F	150×10^{10} Formula (3)	8 [55]	Initiating in the weld at the faying surface, extending into the weld	8.4

8.5 Description 8.5 deals only with mechanically fastened details not pertinent to D1.1.

[a] AWS D1.1/D1.1M deals only with welded details. To maintain consistency and to facilitate cross referencing with other governing specifications, Section 2—Connected Material in Mechanically Fastened Joints, and Description 8.5 are not used in this table.

[b] "Attachment" as used herein is defined as any steel detail welded to a member, which causes a deviation in the stress flow in the member and thus reduces the fatigue resistance. The reduction is due to the presence of the attachment, not due to the loading on the attachment.

Source: Text adapted and illustrations reprinted, with permission, from American Institute of Steel Construction, Inc., 2015. *Specification for Structural Steel Buildings*, Illinois: American Institute of Steel Construction, *Test and Figures from Table A-31*.

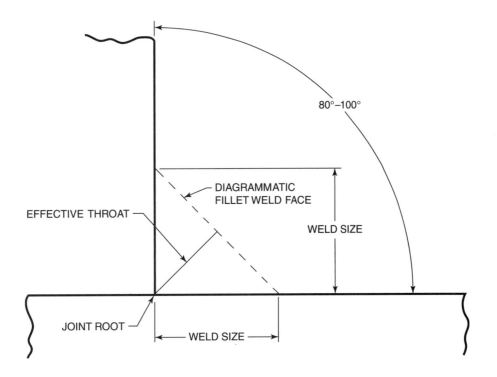

Figure 4.1—Fillet Weld (see 4.4.2.6)

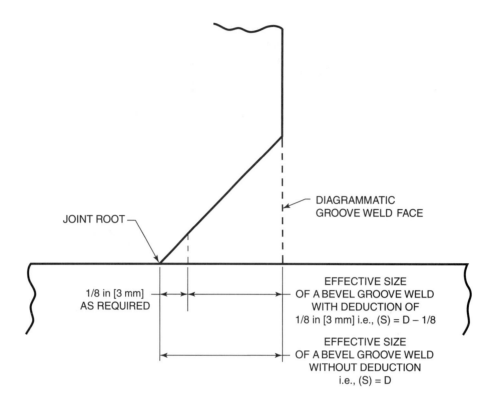

Figure 4.2—Unreinforced Bevel Groove Weld (see 4.4.2.7)

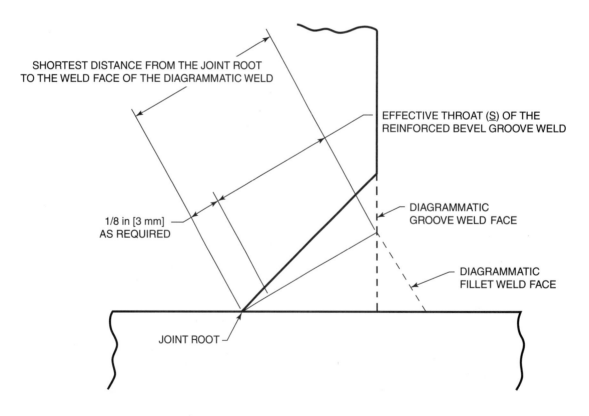

Figure 4.3—Bevel Groove Weld with Reinforcing Fillet Weld (see 4.4.2.7)

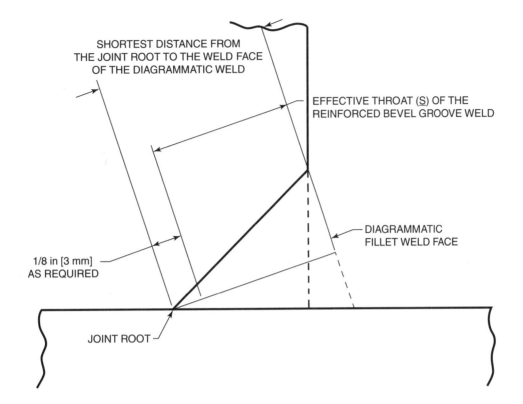

Figure 4.4—Bevel Groove Weld with Reinforcing Fillet Weld (see 4.4.2.7)

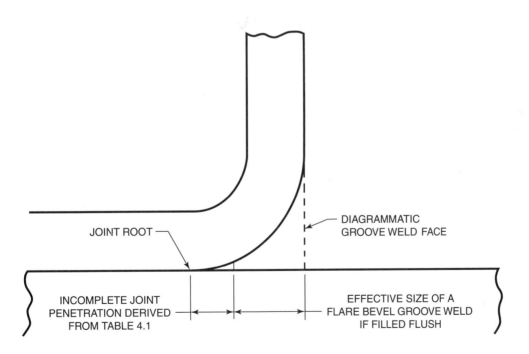

Figure 4.5—Unreinforced Flare Bevel Groove Weld (see 4.4.2.7)

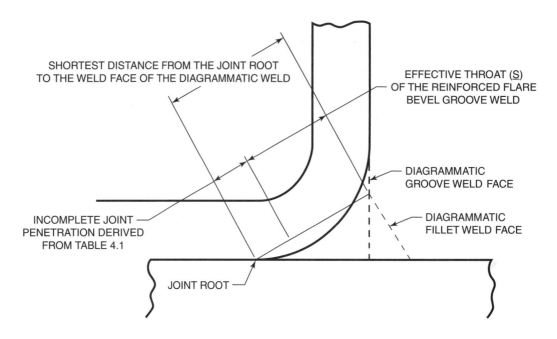

Figure 4.6—Flare Bevel Groove Weld with Reinforcing Fillet Weld (see 4.4.2.7)

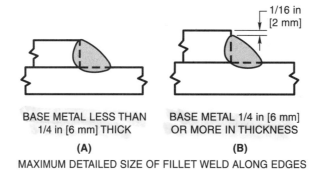

Figure 4.7—Maximum Fillet Weld Size Along Edges in Lap Joints (see 4.4.2.9)

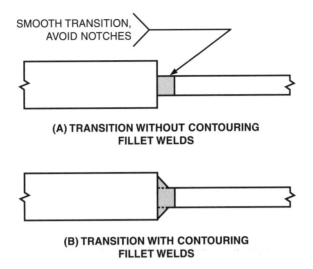

Figure 4.8—Transition of Thickness (Statically Loaded Nontubular) (see 4.7.5 and 4.8.1)

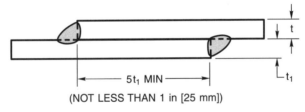

Note: t = thicker member, t_1 = thinner member.

Figure 4.9—Transversely Loaded Fillet Welds (see 4.9.1.1 and 4.9.1.2)

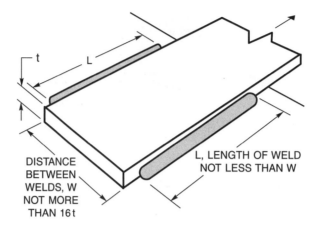

Figure 4.10—Minimum Length of Longitudinal Fillet Welds at End of Plate or Flat Bar Members (see 4.9.2)

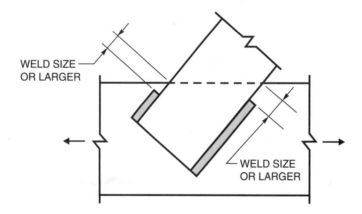

Figure 4.11—Termination of Welds Near Edges Subject to Tension (see 4.9.3.2)

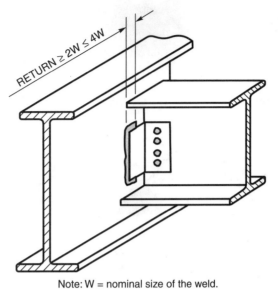

Note: W = nominal size of the weld.

Figure 4.12—End Return at Flexible Connections (see 4.9.3.3 and 4.17.6)

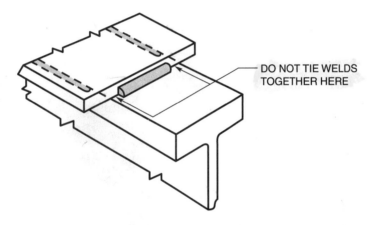

Figure 4.13—Fillet Welds on Opposite Sides of a Common Plane (see 4.9.3.5)

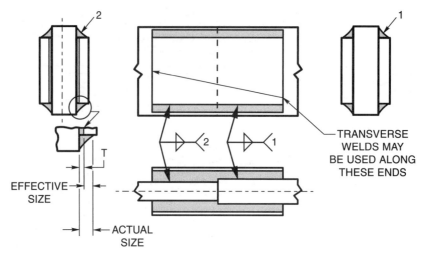

Figure 4.14—Thin Filler Plates in Splice Joint (see 4.11.1)

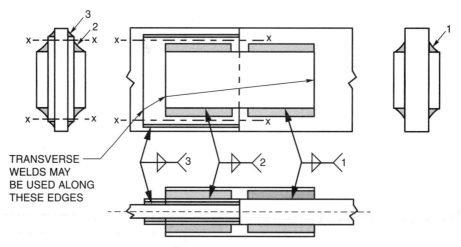

Figure 4.15—Thick Filler Plates in Splice Joint (see 4.11.2)

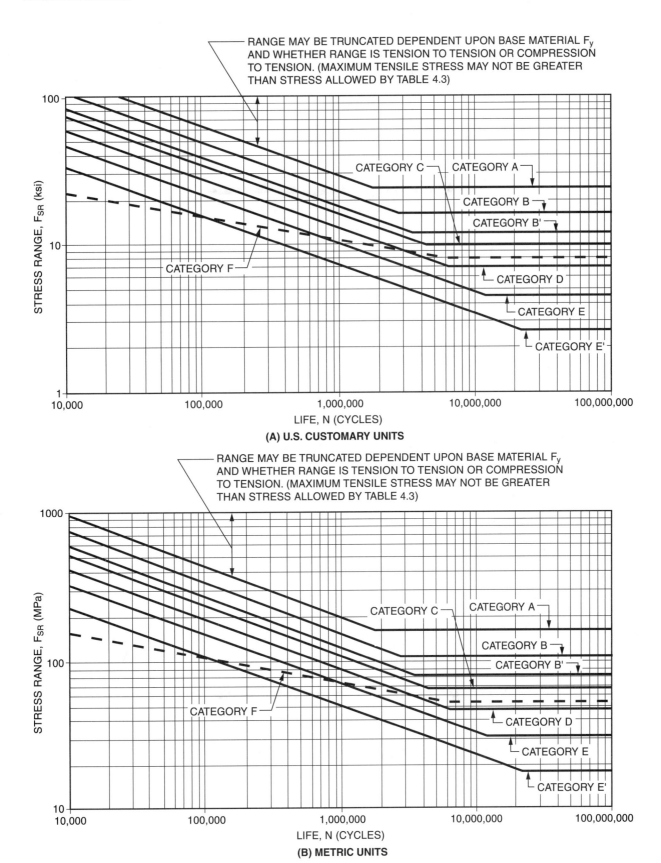

Figure 4.16—Allowable Stress Range for Cyclically Applied Load (Fatigue) in Nontubular Connections (Graphical Plot of Table 4.5) (see 4.16.2)

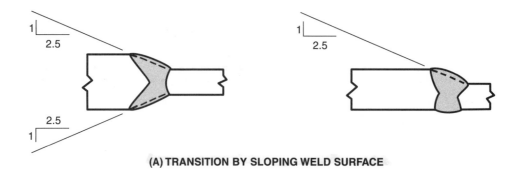

(A) TRANSITION BY SLOPING WELD SURFACE

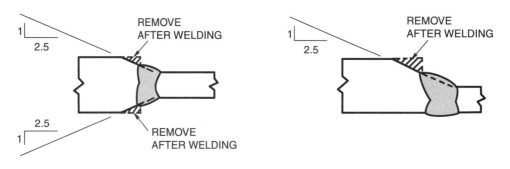

(B) TRANSITION BY SLOPING WELD SURFACE AND CHAMFERING

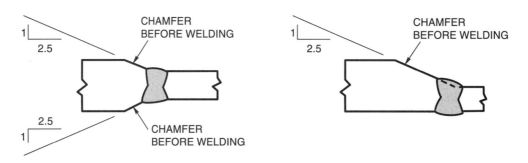

(C) TRANSITION BY CHAMFERING THICKER PART

CENTERLINE ALIGNMENT
(PARTICULARLY APPLICABLE
TO WEB PLATES)

OFFSET ALIGNMENT
(PARTICULARLY APPLICABLE
TO FLANGE PLATES)

Figure 4.17—Transition of Butt Joints in Parts of Unequal Thickness (Cyclically Loaded Nontubular) (see 4.17.1.1)

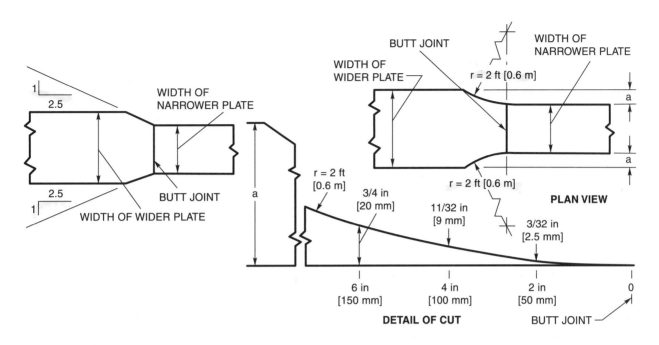

Figure 4.18—Transition of Width (Cyclically Loaded Nontubular) (see 4.17.1.2)

5. Prequalification of WPSs

5.1 Scope

This clause contains requirements for prequalified Welding Procedure Specifications (WPSs). These WPSs are exempt from the requirements for testing required for qualification of WPSs per Clause 6.

It is divided into eight parts as follows:

Part A – WPS Development

Part B – Base Metal

Part C – Weld Joints

Part D – Welding Processes

Part E – Filler Metals and Shielding Gases

Part F – Preheat and Interpass Temperature Requirements

Part G – WPS Requirements

Part H – Post Weld Heat Treatment

Prequalification of WPSs (Welding Procedure Specifications) shall be defined as exempt from the WPS qualification testing required in Clause 6. All prequalified WPSs shall be written. In order for a WPS to be prequalified, conformance with all of the applicable requirements of Clause 5 shall be required. WPSs that do not conform to the requirements of Clause 5 may be qualified by tests in conformance with Clause 6. For convenience, Annex K lists provisions to be included in a prequalified WPS, and which should be addressed in the fabricator's or Contractor's welding program.

Welders, welding operators and tack welders that use prequalified WPSs shall be qualified in conformance with Clause 6, Part C or Clause 10, Part D for tubulars.

Part A
WPS Development

5.2 General WPS Requirements

All the requirements of Table 5.1 shall be met for prequalified WPSs.

5.2.1 All prequalified WPSs to be used shall be prepared by the manufacturer, fabricator, or Contractor as written prequalified WPSs. The written WPS may follow any convenient format (see Annex J for examples). The welding parameters set forth in Table 5.2 shall be specified on the written WPS, and for variables with limits, within the range shown. Changes to the essential variables beyond those permitted by Table 5.2 shall require a new or revised prequalified WPS, or shall require that the WPS be qualified by test in accordance with Clause 6.

5.2.2 Combination of WPSs. A combination of qualified and prequalified WPSs may be used without qualification of the combination, provided the limitation of essential variables applicable to each process is observed.

Part B
Base Metal

5.3 Base Metal

Only base metals listed in Table 5.3 may be used in prequalified WPSs. (For the qualification of listed base metals, and for base metals not listed in Table 5.3, see 6.2.1.)

5.3.1 Engineer's Approval for Auxiliary Attachments. As an alternative to WPS qualification, unlisted materials for auxiliary attachments which fall within the chemical composition range of a steel listed in Table 5.3 may be used in a prequalified WPS when approved by the Engineer. The filler metal of Table 5.4 and minimum preheat shall be in conformance with 5.7, based upon the similar material strength and chemical composition.

Part C
Weld Joints

5.4 Weld Joints

5.4.1 Complete Joint Penetration (CJP) Groove Weld Details.

CJP groove welds which may be used without performing the WPS qualification test described in Clause 6 shall be as detailed in Figure 5.1 and are subject to the limitations described in 5.4.1.1.

5.4.1.1 Joint Dimensions. Dimensions of groove welds specified in 5.4.1 may vary on design or detail drawings within the limits or tolerances shown in the "As Detailed" column in Figure 5.1.

5.4.1.2 Backing. Prequalified CJP groove welds made from one side only, except as allowed for tubular structures, shall have steel backing.

5.4.1.3 Prequalified CJP groove welds detailed without steel backing or spacers may use backing other than steel as listed in 7.9.3 when the following conditions are met:

(1) The backing is removed after welding, and,

(2) The back side of the weld is backgouged to sound metal and backwelded.

Welding procedures for joints welded with backing other than steel in which the weld is to be left in the as welded condition without backgouging and welding from the other side are not prequalified.

5.4.1.4 Double-Sided Groove Preparation. J and U-grooves and the other side of partially welded double-V and double-bevel grooves may be prepared before or after assembly. After backgouging, the other side of partially welded double-V or double-bevel joints should resemble a prequalified U or J-joint configuration at the joint root.

5.4.1.5 GMAW/FCAW in SMAW Joints. Groove preparations detailed for prequalified SMAW joints may be used for prequalified GMAW or FCAW.

5.4.1.6 Corner Joint Preparation. For corner joints, the outside groove preparation may be in either or both members, provided the basic groove configuration is not changed and adequate edge distance is maintained to support the welding operations without excessive melting.

5.4.1.7 Root Openings. Joint root openings may vary as noted in 5.4.1.1 and 5.4.1.8. However, for automatic or mechanized welding using FCAW, GMAW, and SAW processes, the maximum root opening variation (minimum to maximum opening as fit-up) may not exceed 1/8 in [3 mm]. Variations greater than 1/8 in [3 mm] shall be locally corrected prior to automatic or mechanized welding.

5.4.1.8 Fit-up Tolerances. Fit-up tolerances of Figure 5.1 may be applied to the dimensions shown on the detail drawing.

5.4.1.9 J and U-Groove Preparation. J and U-grooves may be prepared before or after assembly.

5.4.2 Partial Joint Penetration (PJP) Groove Weld Details. WPSs for PJP welds which may be used without performing the WPS qualification test described in Clause 6 shall be "As Detailed" in Figure 5.2 and are subject to the limitations described in 5.4.2.

5.4.2.1 Definition. Except as provided in 10.10.2 and Figure 5.1 (B-L1-S), groove welds without steel backing, welded from one side, and groove welds welded from both sides, but without backgouging, are considered PJP groove welds.

5.4.2.2 Joint Dimensions. Dimensions of groove welds specified in 5.4.2 may vary on design or detail drawings within the limits of tolerances shown in the "As Detailed" column in Figure 5.2.

5.4.2.3 Prequalified Weld Size. The weld size (S) of a prequalified PJP groove shall be as shown in Figure 5.2 for the particular welding process, joint designation, groove angle, and welding position proposed for use in welding fabrication.

(1) The minimum weld size of PJP single or double V, bevel-, J-, and U-groove welds, types 2 through 9, shall be as shown in Table 5.5. The base metal thickness shall be sufficient to incorporate the requirements of the joint details selected, conforming to the variances outlined in 5.4.2.2 and the requirements of Table 5.5.

(2) The maximum base metal thickness shall not be limited.

(3) For PJP square groove weld B-P1 and flare-bevel groove welds BTC-P10 and B-P11, minimum weld sizes shall be calculated from Figure 5.2.

(4) Shop or working drawings shall specify the design groove depth "D" applicable for the weld size "(S)" required per 5.4.2.2. (Note that this requirement shall not apply to the B-P1, BTC-P10, and B-P11 details.)

5.4.2.4 GMAW/FCAW in SMAW Joints. Groove preparations detailed for prequalified SMAW joints may be used for prequalified GMAW or FCAW.

5.4.2.5 Corner Joint Preparation. For corner joints, the outside groove preparation may be in either or both members, provided the basic groove configuration is not changed and adequate edge distance is maintained to support the welding operations without excessive melting.

5.4.2.6 Root Openings. Joint root openings may vary as noted in 5.4.2.2 and 5.4.2.7. However, for automatic or mechanized welding using GMAW, FCAW, and SAW processes, the maximum root opening variation (minimum to maximum opening as fit-up) may not exceed 1/8 in [3 mm]. Variations greater than 1/8 in [3 mm] shall be locally corrected prior to automatic or mechanized welding.

5.4.2.7 Fit-up Tolerances. Fit-up tolerances of Figure 5.2 may be applied to the dimensions shown on the detail drawing. However, the use of fit-up tolerances does not exempt the user from meeting the minimum weld size requirements of 5.4.2.3(1).

5.4.2.8 J and U-Groove Preparation. J and U-grooves may be prepared before or after assembly.

5.4.3 Fillet Weld Details. See Table 7.7 for minimum fillet weld sizes and Figure 5.3 for prequalified fillet weld joint details.

5.4.3.1 Details (Nontubular). See Figures 4.1, 4.7, and 4.9 for the limitations for prequalified fillet welds.

5.4.3.2 Skewed T-Joints. Skewed T-joints shall be in conformance with Figures 5.4 and 5.5.

5.4.3.3 Dihedral Angle Limitations. The obtuse side of skewed T-joints with dihedral angles greater than 100° shall be prepared as shown in Figure 5.4, Detail C, to allow placement of a weld of the required size. The amount of machining or grinding, etc., of Figure 5.4, Detail C, should not be more than that required to achieve the required weld size (W).

5.4.3.4 Minimum Weld Size for Skewed T-Joints. For skewed T-joints, the minimum weld size for Details A, B, and C in Figure 5.4 shall be in conformance with Table 7.7.

5.4.4 Plug and Slot Weld Requirements The details of plug and slot welds made by the SMAW, GMAW (except GMAW-S), or FCAW processes are described in 4.4.5.1, 4.4.5.2, 4.4.5.4, and 4.10 and they may be used without performing the WPS qualification described in Clause 6, provided the technique provisions of 7.24 are met.

Part D
Welding Processes

5.5 Welding Processes

5.5.1 Prequalified Processes. SMAW, SAW, GMAW (except GMAW-S), and FCAW WPSs which conform to all of the provisions of Clause 5 shall be deemed as prequalified and are therefore approved for use without performing WPS qualification tests for the process. For WPS prequalification, conformance with all of the applicable provisions of Clause 5 shall be required (see 5.1).

5.5.2 Code Approved Processes. ESW, EGW, GTAW, and GMAW-S welding may be used, provided the WPSs are qualified in conformance with the requirements of Clause 6. Note that the essential variable limitations in Table 6.5 for GMAW shall also apply to GMAW-S.

5.5.3 Other Welding Processes. Other welding processes not covered by 5.5.1 and 5.5.2 may be used, provided the WPSs are qualified by applicable tests as described in Clause 6.

5.5.4 GMAW and FCAW Power Sources. GMAW and FCAW that is done with prequalified WPSs shall be performed using constant voltage (CV) power supplies.

5.5.5 Common Requirements for GMAW Root Pass followed by Parallel Electrode and Multiple Electrode SAW. Welds may also be made in the root of groove or fillet welds using GMAW, followed by parallel or multiple electrode submerged arcs, provided that:

(1) The GMAW conforms to the requirements of this section, and

(2) The spacing between the GMAW arc and the following SAW arc does not exceed 15 in [380 mm].

Part E
Filler Metals and Shielding Gases

5.6 Filler Metal and Shielding Gas

Only filler metals listed in Table 5.4 may be used in prequalified WPSs. For the qualification of listed filler metals, and for filler metals not listed in Table 5.4, see 6.2.1.

5.6.1 Filler Metal Matching. The base metal/filler metal strength relationships below shall be used in conjunction with Tables 5.3 and 5.4 to determine whether matching or undermatching filler metals are required.

Relationship	Base Metal(s)	Filler Metal Strength Relationship Required
Matching	Any steel to itself or any steel to another in the same group	Any filler metal listed in the same group
	Any steel in one group to any steel in another	Any filler metal listed for either strength group. SMAW electrodes shall be the low-hydrogen classification
Undermatching	Any steel to any steel in any group	Any filler metal listed in a strength group below the lower strength group. SMAW electrodes shall be the low-hydrogen classification

Note: See Table 4.3 or 10.2 to determine the filler metal strength requirements to match or undermatch base metal strength.

5.6.2 Weathering Steel Requirements. For exposed, bare, unpainted applications of weathering steel requiring weld metal with atmospheric corrosion resistance and coloring characteristics similar to that of the base metal, the electrode or electrode-flux combination shall conform to Table 5.6.

The exceptions to this requirement are as follows:

5.6.2.1 Single-Pass Groove Welds. Groove welds made with a single pass or a single pass on each side may be made using any of the filler metals for Group II base metals in Table 5.4.

5.6.2.2 Single-Pass Fillet Welds. Single-pass fillet welds up to the following sizes may be made using any of the filler metals for Group II base metals listed in Table 5.4:

SMAW	1/4 in [6 mm]
SAW	5/16 in [8 mm]
GMAW/FCAW	5/16 in [8 mm]

5.6.3 Shielding Gas. Shielding gases for GMAW and FCAW-G shall conform to AWS A5.32M/A5.32 and one of the following:

(1) The shielding gas shall be that used for electrode classification per the applicable AWS A5 specifications, AWS A5.18/A5.18M, A5.20/A5.20M, A5.28/A5.28M, or A5.29/A5.29M.

(2) For AWS A5.36/A5.36M fixed classifications of carbon steel gas shielded FCAW and GMAW, and low-alloy steel FCAW qualified with M21 shielding gas, they shall be limited to the mixed shielding gas requirements of AWS A5.18/A5.18M, A5.20/A5.20M, or A5.29/A5.29M, M21-ArC-20/25(SG-AC-20-25).

(3) The classification shielding gas for all AWS A5.36/A5.36M open classifications shall be limited to the shielding gas designator used for classification(s) and not the range of the shielding gas classification.

(4) For electrodes classified per AWS A5.18/A5.18M, Table 5.7 provides acceptable shielding gases or gas mixtures for production welding.

(5) The electrode/shielding gas combination shall have been tested in accordance with the applicable A5 filler metal specification. The tests shall demonstrate that the electrode/shielding gas combination is capable of meeting all the mechanical and chemical property and NDT requirements for the electrode classification. For FCAW-G and GMAW composite (metal cored) electrodes, tests shall be performed for each electrode manufacturer's brand and trade name to be used. Testing shall be performed by the filler metal manufacturer or gas producer. For FCAW-G, the filler metal shall have been classified by the filler metal manufacturer as an "–M" (i.e. mixed gas) product.

Part F
Preheat and Interpass Temperature Requirements

5.7 Preheat and Interpass Temperature Requirements

5.7.1 Minimum Preheat and Interpass Temperature Requirements. Table 5.8 shall be used to determine the minimum preheat and interpass temperatures for steels listed in the code.

5.7.2 Base Metal/Thickness Combination. The minimum preheat or interpass temperature applied to a joint composed of base metals with different minimum preheats from Table 5.8 (based on category and thickness) shall be the highest of these minimum preheats.

5.7.3 Alternate SAW Preheat and Interpass Temperatures. Preheat and interpass temperatures for parallel or multiple electrode SAW shall be selected in conformance with Table 5.8. For single-pass groove or fillet welds, for combinations of metals being welded and the heat input involved, and with the approval of the Engineer, preheat and interpass temperatures may be established which are sufficient to reduce the hardness in the HAZs of the base metal to less than 225 Vickers hardness number for steel having a minimum specified tensile strength not exceeding 60 ksi [415 MPa], and 280 Vickers hardness number for steel having a minimum specified tensile strength greater than 60 ksi [415 MPa], but not exceeding 70 ksi [485 MPa].

The Vickers hardness number shall be determined in conformance with ASTM E92. If another method of hardness is to be used, the equivalent hardness number shall be determined from ASTM E140, and testing shall be performed according to the applicable ASTM specification.

5.7.3.1 Hardness Requirements. Hardness determination of the HAZ shall be made on the following:

(1) Initial macroetch cross sections of a sample test specimen.

(2) The surface of the member during the progress of the work. The surface shall be ground prior to hardness testing:

(a) The frequency of such HAZ testing shall be at least one test area per weldment of the thicker metal involved in a joint of each 50 ft [15 m] of groove welds or pair of fillet welds.

(b) These hardness determinations may be discontinued after the procedure has been established and the discontinuation is approved by the Engineer.

Part G
WPS Requirements

5.8 WPS Requirements

5.8.1 General WPS Requirements. All the requirements of Table 5.1 shall be met for prequalified WPSs.

5.8.1.1 Vertical-Up Welding Requirements. The progression for all passes in vertical position welding shall be upward, with the following exceptions:

(1) Undercut may be repaired vertically downwards when preheat is in conformance with Table 5.8, but not lower than 70°F [20°C].

(2) When tubular products are welded, the progression of vertical welding may be upwards or downwards, but only in the direction(s) for which the welder is qualified.

5.8.2 Limitation of Variables. Table 5.1 lists WPS variable requirements and limitations by position, weld type, and process.

5.8.2.1 Width/Depth Pass Limitation. Neither the depth nor the maximum width in the cross section of weld metal deposited in each weld pass shall exceed the width at the surface of the weld pass (see Figure 5.6).

Part H
Postweld Heat Treatment

5.9 Postweld Heat Treatment

Postweld heat treatment (PWHT) shall be prequalified provided that it shall be approved by the Engineer and the following conditions shall be met:

(1) The specified minimum yield strength of the base metal shall not exceed 50 ksi [345 MPa].

(2) The base metal shall not be manufactured by quenching and tempering (Q&T), quenching and self-tempering (Q&ST), thermo-mechanical controlled processing (TMCP) or where cold working is used to achieve higher mechanical properties (e.g., certain grades of ASTM A500 tubing).

(3) There shall be no requirements for notch toughness testing of the base metal, HAZ, or weld metal.

(4) There shall be data available demonstrating that the weld metal shall have adequate strength and ductility in the PWHT condition (e.g., as can be found in the relevant AWS A5.X filler metal specification and classification or from the filler metal manufacturer).

(5) PWHT shall be conducted in conformance with 7.8.

Table 5.1
Prequalified WPS Requirements[a] (see 5.2)

Variable	Position	Weld Type	SMAW	SAW[b] Single	SAW[b] Parallel	SAW[b] Multiple	GMAW/ FCAW[c]
Maximum Electrode Diameter	Flat	Fillet[d]	5/16 in [8.0 mm]	1/4 in [6.4 mm]	1/4 in [6.4 mm]	1/4 in [6.4 mm]	1/8 in [3.2 mm]
		Groove[d]	1/4 in [6.4 mm]				
		Root pass	3/16 in [4.8 mm]				
	Horizontal	Fillet	1/4 in [6.4 mm]	1/4 in [6.4 mm]	1/4 in [6.4 mm]	1/4 in [6.4 mm]	1/8 in [3.2 mm]
		Groove	3/16 in [4.8 mm]	Requires WPS Qualification Test			
	Vertical	All	3/16 in [4.8 mm][e]				3/32 in [2.4 mm]
	Overhead	All	3/16 in [4.8 mm][e]				5/64 in [2.0 mm]
Maximum Current	All	Fillet	Within the range of operation recommended by the filler metal manufacturer	1000 A	1200A	Unlimited	Within the range of operation recommended by the filler metal manufacturer
	All	Groove weld root pass with opening		600A	700A		
		Groove weld root pass without opening			900A		
		Groove weld fill passes			1200A		
		Groove weld cap pass		Unlimited			
Maximum Root Pass Thickness[b]	Flat	All	3/8 in [10 mm]	Unlimited			3/8 in [10 mm]
	Horizontal		5/16 in [8 mm]				5/16 in [8 mm]
	Vertical		1/2 in [12 mm]				1/2 in [12 mm]
	Overhead		5/16 in [8 mm]				5/16 in [8 mm]
Maximum Fill Pass Thickness	All	All	3/16 in [5 mm]	1/4 in [6 mm]	Unlimited		1/4 in [6 mm]
Maximum Single Pass Fillet Weld Size[f]	Flat	Fillet	3/8 in [10 mm]	Unlimited			1/2 in [12 mm]
	Horizontal		5/16 in [8 mm]	5/16 in [8 mm]	5/16 in [8 mm]	1/2 in [12 mm]	3/8 in [10 mm]
	Vertical		1/2 in [12 mm]				1/2 in [12 mm]
	Overhead		5/16 in [8 mm]				5/16 in [8 mm]
Maximum Single Pass Layer Width	All (for GMAW/ FCAW) F & H (for SAW)	Root opening > 1/2 in [12 mm]		Split layers	Laterally displaced electrodes or split layer	Split layers	Split layers
		Any layer of width, w		Split layers if w > 5/8 in [16 mm]	Split layers with tandem electrodes if w > 5/8 in [16 mm]	Split layers if w > 1 in [25 mm]	(Footnote g)

[a] Shaded area indicates nonapplicability.
[b] See 5.8.2.1 for width–to–depth limitations.
[c] GMAW-S shall not be prequalified.
[d] Except root passes.
[e] 5/32 in [4.0 mm] for EXX14 and low–hydrogen electrodes.
[f] See 5.6.2 for requirements for welding unpainted and exposed ASTM A588.
[g] In the F, H, or OH positions for nontubulars, split layers when the layer width, w > 5/8 in [16 mm]. In the vertical position for nontubulars or the flat, horizontal, vertical, and overhead positions for tubulars, split layers when the width, w > 1 in [25 mm].

Table 5.2
Essential Variables for Prequalified WPSs (see 5.2.1)

Variables that must be included in a Prequalified WPS

(1) Welding Process(es)[a]
(2) Welding Position(s)
(3) Base Metal Group Number(s) (See Table 5.3)
(4) Base Metal Preheat Category(s) (See Table 5.4)
(5) Filler Metal Classification (SMAW, GMAW, FCAW)
(6) Filler Metal/Flux Classification (SAW)
(7) Nominal Electrode Diameter
(8) Number of Electrodes (SAW)
(9) Electrode Spacing and Orientation (SAW)
(10) Amperage (SAW, FCAW, GMAW)
(11) Voltage (SAW, FCAW, GMAW)

(12) Mode of Transfer (GMAW)
(13) Type of Current (AC or DC)
(14) Current Polarity (AC, DCEN, DCEP)
(15) Wire Feed Speed (SAW, FCAW, GMAW)
(16) Travel Speed
(17) Nominal Shielding Gas Composition (FCAW-G, GMAW)
(18) Shielding Gas Flow Rate (FCAW-G, GMAW)
(19) Weld Type (Fillet, CJP, PJP, Plug, Slot)
(20) Groove Weld Details
(21) Postweld Heat Treatment

Variable Tolerances for Prequalified WPSs

Variable	Allowable tolerance
(22) Amperage (SAW, FCAW, GMAW)	+ or – 10%
(23) Voltage (SAW, FCAW, GMAW)	+ or – 15%
(24) Wire Feed Speed (if not amperage controlled) (SAW, FCAW, GMAW)	+ or – 10%
(25) Travel Speed (SAW, FCAW, GMAW)	+ or – 25%
(26) Shielding Gas Flow Rate (FCAW-G, GMAW)	> 50%, if increase or > 25%, if decrease
(27) Change in the longitudinal spacing of arcs (SAW)	> 10% or 1/8 in [3 mm], whichever is greater
(28) Lateral spacing of arcs (SAW)	> 10% or 1/8 in [3 mm], whichever is greater
(29) The angular orientation of parallel electrodes (SAW)	+ or – 10%
(30) The angle parallel to the direction of travel of the electrode for mechanized or automatic (SAW)	+ or – 10%
(31) The angle of electrode normal to the direction of travel for mechanized or automatic (SAW)	+ or – 10%

[a] A separate WPS shall be required when this variable is changed.

Table 5.3
Approved Base Metals for Prequalified WPSs (see 5.3)

GROUP	Steel Specification	Steel Specification Requirements	Minimum Yield Point/Strength ksi	Minimum Yield Point/Strength MPa	Tensile Range ksi	Tensile Range MPa
I	ASTM A36	≤ 3/4 in [20 mm]	36	250	58–80	400–550
	ASTM A53	Grade B	35	240	60 min.	415 min.
	ASTM A106	Grade B	35	240	60 min.	415 min.
	ASTM A131	Grades A, B, D, E	34	235	58–75	400–520
	ASTM A139	Grade B	35	240	60 min.	415 min.
	ASTM A381	Grade Y35	35	240	60 min.	415 min.
	ASTM A500 (Square/Rectangular)	Grade A	39	270	45 min.	310 min.
		Grade B	46	315	58 min.	400 min.
		Grade C	50	345	62 min.	425 min.
	ASTM A500 (Round)	Grade A	33	230	45 min.	310 min.
		Grade B	42	290	58 min.	400 min.
		Grade C	46	315	62 min.	425 min.
	ASTM A501	Grade A	36	250	58 min.	400 min.
	ASTM A516	Grade 55	30	205	55–75	380–515
		Grade 60	32	220	60–80	415–550
	ASTM A524	Grade I	35	240	60–85	415–586
		Grade II	30	205	55–80	380–550
	ASTM A573	Grade 65	35	240	65–77	450–530
		Grade 58	32	220	58–71	400–490
	ASTM A709	Grade 36 ≤ 3/4 in [20 mm]	36	250	58–80	400–550
	ASTM A1008 SS	Grade 30	30	205	45 min.	310 min.
		Grade 33 Type 1	33	230	48 min.	330 min.
		Grade 40 Type 1	40	275	52 min.	360 min.
	ASTM A1011 SS	Grade 30	30	205	49 min.	340 min.
		Grade 33	33	230	52 min.	360 min.
		Grade 36 Type 1	36	250	53 min.	365 min.
		Grade 40	40	275	55 min.	380 min.
		Grade 45 Type 1	45	310	60 min.	410 min.
	ASTM A1018 SS	Grade 30	30	205	49 min.	340 min.
		Grade 33	33	230	52 min.	360 min.
		Grade 36	36	250	53 min.	365 min.
		Grade 40	40	275	55 min.	380 min.
	API 5L	Grade B	35	241	60 min.	414 min.
		Grade X42	42	290	60 min.	414 min.
	ABS	Grades A, B, D, E[b]	34	235	58–75	400–520

(Continued)

Table 5.3 (Continued)
Approved Base Metals for Prequalified WPSs (see 5.3)

GROUP	Steel Specification	Steel Specification Requirements	Minimum Yield Point/Strength		Tensile Range	
			ksi	MPa	ksi	MPa
II	ASTM A36	All thicknesses	36	250	58–80	400–550
	ASTM A131	Grades AH32, DH32, EH32	46	315	64–85	440–590
		Grades AH36, DH36, EH36	51	355	71–90	490–620
	ASTM A501	Grade B	50	345	70 min.	485 min.
	ASTM A516	Grade 65	35	240	65–85	450–585
		Grade 70	38	260	70–90	485–620
	ASTM A529	Grade 50	50	345	65–100	450–690
		Grade 55	55	380	70–100	485–690
	ASTM A537 Class 1	≤ 2 ½ in [≤ 65 mm]	50	345	70–90	485–620
		> 2 ½ [65 mm] ≤ 4 in [100 mm]	45	310	65–85	450–585
	ASTM A572	Grade 42	42	290	60 min.	415 min.
		Grade 50	50	345	65 min.	450 min.
		Grade 55	55	380	70 min.	485 min.
	ASTM A588[b]	≤ 4 in [100 mm]	50	345	70 min.	485 min.
		> 4 in [100 mm] ≤ 5 in [125 mm]	46	315	67 min.	460 min.
		> 5 in [125 mm] ≤ 8 in [200 mm]	42	290	63 min.	435 min.
		All Shapes	50	345	70 min.	485 min.
	ASTM A595	Grade A	55	380	65 min.	450 min.
		Grades B and C	60	410	70 min.	480 min.
	ASTM A606[b]	Cold-rolled Grade 45	45	310	65 min.	450 min.
		Hot-rolled Grade 50 (AR)	50	340	70 min.	480 min.
		Hot-rolled Grade 50 (A or N)	45	310	65 min.	450 min.
	ASTM A618	Grades Ib, II wall ≤ ¾ in [19 mm]	50	345	70 min.	485 min.
		Grades Ib, II wall > ¾ in ≤ 1-1/2 in [> 19 mm ≤ 38 mm]	46	315	67 min.	460 min.
		Grade III	50	345	65 min.	450 min.
	ASTM A633	Grade A	42	290	63–83	430–570
		Grades C, D ≥ 2-1/2 in [65 mm]	50	345	70–90	485–620
	ASTM A709	Grade 36 Plates ≤ 4 in [100 mm]	36	250	58–80	400–550
		Grade 36 Shapes ≤ 3 in [75 mm]	36	250	58–80	400–550
		Grade 36 Shapes > 3 in [75 mm]	36	250	58 min.	400 min.
		Grade 50	50	345	65 min.	450 min.
		Grade 50W[b]	50	345	70 min.	485 min.
		Grade 50S	50–65	345–450	65 min.	450 min.
		Grade HPS 50W[b]	50	345	70 min.	485 min.
	ASTM A710	Grade A, Class 2 > 2 in ≤ 4 in [> 50 mm ≤ 100 mm]	55	380	65 min.	450 min.
		> 4 in [100 mm]	50	345	60 min.	415 min.
	ASTM A847		50	345	70 min.	485 min.
	ASTM A913	Grade 50	50	345	65 min.	450 min.
	ASTM A992		50–65	345–450	65 min.	450 min.
	ASTM A1008 HSLAS	Grade 45 Class 1	45	310	60 min.	410 min.
		Grade 45 Class 2	45	310	55 min.	380 min.
		Grade 50 Class 1	50	340	65 min.	450 min.
		Grade 50 Class 2	50	340	60 min.	410 min.
		Grade 55 Class 1	55	380	70 min.	480 min.
		Grade 55 Class 2	55	380	65 min.	450 min.

(Continued)

Table 5.3 (Continued)
Approved Base Metals for Prequalified WPSs (see 5.3)

GROUP	Steel Specification Requirements		Minimum Yield Point/Strength		Tensile Range	
	Steel Specification		ksi	MPa	ksi	MPa
II (cont'd)	ASTM A1008 HSLAS-F	Grade 50	50	340	60 min.	410 min.
	ASTM A1011 HSLAS	Grade 45 Class 1	45	310	60 min.	410 min.
		Grade 45 Class 2	45	310	55 min.	380 min.
		Grade 50 Class 1	50	340	65 min.	450 min.
		Grade 50 Class 2	50	340	60 min.	410 min.
		Grade 55 Class 1	55	380	70 min.	480 min.
		Grade 55 Class 2	55	380	65 min.	450 min.
	ASTM A1011 HSLAS-F	Grade 50	50	340	60 min.	410 min.
	ASTM A1011 SS	Grade 50	50	340	65 min.	450 min.
		Grade 55	55	380	70 min.	480 min.
	ASTM A1018 HSLAS	Grade 45 Class 1	45	310	60 min.	410 min.
		Grade 45 Class 2	45	310	55 min.	380 min.
		Grade 50 Class 1	50	340	65 min.	450 min.
		Grade 50 Class 2	50	340	60 min.	410 min.
		Grade 55 Class 1	55	380	70 min.	480 min.
		Grade 55 Class 2	55	380	65 min.	450 min.
	ASTM A1018 HSLAS-F	Grade 50	50	340	60 min.	410 min.
	ASTM A1066	Grade 50	50	345	65 min.	450 min.
	ASTM A1085		50–70	345–485	65 min.	450 min.
	API 2H	Grade 42	42	289	62–82	427–565
		Grade 50	50	345	70–90	483–620
	API 2MT1	Grade 50	50	345	65–90	488–620
	API 2W	Grade 42	42–67	290–462	62 min.	427 min.
		Grade 50	50–75	345–517	65 min.	448 min.
		Grade 50T	50–80	345–552	70 min.	483 min.
	API 2Y	Grade 42	42–67	290–462	62 min.	427 min.
		Grade 50	50–75	345–517	65 min.	448 min.
		Grade 50T	50–80	345–552	70 min.	483 min.
	API 5L	Grade X52	52	359	66 min.	455 min.
	ABS	Grades AH32, DH32, EH32	46	315	64–85	440–590
		Grades AH36, DH36, EH36[b]	51	355	71–90	490–620
III	API 2W	Grade 60	60–90	414–621	75 min.	517 min.
	API 2Y	Grade 60	60–90	414–621	75 min.	517 min.
	ASTM A537 Class 2[b]	≤ 2 ½ in [65 mm]	60	415	80–100	550–690
		> 2 ½ in [65 mm] ≤ 4 in [100 mm]	55	380	75–95	515–655
		> 4 in [100 mm] ≤ 6 in [150 mm]	46	315	70–90	485–620
	ASTM A572	Grade 60	60	415	75 min.	520 min.
		Grade 65	65	450	80 min.	550 min.
	ASTM A633 Grade E[b]	≤ 4 in [100 mm]	60	415	80–100	550–690
		> 4 in [100 mm] ≤ 6 in [150 mm]	55	380	75–95	515–655
	ASTM A710	Grade A, Class 2 ≤ 1 in [20 mm]	65	450	72 min.	495 min.
		Grade A, Class 2 > 1 in ≤ 2 in [> 25 mm ≤ 50 mm]	60	415	72 min.	495 min.
		Grade A, Class 3 > 2 in ≤ 4 in [> 50 mm ≤ 100 mm]	65	450	75 min.	515 min.
		Grade A, Class 3 > 4 in [100 mm]	60	415	70 min.	485 min.
	ASTM A913[a]	Grade 60	60	415	75 min.	520 min.
		Grade 65	65	450	80 min.	550 min.

(Continued)

Table 5.3 (Continued)
Approved Base Metals for Prequalified WPSs (see 5.3)

GROUP	Steel Specification		Minimum Yield Point/Strength		Tensile Range	
			ksi	MPa	ksi	MPa
III (cont'd)	ASTM A1018 HSLAS	Grade 60 Class 2	60	410	70 min.	480 min.
		Grade 70 Class 2	70	480	80 min.	550 min.
	ASTM A1018 HSLAS-F	Grade 60 Class 2	60	410	70 min.	480 min.
		Grade 70 Class 2	70	480	80 min.	550 min.
	ASTM A1066	Grade 60	60	415	75 min.	520 min.
		Grade 65	65	450	80 min.	550 min.
IV	ASTM A709	Grade HPS70W	70	485	85–110	585–760
	ASTM A913[a]	Grade 70	70	485	90 min.	620 min.
	ASTM A1066	Grade 70	70	485	85 min.	585 min.

[a] The heat input limitations of 7.7 shall not apply to ASTM A913 Grades 60, 65, or 70.
[b] Special welding materials and WPS (e.g., E80XX-X low-alloy electrodes) may be required to match the notch toughness of base metal (for applications involving impact loading or low temperature), or for atmospheric corrosion and weathering characteristics (see 5.6.2).

Notes:
1. In joints involving base metals of different groups, either of the following filler metals may be used: (1) that which matches the higher strength base metal, or (2) that which matches the lower strength base metal and produces a low-hydrogen deposit.
2. Match API standard 2B (fabricated tubes) according to steel used.
3. When welds are to be postweld heat treated, the deposited weld metal shall not exceed 0.05% vanadium.
4. See Tables 4.3 and 10.2 for allowable stress requirements for matching filler metal.
5. Filler metal strength properties have been moved to nonmandatory Annex L.
6. AWS A5M (SI Units) electrodes of the same classification may be used in lieu of the AWS A5 (U.S. Customary Units) electrode classification.
7. Any of the electrode classifications for a particular Group (located on the right of Table 5.4) may be used to weld any of the base metals in that Group (located on the left of Table 5.3).

Table 5.4
Filler Metals for Matching Strength for Table 5.3, Groups I, II, III, and IV Metals—SMAW and SAW (see 5.6)

Base Metal Group	AWS Electrode Specification	SMAW		SAW	
		A5.1, Carbon Steel	A5.5[a], Low-Alloy Steel	A5.17, Carbon Steel	A5.23[c], Low-Alloy Steel
I	AWS Electrode Classification	E60XX E70XX	E70XX-X	F6XX-EXXX F6XX-ECXXX F7XX-EXXX F7XX-ECXXX	F7XX-EXXX-XX F7XX-ECXXX-XX
II	AWS Electrode Classification	E7015 E7016 E7018 E7028	E7015-X E7016-X E7018-X	F7XX-EXXX F7XX-ECXXX	F7XX-EXXX-XX F7XX-ECXXX-XX
III	AWS Electrode Classification	N/A	E8015-X E8016-X E8018-X	N/A	F8XX-EXXX-XX F8XX-ECXXX-XX
IV	AWS Electrode Classification	N/A	E9015-X E9016-X E9018-X E9018M	N/A	F9XX-EXXX-XX F9XX-ECXXX-XX

(Continued)

Table 5.4 (Continued)
Filler Metals for Matching Strength for Table 5.3, Group I Metals—FCAW and GMAW Metal Cored (see 5.6)

Base Metal Group	WELDING PROCESS(ES)				
	GMAW		FCAW		Carbon & Low-Alloy Steel GMAW and FCAW
	A5.18, Carbon Steel	A5.28[a], Low-Alloy Steel	A5.20, Carbon Steel	A5.29[a], Low-Alloy Steel	A5.36[c] Open Classification[d] See Note 8 for Annex M
AWS Electrode Specification					
AWS Electrode Classification	ER70S-X E70C-XC E70C-XM (Electrodes with the –GS suffix shall be excluded)	ER70S-XXX E70C-XXX	E7XT-X E7XT-XC E7XT-XM (Electrodes with the –2C, –2M, –3, –10, –13, –14, and –GS suffix shall be excluded and electrodes with the –11 suffix shall be excluded for thicknesses greater than 1/2 in [12 mm])	E6XTX-X E7XTX-X E6XTX-XC E6XTX-XM E7XTX-XC E7XTX-XM	**Carbon Steel GMAW and FCAW** A5.36, Fixed Classification[b] **FCAW Carbon Steel** E7XT-1C E7XT-1M E7XT-5C E7XT-5M E7XT-9C E7XT-9M E7XT-12C E7XT-12M E70T-4 E7XT-6 E7XT-7 E7XT-8 (Flux Cored Electrodes with the T1S, T3S, T10S, T14S, and –GS suffix shall be excluded and electrodes with the T11 suffix shall be excluded for thicknesses greater than 1/2 in [12 mm]) **GMAW-Metal Cored Carbon Steel** E70C-6M (Electrodes with the –GS suffix shall be excluded) (NOTE: A5.36 does not have fixed classifications for other carbon steel metal cored electrodes or for low-alloy steel flux cored or metal cored electrodes)	**FCAW Carbon Steel** E7XTX-XAX-CS1 E7XTX-XAX-CS2 E7XTX-XAX-CS3 (Flux Cored Electrodes with the T1S, T3S, T10S, T14S, and –GS suffix shall be excluded and electrodes with the T11 suffix shall be excluded for thicknesses greater than 1/2 in [12 mm]) **FCAW Low-Alloy Steel** E6XTX-XAX-XXX E7XTX-XAX-XXX **GMAW-Metal Cored Carbon Steel** E7XTX-XAX-CS1 E7XTX-XAX-CS2 (Electrodes with the –GS suffix shall be excluded) **GMAW-Metal Cored Low-Alloy Steel** E7XTX-XAX-XXX
I					

(Continued)

Table 5.4 (Continued)
Filler Metals for Matching Strength for Table 5.3, Group II Metals—FCAW and GMAW Metal Cored (see 5.6)

Base Metal Group		WELDING PROCESS(ES)			
		GMAW		FCAW	Carbon & Low-Alloy Steel GMAW and FCAW
	AWS Electrode Specification	A5.18, Carbon Steel	A5.28[a], Low-Alloy Steel	A5.20, Carbon Steel / A5.29[a], Low-Alloy Steel	A5.36, Open Classification[d] See Note 8 for Annex M / A5.36, Fixed Classification[b]
II	AWS Electrode Classification	ER70S-X E70C-XC E70C-XM (Electrodes with the –GS suffix shall be excluded)	ER70S-XXX E70C-XXX	**FCAW** E7XT-X E7XT-XC E7XT-XM (Electrodes with the –2C, –2M, –3, –10, –13, –14, and –GS suffix shall be excluded and electrodes with the –11 suffix shall be excluded for thicknesses greater than 1/2 in [12 mm]) **Low-Alloy Steel A5.29[a]** E7XTX-X E7XTX-XC E7XTX-XM	**FCAW Carbon Steel** E7XT-1C E7XT-1M E7XT-5C E7XT-5M E7XT-9C E7XT-9M E7XT-12C E7XT-12M E70T-4 E7XT-6 E7XT-7 E7XT-8 (Flux Cored Electrodes with the T1S, T3S, T10S, T14S, and –GS suffix shall be excluded and electrodes with the T11 suffix shall be excluded for thicknesses greater than 1/2 in [12 mm]) **GMAW-Metal Cored Carbon Steel** E70C-6M (Electrodes with the –GS suffix shall be excluded) (NOTE: A5.36 does not have fixed classifications for other carbon steel metal cored electrodes or for low-alloy steel flux cored or metal cored electrodes)	**FCAW Carbon Steel** E7XTX-XAX-CS1 E7XTX-XAX-CS2 E7XTX-XAX-CS3 (Flux Cored Electrodes with the T1S, T3S, T10S, T14S, and –GS suffix shall be excluded and electrodes with the T11 suffix shall be excluded for thicknesses greater than 1/2 in [12 mm]) **FCAW Low-Alloy Steel** E7XTX-AX-XXX E7XTX-XAX-XXX **GMAW-Metal Cored Carbon Steel** E7XTX-XAX-CS1 E7XTX-XAX-CS2 (Electrodes with the –GS suffix shall be excluded) **GMAW-Metal Cored Low-Alloy Steel** E7XTX-XAX-XXX

(Continued)

Table 5.4 (Continued)
Filler Metals for Matching Strength for Table 5.3, Group III and Group IV Metals—FCAW and GMAW Metal Cored (see 5.6)

Base Metal Group		WELDING PROCESS(ES)				
		GMAW		FCAW		Carbon & Low-Alloy Steel GMAW and FCAW
	AWS Electrode Specification	A5.18, Carbon Steel	A5.28[a], Low-Alloy Steel	A5.20, Carbon Steel	A5.29[a], Low-Alloy Steel	A5.36[c] Open Classification[d] See Note 8 for Annex M
				Carbon Steel GMAW and FCAW		
				A5.36, Fixed Classification[b]		
III	AWS Electrode Classification	N/A	ER80S-XXX E80C-XXX	N/A	E8XTX-X E8XTX-XC E8XTX-XM	FCAW Carbon Steel N/A
				FCAW Carbon Steel N/A		FCAW Low-Alloy Steel E8XTX-AX-XXX E8XTX-XAX-XXX
				GMAW-Metal Cored Carbon Steel N/A		GMAW-Metal Cored Carbon Steel N/A
						GMAW-Metal Cored Low-Alloy Steel E8XTX-XAX-XXX
IV	AWS Electrode Classification	N/A	ER90S-XXX E90C-XXX	N/A	E9XTX-X E9XTX-XC E8XTX-XM	FCAW Carbon Steel N/A
				FCAW Carbon Steel N/A		FCAW Low-Alloy Steel E9XTX-AX-XXX E9XTX-XAX-XXX
				GMAW-Metal Cored Carbon Steel N/A		GMAW-Metal Cored Carbon Steel N/A
						GMAW-Metal Cored Low-Alloy Steel E9XTX-XAX-XXX

[a] Filler metals of alloy group B3, B3L, B4, B4L, B5, B5L, B6, B6L, B7, B7L, B8, B8L, B9, E9015-C5L, E9015-D1, E9018-D1, E9018-D3, or any BXH grade in AWS A5.5, A5.23, A5.28, or A5.29 are not prequalified for use in the as-welded condition.
[b] The prequalified argon based shielding gases for carbon and low-alloy steel FCAW and carbon steel GMAW-Metal Core fixed classifications shall be M21-ArC-20-25(SG-AC-20/25), see 5.6.3(2).
[c] Filler metals of alloy group B3, B3L, B4, B4L, B5, B5L, B6, B6L, B7, B7L, B8, B8L, and B9 in AWS A5.36/A5.36M may be "PREQUALIFIED" if classified in the "AS-WELDED" condition.
[d] The prequalified shielding gas for open classification shall be limited to the specific shielding gas for the classification of the electrode and not the range of the shielding gas designator, see 5.6.3(3).

Notes:
1. In joints involving base metals of different groups, either of the following filler metals may be used: (1) that which matches the higher strength base metal, or (2) that which matches the lower strength base metal and produces a low-hydrogen deposit. Preheating shall be in conformance with the requirements applicable to the higher strength group.
2. Match API standard 2B (fabricated tubes) according to steel used.
3. When welds are to be stress-relieved, the deposited weld metal shall not exceed 0.05% vanadium except alloy group B9.
4. See Tables 4.3 and 4.6 for allowable stress requirements for matching filler metal.
5. Filler metal properties have been moved to nonmandatory Annex L.
6. AWS A5M (SI Units) electrodes of the same classification may be used in lieu of the AWS A5 (U.S. Customary Units) electrode classification.
7. Any of the electrode classifications for a particular Group in Table 5.4 may be used to weld any of the base metals in that Group in Table 5.3.
8. AWS A5.36/A5.36M open classifications are listed in Annex U.

Table 5.5
Minimum Prequalified PJP Groove Weld Size (S) (see 5.4.2.3(1))

Base Metal Thickness (T)[a]	Minimum Weld Size[b]	
in [mm]	in	mm
1/8 [3] to 3/16 [5] incl.	1/16	2
Over 3/16 [5] to 1/4 [6] incl.	1/8	3
Over 1/4 [6] to 1/2 [12] incl.	3/16	5
Over 1/2 [12] to 3/4 [20] incl.	1/4	6
Over 3/4 [20] to 1-1/2 [38] incl.	5/16	8
Over 1-1/2 [38] to 2-1/4 [57] incl.	3/8	10
Over 2-1/4 [57] to 6 [150] incl.	1/2	12
Over 6 [150]	5/8	16

[a] For nonlow-hydrogen processes without preheat calculated in conformance with 6.8.4, T equals the thickness of the thicker part joined; single pass welds shall be used. For low-hydrogen processes and nonlow-hydrogen processes established to prevent cracking in conformance with 6.8.4, T equals thickness of the thinner part; single pass requirement does not apply.
[b] Except that the weld size need not exceed the thickness of the thinner part joined.

Table 5.6 (see 5.6.2)
Filler Metal Requirements for Exposed Bare Applications of Weathering Steels

Process	AWS Filler Metal Specification	Approved Electrodes[a]
SMAW	A5.5/A5.5M	All electrodes that deposit weld metal meeting a B2L, C1, C1L, C2, C2L, C3, or WX analysis per A5.5/A5.5M.
SAW	A5.23/A5.23M	All electrode-flux combinations that deposit weld metal with a Ni1, Ni2, Ni3, Ni4, or WX analysis per A5.23/A5.23M.
FCAW	A5.29/A5.29M and A5.36/A5.36M	All electrodes that deposit weld metal with a B2L, K2, Ni1, Ni2, Ni3, Ni4, or WX analysis per A5.29/A5.29M or A5.36/A5.36M.
GMAW	A5.28/A5.28M and A5.36/A5.36M	All electrodes that meet filler metal composition requirements of B2L, G[a], Ni1, Ni2, Ni3, analysis per A5.28/A5.28M or A5.36/A5.36M.

[a] Deposited weld metal shall have a chemical composition the same as that for any one of the weld metals in this table.

Notes:
1. Filler metals shall meet requirements of Table 5.4 in addition to the compositional requirements listed above. The use of the same type of filler metal having next higher tensile strength as listed in AWS filler metal specification may be used.
2. Metal cored electrodes are designated as follows:
   SAW: Insert letter "C" between the letters "E" and "X," e.g., E7AX-ECXXX-Ni1.
   GMAW: Replace the letter "S" with the letter "C," and omit the letter "R," e.g., E80C-Ni1. AWS A5.36/A5.36M composite electrode designation is either a T15 or T16, e.g., E8XT15-XXX-Ni1, E8XT16-XXX-Ni1.

Table 5.7
Prequalified WPS Shielding Gas Options for GMAW Electrodes Conforming to AWS A5.18/A5.18M (see 5.6.3)

Electrode	Shielding Gas	Composition
ER70S-X (except ER70S-G) and E70C-X metal cored electrodes	Ar/CO_2 combinations	Ar 75–90%/CO_2 10–25%
	Ar/O_2 combinations	Ar 95–98%/O_2 2–5%
	100% CO_2	100% CO_2

Table 5.8
Prequalified Minimum Preheat and Interpass Temperature (see 5.7)

CATEGORY	Steel Specification		Welding Process	Thickness of Thickest Part at Point of Welding		Minimum Preheat and Interpass Temperature	
				in	mm	°F	°C
A	ASTM A36		SMAW with other than low-hydrogen electrodes	1/8 to 3/4 incl.	3 to 20 incl.	32[a]	0[a]
	ASTM A53	Grade B					
	ASTM A106	Grade B					
	ASTM A131	Grades A, B, D, E					
	ASTM A139	Grade B					
	ASTM A381	Grade Y35					
	ASTM A500	Grades A, B, C					
	ASTM A501	Grade A					
	ASTM A516	Grades 55, 60		Over 3/4 thru 1-1/2 incl.	Over 20 thru 38 incl.	150	65
	ASTM A524	Grades I, II					
	ASTM A573	Grades 58, 65					
	ASTM A709	Grade 36					
	ASTM A1008 SS	Grade 30		Over 1-1/2 thru 2-1/2 incl.	Over 38 thru 65 incl.	225	110
		Grade 33 Type 1					
		Grade 40 Type 1					
	ASTM A1011 SS	Grades 30, 33		Over 2-1/2	Over 65	300	150
		Grade 36 Type 1					
		Grade 40					
		Grade 45 Type 1					
	ASTM A1018 SS	Grades 30, 33, 36, 40					
	API 5L	Grades B, X42					
	ABS	Grades A, B, D, E					
B	ASTM A36		SMAW with low-hydrogen electrodes, SAW, GMAW, FCAW	1/8 to 3/4 incl.	3 to 20 incl.	32[a]	0[a]
	ASTM A53	Grade B					
	ASTM A106	Grade B					
	ASTM A131	Grades A, B, D, E AH 32, 36 DH 32, 36 EH 32, 36					
	ASTM A139	Grade B					
	ASTM A381	Grade Y35					
	ASTM A500	Grades A, B, C					
	ASTM A501	Grades A, B		Over 3/4 thru 1-1/2 incl.	Over 20 thru 38 incl.	50	10
	ASTM A516	Grades 55, 60, 65, 70					
	ASTM A524	Grades I, II					
	ASTM A529	Grades 50, 55		Over 1-1/2 thru 2-1/2 incl.	Over 38 thru 65 incl.	150	65
	ASTM A537	Classes 1, 2					
	ASTM A572	Grades 42, 50, 55					
	ASTM A573	Grades 58, 65					
	ASTM A588			Over 2-1/2	Over 65	225	110
	ASTM A595	Grades A, B, C					
	ASTM A606						
	ASTM A618	Grades Ib, II, III					
	ASTM A633	Grades A, C, D					
	ASTM A709	Grades 36, 50, 50S, 50W, HPS50W					

(Continued)

Table 5.8 (Continued)
Prequalified Minimum Preheat and Interpass Temperature (see 5.7)

CATEGORY	Steel Specification		Welding Process	Thickness of Thickest Part at Point of Welding		Minimum Preheat and Interpass Temperature	
				in	mm	°F	°C
B (cont'd)	ASTM A710	Grade A, Class 2 >2 in [50 mm]	SMAW with low-hydrogen electrodes, SAW, GMAW, FCAW	1/8 to 3/4 incl.	3 to 20 incl.	50	10
	ASTM A847						
	ASTM A913	Grade 50					
	ASTM A992						
	ASTM A1008 HSLAS	Grade 45 Class 1 Grade 45 Class 2 Grade 50 Class 1 Grade 50 Class 2 Grade 55 Class 1 Grade 55 Class 2					
	ASTM A1008 HSLAS-F	Grade 50					
	ASTM A1011 SS	Grades 50, 55					
	ASTM A1011 HSLAS	Grade 45 Class 1 Grade 45 Class 2 Grade 50 Class 1 Grade 50 Class 2 Grade 55 Class 1 Grade 55 Class 2		Over 3/4 thru 1-1/2 incl.	Over 20 thru 38 incl.	50	10
	ASTM A1011 HSLAS-F	Grade 50					
	ASTM A1018 HSLAS	Grade 45 Class 1 Grade 45 Class 2 Grade 50 Class 1 Grade 50 Class 2 Grade 55 Class 1 Grade 55 Class 2		Over 1-1/2 thru 2-1/2 incl.	Over 38 thru 65 incl.	225	110
				Over 2-1/2	Over 65	300	150
	ASTM A1018 HSLAS-F	Grade 50					
	ASTM A1018 SS	Grades 30, 33, 36, 40					
	ASTM A1066	Grade 50					
	ASTM A1085						
	API 5L	Grades B, X42					
	API Spec. 2H	Grades 42, 50					
	API 2MT1	Grade 50					
	API 2W	Grades 42, 50, 50T					
	API 2Y	Grades 42, 50, 50T					
	ABS	Grades A, B, D, E AH 32, 36 DH 32, 36 EH 32, 36					

(Continued)

Table 5.8 (Continued)
Prequalified Minimum Preheat and Interpass Temperature (see 5.7)

CATEGORY	Steel Specification		Welding Process	Thickness of Thickest Part at Point of Welding		Minimum Preheat and Interpass Temperature	
				in	mm	°F	°C
C	ASTM A572	Grades 60, 65	SMAW with low-hydrogen electrodes, SAW, GMAW, FCAW	1/8 to 3/4 incl.	3 to 20 incl	50	10
	ASTM A633	Grade E					
	ASTM A709[b]	Grade HPS70W					
	ASTM A710	Grade A, Class 2 ≤ 2 in [50 mm]					
	ASTM A710	Grade A, Class 3 > 2 in [50 mm]		Over 3/4 thru 1-1/2 incl.	Over 20 thru 38 incl.	150	65
	ASTM A913	Grades 60, 65, 70					
	ASTM A1018 HSLAS	Grade 60 Class 2					
		Grade 70 Class 2		Over 1-1/2 thru 2-1/2 incl.	Over 38 thru 65 incl.	225	110
	ASTM A1018 HSLAS-F	Grade 60 Class 2					
		Grade 70 Class 2					
	ASTM A1066	Grades 60, 65, 70		Over 2-1/2	Over 65	300	150
	API 2W	Grade 60					
	API 2Y	Grade 60					
	API 5L	Grade X52					
D	ASTM A710	Grade A (All classes)	SMAW, SAW, GMAW, and FCAW with electrodes or electrode-flux combinations capable of depositing weld metal with a maximum diffusible hydrogen content of 8 ml/100 g (H8), when tested according to AWS A4.3.	All thicknesses Over 1/8 in	All thicknesses Over [3 mm]	32[a]	0[a]
	ASTM A913	Grades 50, 60, 65					
E	ASTM A1066	Grades 50, 60, 65	SMAW, SAW, GMAW, and FCAW with electrodes or electrode-flux combinations capable of depositing weld metal with a maximum diffusible hydrogen content of 8 ml/100 g (H8), when tested according to AWS A4.3.	1/8 to 1 incl.	3 to 25 incl.	50	10
				Over 1	Over 25	120	50

[a] When the base metal temperature is below 32°F [0°C], the base metal shall be preheated to a minimum of 70°F [20°C] and the minimum interpass temperature shall be maintained during welding.

[b] For ASTM A709 Grade HPS70W, the maximum preheat and interpass temperatures shall not exceed 400°F [200°C] for thicknesses up to 1-1/2 in [40 mm], inclusive, and 450°F [230°C] for greater thicknesses.

Notes:
1. For modification of preheat requirements for SAW with parallel or multiple electrodes, see 5.7.3.
2. See 7.11 and 7.6 for ambient and base-metal temperature requirements.
3. ASTM A570 and ASTM A607 have been deleted.
4. When welding base metals from different Categories see 5.7.2.

Legend for Figures 5.1 and 5.2

Symbols for joint types
- B — butt joint
- C — corner joint
- T — T-joint
- BC — butt or corner joint
- TC — T- or corner joint
- BTC — butt, T-, or corner joint

Symbols for base metal thickness and penetration
- P — PJP
- L — limited thickness–CJP
- U — unlimited thickness–CJP

Symbol for weld types
- 1 — square-groove
- 2 — single-V-groove
- 3 — double-V-groove
- 4 — single-bevel-groove
- 5 — double-bevel-groove
- 6 — single-U-groove
- 7 — double-U-groove
- 8 — single-J-groove
- 9 — double-J-groove
- 10 — flare-bevel-groove
- 11 — flare-V-groove

Symbols for welding processes if not SMAW
- S — SAW
- G — GMAW
- F — FCAW

Welding processes
- SMAW — shielded metal arc welding
- GMAW — gas metal arc welding
- FCAW — flux cored metal arc welding
- SAW — submerged arc welding

Welding positions
- F — flat
- H — horizontal
- V — vertical
- OH — overhead

Dimensions
- R = Root Opening
- α, β = Groove Angles
- f = Root Face
- r = J- or U-groove Radius
- D, D_1, D_2 = PJP Groove Weld Depth of Groove
- S, S_1, S_2 = PJP Groove Weld Sizes corresponding to S, S_1, S_2, respectively

Joint Designation
The lower case letters, e.g., a, b, c, are used to differentiate between joints that would otherwise have the same joint designation.

Notes for Figures 5.1 and 5.2

[a] Not prequalified for GMAW-S or GTAW.
[b] Joint shall be welded from one side only.
[c] Cyclic load application places restrictions on the use of this detail for butt joints in the flat position (see 4.18.2).
[d] Backgouge root to sound metal before welding second side.
[e] SMAW detailed joints may be used for prequalified GMAW (except GMAW-S) and FCAW.
[f] Minimum weld size (S) as shown in Table 5.5. Depth of groove (D) as specified on drawings.
[g] If fillet welds are used in statically loaded structures to reinforce groove welds in corner and T-joints, these shall be equal to $T_1/4$, but need not exceed 3/8 in [10 mm].
[h] Double-groove welds may have grooves of unequal depth, but the depth of the shallower groove shall be no less than one-fourth of the thickness of the thinner part joined.
[i] Double-groove welds may have grooves of unequal depth, provided these conform to the limitations of Note f. Also, the weld size (S) applies individually to each groove.
[j] The orientation of the two members in the joints may vary from 135° to 180° for butt joints, or 45° to 135° for corner joints, or 45° to 90° for T-joints.
[k] For corner joints, the outside groove preparation may be in either or both members, provided the basic groove configuration is not changed and adequate edge distance is maintained to support the welding operations without excessive edge melting.
[l] Weld size (S) shall be based on joints welded flush.
[m] For flare-V-groove welds and flare-bevel-groove welds to rectangular tubular sections, r shall be as two times the wall thickness.
[n] For flare-V-groove welds to surfaces with different radii r, the smaller r shall be used.
[o] For corner and T-joints the member orientation may vary from 90° to less than or equal to 170° provided the groove angle and root opening are maintained, and the angle between the groove faces and the steel backing is at least 90°. See Figure 5.5.

See Notes on Page 81

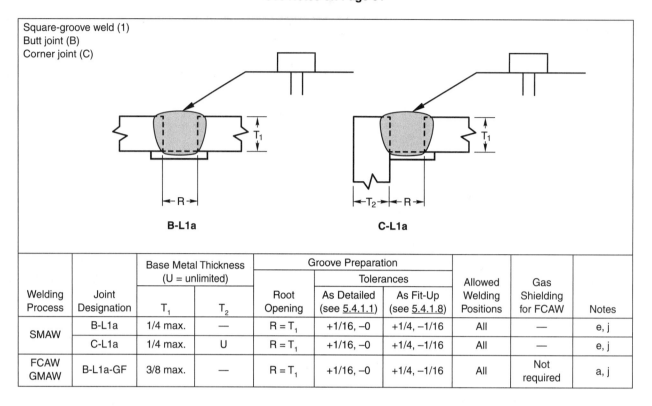

Welding Process	Joint Designation	Base Metal Thickness (U = unlimited)		Groove Preparation			Allowed Welding Positions	Gas Shielding for FCAW	Notes
		T_1	T_2	Root Opening	Tolerances				
					As Detailed (see 5.4.1.1)	As Fit-Up (see 5.4.1.8)			
SMAW	B-L1a	1/4 max.	—	$R = T_1$	+1/16, –0	+1/4, –1/16	All	—	e, j
SMAW	C-L1a	1/4 max.	U	$R = T_1$	+1/16, –0	+1/4, –1/16	All	—	e, j
FCAW GMAW	B-L1a-GF	3/8 max.	—	$R = T_1$	+1/16, –0	+1/4, –1/16	All	Not required	a, j

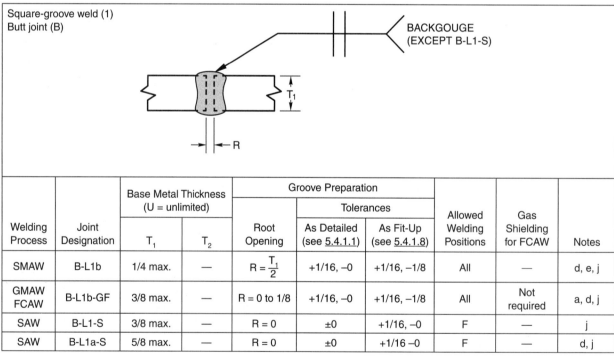

Welding Process	Joint Designation	Base Metal Thickness (U = unlimited)		Groove Preparation			Allowed Welding Positions	Gas Shielding for FCAW	Notes
		T_1	T_2	Root Opening	Tolerances				
					As Detailed (see 5.4.1.1)	As Fit-Up (see 5.4.1.8)			
SMAW	B-L1b	1/4 max.	—	$R = \frac{T_1}{2}$	+1/16, –0	+1/16, –1/8	All	—	d, e, j
GMAW FCAW	B-L1b-GF	3/8 max.	—	R = 0 to 1/8	+1/16, –0	+1/16, –1/8	All	Not required	a, d, j
SAW	B-L1-S	3/8 max.	—	R = 0	±0	+1/16, –0	F	—	j
SAW	B-L1a-S	5/8 max.	—	R = 0	±0	+1/16 –0	F	—	d, j

Figure 5.1—Prequalified CJP Groove Welded Joint Details (See 5.4.1) (Dimensions in Inches)

See Notes on Page 81

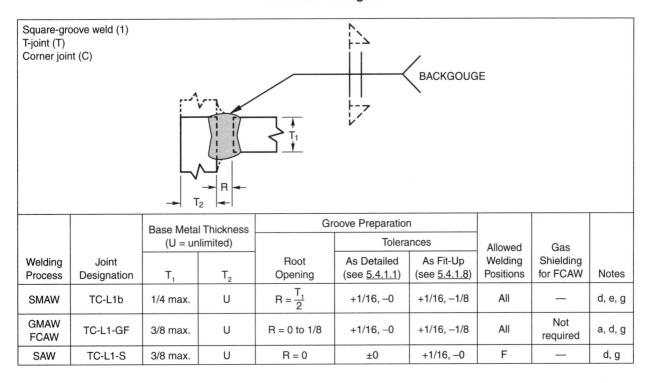

Square-groove weld (1)
T-joint (T)
Corner joint (C)

Welding Process	Joint Designation	Base Metal Thickness (U = unlimited)		Groove Preparation			Allowed Welding Positions	Gas Shielding for FCAW	Notes
		T_1	T_2	Root Opening	Tolerances				
					As Detailed (see 5.4.1.1)	As Fit-Up (see 5.4.1.8)			
SMAW	TC-L1b	1/4 max.	U	$R = \frac{T_1}{2}$	+1/16, –0	+1/16, –1/8	All	—	d, e, g
GMAW FCAW	TC-L1-GF	3/8 max.	U	R = 0 to 1/8	+1/16, –0	+1/16, –1/8	All	Not required	a, d, g
SAW	TC-L1-S	3/8 max.	U	R = 0	±0	+1/16, –0	F	—	d, g

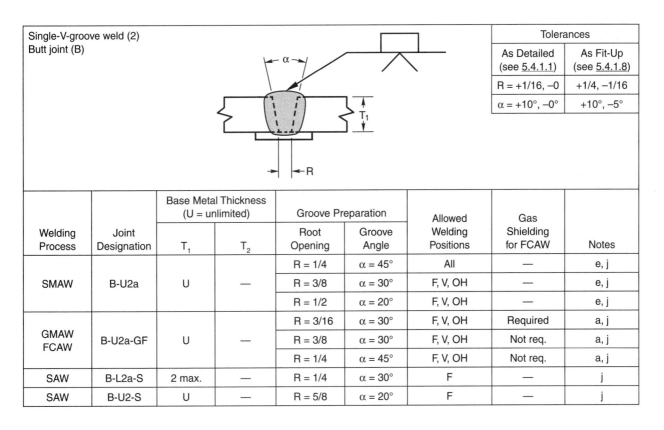

Single-V-groove weld (2)
Butt joint (B)

Tolerances	
As Detailed (see 5.4.1.1)	As Fit-Up (see 5.4.1.8)
R = +1/16, –0	+1/4, –1/16
α = +10°, –0°	+10°, –5°

Welding Process	Joint Designation	Base Metal Thickness (U = unlimited)		Groove Preparation		Allowed Welding Positions	Gas Shielding for FCAW	Notes
		T_1	T_2	Root Opening	Groove Angle			
SMAW	B-U2a	U	—	R = 1/4	α = 45°	All	—	e, j
				R = 3/8	α = 30°	F, V, OH	—	e, j
				R = 1/2	α = 20°	F, V, OH	—	e, j
GMAW FCAW	B-U2a-GF	U	—	R = 3/16	α = 30°	F, V, OH	Required	a, j
				R = 3/8	α = 30°	F, V, OH	Not req.	a, j
				R = 1/4	α = 45°	F, V, OH	Not req.	a, j
SAW	B-L2a-S	2 max.	—	R = 1/4	α = 30°	F	—	j
SAW	B-U2-S	U	—	R = 5/8	α = 20°	F	—	j

Figure 5.1 (Continued)—Prequalified CJP Groove Welded Joint Details (See 5.4.1)
(Dimensions in Inches)

See Notes on Page 81

Single-V-groove weld (2)
Corner joint (C)

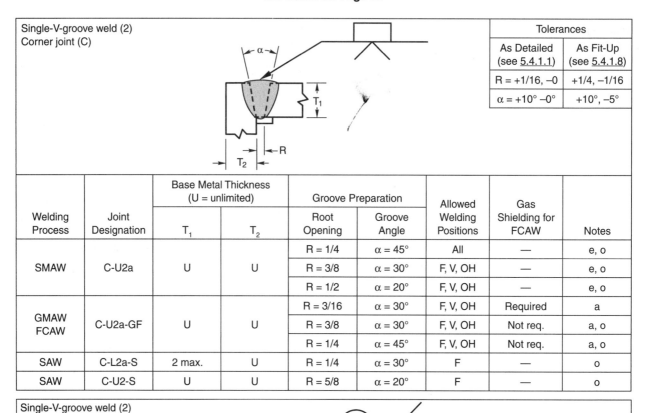

Welding Process	Joint Designation	Base Metal Thickness (U = unlimited)		Groove Preparation		Allowed Welding Positions	Gas Shielding for FCAW	Notes
		T_1	T_2	Root Opening	Groove Angle			
SMAW	C-U2a	U	U	R = 1/4	α = 45°	All	—	e, o
				R = 3/8	α = 30°	F, V, OH	—	e, o
				R = 1/2	α = 20°	F, V, OH	—	e, o
GMAW FCAW	C-U2a-GF	U	U	R = 3/16	α = 30°	F, V, OH	Required	a
				R = 3/8	α = 30°	F, V, OH	Not req.	a, o
				R = 1/4	α = 45°	F, V, OH	Not req.	a, o
SAW	C-L2a-S	2 max.	U	R = 1/4	α = 30°	F	—	o
SAW	C-U2-S	U	U	R = 5/8	α = 20°	F	—	o

Single-V-groove weld (2)
Butt joint (B)

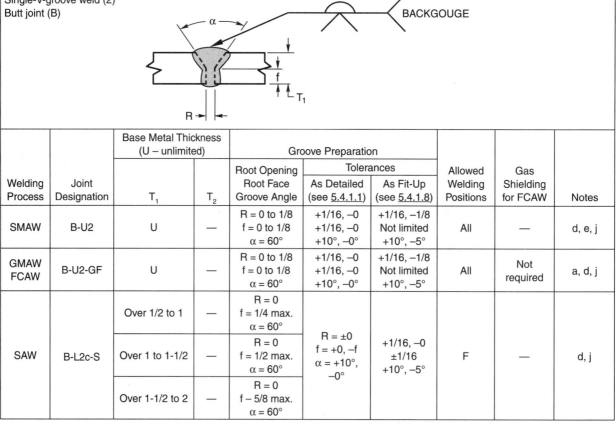

Welding Process	Joint Designation	Base Metal Thickness (U – unlimited)		Groove Preparation			Allowed Welding Positions	Gas Shielding for FCAW	Notes
		T_1	T_2	Root Opening Root Face Groove Angle	Tolerances				
					As Detailed (see 5.4.1.1)	As Fit-Up (see 5.4.1.8)			
SMAW	B-U2	U	—	R = 0 to 1/8 f = 0 to 1/8 α = 60°	+1/16, –0 +1/16, –0 +10°, –0°	+1/16, –1/8 Not limited +10°, –5°	All	—	d, e, j
GMAW FCAW	B-U2-GF	U	—	R = 0 to 1/8 f = 0 to 1/8 α = 60°	+1/16, –0 +1/16, –0 +10°, –0°	+1/16, –1/8 Not limited +10°, –5°	All	Not required	a, d, j
SAW	B-L2c-S	Over 1/2 to 1	—	R = 0 f = 1/4 max. α = 60°	R = ±0 f = +0, –f α = +10°, –0°	+1/16, –0 ±1/16 +10°, –5°	F	—	d, j
		Over 1 to 1-1/2	—	R = 0 f = 1/2 max. α = 60°					
		Over 1-1/2 to 2	—	R = 0 f – 5/8 max. α = 60°					

Figure 5.1 (Continued)—Prequalified CJP Groove Welded Joint Details (See 5.4.1)
(Dimensions in Inches)

See Notes on Page 81

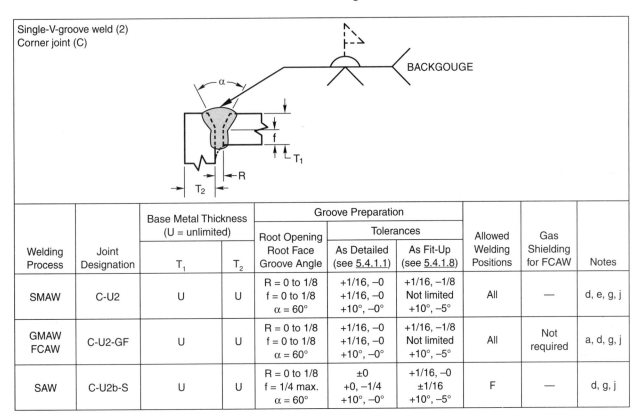

Single-V-groove weld (2)
Corner joint (C)

Welding Process	Joint Designation	Base Metal Thickness (U = unlimited)		Groove Preparation			Allowed Welding Positions	Gas Shielding for FCAW	Notes
		T_1	T_2	Root Opening Root Face Groove Angle	Tolerances As Detailed (see 5.4.1.1)	As Fit-Up (see 5.4.1.8)			
SMAW	C-U2	U	U	R = 0 to 1/8 f = 0 to 1/8 α = 60°	+1/16, –0 +1/16, –0 +10°, –0°	+1/16, –1/8 Not limited +10°, –5°	All	—	d, e, g, j
GMAW FCAW	C-U2-GF	U	U	R = 0 to 1/8 f = 0 to 1/8 α = 60°	+1/16, –0 +1/16, –0 +10°, –0°	+1/16, –1/8 Not limited +10°, –5°	All	Not required	a, d, g, j
SAW	C-U2b-S	U	U	R = 0 to 1/8 f = 1/4 max. α = 60°	±0 +0, –1/4 +10°, –0°	+1/16, –0 ±1/16 +10°, –5°	F	—	d, g, j

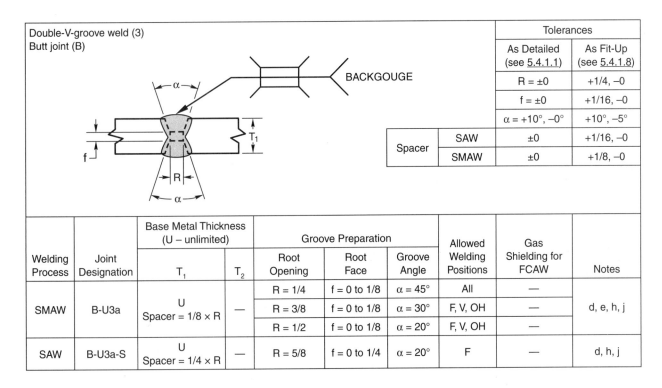

Double-V-groove weld (3)
Butt joint (B)

		Tolerances	
		As Detailed (see 5.4.1.1)	As Fit-Up (see 5.4.1.8)
		R = ±0	+1/4, –0
		f = ±0	+1/16, –0
		α = +10°, –0°	+10°, –5°
Spacer	SAW	±0	+1/16, –0
	SMAW	±0	+1/8, –0

Welding Process	Joint Designation	Base Metal Thickness (U – unlimited)		Groove Preparation			Allowed Welding Positions	Gas Shielding for FCAW	Notes
		T_1	T_2	Root Opening	Root Face	Groove Angle			
SMAW	B-U3a	U Spacer = 1/8 × R	—	R = 1/4 R = 3/8 R = 1/2	f = 0 to 1/8 f = 0 to 1/8 f = 0 to 1/8	α = 45° α = 30° α = 20°	All F, V, OH F, V, OH	— — —	d, e, h, j
SAW	B-U3a-S	U Spacer = 1/4 × R	—	R = 5/8	f = 0 to 1/4	α = 20°	F	—	d, h, j

Figure 5.1 (Continued)—Prequalified CJP Groove Welded Joint Details (See 5.4.1)
(Dimensions in Inches)

See Notes on Page 81

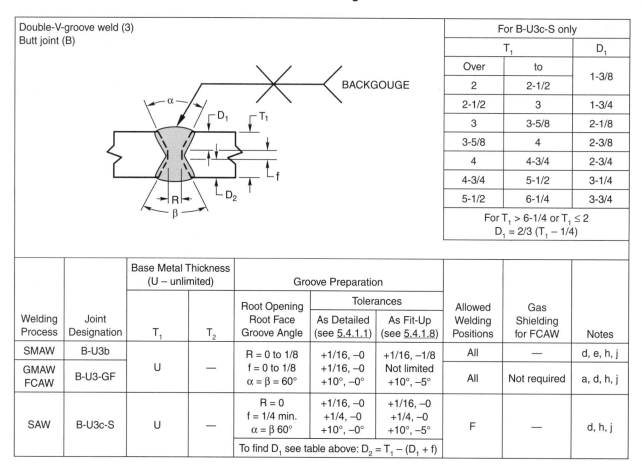

Double-V-groove weld (3)
Butt joint (B)

For B-U3c-S only		
T_1		D_1
Over	to	
2	2-1/2	1-3/8
2-1/2	3	1-3/4
3	3-5/8	2-1/8
3-5/8	4	2-3/8
4	4-3/4	2-3/4
4-3/4	5-1/2	3-1/4
5-1/2	6-1/4	3-3/4
For $T_1 > 6\text{-}1/4$ or $T_1 \leq 2$ $D_1 = 2/3\,(T_1 - 1/4)$		

Welding Process	Joint Designation	Base Metal Thickness (U – unlimited)		Groove Preparation Root Opening Root Face Groove Angle	Tolerances		Allowed Welding Positions	Gas Shielding for FCAW	Notes
		T_1	T_2		As Detailed (see 5.4.1.1)	As Fit-Up (see 5.4.1.8)			
SMAW	B-U3b	U	—	R = 0 to 1/8 f = 0 to 1/8 α = β = 60°	+1/16, –0 +1/16, –0 +10°, –0°	+1/16, –1/8 Not limited +10°, –5°	All	—	d, e, h, j
GMAW FCAW	B-U3-GF						All	Not required	a, d, h, j
SAW	B-U3c-S	U	—	R = 0 f = 1/4 min. α = β 60°	+1/16, –0 +1/4, –0 +10°, –0°	+1/16, –0 +1/4, –0 +10°, –5°	F	—	d, h, j
				To find D_1 see table above: $D_2 = T_1 - (D_1 + f)$					

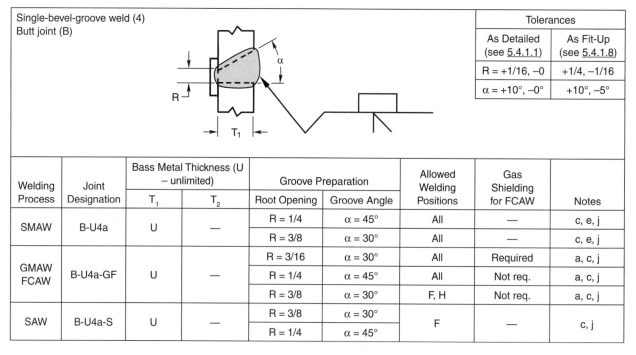

Single-bevel-groove weld (4)
Butt joint (B)

Tolerances	
As Detailed (see 5.4.1.1)	As Fit-Up (see 5.4.1.8)
R = +1/16, –0	+1/4, –1/16
α = +10°, –0°	+10°, –5°

Welding Process	Joint Designation	Bass Metal Thickness (U – unlimited)		Groove Preparation		Allowed Welding Positions	Gas Shielding for FCAW	Notes
		T_1	T_2	Root Opening	Groove Angle			
SMAW	B-U4a	U	—	R = 1/4	α = 45°	All	—	c, e, j
				R = 3/8	α = 30°	All	—	c, e, j
GMAW FCAW	B-U4a-GF	U	—	R = 3/16	α = 30°	All	Required	a, c, j
				R = 1/4	α = 45°	All	Not req.	a, c, j
				R = 3/8	α = 30°	F, H	Not req.	a, c, j
SAW	B-U4a-S	U	—	R = 3/8	α = 30°	F	—	c, j
				R = 1/4	α = 45°			

**Figure 5.1 (Continued)—Prequalified CJP Groove Welded Joint Details (See 5.4.1)
(Dimensions in Inches)**

See Notes on Page 81

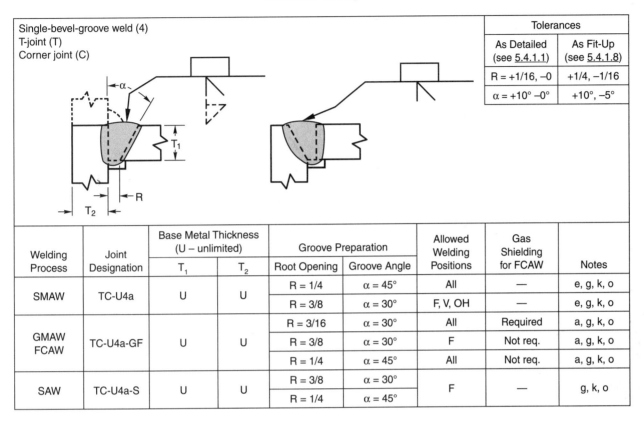

Single-bevel-groove weld (4)
T-joint (T)
Corner joint (C)

				Tolerances	
				As Detailed (see 5.4.1.1)	As Fit-Up (see 5.4.1.8)
				R = +1/16, −0	+1/4, −1/16
				α = +10° −0°	+10°, −5°

Welding Process	Joint Designation	Base Metal Thickness (U – unlimited)		Groove Preparation		Allowed Welding Positions	Gas Shielding for FCAW	Notes
		T₁	T₂	Root Opening	Groove Angle			
SMAW	TC-U4a	U	U	R = 1/4	α = 45°	All	—	e, g, k, o
				R = 3/8	α = 30°	F, V, OH	—	e, g, k, o
GMAW FCAW	TC-U4a-GF	U	U	R = 3/16	α = 30°	All	Required	a, g, k, o
				R = 3/8	α = 30°	F	Not req.	a, g, k, o
				R = 1/4	α = 45°	All	Not req.	a, g, k, o
SAW	TC-U4a-S	U	U	R = 3/8	α = 30°	F	—	g, k, o
				R = 1/4	α = 45°			

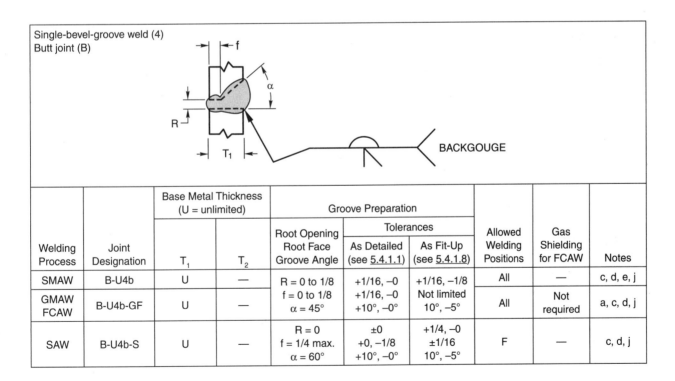

Single-bevel-groove weld (4)
Butt joint (B)

Welding Process	Joint Designation	Base Metal Thickness (U = unlimited)		Groove Preparation			Allowed Welding Positions	Gas Shielding for FCAW	Notes
		T₁	T₂	Root Opening Root Face Groove Angle	Tolerances				
					As Detailed (see 5.4.1.1)	As Fit-Up (see 5.4.1.8)			
SMAW	B-U4b	U	—	R = 0 to 1/8 f = 0 to 1/8 α = 45°	+1/16, −0 +1/16, −0 +10°, −0°	+1/16, −1/8 Not limited 10°, −5°	All	—	c, d, e, j
GMAW FCAW	B-U4b-GF	U	—				All	Not required	a, c, d, j
SAW	B-U4b-S	U	—	R = 0 f = 1/4 max. α = 60°	±0 +0, −1/8 +10°, −0°	+1/4, −0 ±1/16 10°, −5°	F	—	c, d, j

Figure 5.1 (Continued)—Prequalified CJP Groove Welded Joint Details (See 5.4.1)
(Dimensions in Inches)

See Notes on Page 81

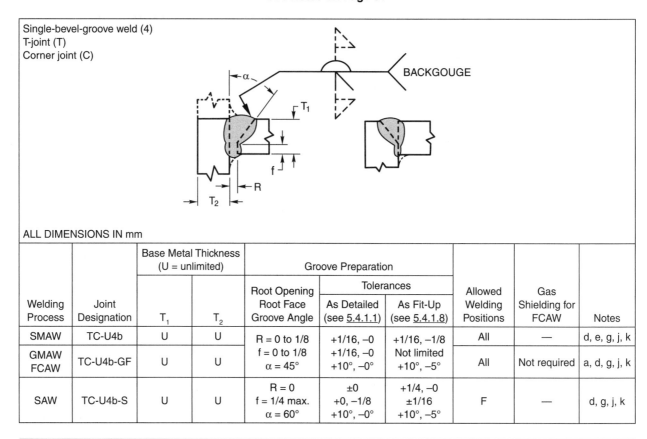

Single-bevel-groove weld (4)
T-joint (T)
Corner joint (C)

ALL DIMENSIONS IN mm

Welding Process	Joint Designation	Base Metal Thickness (U = unlimited)		Groove Preparation			Allowed Welding Positions	Gas Shielding for FCAW	Notes
		T_1	T_2	Root Opening Root Face Groove Angle	Tolerances				
					As Detailed (see 5.4.1.1)	As Fit-Up (see 5.4.1.8)			
SMAW	TC-U4b	U	U	R = 0 to 1/8 f = 0 to 1/8 α = 45°	+1/16, –0 +1/16, –0 +10°, –0°	+1/16, –1/8 Not limited +10°, –5°	All	—	d, e, g, j, k
GMAW FCAW	TC-U4b-GF	U	U				All	Not required	a, d, g, j, k
SAW	TC-U4b-S	U	U	R = 0 f = 1/4 max. α = 60°	±0 +0, –1/8 +10°, –0°	+1/4, –0 ±1/16 +10°, –5°	F	—	d, g, j, k

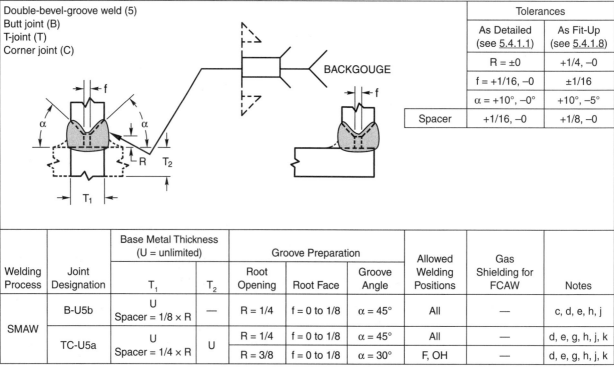

Double-bevel-groove weld (5)
Butt joint (B)
T-joint (T)
Corner joint (C)

	Tolerances	
	As Detailed (see 5.4.1.1)	As Fit-Up (see 5.4.1.8)
R =	±0	+1/4, –0
f =	+1/16, –0	±1/16
α =	+10°, –0°	+10°, –5°
Spacer	+1/16, –0	+1/8, –0

Welding Process	Joint Designation	Base Metal Thickness (U = unlimited)		Groove Preparation			Allowed Welding Positions	Gas Shielding for FCAW	Notes
		T_1	T_2	Root Opening	Root Face	Groove Angle			
SMAW	B-U5b	U Spacer = 1/8 × R	—	R = 1/4	f = 0 to 1/8	α = 45°	All	—	c, d, e, h, j
	TC-U5a	U Spacer = 1/4 × R	U	R = 1/4	f = 0 to 1/8	α = 45°	All	—	d, e, g, h, j, k
				R = 3/8	f = 0 to 1/8	α = 30°	F, OH	—	d, e, g, h, j, k

Figure 5.1 (Continued)—Prequalified CJP Groove Welded Joint Details (See 5.4.1)
(Dimensions in Inches)

See Notes on Page 81

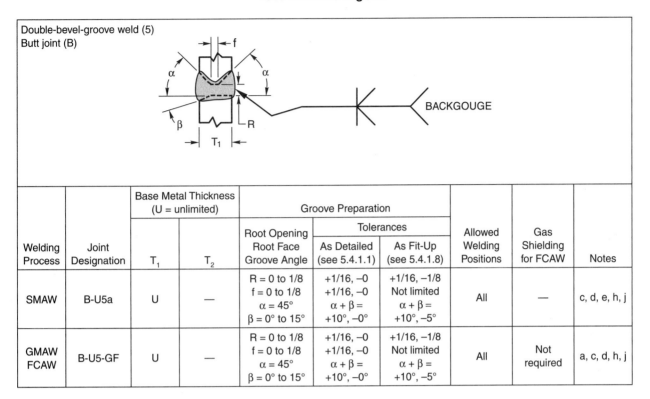

Welding Process	Joint Designation	Base Metal Thickness (U = unlimited)		Groove Preparation			Allowed Welding Positions	Gas Shielding for FCAW	Notes
		T_1	T_2	Root Opening Root Face Groove Angle	Tolerances As Detailed (see 5.4.1.1)	As Fit-Up (see 5.4.1.8)			
SMAW	B-U5a	U	—	R = 0 to 1/8 f = 0 to 1/8 α = 45° β = 0° to 15°	+1/16, –0 +1/16, –0 α + β = +10°, –0°	+1/16, –1/8 Not limited α + β = +10°, –5°	All	—	c, d, e, h, j
GMAW FCAW	B-U5-GF	U	—	R = 0 to 1/8 f = 0 to 1/8 α = 45° β = 0° to 15°	+1/16, –0 +1/16, –0 α + β = +10°, –0°	+1/16, –1/8 Not limited α + β = +10°, –5°	All	Not required	a, c, d, h, j

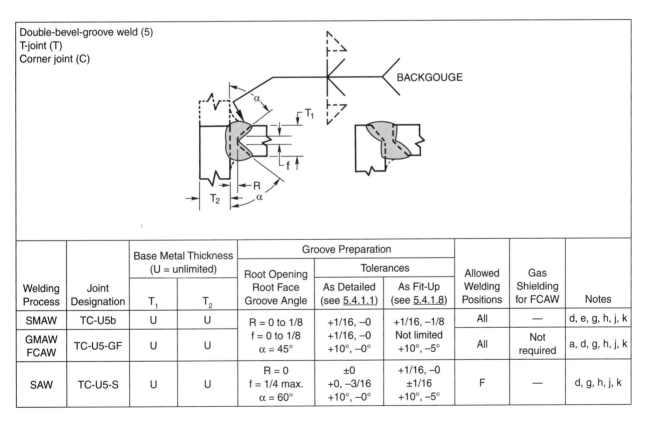

Welding Process	Joint Designation	Base Metal Thickness (U = unlimited)		Groove Preparation			Allowed Welding Positions	Gas Shielding for FCAW	Notes
		T_1	T_2	Root Opening Root Face Groove Angle	Tolerances As Detailed (see 5.4.1.1)	As Fit-Up (see 5.4.1.8)			
SMAW	TC-U5b	U	U	R = 0 to 1/8 f = 0 to 1/8 α = 45°	+1/16, –0 +1/16, –0 +10°, –0°	+1/16, –1/8 Not limited +10°, –5°	All	—	d, e, g, h, j, k
GMAW FCAW	TC-U5-GF	U	U				All	Not required	a, d, g, h, j, k
SAW	TC-U5-S	U	U	R = 0 f = 1/4 max. α = 60°	±0 +0, –3/16 +10°, –0°	+1/16, –0 ±1/16 +10°, –5°	F	—	d, g, h, j, k

Figure 5.1 (Continued)—Prequalified CJP Groove Welded Joint Details (See 5.4.1) (Dimensions in Inches)

See Notes on Page 81

Single-U-groove weld (6)
Butt joint (B)
Corner joint (C)

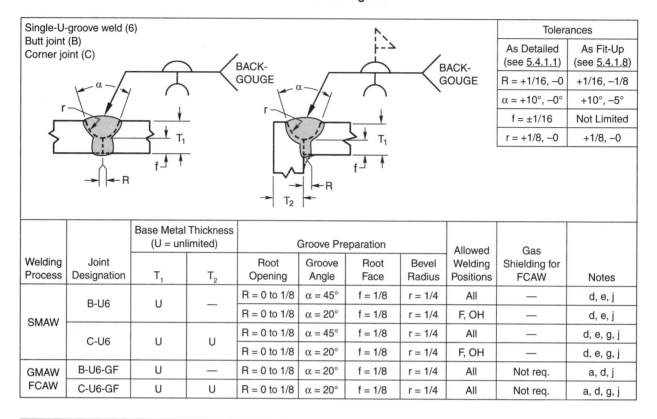

		Tolerances	
		As Detailed (see 5.4.1.1)	As Fit-Up (see 5.4.1.8)
		R = +1/16, –0	+1/16, –1/8
		α = +10°, –0°	+10°, –5°
		f = ±1/16	Not Limited
		r = +1/8, –0	+1/8, –0

Welding Process	Joint Designation	Base Metal Thickness (U = unlimited)		Groove Preparation				Allowed Welding Positions	Gas Shielding for FCAW	Notes
		T_1	T_2	Root Opening	Groove Angle	Root Face	Bevel Radius			
SMAW	B-U6	U	—	R = 0 to 1/8	α = 45°	f = 1/8	r = 1/4	All	—	d, e, j
				R = 0 to 1/8	α = 20°	f = 1/8	r = 1/4	F, OH	—	d, e, j
	C-U6	U	U	R = 0 to 1/8	α = 45°	f = 1/8	r = 1/4	All	—	d, e, g, j
				R = 0 to 1/8	α = 20°	f = 1/8	r = 1/4	F, OH	—	d, e, g, j
GMAW FCAW	B-U6-GF	U	—	R = 0 to 1/8	α = 20°	f = 1/8	r = 1/4	All	Not req.	a, d, j
	C-U6-GF	U	U	R = 0 to 1/8	α = 20°	f = 1/8	r = 1/4	All	Not req.	a, d, g, j

Double-U-groove weld (7)
Butt joint (B)

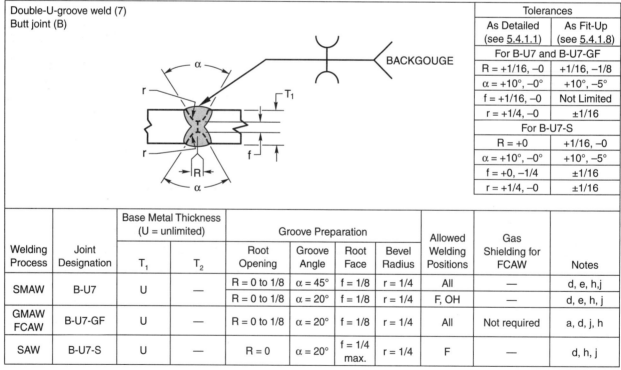

		Tolerances	
		As Detailed (see 5.4.1.1)	As Fit-Up (see 5.4.1.8)
		For B-U7 and B-U7-GF	
		R = +1/16, –0	+1/16, –1/8
		α = +10°, –0°	+10°, –5°
		f = +1/16, –0	Not Limited
		r = +1/4, –0	±1/16
		For B-U7-S	
		R = +0	+1/16, –0
		α = +10°, –0°	+10°, –5°
		f = +0, –1/4	±1/16
		r = +1/4, –0	±1/16

Welding Process	Joint Designation	Base Metal Thickness (U = unlimited)		Groove Preparation				Allowed Welding Positions	Gas Shielding for FCAW	Notes
		T_1	T_2	Root Opening	Groove Angle	Root Face	Bevel Radius			
SMAW	B-U7	U	—	R = 0 to 1/8	α = 45°	f = 1/8	r = 1/4	All	—	d, e, h, j
				R = 0 to 1/8	α = 20°	f = 1/8	r = 1/4	F, OH	—	d, e, h, j
GMAW FCAW	B-U7-GF	U	—	R = 0 to 1/8	α = 20°	f = 1/8	r = 1/4	All	Not required	a, d, j, h
SAW	B-U7-S	U	—	R = 0	α = 20°	f = 1/4 max.	r = 1/4	F	—	d, h, j

Figure 5.1 (Continued)—Prequalified CJP Groove Welded Joint Details (See 5.4.1) (Dimensions in Inches)

See Notes on Page 81

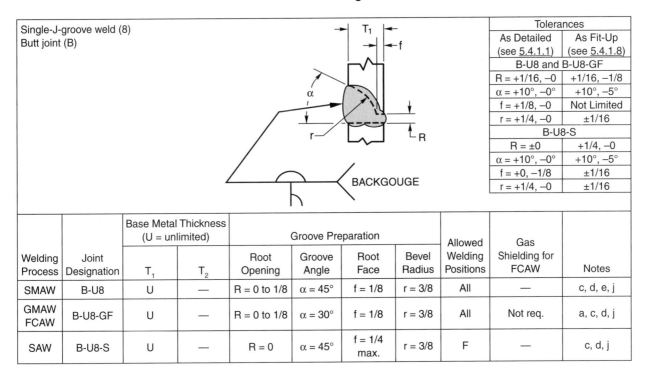

Single-J-groove weld (8)
Butt joint (B)

		Tolerances	
		As Detailed (see 5.4.1.1)	As Fit-Up (see 5.4.1.8)
		B-U8 and B-U8-GF	
		R = +1/16, –0	+1/16, –1/8
		α = +10°, –0°	+10°, –5°
		f = +1/8, –0	Not Limited
		r = +1/4, –0	±1/16
		B-U8-S	
		R = ±0	+1/4, –0
		α = +10°, –0°	+10°, –5°
		f = +0, –1/8	±1/16
		r = +1/4, –0	±1/16

Welding Process	Joint Designation	Base Metal Thickness (U = unlimited)		Groove Preparation				Allowed Welding Positions	Gas Shielding for FCAW	Notes
		T_1	T_2	Root Opening	Groove Angle	Root Face	Bevel Radius			
SMAW	B-U8	U	—	R = 0 to 1/8	α = 45°	f = 1/8	r = 3/8	All	—	c, d, e, j
GMAW FCAW	B-U8-GF	U	—	R = 0 to 1/8	α = 30°	f = 1/8	r = 3/8	All	Not req.	a, c, d, j
SAW	B-U8-S	U	—	R = 0	α = 45°	f = 1/4 max.	r = 3/8	F	—	c, d, j

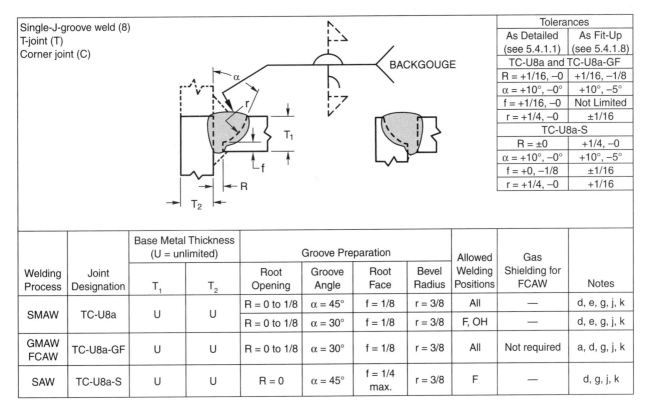

Single-J-groove weld (8)
T-joint (T)
Corner joint (C)

		Tolerances	
		As Detailed (see 5.4.1.1)	As Fit-Up (see 5.4.1.8)
		TC-U8a and TC-U8a-GF	
		R = +1/16, –0	+1/16, –1/8
		α = +10°, –0°	+10°, –5°
		f = +1/16, –0	Not Limited
		r = +1/4, –0	±1/16
		TC-U8a-S	
		R = ±0	+1/4, –0
		α = +10°, –0°	+10°, –5°
		f = +0, –1/8	±1/16
		r = +1/4, –0	+1/16

Welding Process	Joint Designation	Base Metal Thickness (U = unlimited)		Groove Preparation				Allowed Welding Positions	Gas Shielding for FCAW	Notes
		T_1	T_2	Root Opening	Groove Angle	Root Face	Bevel Radius			
SMAW	TC-U8a	U	U	R = 0 to 1/8	α = 45°	f = 1/8	r = 3/8	All	—	d, e, g, j, k
				R = 0 to 1/8	α = 30°	f = 1/8	r = 3/8	F, OH	—	d, e, g, j, k
GMAW FCAW	TC-U8a-GF	U	U	R = 0 to 1/8	α = 30°	f = 1/8	r = 3/8	All	Not required	a, d, g, j, k
SAW	TC-U8a-S	U	U	R = 0	α = 45°	f = 1/4 max.	r = 3/8	F	—	d, g, j, k

Figure 5.1 (Continued)—Prequalified CJP Groove Welded Joint Details (See 5.4.1)
(Dimensions in Inches)

See Notes on Page 81

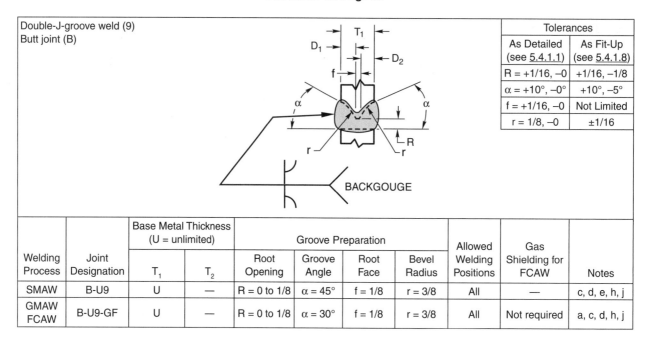

Double-J-groove weld (9)
Butt joint (B)

Tolerances		
	As Detailed (see 5.4.1.1)	As Fit-Up (see 5.4.1.8)
R =	+1/16, –0	+1/16, –1/8
α =	+10°, –0°	+10°, –5°
f =	+1/16, –0	Not Limited
r =	1/8, –0	±1/16

Welding Process	Joint Designation	Base Metal Thickness (U = unlimited)		Groove Preparation				Allowed Welding Positions	Gas Shielding for FCAW	Notes
		T_1	T_2	Root Opening	Groove Angle	Root Face	Bevel Radius			
SMAW	B-U9	U	—	R = 0 to 1/8	α = 45°	f = 1/8	r = 3/8	All	—	c, d, e, h, j
GMAW FCAW	B-U9-GF	U	—	R = 0 to 1/8	α = 30°	f = 1/8	r = 3/8	All	Not required	a, c, d, h, j

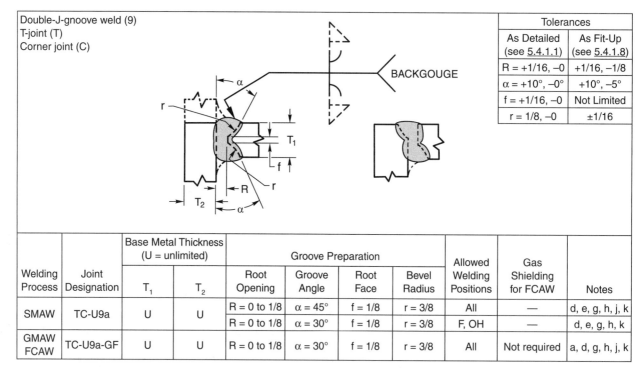

Double-J-groove weld (9)
T-joint (T)
Corner joint (C)

Tolerances		
	As Detailed (see 5.4.1.1)	As Fit-Up (see 5.4.1.8)
R =	+1/16, –0	+1/16, –1/8
α =	+10°, –0°	+10°, –5°
f =	+1/16, –0	Not Limited
r =	1/8, –0	±1/16

Welding Process	Joint Designation	Base Metal Thickness (U = unlimited)		Groove Preparation				Allowed Welding Positions	Gas Shielding for FCAW	Notes
		T_1	T_2	Root Opening	Groove Angle	Root Face	Bevel Radius			
SMAW	TC-U9a	U	U	R = 0 to 1/8	α = 45°	f = 1/8	r = 3/8	All	—	d, e, g, h, j, k
				R = 0 to 1/8	α = 30°	f = 1/8	r = 3/8	F, OH	—	d, e, g, h, k
GMAW FCAW	TC-U9a-GF	U	U	R = 0 to 1/8	α = 30°	f = 1/8	r = 3/8	All	Not required	a, d, g, h, j, k

**Figure 5.1 (Continued)—Prequalified CJP Groove Welded Joint Details (See 5.4.1)
(Dimensions in Inches)**

See Notes on Page 81

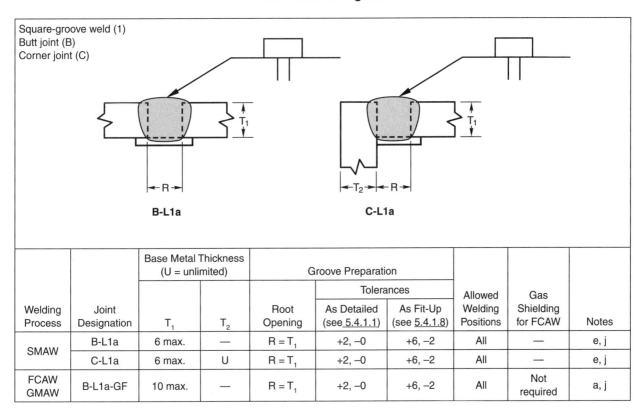

Welding Process	Joint Designation	Base Metal Thickness (U = unlimited)		Root Opening	Groove Preparation		Allowed Welding Positions	Gas Shielding for FCAW	Notes
					Tolerances				
		T_1	T_2		As Detailed (see 5.4.1.1)	As Fit-Up (see 5.4.1.8)			
SMAW	B-L1a	6 max.	—	$R = T_1$	+2, –0	+6, –2	All	—	e, j
SMAW	C-L1a	6 max.	U	$R = T_1$	+2, –0	+6, –2	All	—	e, j
FCAW GMAW	B-L1a-GF	10 max.	—	$R = T_1$	+2, –0	+6, –2	All	Not required	a, j

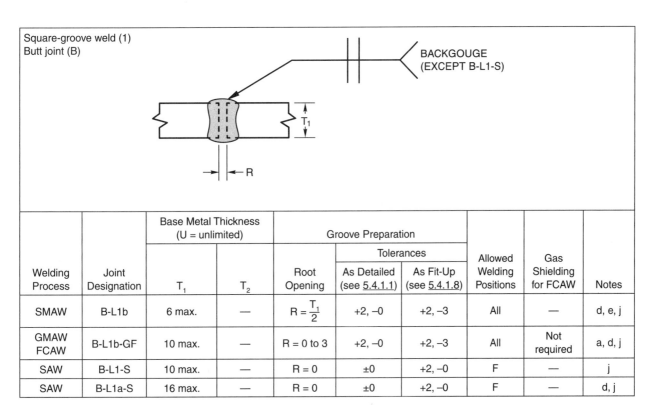

Welding Process	Joint Designation	Base Metal Thickness (U = unlimited)		Root Opening	Groove Preparation		Allowed Welding Positions	Gas Shielding for FCAW	Notes
					Tolerances				
		T_1	T_2		As Detailed (see 5.4.1.1)	As Fit-Up (see 5.4.1.8)			
SMAW	B-L1b	6 max.	—	$R = \frac{T_1}{2}$	+2, –0	+2, –3	All	—	d, e, j
GMAW FCAW	B-L1b-GF	10 max.	—	R = 0 to 3	+2, –0	+2, –3	All	Not required	a, d, j
SAW	B-L1-S	10 max.	—	R = 0	±0	+2, –0	F	—	j
SAW	B-L1a-S	16 max.	—	R = 0	±0	+2, –0	F	—	d, j

Figure 5.1 (Continued)—Prequalified CJP Groove Welded Joint Details (See 5.4.1) (Dimensions in Millimeters)

See Notes on Page 81

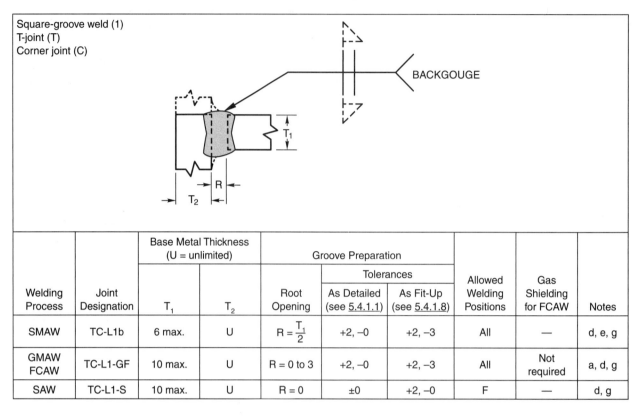

Square-groove weld (1)
T-joint (T)
Corner joint (C)

Welding Process	Joint Designation	Base Metal Thickness (U = unlimited)		Root Opening	Groove Preparation Tolerances		Allowed Welding Positions	Gas Shielding for FCAW	Notes
		T_1	T_2		As Detailed (see 5.4.1.1)	As Fit-Up (see 5.4.1.8)			
SMAW	TC-L1b	6 max.	U	$R = \frac{T_1}{2}$	+2, –0	+2, –3	All	—	d, e, g
GMAW FCAW	TC-L1-GF	10 max.	U	R = 0 to 3	+2, –0	+2, –3	All	Not required	a, d, g
SAW	TC-L1-S	10 max.	U	R = 0	±0	+2, –0	F	—	d, g

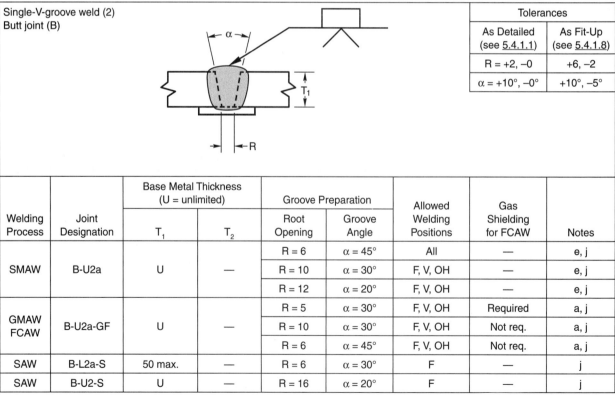

Single-V-groove weld (2)
Butt joint (B)

Welding Process	Joint Designation	Base Metal Thickness (U = unlimited)		Groove Preparation		Allowed Welding Positions	Gas Shielding for FCAW	Notes
		T_1	T_2	Root Opening	Groove Angle			
SMAW	B-U2a	U	—	R = 6	α = 45°	All	—	e, j
				R = 10	α = 30°	F, V, OH	—	e, j
				R = 12	α = 20°	F, V, OH	—	e, j
GMAW FCAW	B-U2a-GF	U	—	R = 5	α = 30°	F, V, OH	Required	a, j
				R = 10	α = 30°	F, V, OH	Not req.	a, j
				R = 6	α = 45°	F, V, OH	Not req.	a, j
SAW	B-L2a-S	50 max.	—	R = 6	α = 30°	F	—	j
SAW	B-U2-S	U	—	R = 16	α = 20°	F	—	j

Tolerances (Single-V-groove):
	As Detailed (see 5.4.1.1)	As Fit-Up (see 5.4.1.8)
R =	+2, –0	+6, –2
α =	+10°, –0°	+10°, –5°

Figure 5.1 (Continued)—Prequalified CJP Groove Welded Joint Details (See 5.4.1) (Dimensions in Millimeters)

See Notes on Page 81

Single-V-groove weld (2)
Corner joint (C)

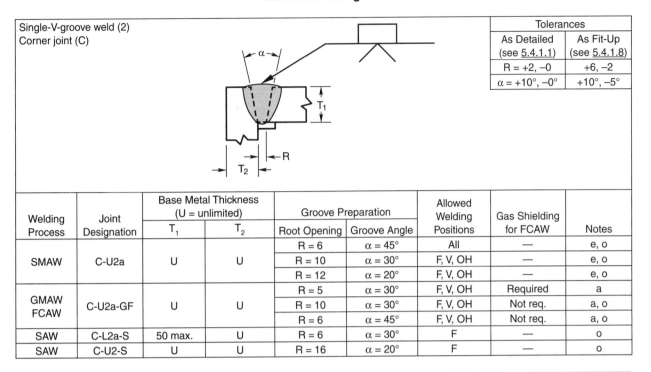

		Tolerances	
		As Detailed (see 5.4.1.1)	As Fit-Up (see 5.4.1.8)
		$R = +2, -0$	$+6, -2$
		$\alpha = +10°, -0°$	$+10°, -5°$

Welding Process	Joint Designation	Base Metal Thickness (U = unlimited)		Groove Preparation		Allowed Welding Positions	Gas Shielding for FCAW	Notes
		T_1	T_2	Root Opening	Groove Angle			
SMAW	C-U2a	U	U	R = 6	$\alpha = 45°$	All	—	e, o
				R = 10	$\alpha = 30°$	F, V, OH	—	e, o
				R = 12	$\alpha = 20°$	F, V, OH	—	e, o
GMAW FCAW	C-U2a-GF	U	U	R = 5	$\alpha = 30°$	F, V, OH	Required	a
				R = 10	$\alpha = 30°$	F, V, OH	Not req.	a, o
				R = 6	$\alpha = 45°$	F, V, OH	Not req.	a, o
SAW	C-L2a-S	50 max.	U	R = 6	$\alpha = 30°$	F	—	o
SAW	C-U2-S	U	U	R = 16	$\alpha = 20°$	F	—	o

Single-V-groove weld (2)
Butt joint (B)

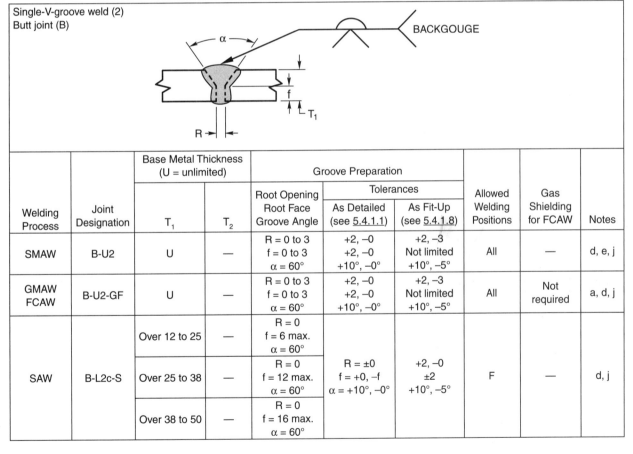

Welding Process	Joint Designation	Base Metal Thickness (U = unlimited)		Groove Preparation			Allowed Welding Positions	Gas Shielding for FCAW	Notes
		T_1	T_2	Root Opening Root Face Groove Angle	Tolerances				
					As Detailed (see 5.4.1.1)	As Fit-Up (see 5.4.1.8)			
SMAW	B-U2	U	—	R = 0 to 3 f = 0 to 3 $\alpha = 60°$	+2, -0 +2, -0 +10°, -0°	+2, -3 Not limited +10°, -5°	All	—	d, e, j
GMAW FCAW	B-U2-GF	U	—	R = 0 to 3 f = 0 to 3 $\alpha = 60°$	+2, -0 +2, -0 +10°, -0°	+2, -3 Not limited +10°, -5°	All	Not required	a, d, j
SAW	B-L2c-S	Over 12 to 25	—	R = 0 f = 6 max. $\alpha = 60°$	R = ±0 f = +0, -f $\alpha = +10°, -0°$	+2, -0 ±2 +10°, -5°	F	—	d, j
		Over 25 to 38	—	R = 0 f = 12 max. $\alpha = 60°$					
		Over 38 to 50	—	R = 0 f = 16 max. $\alpha = 60°$					

Figure 5.1 (Continued)—Prequalified CJP Groove Welded Joint Details (See 5.4.1)
(Dimensions in Millimeters)

See Notes on Page 81

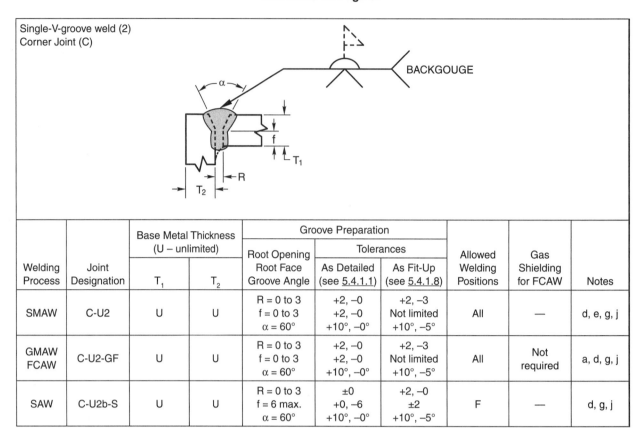

Single-V-groove weld (2)
Corner Joint (C)

Welding Process	Joint Designation	Base Metal Thickness (U – unlimited)		Groove Preparation			Allowed Welding Positions	Gas Shielding for FCAW	Notes
		T_1	T_2	Root Opening Root Face Groove Angle	Tolerances				
					As Detailed (see 5.4.1.1)	As Fit-Up (see 5.4.1.8)			
SMAW	C-U2	U	U	R = 0 to 3 f = 0 to 3 α = 60°	+2, –0 +2, –0 +10°, –0°	+2, –3 Not limited +10°, –5°	All	—	d, e, g, j
GMAW FCAW	C-U2-GF	U	U	R = 0 to 3 f = 0 to 3 α = 60°	+2, –0 +2, –0 +10°, –0°	+2, –3 Not limited +10°, –5°	All	Not required	a, d, g, j
SAW	C-U2b-S	U	U	R = 0 to 3 f = 6 max. α = 60°	±0 +0, –6 +10°, –0°	+2, –0 ±2 +10°, –5°	F	—	d, g, j

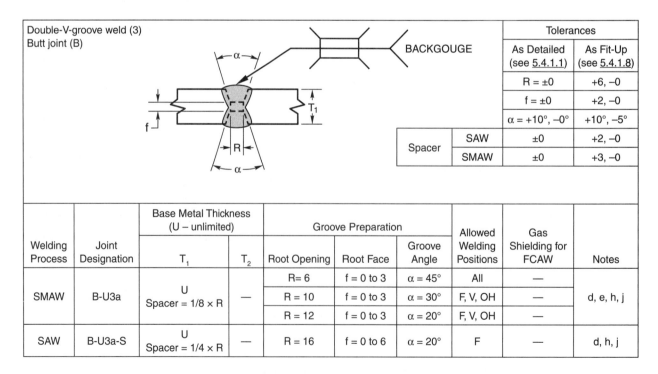

Double-V-groove weld (3)
Butt joint (B)

	Tolerances	
	As Detailed (see 5.4.1.1)	As Fit-Up (see 5.4.1.8)
R =	±0	+6, –0
f =	±0	+2, –0
α =	+10°, –0°	+10°, –5°
Spacer SAW	±0	+2, –0
Spacer SMAW	±0	+3, –0

Welding Process	Joint Designation	Base Metal Thickness (U – unlimited)		Groove Preparation			Allowed Welding Positions	Gas Shielding for FCAW	Notes
		T_1	T_2	Root Opening	Root Face	Groove Angle			
SMAW	B-U3a	U Spacer = 1/8 × R	—	R = 6 R = 10 R = 12	f = 0 to 3 f = 0 to 3 f = 0 to 3	α = 45° α = 30° α = 20°	All F, V, OH F, V, OH	— — —	d, e, h, j
SAW	B-U3a-S	U Spacer = 1/4 × R	—	R = 16	f = 0 to 6	α = 20°	F	—	d, h, j

Figure 5.1 (Continued)—Prequalified CJP Groove Welded Joint Details (See 5.4.1) (Dimensions in Millimeters)

See Notes on Page 81

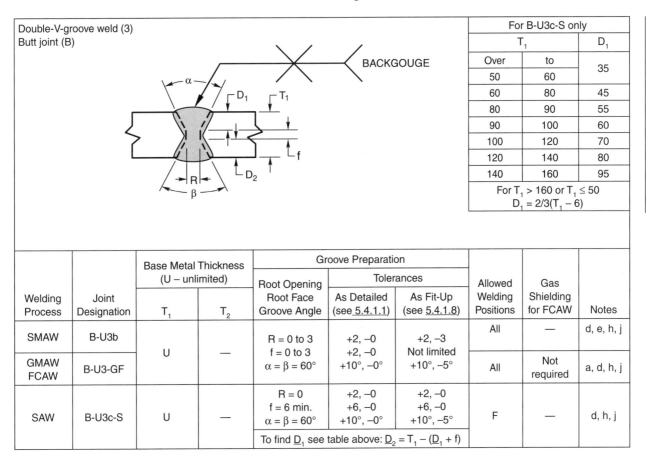

Double-V-groove weld (3)
Butt joint (B)

For B-U3c-S only		
T_1		D_1
Over	to	
50	60	35
60	80	45
80	90	55
90	100	60
100	120	70
120	140	80
140	160	95
For $T_1 > 160$ or $T_1 \leq 50$ $D_1 = 2/3(T_1 - 6)$		

Welding Process	Joint Designation	Base Metal Thickness (U – unlimited)		Groove Preparation			Allowed Welding Positions	Gas Shielding for FCAW	Notes
		T_1	T_2	Root Opening Root Face Groove Angle	Tolerances As Detailed (see 5.4.1.1)	As Fit-Up (see 5.4.1.8)			
SMAW	B-U3b	U	—	R = 0 to 3 f = 0 to 3 $\alpha = \beta = 60°$	+2, –0 +2, –0 +10°, –0°	+2, –3 Not limited +10°, –5°	All	—	d, e, h, j
GMAW FCAW	B-U3-GF						All	Not required	a, d, h, j
SAW	B-U3c-S	U	—	R = 0 f = 6 min. $\alpha = \beta = 60°$	+2, –0 +6, –0 +10°, –0°	+2, –0 +6, –0 +10°, –5°	F	—	d, h, j
				To find D_1 see table above: $D_2 = T_1 - (D_1 + f)$					

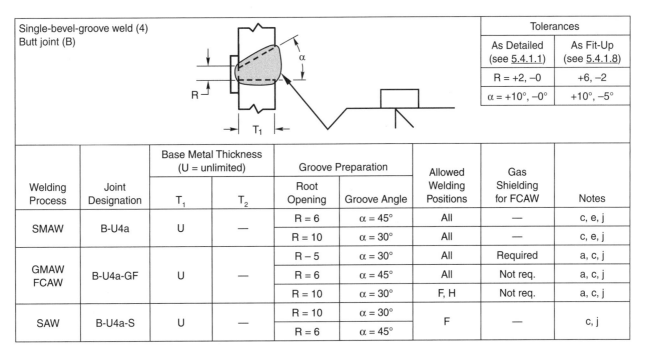

Single-bevel-groove weld (4)
Butt joint (B)

						Tolerances	
						As Detailed (see 5.4.1.1)	As Fit-Up (see 5.4.1.8)
						R = +2, –0	+6, –2
						$\alpha = +10°, –0°$	+10°, –5°

Welding Process	Joint Designation	Base Metal Thickness (U = unlimited)		Groove Preparation		Allowed Welding Positions	Gas Shielding for FCAW	Notes
		T_1	T_2	Root Opening	Groove Angle			
SMAW	B-U4a	U	—	R = 6	$\alpha = 45°$	All	—	c, e, j
				R = 10	$\alpha = 30°$	All	—	c, e, j
GMAW FCAW	B-U4a-GF	U	—	R – 5	$\alpha = 30°$	All	Required	a, c, j
				R = 6	$\alpha = 45°$	All	Not req.	a, c, j
				R = 10	$\alpha = 30°$	F, H	Not req.	a, c, j
SAW	B-U4a-S	U	—	R = 10	$\alpha = 30°$	F	—	c, j
				R = 6	$\alpha = 45°$			

**Figure 5.1 (Continued)—Prequalified CJP Groove Welded Joint Details (See 5.4.1)
(Dimensions in Millimeters)**

See Notes on Page 81

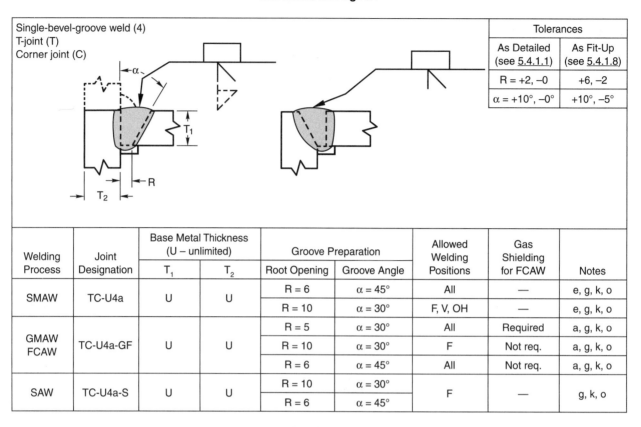

Single-bevel-groove weld (4)
T-joint (T)
Corner joint (C)

Welding Process	Joint Designation	Base Metal Thickness (U – unlimited)		Groove Preparation		Allowed Welding Positions	Gas Shielding for FCAW	Notes
		T_1	T_2	Root Opening	Groove Angle			
SMAW	TC-U4a	U	U	R = 6	α = 45°	All	—	e, g, k, o
				R = 10	α = 30°	F, V, OH	—	e, g, k, o
GMAW FCAW	TC-U4a-GF	U	U	R = 5	α = 30°	All	Required	a, g, k, o
				R = 10	α = 30°	F	Not req.	a, g, k, o
				R = 6	α = 45°	All	Not req.	a, g, k, o
SAW	TC-U4a-S	U	U	R = 10	α = 30°	F	—	g, k, o
				R = 6	α = 45°			

Tolerances:
As Detailed (see 5.4.1.1): R = +2, –0; α = +10°, –0°
As Fit-Up (see 5.4.1.8): +6, –2; +10°, –5°

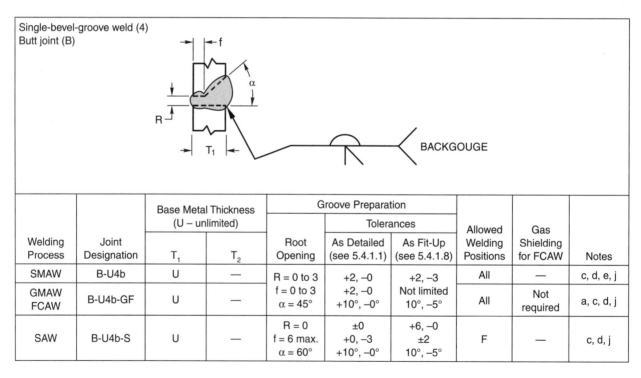

Single-bevel-groove weld (4)
Butt joint (B)

Welding Process	Joint Designation	Base Metal Thickness (U – unlimited)		Groove Preparation			Allowed Welding Positions	Gas Shielding for FCAW	Notes
		T_1	T_2	Root Opening	As Detailed (see 5.4.1.1)	As Fit-Up (see 5.4.1.8)			
SMAW	B-U4b	U	—	R = 0 to 3 f = 0 to 3 α = 45°	+2, –0 +2, –0 +10°, –0°	+2, –3 Not limited 10°, –5°	All	—	c, d, e, j
GMAW FCAW	B-U4b-GF	U	—				All	Not required	a, c, d, j
SAW	B-U4b-S	U	—	R = 0 f = 6 max. α = 60°	±0 +0, –3 +10°, –0°	+6, –0 ±2 10°, –5°	F	—	c, d, j

Figure 5.1 (Continued)—Prequalified CJP Groove Welded Joint Details (See 5.4.1) (Dimensions in Millimeters)

See Notes on Page 81

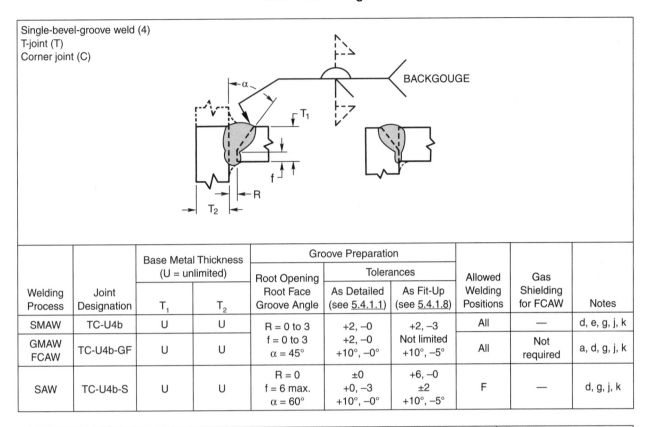

Welding Process	Joint Designation	Base Metal Thickness (U = unlimited)		Groove Preparation			Allowed Welding Positions	Gas Shielding for FCAW	Notes
				Root Opening Root Face Groove Angle	Tolerances				
		T_1	T_2		As Detailed (see 5.4.1.1)	As Fit-Up (see 5.4.1.8)			
SMAW	TC-U4b	U	U	R = 0 to 3 f = 0 to 3 α = 45°	+2, −0 +2, −0 +10°, −0°	+2, −3 Not limited +10°, −5°	All	—	d, e, g, j, k
GMAW FCAW	TC-U4b-GF	U	U				All	Not required	a, d, g, j, k
SAW	TC-U4b-S	U	U	R = 0 f = 6 max. α = 60°	±0 +0, −3 +10°, −0°	+6, −0 ±2 +10°, −5°	F	—	d, g, j, k

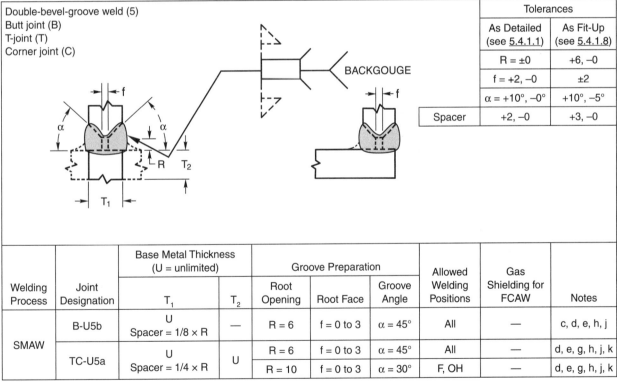

Double-bevel-groove weld (5)
Butt joint (B)
T-joint (T)
Corner joint (C)

	Tolerances	
	As Detailed (see 5.4.1.1)	As Fit-Up (see 5.4.1.8)
R =	±0	+6, −0
f =	+2, −0	±2
α =	+10°, −0°	+10°, −5°
Spacer	+2, −0	+3, −0

Welding Process	Joint Designation	Base Metal Thickness (U = unlimited)		Groove Preparation			Allowed Welding Positions	Gas Shielding for FCAW	Notes
		T_1	T_2	Root Opening	Root Face	Groove Angle			
SMAW	B-U5b	U Spacer = 1/8 × R	—	R = 6	f = 0 to 3	α = 45°	All	—	c, d, e, h, j
	TC-U5a	U Spacer = 1/4 × R	U	R = 6	f = 0 to 3	α = 45°	All	—	d, e, g, h, j, k
				R = 10	f = 0 to 3	α = 30°	F, OH	—	d, e, g, h, j, k

Figure 5.1 (Continued)—Prequalified CJP Groove Welded Joint Details (See 5.4.1)
(Dimensions in Millimeters)

See Notes on Page 81

Double-bevel-groove weld (5)
Butt joint (B)

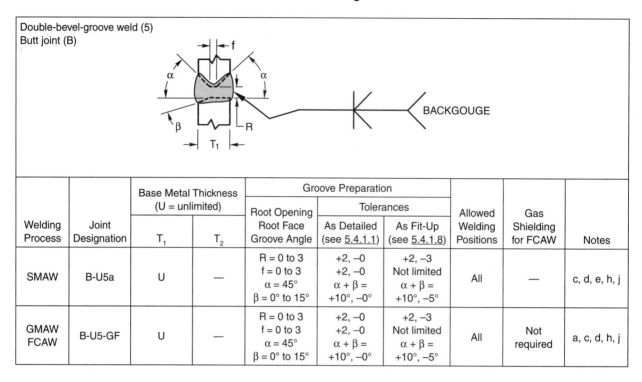

Welding Process	Joint Designation	Base Metal Thickness (U = unlimited)		Groove Preparation			Allowed Welding Positions	Gas Shielding for FCAW	Notes
		T_1	T_2	Root Opening Root Face Groove Angle	Tolerances				
					As Detailed (see 5.4.1.1)	As Fit-Up (see 5.4.1.8)			
SMAW	B-U5a	U	—	R = 0 to 3 f = 0 to 3 α = 45° β = 0° to 15°	+2, –0 +2, –0 α + β = +10°, –0°	+2, –3 Not limited α + β = +10°, –5°	All	—	c, d, e, h, j
GMAW FCAW	B-U5-GF	U	—	R = 0 to 3 f = 0 to 3 α = 45° β = 0° to 15°	+2, –0 +2, –0 α + β = +10°, –0°	+2, –3 Not limited α + β = +10°, –5°	All	Not required	a, c, d, h, j

Double-bevel-groove weld (5)
T-joint (T)
Corner joint (C)

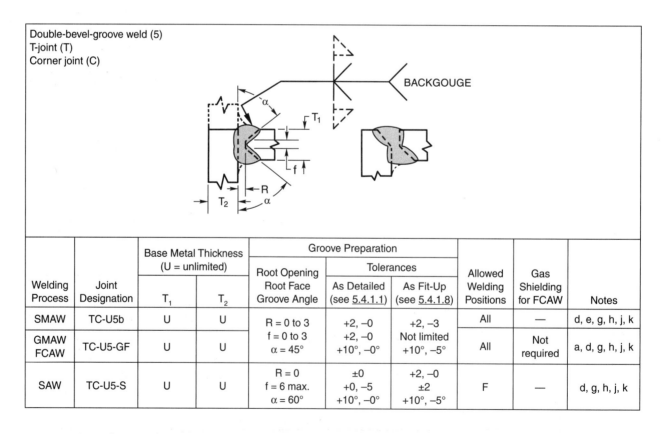

Welding Process	Joint Designation	Base Metal Thickness (U = unlimited)		Groove Preparation			Allowed Welding Positions	Gas Shielding for FCAW	Notes
		T_1	T_2	Root Opening Root Face Groove Angle	Tolerances				
					As Detailed (see 5.4.1.1)	As Fit-Up (see 5.4.1.8)			
SMAW	TC-U5b	U	U	R = 0 to 3 f = 0 to 3 α = 45°	+2, –0 +2, –0 +10°, –0°	+2, –3 Not limited +10°, –5°	All	—	d, e, g, h, j, k
GMAW FCAW	TC-U5-GF	U	U	R = 0 to 3 f = 0 to 3 α = 45°	+2, –0 +2, –0 +10°, –0°	+2, –3 Not limited +10°, –5°	All	Not required	a, d, g, h, j, k
SAW	TC-U5-S	U	U	R = 0 f = 6 max. α = 60°	±0 +0, –5 +10°, –0°	+2, –0 ±2 +10°, –5°	F	—	d, g, h, j, k

Figure 5.1 (Continued)—Prequalified CJP Groove Welded Joint Details (See 5.4.1)
(Dimensions in Millimeters)

See Notes on Page 81

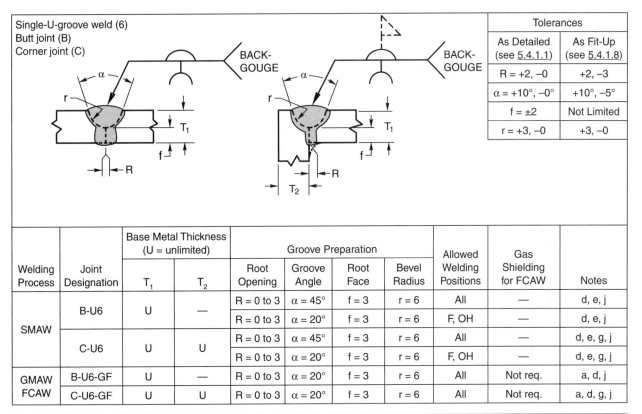

Welding Process	Joint Designation	Base Metal Thickness (U = unlimited)		Groove Preparation				Allowed Welding Positions	Gas Shielding for FCAW	Notes
		T_1	T_2	Root Opening	Groove Angle	Root Face	Bevel Radius			
SMAW	B-U6	U	—	R = 0 to 3	α = 45°	f = 3	r = 6	All	—	d, e, j
				R = 0 to 3	α = 20°	f = 3	r = 6	F, OH	—	d, e, j
	C-U6	U	U	R = 0 to 3	α = 45°	f = 3	r = 6	All	—	d, e, g, j
				R = 0 to 3	α = 20°	f = 3	r = 6	F, OH	—	d, e, g, j
GMAW FCAW	B-U6-GF	U	—	R = 0 to 3	α = 20°	f = 3	r = 6	All	Not req.	a, d, j
	C-U6-GF	U	U	R = 0 to 3	α = 20°	f = 3	r = 6	All	Not req.	a, d, g, j

Single-U-groove weld (6)
Butt joint (B)
Corner joint (C)

Tolerances — As Detailed (see 5.4.1.1): R = +2, –0; α = +10°, –0°; f = ±2; r = +3, –0
As Fit-Up (see 5.4.1.8): +2, –3; +10°, –5°; Not Limited; +3, –0

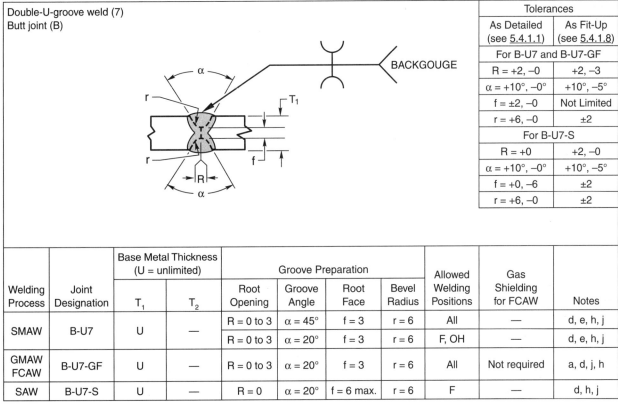

Double-U-groove weld (7)
Butt joint (B)

Tolerances — As Detailed (see 5.4.1.1) / As Fit-Up (see 5.4.1.8)
For B-U7 and B-U7-GF:
R = +2, –0 / +2, –3
α = +10°, –0° / +10°, –5°
f = ±2, –0 / Not Limited
r = +6, –0 / ±2
For B-U7-S:
R = +0 / +2, –0
α = +10°, –0° / +10°, –5°
f = +0, –6 / ±2
r = +6, –0 / ±2

Welding Process	Joint Designation	Base Metal Thickness (U = unlimited)		Groove Preparation				Allowed Welding Positions	Gas Shielding for FCAW	Notes
		T_1	T_2	Root Opening	Groove Angle	Root Face	Bevel Radius			
SMAW	B-U7	U	—	R = 0 to 3	α = 45°	f = 3	r = 6	All	—	d, e, h, j
				R = 0 to 3	α = 20°	f = 3	r = 6	F, OH	—	d, e, h, j
GMAW FCAW	B-U7-GF	U	—	R = 0 to 3	α = 20°	f = 3	r = 6	All	Not required	a, d, j, h
SAW	B-U7-S	U	—	R = 0	α = 20°	f = 6 max.	r = 6	F	—	d, h, j

Figure 5.1 (Continued)—Prequalified CJP Groove Welded Joint Details (See 5.4.1)
(Dimensions in Millimeters)

See Notes on Page 81

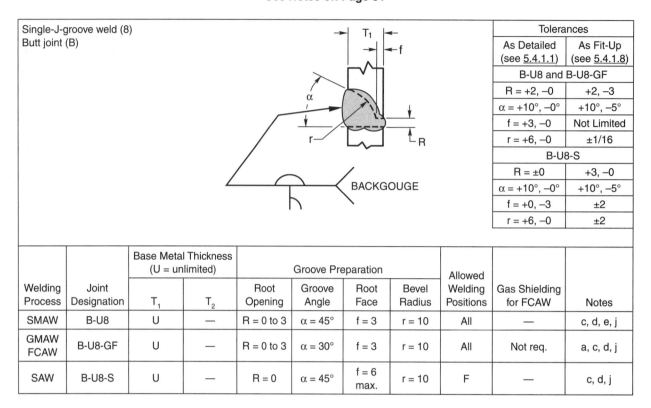

Single-J-groove weld (8)
Butt joint (B)

		Tolerances	
		As Detailed (see 5.4.1.1)	As Fit-Up (see 5.4.1.8)
B-U8 and B-U8-GF			
	R =	+2, –0	+2, –3
	α =	+10°, –0°	+10°, –5°
	f =	+3, –0	Not Limited
	r =	+6, –0	±1/16
B-U8-S			
	R =	±0	+3, –0
	α =	+10°, –0°	+10°, –5°
	f =	+0, –3	±2
	r =	+6, –0	±2

Welding Process	Joint Designation	Base Metal Thickness (U = unlimited)		Groove Preparation				Allowed Welding Positions	Gas Shielding for FCAW	Notes
		T_1	T_2	Root Opening	Groove Angle	Root Face	Bevel Radius			
SMAW	B-U8	U	—	R = 0 to 3	α = 45°	f = 3	r = 10	All	—	c, d, e, j
GMAW FCAW	B-U8-GF	U	—	R = 0 to 3	α = 30°	f = 3	r = 10	All	Not req.	a, c, d, j
SAW	B-U8-S	U	—	R = 0	α = 45°	f = 6 max.	r = 10	F	—	c, d, j

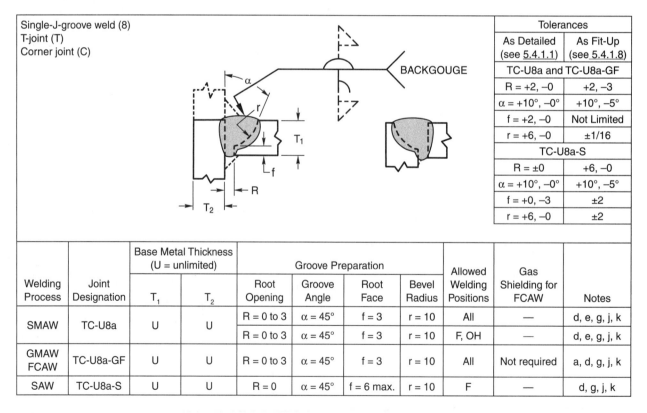

Single-J-groove weld (8)
T-joint (T)
Corner joint (C)

		Tolerances	
		As Detailed (see 5.4.1.1)	As Fit-Up (see 5.4.1.8)
TC-U8a and TC-U8a-GF			
	R =	+2, –0	+2, –3
	α =	+10°, –0°	+10°, –5°
	f =	+2, –0	Not Limited
	r =	+6, –0	±1/16
TC-U8a-S			
	R =	±0	+6, –0
	α =	+10°, –0°	+10°, –5°
	f =	+0, –3	±2
	r =	+6, –0	±2

Welding Process	Joint Designation	Base Metal Thickness (U = unlimited)		Groove Preparation				Allowed Welding Positions	Gas Shielding for FCAW	Notes
		T_1	T_2	Root Opening	Groove Angle	Root Face	Bevel Radius			
SMAW	TC-U8a	U	U	R = 0 to 3	α = 45°	f = 3	r = 10	All	—	d, e, g, j, k
				R = 0 to 3	α = 45°	f = 3	r = 10	F, OH	—	d, e, g, j, k
GMAW FCAW	TC-U8a-GF	U	U	R = 0 to 3	α = 45°	f = 3	r = 10	All	Not required	a, d, g, j, k
SAW	TC-U8a-S	U	U	R = 0	α = 45°	f = 6 max.	r = 10	F	—	d, g, j, k

Figure 5.1 (Continued)—Prequalified CJP Groove Welded Joint Details (See 5.4.1) (Dimensions in Millimeters)

See Notes on Page 81

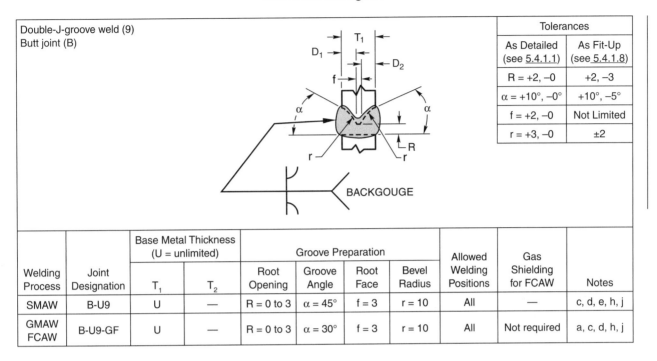

Double-J-groove weld (9)
Butt joint (B)

		Tolerances	
		As Detailed (see 5.4.1.1)	As Fit-Up (see 5.4.1.8)
		R = +2, –0	+2, –3
		α = +10°, –0°	+10°, –5°
		f = +2, –0	Not Limited
		r = +3, –0	±2

Welding Process	Joint Designation	Base Metal Thickness (U = unlimited)		Groove Preparation				Allowed Welding Positions	Gas Shielding for FCAW	Notes
		T_1	T_2	Root Opening	Groove Angle	Root Face	Bevel Radius			
SMAW	B-U9	U	—	R = 0 to 3	α = 45°	f = 3	r = 10	All	—	c, d, e, h, j
GMAW FCAW	B-U9-GF	U	—	R = 0 to 3	α = 30°	f = 3	r = 10	All	Not required	a, c, d, h, j

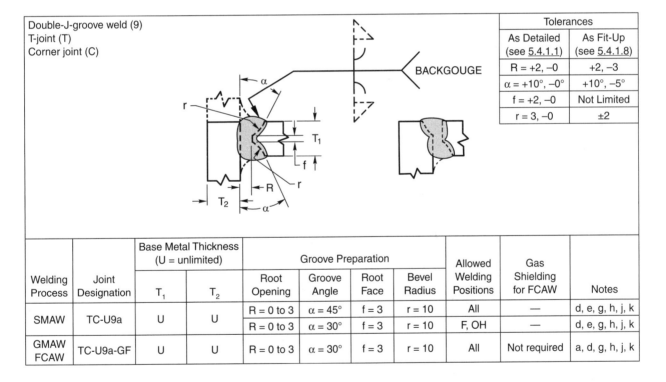

Double-J-groove weld (9)
T-joint (T)
Corner joint (C)

		Tolerances	
		As Detailed (see 5.4.1.1)	As Fit-Up (see 5.4.1.8)
		R = +2, –0	+2, –3
		α = +10°, –0°	+10°, –5°
		f = +2, –0	Not Limited
		r = 3, –0	±2

Welding Process	Joint Designation	Base Metal Thickness (U = unlimited)		Groove Preparation				Allowed Welding Positions	Gas Shielding for FCAW	Notes
		T_1	T_2	Root Opening	Groove Angle	Root Face	Bevel Radius			
SMAW	TC-U9a	U	U	R = 0 to 3	α = 45°	f = 3	r = 10	All	—	d, e, g, h, j, k
				R = 0 to 3	α = 30°	f = 3	r = 10	F, OH	—	d, e, g, h, j, k
GMAW FCAW	TC-U9a-GF	U	U	R = 0 to 3	α = 30°	f = 3	r = 10	All	Not required	a, d, g, h, j, k

Figure 5.1 (Continued)—Prequalified CJP Groove Welded Joint Details (See 5.4.1) (Dimensions in Millimeters)

See Notes on Page 81

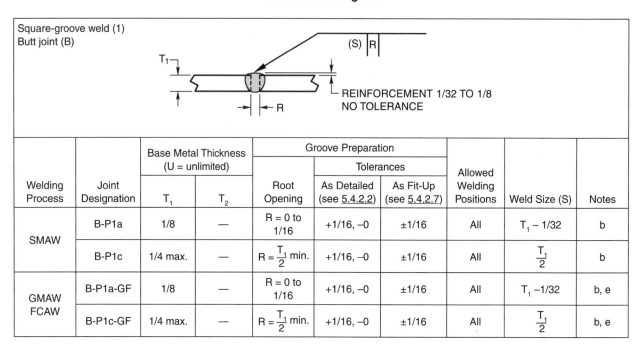

Welding Process	Joint Designation	Base Metal Thickness (U = unlimited)		Groove Preparation			Allowed Welding Positions	Weld Size (S)	Notes
		T_1	T_2	Root Opening	As Detailed (see 5.4.2.2)	As Fit-Up (see 5.4.2.7)			
SMAW	B-P1a	1/8	—	R = 0 to 1/16	+1/16, –0	±1/16	All	$T_1 - 1/32$	b
	B-P1c	1/4 max.	—	$R = \frac{T_1}{2}$ min.	+1/16, –0	±1/16	All	$\frac{T_1}{2}$	b
GMAW FCAW	B-P1a-GF	1/8	—	R = 0 to 1/16	+1/16, –0	±1/16	All	$T_1 - 1/32$	b, e
	B-P1c-GF	1/4 max.	—	$R = \frac{T_1}{2}$ min.	+1/16, –0	±1/16	All	$\frac{T_1}{2}$	b, e

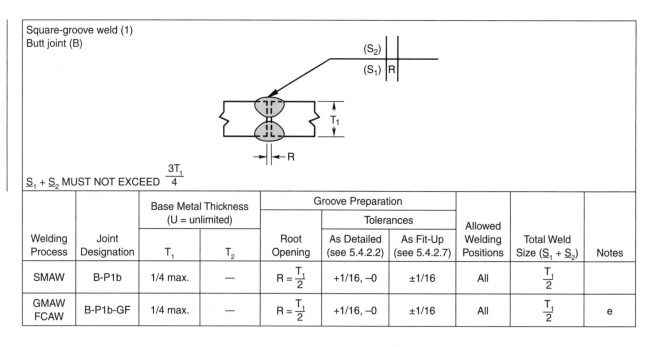

Welding Process	Joint Designation	Base Metal Thickness (U = unlimited)		Groove Preparation			Allowed Welding Positions	Total Weld Size ($S_1 + S_2$)	Notes
		T_1	T_2	Root Opening	As Detailed (see 5.4.2.2)	As Fit-Up (see 5.4.2.7)			
SMAW	B-P1b	1/4 max.	—	$R = \frac{T_1}{2}$	+1/16, –0	±1/16	All	$\frac{T_1}{2}$	
GMAW FCAW	B-P1b-GF	1/4 max.	—	$R = \frac{T_1}{2}$	+1/16, –0	±1/16	All	$\frac{T_1}{2}$	e

Figure 5.2—Prequalified PJP Groove Weld Joint Details (see 5.4.2) (Dimensions in Inches)

See Notes on Page 81

Single-V-groove weld (2)
Butt joint (B)
Corner joint (C)

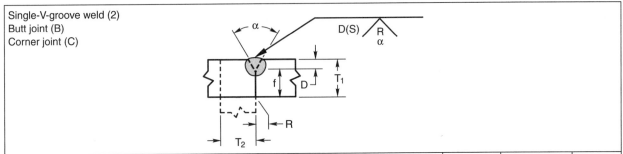

Welding Process	Joint Designation	Base Metal Thickness (U = unlimited)		Groove Preparation			Allowed Welding Positions	Weld Size (S)	Notes
		T_1	T_2	Root Opening Root Face Groove Angle	Tolerances				
					As Detailed (see 5.4.2.2)	As Fit-Up (see 5.4.2.7)			
SMAW	BC-P2	1/4 min.	U	R = 0 f = 1/32 min. α = 60°	+1/16, –0 +U, –0 +10°, –0°	+1/8, –1/16 ±1/16 +10°, –5°	All	$\underline{D}$	b, e, f, j
GMAW FCAW	BC-P2-GF	1/4 min.	U	R = 0 f = 1/8 min. α = 60°	+1/16, –0 +U, –0 +10°, –0°	+1/8, –1/16 ±1/16 +10°, –5°	All	$\underline{D}$	a, b, f, j
SAW	BC-P2-S	7/16 min.	U	R = 0 f = 1/4 min. α = 60°	±0 +U, –0 +10°, –0°	+1/16, –0 ±1/16 +10°, –5°	F	$\underline{D}$	b, f, j

Double-V-groove weld (3)
Butt joint (B)

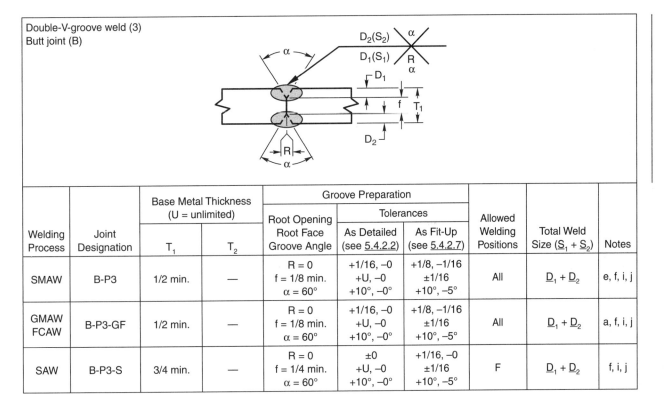

Welding Process	Joint Designation	Base Metal Thickness (U = unlimited)		Groove Preparation			Allowed Welding Positions	Total Weld Size ($\underline{S}_1 + \underline{S}_2$)	Notes
		T_1	T_2	Root Opening Root Face Groove Angle	Tolerances				
					As Detailed (see 5.4.2.2)	As Fit-Up (see 5.4.2.7)			
SMAW	B-P3	1/2 min.	—	R = 0 f = 1/8 min. α = 60°	+1/16, –0 +U, –0 +10°, –0°	+1/8, –1/16 ±1/16 +10°, –5°	All	$\underline{D}_1 + \underline{D}_2$	e, f, i, j
GMAW FCAW	B-P3-GF	1/2 min.	—	R = 0 f = 1/8 min. α = 60°	+1/16, –0 +U, –0 +10°, –0°	+1/8, –1/16 ±1/16 +10°, –5°	All	$\underline{D}_1 + \underline{D}_2$	a, f, i, j
SAW	B-P3-S	3/4 min.	—	R = 0 f = 1/4 min. α = 60°	±0 +U, –0 +10°, –0°	+1/16, –0 ±1/16 +10°, –5°	F	$\underline{D}_1 + \underline{D}_2$	f, i, j

Figure 5.2 (Continued)—Prequalified PJP Groove Welded Joint Details (see 5.4.2) (Dimensions in Inches)

See Notes on Page 81

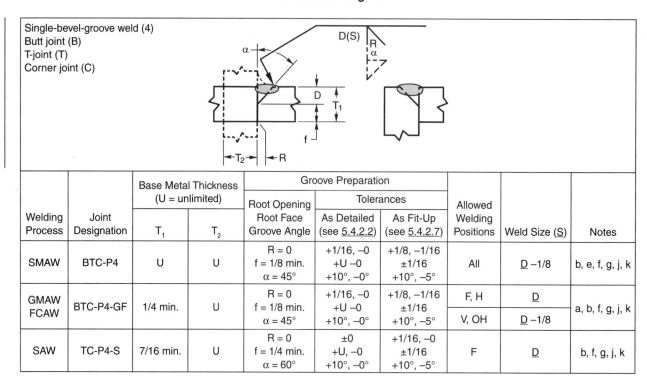

Single-bevel-groove weld (4)
Butt joint (B)
T-joint (T)
Corner joint (C)

Welding Process	Joint Designation	Base Metal Thickness (U = unlimited)		Groove Preparation			Allowed Welding Positions	Weld Size ($\underline{S}$)	Notes
		T_1	T_2	Root Opening Root Face Groove Angle	Tolerances				
					As Detailed (see 5.4.2.2)	As Fit-Up (see 5.4.2.7)			
SMAW	BTC-P4	U	U	R = 0 f = 1/8 min. α = 45°	+1/16, −0 +U −0 +10°, −0°	+1/8, −1/16 ±1/16 +10°, −5°	All	$\underline{D}$ −1/8	b, e, f, g, j, k
GMAW FCAW	BTC-P4-GF	1/4 min.	U	R = 0 f = 1/8 min. α = 45°	+1/16, −0 +U −0 +10°, −0°	+1/8, −1/16 ±1/16 +10°, −5°	F, H	$\underline{D}$	a, b, f, g, j, k
							V, OH	$\underline{D}$ −1/8	
SAW	TC-P4-S	7/16 min.	U	R = 0 f = 1/4 min. α = 60°	±0 +U, −0 +10°, −0°	+1/16, −0 ±1/16 +10°, −5°	F	$\underline{D}$	b, f, g, j, k

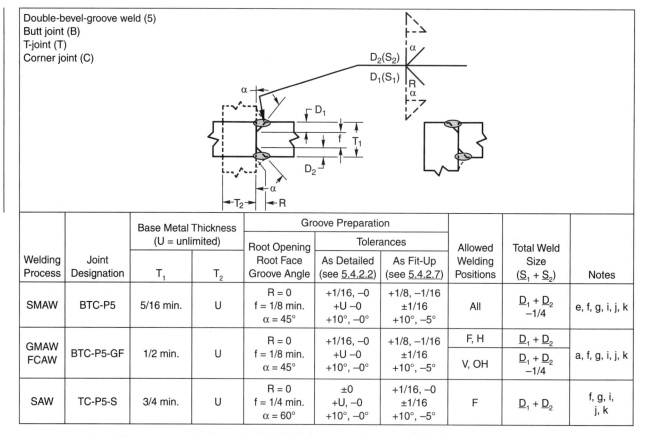

Double-bevel-groove weld (5)
Butt joint (B)
T-joint (T)
Corner joint (C)

Welding Process	Joint Designation	Base Metal Thickness (U = unlimited)		Groove Preparation			Allowed Welding Positions	Total Weld Size ($\underline{S}_1 + \underline{S}_2$)	Notes
		T_1	T_2	Root Opening Root Face Groove Angle	Tolerances				
					As Detailed (see 5.4.2.2)	As Fit-Up (see 5.4.2.7)			
SMAW	BTC-P5	5/16 min.	U	R = 0 f = 1/8 min. α = 45°	+1/16, −0 +U −0 +10°, −0°	+1/8, −1/16 ±1/16 +10°, −5°	All	$\underline{D}_1 + \underline{D}_2$ −1/4	e, f, g, i, j, k
GMAW FCAW	BTC-P5-GF	1/2 min.	U	R = 0 f = 1/8 min. α = 45°	+1/16, −0 +U −0 +10°, −0°	+1/8, −1/16 ±1/16 +10°, −5°	F, H	$\underline{D}_1 + \underline{D}_2$	a, f, g, i, j, k
							V, OH	$\underline{D}_1 + \underline{D}_2$ −1/4	
SAW	TC-P5-S	3/4 min.	U	R = 0 f = 1/4 min. α = 60°	±0 +U, −0 +10°, −0°	+1/16, −0 ±1/16 +10°, −5°	F	$\underline{D}_1 + \underline{D}_2$	f, g, i, j, k

Figure 5.2 (Continued)—Prequalified PJP Groove Welded Joint Details (see 5.4.2)
(Dimensions in Inches)

See Notes on Page 81

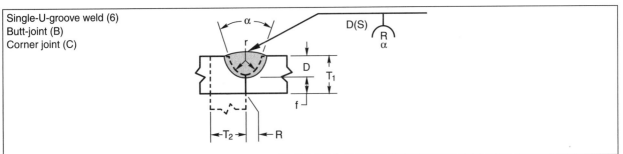

Single-U-groove weld (6)
Butt-joint (B)
Corner joint (C)

Welding Process	Joint Designation	Base Metal Thickness (U = unlimited)		Groove Preparation			Allowed Welding Positions	Weld Size ($\underline{S}$)	Notes
		T_1	T_2	Root Opening Root Face Bevel Radius Groove Angle	Tolerances As Detailed (see 5.4.2.2)	As Fit-Up (see 5.4.2.7)			
SMAW	BC-P6	1/4 min.	U	R = 0 f = 1/32 min. r = 1/4 α = 45°	+1/16, –0 +U, –0 +1/4, –0 +10°, –0°	+1/8, –1/16 ±1/16 ±1/16 +10°, –5°	All	$\underline{D}$	b, e, f, j
GMAW FCAW	BC-P6-GF	1/4 min.	U	R = 0 f = 1/8 min. r = 1/4 α = 20°	+1/16, –0 +U, –0 +1/4, –0 +10°, –0°	+1/8, –1/16 ±1/16 ±1/16 +10°, –5°	All	$\underline{D}$	a, b, f, j
SAW	BC-P6-S	7/16 min.	U	R = 0 f = 1/4 min. r = 1/4 α = 20°	±0 +U, –0 +1/4, –0 +10°, –0°	+1/16, –0 ±1/16 ±1/16 +10°, –5°	F	$\underline{D}$	b, f, j

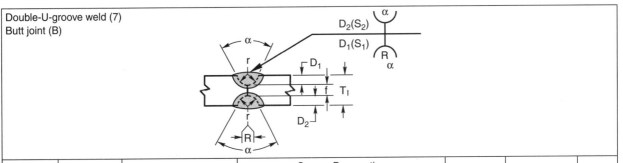

Double-U-groove weld (7)
Butt joint (B)

Welding Process	Joint Designation	Base Metal Thickness (U = unlimited)		Groove Preparation			Allowed Welding Positions	Total Weld Size ($\underline{S}_1 + \underline{S}_2$)	Notes
		T_1	T_2	Root Opening Root Face Bevel Radius Groove Angle	Tolerances As Detailed (see 5.4.2.2)	As Fit-Up (see 5.4.2.7)			
SMAW	B-P7	1/2 min.	—	R = 0 f = 1/8 min. r = 1/4 α = 45°	+1/16, –0 +U, –0 +1/4, –0 +10°, –0°	+1/8, –1/16 ±1/16 ±1/16 +10°, –5°	All	$\underline{D}_1 + \underline{D}_2$	e, f, i, j
GMAW FCAW	B-P7-GF	1/2 min.	—	R = 0 f = 1/8 min. r = 1/4 α = 20°	+1/16, –0 +U, –0 +1/4, –0 +10°, –0°	+1/8, –1/16 ±1/16 ±1/16 +10°, –5°	All	$\underline{D}_1 + \underline{D}_2$	a, f, i, j
SAW	B-P7-S	3/4 min.	—	R = 0 f = 1/4 min. r = 1/4 α = 20°	±0 +U, –0 +1/4, –0 +10°, –0°	+1/16, –0 ±1/16 ±1/16 +10°, –5°	F	$\underline{D}_1 + \underline{D}_2$	f, i, j

**Figure 5.2 (Continued)—Prequalified PJP Groove Welded Joint Details (see 5.4.2)
(Dimensions in Inches)**

See Notes on Page 81

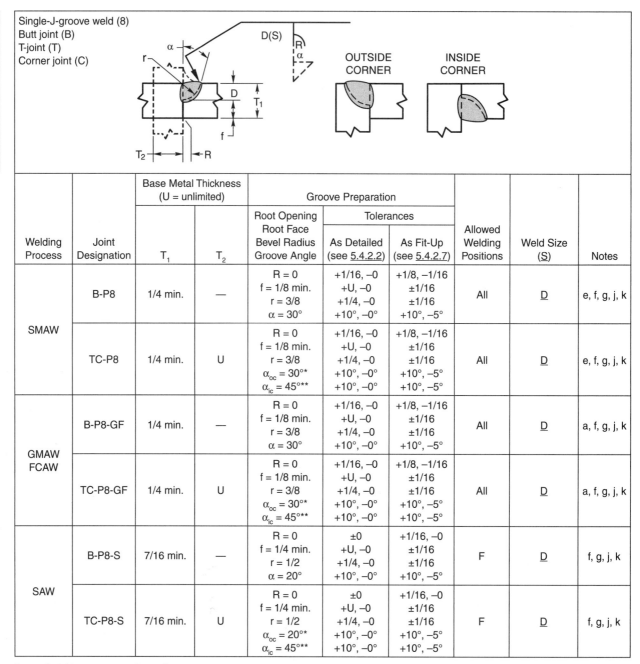

Single-J-groove weld (8)
Butt joint (B)
T-joint (T)
Corner joint (C)

Welding Process	Joint Designation	Base Metal Thickness (U = unlimited)		Groove Preparation			Allowed Welding Positions	Weld Size (S)	Notes
		T_1	T_2	Root Opening Root Face Bevel Radius Groove Angle	Tolerances				
					As Detailed (see 5.4.2.2)	As Fit-Up (see 5.4.2.7)			
SMAW	B-P8	1/4 min.	—	R = 0 f = 1/8 min. r = 3/8 α = 30°	+1/16, −0 +U, −0 +1/4, −0 +10°, −0°	+1/8, −1/16 ±1/16 ±1/16 +10°, −5°	All	D	e, f, g, j, k
	TC-P8	1/4 min.	U	R = 0 f = 1/8 min. r = 3/8 $α_{oc}$ = 30°* $α_{ic}$ = 45°**	+1/16, −0 +U, −0 +1/4, −0 +10°, −0° +10°, −0°	+1/8, −1/16 ±1/16 ±1/16 +10°, −5° +10°, −5°	All	D	e, f, g, j, k
GMAW FCAW	B-P8-GF	1/4 min.	—	R = 0 f = 1/8 min. r = 3/8 α = 30°	+1/16, −0 +U, −0 +1/4, −0 +10°, −0°	+1/8, −1/16 ±1/16 ±1/16 +10°, −5°	All	D	a, f, g, j, k
	TC-P8-GF	1/4 min.	U	R = 0 f = 1/8 min. r = 3/8 $α_{oc}$ = 30°* $α_{ic}$ = 45°**	+1/16, −0 +U, −0 +1/4, −0 +10°, −0° +10°, −0°	+1/8, −1/16 ±1/16 ±1/16 +10°, −5° +10°, −5°	All	D	a, f, g, j, k
SAW	B-P8-S	7/16 min.	—	R = 0 f = 1/4 min. r = 1/2 α = 20°	±0 +U, −0 +1/4, −0 +10°, −0°	+1/16, −0 ±1/16 ±1/16 +10°, −5°	F	D	f, g, j, k
	TC-P8-S	7/16 min.	U	R = 0 f = 1/4 min. r = 1/2 $α_{oc}$ = 20°* $α_{ic}$ = 45°**	±0 +U, −0 +1/4, −0 +10°, −0° +10°, −0°	+1/16, −0 ±1/16 ±1/16 +10°, −5° +10°, −5°	F	D	f, g, j, k

*$α_{oc}$ = Outside corner groove angle.
**$α_{ic}$ = Inside corner groove angle.

Figure 5.2 (Continued)—Prequalified PJP Groove Welded Joint Details (see 5.4.2)
(Dimensions in Inches)

See Notes on Page 81

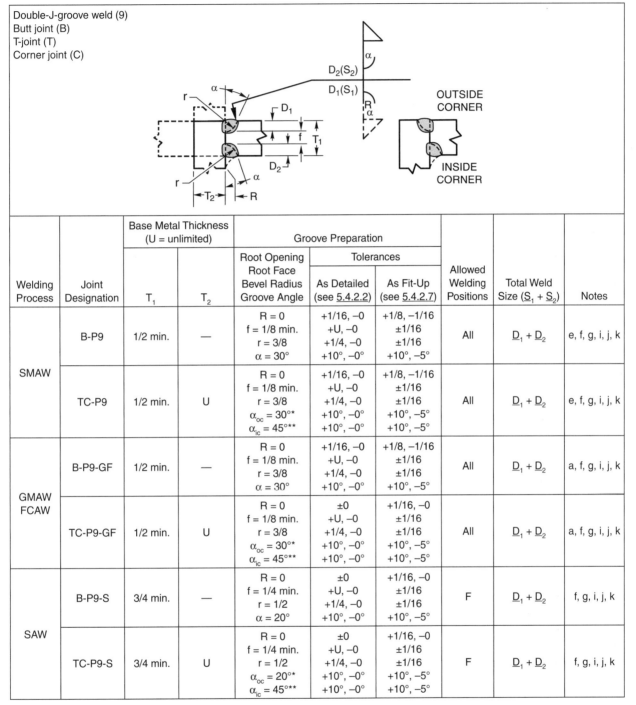

Welding Process	Joint Designation	Base Metal Thickness (U = unlimited)		Groove Preparation			Allowed Welding Positions	Total Weld Size ($\underline{S}_1 + \underline{S}_2$)	Notes
		T_1	T_2	Root Opening Root Face Bevel Radius Groove Angle	Tolerances				
					As Detailed (see 5.4.2.2)	As Fit-Up (see 5.4.2.7)			
SMAW	B-P9	1/2 min.	—	R = 0 f = 1/8 min. r = 3/8 α = 30°	+1/16, –0 +U, –0 +1/4, –0 +10°, –0°	+1/8, –1/16 ±1/16 ±1/16 +10°, –5°	All	$\underline{D}_1 + \underline{D}_2$	e, f, g, i, j, k
	TC-P9	1/2 min.	U	R = 0 f = 1/8 min. r = 3/8 $α_{oc}$ = 30°* $α_{ic}$ = 45°**	+1/16, –0 +U, –0 +1/4, –0 +10°, –0° +10°, –0°	+1/8, –1/16 ±1/16 ±1/16 +10°, –5° +10°, –5°	All	$\underline{D}_1 + \underline{D}_2$	e, f, g, i, j, k
GMAW FCAW	B-P9-GF	1/2 min.	—	R = 0 f = 1/8 min. r = 3/8 α = 30°	+1/16, –0 +U, –0 +1/4, –0 +10°, –0°	+1/8, –1/16 ±1/16 ±1/16 +10°, –5°	All	$\underline{D}_1 + \underline{D}_2$	a, f, g, i, j, k
	TC-P9-GF	1/2 min.	U	R = 0 f = 1/8 min. r = 3/8 $α_{oc}$ = 30°* $α_{ic}$ = 45°**	±0 +U, –0 +1/4, –0 +10°, –0° +10°, –0°	+1/16, –0 ±1/16 ±1/16 +10°, –5° +10°, –5°	All	$\underline{D}_1 + \underline{D}_2$	a, f, g, i, j, k
SAW	B-P9-S	3/4 min.	—	R = 0 f = 1/4 min. r = 1/2 α = 20°	±0 +U, –0 +1/4, –0 +10°, –0°	+1/16, –0 ±1/16 ±1/16 +10°, –5°	F	$\underline{D}_1 + \underline{D}_2$	f, g, i, j, k
	TC-P9-S	3/4 min.	U	R = 0 f = 1/4 min. r = 1/2 $α_{oc}$ = 20°* $α_{ic}$ = 45°**	±0 +U, –0 +1/4, –0 +10°, –0° +10°, –0°	+1/16, –0 ±1/16 ±1/16 +10°, –5° +10°, –5°	F	$\underline{D}_1 + \underline{D}_2$	f, g, i, j, k

*$α_{oc}$ = Outside corner groove angle.
**$α_{ic}$ = Inside corner groove angle.

Figure 5.2 (Continued)—Prequalified PJP Groove Welded Joint Details (see 5.4.2)
(Dimensions in Inches)

See Notes on Page 81

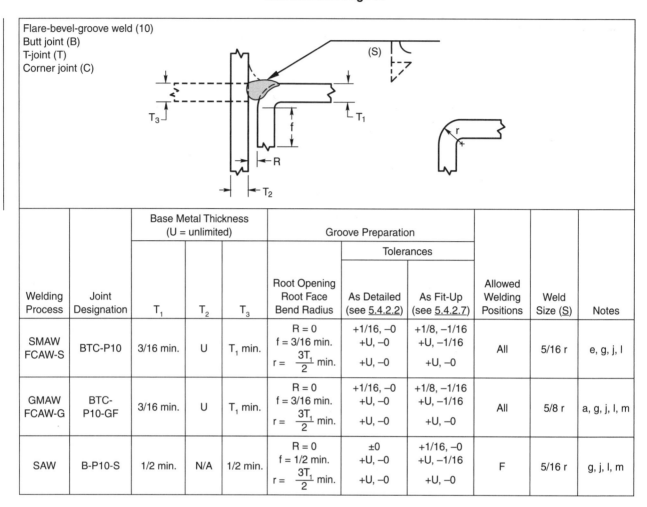

Flare-bevel-groove weld (10)
Butt joint (B)
T-joint (T)
Corner joint (C)

Welding Process	Joint Designation	Base Metal Thickness (U = unlimited)			Groove Preparation			Allowed Welding Positions	Weld Size (S)	Notes
		T_1	T_2	T_3	Root Opening Root Face Bend Radius	Tolerances As Detailed (see 5.4.2.2)	Tolerances As Fit-Up (see 5.4.2.7)			
SMAW FCAW-S	BTC-P10	3/16 min.	U	T_1 min.	R = 0 f = 3/16 min. $r = \frac{3T_1}{2}$ min.	+1/16, –0 +U, –0 +U, –0	+1/8, –1/16 +U, –1/16 +U, –0	All	5/16 r	e, g, j, l
GMAW FCAW-G	BTC-P10-GF	3/16 min.	U	T_1 min.	R = 0 f = 3/16 min. $r = \frac{3T_1}{2}$ min.	+1/16, –0 +U, –0 +U, –0	+1/8, –1/16 +U, –1/16 +U, –0	All	5/8 r	a, g, j, l, m
SAW	B-P10-S	1/2 min.	N/A	1/2 min.	R = 0 f = 1/2 min. $r = \frac{3T_1}{2}$ min.	±0 +U, –0 +U, –0	+1/16, –0 +U, –1/16 +U, –0	F	5/16 r	g, j, l, m

**Figure 5.2 (Continued)—Prequalified PJP Groove Welded Joint Details (see 5.4.2)
(Dimensions in Inches)**

See Notes on Page 81

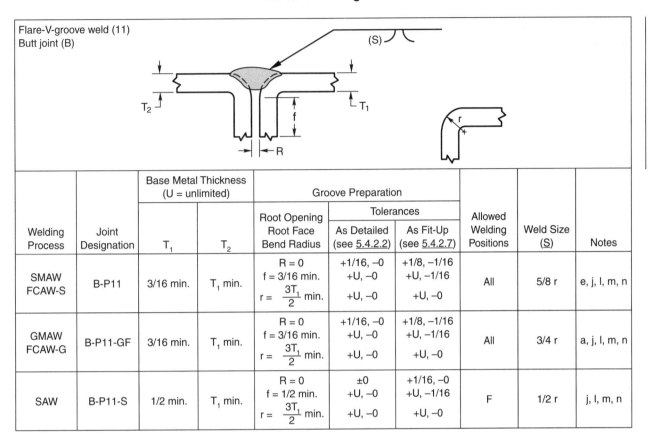

Welding Process	Joint Designation	Base Metal Thickness (U = unlimited)		Groove Preparation			Allowed Welding Positions	Weld Size (S)	Notes
		T_1	T_2	Root Opening Root Face Bend Radius	Tolerances				
					As Detailed (see 5.4.2.2)	As Fit-Up (see 5.4.2.7)			
SMAW FCAW-S	B-P11	3/16 min.	T_1 min.	R = 0 f = 3/16 min. r = $\frac{3T_1}{2}$ min.	+1/16, –0 +U, –0 +U, –0	+1/8, –1/16 +U, –1/16 +U, –0	All	5/8 r	e, j, l, m, n
GMAW FCAW-G	B-P11-GF	3/16 min.	T_1 min.	R = 0 f = 3/16 min. r = $\frac{3T_1}{2}$ min.	+1/16, –0 +U, –0 +U, –0	+1/8, –1/16 +U, –1/16 +U, –0	All	3/4 r	a, j, l, m, n
SAW	B-P11-S	1/2 min.	T_1 min.	R = 0 f = 1/2 min. r = $\frac{3T_1}{2}$ min.	±0 +U, –0 +U, –0	+1/16, –0 +U, –1/16 +U, –0	F	1/2 r	j, l, m, n

Figure 5.2 (Continued)—Prequalified PJP Groove Welded Joint Details (see 5.4.2)
(Dimensions in Inches)

See Notes on Page 81

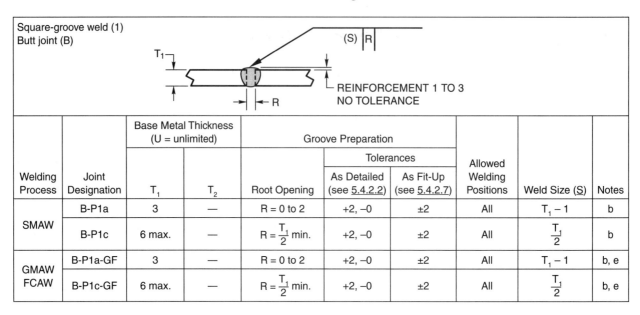

Square-groove weld (1)
Butt joint (B)

REINFORCEMENT 1 TO 3 NO TOLERANCE

Welding Process	Joint Designation	Base Metal Thickness (U = unlimited)		Groove Preparation			Allowed Welding Positions	Weld Size (S)	Notes
		T_1	T_2	Root Opening	Tolerances				
					As Detailed (see 5.4.2.2)	As Fit-Up (see 5.4.2.7)			
SMAW	B-P1a	3	—	R = 0 to 2	+2, –0	±2	All	$T_1 - 1$	b
SMAW	B-P1c	6 max.	—	$R = \frac{T_1}{2}$ min.	+2, –0	±2	All	$\frac{T_1}{2}$	b
GMAW FCAW	B-P1a-GF	3	—	R = 0 to 2	+2, –0	±2	All	$T_1 - 1$	b, e
GMAW FCAW	B-P1c-GF	6 max.	—	$R = \frac{T_1}{2}$ min.	+2, –0	±2	All	$\frac{T_1}{2}$	b, e

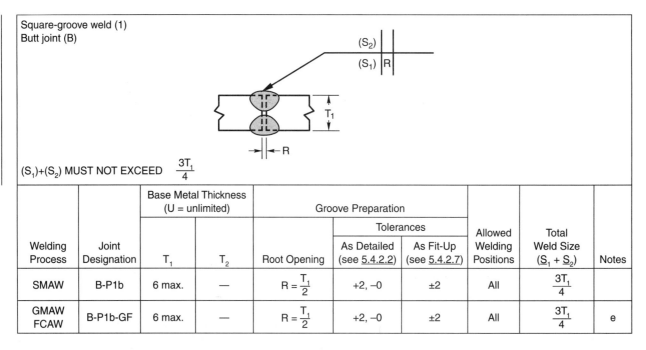

Square-groove weld (1)
Butt joint (B)

$(S_1)+(S_2)$ MUST NOT EXCEED $\frac{3T_1}{4}$

Welding Process	Joint Designation	Base Metal Thickness (U = unlimited)		Groove Preparation			Allowed Welding Positions	Total Weld Size ($S_1 + S_2$)	Notes
		T_1	T_2	Root Opening	Tolerances				
					As Detailed (see 5.4.2.2)	As Fit-Up (see 5.4.2.7)			
SMAW	B-P1b	6 max.	—	$R = \frac{T_1}{2}$	+2, –0	±2	All	$\frac{3T_1}{4}$	
GMAW FCAW	B-P1b-GF	6 max.	—	$R = \frac{T_1}{2}$	+2, –0	±2	All	$\frac{3T_1}{4}$	e

**Figure 5.2 (Continued)—Prequalified PJP Groove Welded Joint Details (See 5.4.2)
(Dimensions in Millimeters)**

See Notes on Page 81

Single-V-groove weld (2)
Butt joint (B)
Corner joint (C)

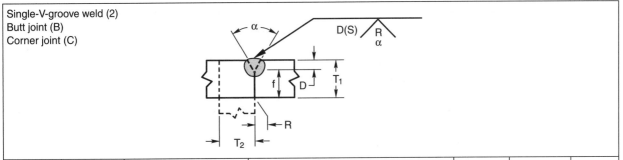

Welding Process	Joint Designation	Base Metal Thickness (U = unlimited)		Groove Preparation			Allowed Welding Positions	Weld Size (S)	Notes
		T_1	T_2	Root Opening Root Face Groove Angle	Tolerances				
					As Detailed (see 5.4.2.2)	As Fit-Up (see 5.4.2.7)			
SMAW	BC-P2	6 min.	U	R = 0 f = 1 min. α = 60°	+2, −0 +U, −0 +10°, −0°	+3, −2 ±2 +10°, −5°	All	D	b, e, f, j
GMAW FCAW	BC-P2-GF	6 min.	U	R = 0 f = 3 min. α = 60°	+2, −0 +U, −0 +10°, −0°	+3, −2 ±2 +10°, −5°	All	D	a, b, f, j
SAW	BC-P2-S	11 min.	U	R = 0 f = 6 min. α = 60°	±0 +U, −0 +10°, −0°	+2, −0 ±2 +10°, −5°	F	D	b, f, j

Double-V-groove weld (3)
Butt joint (B)

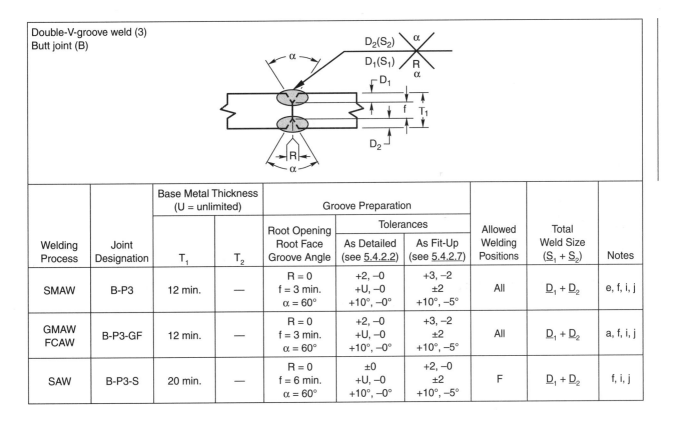

Welding Process	Joint Designation	Base Metal Thickness (U = unlimited)		Groove Preparation			Allowed Welding Positions	Total Weld Size ($S_1 + S_2$)	Notes
		T_1	T_2	Root Opening Root Face Groove Angle	Tolerances				
					As Detailed (see 5.4.2.2)	As Fit-Up (see 5.4.2.7)			
SMAW	B-P3	12 min.	—	R = 0 f = 3 min. α = 60°	+2, −0 +U, −0 +10°, −0°	+3, −2 ±2 +10°, −5°	All	$D_1 + D_2$	e, f, i, j
GMAW FCAW	B-P3-GF	12 min.	—	R = 0 f = 3 min. α = 60°	+2, −0 +U, −0 +10°, −0°	+3, −2 ±2 +10°, −5°	All	$D_1 + D_2$	a, f, i, j
SAW	B-P3-S	20 min.	—	R = 0 f = 6 min. α = 60°	±0 +U, −0 +10°, −0°	+2, −0 ±2 +10°, −5°	F	$D_1 + D_2$	f, i, j

Figure 5.2 (Continued)—Prequalified PJP Groove Welded Joint Details (See 5.4.2)
(Dimensions in Millimeters)

See Notes on Page 81

Single-bevel-groove weld (4)
Butt joint (B)
T-joint (T)
Corner joint (C)

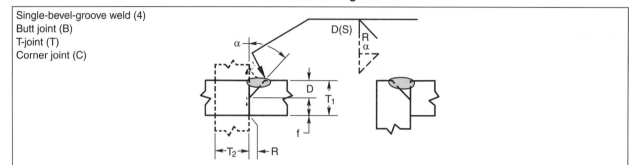

Welding Process	Joint Designation	Base Metal Thickness (U = unlimited)		Groove Preparation			Allowed Welding Positions	Weld Size (S)	Notes
		T_1	T_2	Root Opening Root Face Groove Angle	Tolerances As Detailed (see 5.4.2.2)	As Fit-Up (see 5.4.2.7)			
SMAW	BTC-P4	U	U	R = 0 f = 3 min. α = 45°	+2, –0 +U –0 +10°, –0°	+3, –2 ±2 +10°, –5°	All	D –3	b, e, f, g, j, k
GMAW FCAW	BTC-P4-GF	6 min.	U	R = 0 f = 3 min. α = 45°	+2, –0 +U –0 +10°, –0°	+3, –2 ±2 +10°, –5°	F, H	D	a, b, f, g, j, k
							V, OH	D –3	
SAW	TC-P4-S	11 min.	U	R = 0 f = 6 min. α = 60°	±0 +U, –0 +10°, –0°	+2, –0 ±2 +10°, –5°	F	D	b, f, g, j, k

Double-bevel-groove weld (5)
Butt joint (B)
T-joint (T)
Corner joint (C)

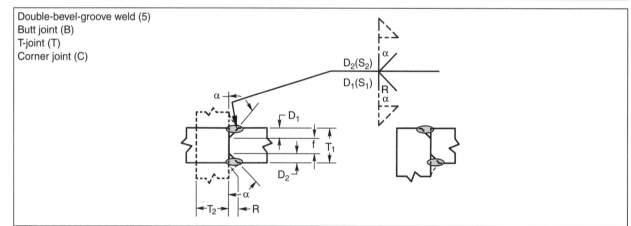

Welding Process	Joint Designation	Base Metal Thickness (U = unlimited)		Groove Preparation			Allowed Welding Positions	Total Weld Size ($S_1 + S_2$)	Notes
		T_1	T_2	Root Opening Root Face Groove Angle	Tolerances As Detailed (see 5.4.2.2)	As Fit-Up (see 5.4.2.7)			
SMAW	BTC-P5	8 min.	U	R = 0 f = 3 min. α = 45°	+2, –0 +U –0 +10°, –0°	+3, –2 ±2 +10°, –5°	All	$D_1 + D_2$ –6	e, f, g, i, j, k
GMAW FCAW	BTC-P5-GF	12 min.	U	R = 0 f = 3 min. α = 45°	+2, –0 +U –0 +10°, –0°	+3, –2 ±2 +10°, –5°	F, H	$D_1 + D_2$	a, f, g, i, j, k
							V, OH	$D_1 + D_2$ –6	
SAW	TC-P5-S	20 min.	U	R = 0 f = 6 min. α = 60°	±0 +U, –0 +10°, –0°	+2, –0 ±2 +10°, –5°	F	$D_1 + D_2$	f, g, i, j, k

**Figure 5.2 (Continued)—Prequalified PJP Groove Welded Joint Details (See 5.4.2)
(Dimensions in Millimeters)**

See Notes on Page 81

Single-U-groove weld (6)
Butt joint (B)
Corner joint (C)

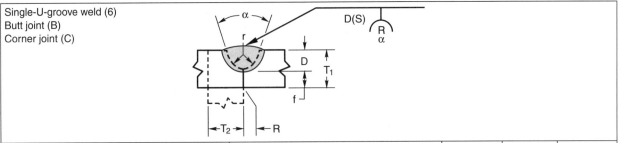

Welding Process	Joint Designation	Base Metal Thickness (U = unlimited)		Groove Preparation			Allowed Welding Positions	Weld Size (S)	Notes
		T_1	T_2	Root Opening Root Face Bevel Radius Groove Angle	Tolerances				
					As Detailed (see 5.4.2.2)	As Fit-Up (see 5.4.2.7)			
SMAW	BC-P6	6 min.	U	R = 0 f = 1 min. r = 6 α = 45°	+2, −0 +U, −0 +6, −0 +10°, −0°	+3, −2 ±2 ±2 +10°, −5°	All	D	b, e, f, j
GMAW FCAW	BC-P6-GF	6 min.	U	R = 0 f = 3 min. r = 6 α = 20°	+2, −0 +U, −0 +6, −0 +10°, −0°	+3, −2 ±2 ±2 +10°, −5°	All	D	a, b, f, j
SAW	BC-P6-S	11 min.	U	R = 0 f = 6 min. r = 6 α = 20°	±0 +U, −0 +6, −0 +10°, −0°	+2, −0 ±2 ±2 +10°, −5°	F	D	b, f, j

Double-U-groove weld (7)
Butt joint (B)

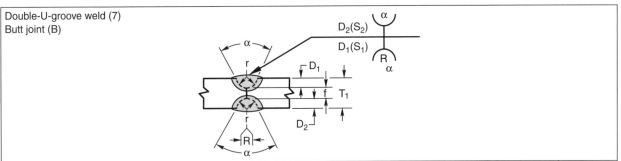

Welding Process	Joint Designation	Base Metal Thickness (U = unlimited)		Groove Preparation			Allowed Welding Positions	Total Weld Size ($S_1 + S_2$)	Notes
		T_1	T_2	Root Opening Root Face Bevel Radius Groove Angle	Tolerances				
					As Detailed (see 5.4.2.2)	As Fit-Up (see 5.4.2.7)			
SMAW	B-P7	12 min.	—	R = 0 f = 3 min. r = 6 α = 45°	+2, −0 +U, −0 +6, −0 +10°, −0°	+3, −2 ±2 ±2 +10°, −5°	All	$D_1 + D_2$	e, f, i, j
GMAW FCAW	B-P7-GF	12 min.	—	R = 0 f = 3 min. r = 6 α = 20°	+2, −0 +U, −0 +6, −0 +10°, −0°	+3, −2 ±2 ±2 +10°, −5°	All	$D_1 + D_2$	a, f, i, j
SAW	B-P7-S	20 min.	—	R = 0 f = 6 min. r = 6 α = 20°	±0 +U, −0 +6, −0 +10°, −0°	+2, −0 ±2 ±2 +10°, −5°	F	$D_1 + D_2$	f, i, j

Figure 5.2 (Continued)—Prequalified PJP Groove Welded Joint Details (See 5.4.2) (Dimensions in Millimeters)

See Notes on Page 81

Single-J-groove weld (8)
Butt joint (B)
T-joint (T)
Corner joint (C)

Welding Process	Joint Designation	Base Metal Thickness (U = unlimited)		Groove Preparation			Allowed Welding Positions	Weld Size (S)	Notes
		T_1	T_2	Root Opening Root Face Bevel Radius Groove Angle	Tolerances				
					As Detailed (see 5.4.2.2)	As Fit-Up (see 5.4.2.7)			
SMAW	B-P8	6 min.	—	R = 0 f = 3 min. r = 10 α = 30°	+2, −0 +U, −0 +6, −0 +10°, −0°	+3, −2 ±2 ±2 +10°, −5°	All	D	e, f, g, j, k
	TC-P8	6 min.	U	R = 0 f = 3 min. r = 10 $α_{oc}$ = 30°* $α_{ic}$ = 45°**	+2, −0 +U, −0 +6, −0 +10°, −0° +10°, −0°	+3, −2 ±2 ±2 +10°, −5° +10°, −5°	All	D	e, f, g, j, k
GMAW FCAW	B-P8-GF	6 min.	—	R = 0 f = 3 min. r = 10 α = 30°	+2, −0 +U, −0 +6, −0 +10°, −0°	+3, −2 ±2 ±2 +10°, −5°	All	D	a, f, g, j, k
	TC-P8-GF	6 min.	U	R = 0 f = 3 min. r = 10 $α_{oc}$ = 30°* $α_{ic}$ = 45°**	+2, −0 +U, −0 +6, −0 +10°, −0° +10°, −0°	+3, −2 ±2 ±2 +10°, −5° +10°, −5°	All	D	a, f, g, j, k
SAW	B-P8-S	11 min.	—	R = 0 f = 6 min. r = 12 α = 20°	±0 +U, −0 +6, −0 +10°, −0°	+2, −0 ±2 ±2 +10°, −5°	F	D	f, g, j, k
	TC-P8-S	11 min.	U	R = 0 f = 6 min. r = 12 $α_{oc}$ = 20°* $α_{ic}$ = 45°**	±0 +U, −0 +6, −0 +10°, −0° +10°, −0°	+2, −0 ±2 ±2 +10°, −5° +10°, −5°	F	D	f, g, j, k

*$α_{oc}$ = Outside corner groove angle.
**$α_{ic}$ = Inside corner groove angle.

Figure 5.2 (Continued)—Prequalified PJP Groove Welded Joint Details (See 5.4.2)
(Dimensions in Millimeters)

See Notes on Page 81

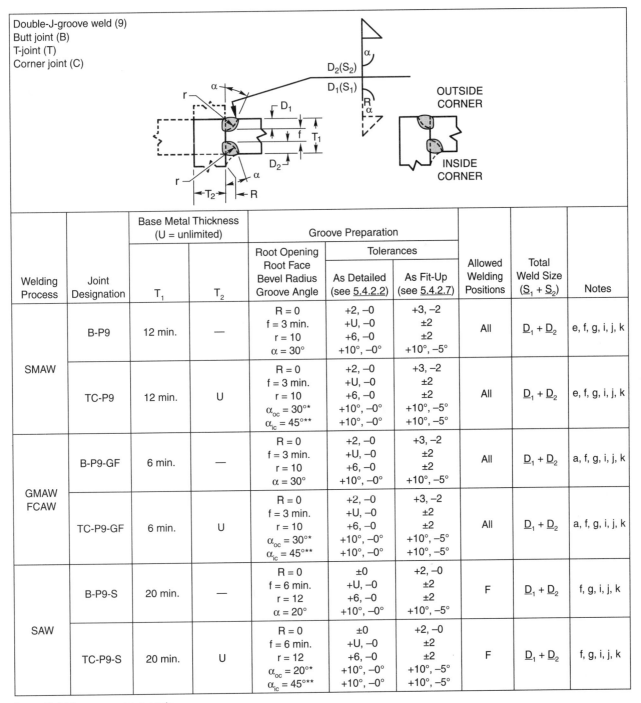

Welding Process	Joint Designation	Base Metal Thickness (U = unlimited)		Groove Preparation			Allowed Welding Positions	Total Weld Size ($\underline{S}_1 + \underline{S}_2$)	Notes
		T_1	T_2	Root Opening Root Face Bevel Radius Groove Angle	Tolerances				
					As Detailed (see 5.4.2.2)	As Fit-Up (see 5.4.2.7)			
SMAW	B-P9	12 min.	—	R = 0 f = 3 min. r = 10 α = 30°	+2, –0 +U, –0 +6, –0 +10°, –0°	+3, –2 ±2 ±2 +10°, –5°	All	$\underline{D}_1 + \underline{D}_2$	e, f, g, i, j, k
	TC-P9	12 min.	U	R = 0 f = 3 min. r = 10 $α_{oc}$ = 30°* $α_{ic}$ = 45°**	+2, –0 +U, –0 +6, –0 +10°, –0° +10°, –0°	+3, –2 ±2 ±2 +10°, –5° +10°, –5°	All	$\underline{D}_1 + \underline{D}_2$	e, f, g, i, j, k
GMAW FCAW	B-P9-GF	6 min.	—	R = 0 f = 3 min. r = 10 α = 30°	+2, –0 +U, –0 +6, –0 +10°, –0°	+3, –2 ±2 ±2 +10°, –5°	All	$\underline{D}_1 + \underline{D}_2$	a, f, g, i, j, k
	TC-P9-GF	6 min.	U	R = 0 f = 3 min. r = 10 $α_{oc}$ = 30°* $α_{ic}$ = 45°**	+2, –0 +U, –0 +6, –0 +10°, –0° +10°, –0°	+3, –2 ±2 ±2 +10°, –5° +10°, –5°	All	$\underline{D}_1 + \underline{D}_2$	a, f, g, i, j, k
SAW	B-P9-S	20 min.	—	R = 0 f = 6 min. r = 12 α = 20°	±0 +U, –0 +6, –0 +10°, –0°	+2, –0 ±2 ±2 +10°, –5°	F	$\underline{D}_1 + \underline{D}_2$	f, g, i, j, k
	TC-P9-S	20 min.	U	R = 0 f = 6 min. r = 12 $α_{oc}$ = 20°* $α_{ic}$ = 45°**	±0 +U, –0 +6, –0 +10°, –0° +10°, –0°	+2, –0 ±2 ±2 +10°, –5° +10°, –5°	F	$\underline{D}_1 + \underline{D}_2$	f, g, i, j, k

*$α_{oc}$ = Outside corner groove angle.
**$α_{ic}$ = Inside corner groove angle.

Figure 5.2 (Continued)—Prequalified PJP Groove Welded Joint Details (See 5.4.2)
(Dimensions in Millimeters)

See Notes on Page 81

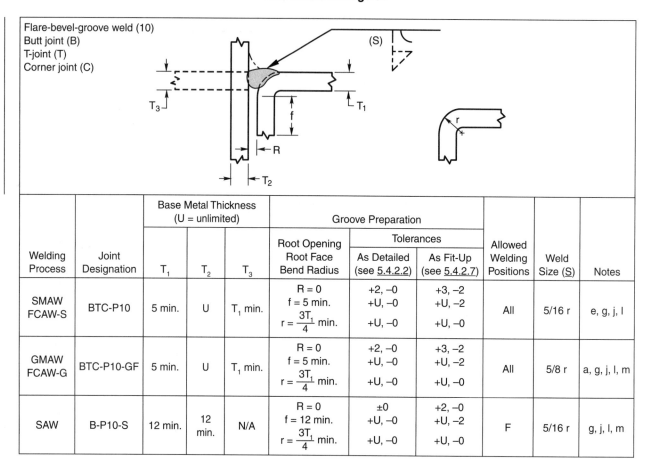

Welding Process	Joint Designation	Base Metal Thickness (U = unlimited)			Groove Preparation			Allowed Welding Positions	Weld Size (S)	Notes
		T_1	T_2	T_3	Root Opening Root Face Bend Radius	Tolerances As Detailed (see 5.4.2.2)	As Fit-Up (see 5.4.2.7)			
SMAW FCAW-S	BTC-P10	5 min.	U	T_1 min.	R = 0 f = 5 min. $r = \frac{3T_1}{4}$ min.	+2, –0 +U, –0 +U, –0	+3, –2 +U, –2 +U, –0	All	5/16 r	e, g, j, l
GMAW FCAW-G	BTC-P10-GF	5 min.	U	T_1 min.	R = 0 f = 5 min. $r = \frac{3T_1}{4}$ min.	+2, –0 +U, –0 +U, –0	+3, –2 +U, –2 +U, –0	All	5/8 r	a, g, j, l, m
SAW	B-P10-S	12 min.	12 min.	N/A	R = 0 f = 12 min. $r = \frac{3T_1}{4}$ min.	±0 +U, –0 +U, –0	+2, –0 +U, –2 +U, –0	F	5/16 r	g, j, l, m

**Figure 5.2 (Continued)—Prequalified PJP Groove Welded Joint Details (See 5.4.2)
(Dimensions in Millimeters)**

See Notes on Page 81

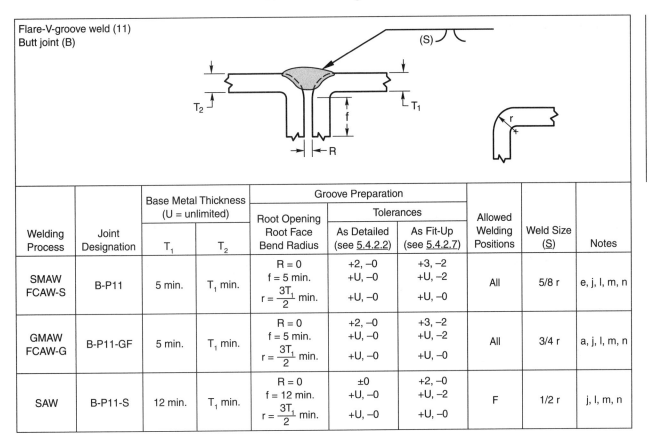

Welding Process	Joint Designation	Base Metal Thickness (U = unlimited)		Groove Preparation			Allowed Welding Positions	Weld Size (S)	Notes
		T_1	T_2	Root Opening Root Face Bend Radius	Tolerances				
					As Detailed (see 5.4.2.2)	As Fit-Up (see 5.4.2.7)			
SMAW FCAW-S	B-P11	5 min.	T_1 min.	R = 0 f = 5 min. $r = \frac{3T_1}{2}$ min.	+2, –0 +U, –0 +U, –0	+3, –2 +U, –2 +U, –0	All	5/8 r	e, j, l, m, n
GMAW FCAW-G	B-P11-GF	5 min.	T_1 min.	R = 0 f = 5 min. $r = \frac{3T_1}{2}$ min.	+2, –0 +U, –0 +U, –0	+3, –2 +U, –2 +U, –0	All	3/4 r	a, j, l, m, n
SAW	B-P11-S	12 min.	T_1 min.	R = 0 f = 12 min. $r = \frac{3T_1}{2}$ min.	±0 +U, –0 +U, –0	+2, –0 +U, –2 +U, –0	F	1/2 r	j, l, m, n

Figure 5.2 (Continued)—Prequalified PJP Groove Welded Joint Details (See 5.4.2) (Dimensions in Millimeters)

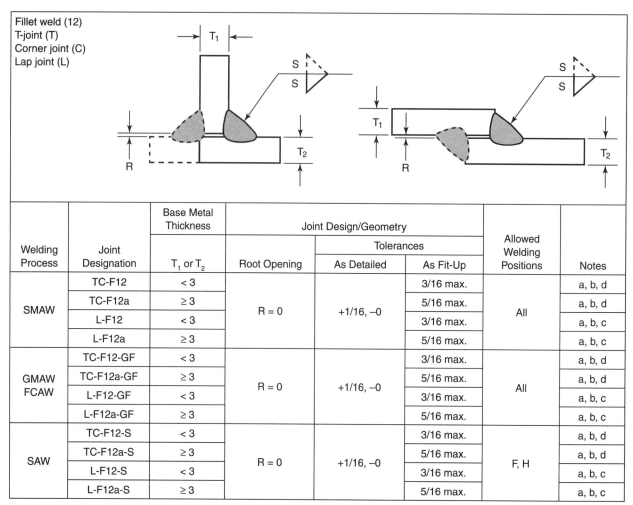

Welding Process	Joint Designation	Base Metal Thickness T_1 or T_2	Joint Design/Geometry Root Opening	Tolerances As Detailed	Tolerances As Fit-Up	Allowed Welding Positions	Notes
SMAW	TC-F12	< 3	R = 0	+1/16, –0	3/16 max.	All	a, b, d
SMAW	TC-F12a	≥ 3	R = 0	+1/16, –0	5/16 max.	All	a, b, d
SMAW	L-F12	< 3	R = 0	+1/16, –0	3/16 max.	All	a, b, c
SMAW	L-F12a	≥ 3	R = 0	+1/16, –0	5/16 max.	All	a, b, c
GMAW FCAW	TC-F12-GF	< 3	R = 0	+1/16, –0	3/16 max.	All	a, b, d
GMAW FCAW	TC-F12a-GF	≥ 3	R = 0	+1/16, –0	5/16 max.	All	a, b, d
GMAW FCAW	L-F12-GF	< 3	R = 0	+1/16, –0	3/16 max.	All	a, b, c
GMAW FCAW	L-F12a-GF	≥ 3	R = 0	+1/16, –0	5/16 max.	All	a, b, c
SAW	TC-F12-S	< 3	R = 0	+1/16, –0	3/16 max.	F, H	a, b, d
SAW	TC-F12a-S	≥ 3	R = 0	+1/16, –0	5/16 max.	F, H	a, b, d
SAW	L-F12-S	< 3	R = 0	+1/16, –0	3/16 max.	F, H	a, b, c
SAW	L-F12a-S	≥ 3	R = 0	+1/16, –0	5/16 max.	F, H	a, b, c

Notes for Figure 5.3
[a] Fillet weld size ("S"). See 4.4.2.8 and Clause 7.13 for minimum fillet weld sizes. See Table 5.1 for maximum single pass size.
[b] See 7.21.1 for additional fillet weld assembly requirements or exceptions.
[c] See 4.4.2.9 for maximum weld size in lap joints.
[d] Perpendicularity of the members shall be within ±10°.

Figure 5.3—Prequalified Fillet Weld Joint Details (Dimensions in Inches) (See 5.4.3)

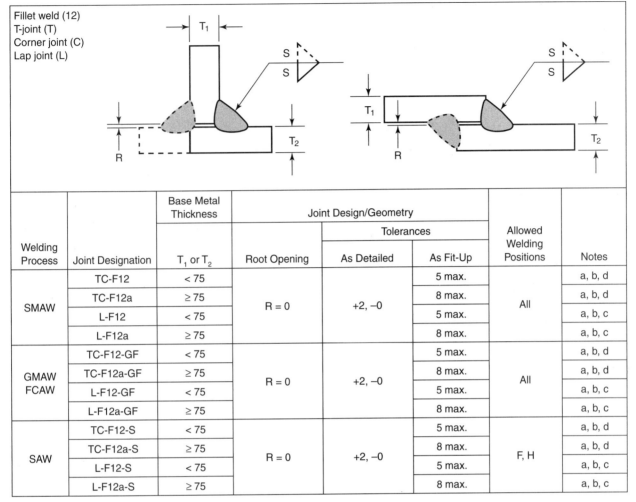

Welding Process	Joint Designation	Base Metal Thickness T_1 or T_2	Joint Design/Geometry			Allowed Welding Positions	Notes
			Root Opening	Tolerances As Detailed	Tolerances As Fit-Up		
SMAW	TC-F12	< 75	R = 0	+2, –0	5 max.	All	a, b, d
	TC-F12a	≥ 75			8 max.		a, b, d
	L-F12	< 75			5 max.		a, b, c
	L-F12a	≥ 75			8 max.		a, b, c
GMAW FCAW	TC-F12-GF	< 75	R = 0	+2, –0	5 max.	All	a, b, d
	TC-F12a-GF	≥ 75			8 max.		a, b, d
	L-F12-GF	< 75			5 max.		a, b, c
	L-F12a-GF	≥ 75			8 max.		a, b, c
SAW	TC-F12-S	< 75	R = 0	+2, –0	5 max.	F, H	a, b, d
	TC-F12a-S	≥ 75			8 max.		a, b, d
	L-F12-S	< 75			5 max.		a, b, c
	L-F12a-S	≥ 75			8 max.		a, b, c

Notes for Figure 5.3

a Fillet weld size ("S"). See 4.4.2.8 and Clause 7.13 for minimum fillet weld sizes. See Table 5.1 for maximum single pass size.
b See 7.21.1 for additional fillet weld assembly requirements or exceptions.
c See 4.4.2.9 for maximum weld size in lap joints.
d Perpendicularity of the members shall be within ±10°.

Figure 5.3 (Continued)—Prequalified Fillet Weld Joint Details (Dimensions in Millimeters) (See 5.4.3)

CLAUSE 5. PREQUALIFICATION OF WPSs   AWS D1.1/D1.1M:2020

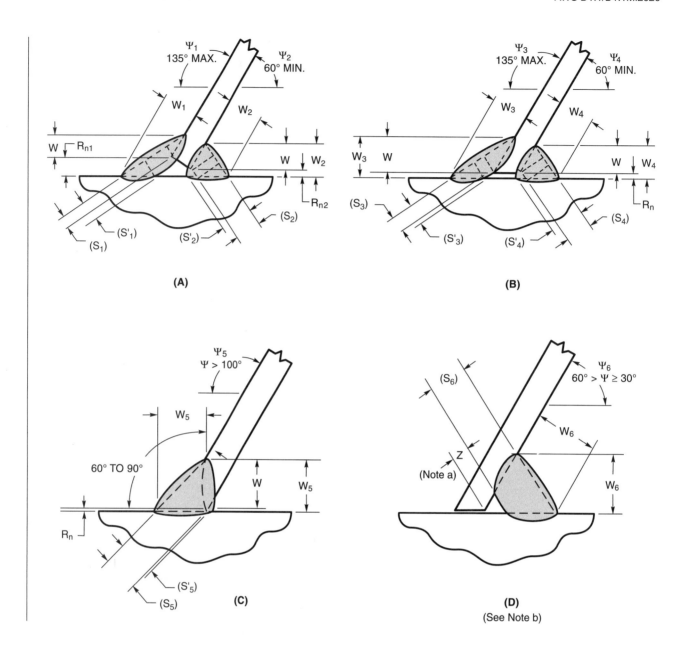

[a] Detail (D). Apply Z loss dimension of Table 4.2 to determine effective throat.
[b] Detail (D) shall not be prequalified for under 30°. For welder qualifications, see Table 6.10.

Notes:
1. (S_n), (S'_n) = Weld size dependent on magnitude of root opening (R_n) (see 7.21.1). (n) represents 1 through 5.
2. t = thickness of thinner part
3. Not prequalified for GMAW-S or GTAW.

Figure 5.4—Prequalified Skewed T-Joint Joint Details (Nontubular) (See 5.4.3.2)

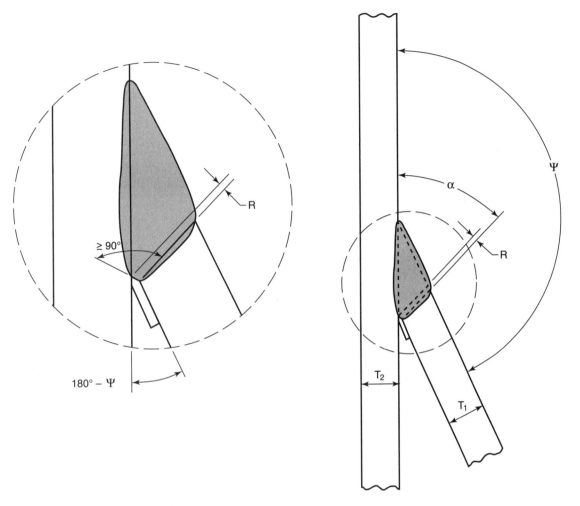

Note: $90° \leq \Psi \leq 170°$.

**Figure 5.5—Prequalified CJP Groove, T-, and Corner Joint
(See Notes for Figures 5.1 and 5.2, Note o)**

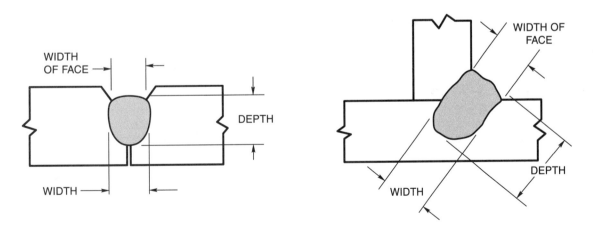

**Figure 5.6—Weld Bead in which Depth and Width Exceed the Width
of the Weld Face (see 5.8.2.1)**

6. Qualification

6.1 Scope

This clause contains the requirements for qualification testing of welding procedure specifications (WPSs) and welding personnel. It is divided into four parts as follows:

Part A—General Requirements. This part covers general requirements of both WPS and welding personnel performance requirements.

Part B—Welding Procedure Specification (WPS) Qualification. This part covers the qualification of a WPS that is not classified as prequalified in conformance with Clause 5.

Part C—Performance Qualification. This part covers the performance qualification tests required by the code to determine a welder's, welding operator's, or tack welder's ability to produce sound welds.

Part D—Requirements for CVN Toughness Testing. This part covers general requirements and procedures for CVN testing when specified by the contract documents.

Part A
General Requirements

6.2 General

The requirements for qualification testing of WPSs and welding personnel (defined as welders, welding operators, and tack welders) are described in this section.

6.2.1 Welding Procedure Specification (WPS). Except for prequalified WPSs conforming to the requirements of Clause 5, a WPS for use in production welding shall be qualified in conformance with Clause 6, Part B. Properly documented evidence of previous WPS qualification may be used.

6.2.1.1 Qualification Responsibility. Each manufacturer or Contractor shall conduct the tests required by this code to qualify the WPS. Properly documented WPSs qualified under the provisions of this code by a company that later has a name change due to voluntary action or consolidation with a parent company may utilize the new name on its WPS documents while maintaining the supporting PQR qualification records with the old company name.

6.2.1.2 WPS Qualification to Other Standards. The acceptability of qualification to other standards is the Engineer's responsibility, to be exercised based upon either the specific structure or service conditions, or both. AWS B2.1-X-XXX Series on Standard Welding Procedure Specifications may, in this manner, be accepted for use in this code.

6.2.1.3 CVN Test Requirements. When required by contract documents, CVN tests shall be included in the WPS qualification. The CVN tests, requirements, and procedure shall be in conformance with the provisions of Part D of this section, or as specified in the contract documents.

6.2.2 Performance Qualification of Welding Personnel. Welders, welding operators and tack welders to be employed to weld under this code, and using the SMAW, SAW, GMAW, GTAW, FCAW, ESW, or EGW processes, shall have been qualified by the applicable tests as described in Part C of this clause.

6.2.2.1 Previous Performance Qualification. Previous performance qualification tests of welders, welding operators, and tack welders that are properly documented are acceptable with the approval of the Engineer. The acceptability of performance qualification to other standards is the Engineer's responsibility, to be exercised based upon either the specific structure or service conditions, or both.

6.2.2.2 Qualification Responsibility. Each manufacturer or Contractor shall be responsible for the qualification of welders, welding operators and tack welders, whether the qualification is conducted by the manufacturer, Contractor, or an independent testing agency.

6.2.3 Period of Effectiveness

6.2.3.1 Welders and Welding Operators. The welder's or welding operator's qualification as specified in this code shall be considered as remaining in effect indefinitely unless:

(1) the welder is not engaged in a given process of welding for which the welder or welding operator is qualified for a period exceeding six months, or

(2) there is some specific reason to question a welder's or welding operator's ability (see 6.25.1).

6.2.3.2 Tack Welders. A tack welder who passes the test described in Part C or those tests required for welder qualification shall be considered eligible to perform tack welding indefinitely in the positions and with the process for which the tack welder is qualified unless there is some specific reason to question the tack welder's ability (see 6.25.2).

6.3 Common Requirements for WPS and Welding Personnel Performance Qualification

6.3.1 Qualification to Earlier Editions. Qualifications performed to and having met the requirements of earlier editions of AWS D1.1 or AWS D1.0 or AWS D2.0 while those editions were in effect are valid and may be used. The use of earlier editions shall be prohibited for new qualifications in lieu of the current editions, unless the specific early edition is specified in the contract documents.

6.3.2 Aging. When allowed by the filler metal specification applicable to weld metal being tested, fully welded qualification test specimens may be aged at 200°F to 220°F [95°C to 105°C] for 48 ± 2 hours.

6.3.3 Records. Records of the test results shall be kept by the manufacturer or Contractor and shall be made available to those authorized to examine them.

6.3.4 Positions of Welds. All welds shall be classified as flat (F), horizontal (H), vertical (V), or overhead (OH), in conformance with the definitions shown in Figure 6.1, Figure 6.2, and 10.11.1.

Test assembly positions are shown in:

(1) Figure 6.3 (groove welds in plate)

(2) Figure 6.4 (fillet welds on plate)

Part B
Welding Procedure Specification (WPS) Qualification

6.4 Production Welding Positions Qualified

The production welding positions qualified by a plate test shall conform to the requirements of Clause 6 and Table 6.1. The production welding positions qualified by a tubular test shall conform to the requirements of Clause 10 and Table 10.8.

6.5 Type of Qualification Tests

The type and number of qualification tests required to qualify a WPS for a given thickness, diameter, or both, shall conform to Table 6.2 (CJP), Table 6.3 (PJP) or Table 6.4 (fillet). Details on the individual NDT and mechanical test requirements are found in the following subclauses:

(1) Visual Inspection (see 6.10.1)

(2) NDT (see 6.10.2)

(3) Face, root and side bend (see 6.10.3.1)

(4) Reduced Section Tension (see 6.10.3.4)

(5) All-Weld-Metal Tension (see 6.10.3.6)

(6) Macroetch (see 6.10.4)

6.6 Weld Types for WPS Qualification

For the purpose of WPS qualification, weld types shall be classified as follows:

(1) CJP Groove Welds for Nontubular Connections (see 6.11)

(2) PJP Groove Welds for Nontubular Connections (see 6.12)

(3) Fillet Welds (see 6.13)

(4) CJP Groove Welds for Tubular Connections (see 10.14)

(5) PJP Groove Welds for Tubular T-, Y-, and K-connections and Butt Joints (see 10.15)

(6) Plug and Slot Welds (see 6.14)

6.7 Preparation of WPS

The manufacturer or Contractor shall prepare a written WPS that specifies all of the applicable essential variables referenced in 6.8. The specific values for these WPS variables shall be obtained from the procedure qualification record (PQR), which shall serve as written confirmation of a successful WPS qualification.

6.8 Essential Variables

6.8.1 SMAW, SAW, GMAW, GTAW, and FCAW. Changes beyond the limitations of PQR essential variables for the SMAW, SAW, GMAW, GTAW, and FCAW processes shown in Table 6.5 and Table 6.7 (when CVN testing is specified) shall require requalification of the WPS (see 6.2.1.3).

6.8.2 ESW and EGW. See Table 6.6 for the PQR essential variable changes requiring WPS requalification for the EGW and ESW processes. Supplementary essential variables (when CVN testing is specified) are shown in Table 6.7.

6.8.3 Base Metal Qualification. Procedure qualification tests using base metals listed in Table 5.3 shall also qualify WPSs using base metals from other groups as specified in Table 6.8. WPSs for base metals not listed in Table 5.3 or Table 6.9 shall be qualified in conformance with Clause 6.

WPSs with steels listed in Table 6.9 shall also qualify Table 5.3 or Table 6.9, steels in conformance with Table 6.8. Table 6.9 also contains recommendations for matching strength filler metal and minimum preheat and interpass temperatures for the materials in the table.

6.8.4 Preheat and Interpass Temperature. The minimum preheat and interpass temperature should be established on the basis of steel composition as shown in Table 5.8. Alternatively, recognized methods of prediction or guidelines such as those provided in Annex B, or other methods may be used. Preheat and interpass temperatures lower than required per Table 5.8 or calculated per Annex B may be used provided they are approved by the Engineer and qualified by WPS testing.

The methods of Annex B are based on laboratory cracking tests and may predict preheat temperatures higher than the minimum temperature shown in Table 5.8. Annex B may be of value in identifying situations where the risk of cracking is increased due to composition, restraint, hydrogen level, or lower welding heat input where higher preheat may be warranted. Alternatively, Annex B may assist in defining conditions under which hydrogen cracking is unlikely and where the minimum requirements of Table 5.8 may be safely relaxed.

6.8.5 Heat Input. When CVN testing is required by the contract documents, heat input shall be calculated as follows:

(1) When non-waveform-controlled welding is to be used, heat input shall be calculated by any one of the methods shown in 6.8.5.1 equations (1) through (3).

(2) When waveform-controlled welding is to be used, heat input shall be calculated by either of the methods shown in 6.8.5.1 equation (2) or (3).

6.8.5.1 Heat Input Calculation Methods. The following equations shall be applicable to the calculation of heat input. Selection of an equation will depend on the measurement capacity of the welding equipment and whether waveform controlled welding is being used.

$$\text{Heat input (Joules/in [Joules/mm])} = \frac{A \times Volts \times 60}{TS\left(\frac{in}{min}\right)\left[\frac{mm}{min}\right]} \quad (1)$$

where:

A = Current, amps
Volts = Voltage
TS = Travel Speed, in/min [mm/min]

$$\text{Heat input (J/in [J/mm])} = \frac{TIE}{L} \quad (2)$$

where:

TIE = Total Instantaneous Energy, J (measured by the power source)

L = Weld bead length, in [mm]

$$\text{Heat input (J/in [J/mm])} = \frac{(AIP \times T)}{L} \quad (3)$$

where:

AIP = Average Instantaneous Power, W(J/s) (measured at the power source)

T = arc time, s
L = weld bead length, in [mm]

6.8.5.2 Maximum Heat Input for Multi-Position WPSs. The maximum heat input for a multiple position WPS shall be established by the PQR with the highest heat input.

6.8.5.3 Measurement of Total Instantaneous Energy or Power. Welding systems, comprised of interconnected power sources and wire feed controllers, shall display total instantaneous energy or average instantaneous power using one of the following:

(1) meters or displays incorporated in the power source/wire feed controller,

(2) external meters with high frequency sampling capable of determining and displaying total instantaneous energy or average instantaneous power, or

(3) upgrades or modifications to the welding equipment to facilitate determination and display total instantaneous energy or average instantaneous power.

6.9 WPS Requirements for Production Welding Using Existing Non-Waveform or Waveform WPSs

6.9.1 WPSs Qualified with Non-Waveform-Controlled Power Sources. WPSs qualified using non-waveform-controlled welding and heat input determined by 6.8.5.1 equation (1) shall remain qualified for use in production welding using waveform-controlled welding equipment, provided the WPS is revised to require heat input determination for production welds using the heat input calculation methods of either 6.8.5.1 equation (2) or (3).

6.9.2 Production Welds Made with Non-Waveform Equipment Using WPSs Qualified with Waveform Equipment. The heat input of production welds determined by 6.8.5.1 equation (1) shall not exceed the heat input limits of the WPS as determined from 6.8.5.1 equation (2) or (3).

6.10 Methods of Testing and Acceptance Criteria for WPS Qualification

The welded test assemblies conforming to 6.10.2 shall have test specimens prepared by cutting the test plate as shown in Figures 6.5 through 6.7, whichever is applicable. The test specimens shall be prepared for testing in conformance with Figures 6.8, 6.9, 6.10, and 6.14, as applicable.

6.10.1 Visual Inspection of Welds. The visual acceptance criteria for qualification of groove and fillet welds (excluding weld tabs) shall conform to the following requirements, as applicable:

6.10.1.1 Visual Inspection of Groove Welds. Groove welds shall meet the following requirements:

(1) Any crack shall be unacceptable, regardless of size.

(2) All craters shall be filled to the full cross section of the weld.

(3) Weld reinforcement shall not exceed 1/8 in [3 mm]. The weld profile shall conform to Figure 7.4 and shall have complete fusion.

(4) Undercut shall not exceed 1/32 in [1 mm].

(5) The weld root for CJP grooves shall be inspected, and shall not have any cracks, incomplete fusion, or inadequate joint penetration.

(6) For CJP grooves welded from one side without backing, root concavity or melt-through shall conform to the following:

  (a) The maximum root concavity shall be 1/16 in [2 mm], provided the total weld thickness is equal to or greater than that of the base metal.

  (b) The maximum melt-through shall be 1/8 in [3 mm].

6.10.1.2 Visual Inspection of Fillet Welds. Fillet welds shall meet the following requirements:

(1) Any crack shall be unacceptable, regardless of size.

(2) All craters shall be filled to the full cross section of the weld.

(3) The fillet weld leg sizes shall not be less than the required leg sizes.

(4) The weld profile shall meet the requirements of Figure 7.4.

(5) Base metal undercut shall not exceed 1/32 in [1 mm].

6.10.2 NDT. Before preparing mechanical test specimens, the qualification test plate, pipe, or tubing shall be nondestructively tested for soundness as follows:

6.10.2.1 RT or UT. Either RT or UT shall be used. The entire length of the weld in test plates, except the discard lengths at each end, shall be examined in conformance with Clause 8, Part E or F, and Clause 10, Part F for tubulars.

6.10.2.2 RT or UT Acceptance Criteria. For acceptable qualification, the weld, as revealed by RT or UT, shall conform to the requirements of Clause 8, Part C or Clause 10, Part F for tubulars.

6.10.3 Mechanical Testing. Mechanical testing shall be as follows:

6.10.3.1 Root, Face, and Side Bend Specimens (see Figure 6.8 for root and face bends, Figure 6.9 for side bends). Each specimen shall be bent in a bend test jig that meets the requirements shown in Figures 6.11 through 6.13 or is substantially in conformance with those figures, provided the maximum bend radius is not exceeded. Any convenient means may be used to move the plunger member with relation to the die member.

The specimen shall be placed on the die member of the jig with the weld at mid-span. Face bend specimens shall be placed with the face of the weld directed toward the gap. Root bend and fillet weld soundness specimens shall be placed with the root of the weld directed toward the gap. Side bend specimens shall be placed with that side showing the greater discontinuity, if any, directed toward the gap.

The plunger shall force the specimen into the die until the specimen becomes U-shaped. The weld and HAZs shall be centered and completely within the bent portion of the specimen after testing. When using the wraparound jig, the specimen shall be firmly clamped on one end so that there is no sliding of the specimen during the bending operation. The weld and HAZs shall be completely in the bent portion of the specimen after testing. Test specimens shall be removed from the jig when the outer roll has been moved 180° from the starting point.

6.10.3.2 Longitudinal Bend Specimens. When material combinations differ markedly in mechanical bending properties, as between two base materials or between the weld metal and the base metal, longitudinal bend tests (face and root) may be used in lieu of the transverse face and root bend tests. The welded test assemblies conforming to 6.10.2 shall have test specimens prepared by cutting the test plate as shown in Figure 6.6 or 6.7, whichever is applicable. The test specimens for the longitudinal bend test shall be prepared for testing as shown in Figure 6.8.

6.10.3.3 Acceptance Criteria for Bend Tests. The convex surface of the bend test specimen shall be visually examined for surface discontinuities. For acceptance, the surface shall contain no discontinuities exceeding the following dimensions:

(1) 1/8 in [3 mm] measured in any direction on the surface

(2) 3/8 in [10 mm]—the sum of the greatest dimensions of all discontinuities exceeding 1/32 in [1 mm], but less than or equal to 1/8 in [3 mm]

(3) 1/4 in [6 mm]—the maximum corner crack, except when that corner crack results from visible slag inclusion or other fusion type discontinuity, then the 1/8 in [3 mm] maximum shall apply.

Specimens with corner cracks exceeding 1/4 in [6 mm] with no evidence of slag inclusions or other fusion type discontinuity shall be disregarded, and a replacement test specimen from the original weldment shall be tested.

6.10.3.4 Reduced-Section Tension Specimens (see Figure 6.10). Before testing, the least width and corresponding thickness of the reduced section shall be measured. The specimen shall be ruptured under tensile load, and the maximum load shall be determined. The cross-sectional area shall be obtained by multiplying the width by the thickness. The tensile strength shall be obtained by dividing the maximum load by the cross-sectional area.

6.10.3.5 Acceptance Criteria for Reduced-Section Tension Test. The tensile strength shall be no less than the minimum of the specified tensile range of the base metal used.

6.10.3.6 All-Weld-Metal Tension Specimen (see Figure 6.14). The test specimen shall be tested in conformance with ASTM A370, *Mechanical Testing of Steel Products*.

6.10.4 Macroetch Test. The weld test specimens shall be prepared with a finish suitable for macroetch examination. A suitable solution shall be used for etching to give a clear definition of the weld.

6.10.4.1 Acceptance Criteria for Macroetch Test. For acceptable qualification, the test specimen, when inspected visually, shall conform to the following requirements:

(1) PJP groove welds; the actual weld size shall be equal to or greater than the specified weld size, (S).

(2) Fillet welds shall have fusion to the root of the joint, but not necessarily beyond.

(3) Minimum leg size shall meet the specified fillet weld size.

(4) The PJP groove welds and fillet welds shall have the following:

    (a) no cracks

    (b) thorough fusion between adjacent layers of weld metal and between weld metal and base metal

    (c) weld profiles conforming to specified detail, but with none of the variations prohibited in 7.23

    (d) no undercut exceeding 1/32 in [1 mm]

6.10.5 Retest. If any one test specimen, of all those tested, fails to meet the test requirements of Clause 6.10 the test coupon shall be considered as failed. Retests may be performed on two additional test specimens according to one of the following alternatives. The results of both test specimens shall meet the test requirements.

(1) If adequate material from the original welded test coupon exists, the additional test specimens shall be removed as close as practicable to the original test specimen location.

(2) If adequate material from the original welded test coupon does not exist, a new test coupon shall be prepared following the original PQR as closely as practicable. The new welded test coupon need only be of sufficient length to provide the required two test specimens. If the new test specimens pass, the essential and supplementary essential variables shall be documented on the PQR.

For material over 1-1/2 in [38 mm] thick when multiple test specimens are required to represent the complete thickness, failure of a test specimen representing a portion of the thickness shall require testing of all test specimens representing the complete thickness from two additional locations in the test coupon.

6.11 CJP Groove Welds

See Table 6.2(1) for the requirements for qualifying a WPS of a CJP weld on nontubular connections. See Figures 6.5–6.7 for the appropriate test plate.

6.11.1.1 Corner or T-Joints. Test specimens for groove welds in corner or T-joints shall be butt joints having the same groove configuration as the corner or T-joint to be used on construction, except the depth of groove need not exceed 1 in [25 mm].

6.12 PJP Groove Welds

Qualification of a PJP groove weld WPS shall be by one of the following methods:

(1) Use of a CJP WPS qualified by testing to support the qualification of a PJP WPS using any Figure 5.2 joint detail (see 6.12.1)

(2) Use of a CJP WPS qualified by testing to support the qualification of a PJP WPS using a joint detail not shown in Figure 5.2 (see 6.12.2)

(3) Qualification of a PJP WPS not supported by a CJP WPS (see 6.12.3)

(4) Qualifying flare-groove welds (see 6.12.4)

Any PJP qualification shall also qualify any fillet weld size on any thickness.

6.12.1 PJP WPS Qualification: Method 1. Qualification of a CJP WPS in accordance with Clause 6 shall qualify PJP groove welds conforming to Figure 5.2 provided the essential variables for the qualified CJP WPS are within the limits listed in Tables 6.5 and 6.6 (when applicable).

6.12.2 PJP WPS Qualification: Method 2. When a CJP WPS has been qualified and is used to support a PJP groove weld detail not shown in Figure 5.2, the qualified PJP weld size shall be determined as shown below:

6.12.2.1 A test plate shall be welded using the minimum root opening and the minimum groove angle to be listed on the WPS. The test plate shall conform to Figure 6.29 (Detail A). Any steel base metal may be used.

6.12.2.2 Three macroetch cross section test specimens shall be prepared and visually examined to demonstrate that the specified weld size is met or exceeded.

6.12.2.3 The maximum weld size of the qualified joint shall be the minimum value of the three macroetch specimens.

6.12.2.4 The qualified WPS shall be prepared, to specify the maximum designed PJP weld size in relation to the depth of groove determined in 6.12.2.3 and the minimum root opening, minimum groove angle, and groove depth qualified.

6.12.3 PJP WPS Qualification: Method 3. Alternately, a PJP WPS shall be qualified and tested as required in Table 6.3. If a PJP bevel- or J- groove weld is to be used for T-joints or inside corner joints, the test joint shall have a temporary restrictive plate in the plane of the square face to simulate the restricted joint configuration. This restrictive plate shall be removed prior to cutting mechanical test specimens.

6.12.4 Qualification of a PJP Flare-Groove Weld WPS. The effective weld size qualified shall be determined by the following:

6.12.4.1 A test plate shall be welded in conformance with Figure 6.29 Details B or C, as applicable. Weld parameters must be within those specified on the CJP WPS. Any steel base metal chemical composition may be used.

6.12.4.2 A minimum of three macroetch cross section test specimens, cut normal to the weld axis, shall be prepared and visually examined to verify that the specified weld size is met or exceeded. The sections shall be taken from the midlength and near the ends of the weld as shown in Figure 6.29 Details B or C.

6.12.4.3 The maximum weld size qualified is the minimum weld size of the three cross sections from 6.12.4.2 above. The minimum radius qualified is that tested.

6.13 Fillet Welds

6.13.1 Type and Number of Specimens. Except as permitted elsewhere in Clause 6, the type and number of specimens that shall be tested to qualify a single-pass fillet weld and/or multiple-pass fillet weld WPS are shown in Table 6.4. Qualification testing may be for either a single-pass fillet weld or multiple-pass fillet weld or both.

6.13.2 Fillet Weld Test. A fillet welded T-joint, as shown in Figure 6.15 for plate or Figure 10.16 for pipe (Detail A or Detail B), shall be made for each WPS and position to be used in construction. Testing is required for the maximum size single-pass fillet weld and the minimum size multiple-pass fillet weld used in construction. These two fillet weld tests may be combined in a single test weldment or assembly or individually qualified as standalone qualifications. Each weldment shall be cut perpendicular to the direction of welding at locations shown in Figure 6.15 or Figure 10.16 as applicable. Specimens representing one face of each cut shall constitute a macroetch test specimen and shall be tested in conformance with 6.10.4.

6.13.3 Consumables Verification Test

6.13.3.1 When a Test is Required. A consumables verification test is required when:

(1) The welding consumable does not conform to the prequalified provisions of Clause 5, and

(2) The WPS using the proposed consumable has not been qualified in accordance with 6.11 or 6.12.

6.13.3.2 Test Plate Weld. The test plate shall be welded as follows:

(1) The test plate shall have the groove configuration shown in Figure 6.16 (Figure 6.17 for SAW), with steel backing.

(2) The plate shall be welded in the 1G (flat) position.

(3) The plate length shall be adequate to provide the test specimens required and oriented as shown in Figure 6.18.

(4) The welding test conditions of current, voltage, travel speed, and gas flow shall approximate those to be used in making production fillet welds as closely as practical.

These conditions establish the WPS from which, when production fillet welds are made, changes in essential variables will be measured in conformance with 6.8.

6.13.3.3 Test Requirements. The test plate shall be tested as follows:

(1) Two side bend (Figure 6.9) specimens and one all-weld-metal tension (Figure 6.14) test specimen shall be removed from the test plate, as shown in Figure 6.18.

(2) The bend test specimens shall be tested in conformance with 6.10.3.1. Those test results shall conform to the requirements of 6.10.3.3.

(3) The tension test specimen shall be tested in conformance with 6.10.3.6. The test result shall determine the strength level for the welding consumable, which shall conform to the requirements of Table 4.3 or the base metal strength level being welded.

6.14 Plug and Slot Welds

When plug and slot welds are specified, WPS qualification shall be in conformance with 6.22.3.

6.15 Welding Processes Requiring Qualification

6.15.1 GTAW, GMAW-S, ESW, and EGW. GTAW, GMAW-S, ESW, and EGW may be used, provided the WPSs are qualified in conformance with the requirements of Clause 6.

6.15.1.1 WPS Requirement (GMAW-S). Prior to use, the Contractor shall prepare a WPS(s) and qualify each WPS in accordance with the requirements of Clause 6. The essential variable limitations in Table 6.5 for GMAW shall also apply to GMAW-S.

6.15.1.2 WPS Requirement (GTAW). Prior to use, the Contractor shall prepare a WPS(s) and qualify each WPS in conformance with the requirements of Clause 6.

6.15.1.3 WPS Requirements (ESW/EGW).

(1) Prior to use, the Contractor shall prepare and qualify each ESW or EGW WPS to be used according to the requirements in Clause 6. The WPS shall include the joint details, filler metal type and diameter, amperage, voltage (type and polarity), speed of vertical travel if not an automatic function of arc length or deposition rate, oscillation (traverse speed, length, and dwell time), type of shielding including flow rate and dew point of gas or type of flux, type of molding shoe, PWHT if used, and other pertinent information.

(2) **All-Weld-Metal Tension Test Requirements.** Prior to use, the Contractor shall demonstrate by the test described in Clause 6, that each combination of shielding and filler metal will produce weld metal having the mechanical properties specified in the latest edition of AWS A5.25, *Specification for Carbon and Low Alloy Steel Electrodes and Fluxes for Electroslag Welding*, or the latest edition of AWS A5.26, *Specification for Carbon and Low Alloy Steel Electrodes for Electrogas Welding*, as applicable, when welded in conformance with the WPS.

(3) **Previous Qualification.** WPSs that have been previously qualified may be used, providing there is proper documentation, and the WPS is approved by the Engineer.

6.15.2 Other Welding Processes. Other welding processes not listed in 5.5.1 or 6.15.1 may be used, provided the WPSs are qualified by testing. The limitation of essential variables applicable to each welding process shall be established by the Contractor developing the WPS and approved by the Engineer. Essential variable ranges shall be based on documented evidence of experience with the process, or a series of tests shall be conducted to establish essential variable limits. Any change in essential variables outside the range so established shall require requalification.

Part C
Performance Qualification

6.16 General

The performance qualification tests required by this code are specifically devised tests to determine a welder's, welding operators, or tack welder's ability to produce sound welds. The qualification tests are not intended to be used as guides for welding or tack welding during actual construction. The latter shall be performed in conformance with a WPS.

6.16.1 Production Welding Positions Qualified

6.16.1.1 Welders and Welding Operators. The qualified production welding positions qualified by a plate test for welders and welding operators shall be in conformance with Table 6.10. The qualified production welding positions qualified by a tubular test for welders and welding operators shall be in conformance with Clause 10 and Table 10.12.

6.16.1.2 Tack Welders. A tack welder shall be qualified by one test plate in each position in which the tack welding is to be performed.

6.16.2 Production Thicknesses and Diameters Qualified

6.16.2.1 Welders or Welding Operators. The range of qualified production welding thicknesses and diameters for which a welder or welding operator is qualified for shall be in conformance with Table 6.11.

6.16.2.2 Tack Welders. Tack welder qualification shall qualify for thicknesses greater than or equal to 1/8 in [3 mm].

6.16.3 Welder and Welding Operator Qualification Through WPS Qualification. A welder or welding operator may also be qualified by welding a satisfactory WPS qualification test plate, pipe or tubing that meets the requirements of 6.10. The welder or welding operator is thereby qualified in conformance with 6.16.1 and 6.16.2.

6.17 Type of Qualification Tests Required

6.17.1 Welders and Welding Operators. The type and number of qualification tests required for welders or welding operators shall conform to Table 6.11. Details on the individual NDT and mechanical test requirements are found in the following subclauses:

(1) Visual Inspection (see 6.10.1) (use WPS requirements)

(2) Face, root, and side bend (see 6.10.3.1) (use WPS requirements)

(3) Macroetch (see 6.23.2)

(4) Fillet Weld Break (see 6.23.4)

6.17.1.1 Substitution of RT for Guided Bend Tests. Except for joints welded by GMAW-S, radiographic examination of a welder or welding operator qualification test plate or test pipe may be made in lieu of bend tests described in 6.17.1(2) (see 6.23.3 for RT requirements).

In lieu of mechanical testing or RT of the qualification test assemblies, a welding operator may be qualified by RT of the initial 15 in [380 mm] of a production groove weld. The material thickness range qualified shall be that shown in Table 6.11.

6.17.1.2 Guided Bend Tests. Mechanical test specimens shall be prepared by cutting the test plate, pipe, or tubing as shown in Figures 6.16, 6.19, 6.20, 6.21, 6.22, and 10.23 for welder qualification or Figure 6.17, 6.22, or 6.24 for welding operator qualification, whichever is applicable. These specimens shall be approximately rectangular in cross section, and be prepared for testing in conformance with Figure 6.8, 6.9, 6.10, or 6.14, whichever is applicable.

6.17.2 Tack Welders. The tack welder shall make a 1/4 in [6 mm] maximum size tack weld approximately 2 in [50 mm] long on the fillet-weld-break specimen as shown in Figure 6.27.

6.17.2.1 Extent of Qualification. A tack welder who passes the fillet weld break test shall be qualified to tack weld all types of joints (except CJP groove welds, welded from one side without backing; e.g., butt joints and T-, Y-, and K-connections) for the process and in the position in which the tack welder is qualified. Tack welds in the foregoing exception shall be performed by welders fully qualified for the process and in the position in which the welding is to be done.

6.18 Weld Types for Welder and Welding Operator Performance Qualification

For the purpose of welder and welding operator qualification, weld types shall be classified as follows:

(1) CJP Groove Welds for Nontubular Connections (see 6.21)

(2) PJP Groove Welds for Nontubular Connections (see 6.22.1)

(3) Fillet Welds for Nontubular Connections (see 6.22.2)

(4) CJP Groove Welds for Tubular Connections (see 10.18)

(5) PJP Groove Welds for Tubular Connections (see 10.19)

(6) Fillet Welds for Tubular Connections (see 10.20)

(7) Plug and Slot Welds for Tubular and Nontubular Connections (see 6.22.3)

6.19 Preparation of Performance Qualification Forms

The welding personnel shall follow a WPS applicable to the qualification test required. All of the WPS essential variable limitations of 6.8 shall apply, in addition to the performance essential variables of 6.20. The Welding Performance Qualification Record (WPQR) shall serve as written verification and shall list all of the applicable essential variables of Table 6.12. Suggested forms are found in Annex J.

6.20 Essential Variables

Changes beyond the limitation of essential variables for welders, welding operators, or tack welders shown in Table 6.12 shall require requalification.

6.21 CJP Groove Welds for Nontubular Connections

See Table 6.10 for the position requirements for welder or welding operator qualification on nontubular connections. Note that qualification on joints with backing qualifies for welding production joints that are backgouged and welded from the second side.

6.21.1 Welder Qualification Plates. The following figure numbers apply to the position and thickness requirements for welders.

(1) Figure 6.16—All Positions—Unlimited Thickness

(2) Figure 6.19—Horizontal Position—Unlimited Thickness

(3) Figure 6.20—All Positions—Limited Thickness

(4) Figure 6.21—Horizontal Position—Limited Thickness

6.21.2 Welding Operator Qualification Test Plates

6.21.2.1 For Other than EGW, ESW, and Plug Welds. The qualification test plate for a welding operator not using EGW or ESW or plug welding shall conform to Figure 6.17. This shall qualify a welding operator for groove and fillet welding in material of unlimited thickness for the process and position tested.

6.21.2.2 For ESW and EGW. The qualification test plate for an ESW or EGW welding operator shall consist of welding a joint of the maximum thickness of material to be used in construction, but the thickness of the material of the test weld need not exceed 1-1/2 in [38 mm] (see Figure 6.24). If a 1-1/2 in [38 mm] thick test weld is made, no test need be made for a lesser thickness. The test shall qualify the welding operator for groove and fillet welds in material of unlimited thickness for this process and test position.

6.22 Extent of Qualification

6.22.1 PJP Groove Welds for Nontubular Connections. Qualification for CJP groove welds shall qualify for all PJP groove welds.

6.22.2 Fillet Welds for Nontubular Connections. Qualification of CJP groove welds shall qualify for fillet welds. However, where only fillet weld qualification is required, see Table 6.11.

6.22.3 Plug and Slot Welds. Qualification for CJP groove welds on tubular or nontubular connections shall qualify for all plug and slot welds. See Table 6.10 for plug and slot weld qualification only. The joint shall consist of a 3/4 in [20 mm] diameter hole in a 3/8 in [10 mm] thick plate with a 3/8 in [10 mm] minimum thickness backing plate (see Figure 6.26).

6.23 Methods of Testing and Acceptance Criteria for Welder and Welding Operator Qualification

6.23.1 Visual Inspection. See 6.10.1 for acceptance criteria.

6.23.2 Macroetch Test. The test specimens shall be prepared with a finish suitable for macroetch examination. A suitable solution shall be used for etching to give a clear definition of the weld.

6.23.2.1 Plug and Fillet Weld Macroetch Tests.

The face of the macroetch shall be smooth for etching.

(1) The plug weld macroetch tests shall be cut from the test joints per:

(a) Welder Qualification—Figure 6.26

(b) Welding Operator Qualification—Figure 6.26

(2) The fillet weld macroetch tests shall be cut from the test joints per:

(a) Welder Qualification—Figure 6.25

(b) Welding Operator Qualification—Figure 6.25

6.23.2.2 Macroetch Test Acceptance Criteria. For acceptable qualification, the test specimen, when inspected visually, shall conform to the following requirements:

(1) Fillet welds shall have fusion to the root of the joint but not necessarily beyond.

(2) Minimum leg size shall meet the specified fillet weld size.

(3) Plug welds shall have:

   (a) No cracks

   (b) Thorough fusion to backing and to sides of the hole

   (c) No visible slag in excess of 1/4 in [6 mm] total accumulated length

6.23.3 RT. If RT is used in lieu of the prescribed bend tests, the weld reinforcement need not to be ground or otherwise smoothed for inspection unless its surface irregularities or juncture with the base metal would cause objectionable weld discontinuities to be obscured in the radiograph. If the backing is removed for RT, the root shall be ground flush (see 7.23.3.1) with the base metal.

6.23.3.1 RT Test Procedure and Technique. The RT procedure and technique shall be in conformance with the requirements of Clause 8, Part E and Clause 10, Part F for tubulars. For welder qualification, exclude 1-1/4 in [32 mm] at each end of the weld from evaluation in the plate test; for welding operator qualification exclude 3 in [75 mm] at each end of the test plate length.

6.23.3.2 RT Acceptance Criteria. For acceptable qualification, the weld, as revealed by the radiograph, shall conform to the requirements of 8.12.2, except that 8.12.2.2 shall not apply.

6.23.4 Fillet Weld Break Test. The entire length of the fillet weld shall be examined visually, and then a 6 in [150 mm] long specimen (see Figure 6.25) or a quarter-section of the pipe fillet weld assembly shall be loaded in such a way that the root of the weld is in tension. At least one welding start and stop shall be located within the test specimen. The load shall be increased or repeated until the specimen fractures or bends flat upon itself.

6.23.4.1 Acceptance Criteria for Fillet Weld Break Test. To pass the visual examination prior to the break test, the weld shall present a reasonably uniform appearance and shall be free of overlap, cracks, and undercut in excess of the requirements of 8.9. There shall be no porosity visible on the weld surface.

The broken specimen shall pass if:

(1) The specimen bends flat upon itself, or

(2) The fillet weld, if fractured, has a fracture surface showing complete fusion to the root of the joint with no inclusion or porosity larger than 3/32 in [2.5 mm] in greatest dimension, and

(3) The sum of the greatest dimensions of all inclusions and porosity shall not exceed 3/8 in [10 mm] in the 6 in [150 mm] long specimen.

6.23.5 Root, Face, and Side Bend Specimens. See 6.10.3.3 for acceptance criteria.

6.24 Method of Testing and Acceptance Criteria for Tack Welder Qualification

A force shall be applied to the specimen as shown in Figure 6.23 until rupture occurs. The force may be applied by any convenient means. The surface of the weld and of the fracture shall be examined visually for defects.

6.24.1 Visual Acceptance Criteria. The tack weld shall present a reasonably uniform appearance and shall be free of overlap, cracks, and undercut exceeding 1/32 in [1 mm]. There shall be no porosity visible on the surface of the tack weld.

6.24.2 Destructive Testing Acceptance Criteria. The fractured surface of the tack weld shall show fusion to the root, but not necessarily beyond, and shall exhibit no incomplete fusion to the base metals or any inclusion or porosity larger than 3/32 in [2.5 mm] in greatest dimension.

6.25 Retest

When a welder, welding operator or tack welder either fails a qualification test, or if there is specific reason to question their welding abilities or period of effectiveness has lapsed, the following shall apply:

6.25.1 Welder and Welding Operator Retest Requirements

6.25.1.1 Immediate Retest. An immediate retest may be made consisting of two welds of each type and position that the welder or welding operator failed. All retest specimens shall meet all of the specified requirements.

6.25.1.2 Retest After Further Training or Practice. A retest may be made, provided there is evidence that the welder or welding operator has had further training or practice. A complete retest of the types and positions failed or in question shall be made.

6.25.1.3 Retest After Lapse of Qualification Period of Effectiveness. When a welder's or welding operator's qualification period of effectiveness has lapsed, a requalification test shall be required. Welders have the option of using a test thickness of 3/8 in [10 mm] to qualify any production welding thickness greater than or equal to 1/8 in [3 mm].

6.25.1.4 Exception—Failure of a Requalification Retest. No immediate retest shall be allowed after failure of a requalification retest. A retest shall be allowed only after further training and practice per 6.25.1.2.

6.25.2 Tack Welder Retest Requirements

6.25.2.1 Retest without Additional Training. In case of failure to pass the test requirements, the tack welder may make one retest without additional training.

6.25.2.2 Retest After Further Training or Practice. A retest may be made, provided the tack welder has had further training or practice. A complete retest shall be required.

Part D
Requirements for CVN *Toughness* Testing

6.26 General: CVN Testing

6.26.1 Application. The CVN test requirements and test procedures in this section shall apply when CVN testing is specified in contract documents or required by this code [see 6.2.1.3 and 7.25.5(3)(d)].

6.26.1.1 Combination of WPSs. Except as required by 6.26.1.2, multiple WPSs, each of which have been qualified with CVN testing, are permitted to be used in a single joint without further qualification testing. WPSs that have been qualified without CVN tests are permitted to have CVN tests conducted on a test coupon welded using the WPS essential variables and then be used with WPSs qualified with CVN testing to deposit weld metal in a single joint.

6.26.1.2 FCAW-S. When welding processes other than FCAW-S are used to deposit weld metal over FCAW-S in a single joint, an additional set of CVN tests shall be conducted in compliance with 6.28.

6.26.2 Test Standards. The CVN test specimens shall be machined and tested in conformance with ASTM E23, *Standard Methods for Notched Bar Impact Testing of Metallic Materials*, for Type A Charpy (simple beam) Impact Specimen, ASTM A370, *Standard Test Method and Definitions for Mechanical Testing of Steel Products*, or AWS B4.0, *Standard Methods for Mechanical Testing of Welds*.

6.26.3 Test Requirements. When WPSs are required to be supported by PQRs with CVN testing, the following shall be required:

(1) A WPS shall be qualified with testing that includes CVN tests, or

(2) If a WPS qualified in accordance with the requirements of Clause 6 exists except that the supporting PQR does not list CVN test results, a test coupon shall be prepared with the WPS parameters such that the heat input does not exceed the heat input of the existing WPS and CVN test specimens can be extracted from the test plate and tested.

If option (1) is used, a new WPS shall be written using the PQR within the limits of Tables 6.1, 6.2, and 6.5, plus those PQR supplementary essential variables applicable to CVN testing (Table 6.7). CVN test specimens shall be removed from the test coupon as shown in one of the test coupon figures (Figures 6.5, 6.6 and 6.7 for plate, Figures 10.14 and 10.15 for tubulars).

If option (2) is used, the existing WPS parameters shall be used to weld the test coupon and the heat input of the existing WPS shall not be exceeded. The coupon shall meet the requirements for WPS qualification (Part B) except mechanical tensile and bend specimen testing shall not be required. The original WPS shall be revised to accommodate the PQR supplementary essential variables applicable to CVN testing (Table 6.7) and the PQR essential variables (Table 6.5).

Option (2) shall not apply to prequalified WPSs.

6.27 CVN Tests

6.27.1 Test Locations. Test specimen locations for CVN testing shall be at the weld metal centerline and in the HAZ when specified.

Unless the alternative locations are specified, the specimen locations within the zone being tested and the joint detail used shall comply with Figure 6.28.

6.27.2 Number of Specimens. Three CVN specimens shall be removed from each test coupon. Alternatively, five CVN specimens may be removed from each test coupon and the specimens with the highest and lowest CVN values shall be discarded prior to determining acceptance.

6.27.3 Specimen Size. Full Size (10 mm x 10 mm) specimens shall be used when test coupon material is 7/16 in [11 mm] or thicker and when the overall test coupon geometry permits. When test coupon is less than 7/16 in [11 mm] or when the coupon geometry prevents full size specimens, sub-sized specimens shall be used. When sub-sized specimens are used for CVN qualification, they shall be made to one of the standard sizes shown in Table 6.14.

The largest possible standard sub-size specimen size shall be machined from the test coupon.

6.27.4 Procedure for Locating the Notch. The CVN test specimens shall conform to the following:

(1) The specimens shall be machined from the test coupon at the appropriate depth as shown in Figure 6.28. The specimens should be made slightly over length to allow for exact positioning of the notch.

(2) The CVN test specimen bars shall be etched with a mild etchant such as 5% nital to reveal the location of the weld fusion line and HAZs.

(3) The longitudinal centerline of the specimens shall be transverse to the weld axis. The CVN notch shall be perpendicular (normal) to the surface of the test coupon.

(4) The centerline of the notch shall be located in the specimens, as shown in Figure 6.28 for the weld joint type being qualified.

Testing may be required at alternative locations to those shown in Figure 6.28 when required as described in 6.26.1.2.

6.27.5 CVN Test Temperature. CVN test specimen testing temperature shall be as specified in contract documents except that for sub-sized specimens, the testing temperature shall be modified in accordance with 6.27.6.

6.27.6 Use of Sub-size CVN Specimens

6.27.6.1 Test Coupon Thickness 7/16 in [11 mm] or greater. When sub-sized specimens are required due to geometry and when the width of the specimens across the notch is less than 80% of the base metal thickness, the test temperature shall be reduced in conformance with Table 6.14. No temperature reduction is required if the width of the specimen across the notch is 80% or more of the base metal thickness. (See C-6.27.6.1 for example calculation.)

6.27.6.2 Test Coupon Thickness less than 7/16 in [11 mm]. When sub-sized specimens are required due to the thickness of the test coupon and the width of the specimen across the notch is less than 80% of the test coupon thickness, the test temperature shall be reduced by an amount equal to the difference (referring to Table 6.14) between the temperature reduction corresponding to the test coupon thickness and the temperature reduction corresponding to the CVN specimen width actually tested. No temperature reduction is required if the width of the specimen across the notch is 80% or more of the base metal thickness. (See C-6.27.6.2 for example calculations.)

6.27.7 Acceptance Criteria. The average energy and minimum single-specimen energy shall be specified by the engineer per 4.3.2 and 1.5.1(5).

The reduction in the minimum acceptance energy values for sub-sized specimens shall be determined in conformance with Table 6.15.

6.27.8 Retests. When the requirements of 6.27.7 are not met, one retest may be performed. The retest shall consist of an additional three (3) CVN specimens removed from the same test coupon as the failed test specimens. The energy value of each CVN specimen shall meet the minimum specified average acceptance criteria [see 6.27.7].

If there is insufficient coupon material remaining to remove the three retest CVN specimens, a complete test coupon shall be prepared and all NDT, mechanical, and CVN tests required by Part B of this clause shall be performed.

6.28 Combining FCAW-S with Other Welding Processes in a Single Joint

This subclause provides testing procedures used to determine the suitability of combining FCAW-S with other welding processes in a single joint.

6.28.1 Filler Metal Variables. Filler metal essential variables for the intermix CVN testing shall be as summarized in Table 6.16. Changes in these essential variables shall require an additional test.

6.28.2 Test Plate Details. A single test plate of ASTM A36, A572 Grade 50, or A992 shall be used to evaluate E70 [E49] filler metal combinations, and ASTM A572 Grade 65 or A913 Grade 65 shall be used to evaluate E80 [E55] filler metal combinations. ASTM A913 Grade 70 shall be used to evaluate E90 [E62] filler metal combinations. The test plate shall be 3/4 in [20 mm] thick, with either a 5/8 in [16 mm] root opening with a 20° included groove angle, or a 1/2 in [12 mm] root opening with a 45° included groove angle. The test plate and specimens shall be as shown in Figure 6.30.

Alternatively, a PQR test plate may be used wherein CVN test specimens have been taken from the intermixed zone

Regardless of the testing method used, the testing shall demonstrate that the acceptance criteria of 6.28 are met.

6.28.3 Welding of Test Plate. The sequence of placement of weld metals shall be the same as that to be employed in production. The first material shall be known as the substrate/root material, and the subsequent material shall be known as the fill material. Approximately one-third the thickness of the test joint shall be welded with the substrate/root material. The balance of the joint shall be welded with the fill material.

6.28.4 Test Specimens Required. Five or ten CVN test specimens shall be made from the test plate, depending on the required number of tests. CVN specimens shall be prepared in accordance with AWS B4.0, *Standard Methods for Mechanical Testing of Welds*.

6.28.5. CVN Specimen Location. The CVN impact bar shall be located as follows:

(1) Transverse specimens from which CVN bars are to be machined shall be etched to reveal the cross section of the weld.

(2) A line shall be scribed on the etched cross section, at the interface of the two welding process deposits (see Figure 6.31).

(3) The CVN specimen shall be taken from primarily material deposited by the second process. The interface location shall be included in the specimen, with the edge of the specimen within 1/16 in [1.5 mm] of the interface location (see Figure 6.32).

6.28.6. Acceptance Criteria. The lowest and highest values obtained from the five test specimens shall be disregarded. Two of the remaining three values shall equal or exceed the specified minimum average absorbed energy at the testing temperature. One of the three may be lower, but not lower than 5 ft·lbf [7 J] below the required absorbed energy. The average of the three shall not be less than the minimum required absorbed energy.

6.29 Reporting

All CVN test results required by this code or contract documents, including all test results from either three-piece or five-piece sets and all retest specimens, shall be reported on the PQR. If CVN testing is performed to obtain results for an existing WPS qualified in accordance with Part B of this clause, the original PQR shall be amended to show the CVN test values.

Table 6.1
WPS Qualification—Production Welding Positions Qualified by Plate, Pipe, and Box Tube Tests (see 6.4)

Qualification Test			Production Plate Welding Qualified			Production Pipe Welding Qualified					Production Box Tube Welding Qualified				
			Groove	Groove	Fillet[b]	Butt Joint[a]		T-, Y-, K-Connections		Fillet[b]	Butt Joint		T-, Y-, K-Connections	Fillet[b]
Weld Type		Test Positions	CJP	PJP		CJP	PJP	CJP	PJP		CJP	PJP	CJP	PJP	
CJP Groove[c]		1G	F	F	F	F	F			F	F	F			F
		2G	F, H	F, H	F, H	F, H	F, H			F, H	F, H	F, H			F, H
		3G	V	V	V	V	V			V	V	V			V
		4G	OH	OH	OH	OH	OH			OH	OH	OH			OH
P L A T E	Fillet[a,c]	1F			F					F					F
		2F			F, H					F, H					F, H
		3F			V					V					V
		4F			OH					OH					OH
	Plug/Slot		Qualifies Plug/Slot Welding for Only the Positions Tested												

CJP—Complete Joint Penetration PJP—Partial Joint Penetration

[a] Qualifies for circumferential welds in pipes equal to or greater than 24 in [600 mm] nominal outer diameter.
[b] Fillet welds in production T-, Y-, or K-connections shall conform to Figure 10.5. WPS qualification shall conform to 6.13.
[c] Qualifies for a welding axis with an essentially straight line, including welding along a line parallel to the axis of circular pipe.

Table 6.2
WPS Qualification—CJP Groove Welds: Number and Type of Test Specimens and Range of Thickness Qualified (see 6.5)

1. Tests on Plate[a]

Nominal Plate Thickness (T) Tested, in [mm]	Number of Specimens				Nominal Base Metal Thickness Qualified, in [mm]	
	Reduced Section Tension (see Fig. 6.10)	Root Bend (see Fig. 6.8)	Face Bend (see Fig. 6.8)	Side Bend (see Fig. 6.9)	Min.	Max.[b]
1/8 ≤ T ≤ 3/8 [3 ≤ T ≤ 10]	2	2	2	(Footnote d)	1/8	2T
3/8 < T < 1 [10 < T < 25]	2	—	—	4	1/8	2T
1 and over [25 and over]	2	—	—	4	1/8	Unlimited

2. Tests on ESW and EGW[c]

Nominal Plate Thickness Tested	Number of Specimens			Nominal Base Metal Thickness Qualified	
	Reduced Section Tension (see Fig. 6.10)	All-Weld-Metal Tension (see Fig. 6.14)	Side Bend (see Fig. 6.9)	Min.	Max.
T	2	1	4	0.5T	1.1T

[a] See Figures 6.6 and 6.7 for test plate requirements.
[b] For square groove welds that are qualified without backgouging, the maximum thickness qualified is limited to the test thickness.
[c] See Figure 6.5 for test plate requirements.
[d] For 3/8 in [10 mm] plate thickness, a side-bend test may be substituted for each of the required face- and root-bend tests.

Table 6.3
WPS Qualification—PJP Groove Welds: Number and Type of Test Specimens and Range of Thickness Qualified (see 6.12.3)

Test Groove Depth, D in [mm]	Number of Specimens					Qualification Ranges		
	Macroetch for Weld Size (S) 6.12.3	Reduced-Section Tension (see Fig. 6.10)	Root Bend (see Fig. 6.8)	Face Bend (see Fig. 6.8)	Side Bend (see Fig. 6.9)	Max Groove Depth	Nominal Plate Thickness, in [mm]	
							Min.	Max.
1/8 ≤ D ≤ 3/8 [3 ≤ D ≤ 10]	3	2	2	2	—	D	1/8 [3]	2T[a]
3/8 < D ≤ 1 [10 < D ≤ 25]	3	2	—	—	4	D	1/8 [3]	Unlimited

[a] T is the thickness of the plate used in the test assembly

Notes:
1. Remove macroetch test specimens to determine weld size prior to removing material to prepare mechanical test specimens.
2. Remove excess material thickness from root side of test plate to the thickness of the weld size determined by the macroetch testing before preparing mechanical bend and tensile specimens.

Table 6.4
WPS Qualification—Fillet Welds: Number and Type of Test Specimens and Range of Thickness Qualified (see 6.13)

Test Specimen	Fillet Size	Number of Welds per WPS	Test Specimens Required[a]			Size Qualified	
			Macroetch 6.13.1 6.10.4	All-Weld-Metal Tension (see Figure 6.14)	Side Bend (see Figure 6.9)	Plate/Pipe Thickness[b]	Fillet Size
Plate T-test (Figure 6.15)	Single pass, max. size to be used in construction	1 in each position to be used	3 faces	—	—	Unlimited	Max. tested single pass and smaller
	Multiple pass, min. size to be used in construction	1 in each position to be used	3 faces	—	—	Unlimited	Min. tested multiple pass and larger
Consumables Verification Test (Figure 6.18)	—	1 in 1G position	—	1	2	Qualifies welding consumables to be used in T-test above	

[a] All welded test plates shall be visually inspected per 6.10.1.
[b] The minimum thickness qualified shall be 1/8 in [3 mm].

Table 6.5
PQR Essential Variable Changes Requiring WPS Requalification for SMAW, SAW, GMAW, FCAW, and GTAW (see 6.8.1)

Essential Variable Changes to PQR Requiring Requalification	Process				
	SMAW	SAW	GMAW	FCAW	GTAW
Filler Metal					
(1) Increase in filler metal classification strength	X		X	X	
(2) Change from low-hydrogen to nonlow-hydrogen SMAW electrode	X				
(3) Change from one electrode or flux-electrode classification to any other electrode or flux-electrode classification[a]		X		X	X
(4) Change to an electrode or flux-electrode classification[b] not covered in:	AWS A5.1 or A5.5	AWS A5.17 or A5.23	AWS A5.18, A5.28, or A5.36	AWS A5.20, A5.29, or A5.36	AWS A5.18 or A5.28
(5) Addition or deletion of filler metal					X
(6) Change from cold wire feed to hot wire feed or vice versa					X
(7) Addition or deletion of supplemental powdered or granular filler metal or cut wire		X			
(8) Increase in the amount of supplemental powdered or granular filler metal or wire		X			
(9) If the alloy content of the weld metal is largely dependent on supplemental powdered filler metal, any WPS change that results in a weld deposit with the important alloying elements not meeting the WPS chemical composition requirements		X			
(10) Change in nominal filler metal diameter by:	> 1/32 in [0.8 mm] increase	Any increase[c]	Any increase or decrease	Any increase	> 1/16 in [1.6 mm] increase or decrease
(11) Change in number of electrodes		X	X	X	X
Process Parameters					
(12) A change in the amperage for each diameter used by:	To a value not recommended by manufacturer	> 10% increase or decrease	> 10% increase or decrease	> 10% increase or decrease	> 25% increase or decrease
(13) A change in type of current (ac or dc) or polarity (electrode positive or negative for dc current)	X	X	X	X	X
(14) A change in the mode of transfer			X		
(15) A change from CV to CC output			X	X	
(16) A change in the voltage for each diameter used by:		> 7% increase or decrease	> 7% increase or decrease	> 7% increase or decrease	
(17) An increase or decrease in the wire feed speed for each electrode diameter (if not amperage controlled) by:		> 10%	> 10%	> 10%	

(Continued)

Table 6.5 (Continued)
PQR Essential Variable Changes Requiring WPS Requalification for SMAW, SAW, GMAW, FCAW, and GTAW (see 6.8.1)

Essential Variable Changes to PQR Requiring Requalification	Process				
	SMAW	SAW	GMAW	FCAW	GTAW
Process Parameters (Cont'd)					
(18) A change in the travel speed[d] by:		> 15% increase or decrease	> 25% increase or decrease	> 25% increase or decrease	> 50% increase or decrease
Shielding Gas					
(19) A change in shielding gas from a single gas to any other single gas or mixture of gas, or in the specified nominal percentage composition of a gas mixture, or to no gas			X	X	X
(20) A change in total gas flow rate by:			Increase > 50% Decrease > 20%	Increase > 50% Decrease > 20%	Increase > 50% Decrease > 20%
(21) A change from the actual classification shielding gas not covered in:			AWS A5.18, A5.28, or A5.36. For A5.36 fixed and open classifications, variations in the shielding gas classification range shall be limited to the specific shielding gas tested or the designator used for the electrode classification.	AWS A5.20, A5.29, or A5.36. For A5.36 fixed and open classifications, variations in the shielding gas classification range shall be limited to the specific shielding gas tested or the designator used for the electrode classification.	
SAW Parameters					
(22) A change of > 10%, or 1/8 in [3 mm], whichever is greater, in the longitudinal spacing of the arcs		X			
(23) A change of > 10%, or 1/8 in [3 mm], whichever is greater, in the lateral spacing of the arcs		X			
(24) An increase or decrease of more than 10° in the angular orientation of any parallel electrode		X			
(25) For mechanized or automatic SAW; an increase or decrease of more than 3° in the angle of the electrode		X			
(26) For mechanized or automatic SAW, an increase or decrease of more than 5° normal to the direction of travel		X			

(Continued)

Table 6.5 (Continued)
PQR Essential Variable Changes Requiring WPS Requalification for SMAW, SAW, GMAW, FCAW, and GTAW (see 6.8.1)

Essential Variable Changes to PQR Requiring Requalification	Process				
	SMAW	SAW	GMAW	FCAW	GTAW
General					
(27) A change in position not qualified by Table 6.1 or 6.10	X	X	X	X	X
(28) A change in diameter, or thickness, or both, not qualified by Table 6.2 or 10.9	X	X	X	X	X
(29) A change in base metal or combination of base metals not listed in on the PQR or qualified by Table 6.8	X	X	X	X	X
(30) Vertical Welding: For any pass from uphill to downhill or vice versa	X		X	X	X
(31) A change in groove type (e.g., single-V to double-V), except qualification of any CJP groove weld qualifies for any groove detail conforming with the requirements of 5.4.1, 5.4.2, 10.9, or 10.10	X	X	X	X	X
(32) A change in the type of groove to a square groove and vice versa	X	X	X	X	X
(33) A change exceeding the tolerances of 5.4.1, 5.4.2, 7.22.4.1 or 10.9, 10.10, and 10.23.2.1 involving: a) A decrease in the groove angle b) A decrease in the root opening c) An increase in the root face for CJP groove welds	X	X	X	X	X
(34) The omission, but not inclusion, of backing or backgouging	X	X	X	X	X
(35) Decrease from preheat temperature[e] by:	> 25°F [15°C]	> 25°F [15°C]	> 25°F [15°C]	> 25°F [15°C]	> 100°F [55°C]
(36) Decrease from interpass temperature[e] by:	> 25°F [15°C]	> 25°F [15°C]	> 25°F [15°C]	> 25°F [15°C]	> 100°F [55°C]
(37) Addition or deletion of PWHT	X	X	X	X	X

[a] The filler metal strength may be decreased without WPS requalification.

[b] AWS A5M (SI Units) electrodes of the same classification may be used in lieu of the AWS A5 (U.S. Customary Units) electrode classification.

[c] For WPSs using alloy flux, any increase or decrease in the electrode diameter shall require WPS requalification.

[d] Travel speed ranges for all sizes of fillet welds may be determined by the largest single pass fillet weld and the smallest multiple-pass fillet weld qualification tests.

[e] The production welding preheat or interpass temperature may be less than the PQR preheat or interpass temperature provided that the provisions of 7.6 are met, and the base metal temperature shall not be less than the WPS temperature at the time of subsequent welding.

Note: An "x" indicates applicability for the process; a shaded block indicates nonapplicability.

Table 6.6
PQR Essential Variable Changes Requiring WPS Requalification for ESW or EGW (see 6.8.2)

Essential Variable Changes to PQR Requiring Requalification	Requalification by WPS Test	Requalification by RT or UT[a]
Filler Metal		
(1) A "significant" change in filler metal or consumable guide metal composition	X	
Molding Shoes (fixed or moveable)		
(2) A change from metallic to nonmetallic or vice versa		X
(3) A change from fusing to nonfusing or vice versa		X
(4) A reduction in any cross-sectional dimension or area of a solid nonfusing shoe > 25%		X
(5) A change in design from nonfusing solid to water cooled or vice versa	X	
Filler Metal Oscillation		
(6) A change in oscillation traverse speed > 10 ipm [4 mm/s]		X
(7) A change in oscillation traverse dwell time > 2 seconds (except as necessary to compensate for joint opening variations)		X
(8) A change in oscillation traverse length that affects by more than 1/8 in [3 mm], the proximity of filler metal to the molding shoes		X
Filler Metal Supplements		
(9) A change in consumable guide metal core cross-sectional area > 30%	X	
(10) A change in the flux system, i.e., cored, magnetic electrode, external, etc.	X	
(11) A change in flux composition including consumable guide coating	X	
(12) A change in flux burden > 30%		X
Electrode/Filler Metal Diameter		
(13) Increase or decrease in electrode diameter > 1/32 in [1 mm]		X
(14) A change in the number of electrodes used	X	
Electrode Amperage		
(15) An increase or decrease in the amperage > 20%	X	
(16) A change in type of current (ac or dc) or polarity		X
Electrode Arc Voltage		
(17) An increase or decrease in the voltage > 10%		X
Process Characteristics		
(18) A change to a combination with any other welding process	X	
(19) A change from single pass to multi-pass and vice versa	X	
(20) A change from constant current to constant voltage and vice versa		X
Wire Feed Speed		
(21) An increase or decrease in the wire feed speed > 40%	X	
Travel Speed		
(22) An increase or decrease in the travel speed (if not an automatic function of arc length or deposition rate) > 20% (except as necessary to compensate for variation in joint opening)		X

(Continued)

Table 6.6 (Continued)
PQR Essential Variable Changes Requiring WPS Requalification for ESW or EGW (see 6.8.2)

Essential Variable Changes to PQR Requiring Requalification	Requalification by WPS Test	Requalification by RT or UT[a]
Electrode Shielding (EGW only)		
(23) A change in shielding gas composition of any one constituent > 5% of total flow	X	
(24) An increase or decrease in the total shielding flow rate > 25%		X
Welding Position		
(25) A change in vertical position by > 10°		X
Groove Type		
(26) An increase in cross-sectional area (for nonsquare grooves)	X	
(27) A decrease in cross-sectional area (for nonsquare grooves)		X
(28) A change in PQR joint thickness, T outside limits of 0.5T–1.1T	X	
(29) An increase or decrease > 1/4 in [6 mm] in square groove root opening		X
Postweld Heat Treatment		
(30) A change in PWHT	X	

[a] Testing shall be performed in conformance with Clause 8, Parts E or F, and Clause 10, Part F for tubulars, as applicable.

Note: An "x" indicates applicability for the requalification method; a shaded block indicates nonapplicability.

Table 6.7
PQR Supplementary Essential Variable Changes for CVN Testing Applications Requiring WPS Requalification for SMAW, SAW, GMAW, FCAW, GTAW (see 6.8.1), and ESW/EGW (see 6.8.2)

Variable	SMAW	SAW	GMAW	FCAW	GTAW	ESW/EGW
Base Metal						
(1) A change in Group Number	X	X	X	X	X	X
(2)(a) Minimum thickness qualified is T or 5/8 in [16 mm] whichever is less, except if T is less than 1/4 in [6 mm], then the minimum thickness qualified is 1/8 in [3 mm]	X	X	X	X	X	
(2)(b) Minimum thickness qualified is 0.5T						X
Filler Metal						
(3) A change in the AWS A5.X Classification, or to a weld metal or filler metal classification not covered by A5.X specifications. Carbon and low-alloy steel FCAW and GMAW-Metal Cored electrodes previously classified under A5.18, A5.20, A5.28, or A5.29 and reclassified under A5.36 without change of manufacturer or brand name, and meeting all of the previous classification requirements used in PQR/WPS CVN qualification shall be acceptable without requalification.	X	X	X	X	X	X
(4) A change in the Flux/Wire classification		X				X[b]
(5) A change in either the electrode or flux trade name when not classified by an AWS specification		X				X
(6) A change from virgin flux to crushed slag flux		X				X[b]
(7) A change in the manufacturer or the manufacturer's brand name or type of electrode			X[a]	X		
Preheat/Interpass Temperature						
(8) An increase of more than 100°F [56°C] in the maximum preheat or interpass temperature qualified	X	X	X	X	X	X
Postweld Heat Treatment						
(9) A change in the PWHT temperature and/or time ranges. The PQR test shall be subject to 80% of the aggregate times at temperature(s). Total time(s) may be applied in signal or multiple heating cycle(s).	X	X	X	X	X	X
Electrical Characteristics						
(10) An increase in heat input over that qualified (see 6.8.5), except when a grain refining austenitizing heat treatment is applied after welding.	X	X	X	X	X	X
Other Variables						
(11) In the vertical position, a change from stringer to weave	X		X	X	X	
(12) A change from multipass per side to single pass per side	X	X	X	X	X	
(13) A change exceeding ±20% in the oscillation variables for mechanized or automatic welding		X	X	X	X	X

[a] Restriction shall only apply to metal cored electrodes listed in AWS A5.18/A5.18M, *Specification for Carbon Steel Electrodes and Rods for Gas Shielded Arc Welding*, AWS A5.28/A5.28M, *Specification for Low-Alloy Steel Electrodes and Rods for Gas Shielded Arc Welding*, and AWS A5.36/A5.36M:2012, *Specification for Carbon and Low-Alloy Steel Flux Cored Arc Welding and Metal Cored Electrodes for Gas Metal Arc Welding*.
[b] ESW only

Table 6.8
Table 5.3, Table 6.9, and Unlisted Steels Qualified by PQR (see 6.8.3)

PQR Base Metal	WPS Base Metal Group Combinations Allowed by PQR
Any Group I Steel to Any Group I Steel	Any Group I Steel to Any Group I Steel
Any Group II Steel to Any Group II Steel	Any Group I Steel to Any Group I Steel Any Group II Steel to Any Group I Steel Any Group II Steel to Any Group II Steel
Any Specific Group III or Table 6.9 Steel to Any Group I Steel	The Specific PQR Group III or Table 6.9 Steel Tested to Any Group I Steel
Any Specific Group III or Table 6.9 Steel to Any Group II Steel	The Specific PQR Group III or Table 6.9 Steel Tested to Any Group I or Group II Steel
Any Group III Steel to the Same or Any Other Group III Steel or Any Group IV Steel to the Same or Any Other Group IV Steel or Any Table 6.9 Steel to the Same or Any Other Table 6.9 Steel	Steels shall be of the same material specification, grade/type and minimum yield strength as the Steels listed in the PQR
Any Combination of Group III, IV, and Table 6.9 Steels	Only the Specific Combination of Steels listed in the PQR
Any Unlisted Steel to Any Unlisted Steel or Any Steel Listed in Table 5.3 or Table 6.9	Only the Specific Combination of Steels listed in the PQR

Notes:
1. Groups I through IV are found in Table 5.3.
2. When allowed by the steel specification, the yield strength may be reduced with increased metal thickness.

Table 6.9
Code-Approved Base Metals and Filler Metals Requiring Qualification per Clause 6 (see 6.8.3)

Base Metal						Matching Strength Filler Metal			Base Metal Thickness, T		Minimum Preheat and Interpass Temperature	
Specification	Minimum Yield Point/Strength		Tensile Range			Process	AWS Electrode Specification[a]	Electrode Classification	in	mm	°F	°C
	ksi	MPa	ksi	MPa								
ASTM A871 Grade 60	60	415	75 min.	520 min.		SMAW	A5.5	E8015-X E8016-X E8018-X				
Grade 65	65	450	80 min.	550 min.								
						SAW	A5.23	F8XX-EXXX-XX F8XX-ECXXX-XX				
						GMAW	A5.28	ER80S-XXX E80C-XXX				
							A5.36	E8XTX-XAX-XXX				
						FCAW	A5.29	E8XTX-X E8XTX-XC E8XTX-XM	Up to 3/4	Up to 20	50	10
							A5.36	E8XTX-AX-XXXX E8XTX-XAX-XXX	Over 3/4 thru 1-1/2	Over 20 thru 38	125	50
ASTM A514 Over 2-1/2 in [65 mm]	90	620	100–130	690–895		SMAW	A5.5	E10015-X E10016-X E10018-X E10018M	Over 1-1/2 thru 2-1/2	Over 38 thru 65	175	80
ASTM A709 Grade HPS 100W [HPS 690W] Over 2-1/2 in to 4 in [65 mm to 100 mm]	90	620	100–130	690–895		SAW	A5.23	F10XX-EXXXX-XX F10XX-ECXXXX-XX	Over 2-1/2	Over 65	225	110
						GMAW	A5.28	ER100S-XXX E100C-XXX				
							A5.36	E10TX-XAX-XXX				
ASTM A710 Grade A, Class 1 5/16 in [8 mm] and under	85	585	90 min.	620 min.								
Over 5/16 in to 3/4 in incl [8 mm to 20 mm incl]	80	550	90 min.	620 min.		FCAW	A5.29	E10XTX-XC E10XTX-XM				
ASTM A710 Grade A, Class 3 1-1/4 in [30 mm] and under	80	550	85 min.	585 min.			A5.36	E10TX-XAX-XXX				
Over 1-1/4 in to 2 in incl [30 mm to 50 mm incl]	75	515	85 min.	585 min.								

(Continued)

CLAUSE 6. QUALIFICATION   AWS D1.1/D1.1M:2020

Table 6.9 (Continued)
Code-Approved Base Metals and Filler Metals Requiring Qualification per Clause 6 (see 6.8.3)

Base Metal						Matching Strength Filler Metal					
Specification	Minimum Yield Point/Strength		Tensile Range		Process	AWS Electrode Specification[a]	Electrode Classification	Base Metal Thickness, T		Minimum Preheat and Interpass Temperature	
	ksi	MPa	ksi	MPa				in	mm	°F	°C
ASTM A514 2-1/2 in [65 mm] and under	100	690	110–130	760–895	SMAW	A5.5	E11015-X E11016-X E11018-X E11018M				
ASTM A517 2-1/2 in [65 mm] and under	100	690	115–135	795–930	SAW	A5.23	F11XX-EXXX-XX F11XX-ECXXX-XX				
Over 2-1/2 in [65 mm]	90	620	105–135	725–930	GMAW	A5.28 A5.36	ER110S-XXX E110C-XXX E11TX-XAX-XXXX	Up to 3/4	Up to 20	50	10
					FCAW	A5.29 A5.36	E11XTX-XC E11XTX-XM E11TX-XAX-XXX	Over 3/4 thru 1-1/2	Over 20 thru 38	125	50
ASTM A709 Grade HPS 100W [HPS 690W] 2-1/2 in [65 mm] and under	100	690	110–130	760–895				Over 1-1/2 thru 2-1/2	Over 38 thru 65	175	80
								Over 2-1/2	Over 65	225	110
ASTM A1043/A1043M Grade 36 Grade 50	36–52 50–65	250–360 345–450	58 min. 65 min.	400 min. 450 min.	SMAW	A5.1	E7015 E7016 E7018 E7028				
						A5.5	E7015-X E7016-X E7018-X				
					SAW	A5.17 A5.23	F7XX-EXXX F7XX-ECXXX F7XX-EXXX-XX F7XX-ECXXX-XX				

(Continued)

Table 6.9 (Continued)
Code-Approved Base Metals and Filler Metals Requiring Qualification per Clause 6 (see 6.8.3)

Base Metal					Matching Strength Filler Metal			Base Metal Thickness, T		Minimum Preheat and Interpass Temperature	
Specification	Minimum Yield Point/Strength		Tensile Range		Process	AWS Electrode Specification[a]	Electrode Classification	in	mm	°F	°C
	ksi	MPa	ksi	MPa							
ASTM A1043/A1043M											
Grade 36	36–52	250–360	58 min.	400 min.	GMAW	A5.18	ER70S-X E70C-XC E70C-XM (Electrodes with the -GS suffix shall be excluded) (Fixed Classification, carbon steel) E70C-6M				
Grade 50 (Cont'd)	50–65	345–450	65 min.	450 min.		A5.36	(Electrodes with the -GS suffix shall be excluded) (Open Classification, carbon steel) E7XT15-XAX-CS1 E7XT15-XAX-CS2	Up to 3/4	Up to 20	50	10
						A5.36	E7XT16-XAX-CS1 E7XT16-XAX-CS2 (Electrodes with the -GS suffix shall be excluded)	Over 3/4 thru 1-1/2	Over 20 thru 38	125	50
						A5.28	ER70S-XXX E70C-XXX (Open Classification, GMAW-Metal Cored low-alloy steel)	Over 1-1/2 thru 2-1/2	Over 38 thru 65	175	80
						A5.36	E7XT-X E7XT-XC E7XT-XM	Over 2-1/2	Over 65	225	110
					FCAW	A5.20	(Electrodes with the –2C, –2M, –3, –10, –13, –14, and -GS suffix shall be excluded, and electrodes with the –11 suffix shall be excluded for thicknesses greater than 1/2 in [12 mm])				

(Continued)

Table 6.9 (Continued)
Code-Approved Base Metals and Filler Metals Requiring Qualification per Clause 6 (see 6.8.3)

Base Metal						Matching Strength Filler Metal			Base Metal Thickness, T		Minimum Preheat and Interpass Temperature	
Specification	Minimum Yield Point/Strength		Tensile Range			Process	AWS Electrode Specification[a]	Electrode Classification	in	mm	°F	°C
	ksi	MPa	ksi	MPa								
ASTM A1043/A1043M								(Fixed Classification, carbon steel)				
Grade 36	36–52	250–360	58 min.	400 min.								
Grade 50	50–65	345–450	65 min.	450 min.				E7XT-1C				
(Cont'd)								E7XT-1M				
								E7XT-5C	Up to 3/4	Up to 20	50	10
								E7XT-5M				
						FCAW (Cont'd)	A5.36	E7XT-9C	Over 3/4 thru 1-1/2	Over 20 thru 38	125	50
								E7XT-9M				
								E7XT-12C				
								E7XT-12M	Over 1-1/2 thru 2-1/2	Over 38 thru 65	175	80
								E70T-4				
								E7XT-6				
								E7XT-7	Over 2-1/2	Over 65	225	110
								E7XT-8				
								(Flux Cored Electrodes with the T1S, T3S, T10S, and -GS suffix shall be excluded and electrodes with the T11 suffix shall be excluded for thicknesses greater than 1/2 in [12 mm]				

(Continued)

Table 6.9 (Continued)
Code-Approved Base Metals and Filler Metals Requiring Qualification per Clause 6 (see 6.8.3)

Base Metal					Matching Strength Filler Metal			Base Metal Thickness, T		Minimum Preheat and Interpass Temperature	
Specification	Minimum Yield Point/Strength		Tensile Range		Process	AWS Electrode Specification[a]	Electrode Classification	in	mm	°F	°C
	ksi	MPa	ksi	MPa							
ASTM A1043/A1043M	36–52	250–360	58 min.	400 min.			(Open Classification, carbon steel)				
	50–65	345–450	65 min.	450 min.		A5.36	E7XTX-XAX-CS1 E7XTX-XAX-CS2 E7XTX-AX-CS3	Up to 3/4	Up to 20	50	10
							(Flux Cored Electrodes with the T1S, T3S, T10S, and -GS suffix shall be excluded and electrodes with the T11 suffix shall be excluded for thicknesses greater than 1/2 in [12 mm])	Over 3/4 thru 1-1/2	Over 20 thru 38	125	50
					FCAW (Cont'd)	A5.29	E7XTX-X E7XTX-XC E7XTX-XM	Over 1-1/2 thru 2-1/2	Over 38 thru 65	175	80
						A5.36	(Open Classification, FCAW, low-alloy steel) E6XTX-AX-XXX E6XTX-XAX-XXX E7XTX-AX-XXX E7XTX-XAX-XXX	Over 2-1/2	Over 65	225	110

[a] A5.36/A5.36M open classifications are listed in Annex M

Notes:
1. When welds are to be stress relieved, the deposited weld metal shall not exceed 0.05% vanadium (see 7.8).
2. When required by contract or job specifications, deposited weld metal shall have a minimum CVN energy of 20 ft-lbf [27 J] at 0°F [–20°C] as determined using CVN testing in conformance with Clause 6, Part D.
3. For ASTM A514, A517, and A709, Grade HPS 100W [HPS 690W], the maximum preheat and interpass temperature shall not exceed 400°F [200°C] for thicknesses up to 1-1/2 in [38 mm] inclusive, and 450°F [230°C] for greater thickness.
4. Filler metal properties have been moved to informative Annex L, except for AWS A5.36, see Annex M.
5. AWS A5M (SI Units) electrodes of the same classification may be used in lieu of the AWS A5 (U.S. Customary Units) electrode classification.

CLAUSE 6. QUALIFICATION   AWS D1.1/D1.1M:2020

Table 6.10
Welder and Welding Operator Qualification—Production Welding Positions Qualified by Plate Tests (see 6.16.1)[a]

Qualification Test			Production Plate Welding Qualified			Production Pipe Welding Qualified					Production Box Tube Welding Qualified			
			Groove		Fillet[d]	Butt Joint[b]		T-, Y-, K-Connections		Fillet[d]	Butt Joint		T-, Y-, K-Connections	Fillet[d]
Weld Type	Test Positions[c]		CJP	PJP		CJP	PJP	CJP	PJP[b,e]		CJP[f]	PJP	CJP	PJP[e]	
Groove[g]	1G		F	F	F, H	F	F			F	F	F		F	F, H
	2G		F, H	F, H	F, H	F, H	F, H			F, H	F, H	F, H		F, H	F, H
	3G		F, H, V	F, H, V	F, H, V	F, H, V	F, H, V		F, H, V	F, H, V	F, H, V	F, H, V		F, H, V	F, H, V
	4G		F, OH	F, OH	F, H, OH	F, OH	F, OH		F, OH	F, H, OH	F, OH	F, OH		F, OH	F, H, OH
	3G + 4G		All	All	All	All	All		All	All	All	All		All	All
Fillet	1F				F					F					F
	2F				F, H					F, H					F, H
	3F				F, H, V					F, H, V					F, H, V
	4F				F, H, OH					F, H, OH					F, H, OH
	3F + 4F				All					All					All
Plug						Qualifies Plug and Slot Welding for Only the Positions Tested									

CJP—Complete Joint Penetration
PJP—Partial Joint Penetration

[a] The qualification of welding operators for electroslag welding (ESW) or electrogas welding (EGW) shall only apply for the position tested.
[b] Only qualified for pipe equal to or greater than 24 in [600 mm] in diameter with backing, backgouging, or both.
[c] See Figures 6.3 and 6.4.
[d] See 6.22 and 10.19 for dihedral angle restrictions for plate joints and tubular T-, Y-, K-connections.
[e] Not qualified for welds having groove angles less than 30° (see 10.14.4.2).
[f] Not qualified for joints welded from one side without backing, or welded from two sides without backgouging.
[g] Groove weld qualification shall also qualify plug and slot welds for the test positions indicated.

Table 6.11
Welder and Welding Operator Qualification—Number and Type of Specimens and Range of Thickness and Diameter Qualified (Dimensions in Inches) (see 6.16.2.1)

(1) Test on Plate

Production Groove or Plug Welds		Number of Specimens[a]				Qualified Dimensions	
Type of Test Weld (Applicable Figures)	Nominal Thickness of Test Plate (T) in	Face Bend[b] (Fig. 6.8)	Root Bend[b] (Fig. 6.8)	Side Bend[b] (Fig. 6.9)	Macro-etch	Nominal Plate, Pipe or Tube Thickness Qualified, in	
						Min.	Max.
Groove (Fig. 6.20 or 6.21)	3/8	1	1	(Footnote c)	—	1/8	3/4 max.[d]
Groove (Fig. 6.16, 6.17, or 6.19)	3/8 < T < 1	—	—	2	—	1/8	2T max.[d]
Groove (Fig. 6.16, 6.17, or 6.19)	1 or over	—	—	2	—	1/8	Unlimited[d]
Plug (Fig. 6.26)	3/8	—	—	—	2	1/8	Unlimited

Production Fillet Welds (T-joint and Skewed)		Number of Specimens[a]					Qualified Dimensions		Dihedral Angles Qualified[e]	
Type of Test Weld (Applicable Figures)	Nominal Test Plate Thickness, T, in	Fillet Weld Break	Macro-etch	Side Bend[b]	Root Bend[b]	Face Bend[b]	Nominal Plate Thickness Qualified, in			
							Min.	Max.	Min.	Max.
Groove (Fig. 6.20 or 6.21)	3/8	—	—	(Footnote c)	1	1	1/8	Unlimited	30°	Unlimited
Groove (Fig. 6.20 or 6.21)	3/8 < T < 1	—	—	2	—	—	1/8	Unlimited	30°	Unlimited
Groove (Fig. 6.16, 6.17, or 6.19)	≥1	—	—	2	—	—	1/8	Unlimited	30°	Unlimited
Fillet Option 1 (Fig. 6.25)	1/2	1	1	—	—	—	1/8	Unlimited	60°	135°
Fillet Option 2 (Fig. 6.22)	3/8	—	—	—	2	—	1/8	Unlimited	60°	135°
Fillet Option 3 (Fig. 10.16) [Any diam. pipe]	> 1/8	—	1	—	—	—	1/8	Unlimited	30°	Unlimited

(2) Tests on Electroslag and Electrogas Welding

Production Plate Groove Welds		Number of Specimens[a]	Nominal Plate Thickness Qualified, in	
Type of Test Weld	Nominal Plate Thickness Tested, T, in	Side Bend[b] (see Fig. 6.9)	Min.	Max.
Groove (Fig. 6.24)	< 1-1/2	2	1/8	T
	1-1/2	2	1/8	Unlimited

[a] All welds shall be visually inspected (see 6.23.1).
[b] Radiographic examination of the test plate may be made in lieu of the bend tests (see 6.17.1.1).
[c] For 3/8 in plate or wall thickness, a side-bend test may be substituted for each of the required face- and root-bend tests.
[d] Also qualifies for welding any fillet or PJP weld size on any thickness of plate, pipe or tubing.
[e] For dihedral angles < 30°, see 10.18.1; except 6GR test not required.

Table 6.11 (Continued)
Welder and Welding Operator Qualification—Number and Type of Specimens and Range of Thickness and Diameter Qualified (Dimensions in Millimeters) (see 6.16.2.1)

(1) Test on Plate

Production Groove or Plug Welds		Number of Specimens[a]				Qualified Dimensions	
Type of Test Weld (Applicable Figures)	Nominal Thickness of Test Plate, T, mm	Face Bend[b] (Fig. 6.8)	Root Bend[b] (Fig. 6.8)	Side Bend[b] (Fig. 6.9)	Macro-etch	Nominal Plate, Pipe or Tube Thickness Qualified, mm	
						Min.	Max.
Groove (Fig. 6.20 or 6.21)	10	1	1	(Footnote c)	—	3	20 max[d]
Groove (Fig. 6.16, 6.17, or 6.19)	10 < T < 25	—	—	2	—	3	2T max[d]
Groove (Fig. 6.16, 6.17, or 6.19)	25 or over	—	—	2	—	3	Unlimited[d]
Plug (Fig. 6.26)	10	—	—	—	2	3	Unlimited

Production Fillet Welds (T-joint and Skewed)		Number of Specimens[a]					Qualified Dimensions		Dihedral Angles Qualified[e]	
Type of Test Weld (Applicable Figures)	Nominal Test Plate Thickness, T, mm	Fillet Weld Break	Macro-etch	Side Bend[b]	Root Bend[b]	Face Bend[b]	Nominal Plate Thickness Qualified, mm			
							Min.	Max.	Min.	Max.
Groove (Fig. 6.20 or 6.21)	10	—	—	(Footnote c)	1	1	3	Unlimited	30°	Unlimited
Groove (Fig. 6.20 or 6.21)	10 < T < 25	—	—	2	—	—	3	Unlimited	30°	Unlimited
Groove (Fig. 6.16, 6.17, or 6.19)	≥ 25	—	—	2	—	—	3	Unlimited	30°	Unlimited
Fillet Option 1 (Fig. 6.25)	12	1	1	—	—	—	3	Unlimited	60°	135°
Fillet Option 2 (Fig. 6.22)	10	—	—	—	2	—	3	Unlimited	60°	135°
Fillet Option 3 (Fig. 10.16) [Any diam. pipe]	> 3	—	1	—	—	—	3	Unlimited	30°	Unlimited

(2) Tests on Electroslag and Electrogas Welding

Production Plate Groove Welds		Number of Specimens[a]	Nominal Plate Thickness Qualified, mm	
Type of Test Weld	Nominal Plate Thickness Tested, T, mm	Side Bend[b] (see Fig. 6.9)	Min.	Max.
Groove (Fig. 6.24)	< 38	2	3	T
	38	2	3	Unlimited

[a] All welds shall be visually inspected (see 6.23.1).
[b] Radiographic examination of the test plate may be made in lieu of the bend tests (see 6.17.1.1).
[c] For 10 mm plate or wall thickness, a side-bend test may be substituted for each of the required face- and root-bend tests.
[d] Also qualifies for welding any fillet or PJP weld size on any thickness of plate, pipe or tubing.
[e] For dihedral angles < 30°, see 10.18.1; except 6GR test not required.

Table 6.12
Welding Personnel Performance Essential Variable Changes Requiring Requalification (see 6.20)

Essential Variable Changes to WPQR Requiring Requalification	Welding Personnel		
	Welders[a]	Welding Operators[a, b]	Tack Welders
(1) To a process not qualified (GMAW-S is considered a separate process)	X	X	X
(2) To an SMAW electrode with an F-number (see Table 6.13) higher than the WPQR electrode F-number	X		X
(3) To a position not qualified	X	X	X
(4) To a diameter or thickness not qualified	X	X	
(5) To a vertical welding progression not qualified (uphill or downhill)	X		
(6) The omission of backing (if used in the WPQR test)	X	X	
(7) To multiple electrodes (if a single electrode was used in the WPQR test) but not vice versa		X[c]	

[a] Welders qualified for SAW, GMAW, FCAW, or GTAW shall be considered as qualified welding operators in the same process(es) and subject to the welder essential variable limitations.
[b] A groove weld qualifies a slot weld for the WPQR position and the thickness ranges as shown in Table 6.11.
[c] Not for ESW or EGW.

Notes:
1. An "x" indicates applicability for the welding personnel; a shaded area indicates nonapplicability.
2. WPQR = Welding Performance Qualification Record.

Table 6.13
Electrode Classification Groups (see Table 6.12)

Group Designation	AWS Electrode Classification
F4	EXX15, EXX16, EXX18, EXX48, EXX15-X, EXX16-X, EXX18-X
F3	EXX10, EXX11, EXX10-X, EXX11-X
F2	EXX12, EXX13, EXX14, EXX13-X
F1	EXX20, EXX24, EXX27, EXX28, EXX20-X, EXX27-X

Note: The letters "XX" used in the classification designation in this table stand for the various strength levels (60 [415], 70 [485], 80 [550], 90 [620], 100 [690], 110 [760], and 120 [830]) of electrodes.

Table 6.14
CVN Test Temperature Reduction (see 6.27.6)

CVN Bar Size		Temperature Reduction	
	mm	°F	°C
Full-size standard bar	10 × 10	0	0
	10 × 9	0	0
	10 × 8	0	0
3/4 size bar	10 × 7.5	5	3
	10 × 7	8	4
2/3 size bar	10 × 6.7	10	6
	10 × 6	15	8
1/2 size bar	10 × 5	20	11
	10 × 4	30	17
1/3 size bar	10 × 3.3	35	19
	10 × 3	40	22
1/4 size bar	10 × 2.5	50	28

Note:
1. Notch shall be placed on the face of the CVN bar with the lesser dimension.
2. Straight line interpolation for intermediate notched face dimensions is permitted.
3. Temperature reductions account for the 9/5 difference between °C and °F but not the 32° offset and are not exact equivalents (see 1.2)

Table 6.15
Charpy V-Notch Test Acceptance Criteria for Various Sub-Size Specimens (see 6.27.7)

Full Size, 10 × 10 mm		3/4 Size, 10 × 7.5 mm		2/3 Size, 10 × 6.7 mm		1/2 Size, 10 × 5 mm		1/3 Size, 10 × 3.3 mm		1/4 Size, 10 × 2.5 mm	
Ft-lbf	[J]	Ft-lbf	[J]	Ft-lbf	[J]	Ft-lbf	[J]	Ft-lbf	[J]	Ft-lbf	[J]
40[a]	[54]	30	[41]	27	[37]	20	[27]	13	[18]	10	[14]
35	[48]	26	[35]	23	[31]	18	[24]	12	[16]	9	[12]
30	[41]	22	[30]	20	[27]	15	[20]	10	[14]	8	[11]
25	[34]	19	[26]	17	[23]	12	[16]	8	[11]	6	[8]
20	[27]	15	[20]	13	[18]	10	[14]	7	[10]	5	[7]
16	[22]	12	[16]	11	[15]	8	[11]	5	[7]	4	[5]
15	[20]	11	[15]	10	[14]	8	[11]	5	[7]	4	[5]
13	[18]	10	[14]	9	[12]	6	[8]	4	[5]	3	[4]
12	[16]	9	[12]	8	[11]	6	[8]	4	[5]	3	[4]
10	[14]	8	[11]	7	[10]	5	[7]	3	[4]	2	[3]
7	[10]	5	[7]	5	[7]	4	[5]	2	[3]	2	[3]

[a] Table is limited to 40 ft-lbf because the relationship between specimen size and test results has been reported to be non-linear for higher values.

Note:
1. Straight line interpolation for intermediate bar face dimensions is permitted.

Source: Adapted, with permission, from American Society of Mechanical Engineers, *ASTM A370-2017, Standard Test Methods and Definitions for Mechanical Testing of Steel Products*, Table 9

**Table 6.16
Filler Metal Essential Variables—FCAW-S Substrate/Root (see 6.28.1)**

	Substrate/Root	Fill					
	FCAW-S	FCAW-S	FCAW-G	SMAW	GMAW	SAW	Other
AWS Classification	X		X	X	X	X	X
Manufacturer	X		X	X		X	X
Manufacturer's Brand and Trade Name	X		X	X		X	X
Diameter			X	X	X	X	X

Note: An "X" in the column indicates that the essential variable is applicable for the particular welding process and weld type.
Source: Reproduced from AWS D1.8/D1.8M:2016, *Structural Welding Code—Seismic*, Table B.1, American Welding Society.

Tabulation of Positions of Groove Welds			
Position	Diagram Reference	Inclination of Axis	Rotation of Face
Flat	A	0° to 15°	150° to 210°
Horizontal	B	0° to 15°	80° to 150° 210° to 280°
Overhead	C	0° to 80°	0° to 80° 280° to 360°
Vertical	D E	15° to 80° 80° to 90°	80° to 280° 0° to 360°

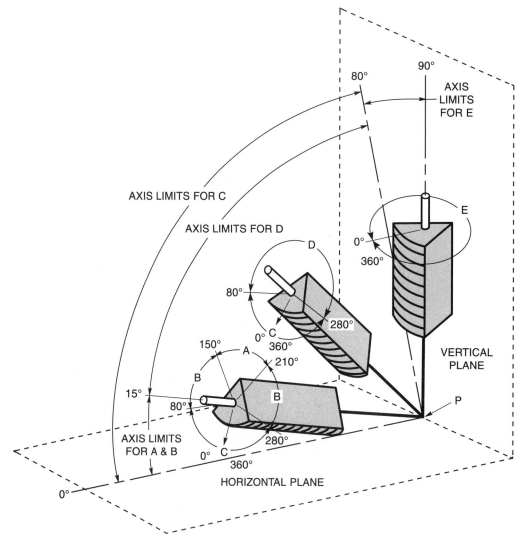

Notes:
1. The horizontal reference plane shall always be taken to lie below the weld under consideration.
2. The inclination of axis shall be measured from the horizontal reference plane toward the vertical reference plane.
3. The angle of rotation of the face shall be determined by a line perpendicular to the theoretical face of the weld which passes through the axis of the weld. The reference position (0°) of rotation of the face invariably points in the direction opposite to that in which the axis angle increases. When looking at point P, the angle of rotation of the face of the weld shall be measured in a clockwise direction from the reference position (0°).

Figure 6.1—Positions of Groove Welds (see 6.3.4)

| Tabulation of Positions of Fillet Welds ||||
Position	Diagram Reference	Inclination of Axis	Rotation of Face
Flat	A	0° to 15°	150° to 210°
Horizontal	B	0° to 15°	125° to 150° 210° to 235°
Overhead	C	0° to 80°	0° to 125° 235° to 360°
Vertical	D E	15° to 80° 80° to 90°	125° to 235° 0° to 360°

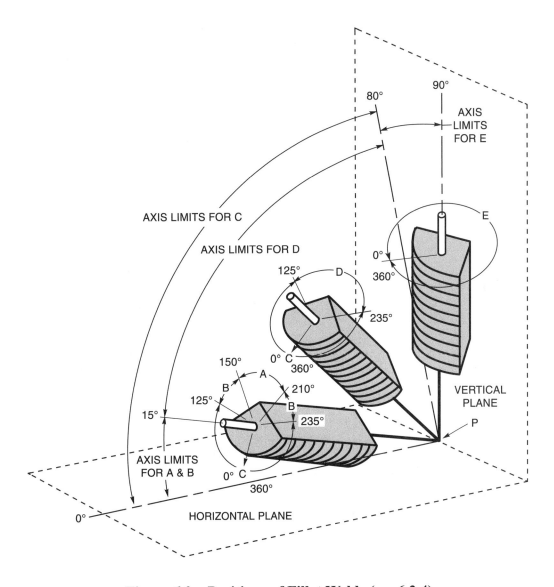

Figure 6.2—Positions of Fillet Welds (see 6.3.4)

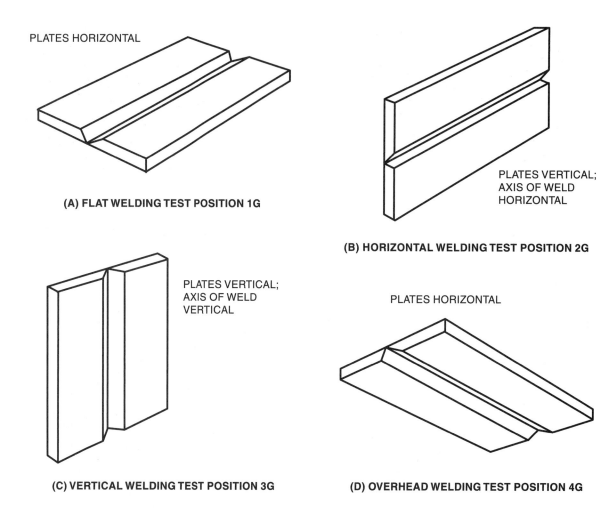

Figure 6.3—Positions Test Plates for Groove Welds (see 6.3.4)

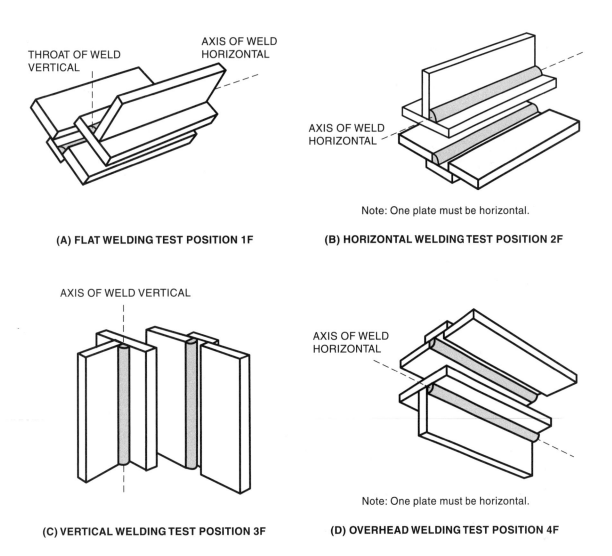

Figure 6.4—Positions Test Plates for Fillet Welds (see 6.3.4)

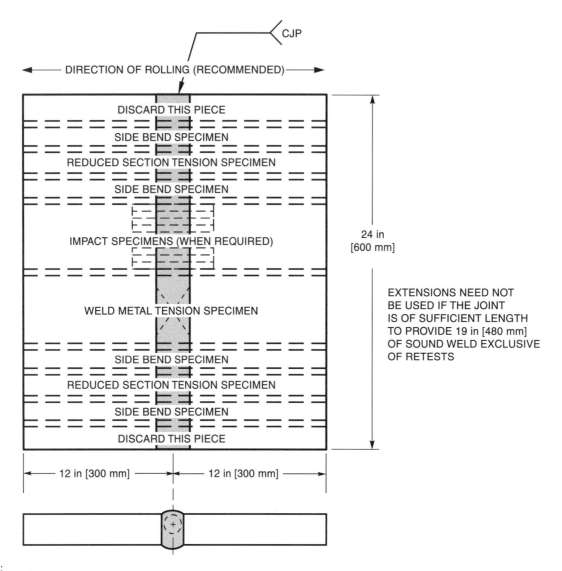

Notes:
1. The groove configuration shown is for illustration only. The groove shape tested shall conform to the production groove shape that is being qualified.
2. When CVN test specimens are required, see Clause 6, Part D for requirements.
3. All dimensions are minimum.

Figure 6.5—Location of Test Specimens on Welded Test Plates—ESW and EGW—WPS Qualification (see 6.10)

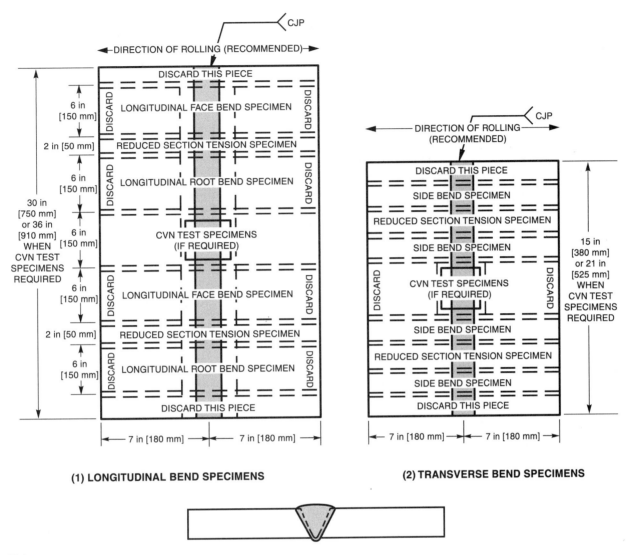

Notes:
1. The groove configuration shown is for illustration only. The groove shape tested shall conform to the production groove shape that is being qualified.
2. When CVN tests are required, the specimens shall be removed from their locations, as shown in Figure 6.28.
3. All dimensions are minimum.

Figure 6.6—Location of Test Specimens on Welded Test Plate Over 3/8 in [10 mm] Thick—WPS Qualification (see 6.10)

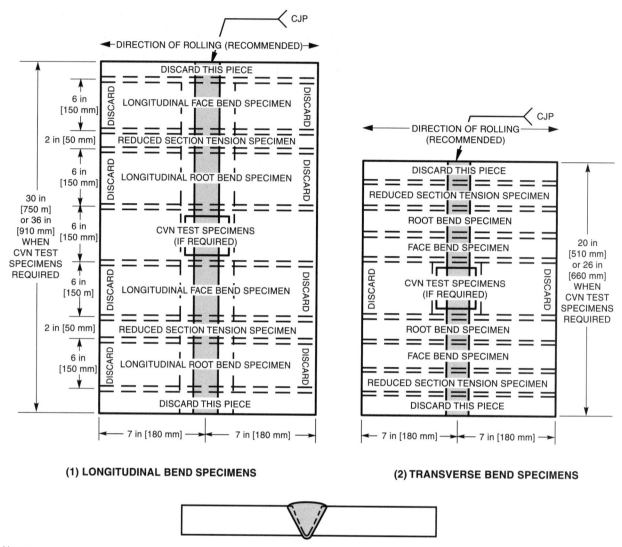

Notes:
1. The groove configuration shown is for illustration only. The groove shape tested shall conform to the production groove shape that is being qualified.
2. When CVN tests are required, the specimens shall be removed from their locations, as shown in Figure 6.28.
3. All dimensions are minimum.
4. For 3/8 in [10 mm] plate, a side-bend test may be substituted for each of the required face- and root-bend tests. See Figure 6.6(2) for plate length and location of specimens.

Figure 6.7—Location of Test Specimens on Welded Test Plate 3/8 in [10 mm] Thick and Under—WPS Qualification (see 6.10)

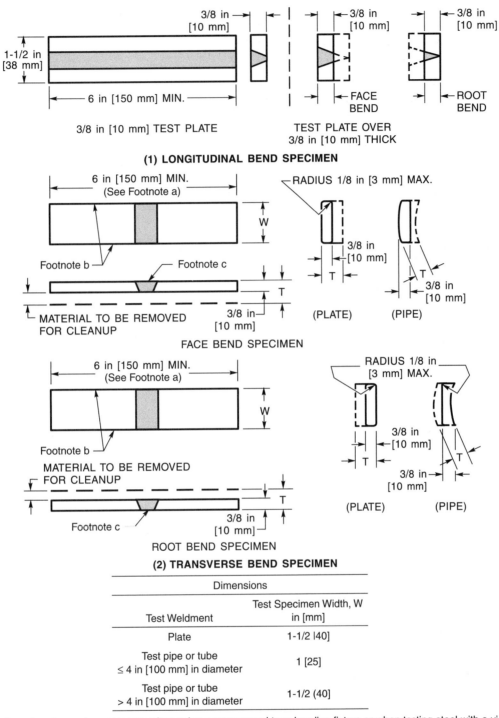

a A longer specimen length may be necessary when using a wraparound type bending fixture or when testing steel with a yield strength of 90 ksi [620 MPa] or more.
b These edges may be thermal cut and may or may not be machined.
c The weld reinforcement and backing, if any, shall be removed flush with the surface of the specimen (see 7.23.3.1 and 7.23.3.2). If a recessed backing is used, this surface may be machined to a depth not exceeding the depth of the recess to remove the backing; in such a case, the thickness of the finished specimen shall be that specified above. Cut surfaces shall be smooth and parallel.

Notes:
1. T = plate or pipe thickness.
2. When the thickness of the test plate is less than 3/8 in [10 mm], the nominal thickness shall be used for face and root bends.

Figure 6.8—Face and Root Bend Specimens (see 6.10.3.1)

CLAUSE 6. QUALIFICATION   AWS D1.1/D1.1M:2020

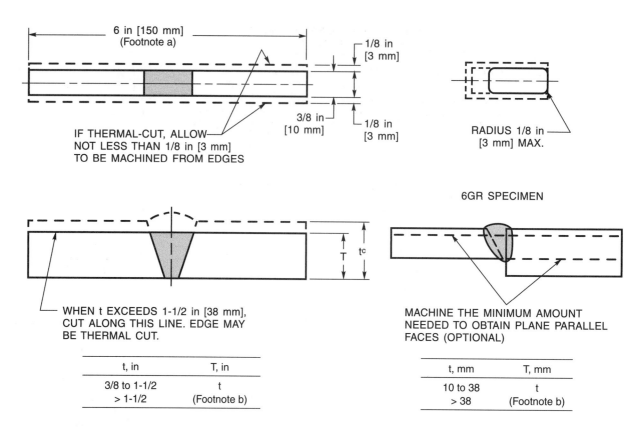

^a A longer specimen length may be necessary when using a wraparound-type bending fixture or when testing steel with a yield strength of 90 ksi [620 MPa] or more.
^b For plates over 1-1/2 in [38 mm] thick, the specimen shall be cut into approximately equal strips with T between 3/4 in [20 mm] and 1-1/2 in [38 mm] and test each strip.
^c t = plate or pipe thickness.

Figure 6.9—Side Bend Specimens (see 6.10.3.1)

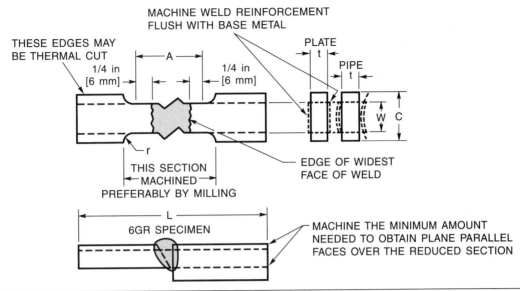

	Dimensions in inches [mm]				
	Test Plate Nominal Thickness, Tp			Test Pipe	
	Tp ≤ 1 in [25 mm]	1 in [25 mm] < Tp < 1-1/2 in [38 mm]	Tp ≥ 1-1/2 in [38 mm]	2 in [50 mm] & 3 in [75 mm] Diameter	6 in [150 mm] & 8 in [200 mm] Diameter or Larger Job Size Pipe
A—Length of reduced section	Widest face of weld + 1/2 in [12 mm], 2-1/4 in [60 mm] min.			Widest face of weld + 1/2 in [12 mm], 2-1/4 in [60 mm] min.	
L—Overall length, min[a]	As required by testing equipment			As required by testing equipment	
W—Width of reduced section[b,c]	3/4 in [20 mm] min.	3/4 in [20 mm] min.	3/4 in [20 mm] min.	1/2 ± 0.01 (12 ± 0.025)	3/4 in [20 mm] min.
C—Width of grip section[c,d]	W + 1/2 in [12 mm] min.	W + 1/2 in [12 mm] min.	W + 1/2 in [12 mm] min.	W + 1/2 in [12 mm] min.	W + 1/2 in [12 mm] min.
t—Specimen thickness[e,f]	Tp	Tp	Tp/n (Note f)	Maximum possible with plane parallel faces within length A	
r—Radius of fillet, min.	1/2 in [12 mm]	1/2 in [12 mm]	1/2 in [12 mm]	1 in [25 mm]	1 in [25 mm]

[a] It is desirable, if possible, to make the length of the grip section large enough to allow the specimen to extend into the grips a distance equal to two-thirds or more of the length of the grips.
[b] The ends of the reduced section shall not differ in width by more than 0.004 in [0.1 mm]. Also, there may be a gradual decrease in width from the ends to the center, but the width of either end shall not be more than 0.015 in [0.38 mm] larger than the width at the center.
[c] Narrower widths (W and C) may be used when necessary. In such cases, the width of the reduced section should be as large as the width of the material being tested allows. If the width of the material is less than W, the sides may be parallel throughout the length of the specimen.
[d] For standard plate-type specimens, the ends of the specimen shall be symmetrical with the center line of the reduced section within 1/4 in [6 mm].
[e] The dimension t is the thickness of the specimen as provided for in the applicable material specifications. The minimum nominal thickness of 1-1/2 in [38 mm] wide specimens shall be 3/16 in [5 mm] except as allowed by the product specification.
[f] For plates over 1-1/2 in [38 mm] thick, specimens may be cut into approximately equal strips. Each strip shall be at least 3/4 in [20 mm] thick. The test results of each strip shall meet the minimum requirements.

Note: Due to limited capacity of some tensile testing machines, alternate specimen dimensions for Table 6.9 steels may be used when approved by the Engineer.

Figure 6.10—Reduced-Section Tension Specimens (see 6.10.3.4)

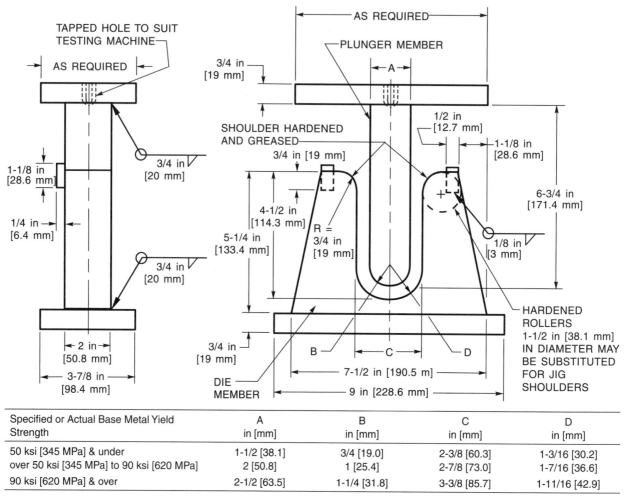

Specified or Actual Base Metal Yield Strength	A in [mm]	B in [mm]	C in [mm]	D in [mm]
50 ksi [345 MPa] & under	1-1/2 [38.1]	3/4 [19.0]	2-3/8 [60.3]	1-3/16 [30.2]
over 50 ksi [345 MPa] to 90 ksi [620 MPa]	2 [50.8]	1 [25.4]	2-7/8 [73.0]	1-7/16 [36.6]
90 ksi [620 MPa] & over	2-1/2 [63.5]	1-1/4 [31.8]	3-3/8 [85.7]	1-11/16 [42.9]

Note: Plunger and interior die surfaces shall be machine-finished.

Figure 6.11—Guided Bend Test Jig (see 6.10.3.1)

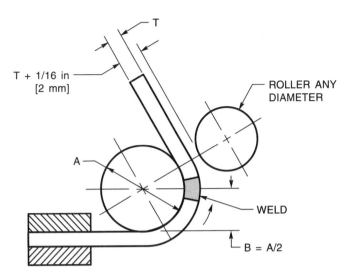

Specified or Actual Base Metal Yield Strength, ksi [MPa]	A in	B in	A mm	B mm
50 [345] & under	1-1/2	3/4	38.1	19.0
over 50 [345] to 90 [620]	2	1	50.8	25.4
90 [620] over	2-1/2	1-1/4	63.5	31.8

Figure 6.12—Alternative Wraparound Guided Bend Test Jig (see 6.10.3.1)

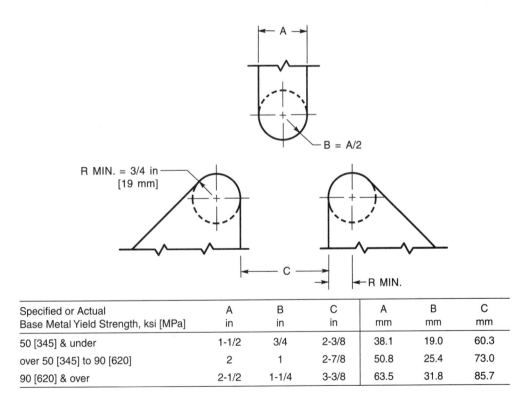

Specified or Actual Base Metal Yield Strength, ksi [MPa]	A in	B in	C in	A mm	B mm	C mm
50 [345] & under	1-1/2	3/4	2-3/8	38.1	19.0	60.3
over 50 [345] to 90 [620]	2	1	2-7/8	50.8	25.4	73.0
90 [620] & over	2-1/2	1-1/4	3-3/8	63.5	31.8	85.7

Figure 6.13—Alternative Roller-Equipped Guided Bend Test Jig for Bottom Ejection of Test Specimen (see 6.10.3.1)

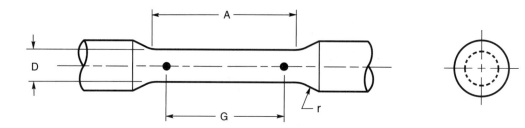

	Dimensions in inches		
	Standard Specimen	Small-Size Specimens Proportional to Standard	
Nominal Diameter	0.500 in Round	0.350 in Round	0.250 in Round
G—Gage length	2.000 ± 0.005	1.400 ± 0.005	1.000 ± 0.005
D—Diameter (Footnote a)	0.500 ± 0.010	0.350 ± 0.007	0.250 ± 0.005
r—Radius of fillet, min.	3/8	1/4	3/16
A—Length of reduced section (Footnote b), min.	2-1/4	1-3/4	1-1/4

	Dimensions (metric version per ASTM E8M)		
	Standard Specimen	Small-Size Specimens Proportional to Standard	
Nominal Diameter	12.5 mm Round	9 mm Round	6 mm Round
G—Gage length	62.5 ± 0.1	45.0 ± 0.1	30.0 ± 0.1
D—Diameter (Footnote a)	12.5 ± 0.2	9.0 ± 0.1	6.0 ± 0.1
r—Radius of fillet, min.	10	8	6
A—Length of reduced section (Footnote b), min.	75	54	36

a The reduced section may have a gradual taper from the ends toward the center, with the ends not more than 1% larger in diameter than the center (controlling dimension).

b If desired, the length of the reduced section may be increased to accommodate an extensometer of any convenient gage length. Reference marks for the measurement of elongation should be spaced at the indicated gage length.

Note: The gage length and fillets shall be as shown, but the ends may be of any form to fit the holders of the testing machine in such a way that the load shall be axial. If the ends are to be held in wedge grips, it is desirable, if possible, to make the length of the grip section great enough to allow the specimen to extend into the grips a distance equal to two-thirds or more of the length of the grips.

Figure 6.14—All-Weld Metal Tension Specimen (see 6.10.3.6)

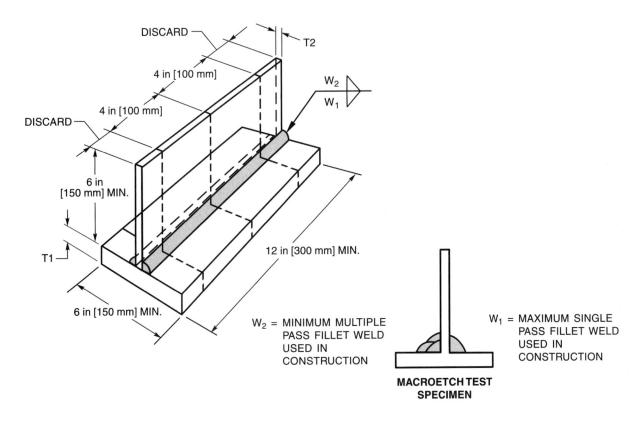

INCHES		
Weld Size	T1 min.	T2 min.
1/8	1/4	3/16
3/16	1/2	3/16
1/4	3/4	1/4
5/16	1	5/16
3/8	1	3/8
1/2	1	1/2
5/8	1	5/8
3/4	1	3/4
> 3/4	1	1

MILLIMETERS		
Weld Size	T1 min.	T2 min.
3	6	5
5	12	5
6	20	6
8	25	8
10	25	10
12	25	12
16	25	16
20	25	20
>20	25	25

Note: Where the maximum plate thickness used in production is less than the value shown above, the maximum thickness of the production pieces may be substituted for T1 and T2.

Figure 6.15—Fillet Weld Soundness Tests for WPS Qualification (see 6.13.2)

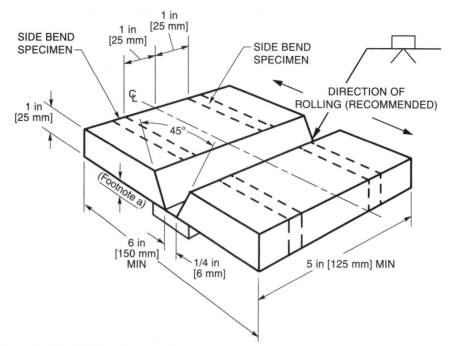

[a] The backing thickness shall be 1/4 in [6 mm] min. to 3/8 in [10 mm] max.; backing width shall be 3 in [75 mm] min. when not removed for RT, otherwise 1 in [25 mm] min.

Note: When RT is used, no tack welds shall be in test area.

Figure 6.16—Test Plate for Unlimited Thickness—Welder Qualification and Fillet Weld Consumable Verification Tests (see 6.13.3.2 and 6.21.1)

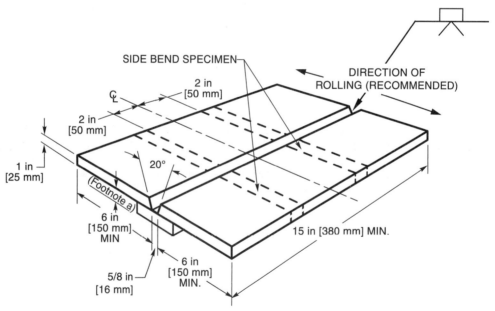

[a] The backing thickness shall be 3/8 in [10 mm] min. to 1/2 in [12 mm] max.; backing width shall be 3 in [75 mm] min. when not removed for RT, otherwise 1-1/2 in [40 mm] min.

Notes:
1. When RT is used, no tack welds shall be in test area.
2. The joint configuration of a qualified WPS may be used in lieu of the groove configuration shown here.

Figure 6.17—Test Plate for Unlimited Thickness—Welder Operator Qualification and Fillet Weld Consumable Verification Tests (see 6.13.3.2 and 6.21.2.2)

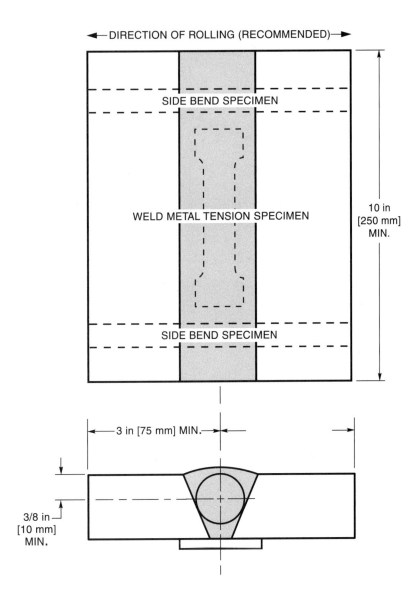

Figure 6.18—Location of Test Specimen on Welded Test Plate 1 in [25 mm] Thick—Consumables Verification for Fillet Weld WPS Qualification (see 6.13.3)

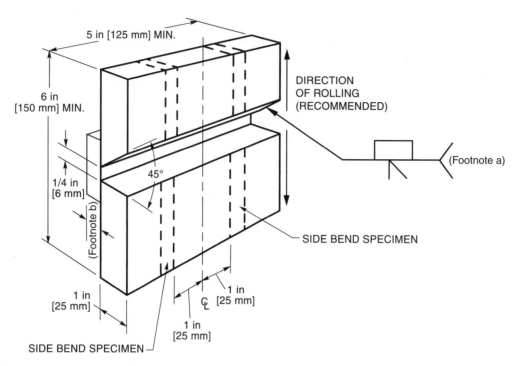

^a When RT is used, no tack welds shall be in test area.
^b The backing thickness shall be 1/4 in [6 mm] min. to 3/8 in [10 mm] max.; backing width shall be 3 in [75 mm] min. when not removed for RT, otherwise 1 in [25 mm] min.

Figure 6.19—Optional Test Plate for Unlimited Thickness—Horizontal Position—Welder Qualification (see 6.21.1)

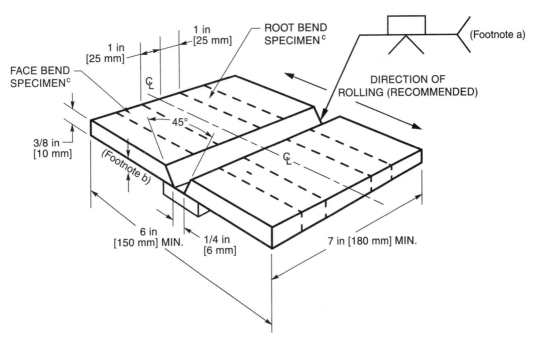

a When RT is used, no tack welds shall be in test area.
b The backing thickness shall be 1/4 in [6 mm] min. to 3/8 in [10 mm] max.; backing width shall be 3 in [75 mm] min. when not removed for RT, otherwise 1 in [25 mm] min.
c For 3/8 in [10 mm] plate, a side-bend test may be substituted for each of the required face- and root-bend tests.

Figure 6.20—Test Plate for Limited Thickness—All Positions—Welder Qualification (see 6.21.1)

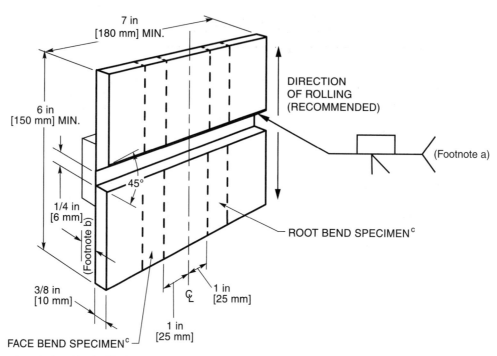

^a When RT is used, no tack welds shall be in test area.
^b The backing thickness shall be 1/4 in [6 mm] min. to 3/8 in [10 mm] max.; backing width shall be 3 in [75 mm] min. when not removed for RT, otherwise 1 in [25 mm] min.
^c For 3/8 in [10 mm] plate, a side-bend test may be substituted for each of the required face- and root-bend tests.

Figure 6.21—Optional Test Plate for Limited Thickness—Horizontal Position—Welder Qualification (see 6.21.1)

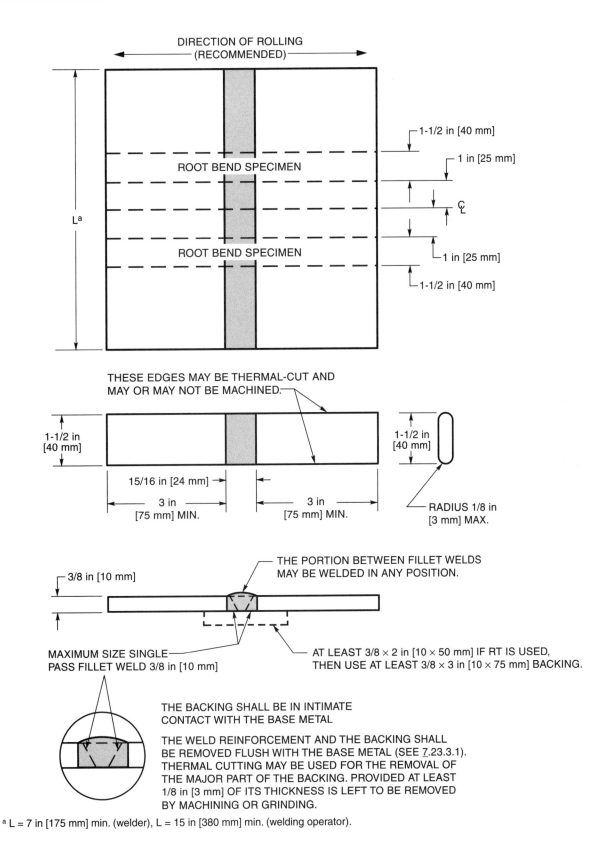

Figure 6.22—Fillet Weld Root Bend Test Plate—Welder or Welding Operator Qualification—Option 2 (see 6.22.2 and Table 6.11 or 10.20 and Table 10.13)

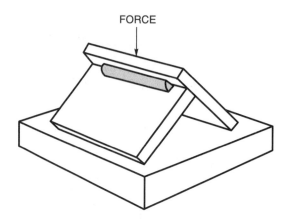

Figure 6.23—Method of Rupturing Specimen—Tack Welder Qualification (See 6.24)

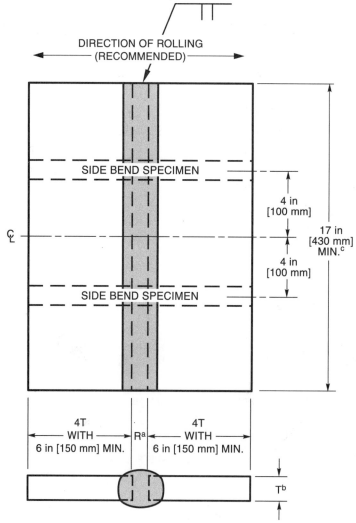

[a] Root opening "R" established by WPS.
[b] T = maximum to be welded in construction but need not exceed 1-1/2 in [38 mm].
[c] Extensions need not be used if joint is of sufficient length to provide 17 in [430 mm] of sound weld.

Figure 6.24—Butt Joint for Welding Operator Qualification—ESW and EGW (See 6.21.2.2)

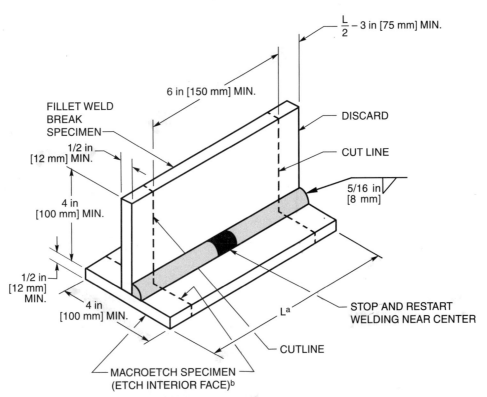

[a] L = 8 in [200 mm] min. welder, 15 in [380 mm] min. (welding operator).
[b] Either end may be used for the required macroetch specimen. The other end may be discarded.

Figure 6.25—Fillet Weld Break and Macroetch Test Plate—Welder or Welding Operator Qualification—Option 1 (See 6.23.2.1)

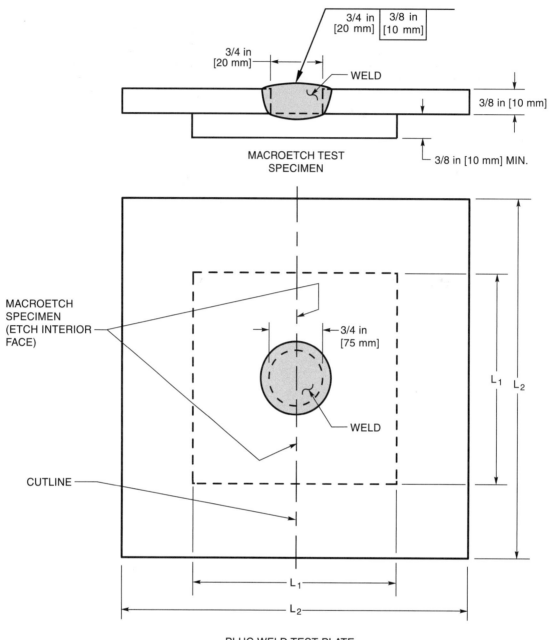

Notes:
1. L1 = 2 in [50 mm] min. (welder), 3 in [75 mm] min. (welding operator);
2. L2 = 3 in [75 mm] min. (welder), 5 in [125 mm] min. (welding operator).

Figure 6.26—Plug Weld Macroetch Test Plate—Welder or Welding Operator Qualification (See 6.14) and WPS Qualification (see 6.22.3)

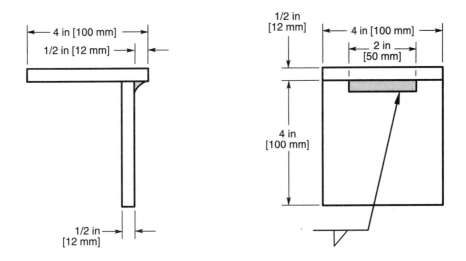

Figure 6.27—Fillet Weld Break Specimen—Tack Welder Qualification (See 6.17.2)

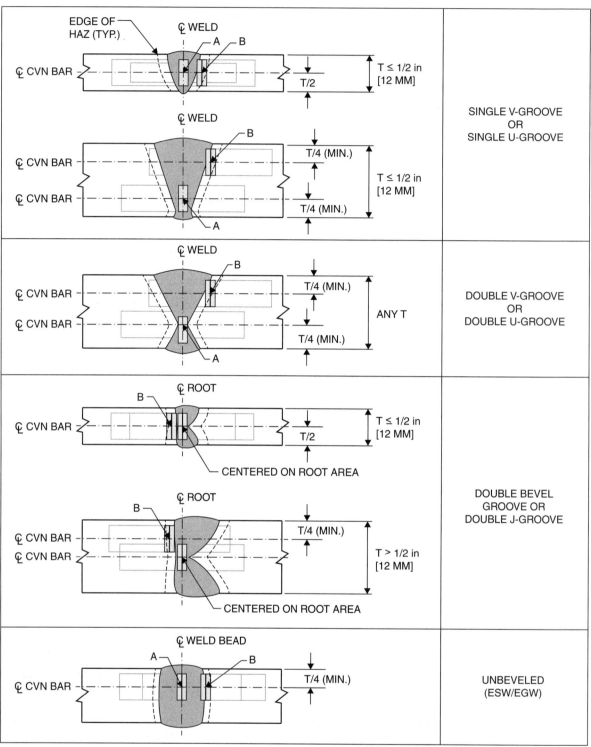

Notes:
1. A = Locate notch on weld centerline for V-, U- and square grooves. Locate notch on root centerline for bevel and J-grooves.
2. B = Locate notch in HAZ when CVNs in the HAZ are specified.
3. The Engineer may specify a notch location a specific distance from the fusion line in lieu of location B.

Figure 6.28—CVN Test Specimen Locations (see 6.27.1)

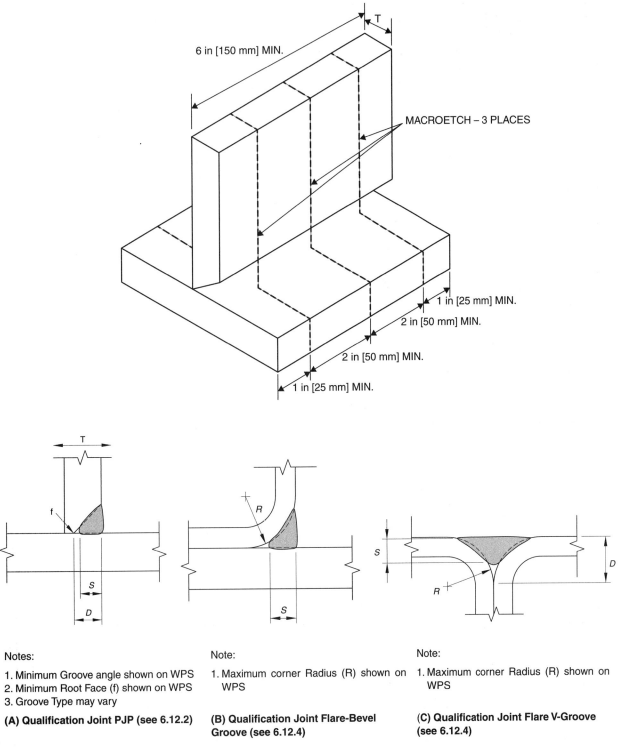

Figure 6.29—Macroetch Test Assemblies for Determination of PJP Weld Size (see 6.12.2 or 6.12.4)

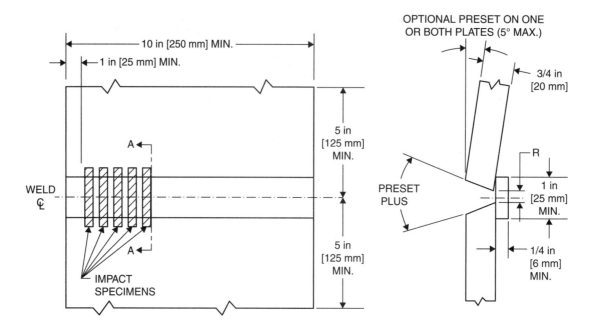

(A) TEST PLATE SHOWING LOCATION OF TEST SPECIMENS

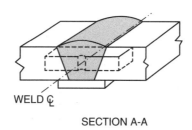

(B) ORIENTATION OF IMPACT SPECIMEN

Note: See Figures 6.31 and 6.32 for positioning of CVN specimen.

	Root Opening (R)	Groove Angle (α)
OPTION 1	1/2 in [12 mm]	45°
OPTION 2	5/8 in [16 mm]	20°

Note: CVN specimen edge to be adjacent to intermix scribe (see Figure 6.32).

Source: Adapted from AWS D1.8/D1.8M:2016, *Structural Welding Code—Seismic*, Table Figure B.1, American Welding Society.

Figure 6.30—Intermix Test Plate (see 6.28.2)

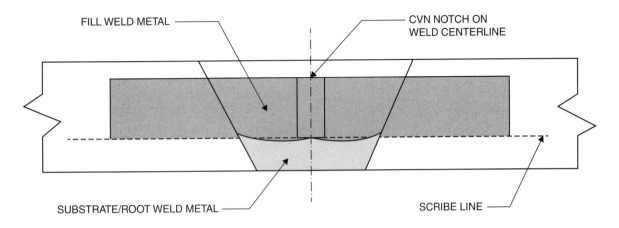

Note: CVN specimen edge is on scribe line.

Source: Adapted from AWS D1.8/D1.8M:2016, *Structural Welding Code—Seismic*, Figure B.2, American Welding Society.

Figure 6.31—Interface Scribe Line Location [see 6.28.5(2)]

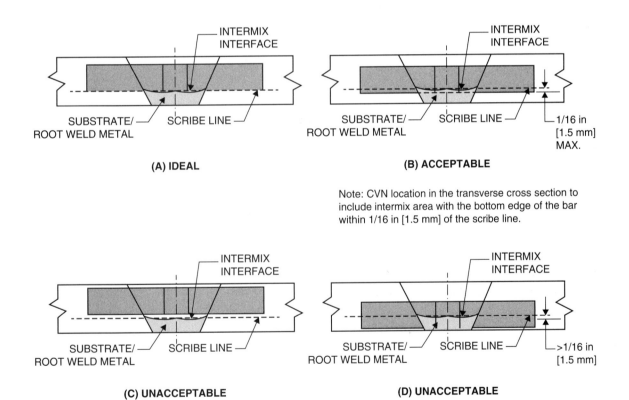

Source: Adapted from AWS D1.8/D1.8M:2016, *Structural Welding Code—Seismic*, Figure B.3, American Welding Society.

Figure 6.32—Intermix CVN Test Specimen Location [see 6.28.5(3)]

7. Fabrication

7.1 Scope

This clause contains requirements for the fabrication and erection of welded assemblies and structures produced by any process acceptable under this code (see 5.5 and 6.15) related to:

(1) Materials – Clauses 7.2 and 7.3

(2) Processes and WPSs – Clauses 7.4, 7.5, 7.6, 7.7, 7.10, 7.11, 7.20, 7.24, 7.26, and 7.29

(3) Weld Details – Clauses 7.9, 7.13, 7.14, 7.15, 7.16, 7.17, 7.18, 7.21, and 7.30

(4) Weld Quality and Repairs – Clauses 7.12, 7.13, 7.23, and 7.25

(5) Member Dimensional Tolerances – Clause 7.22

(6) Post Welding Operations – Clauses 7.3, 7.8, 7.27, 7.28, 7.29, and 7.30

7.2 Base Metal

7.2.1 Specified Base Metal. The contract documents shall designate the specification and classification of base metal to be used. When welding is involved in the structure, approved base metals, listed in Table 5.3 or Table 6.9, should be used wherever possible.

7.2.2 Base Metal for Weld Tabs, Backing, and Spacers

7.2.2.1 Weld Tabs. Weld tabs used in welding shall conform to the following requirements:

(1) When used in welding with an approved steel listed in Table 5.3 or Table 6.9, they may be any of the steels listed in Table 5.3 or Table 6.9.

(2) When used in welding with a steel qualified in conformance with 6.8.3 they may be:

(a) The steel qualified, or

(b) Any steel listed in Table 5.3 or Table 6.9

7.2.2.2 Backing and Shelf Bars. Steel for backing and shelf bars shall conform to the requirements of 7.2.2.1 or ASTM A109 T3 or T4, except that 100 ksi [690 MPa] minimum yield strength steel as backing shall be used only with 100 ksi [690 MPa] minimum yield strength steels.

7.2.2.3 Spacers. Spacers shall be of the same material as the base metal.

7.3 Welding Consumables and Electrode Requirements

7.3.1 General

7.3.1.1 Certification for Electrodes or Electrode Flux Combinations. When requested by the Engineer, the Contractor or fabricator shall furnish certification that the electrode or electrode-flux combination conforms to the requirements of the classification.

7.3.1.2 Suitability of Classification. The classification and size of electrode, arc length, voltage, and amperage shall be suited to the thickness of the material, type of groove, welding positions, and other circumstances attending the work. Welding current shall be within the range recommended by the electrode manufacturer.

7.3.1.3 Shielding Gas. A gas or gas mixture used for shielding shall conform to the requirements of AWS A5.32M/A5.32, *Welding Consumables—Gases and Gas Mixtures for Fusion Welding and Allied Processes*. When requested by the Engineer, the Contractor or fabricator shall furnish the gas manufacturer's certification that the gas or gas mixture conforms to the dew point requirements of AWS A5.32/A5.32M. When mixed at the welding site, suitable meters shall be used for proportioning the gases. Percentage of gases shall conform to the requirements of the WPS.

7.3.1.4 Storage. Welding consumables that have been removed from the original package shall be protected and stored so that the welding properties are not affected.

7.3.1.5 Condition. Electrodes shall be dry and in suitable condition for use.

7.3.2 SMAW Electrodes. Electrodes for SMAW shall conform to the requirements of the latest edition of AWS A5.1/A5.1M, *Specification for Carbon Steel Electrodes for Shielded Metal Arc Welding*, or to the requirements of AWS A5.5/A5.5M, *Specification for Low-Alloy Steel Electrodes for Shielded Metal Arc Welding*.

7.3.2.1 Low-Hydrogen Electrode Storage Conditions. All electrodes having low-hydrogen coverings conforming to AWS A5.1 and AWS A5.5 shall be purchased in hermetically sealed containers or shall be baked by the user in conformance with 7.3.2.4 prior to use. After opening the hermetically sealed container, electrodes not immediately issued for use shall be stored in ovens held at a temperature of at least 250°F [120°C]. Electrodes shall be rebaked no more than once. Electrodes that have been wet shall not be used.

7.3.2.2 Approved Atmospheric Time Periods. After hermetically sealed containers are opened or after electrodes are removed from baking or storage ovens, the electrode exposure to the atmosphere shall not exceed the values shown in column A, Table 7.1, for the specific electrode classification with optional supplemental designators, where applicable. Electrodes exposed to the atmosphere for periods less than those allowed by column A, Table 7.1 may be returned to a holding oven maintained at 250°F [120°C] min.; after a minimum hold period of four hours at 250°F [120°C] min. the electrodes may be reissued.

7.3.2.3 Alternative Atmospheric Exposure Time Periods Established by Tests. The alternative exposure time values shown in column B in Table 7.1 may be used provided testing establishes the maximum allowable time. The testing shall be performed in conformance with AWS A5.5 for each electrode classification and each electrode manufacturer. Such tests shall establish that the maximum moisture content values of AWS A5.5 are not exceeded. Additionally, E70XX or E70XX-X (AWS A5.1 or A5.5) low-hydrogen electrode coverings shall be limited to a maximum moisture content not exceeding 0.4% by weight. These electrodes shall not be used at relative humidity temperature combinations that exceed either the relative humidity or moisture content in the air that prevailed during the testing program.

For proper application of this subclause, see Annex D for the temperature-moisture content chart and its examples. The chart shown in Annex D, or any standard psychometric chart, shall be used in the determination of temperature-relative humidity limits.

7.3.2.4 Baking Electrodes. Electrodes exposed to the atmosphere for periods greater than those allowed in Table 7.1 shall be baked as follows:

(1) All electrodes having low-hydrogen coverings conforming to AWS A5.1 shall be baked for at least two hours between 500°F and 800°F [260°C and 430°C], or

(2) All electrodes having low-hydrogen coverings conforming to AWS A5.5 shall be baked for at least one hour at temperatures between 700°F and 800°F [370°C and 430°C].

All electrodes shall be placed in a suitable oven at a temperature not exceeding one half the final baking temperature for a minimum of one half hour prior to increasing the oven temperature to the final baking temperature. Final baking time shall start after the oven reaches final baking temperature.

7.3.2.5 Low-Hydrogen Electrode Restrictions for ASTM A514 or A517 Steels. When used for welding ASTM A514 or A517 steels, low-hydrogen electrodes shall meet the following requirements, as applicable:

(1) When welding with E90XX-X or higher tensile strength electrodes, the electrode may be used without baking, provided the electrode is furnished in hermetically sealed containers.

(2) When welding with E90XX-X or higher tensile strength electrodes not furnished in hermetically sealed containers, or when welding with E80XX-X or lower tensile strength electrodes whether furnished in hermetically sealed containers or otherwise, the electrodes shall be baked for a minimum of one hour at temperatures between 700°F and 800°F [370°C and 430°C] before being used, except as provided in (3).

(3) When welding with E7018M electrodes, or electrodes with the optional H4R designator, the electrode may be used without baking.

7.3.3 SAW Electrodes and Fluxes. SAW may be performed with one or more single electrodes, one or more parallel electrodes, or combinations of single and parallel electrodes. The spacing between arcs shall be such that the slag cover over the weld metal produced by a leading arc does not cool sufficiently to prevent the proper weld deposit of a following electrode. SAW with multiple electrodes may be used for any groove or fillet weld pass.

7.3.3.1 Electrode-Flux Combination Requirements. The bare electrodes and flux used in combination for SAW of steels shall conform to the requirements in the latest edition of AWS A5.17/A5.17M, *Specification for Carbon Steel Electrodes and Fluxes for Submerged Arc Welding*, or to the requirements of the latest edition of AWS A5.23/A5.23M, *Specification for Low-Alloy Steel Electrodes and Fluxes for Submerged Arc Welding*.

7.3.3.2 Condition of Flux. Flux used for SAW shall be dry and free of contamination from dirt, mill scale, or other foreign material. All flux shall be purchased in packages that can be stored, under normal conditions, for at least six months without such storage affecting its welding characteristics or weld properties. Flux from damaged packages shall be discarded or shall be dried at a minimum temperature of 500°F [260°C] for one hour before use. Flux shall be placed in the dispensing system immediately upon opening a package, or if used from an opened package, the top one inch shall be discarded. Flux that has been wet shall not be used.

7.3.3.3 Flux Reclamation. SAW flux that has not been melted during the welding operation may be reused after recovery by vacuuming, catch pans, sweeping, or other means. The welding fabricator shall have a system for collecting unmelted flux, adding new flux, and welding with the mixture of these two, such that the flux composition and particle size distribution at the weld puddle are relatively constant.

7.3.3.4 Crushed Slag. Crushed slag may be used provided it has its own marking, using the crusher's name and trade designation. In addition, each dry batch or dry blend (lot) of flux, as defined in AWS A5.01M/A5.01, *Welding and Brazing Consumables—Procurement of Filler Metals and Fluxes*, shall be tested in conformance with Schedule I of AWS A5.01M/A5.01 and classified by the Contractor or crusher per AWS A5.17/A5.17M or AWS A5.23/A5.23M, as applicable.

7.3.4 GMAW/FCAW Electrodes. The electrodes for GMAW or FCAW shall conform to the requirements of the following:

(1) AWS A5.18/A5.18M, *Specification for Carbon Steel Electrodes and Rods for Gas Shielded Arc Welding*;

(2) AWS A5.20/A5.20M, *Specification for Carbon Steel Electrodes for Flux Cored Arc Welding*;

(3) AWS A5.28/A5.28M, *Specification for Low-Alloy Steel Electrodes and Rods for Gas Shielded Arc Welding*;

(4) AWS A5.29/A5.29M, *Specification for Low-Alloy Steel Electrodes for Flux Cored Arc Welding*; or

(5) AWS A5.36/A5.36M, *Specification for Carbon and Low-Alloy Steel Flux Cored Electrodes for Flux Cored Arc Welding and Metal Cored Electrodes for Gas Metal Arc Welding*.

7.3.5 GTAW

7.3.5.1 Tungsten Electrodes. Welding current shall be compatible with the diameter and type or classification of electrode. Tungsten electrodes shall be in conformance with AWS A5.12M/A5.12, *Specification for Tungsten and Oxide Dispersed Tungsten Electrodes for Arc Welding and Cutting*.

7.3.5.2 Filler Metal. The filler metal shall conform to all the requirements of the latest edition of AWS A5.18/A5.18M or AWS A5.28/A5.28M or AWS A5.30/A5.30M, *Specification for Consumable Inserts*, as applicable.

7.4 ESW and EGW Processes

7.4.1 Process Limitations. The ESW and EGW processes shall be restricted to use of Table 5.3, Group I, II, and III steels, except that ESW and EGW of ASTM A710 shall not be permitted.

7.4.2 Condition of Electrodes and Guide Tubes. Electrodes and consumable guide tubes shall be dry, clean, and in suitable condition for use.

7.4.3 Condition of Flux. Flux used for ESW shall be dry and free of contamination from dirt, mill scale, or other foreign material. All flux shall be purchased in packages that can be stored, under normal conditions, for at least six months without such storage affecting its welding characteristics or weld properties. Flux from packages damaged in transit or in handling shall be discarded or shall be dried at a minimum temperature of 250°F [120°C] for one hour before use. Flux that has been wet shall not be used.

7.4.4 Weld Starts and Stops. Welds shall be started in such a manner as to allow sufficient heat buildup for complete fusion of the weld metal to the groove faces of the joint. Welds which have been stopped at any point in the weld joint for a sufficient amount of time for the slag or weld pool to begin to solidify may be restarted and completed, provided the completed weld is examined by UT for a minimum of 6 in [150 mm] on either side of the restart and, unless prohibited by joint geometry, also confirmed by RT. All such restart locations shall be recorded and reported to the Engineer.

7.4.5 Preheating. Because of the high-heat input characteristic of these processes, preheating is not normally required. However, no welding shall be performed when the temperature of the base metal at the point of welding is below 32°F [0°C].

7.4.6 Repairs. Welds having discontinuities prohibited by Clause 8, Part C or Clause 10, Part F shall be repaired as allowed by 7.25 utilizing a qualified welding process, or the entire weld shall be removed and replaced.

7.4.7 Weathering Steel Requirements. For ESW and EGW of exposed, bare, unpainted applications of ASTM A588/A588M steel requiring weld metal with atmospheric corrosion resistance and coloring characteristics similar to that of the base metal, the electrode-flux combination shall be in conformance with 6.15.1.3(2), and the filler metal chemical composition shall conform to Table 5.6.

7.5 WPS Variables

The welding variables shall be in conformance with a written WPS (see Annex J, Form J–2, as an example). Each pass will have complete fusion with the adjacent base metal, and such that there will be no depressions or undue undercutting at the toe of the weld. Excessive concavity of initial passes shall be avoided to prevent cracking in the roots of joints under restraint. All welders, welding operators, and tack welders shall be informed in the proper use of the WPS, and the applicable WPS shall be readily available and shall be followed during the performance of welding.

7.6 Preheat and Interpass Temperatures

Base metal shall be preheated, if required, to a temperature not less than the minimum value listed on the WPS (see 5.2.1 for prequalified WPS limitations and Table 6.5 for qualified WPS essential variable limitations).

Preheat and all subsequent minimum interpass temperatures shall be maintained during the welding operation for a distance at least equal to the thickness of the thickest welded part (but not less than 3 in [75 mm]) in all directions from the point of welding.

Minimum interpass temperature requirements shall be considered equal to the preheat requirements, unless otherwise indicated on the WPS.

Preheat and interpass temperature shall be checked just prior to initiating the arc for each pass.

For combinations of base metals, the minimum preheat shall be based on the highest minimum preheat.

7.7 Heat Input Control for Quenched and Tempered Steels

When quenched and tempered steels are welded, the heat input shall be restricted in conjunction with the maximum preheat and interpass temperatures required. Such considerations shall include the additional heat input produced in simultaneous welding on the two sides of a common member. The preceding limitations shall be in conformance with the producer's recommendations. The heat input limitations of this clause shall not apply to ASTM A913/A913M.

7.8 Stress-Relief Heat Treatment

Where required by the contract documents, welded assemblies shall be stress relieved by heat treating. Final machining after stress relieving shall be considered when needed to maintain dimensional tolerances.

7.8.1 Requirements. The stress-relief treatment shall conform to the following requirements:

(1) The temperature of the furnace shall not exceed 600°F [315°C] at the time the welded assembly is placed in it.

(2) Above 600°F [315°C], the rate of heating in °F [°C] per hour shall not exceed 400 [560] divided by the maximum metal thickness of the thicker part, in inches [centimeters], but in no case more than 400°F [220°C] per hour. During the heating period, variations in temperature throughout the portion of the part being heated shall be no greater than 250°F [140°C] within any 15 ft [5 m] interval of length. The rates of heating and cooling need not be less than 100°F [55°C] per hour. However, in all cases, consideration of closed chambers and complex structures may indicate reduced rates of heating and cooling to avoid structural damage due to excessive thermal gradients.

(3) After a maximum temperature of 1100°F [600°C] is reached on quenched and tempered steels, or a mean temperature range between 1100°F and 1200°F [600°C and 650°C] is reached on other steels, the temperature of the assembly shall be held within the specified limits for a time not less than specified in Table 7.2, based on weld thickness. When the specified stress relief is for dimensional stability, the holding time shall be not less than specified in Table 7.2, based on the thickness of the thicker part. During the holding period there shall be no difference greater than 150°F [65°C] between the highest and lowest temperature throughout the portion of the assembly being heated.

(4) Above 600°F [315°C], cooling shall be done in a closed furnace or cooling chamber at a rate no greater than 500°F [260°C] per hour divided by the maximum metal thickness of the thicker part in inches [mm], but in no case more than 500°F [260°C] per hour. From 600°F [315°C], the assembly may be cooled in still air.

7.8.2 Alternative PWHT. Alternatively, when it is impractical to PWHT to the temperature limitations stated in 7.8.1, welded assemblies may be stress-relieved at lower temperatures for longer periods of time, as given in Table 7.3.

7.8.3 Steels Not Recommended for PWHT. Stress relieving of weldments of ASTM A514/A514M, ASTM A517/A517M, ASTM A709/A709M Grade HPS 100W [HPS 690W] and ASTM A710/A710M steels is not generally recommended. Stress relieving may be necessary for those applications where weldments are required to retain dimensional stability during machining or where stress corrosion cracking may be involved, neither condition being unique to weldments involving ASTM A514/A514M, ASTM A517/A517M, ASTM A709/A709M Grade HPS 100W [HPS 690W] and ASTM A710/A710M steels. However, the results of notch toughness tests have shown that PWHT may actually impair weld metal and HAZ toughness, and intergranular cracking may sometimes occur in the grain-coarsened region of the weld HAZ.

7.9 Backing

7.9.1 Attachment of Steel Backing. Steel backing shall conform to the following requirements:

7.9.1.1 Fusion. Groove welds made with steel backing shall have the weld thoroughly fused to the backing.

7.9.1.2 Full-Length Backing. Except as permitted below, steel backing shall be made continuous for the full length of the weld. All joints in steel backing shall be CJP groove weld joints meeting all the requirements of Clause 7 of this code.

For statically loaded applications, backing for welds to the ends of closed sections, such as hollow structural section (HSS), are permitted to be made from one or two pieces with unspliced discontinuities where all of the following conditions are met:

(1) The closed section nominal wall thickness does not exceed 5/8 in [16 mm].

(2) The closed section outside perimeter does not exceed 64 in [1625 mm].

(3) The backing is transverse to the longitudinal axis of the closed section.

(4) The interruption in the backing does not exceed 1/4 in [6 mm].

(5) The weld with discontinuous backing is not closer than the HSS diameter or major cross section dimension from other types of connections.

(6) The interruption in the backing is not located in the corners.

For statically loaded box columns, discontinuous backing is permitted in the CJP welded corners, at field splices and at connection details. Discontinuous backing is permitted in other closed sections where approved by the Engineer.

7.9.1.3 Backing Thickness. Steel backing shall be of sufficient thickness to prevent melt-through.

7.9.1.4 Cyclically Loaded Nontubular Connections. Steel backing in cyclically loaded nontubular structures shall comply with the following:

(1) Steel backing of welds that are transverse to the direction of computed stress shall be removed, and the joints shall be ground or finished smooth except for joints designed in accordance with Table 4.5 (5.5).

(2) Steel backing of welds that are parallel to the direction of stress or are not subject to computed stress need not be removed, unless so specified by the Engineer.

(3) Where external welds are used to attach longitudinal steel backing that is to remain in place, the welds shall be fillet welds that are continuous for the entire length on both sides of the backing.

7.9.1.5 Statically Loaded Connections. Steel backing for welds in statically loaded structures (tubular and nontubular) need not be welded full length and need not be removed unless specified by the Engineer.

7.9.2 Backing Welds. Backing welds may be used for backing of groove or fillet welds. When SMAW is used, low-hydrogen electrodes shall be used.

7.9.3 Non-Steel Backing. Roots of groove welds may be backed by copper, flux, glass tape, ceramic, iron powder, or similar materials to prevent melt-through.

7.10 Welding and Cutting Equipment

All welding and thermal cutting equipment shall be so designed and manufactured, and shall be in such condition, as to enable designated personnel to follow the procedures and attain the results described elsewhere in this code.

7.11 Welding Environment

7.11.1 Maximum Wind Velocity. GMAW, GTAW, EGW, or FCAW-G shall not be done in a draft or wind unless the weld is protected by a shelter. Such shelter shall be of material and shape appropriate to reduce wind velocity in the vicinity of the weld to a maximum of 5 miles per hour [8 kilometers per hour].

7.11.2 Minimum Ambient Temperature. Welding shall not be done

(1) when the ambient temperature is lower than 0°F [–20°C], or

(2) when surfaces are wet or exposed to rain, snow, or

(3) high wind velocities, or

(4) when welding personnel are exposed to inclement conditions.

NOTE: Zero°F does not mean the ambient environmental temperature, but the temperature in the immediate vicinity of the weld. The ambient environmental temperature may be below 0°F [–20°C], but a heated structure or shelter around the area being welded may maintain the temperature adjacent to the weldment at 0°F [–20°C] or higher.

7.12 Conformance with Design

The sizes and lengths of welds shall be no less than those specified by design requirements and detail drawings, except as allowed in Table 8.1 or Table 10.15. The location of welds shall not be changed without approval of the Engineer.

7.13 Minimum Fillet Weld Sizes

The minimum fillet weld size, except for fillet welds used to reinforce groove welds, shall be as shown in Table 7.7. The minimum fillet weld size shall apply in all cases, unless the design drawings specify welds of a larger size.

7.14 Preparation of Base Metal

7.14.1 General. Base metal shall be sufficiently clean to permit welds to be made that will meet the quality requirements of this code.

7.14.2 Mill-Induced Surface Defects. Welds shall not be placed on surfaces that contain fins, tears, cracks, slag, or other base metal defects as defined in the base metal specifications.

7.14.3 Scale and Rust. Loose scale, thick scale, and thick rust shall be removed from the surfaces to be welded, and from surfaces adjacent to the weld. Welds may be made on surfaces that contain mill scale and rust if the mill scale and rust can withstand vigorous hand wire brushing and if the applicable quality requirements of this code are met with the following exception: for girders in cyclically loaded structures, all mill scale shall be removed from the surfaces on which flange-to-web welds are to be made.

7.14.4 Foreign Materials

7.14.4.1 Surfaces to be welded, and surfaces adjacent to the weld, shall be cleaned to remove excessive quantities of the following:

(1) Water

(2) Oil

(3) Grease

(4) Other hydrocarbon based materials

Welding on surfaces containing residual amounts of foreign materials is permitted provided the quality requirements of this code are met.

7.14.4.2 Welds are permitted to be made on surfaces with surface protective coatings or anti-spatter compounds, except those that are prohibited in 7.14.4.1, provided the quality requirements of this code can be met.

7.14.5 Mill-Induced Discontinuities. The limits of acceptability and the repair of visually observed cut surface discontinuities shall be in conformance with Table 7.4, in which the length of discontinuity is the visible long dimension on the cut surface of material and the depth is the distance that the discontinuity extends into the material from the cut surface. All welded repairs shall be in conformance with this code. Removal of the discontinuity may be done from either surface of the base metal. The aggregate length of welding shall not exceed 20% of the length of the plate surface being repaired except with approval of the Engineer.

7.14.5.1 Acceptance Criteria. For discontinuities greater than 1 in [25 mm] in length and depth discovered on cut surfaces, the following procedures shall be observed.

(1) Where discontinuities such as W, X, or Y in Figure 7.1 are observed prior to completing the joint, the size and shape of the discontinuity shall be determined by UT. The area of the discontinuity shall be determined as the area of total loss of back reflection, when tested in conformance with the procedure of ASTM A435/A435M, *Specification for Straight-Beam Ultrasonic Examination of Steel Plates*.

(2) For acceptance of W, X, or Y discontinuities, the area of the discontinuity (or the aggregate area of multiple discontinuities) shall not exceed 4% of the cut material area (length times width) with the following exception: if the length of the discontinuity, or the aggregate width of discontinuities on any transverse section, as measured perpendicular to the cut material length, exceeds 20% of the cut material width, the 4% cut material area shall be reduced by the percentage amount of the width exceeding 20%. (For example, if a discontinuity is 30% of the cut material width, the area of discontinuity cannot exceed 3.6% of the cut material area.) The discontinuity on the cut surface of the cut material shall be removed to a depth of 1 in [25 mm] beyond its intersection with the surface by chipping, gouging, or grinding, and blocked off by welding with a low-hydrogen process in layers not exceeding 1/8 in [3 mm] in thickness for at least the first four layers.

(3) Repair shall not be required if a discontinuity Z, not exceeding the allowable area in 7.14.5.1(2), is discovered after the joint has been completed and is determined to be 1 in [25 mm] or more away from the face of the weld, as measured on the cut base metal surface. If the discontinuity Z is less than 1 in [25 mm] away from the face of the weld, it shall be removed to a distance of 1 in [25 mm] from the fusion zone of the weld by chipping, gouging, or grinding. It shall then be blocked off by welding with a low-hydrogen process in layers not exceeding 1/8 in [3 mm] in thickness for at least the first four layers.

(4) If the area of the discontinuity W, X, Y, or Z exceeds the allowable in 7.14.5.1(2), the cut material or subcomponent shall be rejected and replaced, or repaired at the discretion of the Engineer.

7.14.5.2 Repair. In the repair and determination of limits of mill induced discontinuities visually observed on cut surfaces, the amount of metal removed shall be the minimum necessary to remove the discontinuity or to determine that the limits of Table 7.4 are not exceeded. However, if weld repair is required, sufficient base metal shall be removed to provide access for welding. Cut surfaces may exist at any angle with respect to the rolling direction. All welded repairs of discontinuities shall be made by:

(1) Suitably preparing the repair area

(2) Welding with an approved low-hydrogen process and observing the applicable provisions of this code

(3) Grinding the completed weld smooth and flush (see 7.23.3.1) with the adjacent surface.

NOTE: The requirements of 7.14.5.2 may not be adequate in cases of tensile load applied through the thickness of the material.

7.14.6 Joint Preparation. Machining, thermal cutting, gouging (including plasma arc cutting and gouging), chipping, or grinding may be used for joint preparation, or the removal of unacceptable work or metal, except that oxygen gouging shall only be permitted for use on as-rolled steels.

7.14.7 Material Trimming. For cyclically loaded structures, material thicker than specified in the following list shall be trimmed if and as required to produce a satisfactory welding edge wherever a weld is to carry calculated stress:

(1) Sheared material thicker than 1/2 in [12 mm]

(2) Rolled edges of plates (other than universal mill plates) thicker than 3/8 in [10 mm]

(3) Toes of angles or rolled shapes (other than wide flange sections) thicker than 5/8 in [16 mm]

(4) Universal mill plates or edges of flanges of wide flange sections thicker than 1 in [25 mm]

(5) The preparation for butt joints shall conform to the requirements of the detail drawings

7.14.8 Thermal Cutting Processes. Electric arc cutting and gouging processes (including plasma arc cutting and gouging) and oxyfuel gas cutting processes are recognized under this code for use in preparing, cutting, or trimming materials. The use of these processes shall conform to the applicable requirements of Clause 7.

7.14.8.1 Other Processes. Other thermal cutting and gouging processes may be used under this code, provided the Contractor demonstrates to the Engineer an ability to successfully use the process.

7.14.8.2 Profile Accuracy. Steel and weld metal may be thermally cut, provided a smooth and regular surface free from cracks and notches is secured, and provided that an accurate profile is secured by the use of a mechanical guide. For cyclically loaded structures, freehand thermal cutting shall be done only where approved by the Engineer.

7.14.8.3 Roughness Requirements. In thermal cutting, the equipment shall be so adjusted and manipulated as to avoid cutting beyond (inside) the prescribed lines. The reference standard for evaluation of cut surfaces shall be the surface roughness gauge included in AWS C4.1–77 set, *Criteria for Describing Oxygen-Cut Surfaces and Oxygen Cutting Surface Roughness Gauge*. The roughness of thermal cut surfaces shall be evaluated by visually comparing the cut surface to the roughness represented on the roughness gauge. Surface roughness shall be no greater than that represented by Sample 3, except that for the ends of members not subject to calculated stress, copes in beams with the flange thickness not exceeding 2 in [50 mm], and for materials over 4 in to 8 in [100 mm to 200 mm] thick, surface roughness shall not exceed that represented by Sample 2.

7.14.8.4 Gouge or Notch Limitations. Roughness exceeding these values and notches or gouges not more than 3/16 in [5 mm] deep on otherwise satisfactory surfaces shall be removed by machining or grinding. Notches or gouges exceeding 3/16 in [5 mm] deep may be repaired by grinding if the nominal cross-sectional area is not reduced by more than 2%. Ground or machined surfaces shall be faired to the original surface with a slope not exceeding one in ten. Cut surfaces and adjacent edges shall be left free of slag. In thermal cut surfaces, occasional notches or gouges may, with approval of the Engineer, be repaired by welding.

7.15 Reentrant Corners

Reentrant corners of cut material shall be formed to provide a gradual transition with a radius of not less than 1 in [25 mm] except corners in connection material and beam copes. Adjacent surfaces shall meet without offset or cutting past the point of tangency. The reentrant corners may be formed by thermal cutting, followed by grinding, if necessary, in conformance with the surface requirements of 7.14.8.3.

7.16 Weld Access Holes, Beam Copes, and Connection Material

Weld access holes, beam copes, and cut surfaces in connection materials shall be free of notches. Beam copes and cut surfaces in connection materials shall be free of sharp reentrant corners. Weld access holes shall provide a smooth transition that does not cut past the points of tangency between adjacent surfaces and shall meet the surface requirements of 7.14.8.3.

7.16.1 Weld Access Hole Dimensions. All weld access holes shall have a length from the edge of the weld joint preparation at the inside surface not less than 1–1/2 times the thickness of the material in which the hole is made. The minimum height of the access hole shall be the thickness of the material with the access hole (t_w) but not less than 3/4 in [20 mm] nor does the height need to exceed 2 in [50 mm]. The access hole shall be detailed to provide room for weld backing as needed and provide adequate access for welding.

7.16.1.1 Weld Access Holes in Rolled Sections. The edge of the web shall be sloped or curved from the surface of the flange to the reentrant surface of the access hole. No corner of the weld access hole shall have a radius less than 3/8 in [10 mm].

7.16.1.2 Weld Access Holes in Built-up Sections. For built-up sections where the weld access hole is made after the section is welded, the edge of the web shall be sloped or curved from the surface of the flange to the re-entrant surface of the access hole. No corner of the weld access hole shall have a radius less than 3/8 in [10 mm]. For built-up sections where the access hole is made before the section is welded, the access hole may terminate perpendicular to the flange, providing the weld is terminated at least a distance equal to the weld size away from the access hole. Fillet welds shall not be returned through the access hole (see Figure 7.2).

7.16.2 Shapes to be Galvanized. Weld access holes and beam copes in shapes that are to be galvanized shall be ground to bright metal. If the curved transition portion of weld access holes and beam copes are formed by predrilled or sawed holes, that portion of the access hole or cope need not be ground.

7.16.3 Copes and Access Holes in Heavy Shapes. For rolled shapes with a flange thickness exceeding 2 in [50 mm] and welded sections with plate thickness exceeding 2 in [50 mm]:

(1) If the curved surface of the access hole is thermally cut, a minimum preheat of 150°F [65°C] extending 3 in [75 mm] from the area where the curve is to be cut shall be applied prior to thermal cutting.

(2) The thermally cut surfaces of beam copes and weld access holes shall be ground to bright metal and visually inspected to ensure the surface is free of cracks prior to welding.

7.17 Tack Welds and Construction Aid Welds

7.17.1 General Requirements

(1) Tack welds and construction aid welds shall be made with a qualified or prequalified WPS and by qualified personnel.

(2) Tack welds that are not incorporated in final welds, and construction aid welds that are not removed, shall meet visual inspection requirements before a member is accepted.

7.17.2 Exclusions. Tack welds and construction aid welds are permitted except that:

(1) In tension zones of cyclically loaded structures, there shall be no tack welds not incorporated into the final weld except as permitted by 4.17.2, nor construction aid welds. Locations more than 1/6 of the depth of the web from tension flanges of beams or girders are considered outside the tension zone.

(2) On members made of quenched and tempered steel with specified yield strength greater than 70 ksi [485 MPa], tack welds outside the final weld and construction aid welds shall require the approval of the Engineer.

7.17.3 Removal. At locations other than 7.17.2, tack welds and construction aid welds not incorporated into final welds shall be removed when required by the Engineer.

7.17.4 Additional Tack Weld Requirements

(1) Tack welds incorporated into final welds shall be made with electrodes meeting the requirements of the final welds. These welds shall be cleaned prior to incorporation.

(2) Multipass tack welds shall have cascaded ends or be otherwise prepared for incorporation into the final weld.

(3) Tack welds incorporated into final welds that are qualified with notch toughness or are required to be made with filler metal classified with notch toughness shall be made with compatible filler metals.

7.17.5 Additional Requirements for Tack Welds Incorporated into SAW Welds. The following shall apply in addition to 7.17.4 requirements.

(1) Preheat is not required for single pass tack welds remelted by continuous SAW welds. This is an exception to the qualification requirements of 7.17.1.

(2) Fillet tack welds shall not exceed 3/8 in [10 mm] in size and shall not produce objectionable changes in the appearance of the weld surface.

(3) Tack welds in the roots of joints requiring specific root penetration shall not result in decreased penetration.

(4) Tack welds not conforming to the requirements of (2) and (3) shall be removed or reduced in size by any suitable means before welding.

(5) Tack welds in the root of a joint with steel backing less than 5/16 in [8 mm] thick shall be removed or made continuous for the full length of the joint using SMAW with low-hydrogen electrodes, GMAW, or FCAW-G.

7.18 Camber in Built-Up Members

7.18.1 Camber. Edges of built-up beam and girder webs shall be cut to the prescribed camber with suitable allowance for shrinkage due to cutting and welding. However, moderate variation from the specified camber tolerance may be corrected by a careful application of heat.

7.18.2 Correction. Corrections of errors in camber of quenched and tempered steel shall require approval by the Engineer.

7.19 Splices

7.19.1 Subassembly Splices. All welded subassembly splices in each component part of a cover-plated beam or built-up member shall be made before the component part is welded to other component parts of the member.

7.19.1.1 Shop Splice Location. Shop splices of webs and flanges in built-up girders may be located in a single transverse plane or multiple transverse planes.

7.19.1.2 Cyclically Loaded Splices. For cyclically loaded members, the fatigue stress provisions of the general specifications shall apply.

7.19.2 Member Splices. Long girders or girder sections may be made by welding subassemblies. Splices between sections of rolled beams or built-up girders shall be made in a single transverse plane, when practicable.

7.20 Control of Distortion and Shrinkage

7.20.1 Procedure and Sequence. In assembling and joining parts of a structure or of built-up members and in welding reinforcing parts to members, the procedure and sequence shall be such as will minimize distortion and shrinkage.

7.20.2 Sequencing. Insofar as practicable, all welds shall be made in a sequence that will balance the applied heat of welding while the welding progresses.

7.20.3 Contractor Responsibility. On members or structures where excessive shrinkage or distortion could be expected, the Contractor shall prepare a written welding sequence for that member or structure which meets the quality requirements specified. The welding sequence and distortion control program shall be submitted to the Engineer, for information and comment, before the start of welding on the member or structure in which shrinkage or distortion is likely to affect the adequacy of the member or structure.

7.20.4 Weld Progression. The direction of the general progression in welding on a member shall be from points where the parts are relatively fixed in position with respect to each other toward points having a greater relative freedom of movement.

7.20.5 Minimized Restraint. In assemblies, joints expected to have significant shrinkage should usually be welded before joints expected to have lesser shrinkage. They should also be welded with as little restraint as possible.

7.20.6 Temperature Limitations. In making welds under conditions of severe external shrinkage restraint, once the welding has started, the joint shall not be allowed to cool below the minimum specified preheat until the joint has been completed or sufficient weld has been deposited to ensure freedom from cracking.

7.21 Tolerance of Joint Dimensions

7.21.1 Fillet Weld Assembly. The parts to be joined by fillet welds shall be brought into as close contact as practicable. The root opening shall not exceed 3/16 in [5 mm] except in cases involving either shapes or plates 3 in [75 mm] or greater in thickness if, after straightening and in assembly, the root opening cannot be closed sufficiently to meet this tolerance. In such cases, a maximum root opening of 5/16 in [8 mm] may be used, provided suitable backing is used. Backing may be of flux, glass tape, iron powder, or similar materials, or welds using a low-hydrogen process compatible with the filler metal deposited. If the separation is greater than 1/16 in [2 mm], the legs of the fillet weld shall be increased by the amount of the root opening, or the Contractor shall demonstrate that the required effective throat has been obtained.

7.21.1.1 Faying Surface. The separation between faying surfaces of plug and slot welds, and of butt joints landing on a backing, shall not exceed 1/16 in [2 mm]. Where irregularities in rolled shapes occur after straightening do not allow contact within the above limits, the procedure necessary to bring the material within these limits shall be subject to the approval of the Engineer. The use of filler plates shall be prohibited except as specified on the drawings or as specially approved by the Engineer and made in conformance with 4.11.

7.21.2 PJP Groove Weld Assembly. The parts to be joined by PJP groove welds parallel to the length of the member shall be brought into as close contact as practicable. The root opening between parts shall not exceed 3/16 in [5 mm] except in cases involving rolled shapes or plates 3 in [75 mm] or greater in thickness if, after straightening and in assembly, the root opening cannot be closed sufficiently to meet this tolerance. In such cases, a maximum root opening of 5/16 in [8 mm] may be used, provided suitable backing is used and the final weld meets the requirements for weld size. Tolerances for bearing joints shall be in conformance with the applicable contract specifications.

7.21.3 Butt Joint Alignment. Parts to be joined at butt joints shall be carefully aligned. Where the parts are effectively restrained against bending due to eccentricity in alignment, the offset from the theoretical alignment shall not exceed 10% of the thickness of the thinner part joined, or 1/8 in [3 mm], whichever is smaller. In correcting misalignment in such cases, the parts shall not be drawn in to a greater slope than 1/2 in [12 mm] in 12 in [300 mm]. Measurement of offset shall be based upon the centerline of parts unless otherwise shown on the drawings.

7.21.4 Groove Dimensions

7.21.4.1 Nontubular Cross-Sectional Variations. With the exclusion of ESW and EGW, and with the exception of 7.21.4.2 for root openings in excess of those allowed in Figure 7.3, the dimensions of the cross section of the groove welded joints which vary from those shown on the detail drawings by more than these tolerances shall be referred to the Engineer for approval or correction.

7.21.4.2 Correction. Root openings greater than those allowed in 7.21.4.1, but not greater than twice the thickness of the thinner part or 3/4 in [20 mm], whichever is less, may be corrected by welding to acceptable dimensions prior to joining the parts by welding.

7.21.4.3 Engineer's Approval. Root openings greater than allowed by 7.21.4.2 may be corrected by welding only with the approval of the Engineer.

7.21.5 Gouged Grooves. Grooves produced by gouging shall be in substantial conformance with groove profile dimensions as specified in Figure 5.1 and Figure 5.2 and provisions of 5.4.1 and 5.4.2. Suitable access to the root shall be maintained.

7.21.6 Alignment Methods. Members to be welded shall be brought into correct alignment and held in position by bolts, clamps, wedges, guy lines, struts, and other suitable devices, or by tack welds until welding has been completed. The use of jigs and fixtures is recommended where practicable. Suitable allowances shall be made for warpage and shrinkage.

7.22 Dimensional Tolerance of Welded Structural Members

The dimensions of welded structural members shall conform to the tolerances of (1) the general specifications governing the work, and (2) the special dimensional tolerances in 7.22.1 to 7.22.12. (Note that a tubular column is interpreted as a compression tubular member.)

7.22.1 Straightness of Columns and Trusses. For welded columns and primary truss members, regardless of cross section, the maximum variation in straightness shall be

Lengths of less than 30 ft [9 m]:

$$1/8 \text{ in} \times \frac{[\text{No. of ft of total length}]}{10}$$

$1 \text{ mm} \times [\text{No. of meters of total length}]$

Lengths of 30 ft [9 m] to 45 ft [15 m] = 3/8 in [10 mm]

Lengths over 45 ft [15 m]:

$$3/8 \text{ in} + 1/8 \text{ in} \times \frac{[\text{No. of ft of total length} - 45]}{10}$$

$$10 \text{ mm} + 3 \text{ mm} \times \frac{[\text{No. of meters of total length} - 15]}{3}$$

7.22.2 Beam and Girder Straightness (No Camber Specified). For welded beams or girders, regardless of cross section, where there is no specified camber, the maximum variation in straightness shall be

$$1/8 \text{ in} \times \frac{\text{No. of ft of total length}}{10}$$

$1 \text{ mm} \times \text{No. of meters of total length}$

7.22.3 Beam and Girder Camber (Typical Girder). For welded beams or girders, other than those whose top flange is embedded in concrete without a designed concrete haunch, regardless of cross section, the maximum variation from required camber at shop assembly (for drilling holes for field splices or repairing field welded splices) shall be

at midspan   –0, +1–1/2 in [40 mm] for spans ≥ 100 ft [30 m]
             –0, + 3/4 in [20 mm] for spans < 100 ft [30 m]

at supports, 0 for end supports
             ± 1/8 in [3 mm] for interior supports

at intermediate points, $-0, + \dfrac{4(a)b(1 - a/S)}{S}$

Where

   a = distance in feet (meters) from inspection point to nearest support
   S = span length in feet (meters)
   b = 1–1/2 in [40 mm] for spans ≥ 100 ft [30 m]
   b = 3/4 in [20 mm] for spans < 100 ft [30 m]

See Table 7.5 for tabulated values.

7.22.4 Beam and Girder Camber (without Designed Concrete Haunch). For members whose top flange is embedded in concrete without a designed concrete haunch, the maximum variation from required camber at shop assembly (for drilling holes for field splices or preparing field welded splices) shall be

at midspan, ±3/4 in [20 mm] for spans ≥ 100 ft [30 m]
            ± 3/8 in [10 mm] for spans < 100 ft [30 m]

at supports, 0 for end supports
            ± 1/8 in [3 mm] for interior supports

at intermediate points, $\pm \dfrac{4(a)b(1 - a/S)}{S}$

where a and S are as defined above

b = 3/4 in [20 mm] for spans ≥ 100 ft [30 m]
b = 3/8 in [10 mm] for spans < 100 ft [30 m]

See Table 7.6 for tabulated values.

Regardless of how the camber is shown on the detail drawings, the sign convention for the allowable variation is plus (+) above, and minus (–) below, the detailed camber shape. These provisions also apply to an individual member when no field splices or shop assembly is required. Camber measurements shall be made in the no load condition.

7.22.5 Beam and Girder Sweep. The maximum variation from straightness or specified sweep at the midpoint shall be

$$\pm 1/8 \text{ in} \times \frac{\text{No. of ft of total length}}{10}$$

$$\pm 1 \text{ mm} \times [\text{No. of meters of total length}]$$

provided the member has sufficient lateral flexibility to allow the attachment of diaphragms, cross-frames, lateral bracing, etc., without damaging the structural member or its attachments.

7.22.6 Variation in Web Flatness

7.22.6.1 Measurements. Variations from flatness of girder webs shall be determined by measuring the offset from the actual web centerline to a straight edge whose length is greater than the least panel dimension and placed on a plane parallel to the nominal web plane. Measurements shall be taken prior to erection (see Commentary).

7.22.6.2 Statically Loaded Nontubular Structures. Variations from flatness of webs having a depth, D, and a thickness, t, in panels bounded by stiffeners or flanges, or both, whose least panel dimension is d shall not exceed the following:

Intermediate stiffeners on both sides of web
   where D/t < 150, maximum variation = d/100
   where D/t ≥ 150, maximum variation = d/80

Intermediate stiffeners on one side only of web
   where D/t < 100, maximum variation = d/100
   where D/t ≥ 100, maximum variation = d/67

No intermediate stiffeners
   where D/t ≥ 100, maximum variation = D/150
   (See Annex E for tabulation.)

7.22.6.3 Cyclically Loaded Nontubular Structures. Variation from flatness of webs having a depth, D, and a thickness, t, in panels bounded by stiffeners or flanges, or both, whose least panel dimension is d shall not exceed the following:

Intermediate stiffeners on both sides of web

   Interior girders—
     where D/t < 150—maximum variation = d/115
     where D/t ≥ 150—maximum variation = d/92

   Fascia girders—
     where D/t < 150—maximum variation = d/130
     where D/t ≥ 150—maximum variation = d/105

Intermediate stiffeners on one side only of web

   Interior girders—
     where D/t < 100—maximum variation = d/100
     where D/t ≥ 100—maximum variation = d/67

Fascia girders—
- where D/t < 100—maximum variation = d/120
- where D/t ≥ 100—maximum variation = d/80

No intermediate stiffeners—maximum variation = D/150

(See Annex F for tabulation.)

7.22.6.4 Excessive Distortion. Web distortions of twice the allowable tolerances of 7.22.6.2 or 7.22.6.3 shall be satisfactory when occurring at the end of a girder which has been drilled, or sub-punched and reamed; either during assembly or to a template for a field bolted splice; provided, when the splice plates are bolted, the web assumes the proper dimensional tolerances.

7.22.6.5 Architectural Consideration. If architectural considerations require tolerances more restrictive than described in 7.22.6.2 or 7.22.6.3, specific reference shall be included in the bid documents.

7.22.7 Variation Between Web and Flange Centerlines. For built-up H or I members, the maximum variation between the centerline of the web and the centerline of the flange at contact surface shall not exceed 1/4 in [6 mm].

7.22.8 Flange Warpage and Tilt. For welded beams or girders, the combined warpage and tilt of flange shall be determined by measuring the offset at the toe of the flange from a line normal to the plane of the web through the intersection of the centerline of the web with the outside surface of the flange plate. This offset shall not exceed 1% of the total flange width or 1/4 in [6 mm], whichever is greater, except that welded butt joints of abutting parts shall fulfill the requirements of 7.21.3.

7.22.9 Depth Variation. For welded beams and girders, the maximum allowable variation from specified depth measured at the web centerline shall be

For depths up to 36 in [1 m] incl.	± 1/8 in [3 mm]
For depths over 36 in [1 m] to 72 in [2 m] incl.	± 3/16 in [5 mm]
For depths over 72 in [2 m]	+ 5/16 in [8 mm]
	− 3/16 in [5 mm]

7.22.10 Bearing at Points of Loading. The bearing ends of bearing stiffeners shall be square with the web and shall have at least 75% of the stiffener bearing cross-sectional area in contact with the inner surface of the flanges. The outer surface of the flanges when bearing against a steel base or seat shall fit within 0.010 in [0.25 mm] for 75% of the projected area of web and stiffeners and not more than 1/32 in [1 mm] for the remaining 25% of the projected area. Girders without stiffeners shall bear on the projected area of the web on the outer flange surface within 0.010 in [0.25 mm] and the included angle between web and flange shall not exceed 90° in the bearing length (see Commentary).

7.22.11 Tolerance on Stiffeners

7.22.11.1 Fit of Intermediate Stiffeners. Where tight fit of intermediate stiffeners is specified, it shall be defined as allowing a gap of up to 1/16 in [2 mm] between stiffener and flange.

7.22.11.2 Straightness of Intermediate Stiffeners. The out-of-straightness variation of intermediate stiffeners shall not exceed 1/2 in [12 mm] for girders up to 6 ft [1.8 m] deep, and 3/4 in [20 mm] for girders over 6 ft [1.8 m] deep, with due regard for members which frame into them.

7.22.11.3 Straightness and Location of Bearing Stiffeners. The out-of-straightness variation of bearing stiffeners shall not exceed 1/4 in [6 mm] up to 6 ft [1.8 m] deep or 1/2 in [12 mm] over 6 ft [1.8 m] deep. The actual centerline of the stiffener shall lie within the thickness of the stiffener as measured from the theoretical centerline location.

7.22.12 Other Dimensional Tolerances. Twist of box members and other dimensional tolerances of members not covered by 7.22 shall be individually determined and mutually agreed upon by the Contractor and the Owner with proper regard for erection requirements.

7.23 Weld Profiles

All welds shall meet the visual acceptance criteria of Tables 8.1 or 10.15, and shall be free from cracks, overlaps, and the unacceptable profile discontinuities exhibited in Figure 7.4, Table 7.8, and Table 7.9, except as otherwise allowed in 7.23.1, 7.23.2, and 7.23.3.

7.23.1 Fillet Welds. The faces of fillet welds may be slightly convex, flat, or slightly concave as shown in Figure 7.4 and as allowed by Tables 7.8, 7.9, 8.1, and 10.15.

7.23.2 Exception for Intermittent Fillet Welds. Except for undercut, as allowed by the code, the profile requirements of Figure 7.4 shall not apply to the ends of intermittent fillet welds outside their effective length.

7.23.3 Groove Welds. Groove weld reinforcement shall comply with Tables 7.8 and 7.9 and with the provisions below. Welds shall have a gradual transition to the plane of the base metal surfaces.

7.23.3.1 Flush Surfaces. Welds required to be flush shall be finished so as to not reduce the thicknesses of the thinner base metal or weld metal by more than 1/32 in [1 mm]. Remaining reinforcement shall not exceed 1/32 in [1 mm] in height and shall blend smoothly into the base metal surfaces with transition areas free from undercut. However, all reinforcement shall be removed where the weld forms part of a faying or contact surface.

7.23.3.2 Finish Methods and Values. Where surface finishing is required, surface roughness values (see ASME B46.1) shall not exceed 250 microinches [6.3 micrometers]. Chipping and gouging may be used provided these are followed by grinding or machining. For cyclically loaded structures, finishing shall be parallel to the direction of primary stress, except final roughness of 125 microinches [3.2 micrometers] or less may be finished in any direction.

7.23.4 Shelf Bars. Shelf bars shall conform to the requirements of 7.9.1.1 through 7.9.1.5. Shelf bars may be left in place only for statically loaded members.

7.24 Technique for Plug and Slot Welds

7.24.1 Plug Welds. The technique used to make plug welds when using SMAW, GMAW (except GMAW-S), and FCAW processes shall be as follows:

7.24.1.1 Flat Position. For welds to be made in the flat position, each pass shall be deposited around the root of the joint and then deposited along a spiral path to the center of the hole, fusing and depositing a layer of weld metal in the root and bottom of the joint. The arc shall then be moved to the periphery of the hole and the procedure repeated, fusing and depositing successive layers to fill the hole to the required depth. The slag covering the weld metal should be kept molten until the weld is finished. If the arc is broken or the slag is allowed to cool, the slag must be completely removed before restarting the weld.

7.24.1.2 Vertical Position. For welds to be made in the vertical position, the arc is started at the root of the joint at the lower side of the hole and is carried upward, fusing into the face of the inner plate and to the side of the hole. The arc is stopped at the top of the hole, the slag is cleaned off, and the process is repeated on the opposite side of the hole. After cleaning slag from the weld, other layers should be similarly deposited to fill the hole to the required depth.

7.24.1.3 Overhead Position. For welds to be made in the overhead position, the procedure is the same as for the flat position, except that the slag should be allowed to cool and should be completely removed after depositing each successive bead until the hole is filled to the required depth.

7.24.2 Slot Welds. Slot welds shall be made using techniques similar to those specified in 7.24.1 for plug welds, except that if the length of the slot exceeds three times the width, or if the slot extends to the edge of the part, the technique requirements of 7.24.1.3 shall apply.

7.25 Repairs

The removal of weld metal or portions of the base metal may be done by machining, grinding, chipping, or gouging. It shall be done in such a manner that the adjacent weld metal or base metal is not nicked or gouged. Oxygen gouging shall only be permitted for use on as-rolled steels. Unacceptable portions of the weld shall be removed without substantial removal of the base metal. The surfaces shall be cleaned thoroughly before welding. Weld metal shall be deposited to compensate for any deficiency in size.

7.25.1 Contractor Options. The Contractor has the option of either repairing an unacceptable weld or removing and replacing the entire weld, except as modified by 7.25.3. The repaired or replaced weld shall be retested by the method originally used, and the same technique and quality acceptance criteria shall be applied. If the Contractor elects to repair the weld, it shall be corrected as follows:

7.25.1.1 Overlap, Excessive Convexity, or Excessive Reinforcement. Excessive weld metal shall be removed.

7.25.1.2 Excessive Concavity of Weld or Crater, Undersize Welds, Undercutting. The surfaces shall be prepared (see 7.29) and additional weld metal deposited.

7.25.1.3 Incomplete Fusion, Excessive Weld Porosity, or Slag Inclusions. Unacceptable portions shall be removed (see 7.25) and rewelded.

7.25.1.4 Cracks in Weld or Base Metal. The extent of the crack shall be ascertained by use of acid etching, MT, PT, or other equally positive means; the crack and sound metal 2 in [50 mm] beyond each end of the crack shall be removed, and rewelded.

7.25.2 Localized Heat Repair Temperature Limitations. Members distorted by welding shall be straightened by mechanical means or by application of a limited amount of localized heat. The temperature of heated areas as measured by approved methods shall not exceed 1100°F [600°C] for quenched and tempered steel nor 1200°F [650°C] for other steels. The part to be heated for straightening shall be substantially free of stress and from external forces, except those stresses resulting from a mechanical straightening method used in conjunction with the application of heat.

7.25.3 Engineer's Approval. Prior approval of the Engineer shall be obtained for repairs to base metal (other than those required by 7.14), repair of major or delayed cracks, repairs to ESW and EGW with internal defects, or for a revised design to compensate for deficiencies. The Engineer shall be notified before welded members are cut apart.

7.25.4 Inaccessibility of Unacceptable Welds. If, after an unacceptable weld has been made, work is performed which has rendered that weld inaccessible or has created new conditions that make correction of the unacceptable weld dangerous or ineffectual, then the original conditions shall be restored by removing welds or members, or both, before the corrections are made. If this is not done, the deficiency shall be compensated for by additional work performed conforming to an approved revised design.

7.25.5 Welded Restoration of Base Metal with Mislocated Holes. Mislocated holes may be left open or filled with bolts except when welding is necessary to fulfill contract requirements or when required by the Engineer. Base metal with mislocated holes may be restored by welding provided the Contractor prepares and follows a qualified or prequalified WPS and meets the requirements of (1) through (4) below. The groove geometry shall be considered prequalified.:

(1) Base metal not in tension zones of cyclically loaded members shall be restored by welding, provided the repair weld soundness is verified by the same NDT process specified in the contract documents for groove welds. If the contract documents did not originally specify NDT, the restored base metal shall be evaluated using a method and acceptance criteria specified by the Engineer.

(2) Base metal in tension zones of cyclically loaded members may be restored by welding provided:

(a) The Engineer approves repair by welding and the WPS for the repair.

(b) The WPS is followed in the work and the soundness of the restored base metal is verified by the NDT method(s) specified in the contract documents for examination of tension groove welds or as approved by the Engineer.

(3) In addition to the requirements of (1) and (2), when holes in quenched and tempered base metals are restored by welding:

(a) Appropriate filler metal, heat input, and PWHT (when PWHT is required) shall be used.

(b) Sample welds shall be made using the repair WPS.

(c) RT of the sample welds shall verify that weld soundness conforms to the requirements of 8.12.2.1.

(d) One reduced section tension test (weld metal); two side bend tests (weld metal); and three CVN tests of the HAZ (coarse grained area) removed from sample welds shall be used to demonstrate that the mechanical properties of the repaired area conform to the specified requirements of the base metal (see Clause 6, Part D for CVN testing requirements).

(4) Weld surfaces shall be finished as specified in 7.23.3.1.

7.26 Peening

Peening may be used on intermediate weld layers for control of shrinkage stresses in thick welds to prevent cracking or distortion, or both. No peening shall be done on the root or surface layer of the weld or the base metal at the edges of the weld except as provided in 10.2.3.6(3) for tubulars. Care should be taken to prevent overlapping or cracking of the weld or base metal.

7.26.1 Tools. The use of manual slag hammers, chisels, and lightweight vibrating tools for the removal of slag and spatter is allowed and shall not be considered peening.

7.27 Caulking

Caulking shall be defined as plastic deformation of weld and base metal surfaces by mechanical means to seal or obscure discontinuities. Caulking shall be prohibited for base metals with minimum specified yield strength greater than 50 ksi [345 MPa].

For base metals with minimum specified yield strength of 50 ksi [345 MPa] or less, caulking may be used, provided:

(1) all inspections have been completed and accepted

(2) caulking is necessary to prevent coating failures

(3) the technique and limitations on caulking are approved by the Engineer

7.28 Arc Strikes

Arc strikes outside the area of permanent welds should be avoided on any base metal. Cracks or blemishes caused by arc strikes shall be ground to a smooth contour and checked to ensure soundness.

7.29 Weld Cleaning

7.29.1 In-Process Cleaning. Before welding over previously deposited weld metal, all slag shall be removed and the weld and adjacent base metal shall be cleaned by brushing or other suitable means. This requirement shall apply not only to successive layers but also to successive beads and to the crater area when welding is resumed after any interruption. It shall not, however, restrict the welding of plug and slot welds in conformance with 7.24.

7.29.2 Cleaning of Completed Welds. Slag shall be removed from all completed welds, and the weld and adjacent base metal shall be cleaned by brushing or other suitable means. Tightly adherent spatter remaining after the cleaning operation is acceptable, unless its removal is required for the purpose of NDT. Welded joints shall not be painted until after welding has been completed and the weld accepted.

7.30 Weld Tabs (See 7.2.2)

7.30.1 Use of Weld Tabs. Welds shall be terminated at the end of a joint in a manner that will ensure sound welds. Whenever necessary, this shall be done by use of weld tabs aligned in such a manner to provide an extension of the joint preparation.

7.30.2 Removal of Weld Tabs for Statically Loaded Nontubular Structures. For statically loaded nontubular structures, weld tabs need not be removed unless required by the Engineer.

7.30.3 Removal of Weld Tabs for Cyclically Loaded Nontubular Structures. For cyclically loaded nontubular structures, weld tabs shall be removed upon completion and cooling of the weld, and the ends of the weld shall be made smooth and flush with the edges of abutting parts.

7.30.4 Ends of Welded Butt Joints. Ends of welded butt joints required to be flush shall be finished so as not to reduce the width beyond the detailed width or the actual width furnished, whichever is greater, by more than 1/8 in [3 mm] or so as not to leave reinforcement at each end that exceeds 1/8 in [3 mm]. Ends of welded butt joints shall be faired at a slope not to exceed 1 in 10.

Table 7.1
Allowable Atmospheric Exposure of Low-Hydrogen Electrodes
(See 7.3.2.2 and 7.3.2.3)

Electrode	Column A (hours)	Column B (hours)
A5.1		
E70XX	4 max.	
E70XXR	9 max.	Over 4 to 10 max.
E70XXHZR	9 max.	
E7018M	9 max.	
A5.5		
E70XX-X	4 max.	Over 4 to 10 max.
E80XX-X	2 max.	Over 2 to 10 max.
E90XX-X	1 max.	Over 1 to 5 max.
E100XX-X	1/2 max.	Over 1/2 to 4 max.
E110XX-X	1/2 max.	Over 1/2 to 4 max.

Notes:
1. Column A: Electrodes exposed to atmosphere for longer periods than shown shall be baked before use.
2. Column B: Electrodes exposed to atmosphere for longer periods than those established by testing shall be baked before use.
3. Electrodes shall be issued and held in quivers, or other small open containers. Heated containers are not mandatory.
4. The optional supplemental designator, R, designates a low-hydrogen electrode which has been tested for covering moisture content after exposure to a moist environment for 9 hours and has met the maximum level allowed in AWS A5.1/A5.1M, *Specification for Carbon Steel Electrodes for Shielded Metal Arc Welding*.

Table 7.2
Minimum Holding Time (See 7.8.1)

1/4 in [6 mm] or Less	Over 1/4 in [6 mm] Through 2 in [50 mm]	Over 2 in [50 mm]
15 min.	15 min. for each 1/4 in [6 mm] or fraction thereof	2 hrs plus 15 min. for each additional in [25 mm] or fraction thereof over 2 in [50 mm]

Table 7.3
Alternate Stress-Relief Heat Treatment (See 7.8.2)

Decrease in Temperature below Minimum Specified Temperature		Minimum Holding Time at Decreased Temperature, Hours per Inch [25 mm] of Thickness
Δ°F	Δ°C	
50	30	2
100	60	4
150	90	10
200	120	20

Table 7.4
Limits on Acceptability and Repair of Mill Induced Laminar Discontinuities in Cut Surfaces (See 7.14.5)

Description of Discontinuity	Repair Required
Any discontinuity 1 in [25 mm] in length or less	None, need not be explored
Any discontinuity over 1 in [25 mm] in length and 1/8 in [3 mm] maximum depth	None, but the depth should be explored[a]
Any discontinuity over 1 in [25 mm] in length with depth over 1/8 in [3 mm] but not greater than 1/4 in [6 mm]	Remove, need not weld
Any discontinuity over 1 in [25 mm] in length with depth over 1/4 in [6 mm] but not greater than 1 in [25 mm]	Completely remove and weld
Any discontinuity over 1 in [25 mm] in length with depth greater than 1 in [25 mm]	See 7.14.5.1

[a] A spot check of 10% of the discontinuities on the cut surface in question should be explored by grinding to determine depth. If the depth of any one of the discontinuities explored exceeds 1/8 in [3 mm], then all of the discontinuities over 1 in [25 mm] in length remaining on that cut surface shall be explored by grinding to determine depth. If none of the discontinuities explored in the 10% spot check have a depth exceeding 1/8 in [3 mm], then the remainder of the discontinuities on that cut surface need not be explored.

Table 7.5
Camber Tolerance for Typical Girder (See 7.22.3)

	Camber Tolerance (in inches)				
Span \ a/S	0.1	0.2	0.3	0.4	0.5
≥ 100 ft	9/16	1	1–1/4	1–7/16	1–1/2
< 100 ft	1/4	1/2	5/8	3/4	3/4
	Camber Tolerance (in millimeters)				
Span \ a/S	0.1	0.2	0.3	0.4	0.5
≥ 30 m	14	25	34	38	40
< 30 m	7	13	17	19	20

Table 7.6
Camber Tolerance for Girders without a Designed Concrete Haunch (See 7.22.4)

	Camber Tolerance (in inches)				
a/S \ Span	0.1	0.2	0.3	0.4	0.5
≥ 100 ft	1/4	1/2	5/8	3/4	3/4
< 100 ft	1/8	1/4	5/16	3/8	3/8
	Camber Tolerance (in millimeters)				
a/S \ Span	0.1	0.2	0.3	0.4	0.5
≥ 30 m	7	13	17	19	20
< 30 m	4	6	8	10	10

Table 7.7
Minimum Fillet Weld Sizes (See 7.13)

Base Metal Thickness (T)[a]		Minimum Size of Fillet Weld[b]	
in	mm	in	mm
T ≤ 1/4	T ≤ 6	1/8[c]	3[c]
1/4 < T ≤ 1/2	6 < T ≤ 12	3/16	5
1/2 < T ≤ 3/4	12 < T ≤ 20	1/4	6
3/4 < T	20 < T	5/16	8

[a] For nonlow-hydrogen processes without preheat calculated in conformance with 6.8.4, T equals thickness of the thicker part joined; single-pass welds shall be used.
For nonlow-hydrogen processes using procedures established to prevent cracking in conformance with 6.8.4 and for low-hydrogen processes, T equals thickness of the thinner part joined; single-pass requirement shall not apply.
[b] Except that the weld size need not exceed the thickness of the thinner part joined.
[c] Minimum size for cyclically loaded structures shall be 3/16 in [5 mm].

Table 7.8
Weld Profiles[a] (see 7.23)

Weld Type	Joint Type					
	Butt	Corner-Inside	Corner-Outside	T-Joint	Lap	Butt with Shelf Bar
Groove (CJP or PJP)	Figure 7.4A	Figure 7.4B[b]	Figure 7.4C	Figure 7.4D[b]	N/A	Figure 7.4G
	Schedule A	Schedule B	Schedule A	Schedule B	N/A	See Footnote c
Fillet	N/A	Figure 7.4E	Figure 7.4F	Figure 7.4E	Figure 7.4E	N/A
	N/A	Schedule C	Schedule C or D[d]	Schedule C	Schedule C	N/A

[a] Schedules A through D are given in Table 7.9.
[b] For reinforcing fillet welds required by design, the profile restrictions apply to each groove and fillet, separately.
[c] Welds made using shelf bars and welds made in the horizontal position between vertical bars of unequal thickness are exempt from R and C limitations. See Figures 7.4G and 7.4H for typical details.
[d] See Figure 7.4F for a description of where Schedule C and D apply.

Table 7.9
Weld Profile Schedules (see 7.23)

Schedule A	(t = thickness of thicker plate joined for CJP; t = weld size for PJP)			
	t	R min.	R max.	
	≤ 1 in [25 mm]	0	1/8 in [3 mm]	
	> 1 in [25 mm], ≤ 2 in [50 mm]	0	3/16 in [5 mm]	
	> 2 in [50 mm]	0	1/4 in [6 mm][a]	
Schedule B	(t = thickness of thicker plate joined for CJP; t = weld size for PJP; C = allowable convexity or concavity)			
	t	R min.	R. max.	C max[b]
	< 1 in [25 mm]	0	unlimited	1/8 in [3 mm]
	≥ 1 in [25 mm]	0	unlimited	3/16 in [5 mm]
Schedule C	(W = width of weld face or individual surface bead; C = allowable convexity)			
	W			C max[b]
	≤ 5/16 in [8 mm]			1/16 in [2 mm]
	>5/16 in [8 mm], <1 in [25 mm]			1/8 in [3 mm]
	≥ 1 in [25 mm]			3/16 in [5 mm]
Schedule D	(t = thickness of thinner of the exposed edge dimensions; C = allowable convexity; See Figure 7.4F)			
	t			C max[b]
	any value of t			t/2

[a] For cyclically loaded structures, R max. for materials > 2 in [50 mm] thick is 3/16 in [5 mm].
[b] There is no restriction on concavity as long as minimum weld size (considering both leg and throat) is achieved.

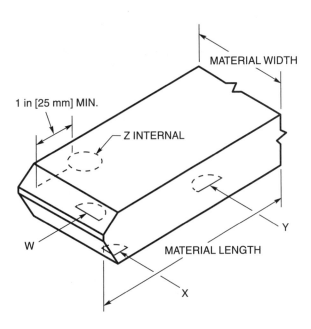

Figure 7.1—Edge Discontinuities in Cut Material (see 7.14.5.1)

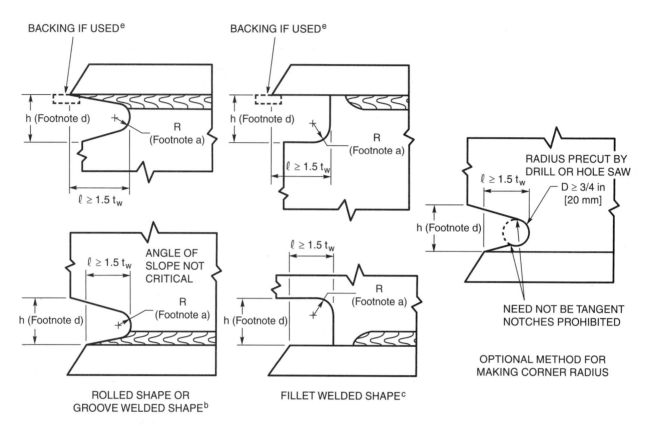

a Radius shall provide smooth notch-free transition; R ≥ 3/8 in [10 mm] (Typical 1/2 in [12 mm]).
b Access hole made after welding web to flange.
c Access hole made before welding web to flange. The web to flange weld shall not be returned through hole.
d h_{min} = 3/4 in [20 mm] or t_w (web thickness), whichever is greater, h_{min} need not exceed 2 in [50 mm].
e These are typical details for joints welded from one side against steel backing. Alternative joint designs should be considered.

Figure 7.2—Weld Access Hole Geometry (see 7.16.1.2)

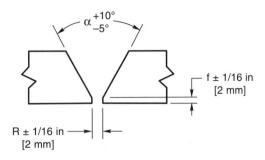

(A) GROOVE WELD WITHOUT BACKING—ROOT NOT BACKGOUGED

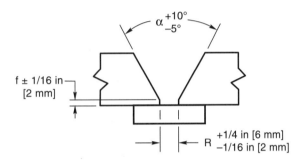

(B) GROOVE WELD WITH BACKING—ROOT NOT BACKGOUGED

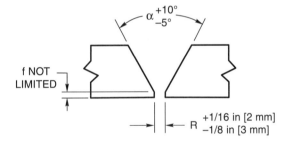

(C) GROOVE WELD WITHOUT BACKING—ROOT BACKGOUGED

	Root Not Backgouged		Root Backgouged	
	in	mm	in	mm
(1) Root face of joint	±1/16	2	Not Limited	
(2) Root opening of joints without backing	±1/16	2	+1/16 −1/8	2 3
Root opening of joint with backing	+1/4 −1/16	6 2	Not applicable	
(3) Groove angle of joint	+10° −5°		+10° −5°	

Note: See 10.23.2.1 for tolerances for CJP tubular groove welds made from one side without backing.

Figure 7.3—Workmanship Tolerances in Assembly of Groove Welded Joints (see 7.21.4.1)

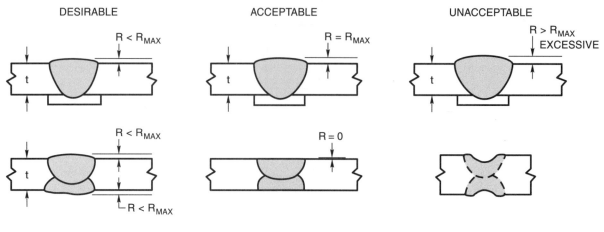

(A) WELD PROFILES FOR BUTT JOINTS

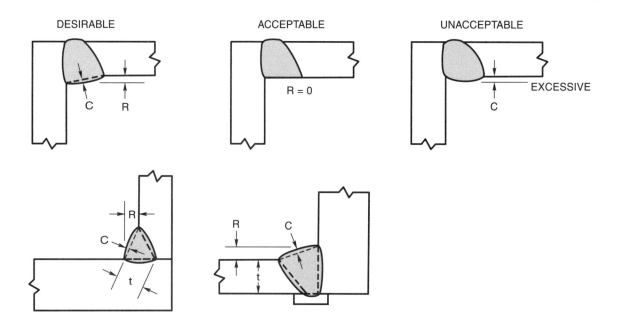

(B) GROOVE WELD PROFILES INSIDE CORNER JOINTS

Figure 7.4—Requirements for Weld Profiles (see Tables 7.8 and 7.9)

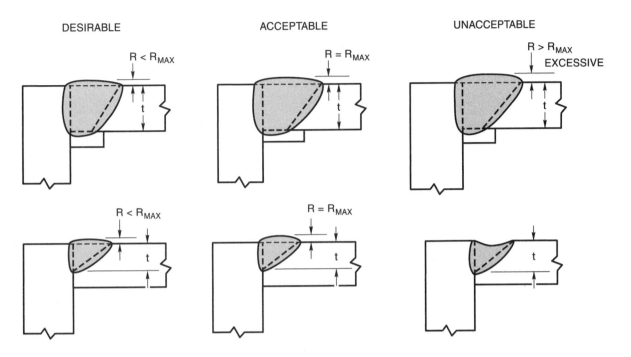

(C) GROOVE WELD PROFILES OUTSIDE CORNER JOINTS

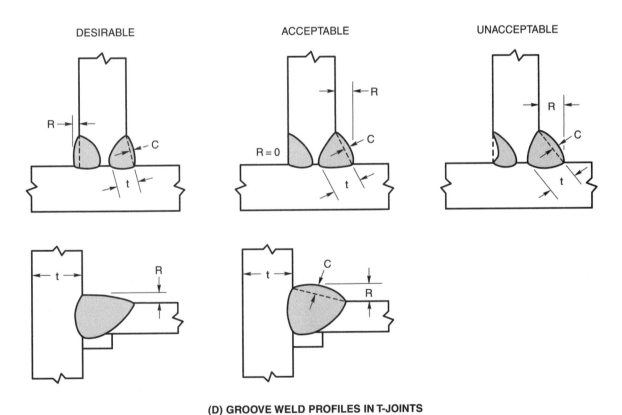

(D) GROOVE WELD PROFILES IN T-JOINTS

Figure 7.4 (Continued)—Requirements for Weld Profiles (see Tables 7.8 and 7.9)

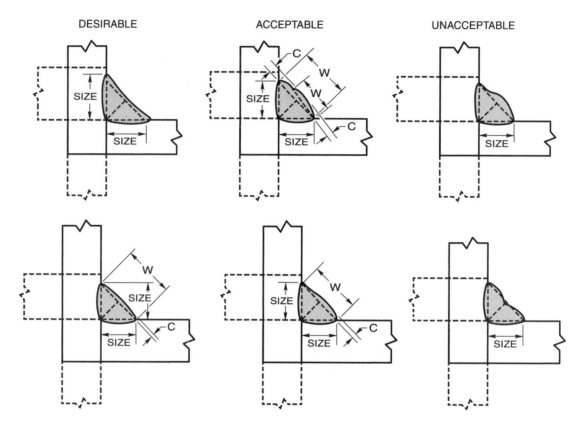

(E) FILLET WELD PROFILES FOR INSIDE CORNER JOINTS, LAP JOINTS, AND T-JOINTS

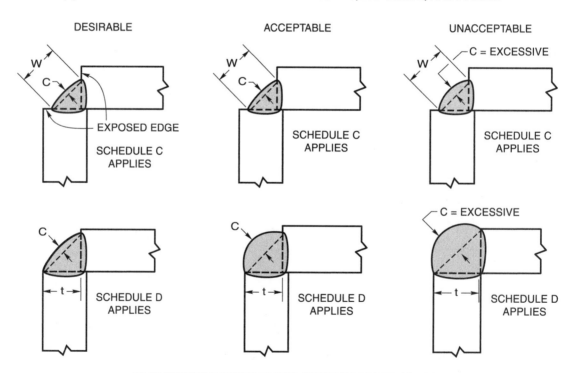

(F) FILLET WELD PROFILES FOR OUTSIDE CORNER JOINTS

Figure 7.4 (Continued)—Requirements for Weld Profiles (see Tables 7.8 and 7.9)

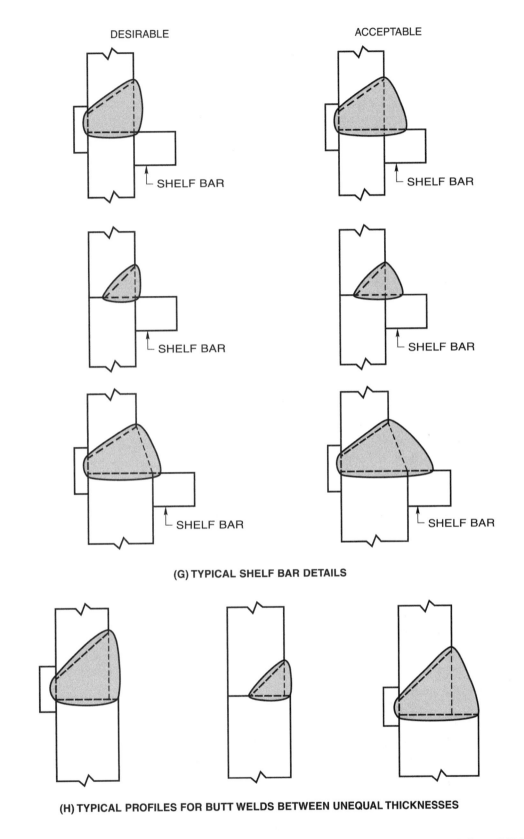

Figure 7.4 (Continued)—Requirements for Weld Profiles (see Tables 7.8 and 7.9)

8. Inspection

Part A
General Requirements

8.1 Scope

This clause contains requirements for inspection and nondestructive testing. It is divided into seven parts as follows:

Part A – General Requirements

Part B – Contractor Responsibilities

Part C – Acceptance Criteria

Part D – NDT Procedures

Part E – Radiographic Testing (RT)

Part F – Ultrasonic Testing (UT) of Groove Welds

Part G – Other Examination Methods

8.1.1 Information Furnished to Bidders. When NDT other than visual is to be required, it shall be so stated in the information furnished to the bidders. This information shall designate the categories of welds to be examined, the extent of examination of each category, and the method or methods of testing.

8.1.2 Inspection and Contract Stipulations. For the purpose of this code, fabrication/erection inspection and testing, and verification inspection and testing shall be separate functions.

8.1.2.1 Contractor's Inspection. This type of inspection and testing shall be performed as necessary prior to assembly, during assembly, during welding, and after welding to ensure that materials and workmanship meet the requirements of the contract documents. Fabrication/erection inspection and testing shall be the responsibilities of the Contractor unless otherwise provided in the contract documents.

8.1.2.2 Verification Inspection. This type of inspection and testing shall be performed and their results reported to the Owner and Contractor in a timely manner to avoid delays in the work. Verification inspection and testing are the prerogatives of the Owner who may perform this function or, when provided in the contract, waive independent verification, or stipulate that both inspection and verification shall be performed by the Contractor.

8.1.3 Definition of Inspector Categories

8.1.3.1 Contractor's Inspector. This inspector is the duly designated person who acts for, and in behalf of, the Contractor on all inspection and quality matters within the scope of the contract documents.

8.1.3.2 Verification Inspector. This inspector is the duly designated person who acts for, and in behalf of, the Owner or Engineer on all inspection and quality matters within the scope of the contract documents.

8.1.3.3 Inspector(s). When the term inspector is used without further qualification as to the specific inspector category described above, it applies equally to inspection and verification within the limits of responsibility described in 8.1.2.

8.1.4 Qualification of Inspection Personnel

8.1.4.1 Engineer's Responsibilities. If the Engineer requires a specific basis of inspection personnel qualification other than those listed in 8.1.4.2, the basis shall be designated in the contract documents.

8.1.4.2 Basis for Qualification of Welding Inspectors. Inspectors responsible for acceptance or rejection of material and workmanship on the basis of visual inspection shall be qualified.

The acceptable qualification basis shall be one of the following:

(1) Current or previous certification as an AWS Certified Welding Inspector (CWI) or Senior Certified Welding Inspector (SCWI) in conformance with the requirements of AWS QC1, *Standard for AWS Certification of Welding Inspectors*,

(2) Current or previous certification as a Level 2 or Level 3 Welding Inspector in conformance with the requirements of Canadian Standards Association (CSA) Standard W178.2, *Certification of Welding Inspectors*,

(3) Current or previous qualification as a Welding Inspector (WI) or Senior Welding Inspector (SWI) in conformance with the requirements of AWS B5.1, *Specification for the Qualification of Welding Inspectors*,

(4) Current or previous qualification as an ASNT SNT-TC–1A-VT Level II in conformance with the requirements of ASNT Recommended Practice No. *SNT-TC–1A, Personnel Qualification and Certification in Nondestructive Testing*, or ANSI/ASNT CP–189 *ASNT Standard for Qualification and Certification of Nondestructive Personnel* or

(5) An individual who, by training or experience, or both, in metals fabrication, inspection and testing, is competent to perform inspection of the work.

8.1.4.3 Alternative Inspector Qualifications.

The basis of alternative qualification for inspection personnel shall be specified in contract documents if different than the requirements in 8.1.4.2. When requested by the Engineer, or an authorized representative of the owner, documentation of the specified alternative qualification of inspection personnel shall be submitted for verification and approval.

8.1.4.4 Assistant Inspector. The Inspector may be supported by Assistant Inspectors who may perform specific inspection functions under the supervision of the Inspector. The work of Assistant Inspectors shall be regularly monitored by the Inspector.

8.1.4.5 Basis for Qualification of Assistant Inspectors. Assistant Inspectors shall be qualified to perform the specific functions to which they are assigned.

The acceptable qualification basis shall be one of the following:

(1) Current or previous certification as an AWS Certified Associate Welding Inspector (CAWI) or higher in conformance with the requirements of AWS QC1, *Standard for AWS Certification of Welding Inspectors*,

(2) Current or previous certification as a Level 1 Welding Inspector or higher in conformance with the requirements of Canadian Standards Association (CSA) W178.2, *Certification of Welding Inspectors*,

(3) Current or previous qualification as an Associate Welding Inspector (AWI) or higher in conformance with the requirements of AWS B5.1, *Specification for the Qualification of Welding Inspectors*,

(4) Current or previous qualification as an ASNT SNT-TC–1A-VT Level I in conformance with the requirements of ASNT Recommended Practice No. *SNT-TC–1A, Personnel Qualification and Certification in Nondestructive Testing*, or ANSI/ASNT CP–189, *ASNT Standard for Qualification and Certification of Nondestructive Personnel* or

(5) An individual who is qualified by training and experience to perform the specific functions they are assigned.

8.1.4.6 Eye Examination. Inspectors and Assistant Inspectors, and personnel performing NDT shall have passed an eye examination to prove natural or corrected near vision acuity, in at least one eye, of Jaeger J–2 at a minimum distance of 12 in [300 mm]. Alternatively, an equivalent test as determined by ophthalmic medical personnel is acceptable. Vision tests are required every three years, or less if necessary, to demonstrate adequacy.

8.1.4.7 Term of Effectiveness. The qualification of an Inspector shall remain in effect indefinitely, provided the Inspector remains active in inspection of welded steel fabrication and their eye examinations are current, unless there is specific reason to question the Inspector's ability.

8.1.5 Inspector Responsibility. The Inspector shall ascertain that all fabrication and erection by welding is performed in conformance with the requirements of the contract documents.

8.1.6 Items to be Furnished to the Inspector. The Inspector shall be furnished complete detailed drawings showing the size, length, type, and location of all welds to be made. The Inspector shall also be furnished the portion of the contract documents that describes material and quality requirements for the products to be fabricated or erected, or both.

8.1.7 Inspector Notification. The Inspector shall be notified in advance of the start of operations subject to inspection and verification.

8.2 Inspection of Materials and Equipment

The Contractor's Inspector shall ensure that only materials and equipment conforming to the requirements of this code shall be used.

8.3 Inspection of WPSs

8.3.1 Prequalified WPS. The Contractor's Inspector shall ensure that all prequalified WPSs to be used for the work conform with the requirements of Clauses 5, 7, 10 (if tubular), and the contract documents.

8.3.2 WPSs Qualified by Test. The Contractor's Inspector shall ensure that all WPSs qualified by test conform with the requirements of Clauses 6, 7, 10 (if tubular), and the contract documents.

8.3.3 WPSs in Production. The Contractor's Inspector shall ensure that all welding operations are performed in conformance with WPSs that meet the requirements of this code and the contract documents.

8.4 Inspection of Welder, Welding Operator, and Tack Welder Qualifications

8.4.1 Determination of Qualification. The Inspector shall allow welding to be performed only by welders, welding operators, and tack welders who are qualified in conformance with the requirements of Clause 6, or Clause 10 for tubulars, or shall ensure that each welder, welding operator, or tack welder has previously demonstrated such qualification under other acceptable supervision and approved by the Engineer in conformance with 6.2.2.1.

8.4.2 Retesting Based on Quality of Work. When the quality of a qualified welder's, welding operator's, or tack welder's work appears to be below the requirements of this code, the Inspector may require that the welder, welding operator, or tack welder demonstrate an ability to produce sound welds of the type that has not met requirements by means of a simple test, such as the fillet weld break test, or by requiring complete requalification in conformance with Clause 6, or Clause 10 for tubulars.

8.4.3 Retesting Based on Qualification Expiration. The Inspector shall require requalification of any welder, welding operator, or tack welder who has not used the process (for which they are qualified) for a period exceeding six months (see 6.2.3.1).

8.5 Inspection of Work and Records

8.5.1 Size, Length, and Location of Welds. The Inspector shall ensure that the size, length, and location of all welds conform to the requirements of this code and to the detail drawings and that no unspecified welds have been added without the approval of the Engineer.

8.5.2 Scope of Examinations. The Inspector shall, at suitable intervals, observe joint preparation, assembly practice, and the welding techniques, and performance of each welder, welding operator, and tack welder to ensure that the applicable requirements of this code are met.

8.5.3 Extent of Examination. The Inspector shall examine the work to ensure that it meets the requirements of this code. Other acceptance criteria, different from those described in the code, may be used when approved by the Engineer. Size and contour of welds shall be measured with suitable gages. Visual inspection for cracks in welds and base metal and other discontinuities should be aided by a strong light, magnifiers, or such other devices as may be found helpful.

8.5.4 Inspector Identification of Inspections Performed. Inspectors shall identify with a distinguishing mark or other recording methods all parts or joints that they have inspected and accepted. Any recording method which is mutually agreeable may be used. Die stamping of cyclically loaded members without the approval of the Engineer shall be prohibited.

8.5.5 Maintenance of Records. The Inspector shall keep a record of qualifications of all welders, welding operators, and tack welders; all WPS qualifications or other tests that are made; and such other information as may be required.

Part B
Contractor Responsibilities

8.6 Obligations of the Contractor

8.6.1 Contractor Responsibilities. The Contractor shall be responsible for visual inspection and necessary correction of all deficiencies in materials and workmanship in conformance with the requirements of this code.

8.6.2 Inspector Requests. The Contractor shall comply with all requests of the Inspector(s) to correct deficiencies in materials and workmanship as provided in the contract documents.

8.6.3 Engineering Judgment. In the event that faulty welding, or its removal for rewelding, damages the base metal so that in the judgment of the Engineer its retention is not in conformance with the intent of the contract documents, the Contractor shall remove and replace the damaged base metal or shall compensate for the deficiency in a manner approved by the Engineer.

8.6.4 Specified NDT Other than Visual. When NDT other than visual inspection is specified in the information furnished to bidders, it shall be the Contractor's responsibility to ensure that all specified welds shall meet the quality requirements of Clause 8, Part C or Clause 10, Part F for tubulars, whichever is applicable.

8.6.5 Nonspecified NDT Other than Visual. If NDT other than visual inspection is not specified in the original contract agreement but is subsequently requested by the Owner, the Contractor shall perform any requested testing or shall allow any testing to be performed in conformance with 8.14. The Owner shall be responsible for all associated costs including handling, surface preparation, NDT, and repair of discontinuities other than those described in 8.9, whichever is applicable, at rates mutually agreeable between Owner and Contractor. However, if such testing should disclose an attempt to defraud or gross nonconformance to this code, repair work shall be done at the Contractor's expense.

Part C
Acceptance Criteria

8.7 Scope

Acceptance criteria for visual and NDT inspection of statically and cyclically loaded nontubular connections are described in Part C. The extent of examination and the acceptance criteria shall be specified in the contract documents in the information furnished to the bidder.

8.8 Engineer's Approval for Alternate Acceptance Criteria

The fundamental premise of the code is to provide general stipulations applicable to most situations. Acceptance criteria for production welds different from those described in the code may be used for a particular application, provided they are suitably documented by the proposer and approved by the Engineer. These alternate acceptance criteria may be based upon evaluation of suitability for service using past experience, experimental evidence or engineering analysis considering material type, service load effects, and environmental factors.

8.9 Visual Inspection

All welds shall be visually inspected and shall be acceptable if the criteria of Table 8.1, or Table 10.15 (if tubular) are satisfied.

8.10 Penetrant Testing (PT) and Magnetic Particle Testing (MT)

Welds that are subject to PT and MT, in addition to visual inspection, shall be evaluated on the basis of the acceptance criteria for visual inspection. The testing shall be performed in conformance with 8.14.4 or 8.14.5, whichever is applicable.

8.11 Nondestructive Testing (NDT)

Except as provided for in 10.28 for tubulars, all NDT methods including equipment requirements and qualifications, personnel qualifications, and operating methods shall be in conformance with Clause 8, Inspection. Acceptance criteria shall be as described in this section. Welds subject to NDT shall have been found acceptable by visual inspection in conformance with 8.9.

For welds subject to NDT in conformance with 8.10, 8.11, 10.25.2, and 10.26.1, the testing may begin immediately after the completed welds have cooled to ambient temperature. Acceptance criteria for ASTM A514, A517, and A709 Grade HPS 100W [690W] steels shall be based on NDT performed not less than 48 hours after completion of the welds.

8.12 Radiographic Testing (RT)

Welds shown by RT that do not meet the requirements of Part C, or alternate acceptance criteria per 8.8, shall be repaired in conformance with 7.25. Discontinuities other than cracks shall be evaluated on the basis of being either elongated or rounded. Regardless of the type of discontinuity, an elongated discontinuity shall be defined as one in which its length exceeds three times its width. A rounded discontinuity shall be defined as one in which its length is three times its width or less and may be round or irregular and may have tails.

8.12.1 Discontinuity Acceptance Criteria for Statically Loaded Nontubular Connections. Welds that are subject to RT in addition to visual inspection shall have no cracks and shall be unacceptable if the RT shows any discontinuities exceeding the following limitations. The limitations given by Figure 8.1 for 1–1/8 in [30 mm] weld size (S) shall apply to all weld sizes greater than 1-1/8 in [30 mm].

(1) Elongated discontinuities exceeding the maximum size of Figure 8.1.

(2) Discontinuities closer than the minimum clearance allowance of Figure 8.1.

(3) Rounded discontinuities greater than a maximum size of S/3, not to exceed 1/4 in [6 mm]. However, when S is greater than 2 in [50 mm], the maximum rounded indication may be 3/8 in [10 mm]. The minimum clearance of rounded discontinuities greater than or equal to 3/32 in [2.5 mm] to an acceptable elongated or rounded discontinuity or to an edge or end of an intersecting weld shall be three times the greatest dimension of the larger of the discontinuities being considered.

(4) At the intersection of a weld with another weld or a free edge (i.e., an edge beyond which no material extension exists), acceptable discontinuities shall conform to the limitations of Figure 8.1, Cases I–IV.

(5) Isolated discontinuities such as a cluster of rounded indications, having a sum of their greatest dimensions exceeding the maximum size single discontinuity allowed in Figure 8.1. The minimum clearance to another cluster or an elongated or rounded discontinuity or to an edge or end of an intersecting weld shall be three times the greatest dimension of the larger of the discontinuities being considered.

(6) The sum of individual discontinuities each having a greater dimension of less than 3/32 in [2.5 mm] shall not exceed 2S/3 or 3/8 in [10 mm], whichever is less, in any linear 1 in [25 mm] of weld. This requirement is independent of (1), (2), and (3) above.

(7) In-line discontinuities, where the sum of the greatest dimensions exceeds S in any length of 6S. When the length of the weld being examined is less than 6S, the allowable sum of the greatest dimensions shall be proportionally less.

8.12.2 Discontinuity Acceptance Criteria for Cyclically Loaded Nontubular Connections. Welds that are subject to RT in addition to visual inspection shall have no cracks and shall be unacceptable if the RT shows any of the types of discontinuities described in 8.12.2.1, 8.12.2.2, or 8.12.2.3. The limitations given by Figures 8.2 and 8.3 for 1–1/2 in [38 mm] weld size (S) shall apply to all weld sizes greater than 1–1/2 in [38 mm].

8.12.2.1 Cyclically Loaded Nontubular Connections in Tension

(1) Discontinuities exceeding the maximum size of Figure 8.2.

(2) Discontinuities closer than the minimum clearance allowance of Figure 8.2.

(3) At the intersection of a weld with another weld or a free edge (i.e., an edge beyond which no material extension exists), acceptable discontinuities shall conform to the limitations of Figure 8.2, Cases I–IV.

(4) Isolated discontinuities such as a cluster of rounded indications, having a sum of their greatest dimensions exceeding the maximum size single discontinuity allowed in Figure 8.2. The minimum clearance to another cluster or an elongated or rounded discontinuity or to an edge or end of an intersecting weld shall be three times the greatest dimension of the larger of the discontinuities being considered.

(5) The sum of individual discontinuities each having a greater dimension of less than 3/32 in [2.5 mm] shall not exceed 2S/3 or 3/8 in [10 mm], whichever is less, in any linear 1 in [25 mm] of weld. This requirement is independent of (1), (2), and (3) above.

(6) In-line discontinuities, where the sum of the greatest dimensions exceeds S in any length of 6S. When the length of the weld being examined is less than 6S, the allowable sum of the greatest dimensions shall be proportionally less.

8.12.2.2 Cyclically Loaded Nontubular Connections in Compression

(1) Discontinuities exceeding the maximum size of Figure 8.3.

(2) Discontinuities closer than the minimum clearance allowance of Figure 8.3.

(3) At the intersection of a weld with another weld or a free edge (i.e., an edge beyond which no material extension exists), acceptable discontinuities shall conform to the limitations of Figure 8.3, Cases I–V.

(4) Isolated discontinuities such as a cluster of rounded indications, having a sum of their greatest dimensions exceeding the maximum size single discontinuity allowed in Figure 8.3. The minimum clearance to another cluster or an elongated or rounded discontinuity or to an edge or end of an intersecting weld shall be three times the greatest dimension of the larger of the discontinuities being considered.

(5) The sum of individual discontinuities each having a greater dimension of less than 3/32 in [2.5 mm] shall not exceed 2S/3 or 3/8 in [10 mm], whichever is less, in any linear 1 in [25 mm] of weld. This requirement is independent of (1), (2), and (3) above.

(6) In-line discontinuities, where the sum of the greatest dimensions exceeds S in any length of 6S. When the length of the weld being examined is less than 6S, the allowable sum of the greatest dimensions shall be proportionally less.

8.12.2.3 Discontinuities Less than 1/16 in [2 mm].
In addition to the requirements of 8.12.2.1 and 8.12.2.2, discontinuities having a greatest dimension of less than 1/16 in [2 mm] shall be unacceptable if the sum of their greatest dimensions exceeds 3/8 in [10 mm] in any linear inch of weld.

8.13 Ultrasonic Testing (UT)

8.13.1 Acceptance Criteria for Statically Loaded Nontubular Connections.
The acceptance criteria for welds subject to UT in addition to visual inspection shall meet the requirements of Table 8.2. For CJP web-to-flange welds, acceptance of discontinuities detected by scanning movements other than scanning pattern 'E' (see 8.30.2.2) may be based on weld thickness equal to the actual web thickness plus 1 in [25 mm].

Discontinuities detected by scanning pattern 'E' shall be evaluated to the criteria of Table 8.2 for the actual web thickness. When CJP web-to-flange welds are subject to calculated tensile stress normal to the weld, they should be so designated on the design drawing and shall conform to the requirements of Table 8.2. Ultrasonically tested welds are evaluated on the basis of a discontinuity reflecting ultrasound in proportion to its effect on the integrity of the weld. Indications of discontinuities that remain on the display as the search unit is moved towards and away from the discontinuity (scanning movement "B") may be indicative of planar discontinuities with significant through-throat dimension.

Since the major reflecting surface of the most critical discontinuities is oriented a minimum of 20° (for a 70° search unit) to 45° (for a 45° search unit) from perpendicular to the sound beam, amplitude evaluation (dB rating) does not allow reliable disposition. When indications exhibiting these planar characteristics are present at scanning sensitivity, a more detailed evaluation of the discontinuity by other means shall be required (e.g., alternate UT techniques, RT, grinding or gouging for visual inspection, etc.).

8.13.2 Acceptance Criteria for Cyclically Loaded Nontubular Connections.
The acceptance criteria for welds subject to UT in addition to visual inspection shall meet the following requirements:

(1) Welds subject to tensile stress under any condition of loading shall conform to the requirements of Table 8.3.

(2) Welds subject to compressive stress shall conform to the requirements of Table 8.2.

8.13.2.1 Indications. Ultrasonically tested welds are evaluated on the basis of a discontinuity reflecting ultrasound in proportion to its effect on the integrity of the weld. Indications of discontinuities that remain on the display as the search unit is moved towards and away from the discontinuity (scanning movement "B") may be indicative of planar discontinuities with significant through throat dimension. As the orientation of such discontinuities, relative to the sound beam, deviates from the perpendicular, dB ratings which do not allow direct, reliable evaluation of the welded joint integrity may result. When indications that exhibit these planar characteristics are present at scanning sensitivity, a more detailed evaluation of the discontinuity by other means may be required (e.g., alternate UT techniques, RT, grinding, or gouging for visual inspection, etc.).

8.13.2.2 Scanning. CJP web-to-flange welds shall conform to the requirements of Table 8.2, and acceptance for discontinuities detected by scanning movements other than scanning pattern 'E' (see 8.30.2.2) may be based on a weld thickness equal to the actual web thickness plus 1 in [25 mm]. Discontinuities detected by scanning pattern 'S' shall be evaluated to the criteria of 8.13.2 for the actual web thickness. When such web-to-flange welds are subject to calculated tensile stress normal to the weld, they shall be so designated on design drawings and shall conform to the requirements of Table 8.3.

Part D
NDT Procedures

8.14 Procedures

The NDT procedures as described in this code have been in use for many years and provide reasonable assurance of weld integrity; however, it appears that some users of the code incorrectly consider each method capable of detecting all unacceptable discontinuities. Users of the code should become familiar with all the limitations of NDT methods to be used, particularly the inability to detect and characterize planar discontinuities with specific orientations. (The limitations and complementary use of each method are explained in the latest edition of AWS B1.10, *Guide for Nondestructive Examination of Welds*.)

8.14.1 RT. When RT is used, the procedure and technique shall be in conformance with Part E of this clause or Clause 10, Part F and 10.27 for tubulars.

8.14.2 Radiation Imaging Systems. When examination is performed using radiation imaging systems, the procedures and techniques shall be in conformance with Part G of this clause.

8.14.3 UT. When UT is used, the procedure and technique shall be in conformance with Part F of this clause.

8.14.4 MT. When MT is used, the procedure and technique shall be in conformance with ASTM E709, and the standard of acceptance shall be in conformance with Clause 8, Part C, of this code.

8.14.5 PT. For detecting discontinuities that are open to the surface, PT may be used. The standard methods set forth in ASTM E165 shall be used for PT inspection, and the standards of acceptance shall be in conformance with Clause 8, Part C, of this code.

8.14.6 Personnel Qualification

8.14.6.1 ASNT Requirements. Personnel performing NDT other than visual shall be qualified in conformance with the current edition of the American Society for Nondestructive Testing Recommended Practice No. SNT-TC–1A. Individuals who perform NDT shall be qualified for:

(1) NDT Level II, or

(2) NDT Level I working under the NDT Level II

8.14.6.2 Certification. Certification of Level I and Level II individuals shall be performed by a Level III individual who has:

(1) been certified by the American Society for Nondestructive Testing, or

(2) the education, training, experience, and has successfully passed the written examination described in SNT-TC–1A.

8.14.6.3 Exemption of QC1 Requirements. Personnel performing NDT under the provisions of 8.14.6 need not be qualified and certified under the provisions of AWS QC1.

8.15 Extent of Testing

Information furnished to the bidders shall clearly identify the extent of NDT (types, categories, or location) of welds to be tested.

8.15.1 Full Testing. Weld joints requiring testing by contract specification shall be tested for their full length, unless partial or spot testing is specified.

8.15.2 Partial Testing. When partial testing is specified, the location and lengths of welds or categories of weld to be tested shall be clearly designated in the contract documents.

8.15.3 Spot Testing. When spot testing is specified, the number of spots in each designated category of welded joint to be tested in a stated length of weld or a designated segment of weld shall be included in the information furnished to the bidders. Each spot test shall cover at least 4 in [100 mm] of the weld length. When spot testing reveals indications of unacceptable discontinuities that require repair, the extent of those discontinuities shall be explored. Two additional spots in the same segment of weld joint shall be taken at locations away from the original spot. The location of the additional spots shall be agreed upon between the Contractor and the Verification Inspector.

When either of the two additional spots shows defects that require repair, the entire segment of weld represented by the original spot shall be completely tested. If the weld involves more than one segment, two additional spots in each segment shall be tested at locations agreed upon by the Contractor and the Verification Inspector, subject to the foregoing interpretation.

8.15.4 Relevant Information. NDT personnel shall, prior to testing, be furnished or have access to relevant information regarding weld joint geometries, material thicknesses, and welding processes used in making the weldment. NDT personnel shall be apprised of any subsequent repairs to the weld.

Part E
Radiographic Testing (RT)

8.16 RT of Groove Welds in Butt Joints

8.16.1 Procedures and Standards. The procedures and standards set forth in Part E shall govern radiographic testing (RT) when such inspection is required by the contract documents as provided in 8.15. The requirements described herein are specifically for testing groove welds in butt joints in plate, shapes, and bars by X-ray or gamma-ray sources. The methods shall conform to ASTM E94, *Standard Guide for Radiographic Examination using Industrial Radiographic Film*; and ASTM E1032, *Radiographic Examination of Weldments using Industrial X-Ray Film*.

Digital Image archival shall be in accordance with ASTM E2339, *Standard Practice for Digital Imaging and Communication in Nondestructive Evaluation (DICONDE)*.

8.16.1.1 Digital Radiography. The digital radiography method shall be either computed radiography (CR) using a photostimulable phosphor plate, more commonly known as a storage phosphor imaging plate (SPIP), or direct radiography (DR) using a Digital Detector Array (DDA).

(1) CR shall comply with ASTM E2033, *Standard Practice for Radiographic Examination using Computed Radiology (Photostimulable Luminescence Method)* and ASTM E2445, *Standard Practice for Performance Evaluation and Long-Term Stability of Computed Radiology Systems.*

(2) DR shall comply with ASTM E2698, *Standard Practice for Radiological Examination Using Digital Detector Arrays* and ASTM E2737, *Standard Practice for Digital Detector Array Performance Evaluation and Long-Term Stability.*

8.16.2 Variations. Variations in testing procedures, equipment, and acceptance standards may be used upon agreement between the Contractor and the Owner. Such variations include, but are not limited to, the following:

(1) RT of fillet welds and all welds in T and corner joints;

(2) changes in source-to-film distance;

(3) unusual application of film;

(4) use of SPIPS and DDAs;

(5) unusual hole-type or wire-type image quality indicators (IQI) applications (including film side IQI);

(6) and for RT of thicknesses greater than 6 in [150 mm];

(7) film types;

(8) density or brightness requirements;

(9) variations in exposure, film development or image processing; and

(10) viewing techniques.

8.17 RT Procedures

8.17.1 Procedure. Radiographic film and digital images shall be made using a single source of either X-ray or gamma-ray radiation. The radiographic film or digital image sensitivity shall be judged based on hole-type or wire IQI. Radiographic technique and equipment shall provide sufficient sensitivity to clearly delineate the required hole-type IQIs and the essential holes or wires as described in 8.17.7 and 10.27.1; Tables 8.4, 8.5, 10.16, and 10.17; and Figures 8.4 and 8.5. Identifying letters and numbers shall show clearly in the radiographic film or digital image.

8.17.2 Safety Requirements. RT shall be performed in conformance with all applicable safety requirements.

8.17.3 Removal of Reinforcement. When the contract documents require the removal of weld reinforcement, the welds shall be prepared as described in 7.23.3.1. Other weld surfaces need not be ground or otherwise smoothed unless surface irregularities or the junction between weld and base metal may cause objectionable weld discontinuities to be obscured in the radiographic film or digital image.

8.17.3.1 Tabs. Weld tabs shall be removed prior to RT unless otherwise approved by the Engineer.

8.17.3.2 Steel Backing. When required by 7.9.1.4 or other provisions of the contract documents, steel backing shall be removed, and the surface shall be finished flush by grinding prior to RT. Grinding shall be as described in 7.23.3.1.

8.17.3.3 Reinforcement. When weld reinforcement or backing, or both, is not removed, or wire IQI alternate placement is not used, steel shims which extend at least 1/8 in [3 mm] beyond three sides of the required hole-type IQI or wire IQI shall be placed under the hole-type IQI or wire IQI, so that the total thickness of steel between the hole-type IQI and the film is approximately equal to the average thickness of the weld measured through its reinforcement and backing.

8.17.4 Radiographic Film. Radiographic film shall be as described in ASTM E94. Lead foil screens shall be used as described in ASTM E94. Fluorescent screens shall be prohibited.

8.17.5 Technique. Radiographs shall be made with a single source of radiation centered as near as practicable with respect to the length and width of that portion of the weld being examined.

8.17.5.1 Geometric Unsharpness. Gamma ray or X-ray sources, regardless of size, shall be capable of meeting the geometric unsharpness limitation as defined in 8.17.5.1.1. The calculation of geometric unsharpness below shall determine the minimum source-to-film distance:

$U_g = Fd/D$

where:

U_g = geometric unsharpness
F = source size – the maximum projected dimension of the radiating source (or effective focal spot) in the plane perpendicular to the distance D from the source side of the subject being radiographed.
D = distance from source of radiation to subject being radiographed.
d = distance from source side of subject to the film.

8.17.5.1.1 Geometric Unsharpness Limitation. (D) and (d) shall be determined at the approximate center of the area of interest. Material thickness for determining allowable geometric unsharpness shall be the thickness upon which the IQI selection is based. The maximum values for geometric unsharpness are as follows:

Material Thickness, in [mm]	Ug Maximum, in [mm]
Under 2 [50]	0.020 [0.5]
2 through 3 [50 – 75]	0.030 [0.75]
Over 3 through 4 [75 – 100]	0.040 [1]
Greater than 4 [100]	0.070 [1.75]

8.17.5.2 Source-to-Subject Distance. The source to-subject distance shall not be less than the total length of film being exposed in a single plane. This provision shall not apply to panoramic exposures.

8.17.5.3 Source-to-Subject Distance Limitations. The source-to-subject distance shall not be less than seven times the thickness of weld plus reinforcement and backing, if any, nor such that the inspecting radiation shall penetrate any portion of the weld represented in the radiograph at an angle greater than 26–1/2° from a line normal to the weld surface.

8.17.6 Sources. X-ray units, 600 kVp maximum, and iridium 192 may be used as a source for all RT provided they have adequate penetrating ability. Cobalt 60 shall only be used as a radiographic source when the steel being radiographed exceeds 2–1/2 in [65 mm] in thickness. Other radiographic sources may be used with the approval of the Engineer.

8.17.7 IQI Selection and Placement. IQIs shall be selected and placed on the weldment in the area of interest being radiographed as shown in Table 8.6. Steel backing shall not be considered part of the weld or weld reinforcement in IQI selection.

8.17.8 Technique. Welded joints shall be radiographed and the film indexed by methods that will provide complete and continuous inspection of the joint within the limits specified to be examined. Joint limits shall show clearly in the radiographs. Short film, short screens, excessive undercut by scattered radiation, or any other process that obscures portions of the total weld length shall render the radiograph unacceptable.

8.17.8.1 Film or Plate Length. Film, SPIPs, and DDAs shall have sufficient length and shall be placed to provide at least 1/2 in [12 mm] of imaging data beyond the projected edge of the weld.

8.17.8.2 Overlapping Film. Welds longer than 14 in [350 mm] may be radiographed by overlapping film cassettes and making a single exposure, or by using single film cassettes with either SPIPs or conventional film, or DDAs, and making separate exposures. The provisions of 8.17.5 shall apply.

8.17.8.3 Backscatter. To check for backscatter radiation, a lead symbol "B", 1/2 in [12 mm] high, 1/16 in [2 mm] thick shall be attached to the back of each film cassette. If the "B" image appears on the radiograph, the radiograph shall be considered unacceptable.

8.17.9 Film, SPIP, and DDA Width. Film, SPIP, and DDA widths shall be sufficient to depict all portions of the weld joint, including the HAZs, and shall provide sufficient additional space for the required hole-type IQIs or wire IQI and film identification without infringing upon the area of interest.

8.17.10 Quality of Radiographic Film and Digital Images. All radiographic film and digital images shall be free from imperfections and artifacts to the extent that they cannot mask or be confused with the image of any discontinuity in the area of interest in the radiograph.

8.17.10.1 Radiographic Film and Digital Image Artifacts.

Imperfections include, but are not limited to the following as applicable:

(1) fogging

(2) processing defects such as streaks, water marks, or chemical stains

(3) scratches, finger marks, crimps, dirtiness, static marks, smudges, or tears

(4) loss of detail due to poor screen-to-film contact

(5) false indications due to defective screens, SPIP, DDA, or internal faults

(6) artifacts due to non-functional pixels

8.17.11 Density Limitations.

8.17.11.1 Radiographic Film. The transmitted film density through the radiographic image of the body of the required hole-type IQI(s) and the area of interest shall be 1.8 minimum for single film viewing for radiographs made with an X-ray source and 2.0 minimum for radiographs made with a gamma-ray source. For composite viewing of double film exposures, the minimum density shall be 2.6. Each radiograph of a composite set shall have a minimum density of 1.3. The maximum density shall be 4.0 for either single or composite viewing.

8.17.11.1.1 H & D Density. The density measured shall be H & D density (radiographic density), which is a measure of film blackening, expressed as:

$$D = \log I_o/I$$

where:

D = H & D (radiographic) density
I_o = light intensity on the film, and
I = light transmitted through the film.

8.17.11.1.2 Transitions. When weld transitions in thickness are radiographed and the ratio of the thickness of the thicker section to the thickness of the thinner section is 3 or greater, radiographs should be exposed to produce single film densities of 3.0 to 4.0 in the thinner section. When this is done, the minimum density requirements of 8.17.11.1 shall be waived unless otherwise provided in the contract documents.

8.17.11.2 Digital Image Sensitivity Range. The contrast and brightness range that demonstrates the required sensitivity shall be considered valid contrast and brightness values for interpretation. When multiple IQIs are used to cover different thickness ranges, the contrast and brightness range that demonstrates the required IQI image of each IQI shall be determined. Intervening thicknesses may be interpreted using the overlapping portions of the determined contrast and brightness ranges. When there is no overlap, additional IQI(s) shall be used.

8.17.12 Identification Marks. A radiograph identification mark and two location identification marks shall be placed on the steel at each radiograph location. A corresponding radiograph identification mark and two location identification marks, all of which shall show in the image, shall be provided by placing lead numbers or letters, or both, over each of the identical identification and location marks made on the steel to provide a means for matching the image to the weld or base metal. Additional identification information may be pre-printed no less than 3/4 in [20 mm] from the edge of the weld or shall be produced on the radiograph by placing lead figures on the steel. Information required to show on the radiograph shall include the Owner's contract identification, initials of the RT company, initials of the fabricator, the fabricator shop order number, the radiographic identification mark, the date, and the weld repair number, if applicable. For digital images, the additional information may be added as text on the processed image. When such text information is added, the software locking tool shall be enabled to prevent subsequent editing of the information.

8.17.13 Edge Blocks. Edge blocks shall be used when radiographing butt joints greater than 1/2 in [12 mm] thickness. The edge blocks shall have a length sufficient to extend beyond each side of the weld centerline for a minimum distance equal to the weld thickness, but no less than 2 in [50 mm], and shall have a thickness equal to or greater than the thickness of the weld. The minimum width of the edge blocks shall be equal to half the weld thickness, but not less than 1 in [25 mm]. The edge blocks shall be centered on the weld against the plate being radiographed, allowing no more than 1/16 in [2 mm] gap for the minimum specified length of the edge blocks. Edge blocks shall be made of radiographically clean steel and the surface shall have a finish of ANSI 125 μin [3 μm] or smoother (see Figure 8.10).

8.17.14 Linear Reference Comparators. When using SPIPs or DDAs, a measuring scale or an object of known dimension shall be used to serve as a linear reference comparator. The comparator shall be attached to the SPIP holder or DDA prior to exposure. As an alternative, when using SPIPs a transparent scale with opaque gradations may be placed on the SPIP prior to processing. In any case the reference comparator shall not interfere with interpretation of the image.

8.18 Examination, Report, and Disposition of Radiographs

8.18.1 Equipment Provided by Contractor.

8.18.1.1 Film Radiography. The Contractor shall provide a suitable variable intensity illuminator (viewer) with spot review or masked spot review capability. The viewer shall incorporate a means for adjusting the size of the spot under examination. The viewer shall have sufficient capacity to properly illuminate radiographs with an H & D density of 4.0. Film review shall be done in an area of subdued light.

8.18.1.2 Digital Radiography. The work station monitor, for evaluating images, shall have a display resolution with a pixel count which is equal to or greater than the pixel count of the direct imaging plate.

8.18.2 Reports. Before a weld subject to RT by the Contractor for the Owner is accepted, all of its radiographic film and digital images, including any that show unacceptable quality prior to repair, and a report interpreting them shall be submitted to the Verification Inspector.

8.18.3 Radiographic Film or Digital Image Retention. A full set of radiographic film or digital images, including any that show unacceptable quality prior to repair, shall be delivered to the Owner upon completion of the work. The Contractor's obligation to retain radiographic film or digital images shall cease:

(1) upon delivery of this full set to the Owner, or

(2) one full year after the completion of the Contractor's work, provided the Owner is given prior written notice.

8.18.4 Radiographic Film or Digital Image Archival

8.18.4.1 Radiographic Film. Radiographic film shall be stored in accordance with ASTM E1254, *Standard Guide for Storage of Radiographs and Unexposed Industrial Radiographic Films.*

8.18.4.2 Radiographic Digital Images. Radiographic digital images shall be archived using a reproducible electronic medium. Data file format shall comply with ASTM E2699, *Standard Practice for Digital Imaging and Communication in Nondestructive Evaluation (DICONDE) for Digital Radiographic (DR) Test Methods.* The image archival method shall be documented and proven at system installation. This shall include the image file nomenclature to enable the retrieval of images at a later date. Archived image files shall maintain the bit depth and spatial resolution of the original image. Image data compression is not allowed. The initial image presented by the SPIP or DDA system shall be preserved (stored) without altering the original spatial resolution and pixel intensity. The final image used for disposition shall also be archived when additional image processing is applied (excluding window/level and digital image zoom) to achieve the required image quality level. Annotations made to the image shall be stored in a manner which will not mask or hide diagnostic areas of the image.

Part F
Ultrasonic Testing (UT) of Groove Welds

8.19 General

8.19.1 Procedures and Standards. The procedures and standards set forth in Part F shall govern the UT of groove welds and HAZs between the thicknesses of 5/16 in and 8 in [8 mm and 200 mm] inclusive, when such testing is required by 8.14 of this code. For thicknesses less than 5/16 in [8 mm] or greater than 8 in [200 mm], testing shall be performed in conformance with Annex O. These procedures and standards shall be prohibited for testing tube-to-tube T-, Y-, or K-connections.

8.19.2 Variations. Annex O and Annex H specify alternative ultrasonic techniques for performing conventional UT and phased array UT examination of groove welds. Other variations in testing procedures, equipment, and acceptance standards not included in Part F of Clause 8 may be used with the approval of the Engineer. Such variations include other thicknesses, weld geometries, transducer sizes, frequencies, couplant, painted surfaces, testing techniques, etc. Such approved variations shall be recorded in the inspection records.

8.19.3 Piping Porosity. To detect possible piping porosity, RT is recommended to supplement UT of ESW or EGW welds.

8.19.4 Base Metal. These procedures are not intended to be employed for the procurement testing of base metals. However, welding related discontinuities (cracking, lamellar tearing, delaminations, etc.) in the adjacent base metal which would not be acceptable under the provisions of this code shall be reported to the Engineer for disposition.

8.20 Qualification Requirements

In satisfying the requirements of 8.14.6, the qualification of the UT operator shall include a specific and practical examination which shall be based on the requirements of this code. This examination shall require the UT operator to demonstrate the ability to apply the rules of this code in the accurate detection and disposition of discontinuities.

8.21 UT Equipment

8.21.1 Equipment Requirements. The UT instrument shall be the pulse echo type suitable for use with transducers oscillating at frequencies between 1 and 6 megahertz. The display shall be an "A" scan rectified video trace. The requirements shall be as shown in Table 8.8.

8.21.2 Horizontal Linearity. The horizontal linearity of the test instrument shall be qualified over the full soundpath distance to be used in testing in conformance with 8.28.1.

8.21.3 Requirements for Test Instruments. Test instruments shall include internal stabilization so that after warm-up, no variation in response greater than ± 1 dB occurs with a supply voltage change of 15% nominal or, in the case of a battery, throughout the charge operating life. There shall be an alarm or meter to signal a drop in battery voltage prior to instrument shutoff due to battery exhaustion.

8.21.4 Calibration of Test Instruments. The test instrument shall have a calibrated gain control (attenuator) adjustable in discrete 1 or 2 dB steps over a range of at least 60 dB. The accuracy of the attenuator settings shall be within plus or minus 1 dB. The procedure for qualification shall be as described in 8.23.2 and 8.28.2.

8.21.5 Display Range. The dynamic range of the instrument's display shall be such that a difference of 1 dB of amplitude can be easily detected on the display.

8.21.6 Straight-Beam (Longitudinal Wave) Search Units. Straight-beam (longitudinal wave) search unit transducers shall have an active area of not less than 1/2 in^2 [323 mm^2] nor more than 1 in^2 [645 mm^2]. The transducer shall be round or square. Transducers shall be capable of resolving the three reflections as described in 8.27.1.3.

8.21.7 Angle-Beam Search Units. Angle-beam search units shall consist of a transducer and an angle wedge. The unit may be comprised of the two separate elements or may be an integral unit.

8.21.7.1 Frequency. The transducer frequency shall be between 2 and 2.5 MHz, inclusive.

8.21.7.2 Transducer Dimensions. The transducer crystal shall be square or rectangular in shape and may vary from 5/8 in to 1 in [15 mm to 25 mm] in width and from 5/8 in to 13/16 in [15 mm to 20 mm] in height (see Figure 8.11). The maximum width to height ratio shall be 1.2 to 1.0, and the minimum width-to-height ratio shall be 1.0 to 1.0.

8.21.7.3 Angles. The search unit shall produce a sound beam in the material being tested within plus or minus 2° of one of the following proper angles: 70°, 60°, or 45°, as described in 8.27.2.2.

8.21.7.4 Marking. Each search unit shall be marked to clearly indicate the frequency of the transducer, nominal angle of refraction, and index point. The index point location procedure is described in 8.27.2.1.

8.21.7.5 Internal Reflections. Maximum allowable internal reflections from the search unit shall be as described in 8.23.3 and 8.28.3.

8.21.7.6 Edge Distance. The dimensions of the search unit shall be such that the distance from the leading edge of the search unit to the index point shall not exceed 1 in [25 mm].

8.21.7.7 IIW Type Block. The qualification procedure using the IIW reference block or other IIW type block shall be in conformance with 8.27.2.6 and as shown in Figure 8.12.

8.22 Reference Standards

8.22.1 IIW Standard. Any of the International Institute of Welding (IIW) type UT reference blocks may be used as the standard for both distance and sensitivity calibration, provided the block includes the 0.060 in [1.5 mm] diameter hole as shown in Figure 8.13 and distance, resolution, and angle verification features of Figure 8.16 (positions A through G). IIW type blocks shall conform to ASTM E164. Other portable blocks may be used, provided the reference level sensitivity for instrument/ search unit combination is adjusted to be the equivalent of that achieved with the IIW type block (see Annex G, for examples).

8.22.2 Prohibited Reflectors. The use of a "corner" reflector for calibration purposes shall be prohibited.

8.22.3 Resolution Requirements. The combination of search unit and instrument shall resolve three holes in the RC resolution reference test block shown in Figure 8.14. The search unit position is described in 8.27.2.5. The resolution shall be evaluated with the instrument controls set at normal test settings and with indications from the holes brought to midscreen height. Resolution shall be sufficient to distinguish at least the peaks of indications from the three holes. Use of the RC resolution reference block for calibration shall be prohibited. Each combination of instrument search unit (shoe and transducer) shall be checked prior to its initial use. This equipment verification shall be done initially with each search unit and UT unit combination. The verification need not be done again provided documentation is maintained that records the following items:

(1) UT machine's make, model and serial number

(2) Search unit's manufacturer, type, size, angle, and serial number

(3) Date of verification and technician's name

8.23 Equipment Qualification

8.23.1 Horizontal Linearity. The horizontal linearity of the test instrument shall be requalified at two-month intervals in each of the distance ranges that the instrument will be used. The qualification procedure shall be in conformance with 8.28.1 (see Annex G, for alternative method).

8.23.2 Gain Control. The instrument's gain control (attenuator) shall meet the requirements of 8.21.4 and shall be checked for correct calibration at two month intervals in conformance with 8.28.2. Alternative methods may be used for calibrated gain control (attenuator) qualification if proven at least equivalent with 8.28.2.

8.23.3 Internal Reflections. Maximum internal reflections from each search unit shall be verified at a maximum time interval of 40 hours of instrument use in conformance with 8.28.3.

8.23.4 Calibration of Angle-Beam Search Units. With the use of an approved calibration block, each angle beam search unit shall be checked after each eight hours of use to determine that the contact face is flat, that the sound entry point is correct, and that the beam angle is within the allowed plus or minus 2° tolerance in conformance with 8.27.2.1 and 8.27.2.2. Search units which do not meet these requirements shall be corrected or replaced.

8.24 Calibration for Testing

8.24.1 Position of Reject Control. All calibrations and tests shall be made with the reject (clipping or suppression) control turned off. Use of the reject (clipping or suppression) control may alter the amplitude linearity of the instrument and invalidate test results.

8.24.2 Technique. Calibration for sensitivity and horizontal sweep (distance) shall be made by the UT operator just prior to and at the location of the first weld tested. After that, requirements of 8.24.3 recalibration shall apply.

8.24.3 Recalibration. Recalibration shall be made after a change of operators, each two-hour maximum time interval, or when the electrical circuitry is disturbed in any way which includes the following:

(1) Transducer change

(2) Battery change

(3) Electrical outlet change

(4) Coaxial cable change

(5) Power outage (failure)

8.24.4 Straight-Beam Testing of Base Metal. Calibration for straight-beam testing of base metal shall be made with the search unit applied to Face A of the base metal and performed as follows:

8.24.4.1 Sweep. The horizontal sweep shall be adjusted for distance calibration to present the equivalent of at least two plate thicknesses on the display.

8.24.4.2 Sensitivity. The sensitivity shall be adjusted at a location free of indications so that the first back reflection from the far side of the plate will be 50% to 75% of full screen height.

8.24.5 Calibration for Angle-Beam Testing. Calibration for angle-beam testing shall be performed as follows (see Annex G, G2.4 for alternative method).

8.24.5.1 Horizontal Sweep. The horizontal sweep shall be adjusted to represent the actual sound-path distance by using the IIW type block or alternative blocks as described in 8.22.1. The distance calibration shall be made using either the 5 in [125 mm] scale or 10 in [250 mm] scale on the display, whichever is appropriate. If, however, the joint configuration or thickness prevents full examination of the weld at either of these settings, the distance calibration shall be made using 15 in or 20 in [400 mm or 500 mm] scale as required. The search unit position is described in 8.27.2.3.

NOTE: The horizontal location of all screen indications is based on the location at which the left side of the trace deflection breaks the horizontal base line.

8.24.5.2 Zero Reference Level. The zero reference level sensitivity used for discontinuity evaluation ("b" on the ultrasonic test report, Annex P, Form P–11) shall be attained by adjusting the calibrated gain control (attenuator) of the discontinuity detector, meeting the requirements of 8.21, so that a maximized horizontal trace deflection (adjusted to horizontal reference line height with calibrated gain control [attenuator]) results on the display between 40% and 60% screen height, in conformance with 8.27.2.4.

8.25 Testing Procedures

8.25.1 "X" Line. An "X" line for discontinuity location shall be marked on the test face of the weldment in a direction parallel to the weld axis. The location distance perpendicular to the weld axis shall be based on the dimensional figures on the detail drawing and usually falls on the centerline of the butt joint welds, and always falls on the near face of the connecting member of T and corner joint welds (the face opposite Face C).

8.25.2 "Y" Line. A "Y" accompanied with a weld identification number shall be clearly marked on the base metal adjacent to the weld that is subject to UT. This marking is used for the following purposes:

(1) Weld identification

(2) Identification of Face A

(3) Distance measurements and direction (+ or –) from the "X" line

(4) Location measurement from weld ends or edges

8.25.3 Cleanliness. All surfaces to which a search unit is applied shall be free of weld spatter, dirt, grease, oil (other than that used as a couplant), paint, and loose scale and shall have a contour allowing intimate coupling.

8.25.4 Couplants. A couplant material shall be used between the search unit and the test material. The couplant shall be either glycerin or cellulose gum and water mixture of a suitable consistency. A wetting agent may be added if needed. Light machine oil may be used for couplant on calibration blocks.

8.25.5 Extent of Testing. The entire base metal through which ultrasound must travel to test the weld shall be tested for laminar reflectors using a straight-beam search unit conforming to the requirements of 8.21.6 and calibrated in conformance with 8.24.4. If any area of base metal exhibits total loss of back reflection or an indication equal to or greater than the original back reflection height is located in a position that will interfere with the normal weld scanning procedure, its size, location, and depth from the Face A shall be determined and reported on the UT report, and an alternate weld scanning procedure shall be used.

8.25.5.1 Reflector Size. The reflector size evaluation procedure shall be in conformance with 8.29.1.

8.25.5.2 Inaccessibility. If part of a weld is inaccessible to testing in conformance with the requirements of Table 8.7, due to laminar content recorded in conformance with 8.25.5, the testing shall be conducted using one or more of the following alternative procedures as necessary to attain full weld coverage:

(1) Weld surface(s) shall be ground flush in conformance with 7.23.3.1.

(2) Testing from Faces A and B shall be performed.

(3) Other search unit angles shall be used.

8.25.6 Testing of Welds. Welds shall be tested using an angle beam search unit conforming to the requirements of 8.21.7 with the instrument calibrated in conformance with 8.24.5 using the angle as shown in Table 8.7. Following calibration and during testing, the only instrument adjustment allowed is the sensitivity level adjustment with the calibrated gain control (attenuator). The reject (clipping or suppression) control shall be turned off. Sensitivity shall be increased from the reference level for weld scanning in conformance with Table 8.2 or Table 8.3, as applicable.

8.25.6.1 Scanning. The testing angle and scanning procedure shall be in conformance with those shown in Table 8.7.

8.25.6.2 Butt Joints. All butt joint welds shall be tested from each side of the weld axis. Corner and T-joint welds shall be primarily tested from one side of the weld axis only. All welds shall be tested using the applicable scanning pattern or patterns shown in Figure 8.15 as necessary to detect both longitudinal and transverse discontinuities. It is intended that, as a minimum, all welds be tested by passing sound through the entire volume of the weld and the HAZ in two crossing directions, wherever practical.

8.25.6.3 Maximum Indication. When a discontinuity indication appears on the screen, the maximum attainable indication from the discontinuity shall be adjusted to produce a horizontal reference level trace deflection on the display. This adjustment shall be made with the calibrated gain control (attenuator), and the instrument reading in decibels shall be used as the "Indication Level, a," for calculating the "Indication Rating, d," as shown on the test report (Annex P, Form P–11).

8.25.6.4 Attenuation Factor. The "Attenuation Factor, (c)," on the test report shall be attained by subtracting 1 in [25 mm] from the sound-path distance and multiplying the remainder by 2 for U.S. Customary Units or by 0.08 for SI Units.

The factor (c) shall be rounded to the closest significant decimal place (0.1). Values less than 0.05 shall be reduced to the lower 0.1 and those of 0.05 or greater increased to the higher 0.1.

8.25.6.5 Indication Rating. The "Indication Rating, (d)," in the UT Report, Annex P, Form P–11, represents the algebraic difference in decibels between the indication level and the reference level with correction for attenuation as indicated in the following expressions:

Instruments with gain in dB: $a - b - c = d$

Instruments with attenuation in dB: $b - a - c = d$

The indication rating shall be rounded to the nearest whole number (1 dB) value. Resulting decimal values less than 0.5 dB shall be rounded down and those of 0.5 dB or greater shall be rounded up.

8.25.7 Length of Discontinuities. The length of discontinuities shall be determined in conformance with the procedure described in 8.29.2.

8.25.8 Basis for Acceptance or Rejection. Each weld discontinuity shall be accepted or rejected on the basis of its indication rating and its length, in conformance with Table 8.2 for statically loaded structures or Table 8.3 for cyclically loaded structures, whichever is applicable. Only those discontinuities which are unacceptable need be recorded on the test report, except that for welds designated in the contract documents as being "Fracture Critical," acceptable ratings that are within 6 dB, inclusive, of the minimum unacceptable rating shall be recorded on the test report.

8.25.9 Identification of Rejected Area. Each unacceptable discontinuity shall be indicated on the weld by a mark directly over the discontinuity for its entire length. The depth from the surface and indication rating shall be noted on nearby base metal.

8.25.10 Repair. Welds found unacceptable by UT shall be repaired by methods allowed by 7.25 of this code. Repaired areas shall be retested ultrasonically with results tabulated on the original form (if available) or additional report forms.

8.25.11 Retest Reports. Evaluation of retested repaired weld areas shall be tabulated on a new line on the report form. If the original report form is used, an R1, R2, . . . Rn shall prefix the indication number. If additional report forms are used, the R number shall prefix the report number.

8.25.12 Steel Backing. UT of CJP groove welds with steel backing shall be performed with a UT procedure that recognizes potential reflectors created by the base metal-backing interface (see Commentary C–8.25.12 for additional guidance scanning groove welds containing steel backing).

8.26 Preparation and Disposition of Reports

8.26.1 Content of Reports. A report form which clearly identifies the work and the area of inspection shall be completed by the UT operator at the time of inspection. The report form for welds that are acceptable need only contain sufficient information to identify the weld, the operator (signature), and the acceptability of the weld. An example of such a form is shown in Annex P, Form P–11.

8.26.2 Prior Inspection Reports. Before a weld subject to UT by the Contractor for the Owner is accepted, all report forms pertaining to the weld, including any that show unacceptable quality prior to repair, shall be submitted to the Inspector.

8.26.3 Completed Reports. A full set of completed report forms of welds subject to UT by the Contractor for the Owner, including any that show unacceptable quality prior to repair, shall be delivered to the Owner upon completion of the work. The Contractor's obligation to retain UT reports shall cease:

(1) upon delivery of this full set to the Owner, or

(2) one full year after completion of the Contractor's work, provided that the Owner is given prior written notice.

8.27 Calibration of the UT Unit with IIW Type or Other Approved Reference Blocks (Annex G)

See 8.22 and Figures 8.13, 8.14, and 8.16.

8.27.1 Longitudinal Mode

8.27.1.1 Distance Calibration. See Annex G, G1 for alternative method.

(1) The transducer shall be set in position G on the IIW type block.

(2) The instrument shall be adjusted to produce indications at 1 in [25 mm on a metric block], 2 in [50 mm on a metric block], 3 in [75 mm on a metric block], 4 in [100 mm on a metric block], etc., on the display.

8.27.1.2 Amplitude. See Annex G, G1.2 for alternative method.

(1) The transducer shall be set in position G on the IIW type block.

(2) The gain shall be adjusted until the maximized indication from first back reflection attains 50% to 75% screen height.

8.27.1.3 Resolution

(1) The transducer shall be set in position F on the IIW type block.

(2) Transducer and instrument shall resolve all three distances.

8.27.1.4 Horizontal Linearity Qualification. Qualification procedure shall be per 8.23.1.

8.27.1.5 Gain Control (Attenuation) Qualification. The qualification procedure shall be in conformance with 8.23.2 or an alternative method, in conformance with 8.23.2, shall be used.

8.27.2 Shear Wave Mode (Transverse)

8.27.2.1 Index Point. The transducer sound entry point (index point) shall be located or checked by the following procedure:

(1) The transducer shall be set in position D on the IIW type block.

(2) The transducer shall be moved until the signal from the radius is maximized. The point on the transducer which aligns with the radius line on the calibration block is the point of sound entry (see Annex G, G2.1 for alternative method).

8.27.2.2 Angle. The transducer sound-path angle shall be checked or determined by one of the following procedures:

(1) The transducer shall be set in position B on IIW type block for angles 40° through 60°, or in position C on IIW type block for angles 60° through 70° (see Figure 8.16).

(2) For the selected angle, the transducer shall be moved back and forth over the line indicative of the transducer angle until the signal from the radius is maximized. The sound entry point on the transducer shall be compared with the angle mark on the calibration block (tolerance ± 2°) (see Annex G, G2.2 for alternative methods).

8.27.2.3 Distance Calibration Procedure. The transducer shall be set in position D on an IIW type block (any angle). The instrument shall then be adjusted to attain one indication at 4 in [100 mm on a metric block] and a second indication at 8 in [200 mm on a metric block] or 9 in [225 mm on a metric block] (see Annex G, G2.3 for alternative methods).

8.27.2.4 Amplitude or Sensitivity Calibration Procedure. The transducer shall be set in position A on the IIW type block (any angle). The maximized signal shall then be adjusted from the 0.060 in [1.59 mm] hole to attain a horizontal reference-line height indication (see Annex G, G2.4 for alternative method). The maximum decibel reading obtained shall be used as the "Reference Level, b" reading on the Test Report sheet (Annex P, Form P–11) in conformance with 8.22.1.

8.27.2.5 Resolution

(1) The transducer shall be set on resolution block RC position Q for 70° angle, position R for 60° angle, or position S for 45° angle.

(2) Transducer and instrument shall resolve the three test holes, at least to the extent of distinguishing the peaks of the indications from the three holes.

8.27.2.6 Approach Distance of Search Unit. The minimum allowable distance between the toe of the search unit and the edge of IIW type block shall be as follows (see Figure 8.12):

for 70° transducer	X = 2 in [50 mm]
for 60° transducer	X = 1–7/16 in [37 mm]
for 45° transducer	X = 1 in [25 mm]

8.28 Equipment Qualification Procedures

8.28.1 Horizontal Linearity Procedure. *NOTE: Since this qualification procedure is performed with a straight-beam search unit which produces longitudinal waves with a sound velocity of almost double that of shear waves, it is necessary to double the shear wave distance ranges to be used in applying this procedure.* Example: The use of a 10 in [250 mm] screen calibration in shear wave would require a 20 in [500 mm] screen calibration for this qualification procedure.

The following procedure shall be used for instrument qualification (see Annex G, G3, for alternative method):

(1) A straight-beam search unit shall be coupled meeting the requirements of 8.21.6 to the IIW type block or DS block in Position G, T, or U (see Figure 8.16) as necessary to attain five back reflections in the qualification range being certified (see Figure 8.16).

(2) The first and fifth back reflections shall be adjusted to their proper locations with use of the distance calibration and zero delay adjustments.

(3) Each indication shall be adjusted to reference level with the gain or attenuation control for horizontal location examination.

(4) Each intermediate trace deflection location shall be correct within 2% of the screen width 8.28.2.

8.28.2 dB Accuracy

8.28.2.1 Procedure. *NOTE: In order to attain the required accuracy (± 1%) in reading the indication height, the display shall be graduated vertically at 2% intervals, or 2.5% for instruments with digital amplitude readout, at horizontal mid-screen height. These graduations shall be placed on the display between 60% and 100% of screen height. This may be accomplished with use of a graduated transparent screen overlay. If this overlay is applied as a permanent part of the UT unit, care should be taken that the overlay does not obscure normal testing displays.*

(1) A straight-beam search unit shall be coupled, meeting the requirements of 8.21.6 to the DS block shown in Figure 8.14 and position "T," Figure 8.16.

(2) The distance calibration shall be adjusted so that the first 2 in [50 mm] back reflection indication (hereafter called *the indication*) is at horizontal mid-screen.

(3) The calibrated gain or attenuation control shall be adjusted so that the indication is exactly at or slightly above 40% screen height.

(4) The search unit shall be moved toward position U, see Figure 8.16, until the indication is at exactly 40% screen height.

(5) The sound amplitude shall be increased 6 dB with the calibrated gain or attenuation control. The indication level theoretically should be exactly at 80% screen height.

(6) The dB reading shall be recorded under "a" and actual % screen height under "b" from step 5 on the certification report (Annex P, Form P–8), Line 1.

(7) The search unit shall be moved further toward position U, Figure 8.16, until the indication is at exactly 40% screen height.

(8) Step 5 shall be repeated.

(9) Step 6 shall be repeated; except, information should be applied to the next consecutive line on Annex P, Form P–8.

(10) Steps 7, 8, and 9 shall be repeated consecutively until the full range of the gain control (attenuator) is reached (60 dB minimum).

(11) The information from Rows "a" and "b" shall be applied to equation 8.28.2.2 or the nomograph described in 8.28.2.3 to calculate the corrected dB.

(12) Corrected dB from step 11 to Row "c" shall be applied.

(13) Row "c" value shall be subtracted from Row "a" value and the difference in Row "d," dB error shall be applied.

NOTE: These values may be either positive or negative and so noted. Examples of Application of Forms P–8, P–9, and P–10 are found in Annex P.

(14) Information shall be tabulated on a form, including minimum equivalent information as displayed on Form P–8, and the unit evaluated in conformance with instructions shown on that form.

(15) Form P–9 provides a relatively simple means of evaluating data from item (14). Instructions for this evaluation are given in items (16) through (18).

(16) The dB information from Row "e" (Form P–8) shall be applied vertically and dB reading from Row "a" (Form P–8) horizontally as X and Y coordinates for plotting a dB curve on Form P–9.

(17) The longest horizontal length, as represented by the dB reading difference, which can be inscribed in a rectangle representing 2 dB in height, denotes the dB range in which the equipment meets the code requirements. The minimum allowable range is 60 dB.

(18) Equipment that does not meet this minimum requirement may be used, provided correction factors are developed and used for discontinuity evaluation outside the instrument acceptable linearity range, or the weld testing and discontinuity evaluation is kept within the acceptable vertical linearity range of the equipment.

NOTE: The dB error figures (Row "d") may be used as correction factor figures.

8.28.2.2 Decibel Equation. The following equation shall be used to calculate decibels:

$$dB_2 - dB_1 = 20 \times \text{Log} \frac{\%_2}{\%_1}$$

$$dB_2 = 20 \times \text{Log} \frac{\%_2}{\%_1} + dB_1$$

As related to Annex P, From P-8

dB_1 = Row "a"
dB_2 = Row "c"
$\%_1$ = Row "b"
$\%_2$ = Defined on From P-8

8.28.2.3 Annex P. The following notes apply to the use of the nomograph in Annex P Form P–10:

(1) Rows a, b, c, d, and e are on certification sheet, Annex P, Form P–8.

(2) The A, B, and C scales are on the nomograph, Annex P, Form P–10.

(3) The zero points on the C scale shall be prefixed by adding the necessary value to correspond with the instrument settings; i.e., 0, 10, 20, 30, etc.

8.28.2.4 Procedure. The following procedures shall apply to the use of the nomograph in Annex P, Form P–10:

(1) A straight line between the decibel reading from Row "a" applied to the C scale and the corresponding percentage from Row "b" applied to the A scale shall be extended.

(2) The point where the straight line from step 1 crosses the pivot line B as a pivot point for a second straight line shall be used.

(3) A second straight line from the average % point on the A scale through the pivot point developed in step 2 and on to the dB scale C shall be extended.

(4) This point on the C scale is indicative of the corrected dB for use in Row "c."

8.28.2.5 Nomograph. For an example of the use of the nomograph, see Annex P, Form P–10.

8.28.3 Internal Reflections Procedure

(1) Calibrate the equipment in conformance with 8.24.5.

(2) Remove the search unit from the calibration block without changing any other equipment adjustments.

(3) Increase the calibrated gain or attenuation 20 dB more sensitive than reference level.

(4) The screen area beyond 1/2 in [12 mm] sound path and above reference level height shall be free of any indication.

8.29 Discontinuity Size Evaluation Procedures

8.29.1 Straight-Beam (Longitudinal) Testing. The size of lamellar discontinuities is not always easily determined, especially those that are smaller than the transducer size. When the discontinuity is larger than the transducer, a full loss of back reflection will occur and a 6 dB loss of amplitude and measurement to the centerline of the transducer is usually reliable for determining discontinuity edges. However, the approximate size evaluation of those reflectors, which are smaller than the transducer, shall be made by beginning outside of the discontinuity with equipment calibrated in conformance with 8.24.4 and moving the transducer toward the area of discontinuity until an indication on the display begins to form. The leading edge of the search unit at this point is indicative of the edge of the discontinuity.

8.29.2 Angle-Beam (Shear) Testing. The following procedure shall be used to determine lengths of indications which have dB ratings more serious than for a Class D indication. The length of such indication shall be determined by measuring the distance between the transducer centerline locations where the indication rating amplitude drops 50% (6 dB) below the rating for the applicable discontinuity classification. This length shall be recorded under "discontinuity length" on the test report. Where warranted by discontinuity amplitude, this procedure shall be repeated to determine the length of Class A, B, and C discontinuities.

8.30 Scanning Patterns (See Figure 8.15)

8.30.1 Longitudinal Discontinuities

8.30.1.1 Scanning Movement A. Rotation angle a = 10°.

8.30.1.2 Scanning Movement B. Scanning distance b shall be such that the section of weld being tested is covered.

8.30.1.3 Scanning Movement C. Progression distance c shall be approximately one-half the transducer width.

NOTE: Movements A, B, and C may be combined into one scanning pattern.

8.30.2 Transverse Discontinuities

8.30.2.1 Ground Welds. Scanning pattern D shall be used when welds are ground flush.

8.30.2.2 Unground Welds. Scanning pattern E shall be used when the weld reinforcement is not ground flush. Scanning angle e = 15° max.

NOTE: The scanning pattern shall cover the full weld section.

8.30.3 ESW or EGW Welds (Additional Scanning Pattern). Scanning Pattern E Search unit rotation angle e between 45° and 60°.

NOTE: The scanning pattern shall cover the full weld section.

8.31 Examples of dB Accuracy Certification

Annex P shows examples of the use of Forms P–8, P–9, and P–10 for the solution to a typical application of 8.28.2.

Part G
Other Examination Methods

8.32 General Requirements

This part contains NDT methods not addressed in Parts D, E, or Part F of Clause 8, or Clause 10, Part F for tubulars of this code. The NDT methods set forth in part G may be used as an alternative to the methods outlined in Parts D, E, or F of Clause 8, or Clause 10, Part F for tubulars, providing procedures, qualification criteria for procedures and personnel, and acceptance criteria are documented in writing and approved by the Engineer.

8.33 Radiation Imaging Systems

Examination of welds may be performed using ionizing radiation methods other than RT, such as electronic imaging, including real-time imaging systems. Sensitivity of such examination as seen on the monitoring equipment (when used for acceptance and rejection) and the recording medium shall be no less than that required for RT.

8.33.1 Procedures. Written procedures shall contain the following essential variables:

(1) Equipment identification including manufacturer, make, model, and serial number,

(2) Radiation and imaging control setting for each combination of variables established herein,

(3) Weld thickness ranges,

(4) Weld joint types,

(5) Scanning speed,

(6) Radiation source to weld distance,

(7) Image conversion screen to weld distance,

(8) Angle of X-rays through the weld (from normal),

(9) IQI location (source side or screen side),

(10) Type of recording medium (video recording, photographic still film, photographic movie film, or other acceptable mediums),

(11) Computer enhancement (if used),

(12) Width of radiation beam,

(13) Indication characterization protocol and acceptance criteria, if different from this code.

8.33.2 IQI. The wire-type IQI, as described in Part E, shall be used. IQI placement shall be as specified in Part E of this clause, or Clause 10, Part F for tubulars for static examination. For in-motion examination, placement shall be as follows:

(1) Two IQIs positioned at each end of area of interest and tracked with the run,

(2) One IQI at each end of the run and positioned at a distance no greater than 10 ft. [3 m] between any two IQIs during the run.

8.34 Advanced Ultrasonic Systems

Advanced Ultrasonic Systems includes, but are not limited to, multiple probe, multi-channel systems, automated inspection, time-of-flight diffraction (TOFD), and phased array systems.

8.34.1 Procedures. Written procedures shall contain the following essential variables:

(1) Equipment identification including manufacturer, make, model, and serial numbers;

(2) Type of probes, including size, shape, angle, and frequency (MHz);

(3) Scanning control settings for each combination of variables established herein;

(4) Setup and calibration procedure for equipment and probes using industry standards or workmanship samples;

(5) Weld thickness ranges;

(6) Weld joint type;

(7) Scanning speeds;

(8) Number of probes;

(9) Scanning angle;

(10) Type of scan (A, B, C, other);

(11) Type of recording medium (video recording, computer assisted, or other acceptable mediums);

(12) Computer based enhancement (if used);

(13) Identification of computer software (if used); and

(14) Indication characterization protocol and acceptance criteria, if different from this code.

8.35 Additional Requirements

8.35.1 Procedure Qualification. Procedures shall be qualified by testing the NDT method (system) and recording medium to establish and record all essential variables and conditions. Qualification testing shall consist of determining that each combination of the essential variables or ranges of variables can provide the minimum required sensitivity. Test results shall be recorded on the recording medium that is to be used for production examination. Procedures shall be approved by an individual qualified as ASNT SNT-TC–1A, Level III (see 8.35.2).

8.35.2 Personnel Qualifications. In addition to the personnel qualifications of 8.14.6, the following shall apply.

(1) Level III—shall have minimum of six months experience using the same or similar equipment and procedures for examination of welds in structural or piping metallic materials.

(2) Levels I and II—shall be certified by the Level III above and have a minimum of three months experience using the same or similar equipment and procedures for examination of welds in structural or piping metallic materials. Qualification shall consist of written and practical examinations for demonstrating capability to use the equipment and procedures to be used for production examination.

8.35.3 Image Enhancement. Computer enhancement of the recording images shall be acceptable for improving the recorded image and obtaining additional information, providing required minimum sensitivity and accuracy of characterizing discontinuities are maintained. Computer enhanced images shall be clearly marked that enhancement was used and enhancement procedures identified.

8.35.4 Records—Radiation Imaging Examinations. Examinations, which are used for acceptance or rejection of welds, shall be recorded on an acceptable medium. The record shall be in-motion or static, whichever is used to accept or reject the welds. A written record shall be included with the recorded images giving the following information as a minimum:

(1) Identification and description of welds examined

(2) Procedure(s) used

(3) Equipment used

(4) Location of the welds within the recorder medium

(5) Results, including a list of unacceptable welds and repairs and their locations within the recorded medium.

Table 8.1
Visual Inspection Acceptance Criteria (see 8.9)

Discontinuity Category and Inspection Criteria	Statically Loaded Nontubular Connections	Cyclically Loaded Nontubular Connections
(1) Crack Prohibition Any crack shall be unacceptable, regardless of size or location.	X	X
(2) Weld/Base Metal Fusion Complete fusion shall exist between adjacent layers of weld metal and between weld metal and base metal.	X	X
(3) Crater Cross Section All craters shall be filled to provide the specified weld size, except for the ends of intermittent fillet welds outside of their effective length.	X	X
(4) Weld Profiles Weld profiles shall be in conformance with 7.23.	X	X
(5) Time of Inspection Visual inspection of welds in all steels may begin immediately after the completed welds have cooled to ambient temperature. Acceptance criteria for ASTM A514, A517, and A709 Grade HPS 100W [HPS 690W] steels shall be based on visual inspection performed not less than 48 hours after completion of the weld.	X	X

(6) Undersized Welds
The size of a fillet weld in any continuous weld may be less than the specified nominal size (L) without correction by the following amounts (U):
    L,   U,
specified nominal weld size, in [mm]   allowable decrease from L, in [mm]
    ≤ 3/16 [5]   ≤ 1/16 [2]
    1/4 [6]   ≤ 3/32 [2.5]
    ≥ 5/16 [8]   ≤ 1/8 [3]
In all cases, the undersize portion of the weld shall not exceed 10% of the weld length. On web-to-flange welds on girders, underrun shall be prohibited at the ends for a length equal to twice the width of the flange.
	X	X

(7) Undercut (A) For material less than 1 in [25 mm] thick, undercut shall not exceed 1/32 in [1 mm], with the following exception: undercut shall not exceed 1/16 in [2 mm] for any accumulated length up to 2 in [50 mm] in any 12 in [300 mm]. For material equal to or greater than 1 in [25 mm] thick, undercut shall not exceed 1/16 in [2 mm] for any length of weld.	X	
(B) In primary members, undercut shall be no more than 0.01 in [0.25 mm] deep when the weld is transverse to tensile stress under any design loading condition. Undercut shall be no more than 1/32 in [1 mm] deep for all other cases.		X
(8) Porosity (A) CJP groove welds in butt joints transverse to the direction of computed tensile stress shall have no visible piping porosity. For all other groove welds and for fillet welds, the sum of the visible piping porosity 1/32 in [1 mm] or greater in diameter shall not exceed 3/8 in [10 mm] in any linear inch of weld and shall not exceed 3/4 in [20 mm] in any 12 in [300 mm] length of weld.	X	
(B) The frequency of piping porosity in fillet welds shall not exceed one in each 4 in [100 mm] of weld length and the maximum diameter shall not exceed 3/32 in [2.5 mm]. Exception: for fillet welds connecting stiffeners to web, the sum of the diameters of piping porosity shall not exceed 3/8 in [10 mm] in any linear inch of weld and shall not exceed 3/4 in [20 mm] in any 12 in [300 mm] length of weld.		X
(C) CJP groove welds in butt joints transverse to the direction of computed tensile stress shall have no piping porosity. For all other groove welds, the frequency of piping porosity shall not exceed one in 4 in [100 mm] of length and the maximum diameter shall not exceed 3/32 in [2.5 mm].		X

Note: An "X" indicates applicability for the connection type; a shaded area indicates non-applicability.

Table 8.2
UT Acceptance-Rejection Criteria (Statically Loaded Nontubular Connections and Cyclically Loaded Nontubular Connections in Compression) (see 8.13.1, 8.13.2(2), and C–8.25.6)

Discontinuity Severity Class	Weld Size[a] in inches [mm] and Search Unit Angle											
	5/16 through 3/4 [8–20]	> 3/4 through 1-1/2 [20–38]	> 1-1/2 through 2-1/2 [38–65]			> 2-1/2 through 4 [65–100]			> 4 through 8 [100–200]		
	70°	70°	70°	60°	45°	70°	60°	45°	70°	60°	45°
Class A	+5 & lower	+2 & lower	–2 & lower	+1 & lower	+3 & lower	–5 & lower	–2 & lower	0 & lower	–7 & lower	–4 & lower	–1 & lower
Class B	+6	+3	–1 0	+2 +3	+4 +5	–4 –3	–1 0	+1 +2	–6 –5	–3 –2	0 +1
Class C	+7	+4	+1 +2	+4 +5	+6 +7	–2 to +2	+1 +2	+3 +4	–4 to +2	–1 to +2	+2 +3
Class D	+8 & up	+5 & up	+3 & up	+6 & up	+8 & up	+3 & up	+3 & up	+5 & up	+3 & up	+3 & up	+4 & up

[a] Weld size in butt joints shall be the nominal thickness of the thinner of the two parts being joined.

Notes:
1. Class B and C discontinuities shall be separated by at least 2L, L being the length of the longer discontinuity, except that when two or more such discontinuities are not separated by at least 2L, but the combined length of discontinuities and their separation distance is equal to or less than the maximum allowable length under the provisions of Class B or C, the discontinuity shall be considered a single acceptable discontinuity.
2. Class B and C discontinuities shall not begin at a distance less than 2L from weld ends carrying primary tensile stress, L being the discontinuity length.
3. Discontinuities detected at "scanning level" in the root face area of CJP double groove weld joints shall be evaluated using an indication rating 4 dB more sensitive than described in 8.25.6.5 when such welds are designated as "tension welds" on the drawing (subtract 4 dB from the indication rating "d"). This shall not apply if the weld joint is backgouged to sound metal to remove the root face and MT used to verify that the root face has been removed.
4. ESW or EGW: Discontinuities detected at "scanning level" that exceed 2 in [50 mm] in length shall be suspected as being piping porosity and shall be further evaluated with radiography.
5. For indications that remain on the display as the search unit is moved, refer to 8.13.1.

Class A (large discontinuities)
Any indication in this category shall be rejected (regardless of length).

Class B (medium discontinuities)
Any indication in this category having a length greater than 3/4 in [20 mm] shall be rejected.

Class C (small discontinuities)
Any indication in this category having a length greater than 2 in [50 mm] shall be rejected.

Class D (minor discontinuities)
Any indication in this category shall be accepted regardless of length or location in the weld.

Scanning Levels	
Sound path[b] in inches [mm]	Above Zero Reference, dB
through 2-1/2 [65 mm]	14
> 2-1/2 through 5 [65–125 mm]	19
> 5 through 10 [125–250 mm]	29
> 10 through 15 [250–380 mm]	39

[b] This column refers to sound path distance; NOT material thickness.

Table 8.3
UT Acceptance-Rejection Criteria (Cyclically Loaded Nontubular Connections in Tension) (see 8.13.2 and C–8.25.6)

Discontinuity Severity Class	Weld Size[a] in inches [mm] and Search Unit Angle										
	5/16 through 3/4 [8–20]	> 3/4 through 1–1/2 [20–38]	> 1–1/2 through 2–1/2 [38–65]			> 2–1/2 through 4 [65–100]			> 4 through 8 [100–200]		
	70°	70°	70°	60°	45°	70°	60°	45°	70°	60°	45°
Class A	+10 & lower	+8 & lower	+4 & lower	+7 & lower	+9 & lower	+1 & lower	+4 & lower	+6 & lower	–2 & lower	+1 & lower	+3 & lower
Class B	+11	+9	+5 +6	+8 +9	+10 +11	+2 +3	+5 +6	+7 +8	–1 0	+2 +3	+4 +5
Class C	+12	+10	+7 +8	+10 +11	+12 +13	+4 +5	+7 +8	+9 +10	+1 +2	+4 +5	+6 +7
Class D	+13 & up	+11 & up	+9 & up	+12 & up	+14 & up	+6 & up	+9 & up	+11 & up	+3 & up	+6 & up	+8 & up

[a] Weld size in butt joints shall be the nominal thickness of the thinner of the two parts being joined.

Notes:
1. Class B and C discontinuities shall be separated by at least 2L, L being the length of the longer discontinuity, except that when two or more such discontinuities are not separated by at least 2L, but the combined length of discontinuities and their separation distance is equal to or less than the maximum allowable length under the provisions of Class B or C, the discontinuity shall be considered a single acceptable discontinuity.
2. Class B and C discontinuities shall not begin at a distance less than 2L from weld ends carrying primary tensile stress, L being the discontinuity length.
3. Discontinuities detected at "scanning level" in the root face area of CJP double groove weld joints shall be evaluated using an indication rating 4 dB more sensitive than described in 8.25.6.5 when such welds are designated as "tension welds" on the drawing (subtract 4 dB from the indication rating "d"). This shall not apply if the weld joint is backgouged to sound metal to remove the root face and MT used to verify that the root face has been removed.
4. For indications that remain on the display as the search unit is moved, refer to 8.13.2.1.

Class A (large discontinuities)
Any indication in this category shall be rejected (regardless of length).

Class B (medium discontinuities)
Any indication in this category having a length greater than 3/4 in [20 mm] shall be rejected.

Class C (small discontinuities)
Any indication in this category having a length greater than 2 in [50 mm] in the middle half or 3/4 in [20 mm] length in the top or bottom quarter of weld thickness shall be rejected.

Class D (minor discontinuities)
Any indication in this category shall be accepted regardless of length or location in the weld.

Scanning Levels	
Sound path[b] in [mm]	Above Zero Reference, dB
through 2–1/2 [65 mm]	20
> 2–1/2 through 5 [65–125 mm]	25
> 5 through 10 [125–250 mm]	35
> 10 through 15 [250–380 mm]	45

[b] This column refers to sound path distance; NOT material thickness.

Table 8.4
Hole-Type IQI Requirements (see 8.17.1)

Nominal Material Thickness Range, in	Nominal Material Thickness Range, mm	Source Side Designation	Essential Hole
Up to 0.25 incl.	Up to 6 incl.	10	4T
Over 0.25 to 0.375	Over 6 through 10	12	4T
Over 0.375 to 0.50	Over 10 through 12	15	4T
Over 0.50 to 0.625	Over 12 through 16	15	4T
Over 0.625 to 0.75	Over 16 through 20	17	4T
Over 0.75 to 0.875	Over 20 through 22	20	4T
Over 0.875 to 1.00	Over 22 through 25	20	4T
Over 1.00 to 1.25	Over 25 through 32	25	4T
Over 1.25 to 1.50	Over 32 through 38	30	2T
Over 1.50 to 2.00	Over 38 through 50	35	2T
Over 2.00 to 2.50	Over 50 through 65	40	2T
Over 2.50 to 3.00	Over 65 through 75	45	2T
Over 3.00 to 4.00	Over 75 through 100	50	2T
Over 4.00 to 6.00	Over 100 through 150	60	2T
Over 6.00 to 8.00	Over 150 through 200	80	2T

Table 8.5
Wire IQI Requirements (see 8.17.1)

Nominal Material Thickness Range, in	Nominal Material Thickness Range, mm	Source Side Essential Wire Identity
Up to 0.25 incl.	Up to 6 incl.	6
Over 0.25 to 0.375	Over 6 to 10	7
Over 0.375 to 0.625	Over 10 to 16	8
Over 0.625 to 0.75	Over 16 to 20	9
Over 0.75 to 1.50	Over 20 to 38	10
Over 1.50 to 2.00	Over 38 to 50	11
Over 2.00 to 2.50	Over 50 to 65	12
Over 2.50 to 4.00	Over 65 to 100	13
Over 4.00 to 6.00	Over 100 to 150	14
Over 6.00 to 8.00	Over 150 to 200	16

Table 8.6
IQI Selection and Placement (see 8.17.7)

IQI Types	Equal T ≥ 10 in [250 mm] L		Equal T < 10 in [250 mm] L		Unequal T ≥ 10 in [250 mm] L		Unequal T < 10 in [250 mm] L	
	Hole	Wire	Hole	Wire	Hole	Wire	Hole	Wire
Number of IQIs Nontubular	2	2	1	1	3	2	2	1
ASTM Standard Selection—	E1025	E747	E1025	E747	E1025	E747	E1025	E747
Table	8.4	8.5	8.4	8.5	8.4	8.5	8.4	8.5
Figures	8.6		8.7		8.8		8.9	

T = Nominal base metal thickness (T1 and T2 of Figures).
L = Weld Length in area of interest of each radiograph.
Note: T may be increased to provide for the thickness of allowable weld reinforcement provided shims are used under hole IQIs per 8.17.3.3.

Table 8.7
Testing Angle (see 8.25)

	Procedure Chart																	
	Material Thickness, in [mm]																	
Application	5/16 [8] to 1–1/2 [38]		> 1–1/2 [38] to 1–3/4 [45]		> 1–3/4 [45] to 2–1/2 [65]		> 2–1/2 [65] to 3–1/2 [90]		> 3–1/2 [90] to 4–1/2 [110]		> 4–1/2 [110] to 5 [130]		> 5 [130] to 6–1/2 [160]		> 6–1/2 [160] to 7 [180]		>7 [180] to 8 [200]	
Butt Joint	1	O	1	*	1G or 4	* F	1G or 5	* F	6 or 7	* F	8 or 10	* F	9 or 11	* F	12 or 13	* F	12	* F
T-Joint	1	O	1	F or XF	4	F or XF	5	F or XF	7	F or XF	10	F or XF	11	F or XF	13	F or XF	—	—
Corner Joint	1	O	1	F or XF	1G or 4	F or XF	1G or 5	F or XF	6 or 7	F or XF	8 or 10	F or XF	9 or 11	F or XF	13 or 14	F or XF	—	—
ESW/EGW Welds	1	O	1	O	1G or 4	1**	1G or 5	P1 or P3	6 or 7	P3	11 or 15	P3	11 or 15	P3	11 or 15	P3	11 or 15**	P3

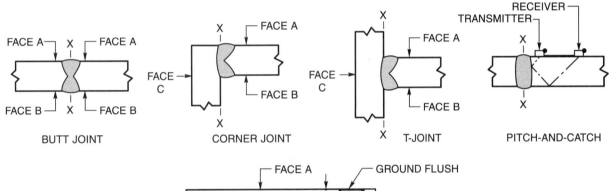

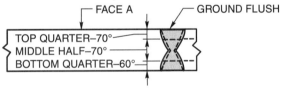

Notes:
1. Where possible, all examinations shall be made from Face A and in Leg 1, unless otherwise specified in this table.
2. Root areas of single groove weld joints that have backing not requiring removal by contract, shall be tested in Leg 1, where possible, with Face A being that opposite the backing. (Grinding of the weld face or testing from additional weld faces may be necessary to permit complete scanning of the weld root.)
3. Examination in Leg II or III shall be made only to satisfy provisions of this table or when necessary to test weld areas made inaccessible by an unground weld surface, or interference with other portions of the weldment, or to meet the requirements of 8.25.6.2.
4. A maximum of Leg III shall be used only where thickness or geometry prevents scanning of complete weld areas and HAZs in Leg I or Leg II.
5. On tension welds in cyclically loaded structures, the top quarter of thickness shall be tested with the final leg of sound progressing from Face B toward Face A, the bottom quarter of thickness shall be tested with the final leg of sound progressing from Face A toward Face B; i.e., the top quarter of thickness shall be tested either from Face A in Leg II or from Face B in Leg I at the Contractor's option, unless otherwise specified in the contract documents.
6. The weld face indicated shall be ground flush before using procedure 1G, 6, 8, 9, 12, 14, or 15. Face A for both connected members shall be in the same plane.

(See Legend on next page)

Table 8.7 (Continued)
Testing Angle (see 8.25)

Legend:
X — Check from Face C.
G — Grind weld face flush.
O — Not required.
* — Required only where display reference height indication of discontinuity is noted at the weld metal-base metal interface while searching at scanning level with primary procedures selected from first column.
** — Use 15 in [400 mm] or 20 in [500 mm] screen distance calibration.
P — Pitch and catch shall be conducted for further discontinuity evaluation in only the middle half of the material thickness with only 45° or 70° transducers of equal specification, both facing the weld. (Transducers must be held in a fixture to control positioning—see sketch.) Amplitude calibration for pitch and catch is normally made by calibrating a single search unit. When switching to dual search units for pitch and catch inspection, there should be assurance that this calibration does not change as a result of instrument variables.
F — Weld metal-base metal interface indications shall be further evaluated with either 70°, 60°, or 45° transducer—whichever sound path is nearest to being perpendicular to the suspected fusion surface.
Face A — the face of the material from which the initial scanning is done (on T-and corner joints, follow above sketches).
Face B — opposite Face A (same plate).
Face C — the face opposite the weld on the connecting member or a T-or corner joint.

Procedure Legend

No.	Area of Weld Thickness		
	Top Quarter	Middle Half	Bottom Quarter
1	70°	70°	70°
2	60°	60°	60°
3	45°	45°	45°
4	60°	70°	70°
5	45°	70°	70°
6	70° G Face A	70°	60°
7	60° Face B	70°	60°
8	70° G Face A	60°	60°
9	70° G Face A	60°	45°
10	60° Face B	60°	60°
11	45° Face B	70°**	45°
12	70° G Face A	45°	70° G Face B
13	45° Face B	45°	45°
14	70° G Face A	45°	45°
15	70° G Face A	70° A Face B	70° G Face B

Table 8.8
UT Equipment Qualification and Calibration Requirements (see 8.21.1)

		Type of Qualification or Calibration Activity		Minimum Frequency	
		Description	Code Clause	Minimum Frequency	Code Clause
Equipment Qualification Procedures	Instruments	Horizontal Linearity	8.28.1	2 months	8.23.1
		Gain Control/dB Accuracy	8.28.2	2 months	8.23.2
	Search Units	Internal Reflections	8.28.3	40 hours of use[a]	8.23.3
		Angle Beam Search Units (Index Point, Angle)	8.27.2.1 8.27.2.2	8 hours of use[a]	8.23.4
	Instrument/Search Unit Combinations	Resolution (Angle Beam)	8.22.3 8.27.2.5	Prior to initial use[b]	8.22.3
		Resolution (Straight Beam)	8.27.1.3	Prior to initial use[b]	8.22.3
Calibration for Testing	Straight Beam (for Base Material Testing)	Range	8.24.4.1 or 8.27.1.1	Just prior to and at the location of the first weld tested[c]	8.24.2
		Sensitivity	8.24.4.2 or 8.27.1.2		
	Angle Beams	Range	8.24.5.1 8.27.2.3		
		Sensitivity	8.24.5.2 8.27.2.4		
		Index Point	8.27.2.1		
		Angle	8.27.2.2		
	Straight Beam and Angle Beam	Recalibration	8.24.3	2 hours[d]	8.24.3

[a] Must be performed for each search unit.
[b] Must be performed for each combination of search unit (transducer and shoe) and instrument prior to initial use.
[c] After the requirements of 8.24.2 are met the recalibration requirements of 8.24.3 shall apply.
[d] Or when electrical circuitry is disturbed in any way which includes the following:
   (1) Transducer Change
   (2) Battery Change
   (3) Electrical outlet change
   (4) Coaxial cable change
   (5) Power outage (failure)

Legend for Figures 8.1, 8.2, and 8.3

Dimensions of Discontinuities

B = Maximum allowed dimension of a radiographed discontinuity.
L = Largest dimension of a radiographed discontinuity.
L' = Largest dimension of adjacent discontinuities.
C = Minimum clearance measured along the longitudinal axis of the weld between edges of porosity or fusion-type discontinuities (larger of adjacent discontinuities governs), or to an edge or an end of an intersecting weld.
C1 = Minimum allowed distance between the nearest discontinuity to the free edge of a plate or tubular, or the intersection of a longitudinal weld with a girth weld, measured parallel to the longitudinal weld axis.
W = Smallest dimension of either of adjacent discontinuities.

Material Dimensions

S = Weld size
T = Plate or pipe thickness for CJP groove welds

Definitions of Discontinuities

- An elongated discontinuity shall have the largest dimension (L) exceed 3 times the smallest dimension.
- A rounded discontinuity shall have the largest dimension (L) less than or equal to 3 times the smallest dimension.
- A cluster shall be defined as a group of nonaligned, acceptably-sized, individual adjacent discontinuities with spacing less than the minimum allowed (C) for the largest individual adjacent discontinuity (L'), but with the sum of the greatest dimensions (L) of all discontinuities in the cluster equal to or less than the maximum allowable individual discontinuity size (B). Such clusters shall be considered as individual discontinuities of size L for the purpose of assessing minimum spacing.
- Aligned discontinuities shall have the major axes of each discontinuity approximately aligned.

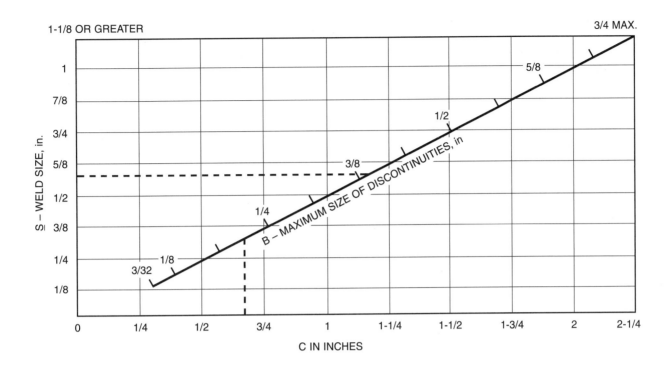

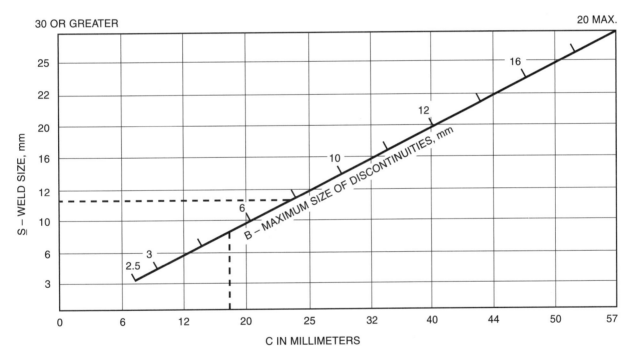

Notes:
1. To determine the maximum size of discontinuity allowed in any joint or weld size, project $\underline{S}$ horizontally to B.
2. To determine the minimum clearance allowed between edges of discontinuities of any size greater than or equal to 3/32 in [2.5 mm], project B vertically to C.
3. See Legend on page 246 for definitions.

Figure 8.1—Discontinuity Acceptance Criteria for Statically Loaded Nontubular and Statically or Cyclically Loaded Tubular Connections (see 8.12.1 and 10.25.2 for tubulars)

KEY FOR FIGURE 8.1, CASES I, II, III, AND IV

DISCONTINUITY A = ROUNDED OR ELONGATED DISCONTINUITY LOCATED IN WELD A
DISCONTINUITY B = ROUNDED OR ELONGATED DISCONTINUITY LOCATED IN WELD B
L AND W = LARGEST AND SMALLEST DIMENSIONS, RESPECTIVELY, OF DISCONTINUITY A
L' AND W' = LARGEST AND SMALLEST DIMENSIONS, RESPECTIVELY, OF DISCONTINUITY B
S = WELD SIZE
C_1 = SHORTEST DISTANCE PARALLEL TO THE WELD A AXIS, BETWEEN THE NEAREST DISCONTINUITY EDGES

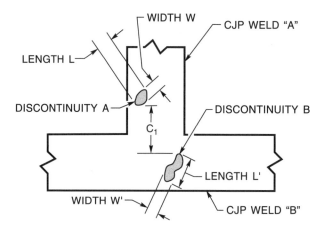

CASE I DISCONTINUITY LIMITATIONS[a]

DISCONTINUITY DIMENSION	LIMITATIONS	CONDITIONS
L	< S/3, ≤ 1/4 in [6 mm]	S ≤ 2 in [50 mm]
	≤ 3/8 in [10 mm]	S > 2 in [50 mm]
C_1	≥ 3L	(A) ONE DISCONTINUITY ROUNDED, THE OTHER ROUNDED OR ELONGATED[a] (B) L ≥ 3/32 in [2.5 mm]

[a] The elongated discontinuity may be located in either weld "A" or "B." For the purposes of this illustration the elongated discontinuity "B" was located in weld "B."

Case I—Discontinuity at Weld Intersection

Figure 8.1 (Continued)—Discontinuity Acceptance Criteria for Statically Loaded Nontubular and Statically or Cyclically Loaded Tubular Connections (see 8.12.1 and 10.25.2 for tubulars)

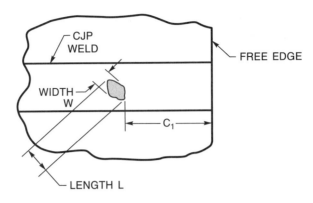

CASE II DISCONTINUITY LIMITATIONS

DISCONTINUITY DIMENSION	LIMITATIONS	CONDITIONS
L	< $S/3$, ≤ 1/4 in [6 mm]	S ≤ 2 in [50 mm]
L	≤ 3/8 in [10 mm]	S > 2 in [50 mm]
C_1	≥ 3L	L ≥ 3/32 in [2.5 mm]

Case II—Discontinuity at Free Edge of CJP Groove Weld

Figure 8.1 (Continued)—Discontinuity Acceptance Criteria for Statically Loaded Nontubular and Statically or Cyclically Loaded Tubular Connections (see 8.12.1 and 10.25.2 for tubulars)

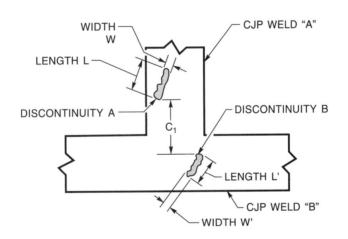

CASE III DISCONTINUITY LIMITATIONS

DISCONTINUITY DIMENSION	LIMITATIONS	CONDITIONS
L	≤ $S/3$	L/W > 3
C_1	> 3L OR 2S, WHICHEVER IS GREATER	L ≥ 3/32 in [2.5 mm]

Case III—Discontinuity at Weld Intersection

Figure 8.1 (Continued)—Discontinuity Acceptance Criteria for Statically Loaded Nontubular and Statically or Cyclically Loaded Tubular Connections (see 8.12.1 and 10.25.2 for tubulars)

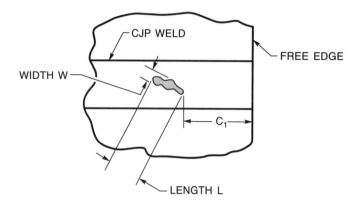

CASE IV DISCONTINUITY LIMITATIONS

DISCONTINUITY DIMENSION	LIMITATIONS	CONDITIONS
L	≤ S/3	L/W > 3
C_1	≥ 3L OR 2S, WHICHEVER IS GREATER	L ≥ 3/32 in [2.5 mm]

Case IV—Discontinuity at Free Edge of CJP Groove Weld

Figure 8.1 (Continued)—Discontinuity Acceptance Criteria for Statically Loaded Nontubular and Statically or Cyclically Loaded Tubular Connections (see 8.12.1 and 10.25.2 for tubulars)

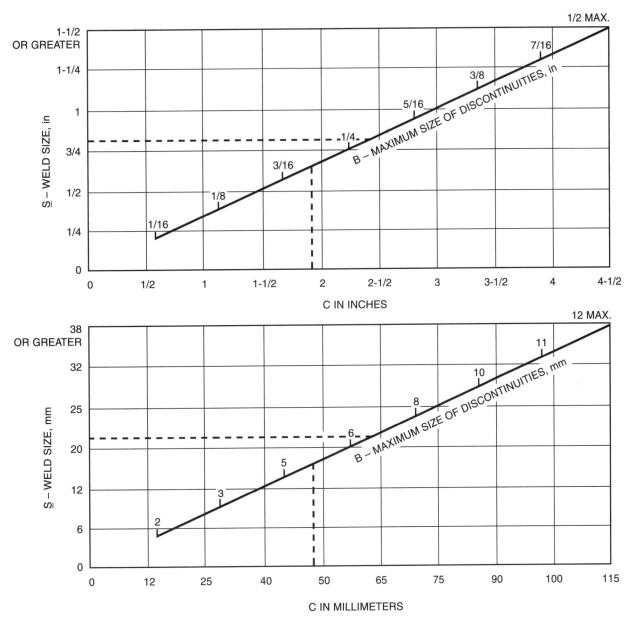

Notes:
1. To determine the maximum size of discontinuity allowed in any joint or weld size, project S horizontally to B.
2. To determine the minimum clearance allowed between edges of discontinuities of any size, project B vertically to C.
3. See Legend on page 246 for definitions.

Figure 8.2—Discontinuity Acceptance Criteria for Cyclically Loaded Nontubular Connections in Tension (Limitations of Porosity and Fusion Discontinuities) (see 8.12.2.1)

KEY FOR FIGURE 8.2, CASES I, II, III AND IV

DISCONTINUITY A = ROUNDED OR ELONGATED DISCONTINUITY LOCATED IN WELD A
DISCONTINUITY B = ROUNDED OR ELONGATED DISCONTINUITY LOCATED IN WELD B
L AND W = LARGEST AND SMALLEST DIMENSIONS, RESPECTIVELY, OF DISCONTINUITY A
L' AND W' = LARGEST AND SMALLEST DIMENSIONS, RESPECTIVELY, OF DISCONTINUITY B
S = WELD SIZE
C_1 = SHORTEST DISTANCE PARALLEL TO THE WELD A AXIS, BETWEEN THE NEAREST DISCONTINUITY EDGES

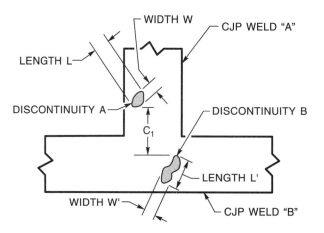

CASE I DISCONTINUITY LIMITATIONS[a]

DISCONTINUITY DIMENSION	LIMITATIONS	CONDITIONS
L	SEE FIGURE 8.2 GRAPH (B DIMENSION)	L ≥ 1/16 in [2 mm]
C_1	SEE FIGURE 8.2 GRAPH (C DIMENSION)	—

[a] The elongated discontinuity may be located in either weld "A" or "B." For the purposes of this illustration the elongated discontinuity "B" was located in weld "B."

Case I—Discontinuity at Weld Intersection

Figure 8.2 (Continued)—Discontinuity Acceptance Criteria for Cyclically Loaded Nontubular Connections in Tension (Limitations of Porosity and Fusion Discontinuities) (see 8.12.2.1)

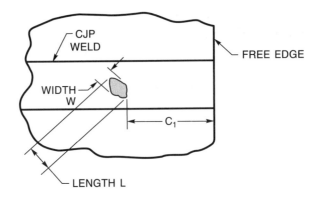

CASE II DISCONTINUITY LIMITATIONS

DISCONTINUITY DIMENSION	LIMITATIONS	CONDITIONS
L	SEE FIGURE 8.2 GRAPH (B DIMENSION)	L ≥ 1/16 in [2 mm]
C_1	SEE FIGURE 8.2 GRAPH (C DIMENSION)	—

Case II—Discontinuity at Free Edge of CJP Groove Weld

Figure 8.2 (Continued)—Discontinuity Acceptance Criteria for Cyclically Loaded Nontubular Connections in Tension (Limitations of Porosity and Fusion Discontinuities) (see 8.12.2.1)

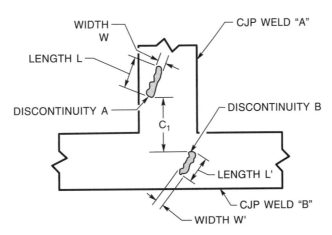

CASE III DISCONTINUITY LIMITATIONS

DISCONTINUITY DIMENSION	LIMITATIONS	CONDITIONS
L	SEE FIGURE 8.2 GRAPH (B DIMENSION)	L ≥ 1/16 in [2 mm]
C_1	SEE FIGURE 8.2 GRAPH (C DIMENSION)	—

Case III—Discontinuity at Weld Intersection

Figure 8.2 (Continued)—Discontinuity Acceptance Criteria for Cyclically Loaded Nontubular Connections in Tension (Limitations of Porosity and Fusion Discontinuities) (see 8.12.2.1)

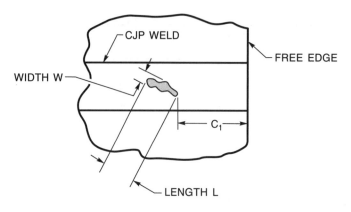

CASE IV DISCONTINUITY LIMITATIONS

DISCONTINUITY DIMENSION	LIMITATIONS	CONDITIONS
L	SEE FIGURE 8.2 GRAPH (B DIMENSION)	L ≥ 1/16 in [2 mm]
C_1	SEE FIGURE 8.2 GRAPH (C DIMENSION)	—

Case IV—Discontinuity at Free Edge of CJP Groove Weld

Figure 8.2 (Continued)—Discontinuity Acceptance Criteria for Cyclically Loaded Nontubular Connections in Tension (Limitations of Porosity and Fusion Discontinuities) (see 8.12.2.1)

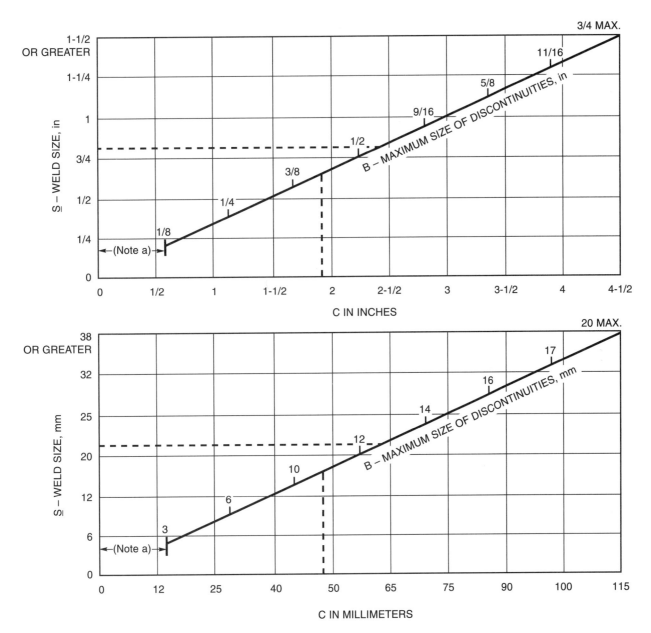

a The maximum size of a discontinuity located within this distance from an edge of plate shall be 1/8 in [3 mm], but a 1/8 in [3 mm] discontinuity shall be 1/4 in [6 mm] or more away from the edge. The sum of discontinuities less than 1/8 in [3 mm] in size and located within this distance from the edge shall not exceed 3/16 in [5 mm]. Discontinuities 1/16 in [2 mm] to less than 1/8 in [3 mm] shall not be restricted in other locations unless they are separated by less than 2 L (L being the length of the larger discontinuity); in which case, the discontinuities shall be measured as one length equal to the total length of the discontinuities and space and evaluated as shown in this figure.

Notes:
1. To determine the maximum size of discontinuity allowed in any joint or weld size, project $\underline{S}$ horizontally to B.
2. To determine the minimum clearance allowed between edges of discontinuities of any size, project B vertically to C.
3. See Legend on page 246 for definitions.

Figure 8.3—Discontinuity Acceptance Criteria for Cyclically Loaded Nontubular Connections in Compression (Limitations of Porosity or Fusion-Type Discontinuities) (see 8.12.2.2)

KEY FOR FIGURE 8.3, CASES I, II, III, IV, AND V

DISCONTINUITY A = ROUNDED OR ELONGATED DISCONTINUITY LOCATED IN WELD A
DISCONTINUITY B = ROUNDED OR ELONGATED DISCONTINUITY LOCATED IN WELD B
L AND W = LARGEST AND SMALLEST DIMENSIONS, RESPECTIVELY, OF DISCONTINUITY A
L' AND W' = LARGEST AND SMALLEST DIMENSIONS, RESPECTIVELY, OF DISCONTINUITY B
S = WELD SIZE
C_1 = SHORTEST DISTANCE PARALLEL TO THE WELD A AXIS, BETWEEN THE NEAREST DISCONTINUITY EDGES

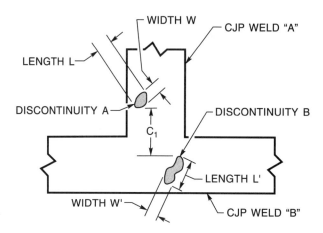

CASE I DISCONTINUITY LIMITATIONS[a]

DISCONTINUITY DIMENSION	LIMITATIONS	CONDITIONS
L	SEE FIGURE 8.3 GRAPH (B DIMENSION)	L ≥ 1/8 in [3 mm]
C_1	SEE FIGURE 8.3 GRAPH (C DIMENSION)	C_1 ≥ 2L or 2L', WHICHEVER IS GREATER

[a] The elongated discontinuity may be located in either weld "A" or "B." For the purposes of this illustration the elongated discontinuity "B" was located in weld "B."

Case I—Discontinuity at Weld Intersection

Figure 8.3 (Continued)—Discontinuity Acceptance Criteria for Cyclically Loaded Nontubular Connections in Compression (Limitations of Porosity or Fusion-Type Discontinuities) (see 8.12.2.2)

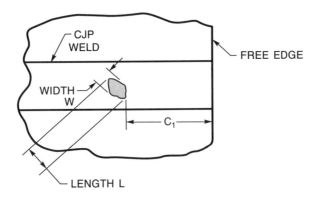

CASE II DISCONTINUITY LIMITATIONS

DISCONTINUITY DIMENSION	LIMITATIONS	CONDITIONS
L	SEE FIGURE 8.3 GRAPH (B DIMENSION)	$L \geq 1/8$ in [3 mm]
C_1	SEE FIGURE 8.3 GRAPH (C DIMENSION)	$C_1 \geq 5/8$ in [16 mm]

Case II—Discontinuity at Free Edge of CJP Groove Weld

Figure 8.3 (Continued)—Discontinuity Acceptance Criteria for Cyclically Loaded Nontubular Connections in Compression (Limitations of Porosity or Fusion-Type Discontinuities) (see 8.12.2.2)

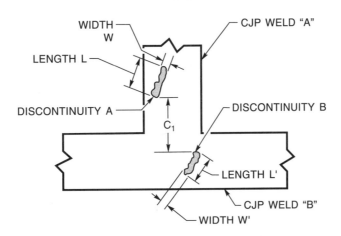

CASE III DISCONTINUITY LIMITATIONS

DISCONTINUITY DIMENSION	LIMITATIONS	CONDITIONS
L	SEE FIGURE 8.3 GRAPH (B DIMENSION)	$L \geq 1/8$ in [3 mm]
C_1	SEE FIGURE 8.3 GRAPH (C DIMENSION)	$C_1 \geq 2L$ or $2L'$, WHICHEVER IS GREATER

Case III—Discontinuity at Weld Intersection

Figure 8.3 (Continued)—Discontinuity Acceptance Criteria for Cyclically Loaded Nontubular Connections in Compression (Limitations of Porosity or Fusion-Type Discontinuities) (see 8.12.2.2)

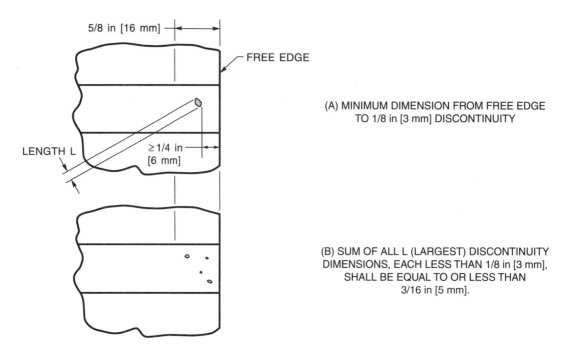

(A) MINIMUM DIMENSION FROM FREE EDGE TO 1/8 in [3 mm] DISCONTINUITY

(B) SUM OF ALL L (LARGEST) DISCONTINUITY DIMENSIONS, EACH LESS THAN 1/8 in [3 mm], SHALL BE EQUAL TO OR LESS THAN 3/16 in [5 mm].

Note: All dimensions between discontinuities ≥ 2L (L being largest of any two)

Case IV—Discontinuities Within 5/8 in [16 mm] of a Free Edge

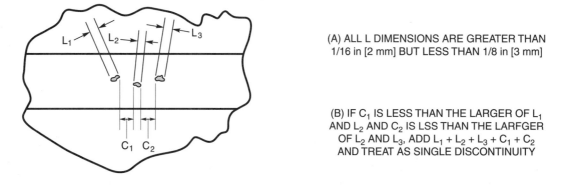

(A) ALL L DIMENSIONS ARE GREATER THAN 1/16 in [2 mm] BUT LESS THAN 1/8 in [3 mm]

(B) IF C_1 IS LESS THAN THE LARGER OF L_1 AND L_2 AND C_2 IS LSS THAN THE LARFGER OF L_2 AND L_3, ADD $L_1 + L_2 + L_3 + C_1 + C_2$ AND TREAT AS SINGLE DISCONTINUITY

Note: The weld shown above is for illustration only. These limitations apply to all locations or intersections. The number of discontinuities is also for illustration only.

Case V—Discontinuities Separated by Less Than 2L Anywhere in Weld (Use Figure 8.3 Graph "B" Dimension for Single Flaw)

Figure 8.3 (Continued)—Discontinuity Acceptance Criteria for Cyclically Loaded Nontubular Connections in Compression (Limitations of Porosity or Fusion-Type Discontinuities) (see 8.12.2.2)

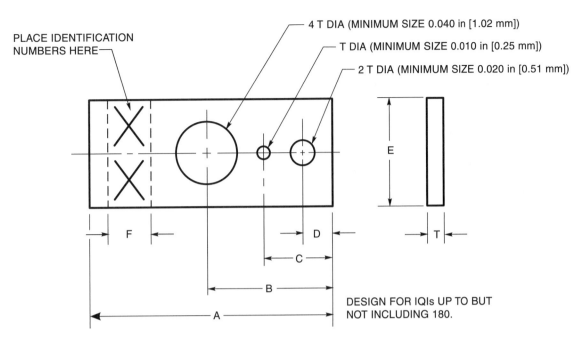

Table of Dimensions of IQI (in)							
Number[a]	A	B	C	D	E	F	IQI Thickness and Hole Diameter Tolerances
5–20	1.500 ± 0.015	0.750 ± 0.015	0.438 ± 0.015	0.250 ± 0.015	0.500 ± 0.015	0.250 ± 0.030	± 0.0005
21–59	1.500 ± 0.015	0.750 ± 0.015	0.438 ± 0.015	0.250 ± 0.015	0.500 ± 0.015	0.250 ± 0.030	± 0.0025
60–179	2.250 ± 0.030	1.375 ± 0.030	0.750 ± 0.015	0.375 ± 0.030	1.000 ± 0.030	0.375 ± 0.030	± 0.005

Table of Dimensions of IQI (mm)							
Number[a]	A	B	C	D	E	F	IQI Thickness and Hole Diameter Tolerances
5–20	38.10 ± 0.38	19.05 ± 0.38	11.13 ± 0.38	6.35 ± 0.38	12.70 ± 0.38	6.35 ± 0.38	± 0.013
21–59	38.10 ± 0.38	19.05 ± 0.38	11.13 ± 0.38	6.35 ± 0.38	12.70 ± 0.38	6.35 ± 0.38	± 0.06
60–179	57.15 ± 0.80	34.92 ± 0.80	19.05 ± 0.80	9.52 ± 0.80	25.40 ± 0.80	9.525 ± 0.80	± 0.13

[a] IQIs No. 5 through 9 are not 1T, 2T, and 4T.

Note: Holes shall be true and normal to the IQI. Do not chamfer.

Figure 8.4—Hole-Type IQI (see 8.17.1 and 10.28.1)
(Reprinted by permission of the ASTM International, copyright.)

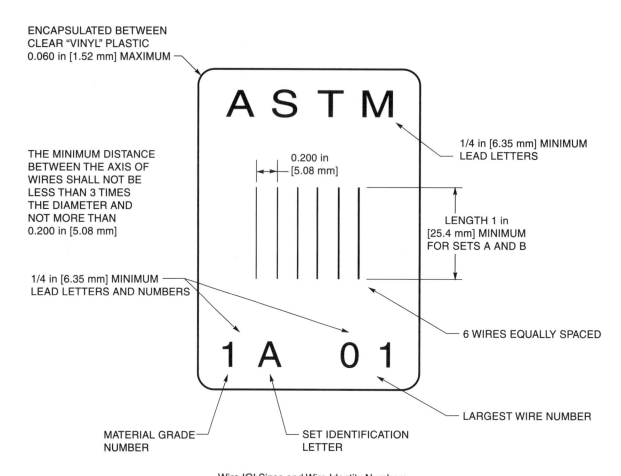

Wire IQI Sizes and Wire Identity Numbers

Set A		Set B		Set C		Set D	
Wire Diameter in [mm]	Wire Identity	Wire Diameter in [mm]	Wire Identity	Wire Diameter in [mm]	Wire Identity	Wire Diameter in [mm]	Wire Identity
0.0032 [0.08]	1	0.010 [0.25]	6	0.032 [0.81]	11	0.10 [2.5]	16
0.004 [0.1]	2	0.013 [0.33]	7	0.040 [1.02]	12	0.125 [3.2]	17
0.005 [0.13]	3	0.016 [0.4]	8	0.050 [1.27]	13	0.160 [4.06]	18
0.0063 [0.16]	4	0.020 [0.51]	9	0.063 [1.6]	14	0.20 [5.1]	19
0.008 [0.2]	5	0.025 [0.64]	10	0.080 [2.03]	15	0.25 [6.4]	20
0.010 [0.25]	6	0.032 [0.81]	11	0.100 [2.5]	16	0.32 [8]	21

Figure 8.5—Wire IQI (see 8.17.1 and 10.27.1)
(Reprinted by permission of the ASTM International, copyright.)

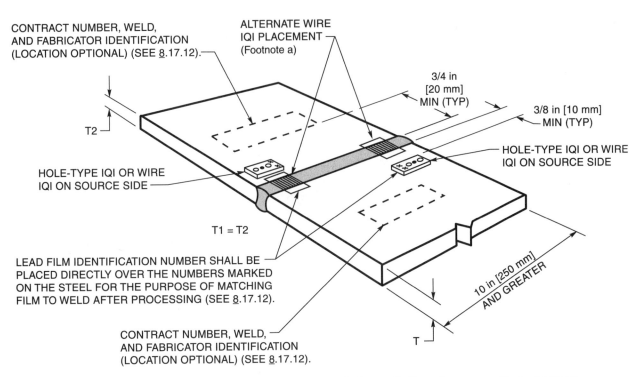

Figure 8.6—RT Identification and Hole-Type or Wire IQI Locations on Approximately Equal Thickness Joints 10 in [250 mm] and Greater in Length (see 8.17.7 and 10.27.2)

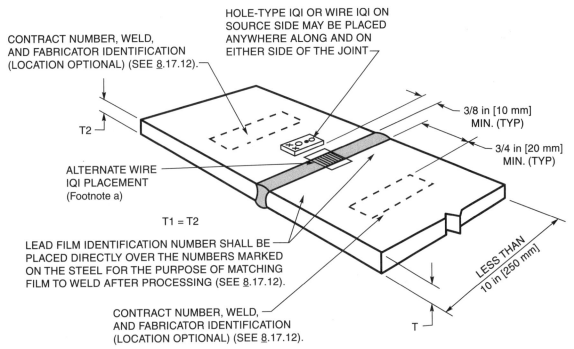

Figure 8.7—RT Identification and Hole-Type or Wire IQI Locations on Approximately Equal Thickness Joints Less than 10 in [250 mm] in Length (see 8.17.7 and 10.27.2)

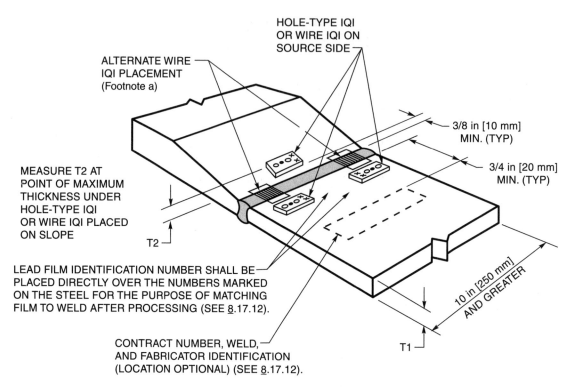

[a] Alternate source side IQI placement allowed for tubular applications and other applications when approved by the Engineer.

Figure 8.8—RT Identification and Hole-Type or Wire IQI Locations on Transition Joints 10 in [250 mm] and Greater in Length (see 8.17.7 and 10.27.2)

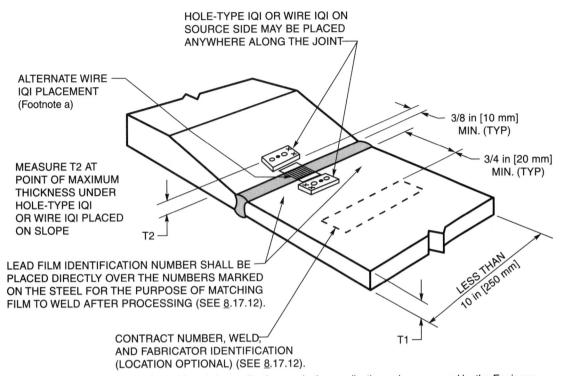

Figure 8.9—RT Identification and Hole-Type or Wire IQI Locations on Transition Joints Less than 10 in [250 mm] in Length (see 8.17.7 and 10.27.2)

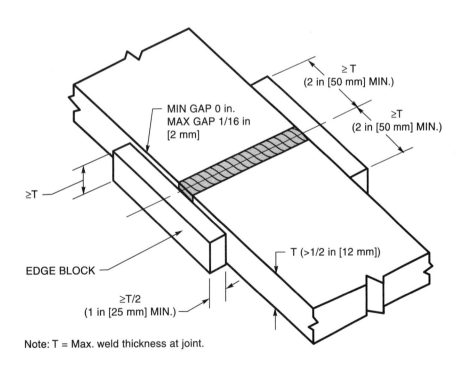

Figure 8.10—RT Edge Blocks (see 8.17.13)

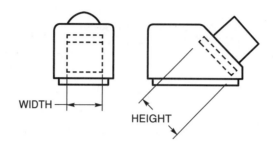

Figure 8.11—Transducer Crystal (see 8.21.7.2)

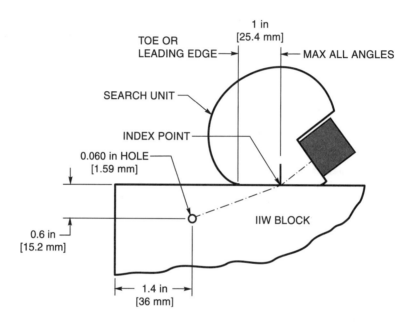

Figure 8.12—Qualification Procedure of Search Unit Using IIW Reference Block (see 8.21.7.7)

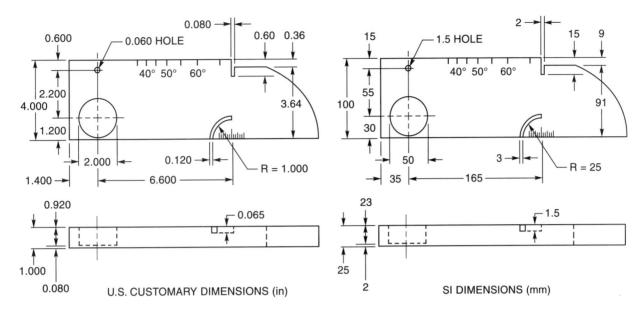

Notes:
1. The dimensional tolerance between all surfaces involved in referencing or calibrating shall be within ±0.005 in [0.13 mm] of detailed dimension.
2. The surface finish of all surfaces to which sound is applied or reflected from shall have a maximum of 125 μin [3.17 μm] r.m.s.
3. All material shall be ASTM A36 or acoustically equivalent.
4. All holes shall have a smooth internal finish and shall be drilled 90° to the material surface.
5. Degree lines and identification markings shall be indented into the material surface so that permanent orientation can be maintained.
6. These notes shall apply to all sketches in Figures 8.13 and 8.14.

Figure 8.13—Typical IIW Type Block (see 8.22.1)

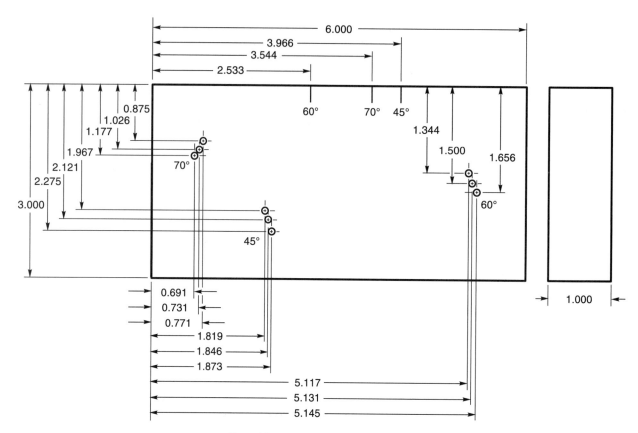
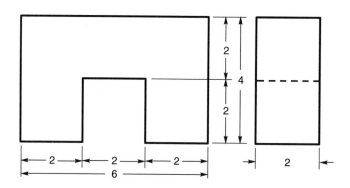

Figure 8.14—Qualification Blocks (see 8.22.3 and 8.27) (Dimensions in Inches)

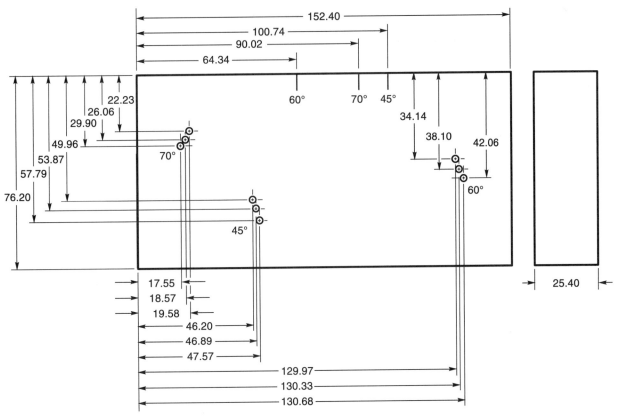

Note: All holes are 1.59 mm in diameter.

RC – RESOLUTION REFERENCE BLOCK

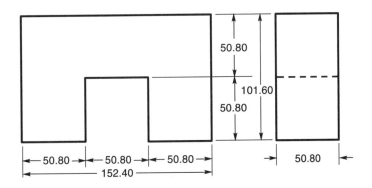

TYPE – DISTANCE AND SENSITIVITY REFERENCE BLOCK

**Figure 8.14 (Continued)—Qualification Blocks (see 8.22.3 and 8.27)
(Dimensions in Millimeters)**

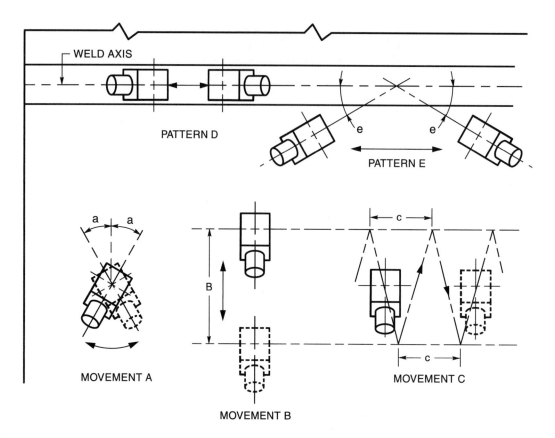

Notes:
1. Testing patterns are all symmetrical around the weld axis with the exception of pattern D, which shall be conducted directly over the weld axis.
2. Testing from both sides of the weld axis shall be made wherever mechanically possible.

Figure 8.15—Plan View of UT Scanning Patterns (see 8.30)

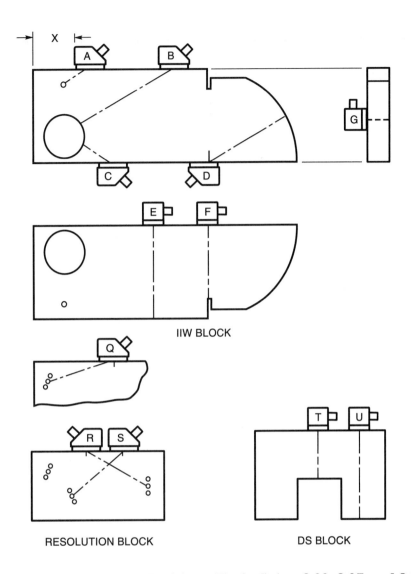

Figure 8.16—Transducer Positions (Typical) (see 8.22, 8.27, and 8.28)

9. Stud Welding

9.1 Scope

Clause 9 contains general requirements (9.2) for welding of steel studs to steel, and stipulates specific requirements for:

(1) Mechanical properties and material of steel studs, and requirements for qualification of stud bases – Clauses 9.3 and 9.9.

(2) Application qualification testing, operator qualification, preproduction testing, and workmanship – Clause 9.6.

(3) Stud welding during production, fabrication/ erection, and inspection – Clauses 9.4, 9.5, and 9.7.

(4) The stud manufacturer's certification of stud base weldability – Clause 9.9.

NOTE: Approved steels; for studs, see 9.2.6; for base metals, see Table 5.3 (Groups I and II). For guidance, see C-9.6.2.

9.2 General Requirements

9.2.1 Stud Design. Studs shall be of suitable design for arc welding to steel members with the use of automatically timed stud welding equipment. The type and size of the stud shall be as specified by the drawings, specifications, or special provisions. For headed-type studs, see Figure 9.1. Alternative head configurations may be used with proof of mechanical and embedment tests confirming full-strength development of the design, and with the approval of the Engineer.

9.2.1.1 Manufacturer's ID. All headed anchor studs and deformed anchor bars shall have a manufacturer's permanent identification.

9.2.2 Arc Shields. An arc shield (ferrule) of heat-resistant ceramic or other suitable material shall be furnished with each stud.

9.2.3 Flux. A suitable deoxidizing and arc stabilizing flux for welding shall be furnished with each stud of 5/16 in [8 mm] diameter or larger. Studs less than 5/16 in [8 mm] in diameter may be furnished with or without flux.

9.2.4 Stud Bases. A stud base, to be qualified, shall have passed the test described in 9.9. Only studs with qualified stud bases shall be used. Qualification of stud bases in conformance with 9.9 shall be at the manufacturer's expense. The arc shield used in production shall be the same as used in qualification tests or as recommended by the manufacturer. When requested by the Engineer, the Contractor shall provide the following information:

(1) A description of the stud and arc shield

(2) Certification from the manufacturer that the stud base is qualified in conformance with 9.9.

(3) Qualification test data

9.2.5 Stud Finish

9.2.5.1 Stud finish shall be produced by heading, rolling, or machining. Finished studs shall be of uniform quality and condition, free of defects that may affect the welding quality, suitability for intended application, or fit of the studs in the specified ceramic arc shields (ferrules). Such defects include laps, fins, seams, cracks, twists, bends, thread defects, discontinuities, or foreign materials (see 9.4.1 and 9.4.2).

9.2.5.2 Headed studs are subject to cracks or bursts in the stud head which are abrupt interruptions of the periphery caused by radial separation of the metal extending from the head inward to the stud shank. These cracks or bursts shall not be the cause for rejection, provided that they do not exceed one half of the distance from the stud head to the stud shank as determined by visual inspection (see Figure C-9.1.) Studs shall be rejected if the cracks or bursts are of a number or width that does not permit the head to fit into the welding tool chuck or cause arcing between the stud head and the chuck affecting chuck life or weld quality.

9.2.6 Stud Material. Studs shall be made from cold drawn bar conforming to the requirements of ASTM A29/A29M-12e1, *Standard Specification for General Requirements for Steel Bars, Carbon and Alloy, Hot-Wrought, Grades 1010 through* 1020, inclusive, either semi-killed or killed aluminum or silicon deoxidation.

9.2.7 Base Metal Thickness. When welding directly to base metal, the base metal shall be no thinner than 1/3 the stud diameter. When welding through deck, the stud diameter shall be no greater than 2.5 times the base material thickness. In no case shall studs be welded through more than two plies of metal decking.

9.3 Mechanical Requirements

9.3.1 Standard Mechanical Requirements. At the manufacturer's option, mechanical properties of studs shall be determined by testing either the steel after cold finishing or the full diameter finished studs. In either case, the studs shall conform to the standard properties shown in Table 9.1.

9.3.2 Testing. Mechanical properties shall be determined in conformance with the applicable sections of ASTM A370, *Mechanical Testing of Steel Products*. A typical test fixture is used, similar to that shown in Figure 9.2.

9.3.3 Engineer's Request. Upon request by the Engineer, the Contractor shall furnish:

(1) The stud manufacturer's certification that the studs, as delivered, conform to the applicable requirements of 9.2 and 9.3.

(2) Certified copies of the stud manufacturer's test reports covering the last completed set of in-plant quality control mechanical tests, required by 9.3 for each diameter delivered.

(3) Certified material test reports (CMTR) from the steel supplier indicating diameter, chemical properties, and grade on each heat number delivered.

9.3.4 Absence of Quality Control Tests. When quality control tests are not available, the Contractor shall furnish a chemical test report conforming to 9.2.6 and a mechanical test report conforming to the requirements of 9.3 for each lot number. Unidentified and untraceable studs shall not be used.

9.3.5 Additional Studs. The Contractor is responsible for furnishing additional studs of each type and size, at the request of the Engineer, for checking the requirements of 9.2 and 9.3. Testing shall be at the Owner's expense.

9.4 Workmanship/Fabrication

9.4.1 Cleanliness. At the time of welding, the studs shall be free from rust, rust pits, scale, oil, moisture, or other deleterious matter that would adversely affect the welding operation.

9.4.2 Coating Restrictions. The stud base shall not be painted, galvanized, or cadmium-plated prior to welding.

9.4.3 Base Metal Preparation. The areas to which the studs are to be welded shall be free of scale, rust, moisture, paint, or other injurious material to the extent necessary to obtain satisfactory welds and prevent objectionable fumes. These areas may be cleaned by wire brushing, scaling, prick-punching, or grinding.

9.4.4 Moisture. The arc shields or ferrules shall be kept dry. Any arc shields which show signs of surface moisture from dew or rain shall be oven dried at 250°F [120°C] for two hours before use.

9.4.5 Spacing Requirements. Longitudinal and lateral spacings of stud shear connectors (Type B) may vary a maximum of 1 in [25 mm] from the location shown in the drawings. The minimum distance from the edge of a stud base to the edge of a flange shall be the diameter of the stud plus 1/8 in [3 mm], but preferably not less than 1-1/2 in [40 mm].

9.4.6 Arc Shield Removal. After welding, arc shields shall be broken free from studs to be embedded in concrete, and, where practical, from all other studs.

9.4.7 Acceptance Criteria. The studs, after welding, shall be free of any discontinuities or substances that would interfere with their intended function and have a full 360° flash. However, nonfusion on the legs of the flash and small shrink fissures shall be acceptable. The fillet weld profiles shown in Figure 7.4 shall not apply to the flash of automatically timed stud welds.

9.5 Technique

9.5.1 Automatic Mechanized Welding. Studs shall be welded with automatically timed stud welding equipment connected to a suitable source of direct current electrode negative power. Welding voltage, current, time, and gun settings for lift and plunge should be set at optimum settings, based on past practice, recommendations of stud and equipment manufacturer, or both. AWS C5.4, *Recommended Practices for Stud Welding*, should also be used for technique guidance.

9.5.2 Multiple Welding Guns. If two or more stud welding guns shall be operated from the same power source, they shall be interlocked so that only one gun can operate at a time, and so that the power source has fully recovered from making one weld before another weld is started.

9.5.3 Movement of Welding Gun. While in operation, the welding gun shall be held in position nominally perpendicular to the base material, and without movement until the weld metal has solidified.

9.5.4 Ambient and Base Metal Temperature Requirements. Welding shall not be done when the base metal temperature is below 0°F [−18°C] or when the surface is wet or exposed to falling rain or snow. When the temperature of the base metal is below 32°F [0°C], one additional stud in each 100 studs welded shall be tested by methods described in 9.7.1.3 and 9.7.1.4, except that the angle of testing shall be approximately 15°. This is in addition to the first two studs tested for each start of a new production period or change in set-up. Set-up includes stud gun, power source, stud diameter, gun lift and plunge, total welding lead length, and changes greater than ± 5% in current (amperage) and time.

9.5.5 Fillet Weld Option. At the option of the Contractor, studs may be attached with fillet welds using a WPS qualified in accordance with Clause 5 or 6.

9.5.5.1 Production Control. Conformance to 9.7 shall not be required for studs attached by fillet welding.

9.5.5.2 Stud Materials. Stud materials shall meet the following requirements:

(1) Studs meeting the requirements of 9.2.6 shall be acceptable for use in a prequalified WPS;

(2) Stud materials shall be considered Group II material for the selection of matching filler metal strengths.

9.5.5.3 Surfaces. Base metal shall be sufficiently clean to permit making welds that meet Clauses 7, 8, and 9.4.3. The cleanliness of the stud shall be in accordance with 9.4.1. Base metal shall be prepared to meet the requirements of 9.4.3 and 7.14.1.

9.5.5.4 Stud Fit. The base of the stud shall be prepared so that the base is in contact with the base metal. When a contact over the entire stud base with the base metal cannot be made, the following apply:

(1) The gap between the base material and the stud shall not exceed 1/16 in [2 mm] at any point, except as permitted below.

(2) If the separation is greater than 1/16 in [2 mm], the legs of the fillet weld shall be increased by the amount of the root opening, or the contractor shall demonstrate that the required effective throat has been obtained.

9.5.5.5 Tack Welds. Tack welds are permitted when required to facilitate the installation. Tack welds shall comply with 7.17.

9.5.5.6 Minimum Weld Size. When fillet welds are used, the minimum size shall be the larger of those required in Table 7.7 or Table 9.2.

9.5.5.7 Preheat. The base metal to which studs are welded shall be preheated in conformance with the requirements of the WPS.

9.5.5.8 Welders. Welders shall be qualified to perform fillet welding per Clause 6 or Clause 10.

9.5.5.9 Repairs of Fillet Welded Studs. Fillet welds that require repair shall be repaired in accordance with 7.25.

9.5.5.10 Visual Inspection. All fillet welded studs and welded repairs shall be visually inspected in conformance with 8.9.

9.6 Stud Application Qualification Requirements

All Welding Procedure Specifications (WPSs) shall be written by the Contractor.

9.6.1 Prequalified WPSs. Studs conforming to 9.2.6 which are shop or field applied in the flat (down-hand) position to a planar and horizontal surface to steel base metals in Groups I or II per Table 5.3 are deemed prequalified by virtue of the manufacturer's stud base qualification when welding is performed within the ranges of the amperage and time of the manufacturer's welding recommendations. The limit of flat position is defined as 0° – 15° slope on the surface to which the stud is applied. For curved surfaces the 15 degree flat position tolerance is measured at the tangent point where the stud is to be applied.

9.6.2 WPSs Qualified by Testing. Examples of stud applications using mechanized welding that require tests of this section are the following:

(1) Studs which are applied on nonplanar surfaces or to a planar surface in the vertical or overhead positions.

(2) Studs which are welded through decking. The tests shall be conducted using the thinnest base material thickness used in construction but no thinner than 1/2 inch, a Group I or II base material, the same base material coating, the thickest decking gauge and coating material used in construction.

(3) Studs welded to other than Groups I or II steels listed in Table 5.3.

9.6.3 Responsibilities for Tests. The Contractor shall be responsible for the performance of these tests. Tests may be performed by the Contractor, the stud manufacturer, or by another testing agency satisfactory to all parties involved.

9.6.4 Preparation of Specimens

9.6.4.1 Test Specimens. To qualify applications involving materials listed in Table 5.3, Groups I and II: specimens may be prepared using ASTM A36 steel base materials or base materials listed in Table 5.3, Groups I and II.

9.6.4.2 Recorded Information. To qualify applications involving materials other than those listed in Table 5.3, Groups I and II, the test specimen base material shall be of the chemical, physical and grade specifications to be used in production.

9.6.5 Number of Specimens. Ten specimens shall be welded consecutively using recommended procedures and settings for each diameter, position, and surface geometry.

9.6.6 Test Required. The ten specimens shall be tested using one or more of the following methods: bending, torquing, or tensioning.

9.6.7 Test Methods

9.6.7.1 Bend Test. Studs shall be tested by alternately bending 30° in opposite directions in a typical test fixture as shown in Figure 9.4 until failure occurs. Alternatively, studs may be bent 90° from their original axis. Type C studs, when bent 90°, shall be bent over a pin with a diameter of 4 times the diameter of the stud. In either case, a stud application shall be considered qualified if the studs are bent 90° and fracture occurs in the plate or shape material or in the shank of the stud and not in the weld.

9.6.7.2 Torque Test. Studs shall be torque tested using a torque test arrangement that is substantially in conformance with Figure 9.3. A stud application shall be considered qualified if all test specimens are torqued to destruction without failure in the weld.

9.6.7.3 Tension Test. Studs shall be tension tested to destruction using any machine capable of supplying the required force. A stud application shall be considered qualified if the test specimens do not fail in the weld.

9.6.8 Application Qualification Test Data. Application Qualification Test Data shall include the following:

(1) Drawings that show shapes and dimensions of studs and arc shields.

(2) A complete description of stud and base materials, and a description (part number) of the arc shield.

(3) Welding position and settings (current, time).

(4) A record, which shall be made for each qualification and shall be available for each contract. A suggested WPS/PQR form for nonprequalified application may be found in Annex J, Form J-8.

9.7 Production Control

9.7.1 Pre-Production Testing

9.7.1.1 Start of Shift. Before production welding with a particular set-up and with a given size and type of stud, and at the beginning of each day's or shift's production, testing shall be performed on the first two studs that are welded. The stud technique may be developed on a piece of material similar to the production member in thickness and properties. If actual production thickness is not available, the thickness may vary ± 25%. All test studs shall be welded in the same general position as required on the production member (flat, vertical, or overhead).

9.7.1.2 Production Member Option. Instead of being welded to separate material, the test studs may be welded on the production member, except when separate plates are required by 9.7.1.5.

9.7.1.3 Flash Requirement. Studs shall exhibit full 360° flash with no evidence of undercut into the stud base.

9.7.1.4 Bend and Torque Test. In addition to visual examination, the test shall consist of bending the studs after they are allowed to cool, to an angle of approximately 30° from their original axes by either striking the studs with a hammer on the unwelded end or placing a pipe or other suitable device over the stud and manually or mechanically bending the stud. Alternatively for threaded studs, the torque test of Figure 9.3 may be used instead of bending.

Testing performed at temperatures below 50°F [10°C], shall be done by continuous slow application of load.

9.7.1.5 Event of Failure. If on visual examination the test studs do not exhibit 360° flash, or if on testing, failure occurs in the weld zone of either stud, the procedure shall be corrected, and two more studs shall be welded to separate material or on the production member and tested in conformance with the provisions of 9.7.1.3 and 9.7.1.4. If either of the second two studs fails, additional welding shall be continued on separate plates until two consecutive studs are tested and found to be satisfactory before any more production studs are welded to the member.

9.7.2 Production Welding.
Once production welding has begun, any changes made to the welding setup, as determined in 9.7.1, shall require that the testing in 9.7.1.3 and 9.7.1.4 be performed prior to resuming production welding.

9.7.3 Repair of Studs.
In production, studs on which a full 360° flash is not obtained may, at the option of the Contractor, be repaired by adding the minimum fillet weld as required by 9.5.5 in place of the missing flash. The repair weld shall extend at least 3/8 in [10 mm] beyond each end of the discontinuity being repaired.

9.7.4 Operator Qualification.
The pre-production test required by 9.7.1, if successful, shall also serve to qualify the stud welding operator. Before any production studs are welded by an operator not involved in the preproduction set-up of 9.7.1, the first two studs welded by the operator shall have been tested in conformance with the provisions of 9.7.1.3 and 9.7.1.4. When the two welded studs have been tested and found satisfactory, the operator may then weld production studs.

9.7.5 Removal Area Repair.
If an unacceptable stud has been removed from a component subjected to tensile stresses, the area from which the stud was removed shall be made smooth and flush. Where in such areas the base metal has been pulled out in the course of stud removal, SMAW with low-hydrogen electrodes in conformance with the requirements of this code shall be used to fill the pockets, and the weld surface shall be flush.

In compression areas of members, if stud failures are confined to shanks or fusion zones of studs, a new stud may be welded adjacent to each unacceptable area in lieu of repair and replacement on the existing weld area (see 9.4.5). If base metal is pulled out during stud removal, the repair provisions shall be the same as for tension areas except that when the depth of discontinuity is the lesser of 1/8 in [3 mm] or 7% of the base metal thickness, the discontinuity may be faired by grinding in lieu of filling with weld metal. Where a replacement stud is to be provided, the base metal repair shall be made prior to welding the replacement stud. Replacement studs (other than threaded type which should be torque tested) shall be tested by bending to an angle of approximately 15° from their original axes. The areas of components exposed to view in completed structures shall be made smooth and flush where a stud has been removed.

9.8 Fabrication and Verification Inspection Requirements

9.8.1 Visual Inspection.
If a visual inspection reveals any stud that does not show a full 360° flash or any stud that has been repaired by welding, such stud shall be bent to an angle of approximately 15° from its original axis. Threaded studs shall be torque tested. The method of bending shall be in conformance with 9.7.1.4. The direction of bending for

studs with less than a 360° flash shall be opposite to the missing portion of the flash. Torque testing shall be in conformance with Figure 9.3.

9.8.2 Additional Tests. The Verification Inspector, where conditions warrant, may select a reasonable number of additional studs to be subjected to the tests described in 9.8.1.

9.8.3 Bent Stud Acceptance Criteria. The bent stud shear connectors (Type B) and deformed anchors (Type C) and other studs to be embedded in concrete (Type A) that show no sign of failure shall be acceptable for use and left in the bent position. When bent studs are required by the contract documents to be straightened, the straightening operation shall be done without heating, and before completion of the production stud welding operation.

9.8.4 Torque Test Acceptance Criteria. Threaded studs (Type A) torque tested to the proof load torque level in Figure 9.3 that show no sign of failure shall be acceptable for use.

9.8.5 Corrective Action. Welded studs not conforming to the requirements of the code shall be repaired or replaced by the Contractor. The Contractor shall revise the welding procedure as necessary to ensure that subsequent stud welding will meet code requirements.

9.8.6 Owner's Option. At the option and the expense of the Owner, the Contractor may be required, at any time, to submit studs of the types used under the contract for a qualification check in conformance with the procedures of 9.9.

9.9 Manufacturers' Stud Base Qualification Requirements

9.9.1 Purpose. The purpose of these requirements is to prescribe tests for the stud manufacturers' certification of stud base weldability.

9.9.2 Responsibility for Tests. The stud manufacturer shall be responsible for the performance of the qualification test. These tests may be performed by a testing agency satisfactory to the Engineer. The agency performing the tests shall submit a certified report to the manufacturer of the studs giving procedures and results for all tests including the information described in 9.9.10.

9.9.3 Extent of Qualification. Qualification of a stud base shall constitute qualification of stud bases with the same geometry, flux, and arc shield, having the same diameter and diameters that are smaller by less than 1/8 in [3 mm]. A stud base qualified with an approved grade of ASTM A29 steel and meets the standard mechanical properties (see 9.3.1) shall constitute qualification for all other approved grades of ASTM A29 steel (see 9.2.6), provided that conformance with all other provisions stated herein shall be achieved.

9.9.4 Duration of Qualification. A size of stud base with arc shield, once qualified, shall be considered qualified until the stud manufacturer makes any change in the stud base geometry, material, flux, or arc shield which affects the welding characteristics.

9.9.5 Preparation of Specimens

9.9.5.1 Materials. Test specimens shall be prepared by welding representative studs to suitable specimen plates of ASTM A36 steel or any of the other materials listed in Table 5.3 or Table 6.9. Studs to be welded through metal decking shall have the weld base qualification testing done by welding through metal decking representative of that used in construction, galvanized per ASTM A653 coating designation G90 for one thickness of deck or G60 for two deck plies. When studs are to be welded through decking, the stud base qualification test shall include decking representative of that to be used in construction. Welding shall be done in the flat position (plate surface horizontal). Tests for threaded studs shall be on blanks (studs without threads).

9.9.5.2 Welding Equipment. Studs shall be welded with power source, welding gun, and automatically controlled equipment as recommended by the stud manufacturer. Welding voltage, current, and time (see 9.9.6) shall be measured and recorded for each specimen. Lift and plunge shall be at the optimum setting as recommended by the manufacturer.

9.9.6 Number of Test Specimens

9.9.6.1 High Current. For studs 7/8 in [22 mm] or less in diameter, 30 test specimens shall be welded consecutively with constant optimum time, but with current 10% above optimum. For studs over 7/8 in [22 mm] diameter, 10 test specimens shall be welded consecutively with constant optimum time. Optimum current and time shall be the midpoint of the range normally recommended by the manufacturer for production welding.

9.9.6.2 Low Current. For studs 7/8 in [22 mm] or less in diameter, 30 test specimens shall be welded consecutively with constant optimum time, but with current 10% below optimum. For studs over 7/8 in [22 mm] diameter, 10 test specimens shall be welded consecutively with constant optimum time, but with current 5% below optimum.

9.9.6.3 Metal Deck. For studs to be welded through metal deck, the range of weld base diameters shall be qualified by welding 10 studs at the optimum current and time as recommended by the manufacturer conforming to the following:

(1) Maximum and minimum diameters welded through one thickness of 16 gage deck, coating designation G90.

(2) Maximum and minimum diameters welded through two plies of 16 gage deck coating designation G60.

(3) Maximum and minimum diameters welded through one thickness of 18 gage G60 deck over one thickness of 16 gage G60 deck.

(4) Maximum and minimum diameters welded through two plies of 18 gage deck, both with G60 coating designation.

The range of diameters from maximum to minimum welded through two plies of 18 gage metal deck with G60 galvanizing shall be qualified for welding through one or two plies of metal deck 18 gage or less in thickness.

9.9.7 Tests

9.9.7.1 Tension Tests. Ten of the specimens welded in conformance with 9.9.6.1 and ten in conformance with 9.9.6.2 shall be subjected to a tension test in a fixture similar to that shown in Figure 9.2, except that studs without heads may be gripped on the unwelded end in the jaws of the tension testing machine. A stud base shall be considered as qualified if all test specimens have a tensile strength equal to or above the minimum described in 9.3.1.

9.9.7.2 Bend Tests (Studs 7/8 in [22 mm] or less in diameter). Twenty of the specimens welded in conformance with 9.9.6.1 and twenty in conformance with 9.9.6.2 shall be bend tested by being bent alternately 30° from their original axes in opposite directions until failure occurs. Studs shall be bent in a bend testing device as shown in Figure 9.4, except that studs less than 1/2 in [12 mm] diameter may be bent using a device as shown in Figure 9.5. A stud base shall be considered as qualified if, on all test specimens, fracture occurs in the plate material or shank of the stud and not in the weld or HAZ. All test specimens for studs over 7/8 in [22 mm] shall only be subjected to tensile tests.

9.9.7.3 Weld through Deck Tests. All welds-through-deck stud specimens shall be tested by bending 30° in opposite directions in a bend testing device as shown in Figure 9.4, or by bend testing 90° from their original axis or tension testing to destruction in a machine capable of supplying the required force. With any test method used, the range of stud diameters from maximum to minimum shall be considered as qualified weld bases for through deck welding if, on all test specimens, fracture occurs in the plate material or shank of the stud and not in the weld or HAZ.

9.9.8 Retests. If failure occurs in a weld or the HAZ in any of the bend test groups of 9.9.7.2 or at less than specified minimum tensile strength of the stud in any of the tension groups in 9.9.7.1, a new test group (described in 9.9.6.1 or 9.9.6.2, as applicable) shall be prepared and tested. If such failures are repeated, the stud base shall fail to qualify.

9.9.9 Acceptance. For a manufacturer's stud base and arc shield combination to be qualified, each stud of each group of 30 studs shall, by test or retest, meet the requirements described in 9.9.7. Qualification of a given diameter of stud base shall be considered qualification for stud bases of the same nominal diameter (see 9.9.3, stud base geometry, material, flux, and arc shield).

9.9.10 Manufacturer's Qualification Test Data. The test data shall include the following:

(1) Drawings showing shapes and dimensions with tolerances of stud, arc shields, and flux;

(2) A complete description of materials used in the studs, including the quantity and type of flux, and a description of the arc shields; and

(3) Certified results of tests.

Table 9.1
Mechanical Property Requirements for Studs (see 9.3.1)

		Type A[a]	Type B[b]	Type C[c]
Tensile Strength	Psi min.	61 000	65 000	80 000
	MPa min.	420	420	552
Yield Strength (0.2% offset)	Psi min.	49 000	51 000	—
	MPa min.	340	350	—
(0.5% offset)	Psi min.	—	—	70 000
	MPa min.	—	—	485
Elongation	% in 2 in. min.	17%	20%	—
	% in 5x dia. min.	14%	15%	—
Reduction of area	% min.	50%	50%	—

[a] Type A studs shall be general purpose of any type and size used for purposes other than shear transfer in composite beam design and construction.
[b] Type B studs shall be studs that are headed, bent, or of other configuration in 3/8 in [10 mm], 1/2 in [12 mm], 5/8 in [16 mm], 3/4 in [20 mm], 7/8 in [22 mm], and 1 in [25 mm] diameter that are used as an essential component in composite beam design and concrete anchorage design.
[c] Type C studs shall be cold-worked deformed steel bars manufactured in conformance with specification ASTM A496 having a nominal diameter equivalent to the diameter of a plain wire having the same weight per foot as the deformed wire. ASTM A496 specifies a maximum diameter of 0.628 in [16 mm] maximum. Any bar supplied above that diameter shall have the same physical characteristics regarding deformations as required by ASTM A496.

Table 9.2
Minimum Fillet Weld Size for Studs (see 9.5.5.6)

(Dimensions in Inches)				
Stud Diameter	1/4 thru 7/16	1/2	5/8, 3/4, 7/8	1
Min. Size Fillet	3/16	1/4	5/16	3/8
(Dimensions in Millimeters)				
Stud Diameter	6 thru 11	12	16, 20, 22	25
Min. Size Fillet	5	6	8	10

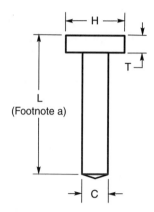

a Manufactured length before welding.

Standard Dimensions, in				
Shank Diameter (C)	Length Tolerances (L)	Head Diameter (H)	Minimum Head Height (T)	
3/8	+0.010 −0.010	±1/16	3/4 ± 1/64	9/32
1/2	+0.010 −0.010	± 1/16	1 ± 1/64	9/32
5/8	+0.010 −0.010	± 1/16	1-1/4 ± 1/64	9/32
3/4	+0.015 −0.015	± 1/16	1-1/4 ± 1/64	3/8
7/8	+0.015 −0.015	± 1/16	1-3/8 ± 1/64	3/8
1	+0.020 −0.020	± 1/16	1-5/8 ± 1/64	1/2
Standard Dimensions, mm				
Shank Diameter (C)	Length Tolerances (L)	Head Diameter (H)	Minimum Head Height (T)	
10	+0.25 −0.25	± 1.6	19 ± 0.40	7.1
13	+0.25 −0.25	± 1.6	25 ± 0.40	7.1
16	+0.25 −0.25	± 1.6	32 ± 0.40	7.1
19	+0.40 −0.40	± 1.6	32 ± 0.40	9.5
22	+0.40 −0.40	± 1.6	35 ± 0.40	9.5
25	+0.40 −0.40	± 1.6	41 ± 0.40	12.7

Figure 9.1—Dimension and Tolerances of Standard-Type Headed Studs (see 9.2.1)

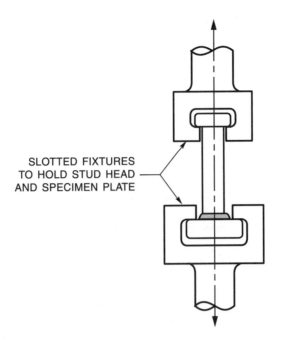

Figure 9.2—Typical Tension Test Fixture (see 9.3.2)

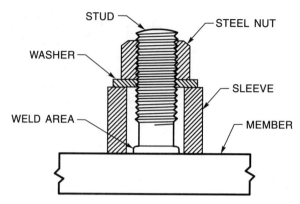

Note: Dimensions of test fixture details should be appropriate to the size of the stud. The threads of the stud shall be clean and free of lubricant other than the residue of cutting/cold forming lubricants in the "as received" condition from the manufacturer.

Required Proof Torque for Testing Threaded Studs[a]								
Nominal Diameter		M.E.T.A.[b]		Thread			Proof Testing Torque[c]	
in	mm	in²	mm²	no./in	pitch-mm	Series	lb-ft	Joule
0.236	M6	0.031	20.1		1.0	ISO-724	5.4	7.4
1/4	6.4	0.036	23.2	28		UNF	6.6	9.0
		0.032	20.6	20		UNC	5.9	7.8
5/16	7.9	0.058	37.4	24		UNF	13.3	18.1
		0.052	33.5	18		UNC	11.9	16.1
0.315	M8	0.057	36.6		1.25	ISO-724	13.2	17.9
3/8	9.5	0.088	56.8	24		UNF	24.3	32.9
		0.078	50.3	16		UNC	21.5	29.2
0.394	M10	0.090	58.0		1.5	ISO-724	26.2	35.5
7/16	11.1	0.118	76.1	20		UNF	37.9	51.4
		0.106	68.4	14		UNC	34.8	47.2
0.472	M12	0.131	84.3		1.75	ISO-724	45.7	61.9
1/2	12.7	0.160	103.2	20		UNF	58.8	79.7
		0.142	91.6	13		UNC	52.2	70.8
0.551	M14	0.178	115.0		2.0	ISO-724	72.7	98.5
9/16	14.3	0.203	131.0	18		UNF	83.9	113.8
		0.182	117.4	12		UNC	75.2	102.0
5/8	15.9	0.255	164.5	18		UNF	117.1	158.8
		0.226	145.8	11		UNC	103.8	140.8
0.630	M16	0.243	157.0		2.0	ISO-724	113.4	153.7
3/4	19.1	0.372	240.0	16		UNF	205.0	278.0
		0.334	215.5	10		UNC	184.1	249.7
0.787	M20	0.380	245.0		2.5	ISO-724	221.2	299.9
0.866	M22	0.470	303.0		2.5	ISO-724	300.9	408.0
7/8	22.2	0.509	328.4	14		UNF	327.3	443.9
		0.462	298.1	9		UNC	297.1	402.9
0.945	M24	0.547	353.0		3.0	ISO-724	382.4	518.5
1	25.4	0.678	437.4	12		UNF	498.3	675.7
		0.606	391.0	8		UNC	445.4	604.0

[a] Torque figures are based on Type A threaded studs with a minimum yield stress of 49 000 psi [340 MPa].
[b] Mean Effective Thread Area (M.E.T.A) shall be defined as the effective stress area based on a mean diameter taken approximately midway between the minor and the pitch diameters.
[c] Values are calculated on a proof testing torque of 0.9 times Nominal Stud Diameter times 0.2 Friction Coefficient Factor times Mean Effective Thread Area times Minimum Yield Stress for unplated studs in the as-received condition. Plating, coatings, or oil/grease deposits will change the Friction Coefficient Factor.

Figure 9.3—Torque Testing Arrangement and Table of Testing Torques (see 9.7.1.4)

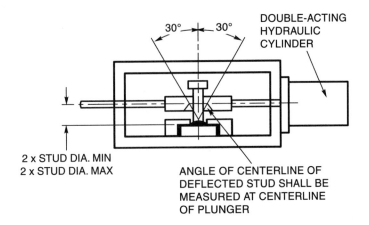

Notes:
1. Fixture holds specimen and stud is bent 30° alternately in opposite directions.
2. Load can be applied with hydraulic cylinder (shown) or fixture adapted for use with tension test machine.

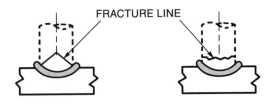

TYPICAL FRACTURES IN SHANK OF STUD

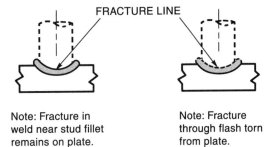

Note: Fracture in weld near stud fillet remains on plate.

Note: Fracture through flash torn from plate.

TYPICAL WELD FAILURES

Figure 9.4—Bend Testing Device (see 9.9.7.2)

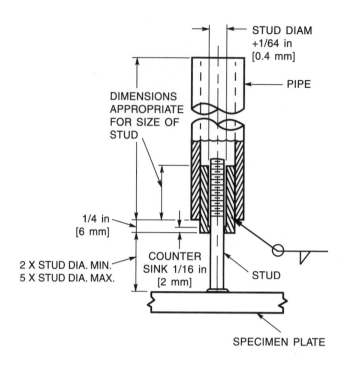

Figure 9.5—Suggested Type of Device for Qualification Testing of Small Studs (see 9.9.7.2)

10. Tubular Structures

10.1 Scope

This Clause supplements Clauses 1–9. The specific requirements of Clause 10 apply only to tubular connections as defined in 10.3 and Clause 3. Such connections may be butt joints between tubulars, T-, Y-, K- connections of tubulars to tubulars, or tubulars welded to flat plates or flat elements of other members. Where this clause refers to rectangular tubulars, square tubulars are included unless noted.

10.1.1 The scope of the term "tubular members" includes:

(1) Circular hollow structural sections (HSS);

(2) Rectangular HSS;

(3) Square HSS;

(4) Fabricated members built from flat plates and or shapes joined together with CJP longitudinal seams acting as chord members designed using the provisions of Clause 10. For design purposes, this clause shall be used with the applicable requirements of Clause 4, Part A. All provisions of Clause 10 apply to both static and cyclic applications, with the exception that the fatigue provisions of 10.2.3, only apply to cyclic applications.

10.1.2 Clause 10 is divided into six parts, as follows:

Part A — **Design of Tubular Connections**

Part B — **Prequalification of Welding Procedure Specifications (WPSs)**

Part C — **Welding Procedure Specification (WPS) Qualification**

Part D — **Performance Qualification**

Part E — **Fabrication**

Part F — **Inspection**

Part A
Design of Tubular Connections

10.2 Design Criteria

10.2.1 Statically Loaded Tubular Connections. Provisions for the design of statically loaded tubular members and connections are given in ANSI/AISC 360, *Specification for Structural Steel Buildings,* with further guidance, in the AISC *Steel Construction Manual,* and AISC Design Guide No. 24, *Hollow Structural Section Connections.* AISC permits member and connection design using either allowable strength design (ASD) or load and resistance factor design (LRFD) formats.

For the preliminary static design of circular tube connections a simple punching shear check for connection efficiency is provided in Annex R.

10.2.2 Welds Stresses. The available stresses in welds shall not exceed those given in Table 10.2, except as modified by 10.5 and ANSI/AISC 360 as it applies to rectangular and square HSS.

10.2.3 Cyclically Loaded Tubular Connections. Tubular connections subjected to repeated applications of stress shall be designed for fatigue in accordance with this Clause in addition to the general provisions of Clause 4, Part C—*Specific Requirements for Design on Nontubular Connections (Cyclically Loaded)*. For stress range and corresponding cycle count below the values of the K2 curve in Figure 10.1 fatigue analysis is not necessary. Other fatigue analysis exceptions shall be approved by the Engineer.

The applicable stress range is the nominal stress range calculated by ordinary beam theory or similar consideration of the total load and moment on a gross section and the resulting axial, shear and bending stress ranges. Fatigue Stress Design Parameters and S-N curves (i.e. Stress vs. Number of Cycles Curves) for various stress categories are included in 10.2.3.3.

10.2.3.1 Stress Range and Member Type. In the design of members and connections subject to repeated variations in live load stress, consideration shall be given to the number of stress cycles, the expected range of stress, and type and location of member or detail.

10.2.3.2 Fatigue Stress Categories. The type and location of material shall be categorized as shown in Tables 10.1 and 10.3.

10.2.3.3 Basic Allowable Stress Limitation. Where the applicable design specification has a fatigue requirement, the maximum stress shall not exceed the basic allowable stress provided elsewhere, and the range of stress at a given number of cycles shall not exceed the values given in Figure 10.1.

10.2.3.4 Cumulative Damage. Where the fatigue environment involves stress ranges of varying magnitude and varying numbers of applications, the cumulative fatigue damage ratio, D, summed over all the various loads, shall not exceed unity, where

$$D = \sum \frac{n}{N}$$

where

n = number of cycles applied at a given stress range

N = number of cycles for which the given stress range would be allowed in Figure 10.1

10.2.3.5 Critical Members. For critical members whose sole failure mode would be catastrophic, D (see 10.2.3.4) shall be limited to a fractional value of 1/3 maximum.

10.2.3.6 Fatigue Behavior Improvement. For the purpose of enhanced fatigue behavior, and where specified in contract documents, the following profile improvements may be undertaken for welds in tubular T-, Y-, or K-connections:

(1) A capping layer may be applied so that the as-welded surface merges smoothly with the adjoining base metal, and approximates the profile shown in Figure 10.11. Notches in the profile shall not be deeper than 0.04 in or 1 mm, relative to a disc having a diameter equal to or greater than the branch member thickness.

(2) The weld surface may be ground to the profile shown in Figure 10.11. Final grinding marks shall be transverse to the weld axis.

(3) The toe of the weld may be peened with a blunt instrument, so as to produce local plastic deformation which smooths the transition between weld and base metal, while inducing a compressive residual stress. Such peening shall always be done after visual inspection, and be followed by MT as described below. Consideration should be given to the possibility of locally degraded notch toughness due to peening.

In order to qualify fatigue categories X_1 and K_1, representative welds (all welds where peening has been applied) shall receive MT for surface and near-surface discontinuities. Any indications which cannot be resolved by light grinding shall be repaired in conformance with 7.25.1.4.

10.2.3.7 Size and Profile Effects. Applicability of welds to the fatigue categories listed below is limited to the following maximum weld size or base metal thicknesses:

C₁   2 in [50 mm] thinner member at transition
C₂   1 in [25 mm] attachment
D   1 in [25 mm] attachment
E   1 in [25 mm] attachment
ET   1.5 in [38 mm] branch
F   0.7 in [18 mm] weld size
FT   1 in [25 mm] weld size

For applications exceeding these limits, consideration should be given to reducing the allowable stresses or improving the weld profile (see Commentary). For T-, Y-, and K-connections, two levels of fatigue performance are provided for in Table 10.4. The designer shall designate when Level I is to apply; in the absence of such designation, and for applications where fatigue is not a consideration, Level II shall be the minimum acceptable standard.

10.3 Identification and Parts of Tubular Connections

Members in tubular structures shall be identified as shown in Figure 10.2.

10.4 Symbols

Symbols used in this clause are as shown in Annex I.

10.5 Weld Design

10.5.1 Weld Effective Lengths. For branch elements (including plates) transverse to the direction of the chord member or transverse to the overlapped branch member axis in an overlapped connection, the design of welds shall involve a consideration of the weld effective length (a proportion of the weld total length), in conformance with 10.5.5 or 10.5.6 for circular, square, and rectangular tubulars, respectively.

10.5.2 Fillet Welds

10.5.2.1 Effective Area. The effective area shall be in conformance with 4.4.2.10. The total length is calculated as the perimeter of the branch member at its intersection with the chord member. For rectangular tubulars, the total weld length shall be calculated as the perimeter of the branch member at its intersection with the chord member or by treating the rectangular branch as a box shape (and ignoring the corners). Weld effective length considerations may be waived if the fillet weld is designed to develop the yield strength of the adjoining branch member wall.

10.5.2.2 Beta Limitation for Prequalified Details. Details for prequalified fillet welds in tubular T-, Y-, and K-connections are described in Figure 10.5. These details shall be limited to $\beta \leq 1/3$ for circular tubular connections with circular chords, and $\beta \leq 0.8$ for tubular connections with rectangular chords. These details are subject to the limitations of 10.8.1. For a rectangular tubular chord member with large corner radii, a smaller limit on β may be required to keep the branch member and the weld on the flat face.

10.5.2.3 Lap Joints. Lap joints of telescoping tubulars may be single pass or multi-pass fillet welded in conformance with Figure 10.3.

10.5.3 Groove Welds. The effective area shall be in conformance with 4.4.1.5. The total weld length is calculated as the perimeter of the branch member at its intersection with the chord member.

10.5.3.1 Prequalified PJP Groove Weld Details. Prequalified PJP groove welds in tubular T-, Y-, or K-connections shall conform to Figure 10.6. The Engineer shall use the figure in conjunction with Table 10.5 to calculate the minimum required weld size. To determine the minimum weld size, the Z loss dimension shall be deducted from the distance between the joint root and the weld surface.

10.5.3.2 Prequalified CJP Groove Weld Details Welded from One Side without Backing in T-, Y-, and K-Connections. See 10.10.2 for the detail options. If fatigue behavior improvement is required, the details selected shall be based on the profile requirements of 10.2.3.6 and Table 10.4. Weld stress calculations shall not be required for CJP welds if all of the following conditions apply:

(1) the base metal configuration satisfies code requirements,

(2) matching or one strength level higher filler metal is used in accordance with the code, and

(3) the attached members are adequately designed to the relevant design standard.

10.5.4 Stresses in PJP and Fillet Welds. The maximum stress at any point in the weld shall not exceed the allowable stress in Table 10.2 or the corresponding local branch stress computed by:

$$f_{weld} = Q_z (t_b/t_w) [(f_a/K_a) + (f_b/K_b)]$$

where

t_b = thickness of branch member

t_w = effective throat of the weld

f_a and f_b = nominal axial and bending stresses in the branch

Q_z, K_a and K_b are factors given in 10.5.5.

10.5.5 Weld Effective Lengths in Circular-to-Circular Tubular Connections. When using fillet welds or PJP welds sized only to resist required load or forces, the minimum required effective throat shall be determined as follows:

$$t_w = (t_b / f_{weld}) [(f_a/K_a) + (f_b/K_b)] Q_z$$

t_w = effective throat of weld

t_b = thickness of branch member

f_{weld} = available weld stress as applicable, for ASD from Table 4.3, or for both LRFD and ASD from AISC 360 Table J2.5

f_a and f_b = required axial and bending stresses in the branch, ASD or LRFD as applicable

$K_a = (1 + 1/\sin \Theta_1)/2$ for branch axial loading

$$K_b = \sqrt{K_{bip}^2 + K_{bop}^2}$$

$K_{bip} = (3 + 1/\sin \Theta_1)/4 \sin \Theta_1$ for branch in-plane bending

$K_{bop} = (1 + 3/\sin \Theta_2)/4 \sin \Theta_2$ for branch out-of-plane bending

where Θ_1 is the in-plane acute angle between the branch and chord member axes and Q_2 is the out-of-plane acute angle between the branch and chord member axes.

Q_z = load amplification factor to account for non-uniform distribution of load (or alternatively a weld effective length factor), as follows:

Electrode minimum specified tensile strength = 60 – 70 ksi

	ASD	LRFD
E60XX and E70XX—	1.35	1.5
Higher strengths—	1.60	1.8

10.5.6 Weld Effective Lengths in Rectangular T-, Y-, and K- Connections. Weld effective lengths and effective section properties, for rectangular T-,Y-, and K-connections shall comply with AISC 360 *Specification for Structural Steel Buildings*, Chapter K5.

10.6 Thickness Transition

Tension butt joints in cyclically loaded axially aligned primary members of different material thicknesses or size shall be made in such a manner that the slope through the transition zone does not exceed 1 in 2–1/2. The transition shall be accomplished by chamfering the thicker part, sloping the weld metal, or by any combination of these methods (see Figure 10.4).

10.7 Material Limitations

10.7.1 Material Limitations for Tubular T-, Y-, K-Connections. Tubular connection materials shall be selected understanding that local stress concentrations typically exist which may lead to local yielding and plastic strains at the design load. Likewise, cyclically loaded tubular connection materials shall be selected considering the additional demands on ductility brought forth by weld discontinuities and potential fatigue cracks.

10.7.2 Tubular Base Metal Notch Toughness

10.7.2.1 CVN Test Requirements. Welded tubular members in tension shall be required to demonstrate CVN test absorbed energy of 20 ft·lbf at 70°F [27 J at 21°C] for base metal thickness of 2 in [50 mm] or greater with a specified minimum yield strength of 40 ksi [280 MPa] or greater.

CVN testing shall be in conformance with ASTM A673 (Frequency H, heat lot). For the purposes of this subclause, a tension member is defined as one having more than 10 ksi [70 MPa] tensile stress due to unfactored design loads.

10.7.2.2 LAST Requirements. Fatigue-controlled tubulars used as the main member or heavy wall joint-can in T-, Y-, and K-connections, shall be required to demonstrate CVN test absorbed energy of 20 ft·lbf [27 J] at the Lowest Anticipated Service Temperature (LAST) for the following conditions:

(1) Base metal thickness of 2 in [50 mm] or greater.

(2) Base metal thickness of 1 in [25 mm] or greater with a specified yield strength of 50 ksi [345 MPa] or greater.

When the LAST is not specified, or the structure is not governed by cyclic or fatigue loading, testing shall be at a temperature not greater than 40°F [4°C]. CVN testing shall normally represent the as-furnished tubulars, and be tested in conformance with ASTM A673 Frequency H (heat lot).

Part B
Prequalification of Welding Procedure Specifications (WPSs)

10.8 Fillet Weld Requirements

10.8.1 Details. For prequalified status, fillet welded tubular connections shall conform to the following provisions:

(1) **Prequalified WPSs.** Fillet welded tubular connections made by SMAW, GMAW, or FCAW processes that may be used without performing WPS qualification tests are detailed in Figure 10.5 (see 10.5.2.2 for limitations). These details may also be used for GMAW-S qualified in conformance with 10.14.4.3.

(2) Details for lap joints are shown in Figure 10.3.

10.9 PJP Requirements

10.9.1 Details. Details for PJP tubular groove welds that are accorded prequalified status shall conform to the following provisions:

(1) PJP tubular groove welds, other than T-, Y-, and K-connections, may be used without performing the WPS qualification tests, when these may be applied and shall meet all of the joint dimension limitations as described in Figure 5.1.

(2) PJP T-, Y-, and K-tubular connections, welded only by the SMAW, GMAW, or FCAW process, may be used without performing the WPS qualification tests, when they may be applied and shall meet all of the joint dimension limitations as described in Figure 10.6. These details may also be used for GMAW-S qualified in conformance with 10.14.4.3.

10.9.1.1 Matched Box Connections. Details for PJP groove welds in these connections, the corner dimensions and the radii of the main tube are shown in Figure 10.6. Fillet welds may be used in toe and heel zones (see Figure 10.5). If the corner dimension or the radius of the main tube, or both, are less than as shown in Figure 10.6, a sample joint of the side detail shall be made and sectioned to verify the weld size. The test weld shall be made in the horizontal position. This requirement may be waived if the branch tube is beveled as shown for CJP groove welds in Figure 10.7.

10.10 CJP Groove Weld Requirements

10.10.1 Butt Joints. For tubular groove welds to be given prequalified status, the following conditions shall apply:

(1) **Prequalified WPSs.** Where welding from both sides or welding from one side with backing is possible, any WPS and groove detail that is appropriately prequalified in conformance with Clause 5 may be used, except that SAW is only prequalified for diameters greater than or equal to 24 in [600 mm]. Welded joint details shall be in conformance with Clause 5.

(2) **Nonprequalified Joint Detail.** There are no prequalified joint details for CJP groove welds in butt joints made from one side without backing (see 10.14.2).

10.10.2 Tubular T-, Y-, and K-Connections. Details for CJP groove welds welded from one side without backing in tubular T-, Y-, and K-connections used in circular tubes are described in this section. The applicable range of Details A, B, C, and D are described in Figures 10.7 and 10.8, and the ranges of local dihedral angles, [Ψ], corresponding to these are described in Table 10.6.

Joint dimensions including groove angles are described in Table 10.7 and Figure 10.9. When selecting a profile (compatible with fatigue category used in design) as a function of thickness, the guidelines of 10.2.3.7 shall be observed. Alternative weld profiles that may be required for thicker sections are described in Figure 10.10. In the absence of special fatigue requirements, these profiles shall be applicable to branch thicknesses exceeding 5/8 in [16 mm].

Improved weld profiles meeting the requirements of 10.2.3.6 and 10.2.3.7 are described in Figure 10.11. In the absence of special fatigue requirements, these profiles shall be applicable to branch thicknesses exceeding 1–1/2 in [38 mm] (not required for static compression loading). Prequalified details for CJP groove welds in tubular T-, Y-, and K-connections utilizing box sections are further described in Figure 10.7. The foregoing details are subject to the limitation of 10.10.1.

NOTE: See the Commentary (C-10.14.4) for engineering guidance in the selection of a suitable profile.

The joint dimensions and groove angles shall not vary from the ranges detailed in Table 10.7 and shown in Figure 10.7 and Figures 10.9–10.11. The root face of joints shall be zero unless dimensioned otherwise. It may be detailed to exceed zero or the specified dimension by not more than 1/16 in [2 mm]. It may not be detailed less than the specified dimensions.

10.10.2.1 Joint Details. Details for CJP groove welds in tubular T-, Y-, and K-connections are described in 10.10.2. These details are prequalified for SMAW and FCAW. These details may also be used for GMAW-S qualified in conformance with 10.14.4.3.

Part C
Welding Procedure Specification (WPS) Qualification

10.11 Common Requirements for WPS and Welding Personnel Performance Qualification

10.11.1 Positions of Welds. All welds shall be classified as flat (F), horizontal (H), vertical (V), and overhead (OH), in conformance with the definitions shown in Figures 6.1 and 6.2.

Test assembly positions are shown in:

(1) Figure 10.12 (groove welds in pipe or tubing)

(2) Figure 10.13 (fillet welds in pipe or tubing)

10.12 Production Welding Positions Qualified

Tubular production welding positions qualified by a tubular test shall conform to the requirements of Table 10.8. Tubular production welding positions qualified by a plate test shall conform to the requirements in Clause 6 and Table 6.1.

10.13 Type of Qualification Tests, Methods of Testing, and Acceptance Criteria for WPS Qualification

The type and number of qualification tests required to qualify a WPS for a given thickness, diameter or both, shall conform to Table 10.9 (CJP), Table 10.10 (PJP), or Table 10.11 (Fillet). Details on the individual NDT and mechanical test requirements are found in 6.5.

The welded test assemblies shall have test specimens prepared by cutting the pipe or tubing as shown in Figure 10.14 or Figure 10.15. Methods of testing and acceptance criteria shall be per 6.10 with the following exceptions:

(1) For T-, Y- and K-connections melt-through is not limited.

(2) For tubulars, the full circumference of the weld shall be RT or UT examined in conformance with Clause 8, Part C and Clause 10, Part F, as applicable.

10.14 CJP Groove Welds for Tubular Connections

CJP groove welds shall be classified as follows:

(1) CJP butt joints with backing or backgouging (see 10.14.1).

(2) CJP butt joints without backing welded from one side only (see 10.14.2).

(3) T-, Y-, K-connections with backing or backgouging (see 10.14.3).

(4) T-, Y-, K-connections without backing welded from one side only (see 10.14.4).

10.14.1 CJP Butt Joints with Backing or Backgouging. A WPS with backing or backgouging shall be qualified using the detail shown in Figure 10.18(A) (with backgouging) or Figure 10.18(B) (with backing).

10.14.2 CJP Butt Joints without Backing Welded from One Side Only. A WPS without backing welded from one side only shall be qualified using the joint detail shown in Figure 10.18(A).

10.14.3 T-, Y-, or K-Connections with Backing or Backgouging. A WPS for tubular T-, Y-, or K-connections with backing or backgouging shall be qualified using:

(1) the appropriate nominal pipe OD selected from Table 10.9, and

(2) the joint detail of Figure 10.18(B), or

(3) for nominal pipe ODs equal to or greater than 24 in [600 mm], a plate qualification in conformance with 6.10 using the joint detail of Figure 10.18(B).

10.14.4 T-, Y-, or K-Connections without Backing Welded from One Side Only. When qualification is required, a WPS for T-, Y-, or K-connections without backing welded from one side only shall require the following:

10.14.4.1 WPSs without Prequalified Status. For a WPS whose essential variables are outside the prequalified range, qualification for CJP tubular groove welds shall require the following:

(1) Qualification in conformance with Figure 10.20 for pipes with outside diameters greater than or equal to 4 in [100 mm] or Figure 10.20 and Figure 10.22 for box tubes. Qualification in conformance with Figure 10.21 for pipes with outside diameters less than 4 in [100 mm] or Figure 10.22 for box tubes.

(2) A Sample Joint or Tubular Mock-Up. The sample joint or tubular mock-up shall provide at least one macroetch test section for each of the following conditions:

(a) The groove combining the greatest groove depth with the smallest groove angle, or combination of grooves to be used: test with welding position vertical.

(b) The narrowest root opening to be used with a 37.5° groove angle: one test welded in the flat position and one test welded in the overhead position.

(c) The widest root opening to be used with a 37.5° groove angle: one test to be welded in the flat position and one test to be welded in the overhead position.

(d) For matched box connections only, the minimum groove angle, corner dimension and corner radius to be used in combination: one test in horizontal position.

(3) The macroetch test specimens required in (1) and (2) above shall be examined for discontinuities and shall have:

(a) No cracks

(b) Thorough fusion between adjacent layers of weld metal and between weld metal and base metal

(c) Weld details conforming to the specified detail but with none of the variations prohibited in 7.23.

(d) No undercut exceeding the values allowed in 8.9.

(e) For porosity 1/32 in [1 mm] or larger, accumulated porosity shall not exceed 1/4 in [6 mm]

(f) No accumulated slag, the sum of the greatest dimension of which shall not exceed 1/4 in [6 mm]

Those specimens not conforming to (a) through (f) shall be considered unacceptable; (b) through (f) not applicable to backup weld.

10.14.4.2 CJP Groove Welds in a T-, Y-, or K-Connection WPS with Dihedral Angles Less than 30°. The sample joint described in 10.14.4.1(2)(a) shall be required. Three macroetch test sections shall be cut from the test specimens, shall conform to the requirements of 10.14.4.1(3) and shall show the required theoretical weld (with due allowance for backup welds to be discounted, as shown in Details C and D of Figures 10.9–10.11) (see Figure 10.19 for test joint details).

10.14.4.3 CJP Groove Welds in a T-, Y-, or K-Connection WPS Using GMAW-S. For T-, Y-, and K-connections, where GMAW-S is used, qualification in conformance with Clause 6 shall be required prior to welding the standard joint configurations detailed in 10.10.2. The joint tested shall incorporate a 37.5° single bevel groove, offset root and restriction ring as shown in Figure 10.20.

10.14.4.4 Weldments Requiring CVN Toughness. WPSs for butt joints (longitudinal or circumferential seams) within 0.5D of attached branch members, in tubular connection joint-cans requiring CVN testing under 10.7.2.2, shall be required to demonstrate weld metal CVN absorbed energy of 20 ft·lbf [27 J] at the LAST (Lowest Anticipated Service Temperature), or at 0°F [-18°C], whichever is lower. If AWS specifications for the welding materials to be used do not encompass this requirement, or if production welding is outside the range covered by prior testing, e.g., tests per AWS filler metal specifications, then weld metal CVN tests shall be made during WPS qualification, as described in Clause 6, Part D.

10.15 PJP and Fillet Welds Tubular T-, Y-, or K-Connections and Butt Joints

When PJP groove welds are specified, in T-, Y-, or K-connections or butt welds, qualification shall be in conformance with Table 10.10. When fillet welds are specified in T-, Y- or K-connections, qualification shall be in conformance with 6.13, Table 10.11 and Figure 10.16.

Part D
Performance Qualification

10.16 Production Welding Positions, Thicknesses, and Diameters Qualified

10.16.1 Welders and Welding Operators. The qualified tubular production welding positions qualified by a tubular test for welders and welding operators shall be in conformance with Table 10.12. The qualified tubular production welding positions qualified by a plate test for welders and welding operators shall be in conformance with Clause 6 and Table 6.10.

For tests on tubulars, the number and type of test specimens and the range of qualified production welding thicknesses and diameters for which a welder or welding operator is qualified shall be in conformance with Table 10.13. Mechanical test specimens shall be prepared by cutting the pipe or tubing as shown in Figure 10.23 and as specified in 6.17.1.2. If tests are performed using plate, the qualification limitations shall be in conformance with Table 6.11.

10.16.2 Tack Welders. Tack welder qualification shall qualify for tubular thickness greater than 1/8 in [3 mm] and all diameters, but does not include CJP butt joints and T-, Y-, and K-connections welded from one side. Tack welds in the foregoing exception shall be performed by welders fully qualified for the process and position in which the welding is to be done.

10.17 Weld Types for Welder and Welding Operator Performance Qualification

For the purpose of welder and welding operator qualification, weld types shall be classified as follows:

(1) CJP Groove Welds for Tubular Connections (see 10.18)

(2) PJP Groove Welds for Tubular Connections (see 10.19)

(3) Fillet Welds for Tubular Connections (see 10.20)

Plate welds parallel to the tubular centerline and tubular longseam welds do not require tubular qualification and may be qualified using Clause 6.

10.18 CJP Groove Welds for Tubular Connections

Welder or welding operator qualification tests shall use the following details:

(1) CJP groove butt joints with backing or backgouging in pipe. Use Figure 10.17(B).

(2) CJP groove butt joints without backing or backgouging. Use Figure 10.17(A).

(3) CJP groove butt joints or T-, Y-, and K-connections with backing in box tubing. Use Figure 10.17(B) in pipe (any diameter), plate or box tubing.

(4) CJP groove T-, Y-, and K-Connections welded from one side with backing in pipe. Use Figure 10.17(B) in pipe of the appropriate diameter.

(5) CJP groove T-, Y-, and K-connections welded from one side without backing in pipe. Use Figure 10.20 for nominal pipe diameter of ≥ 6 in [150 mm] or Figure 10.21 for nominal pipe ≤ 4 in [100 mm].

(6) CJP groove T-, Y-, and K-connection welded from one side without backing or backgouging in box tubing. The options are the following:

(a) Figure 10.20 in pipe (any diameter) or box tubing plus Figure 10.22 in box tubing.

(b) Figure 10.20 in box tubing with macroetch specimens removed from the locations shown in Figure 10.22.

See Table 10.13 for the production ranges of diameter and thickness qualified by the test assembly diameters and thicknesses.

10.18.1 Other Joint Details or WPSs. For joint details, WPSs, or assumed depth of sound welds that are more difficult than those described herein, a test described in 10.14.4.2 shall be performed by each welder in addition to the 6GR tests (see Figures 10.21 or 10.22). The test position shall be vertical.

10.19 PJP Groove Welds for Tubular Connections

Qualification for CJP groove welds on tubular connections shall qualify for all PJP groove welds.

10.20 Fillet Welds for Tubular Connections

See Table 10.13 for fillet weld qualification requirements.

10.21 Methods of Testing and Acceptance Criteria for Welder and Welding Operator Qualification

10.21.1 Macroetch Test for T-, Y-, and K-Connections. The corner macroetch test joint for T-, Y-, and K-connections on box tubing in Figure 10.22 shall have four macroetch test specimens cut from the weld corners at the locations shown in Figure 10.22. One face from each corner specimen shall be smooth for etching. If the welder tested on a 6GR coupon (Figure 10.21) using box tubing, the four required corner macroetch test specimens may be cut from the corners of the 6GR coupon in a manner similar to Figure 10.22. One face from each corner specimen shall be smooth for etching.

10.21.1.1 Macroetch Test Acceptance Criteria. For acceptable qualification, the test specimen, when inspected visually, shall conform to Clause 6 and the following requirements:

(1) Fillet welds and the corner macroetch test joint for T-, Y-, and K-connections on box tubing, Figure 10.22, shall have:

   (a) No cracks

   (b) Thorough fusion between adjacent layers of weld metals and between weld metal and base metal

   (c) Weld profiles conforming to intended detail, but with none of the variations prohibited in 7.23

   (d) No undercut exceeding 1/32 in [1 mm]

   (e) For porosity 1/32 in [1 mm] or larger, accumulated porosity not exceeding 1/4 in [6 mm]

   (f) No accumulated slag, the sum of the greatest dimensions of which shall not exceed 1/4 in [4 mm]

10.21.2 RT Test Procedure and Technique. Welded test pipe or tubing 4 in [100 mm] in diameter or larger shall be examined for a minimum of one-half of the weld perimeter selected to include a sample of all positions welded. For example, a test pipe or tube welded in the 5G, 6G, or 6GR position shall be radiographed from the top centerline to the bottom centerline on either side. Welded test pipe or tubing less than 4 in [100 mm] in diameter shall require 100% RT.

Part E
Fabrication

10.22 Backing

10.22.1 Full-Length Backing. Except as permitted below, steel backing shall be made continuous for the full length of the weld. All joints in the steel backing shall be CJP groove weld joints meeting all the requirements of Clause 7 of this code. For statically loaded applications, backing for welds to the ends of closed sections, such as hollow structural sections (HSS), are permitted to be made from one or two pieces with unspliced discontinuities where all of the following conditions are met:

(1) The closed section nominal wall thickness does not exceed 5/8 in [16 mm].

(2) The closed section outside perimeter does not exceed 64 in [1625 mm].

(3) The backing is transverse to the longitudinal axis of the closed section.

(4) The interruption in the backing does not exceed 1/4 in [6 mm].

(5) The weld with discontinuous backing is not closer than the HSS diameter or major cross section dimension from other types of connections.

(6) The interruption in the backing is not located in the corners.

For statically loaded box columns, discontinuous backing is permitted in the CJP welded corners, at field splices and at connection details. Discontinuous backing is permitted in other closed sections where approved by the Engineer.

NOTE: Commercially available steel backing for pipe and tubing is acceptable, provided there is no evidence of melting on exposed interior surfaces.

10.23 Tolerance of Joint Dimensions

10.23.1 Girth Weld Alignment (Tubular). Abutting parts to be joined by girth welds shall be carefully aligned. No two girth welds shall be located closer than one pipe diameter or 3 ft [1 m], whichever is less. There shall be no more than two girth welds in any 10 ft [3 m] interval of pipe, except as may be agreed upon by the Owner and Contractor. Radial offset of abutting edges of girth seams shall not exceed 0.2t (where t is the thickness of the thinner member) and the maximum allowable shall be 1/4 in [6 mm], provided that any offset exceeding 1/8 in [3 mm] is welded from both sides. However, with the approval of the Engineer, one localized area per girth seam may be offset up to 0.3t with a maximum of 3/8 in [10 mm], provided the localized area is under 8t in length. Filler metal shall be added to this region to provide a 4 to 1 transition and may be added in conjunction with making the weld. Offsets in excess of this shall be corrected as provided in 7.21.3. Longitudinal weld seams of adjoining sections shall be staggered a minimum of 90°, unless closer spacing is agreed upon by the Owner and fabricator.

10.23.2 Groove Dimensions

10.23.2.1 Tubular Cross-Sectional Variations. Variation in cross section dimension of groove welded joints, from those shown on the detailed drawings, shall be in conformance with 7.21.4.1 except:

(1) Tolerances for T-, Y-, and K-connections are included in the ranges given in 10.10.2.

(2) The tolerances shown in Table 10.14 apply to CJP tubular groove welds in butt joints, made from one side only, without backing.

Part F
Inspection

10.24 Visual Inspection

All welds shall be visually inspected and shall be acceptable if the criteria in Table 10.15 are satisfied.

10.25 NDT

10.25.1 Scope. Acceptance criteria for inspection of tubular connections are described in Part C of Clause 8, as applicable, and Part F of this clause. The acceptance criteria shall be specified in the contract documents on information furnished to the bidder.

10.25.2 Tubular Connection Requirements. For CJP groove butt joints welded from one side without backing, the entire length of all completed tubular production welds shall be examined by either RT or UT. The acceptance criteria shall conform to 8.12.1 for RT (see Figure 8.1) or 10.26.1 for UT, as applicable.

10.26 UT

10.26.1 Acceptance Criteria for Tubular Connections. Acceptance criteria for UT shall be as provided in contract documents. Class R or Class X, or both, may be incorporated by reference. Amplitude based acceptance criteria as given by 8.13.1 may also be used for groove welds in butt joints in tubing 24 in [600 mm] in diameter and over, provided all relevant provisions of Clause 8, Part F, are followed. However, these amplitude criteria shall not be applied to tubular T, Y-, and K-connections.

10.26.1.1 Class R (Applicable When UT is Used as an Alternate to RT). All indications having one-half (6 dB) or less amplitude than the standard sensitivity level (with due regard for 10.29.6) shall be disregarded. Indications exceeding the disregard level shall be evaluated as follows:

(1) Isolated random spherical reflectors, with 1 in [25 mm] minimum separation up to the standard sensitivity level shall be accepted. Larger reflectors shall be evaluated as linear reflectors.

(2) Aligned spherical reflectors shall be evaluated as linear reflectors.

(3) Clustered spherical reflectors having a density of more than one per square inch [645 square millimeters] with indications above the disregard levels (projected area normal to the direction of applied stress, averaged over a 6 in [150 mm] length of weld) shall be rejected.

(4) Linear or planar reflectors whose lengths (extent) exceed the limits of Figure 10.24 shall be rejected. Additionally, root reflectors shall not exceed the limits of Class X.

10.26.1.2 Class X (Experience-Based, Fitness for Purpose Criteria Applicable to T-, Y-, and K-Connections in Structures with Notch-Tough Weldments). All indications having half (6 dB) or less amplitude than the standard sensitivity level (with due regard for 10.29.6) shall be disregarded. Indications exceeding the disregard level shall be evaluated as follows:

(1) Spherical reflectors shall be as described in Class R, except that any indications within the following limits for linear or planar shall be acceptable.

(2) Linear or planar reflectors shall be evaluated by means of beam boundary techniques, and those whose dimensions exceeded the limits of Figure 10.25 shall be rejected. The root area shall be defined as that lying within 1/4 in [6 mm] or $t_w/4$, whichever is greater, of the root of the theoretical weld, as shown in Figure 10.9.

10.27 RT Procedures

10.27.1 Procedure. In addition to the requirements of 8.17, IQI selection shall conform to Tables 10.16 and 10.17 and Figures 8.4 and 8.5.

10.27.2 IQI Selection and Placement. IQIs shall be selected and placed on the weldment in the area of interest being radiographed as shown in Table 10.18. When a complete circumferential pipe weld is radiographed with a single exposure and the radiation source is placed at the center of the curvature, at least three equally spaced IQIs shall be used. Steel backing shall not be considered part of the weld or weld reinforcement in IQI selection.

10.28 Supplementary RT Requirements for Tubular Connections

10.28.1 Circumferential Groove Welds in Butt Joints. The technique used to radiograph circumferential butt joints shall be capable of covering the entire circumference. The technique shall preferably be single-wall exposure/ single-wall view. Where accessibility or pipe size prohibits this, the technique may be double-wall exposure/ single-wall view or double-wall exposure/double-wall view.

10.28.1.1 Single-Wall Exposure/Single-Wall View. The source of radiation shall be placed inside the pipe and the film on the outside of the pipe (see Figure 10.26). Panoramic exposure may be made if the source-to-object requirements are satisfied; if not, a minimum of three exposures shall be made. The IQI may be selected and placed on the source side of the pipe. If not practicable, it may be placed on the film side of the pipe.

10.28.1.2 Double-Wall Exposure/Single-Wall View. Where access or geometrical conditions prohibit single-wall exposure, the source may be placed on the outside of the pipe and film on the opposite wall outside the pipe (see Figure 10.27). A minimum of three exposures shall be required to cover the complete circumference. The IQI may be selected and placed on the film side of the pipe.

10.28.1.3 Double-Wall Exposure/Double-Wall View. When the outside diameter of the pipe is 3–1/2 in [90 mm] or less, both the source side and film side weld may be projected onto the film and both walls viewed for acceptance. The source of radiation shall be offset from the pipe by a distance that is at least seven times the outside diameter. The radiation beam shall be offset from the plane of the weld centerline at an angle sufficient to separate the images of the source side and film side welds. There shall be no overlap of the two zones interpreted. A minimum of two exposures 90° to each other shall be required (see Figure 10.28). The weld may also be radiographed by superimposing the two welds, in which case there shall be a minimum of three exposures 60° to each other (see Figure 10.29). In each of these two techniques, the IQI shall be placed on the source side of the pipe.

10.29 UT of Tubular T-, Y-, and K-Connections

10.29.1 Procedure. All UT shall be in conformance with a written procedure which has been prepared or approved by an individual certified as SNT-TC-1A, Level III, and experienced in UT of tubular structures. The procedure shall be based upon the requirements of this clause and Clause 8, Part F, as applicable. The procedure shall contain, as a minimum, the following information regarding the UT method and techniques:

(1) The type of weld joint configuration to be examined (i.e., the applicable range of diameter, thickness, and local dihedral angle). Conventional techniques are generally limited to diameters of 12–3/4 in [325 mm] and larger, thicknesses of 1/2 in [12 mm] and above, and local dihedral angles of 30° or greater. Special techniques for smaller sizes may be used, provided they are qualified as described herein, using the smaller size of application.

(2) Acceptance criteria for each type and size weld

(3) Type(s) of UT instrumentation (make and model)

(4) Transducer (search unit) frequency, size and shape of active area, beam angle, and type of wedge on angle beam probes. Procedures using transducers with frequencies up to 6 MHz, sized down to 1/4 in [6 mm], and of different shape than specified elsewhere, may be used, provided they are qualified as described herein.

(5) Surface preparation and couplant (where used)

(6) Type of calibration test block and reference reflector

(7) Method of calibration and required accuracy for distance (sweep), vertical linearity, beam spread, angle, sensitivity, and resolution

(8) Recalibration interval for each item in (7) above

(9) Method for determining acoustical continuity of base metal (see 10.29.4), and for establishing geometry as a function of local dihedral angle and thickness

(10) Scanning pattern and sensitivity (see 10.29.5)

(11) Transfer correction for surface curvature and roughness (where amplitude methods are used (see 10.29.3)

(12) Methods for determining effective beam angle (in curved material), indexing root area, and discontinuity locations

(13) Method of discontinuity length and height determination

(14) Method of discontinuity verification during excavation and repair

10.29.2 Personnel. In addition to personnel requirements of 8.14.6, when examination of T-, Y-, and K-connections is to be performed, the operator shall be required to demonstrate an ability to apply the special techniques required for such an examination. Practical tests for this purpose shall be performed upon mock-up welds that represent the type of welds to be inspected, including a representative range of dihedral angle and thickness to be encountered in production, using the applicable qualified and approved procedures. Each mock-up shall contain natural or artificial discontinuities that yield UT indications above and below the reject criteria specified in the approved procedure. Performance shall be judged on the basis of the ability of the operator to determine the size and classification of each discontinuity with an accuracy required to accept or reject each weldment and accurately locate the unacceptable discontinuities along the weld and within the cross section of the weld. At least 70% of the unacceptable discontinuities shall be correctly identified as unacceptable. Every discontinuity exceeding its maximum acceptable dimensions by a factor of two, or by an amplitude of 6 dB shall be located and reported.

10.29.3 Calibration. UT equipment qualification and calibration methods shall meet the requirements of the approved procedure and Clause 8, Part F, except as follows:

10.29.3.1 Range. Range (distance) calibration shall include, as a minimum, the entire sound path distance to be used during the specific examination. This may be adjusted to represent either the sound-path travel, surface distance, or equivalent depth below contact surface, displayed along the instrument horizontal scale, as described in the approved procedure.

10.29.3.2 Sensitivity Calibration. Standard sensitivity for examination of production welds using amplitude techniques shall be: basic sensitivity + distant amplitude correction + transfer correction. This calibration shall be performed at least once for each joint to be tested; except that, for repetitive testing of the same size and configuration, the calibration frequency of 8.24.3 may be used.

(1) **Basic Sensitivity**. Reference level screen height obtained using maximum reflection from the 0.060 in [1.5 mm] diameter hole in the IIW type block (or other block which results in the same basic calibration sensitivity) as described in 8.24 (or 8.27).

(2) **Distance Amplitude Correction**. The sensitivity level shall be adjusted to provide for attenuation loss throughout the range of sound path to be used by either distance amplitude correction curves, electronic means, or as described in 8.25.6.4. Where high frequency transducers are used, the greater attenuation shall be taken into account. Transfer correction may be used to accommodate UT through tight layers of paint not exceeding 10 mils [0.25 mm] in thickness.

10.29.4 Base Metal Examination. The entire area subject to UT scanning shall be examined by the longitudinal wave technique to detect laminar reflectors that could interfere with the intended, directed sound wave propagation. All areas containing laminar reflectors shall be marked for identification prior to weld examination and the consequences considered in selection of search unit angles and scanning techniques for examination of the welds in that area. The Engineer shall be notified of base material discontinuities that exceed the limits of 7.14.5.1.

10.29.5 Weld Scanning. Weld scanning of T-, Y-, and K-connections shall be performed from the branch member surface (see Figure 10.30). All examinations shall be made in leg I and II where possible. For initial scanning, the sensitivity shall be increased by 12 dB above that established in 10.29.3 for the maximum sound path. Indication evaluation shall be performed with reference to the standard sensitivity.

10.29.6 Optimum Angle. Indications found in the root areas of groove welds in butt joints and along the fusion face of all welds shall be further evaluated with either 70°, 60°, or 45° search angle, whichever is nearest to being perpendicular to the expected fusion face.

10.29.7 Discontinuity Evaluation. Discontinuities shall be evaluated by use of a combination of beam boundary and amplitude techniques. Sizes shall be given as length and height (depth dimension) or amplitude, as applicable. Amplitude shall be related to "standard calibration." In addition, discontinuities shall be classified as linear or planar versus spherical, by noting changes in amplitude as the transducer is swung in an arc centered on the reflector. The location (position) of discontinuities within the weld cross section, as well as from an established reference point along the weld axis, shall be determined.

10.29.8 Reports

10.29.8.1 Forms. A report form that clearly identifies the work and the area of inspection shall be completed by the UT operator at the time of inspection. A detailed report and sketch showing the location along the weld axis, location within the weld cross section, size (or indication rating), extent, orientation, and classification for each discontinuity shall be completed for each weld in which significant indications are found.

10.29.8.2 Reported Discontinuities. When specified, discontinuities approaching unacceptable size, particularly those about which there is some doubt in their evaluation, shall also be reported.

10.29.8.3 Incomplete Inspection. Areas for which complete inspection was not practicable shall also be noted, along with the reason why the inspection was incomplete.

10.29.8.4 Reference Marks. Unless otherwise specified, the reference position and the location and extent of unacceptable discontinuities shall also be marked physically on the workpiece.

Table 10.1
Fatigue Stress Design Parameters (see 10.2.3.2)

Description	Stress Category	Constant C_f	Threshold F_{TH} ksi [MPa]	Potential Crack Initiation Point	Illustrative Examples
Section 1—Welded Joints Transverse to Direction of Stress					
1.1 Base metal and filler metal in or adjacent to CJP groove welded butt splices with backing left in place. Tack welds inside groove Tack welds outside the groove and not closer than 1/2 in [12 mm] to edge of base metal	D E	22×10^8 11×10^8	7 [48] 4.5 [31]	From the toe of the groove weld or the toe of the weld attaching backing	1.1

Table 10.2
Available Stresses in Tubular Connection Welds (see 10.2.2 and 10.5.4)

Type of Weld	Tubular Application	Kind of Stress	Allowable Stress Design (ASD) Allowable Stress	Load and Resistance Factor Design (LRFD) Resistance Factor Φ	Load and Resistance Factor Design (LRFD) Nominal Strength	Required Filler Metal Strength Level[a]
CJP Groove Weld	Longitudinal butt joints (longitudinal seams)	Tension or compression parallel to axis of the weld[b]	Same as for base metal[c]	0.9	$0.6 F_y$	Filler metal with strength equal to or less than matching filler metal may be used
		Beam or torsional shear	Base metal 0.40 F_y Filler metal 0.3 F_{EXX}	0.9 0.8	$0.6 F_y$ $0.6 F_{EXX}$	
		Compression normal to the effective area[b]		0.9	F_y	
	Circumferential butt joints (girth seams)	Shear on effective area	Same as for base metal	Base metal 0.9 Weld metal 0.8	$0.6 F_y$ $0.6 F_{EXX}$	Matching filler metal shall be used
		Tension normal to the effective area		0.9	F_y	
	Weld joints in structural T-, Y-, or K-connections in structures designed for critical loading such as fatigue, which would normally call for CJP welds	Tension, compression or shear on base metal adjoining weld conforming to detail of Figures 10.7 and 10.9–10.11 (tubular weld made from outside only without backing)	Same as for base metal or as limited by connection geometry	Same as for base metal or as limited by connection geometry		Matching filler metal shall be used
		Tension, compression, or shear on effective area of groove welds, made from both sides or with backing				
Fillet Weld	Longitudinal joints of built-up tubular members	Tension or compression parallel to axis of the weld	Same as for base metal	0.9	F_y	Filler metal with a strength level equal to or less than matching filler metal may be used
		Shear on effective area	$0.30 F^e_{EXX}$	0.75	$0.6 F_{EXX}$	
	Joints in structural T-, Y-, or K-connections in circular lap joints and joints of attachments to tubes	Shear on effective throat regardless of direction of loading (see 10.5)	$0.30 F_{EXX}$ or as limited by connection geometry	0.75	$0.6 F_{EXX}$ or as limited by connection geometry	Filler metal with a strength level equal to or less than matching filler metal may be used

(Continued)

Table 10.2 (Continued)
Available Stresses in Tubular Connection Welds (see 10.2.2 and 10.5.4)

Type of Weld	Tubular Application	Kind of Stress		Allowable Stress Design (ASD) Allowable Stress	Load and Resistance Factor Design (LRFD) Resistance Factor Φ	Load and Resistance Factor Design (LRFD) Nominal Strength	Required Filler Metal Strength Level[a]
Plug and Slot Welds	Shear parallel to faying surfaces (on effective area)			Base metal 0.40 F_y Filler metal 0.3 F_{EXX}	Not Applicable		Filler metal with a strength level equal to or less than matching filler metal may be used
PJP Groove Weld	Longitudinal seam of tubular members	Tension or compression parallel to axis of the weld[b]		Same as for base metal[c]	0.9	F_y	Filler metal with a strength level equal to or less than matching filler metal may be used
	Circumferential and longitudinal joints that transfer loads	Compression normal to the effective area	Joint not designed to bear	0.50 F_{EXX}, except that stress on adjoining base metal shall not exceed 0.60 F_y	0.9	F_y	Filler metal with a strength level equal to or less than matching filler metal may be used
			Joint designed to bear	Same as for base metal			
		Shear on effective area		0.30 F_{EXX}, except that stress on adjoining base metal shall not exceed 0.50 F_y for tension, or 0.40 F_y for shear	0.75	0.6 F_{EXX}	Filler metal with a strength level equal to or less than matching filler metal may be used
		Tension on effective area			Base metal 0.9 Filler metal 0.8	F_y 0.6 F_{EXX}	
	Structural T-, Y-, or K-connection in ordinary structures	Load transfer across the weld as stress on the effective throat (see 10.5)		0.30 F_{EXX} or as limited by connection geometry except that stress on an adjoining base metal shall not exceed 0.50 F_y for tension and compression, nor 0.40 F_y for shear	Base metal 0.9 Filler metal 0.8 or as limited by connection geometry	F_y 0.6 F_{EXX}	Matching filler metal shall be used

[a] For matching filler metal see Table 5.4.
[b] Beam or torsional shear up to 0.30 minimum specified tensile strength of filler metal is allowed, except that shear on adjoining base metal shall not exceed 0.40 F_y (LRFD; see shear).
[c] Groove and fillet welds parallel to the longitudinal axis of tension or compression members, except in connection areas, shall not be considered as transferring stress and hence may take the same stress as that in the base metal, regardless of electrode (filler metal) classification.
[e] Alternatively, see 4.6.4.2 and 4.6.4.3.

Table 10.3
Stress Categories for Type and Location of Material for Circular Sections (see 10.2.3.2)

Stress Category	Situation	Kinds of Stress[a]
A	Plain unwelded pipe	TCBR
B	Pipe with longitudinal seam	TCBR
B	Butt splices, CJP groove welds, ground flush and inspected by RT or UT (Class R)	TCBR
B	Members with continuously welded longitudinal stiffeners	TCBR
C_1	Butt splices, CJP groove welds, as welded	TCBR
C_2	Members with transverse (ring) stiffeners	TCBR
D	Members with miscellaneous attachments such as clips, brackets, etc.	TCBR
D	Cruciform and T-joints with CJP welds (except at tubular connections)	TCBR
DT	Connections designed as a simple T-, Y-, or K-connections with CJP groove welds conforming to Figures 10.9–10.11 (including overlapping connections in which the main member at each intersection meets punching shear requirements) (see Note b)	TCBR in branch member (Note: Main member must be checked separately per category K_1 or K_2)
E	Balanced cruciform and T-joints with PJP groove welds or fillet welds (except at tubular connections)	TCBR in member; weld must also be checked per category F
E	Members where doubler wrap, cover plates, longitudinal stiffeners, gusset plates, etc., terminate (except at tubular connections)	TCBR in member; weld must also be checked per category F
ET	Simple T-, Y-, and K-connections with PJP groove welds or fillet welds; also, complex tubular connections in which the punching shear capacity of the main member cannot carry the entire load and load transfer is accomplished by overlap (negative eccentricity), gusset plates, ring stiffeners, etc. (see Note b)	TCBR in branch member (Note: Main member in simple T-, Y-, or K-connections must be checked separately per category K_1 or K_2; weld must also be checked per category FT)
F	End weld of cover plate or doubler wrap; welds on gusset plates, stiffeners, etc.	Shear in weld
F	Cruciform and T-joints, loaded in tension or bending, having fillet or PJP groove welds (except at tubular connections)	Shear in weld (regardless of direction of loading) (see 10.5)
FT	Simple T-, Y-, or K-connections loaded in tension or bending, having fillet or PJP groove welds	Shear in weld (regardless of direction of loading)
X_2	Intersecting members at simple T-, Y-, and K-connections; any connection whose adequacy is determined by testing an accurately scaled model or by theoretical analysis (e.g., finite element)	Greatest total range of hot spot stress or strain on the outside surface of intersecting members at the toe of the weld joining them—measured after shakedown in model or prototype connection or calculated with best available theory

(Continued)

Table 10.3 (Continued)
Stress Categories for Type and Location of Material for Circular Sections (see 10.2.3.2)

Stress Category	Situation	Kinds of Stress[a]
X_1	As for X_2, profile improved per 10.2.3.6 and 10.2.3.7	As for X_2
X_1	Unreinforced cone-cylinder intersection	Hot-spot stress at angle change; calculate per Note d
K_2	Simple T-, Y-, and K-connections in which the gamma ratio R/t_c of main member does not exceed 24 (see Note c).	Punching shear for main members; calculate per Note e
K_1	As for K_2, profile improved per 10.2.3.6 and 10.2.3.7	

[a] T = tension, C = compression, B = bending, R = reversal—i.e., total range of nominal axial and bending stress.

[b] Empirical curves (Figure 10.1) based on "typical" connection geometries; if actual stress concentration factors or hot spot strains are known, use of curve X_1 or X_2 is preferred.

[c] Empirical curves (Figure 10.1) based on tests with gamma (R/t_c) of 18 to 24; curves on safe side for very heavy chord members (low R/t_c); for chord members (R/t_c greater than 24) reduce allowable stress in proportion to

$$\frac{\text{Allowable fatigue stress}}{\text{Stress from curve K}} \left(\frac{24}{R/t_c}\right)^{0.7}$$

Where actual stress concentration factors or hot-spot strains are known, use of curve X_1 or X_2 is preferred.

[d] Stress concentration factor – $SCF = \dfrac{1}{\cos\bar{\psi}} + 1.17 \tan\bar{\psi}\sqrt{\gamma_b}$

Where

$\bar{\psi}$ = angle change at transition

γ_b = radius to thickness ratio of tube at transition

[e] Cyclic range of punching shear is given by

$$V_p = \tau \sin\theta \left[(\alpha f_a + \sqrt{(0.67 f_{by})^2 + (1.5 f_{bz})^2}) \right]$$

where

t and θ are defined in Figure 10.2, and

f_a = cyclic range of nominal branch member stress for axial load.

f_{by} = cyclic range of in-plane bending stress.

f_{bz} = cyclic range of out-of-plane bending stress.

α is as defined as:		
Chord ovalizing parameter needed for Q_q	$\alpha = 1.0 + 0.7\, g/d_b$ $1.0 \leq \alpha < 1.7$	For axial load in gap K-connections having all members in same plane and loads transverse to main member essentially balanced[f]
	$\alpha = 1.7$ $\alpha = 2.4$	For axial load in T- and Y connections For axial load in cross connections
	$\alpha = 0.67$ $\alpha = 1.5$	For in-plane bending[g] For out-of-plane bending[g]

[f] Gap g is defined in Figures 10.2(E), (F), and (H); d_b is branch diameter.

[g] For combinations of the in-plane bending and out-of-plane bending, use interpolated values of α and λ.

Table 10.4
Fatigue Category Limitations on Weld Size or Thickness and Weld Profile (Tubular Connections) (see 10.2.3.7)

Weld Profile	Level I — Limiting Branch Member Thickness for Categories X_1, X_2, DT in [mm]	Level II — Limiting Branch Member Thickness for Categories X_2, K_2 in [mm]
Standard flat weld profile Figure 10.9	0.375 [10]	0.625 [16]
Profile with toe fillet Figure 10.10	0.625 [16]	1.50 [38] qualified for unlimited thickness for static compression loading
Concave profile, as welded, Figure 10.11 with disk test per 10.2.3.6(1)	1.00 [25]	unlimited
Concave smooth profile Figure 10.11 fully ground per 10.2.3.6(2)	unlimited	—

Table 10.5
Z Loss Dimensions for Calculating Prequalified PJP T-, Y-, and K-Tubular Connection Minimum Weld Sizes (see 10.5.3.1)

Joint Included Angle ϕ	Position of Welding: V or OH			Position of Welding: H or F		
	Process	Z (in)	Z (mm)	Process	Z (in)	Z (mm)
$\phi \geq 60°$	SMAW	0	0	SMAW	0	0
	FCAW-S	0	0	FCAW-S	0	0
	FCAW-G	0	0	FCAW-G	0	0
	GMAW	N/A	N/A	GMAW	0	0
	GMAW-S[a]	0	0	GMAW-S[a]	0	0
$60° > \phi \geq 45°$	SMAW	1/8	3	SMAW	1/8	3
	FCAW-S	1/8	3	FCAW-S	0	0
	FCAW-G	1/8	3	FCAW-G	0	0
	GMAW	N/A	N/A	GMAW	0	0
	GMAW-S[a]	1/8	3	GMAW-S[a]	1/8	3
$45° > \phi \geq 30°$	SMAW	1/4	6	SMAW	1/4	6
	FCAW-S	1/4	6	FCAW-S	1/8	3
	FCAW-G	3/8	10	FCAW-G	1/4	6
	GMAW	N/A	N/A	GMAW	1/4	6
	GMAW-S[a]	3/8	10	GMAW-S[a]	1/4	6

[a] See 10.9.1(2) for qualification requirements for welding prequalified PJP T-, Y-, K-Connections details with GMAW-S.

Table 10.6
Joint Detail Applications for Prequalified CJP T-, Y-, and K-Tubular Connections
(see 10.10.2 and Figure 10.8)

Detail	Applicable Range of Local Dihedral Angle, Ψ
A	180° to 135°
B	150° to 50°
C	75° to 30°
D	40° to 15° } Not prequalified for groove angles under 30°

Notes:
1. The applicable joint detail (A, B, C, or D) for a particular part of the connection shall be determined by the local dihedral angle, Ψ, which changes continuously in progressing around the branch member.
2. The angle and dimensional ranges given in Detail A, B, C, or D include maximum allowable tolerances.
3. See Clause 3 for definition of local dihedral angle.

Table 10.7
Prequalified Joint Dimensions and Groove Angles for CJP Groove Welds in Tubular T-, Y-, and K-Connections Made by SMAW, GMAW-S, and FCAW (see 10.10.2)

	Detail A $\Psi = 180° - 135°$		Detail B $\Psi = 150° - 50°$		Detail C $\Psi = 75° - 30°$[a]		Detail D $\Psi = 40° - 15°$[a]
End preparation (ω) max.			90°[b]		(Footnote b)		
min.			10° or 45° for $\Psi > 105°$		10°		
	FCAW-S SMAW[c]	GMAW-S FCAW-G[d]	FCAW-S SMAW[c]	GMAW-S FCAW-G[d]	(Footnote e)		
						W max.	ϕ
				1/4 in [6 mm] for $\phi > 45°$	FCAW-S SMAW (1)	1/8 in [3 mm] 3/16 in [5 mm]	25°–40° 15°–25°
Fit-up or root opening (R) max.	3/16 in [5 mm]	3/16 in [5 mm]	1/4 in [6 mm]	5/16 in [8 mm] for $\phi \le 45°$			
min.	1/16 in [2 mm] No min. for $\phi > 90°$	1/16 in [2 mm] No min. for $\phi > 120°$	1/16 in [2 mm]	1/16 in [2 mm]	GMAW-S FCAW-G (2)	1/8 in [3 mm] 1/4 in [6 mm] 3/8 in [10 mm] 1/2 in [12 mm]	30°–40° 25°–30° 20°–25° 15°–20°
Joint included angle φ max.	90°		60° for $\Psi \le 105°$		40°; if more use Detail B		
min.	45°		37-1/2°; if less use Detail C		1/2 Ψ		
Completed weld t_w	$\ge t_b$		$\ge t_b$ for $\Psi > 90°$ $\ge t_b$ /sin Ψ for $\Psi < 90°$		$\ge t_b$ /sin Ψ but need not exceed 1.75 t_b		$\ge 2t_b$
L	$\ge t_b$ /sin Ψ but need not exceed 1.75 t_b				Weld may be built up to meet this		

[a] Not prequalified for groove angles (ϕ) under 30°.
[b] Otherwise as needed to obtain required ϕ.
[c] These root details apply to SMAW and FCAW-S.
[d] These root details apply to GMAW-S and FCAW-G.
[e] Initial passes of back-up weld discounted until width of groove (W) is sufficient to assure sound weld.

Notes:
1. For GMAW-S see 10.14.4.3. These details are not intended for GMAW (spray transfer).
2. See Figure 10.9 for minimum standard profile (limited thickness).
3. See Figure 10.10 for alternate toe-fillet profile.
4. See Figure 10.11 for improved profile (see 10.2.3.6 and 10.2.3.7).

Table 10.8
WPS Qualification—Production Welding Positions Qualified by Plate, Pipe, and Box Tube Tests (see 10.12)

Qualification Test		Production Plate Welding Qualified			Production Pipe Welding Qualified					Production Box Tube Welding Qualified				
					Butt Joint		T-, Y-, K- Connections			Butt Joint		T-, Y-, K- Connections		
Weld Type	Test Positions	Groove CJP	Groove PJP	Fillet[a]	CJP	PJP	CJP	PJP	Fillet[a]	CJP	PJP	CJP	PJP	Fillet[a]
CJP Groove	1G Rotated	F	F	F	F[b]	F		F	F	F[b]	F		F	F
	2G	F, H	F, H	F, H	(F, H)[b]	F, H		F, H	F, H	(F, H)[b]	F, H		F, H	F, H
	5G	F, V, OH	F, V, OH	F, V, OH	(F, V, OH)[b]	F, V, OH		F, V, OH	F, V, OH	(F, V, OH)[b]	F, V, OH		F, V, OH	F, V, OH
	(2G + 5G)	All	All	All	All[b]	All	All[c]	All[d]	All	All[b]	All	All[e]	All[d,f]	All
	6G	All	All	All	All[b]	All		All[d]	All	All[b]	All		All[d,f]	All
	6GR	All[g]	All	All	All[g]	All	All[c]	All	All	All[b]	All	All[e]	All	All
Fillet	1F Rotated			F					F					F
	2F			F, H					F, H					F, H
	2F Rotated			F, H					F, H					F, H
	4F			F, H, OH					F, H, OH					F, H, OH
	5F			All					All					All

CJP—Complete Joint Penetration
PJP—Partial Joint Penetration

[a] Fillet welds in production T-, Y-, or K-connections shall conform to Figure 10.5. WPS qualification shall conform to 10.15.
[b] Production butt joint details without backing or backgouging require qualification testing of the joint detail shown in Figure 10.18(A).
[c] For production joints of CJP T-, Y-, and K-connections that conform to either 10.9, 10.10, or 10.11 and Table 10.7, use Figure 10.20 detail for testing. For other production joints, see 10.14.4.1.
[d] For production joints of PJP T-, Y-, and K-connections that conform to Figure 10.6, use either the Figure 10.18(A) or Figure 10.18(B) detail for testing.
[e] For production joints of CJP T-, Y-, and K-connections that conform to Figure 10.7, and Table 10.7, use Figures 10.20 and 10.22 detail for testing, or, alternatively, test the Figure 10.20 joint and cut macroetch specimens from the corner locations shown in Figure 10.22. For other production joints, see 10.14.4.1.
[f] For matched box connections with corner radii less than twice the chord member thickness, see 10.9.1.1.
[g] Limited to prequalified joint details (see 10.9 or 10.10).

Table 10.9
WPS Qualification-CJP Groove Welds: Number and Type of Test Specimens and Range of Thickness and Diameter Qualified (See 10.13) (Dimensions in Inches)

1. Tests on Pipe or Tubing[a,b]

	Nominal Pipe Size or Diam., in	Nominal Wall Thickness, T, in	Number of Specimens				Nominal Diameter[e] of Pipe or Tube Size Qualified, in	Nominal Plate, Pipe or Tube Wall Thickness[c,d] Qualified, in	
			Reduced Section Tension (see Fig. 6.10)	Root Bend (see Fig. 6.8)	Face Bend (see Fig. 6.8)	Side Bend (see Fig. 6.9)		Min.	Max.
Job Size Test Pipes	< 24	1/8 ≤ T ≤ 3/8	2	2	2	(Footnote f)	Test diam. and over	1/8	2T
	< 24	3/8 < T < 3/4	2			4	Test diam. and over	T/2	2T
	< 24	T ≥ 3/4	2			4	Test diam. and over	3/8	Unlimited
	≥ 24	1/8 ≤ T ≤ 3/8	2	2	2	(Footnote f)	Test diam. and over	1/8	2T
	≥ 24	3/8 < T < 3/4	2			4	24 and over	T/2	2T
	≥ 24	T ≥ 3/4	2			4	24 and over	3/8	Unlimited
Standard Test Pipes		2 in Sch. 80 or 3 in Sch. 40	2	2	2		3/4 through 4	1/8	3/4
		6 in Sch. 120 or 8 in Sch. 80	2			4	4 and over	3/16	Unlimited

[a] All pipe or tube welds shall be visually inspected (see 6.10.1) and subject to NDT (see 6.10.2).
[b] See Table 10.8 for the groove details required for qualification of tubular butt and T-, Y-, K-connection joints.
[c] For square groove welds that are qualified without backgouging, the maximum thickness qualified shall be limited to the test thickness.
[d] CJP groove weld qualification on any thickness or diameter shall qualify any size of fillet or PJP groove weld for any thickness or diameter (see 6.12.3).
[e] Qualification with any pipe diameter shall qualify all box section widths and depths.
[f] For 3/8 in wall thickness, a side-bend test may be substituted for each of the required face- and root-bend tests.

Table 10.9 (Continued)
WPS Qualification—CJP Groove Welds: Number and Type of Test Specimens and Range of Thickness and Diameter Qualified (see 10.13) (Dimensions in Millimeters)

1. Tests on Pipe or Tubing[a, b]

	Nominal Pipe Size or Diam., mm	Nominal Wall Thickness, T, mm	Number of Specimens				Nominal Diameter[e] of Pipe or Tube Size Qualified, mm	Nominal Plate, Pipe or Tube Wall Thickness[c,d] Qualified, mm	
			Reduced Section Tension (see Fig. 6.10)	Root Bend (see Fig. 6.8)	Face Bend (see Fig. 6.8)	Side Bend (see Fig. 6.9)		Min.	Max.
Job Size Test Pipes	< 600	3 ≤ T ≤ 10	2	2	2	(Footnote f)	Test diam. and over	3	2T
		10 < T < 20	2			4	Test diam. and over	T/2	2T
		T ≥ 20	2			4	Test diam. and over	10	Unlimited
	≥ 600	3 ≤ T ≤ 10	2	2	2	(Footnote f)	Test diam. and over	3	2T
		10 < T < 20	2			4	600 and over	T/2	2T
		T ≥ 20	2			4	600 and over	10	Unlimited
Standard Test Pipes	DN 50 × 5.5 mm WT or DN 80 × 5.5 mm WT		2	2	2		20 through 100	3	20
	DN 150 × 14.3 mm WT or DN 200 × 12.7 mm WT		2			4	100 and over	5	Unlimited

[a] All pipe or tube welds shall be visually inspected (see 6.10.1) and subject to NDT (see 6.10.2).
[b] See Table 10.8 for the groove details required for qualification of tubular butt and T-, Y-, K-connection joints.
[c] For square groove welds that are qualified without backgouging, the maximum thickness qualified shall be limited to the test thickness.
[d] CJP groove weld qualification on any thickness or diameter shall qualify any size of fillet or PJP groove weld for any thickness or diameter (see 6.12.3).
[e] Qualification with any pipe diameter shall qualify all box section widths and depths.
[f] For 10 mm wall thickness, a side-bend test may be substituted for each of the required face- and root-bend tests.

Table 10.10
WPS Qualification—PJP Groove Welds: Number and Type of Test Specimens and Range of Thickness Qualified (See 10.13)

Test Groove Depth, T in [mm]	Number of Specimens[a,b]					Qualification Ranges[c,d]		
	Macroetch for Weld Size (S) 6.12.3	Reduced-Section Tension (see Fig. 6.10)	Root Bend (see Fig. 6.8)	Face Bend (see Fig. 6.8)	Side Bend (see Fig. 6.9)	Groove Depth	Nominal Pipe or Tubing Thickness, in [mm]	
							Min.	Max.
1/8 ≤ T ≤ 3/8 [3 ≤ T ≤ 10]	3	2	2	2		D	1/8 [3]	2T
3/8 < T ≤ 1 [10 < T ≤ 25]	3	2			4	D	1/8 [3]	Unlimited

[a] One pipe or tubing per position shall be required (see Figure 10.18). Use the production PJP groove detail for qualification. All pipes or tubing shall be visually inspected (see 6.10.1).
[b] If a PJP bevel- or J-groove weld is to be used for T-joints or double-bevel- or double-J-groove weld is to be used for corner joints, the butt joint shall have a temporary restrictive plate in the plane of the square face to simulate a T-joint configuration.
[c] See the pipe diameter qualification requirements of Table 10.9.
[d] Any PJP qualification shall also qualify any fillet weld size on any thickness.

Table 10.11
WPS Qualification—Fillet Welds: Number and Type of Test Specimens and Range of Thickness Qualified (see 10.13)

Test Specimen	Fillet Size	Number of Welds per WPS	Test Specimens Required[a]			Sizes Qualified	
			Macroetch 6.13.1 6.10.4	All-Weld-Metal Tension (see Figure 6.14)	Side Bend (see Figure 6.9)	Pipe Thickness[b]	Fillet Size
Pipe T-test[c] (Figure 10.16)	Single pass, max. size to be used in construction	1 in each position to be used (see Table 10.8)	3 faces (except for 4F & 5F, 4 faces req'd)			Unlimited	Max. tested single pass and smaller
	Multiple pass, min. size to be used in construction	1 in each position to be used (see Table 10.8)	3 faces (except for 4F & 5F, 4 faces req'd)			Unlimited	Min. tested multiple pass and larger

[a] All welded test pipes shall be visually inspected per 6.10.1.
[b] The minimum thickness qualified shall be 1/8 in [3 mm].
[c] See Table 10.9 for pipe diameter qualification.

Table 10.12
Welder and Welding Operator Qualification—Production Welding Positions Qualified by Pipe and Box Tube Tests (see 10.16.1)

Qualification Test			Production Plate Welding Qualified			Production Pipe Welding Qualified					Production Box Tube Welding Qualified				
			Groove		Fillet[b]	Butt Joint		T-, Y-, K-Connections		Fillet[b]	Butt Joint		T-, Y-, K-Connections		Fillet[b]
Weld Type		Test Positions[a]	CJP	PJP		CJP[c]	PJP[c]	CJP[c,d]	PJP[c,d]		CJP	PJP	CJP	PJP[d]	
Groove[e] (Pipe or Box)	T U B U L A R	1G Rotated[f]	F	F	F, H	F	F		F	F, H	F	F		F	F, H
		2G 5G[f]	F, H	F, H	F, H	F, H	F, H		F, H	F, H	F, H	F, H		F, H	F, H
		6G[f]	F, V, OH	F, V, OH	F, V, OH	F, V, OH	F, V, OH		F, V, OH	F, V, OH	F, V, OH	F, V, OH		F, V, OH	F, V, H
		(2G + 5G)[f]	All	All	All	All	All		All	All	All	All		All	All
			All	All	All	All	All		All	All	All	All		All	All
		6GR (Fig. 10.20)	All	All	All	All[g]	All	All	All	All	All[g]	All		All	All
		6GR (Figs. 10.20 & 10.22)	All	All	All	All[g]	All	All	All	All	All[g]	All	All[d,h]	All	All
Pipe Fillet		1F Rotated			F						F					F
		2F			F, H						F, H					F, H
		2F Rotated			F, H						F, H					F, H
		4F			F, H, OH						F, H, OH					F, H, OH
		5F			All						All					All

CJP—Complete Joint Penetration
PJP—Partial Joint Penetration

[a] See Figures 10.12 and 10.13.
[b] See 10.14 for dihedral angle restrictions for tubular T-, Y-, K-connections.
[c] Qualification using box tubing (Figure 10.20) also qualifies welding pipe equal to or greater than 24 in [600 mm] in diameter.
[d] Not qualified for welds having groove angles less than 30° (see 10.14.4.2).
[e] Groove weld qualification shall also qualify plug and slot welds for the test positions indicated.
[f] Qualification for welding production joints without backing or backgouging shall require using the Figure 10.22(A) joint detail. For welding production joints with backing or backgouging, either Figure 10.22(A) or Figure 10.22(B) joint detail may be used for qualification.
[g] Not qualified for joints welded from one side without backing, or welded from two sides without backgouging.
[h] Pipe or box tubing is required for the 6GR qualification (Figure 10.20). If box tubing is used per Figure 10.20, the macroetch test may be performed on the corners of the test specimen (similar to Figure 10.22).

Table 10.13
Welder and Welding Operator Qualification—Number and Type of Specimens and Range of Thickness and Diameter Qualified (Dimensions in Inches) (see 10.16)

Tests on Pipe or Tubing[a]

Production CJP Groove Butt Joints			Number of Specimens[b]						Qualified Dimensions			
			1G and 2G Positions Only			5G, 6G, and 6GR Positions Only			Nominal Pipe or Tube Size Qualified, in		Nominal Plate, Pipe or Tube Wall Thickness[c] Qualified, in	
Type of Test Weld	Nominal Size of Test Pipe, in	Nominal Test Thickness, in	Face Bend[d]	Root Bend[d]	Side Bend[d]	Face Bend[d]	Root Bend[d]	Side Bend[d]	Min.	Max.	Min.	Max.
Groove	≤ 4	Unlimited	1	1	Footnote e	2	2	Footnote e	3/4	4	1/8	3/4
Groove	> 4	≤ 3/8	1	1	Footnote e	2	2	Footnote e	Footnote f	Unlimited	1/8	3/4
Groove	> 4	> 3/8			2			4	Footnote f	Unlimited	3/16	Unlimited

Production T-, Y-, or K-Connection CJP Groove Welds			Number of Specimens[b]		Qualified Dimensions					
					Nominal Pipe or Tube Size Qualified, in		Nominal Wall or Plate Thickness[c] Qualified, in		Dihedral Angles Qualified[g]	
Type of Test Weld	Nominal Size of Test Pipe, in	Nominal Test Thickness, in	Side Bend[d]	Macroetch	Min.	Max.	Min.	Max.	Min.	Max.
Pipe Groove (Fig. 10.20)	≥ 6 O.D.	≥ 1/2	4		4	Unlimited	3/16	Unlimited	30°	Unlimited
Pipe Groove (Fig. 10.21)	< 4 O.D.	≥ 0.203	Footnote h		3/4	< 4	1/8	Unlimited	30°	Unlimited
Box Groove (Fig. 10.22)	Unlimited	≥ 1/2	4	4	Unlimited (Box only)	Unlimited (Box only)	3/16	Unlimited	30°	Unlimited

Production T-, Y-, or K-Connection Fillet Welds			Number of Specimens[b]				Qualified Dimensions					
							Nominal Pipe or Tube Size Qualified, in		Nominal Wall or Plate Thickness Qualified, in		Dihedral Angles Qualified[g]	
Type of Test Weld	Nominal Size of Test Pipe, D	Nominal Test Thickness, in	Fillet Weld Break	Macroetch	Root Bend[d]	Face Bend[d]	Min.	Max.	Min.	Max.	Min.	Max.
5G position (Groove)	Unlimited	≥ 1/8			2[e]	2[e]	Footnote f	Unlimited	1/8[c]	Unlimited[c]	30°	Unlimited
Option 1 —Fillet (Fig. 6.25)[i]		≥ 1/2	1	1			24	Unlimited	1/8	Unlimited	60°	Unlimited
Option 2 —Fillet (Fig. 6.22)[i]		3/8		2			24	Unlimited	1/8	Unlimited	60°	Unlimited
Option 3 —Fillet (Fig. 10.16)	Unlimited	≥ 1/8		1			D	Unlimited	1/8	Unlimited	30°	Unlimited

[a] See Table 10.12 for appropriate groove details.
[b] All welds shall be visually inspected (see 6.23.1).
[c] Also qualifies for welding any fillet or PJP weld size on any thickness of plate, pipe or tubing.
[d] Radiographic examination of the test pipe or tubing may be made in lieu of the bend tests (see 6.17.1.1).
[e] For 3/8 in wall thickness, a side-bend test may be substituted for each of the required face- and root-bend tests.
[f] The minimum pipe size qualified shall be 1/2 the test diameter or 4 in, whichever is greater.
[g] For dihedral angles < 30°, see 10.18.1; except 6GR test not required.
[h] Two root and two face bends.
[i] Two plates required, each subject to the test specimen requirements described. One plate shall be welded in the 3F position and the other in the 4F position.

Table 10.13 (Continued)
Welder and Welding Operator Qualification—Number and Type of Specimens and Range of Thickness and Diameter Qualified (Dimensions in Millimeters) (see 10.16)

Tests on Pipe or Tubing[a]

Production CJP Groove Butt Joints			Number of Specimens[b]						Qualified Dimensions			
			1G and 2G Positions Only			5G, 6G, and 6GR Positions Only			Nominal Pipe or Tube Size Qualified, mm		Nominal Plate, Pipe or Tube Wall Thickness[c] Qualified, mm	
Type of Test Weld	Nominal Size of Test Pipe, mm	Nominal Test Thickness, mm	Face Bend[d]	Root Bend[d]	Side Bend[d]	Face Bend[d]	Root Bend[d]	Side Bend[d]	Min.	Max.	Min.	Max.
Groove	≤ 100	Unlimited	1	1	Footnote e	2	2	Footnote e	20	100	3	20
Groove	> 100	≤ 10	1	1	Footnote e	2	2	Footnote e	Footnote f	Unlimited	3	20
Groove	> 100	> 10			2			4	Footnote f	Unlimited	5	Unlimited

Production T-, Y-, or K-Connection CJP Groove Welds			Number of Specimens[b]		Qualified Dimensions					
					Nominal Pipe or Tube Size Qualified, mm		Nominal Wall or Plate Thickness[c] Qualified, mm		Dihedral Angles Qualified[g]	
Type of Test Weld	Nominal Size of Test Pipe, mm	Nominal Test Thickness, mm	Side Bend[d]	Macroetch	Min.	Max.	Min.	Max.	Min.	Max.
Pipe Groove (Fig. 10.20)	≥ 150 O.D.	≥ 12	4		100	Unlimited	5	Unlimited	30°	Unlimited
Pipe Groove (Fig. 10.21)	< 100 O.D.	≥ 5	Footnote h		20	< 100	3	Unlimited	30°	Unlimited
Box Groove (Fig. 10.22)	Unlimited	≥ 12	4	4	Unlimited (Box only)	Unlimited (Box only)	5	Unlimited	30°	Unlimited

Production T-, Y-, or K-Connection Fillet Welds			Number of Specimens[b]				Qualified Dimensions					
							Nominal Pipe or Tube Size Qualified, mm		Nominal Wall or Plate Thickness Qualified, mm		Dihedral Angles Qualified[g]	
Type of Test Weld	Nominal Size of Test Pipe, D	Nominal Test Thickness, mm	Fillet Weld Break	Macroetch	Root Bend[d]	Face Bend[d]	Min.	Max.	Min.	Max.	Min.	Max.
5G position (Groove)	Unlimited	≥ 3			2[e]	2[e]	Footnote f	Unlimited	3[c]	Unlimited[c]	30°	Unlimited
Option 1 —Fillet (Fig. 10.21)[i]		≥ 12	1	1			600	Unlimited	3	Unlimited	60°	Unlimited
Option 2 —Fillet (Fig. 10.18)[i]		10			2		600	Unlimited	3	Unlimited	60°	Unlimited
Option 3 —Fillet (Fig. 10.16)	Unlimited	≥ 3		1			D	Unlimited	3	Unlimited	30°	Unlimited

[a] See Table 10.12 for appropriate groove details.
[b] All welds shall be visually inspected (see 6.23.1).
[c] Also qualifies for welding any fillet or PJP weld size on any thickness of plate, pipe or tubing.
[d] Radiographic examination of the test pipe or tubing may be made in lieu of the bend tests (see 6.17.1.1).
[e] For 10 mm wall thickness, a side-bend test may be substituted for each of the required face- and root-bend tests.
[f] The minimum pipe size qualified shall be 1/2 the test diameter or 100 mm, whichever is greater.
[g] For dihedral angles < 30°, see 10.18.1; except 6GR test not required.
[h] Two root and two face bends.
[i] Two plates required, each subject to the test specimen requirements described. One plate shall be welded in the 3F position and the other in the 4F position.

Table 10.14
Tubular Root Opening Tolerances, Butt Joints Welded Without Backing (see 10.23.2.1)

	Root Face of Joint		Root Opening of Joints without Steel Backing		Groove Angle of Joint
	in	mm	in	mm	deg
SMAW	± 1/16	± 2	± 1/16	± 2	± 5
GMAW	± 1/32	± 1	± 1/16	± 2	± 5
FCAW	± 1/16	± 2	± 1/16	± 2	± 5

Note: Root openings wider than allowed by the above tolerances, but not greater than the thickness of the thinner part, may be built up by welding to acceptable dimensions prior to the joining of the parts by welding.

Table 10.15
Visual Inspection Acceptance Criteria (see 10.24)

Discontinuity Category and Inspection Criteria	Tubular Connections (All Loads)
(1) Crack Prohibition Any crack shall be unacceptable, regardless of size or location.	X
(2) Weld/Base Metal Fusion Complete fusion shall exist between adjacent layers of weld metal and between weld metal and base metal.	X
(3) Crater Cross Section All craters shall be filled to provide the specified weld size, except for the ends of intermittent fillet welds outside of their effective length.	X
(4) Weld Profiles Weld profiles shall be in conformance with 7.23.	X
(5) Time of Inspection Visual inspection of welds in all steels may begin immediately after the completed welds have cooled to ambient temperature. Acceptance criteria for ASTM A514, A517, and A709 Grade HPS 100W [HPS 690W] steels shall be based on visual inspection performed not less than 48 hours after completion of the weld.	X
(6) Undersized Welds The size of a fillet weld in any continuous weld may be less than the specified nominal size (L) without correction by the following amounts (U): L, specified nominal weld size, in [mm] U, allowable decrease from L, in [mm] $\leq$ 3/16 [5] $\leq$ 1/16 [2] 1/4 [6] $\leq$ 3/32 [2.5] $\geq$ 5/16 [8] $\leq$ 1/8 [3] In all cases, the undersize portion of the weld shall not exceed 10% of the weld length. On web-to-flange welds on girders, underrun shall be prohibited at the ends for a length equal to twice the width of the flange.	X
(7) Undercut (A) For material less than 1 in [25 mm] thick, undercut shall not exceed 1/32 in [1 mm], with the following exception: undercut shall not exceed 1/16 in [2 mm] for any accumulated length up to 2 in [50 mm] in any 12 in [300 mm]. For material equal to or greater than 1 in [25 mm] thick, undercut shall not exceed 1/16 in [2 mm] for any length of weld.	
(B) In primary members, undercut shall be no more than 0.01 in [0.25 mm] deep when the weld is transverse to tensile stress under any design loading condition. Undercut shall be no more than 1/32 in [1 mm] deep for all other cases.	X
(8) Porosity (A) CJP groove welds in butt joints transverse to the direction of computed tensile stress shall have no visible piping porosity. For all other groove welds and for fillet welds, the sum of the visible piping porosity 1/32 in [1 mm] or greater in diameter shall not exceed 3/8 in [10 mm] in any linear inch of weld and shall not exceed 3/4 in [20 mm] in any 12 in [300 mm] length of weld.	
(B) The frequency of piping porosity in fillet welds shall not exceed one in each 4 in [100 mm] of weld length and the maximum diameter shall not exceed 3/32 in [2.5 mm]. Exception: for fillet welds connecting stiffeners to web, the sum of the diameters of piping porosity shall not exceed 3/8 in [10 mm] in any linear inch of weld and shall not exceed 3/4 in [20 mm] in any 12 in [300 mm] length of weld.	X
(C) CJP groove welds in butt joints transverse to the direction of computed tensile stress shall have no piping porosity. For all other groove welds, the frequency of piping porosity shall not exceed one in 4 in [100 mm] of length and the maximum diameter shall not exceed 3/32 in [2.5 mm].	X

Note: An "X" indicates applicability for the connection type; a shaded area indicates non-applicability.

Table 10.16
Hole-Type IQI Requirements (see 10.27.1)

Nominal Material Thickness[a] Range, in	Nominal Material Thickness[a] Range, mm	Film Side Designation	Film Side Essential Hole
Up to 0.25 incl.	Up to 6 incl.	7	4T
Over 0.25 to 0.375	Over 6 through 10	10	4T
Over 0.375 to 0.50	Over 10 through 12	12	4T
Over 0.50 to 0.625	Over 12 through 16	12	4T
Over 0.625 to 0.75	Over 16 through 20	15	4T
Over 0.75 to 0.875	Over 20 through 22	17	4T
Over 0.875 to 1.00	Over 22 through 25	17	4T
Over 1.00 to 1.25	Over 25 through 32	20	4T
Over 1.25 to 1.50	Over 32 through 38	25	2T
Over 1.50 to 2.00	Over 38 through 50	30	2T
Over 2.00 to 2.50	Over 50 through 65	35	2T
Over 2.50 to 3.00	Over 65 through 75	40	2T
Over 3.00 to 4.00	Over 75 through 100	45	2T
Over 4.00 to 6.00	Over 100 through 150	50	2T
Over 6.00 to 8.00	Over 150 through 200	60	2T

[a] Single-wall radiographic thickness.

Table 10.17
Wire IQI Requirements (see 10.27.1)

Nominal Material Thickness[a] Range, in	Nominal Material Thickness[a] Range, mm	Film Side Essential Wire Identity
Up to 0.25 incl.	Up to 6 incl.	5
Over 0.25 to 0.375	Over 6 to 10	6
Over 0.375 to 0.625	Over 10 to 16	7
Over 0.625 to 0.75	Over 16 to 20	8
Over 0.75 to 1.50	Over 20 to 38	9
Over 1.50 to 2.00	Over 38 to 50	10
Over 2.00 to 2.50	Over 50 to 65	11
Over 2.50 to 4.00	Over 65 to 100	12
Over 4.00 to 6.00	Over 100 to 150	13
Over 6.00 to 8.00	Over 150 to 200	14

[a] Single-wall radiographic thickness

Table 10.18
IQI Selection and Placement (see 10.27.2)

IQI Types	Equal T, L ≥ 10 in [250 mm] Hole	Equal T, L ≥ 10 in [250 mm] Wire	Equal T, L < 10 in [250 mm] Hole	Equal T, L < 10 in [250 mm] Wire	Unequal T, L ≥ 10 in [250 mm] Hole	Unequal T, L ≥ 10 in [250 mm] Wire	Unequal T, L < 10 in [250 mm] Hole	Unequal T, L < 10 in [250 mm] Wire
Number of IQIs Pipe Girth	3	3	3	3	3	3	3	3
ASTM Standard Selection— Tables	E1025 10.16	E747 10.17	E1025 10.16	E747 10.17	E1025 10.16	E747 10.17	E1025 10.16	E747 10.17
Figure	8.6		8.7		8.8		8.9	

Notes:
1. T = Nominal base metal thickness (T1 and T2 of Figures).
2. L = Weld Length in area of interest of each radiograph.
3. T may be increased to provide for the thickness of allowable weld reinforcement provided shims are used under hole IQIs per 8.17.3.3.

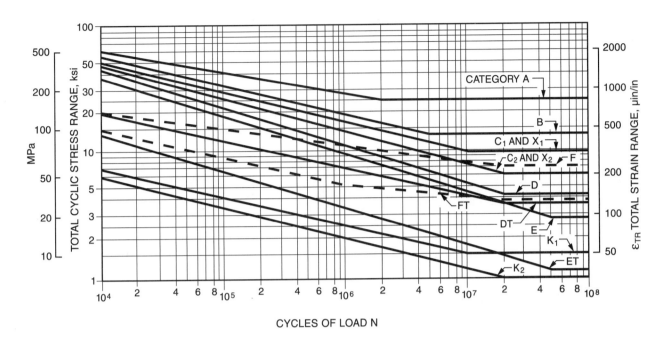

Figure 10.1—Allowable Fatigue Stress and Strain Ranges for Stress Categories (see Table 10.3), Tubular Structures for Atmospheric Service (see 10.2.3.3)

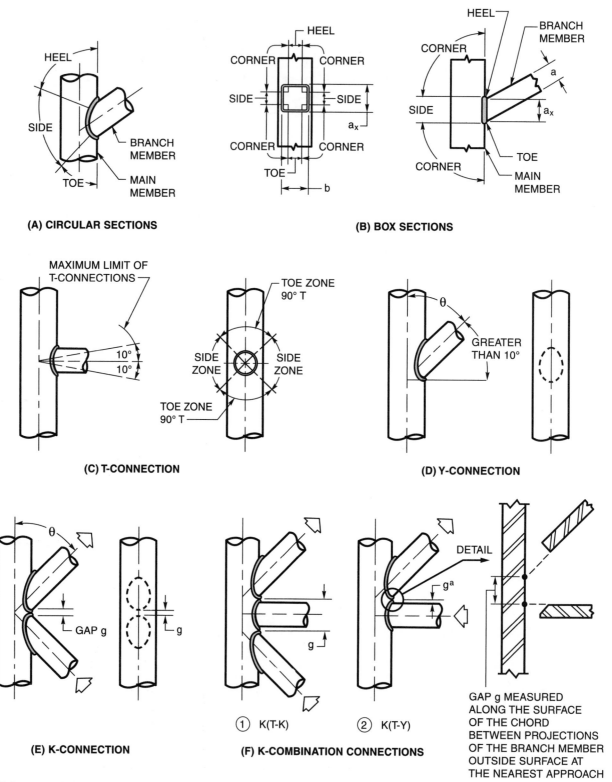

[a] Relevant gap is between braces whose loads are essentially balanced. Type (2) may also be referred to as an N-connection.

Figure 10.2—Parts of a Tubular Connection (see 10.3)

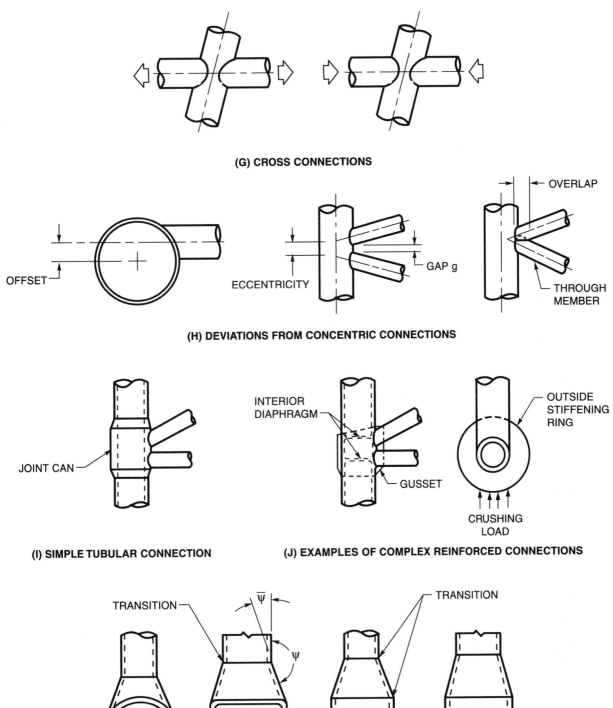

Figure 10.2 (Continued)—Parts of a Tubular Connection (see 10.3)

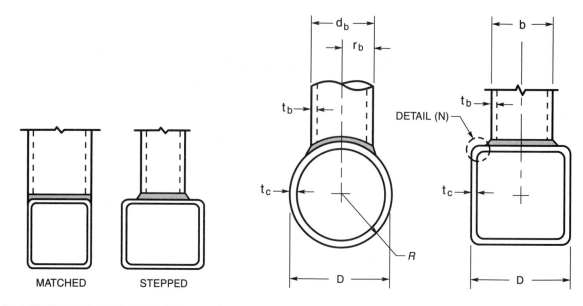

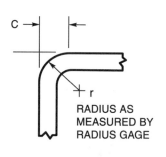

PARAMETER	CIRCULAR SECTIONS	BOX SECTIONS
β	r_b/R OR d_b/D	b/D
η	—	a_x/D
γ	R/t_c	$D/2t_c$
τ	t_b/t_c	t_b/t_c
θ	ANGLE BETWEEN MEMBER CENTERLINES	
ψ	LOCAL DIHEDRAL ANGLE AT A GIVEN POINT ON WELDED JOINT	
C	CORNER DIMENSION AS MEASURED TO THE POINT OF TANGENCY OR CONTACT WITH A 90° SQUARE PLACED ON THE CORNER	

Figure 10.2 (Continued)—Parts of a Tubular Connection (see 10.3)

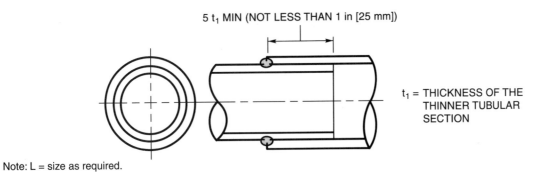

Note: L = size as required.

Figure 10.3—Fillet Welded Lap Joint (Tubular) [see 10.5.2.3 and 10.8.1(2)]

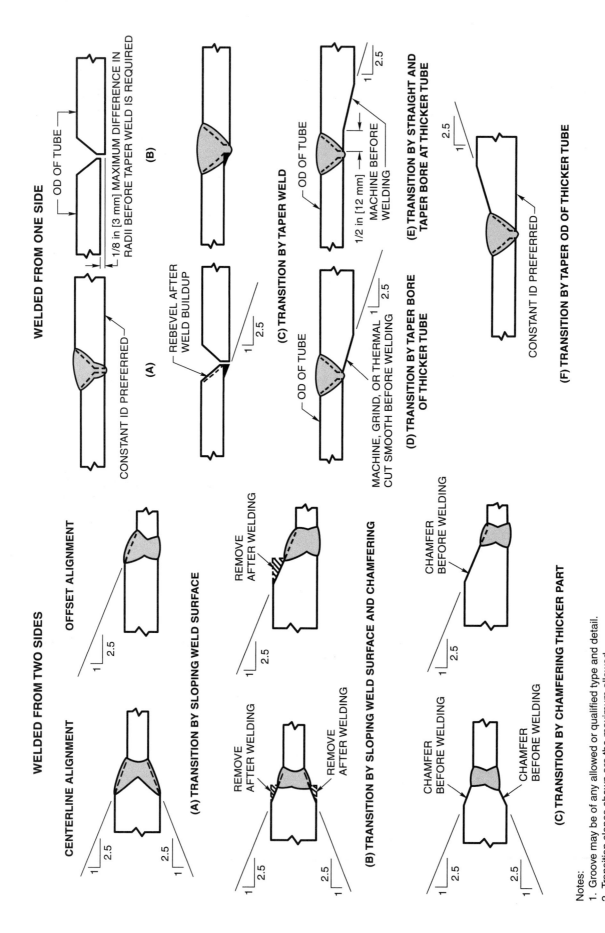

Figure 10.4—Transition of Thickness of Butt Joints in Parts of Unequal Thickness (Tubular) (see 10.6)

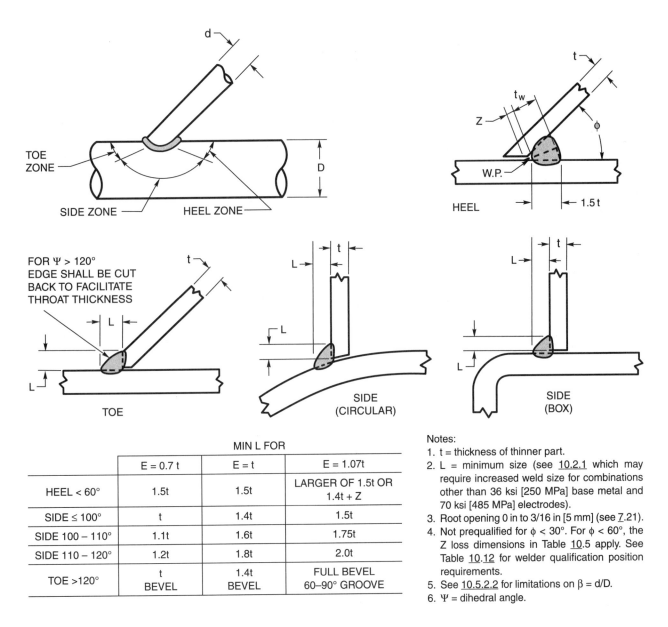

Figure 10.5—Fillet Welded Prequalified Tubular Joints Made by SMAW, GMAW, and FCAW (see 10.8.1)

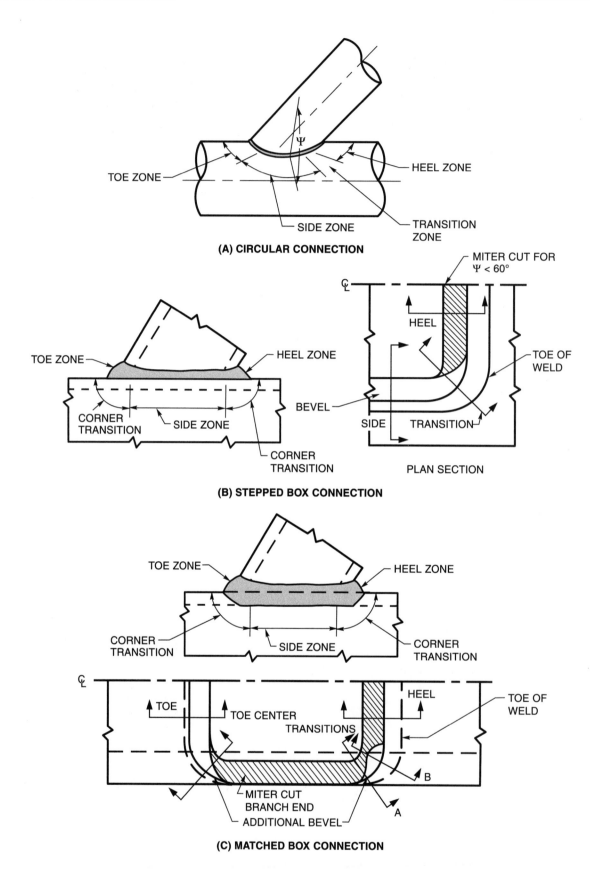

Figure 10.6—Prequalified Joint Details for PJP T-, Y-, and K-Tubular Connections (see 10.9.1)

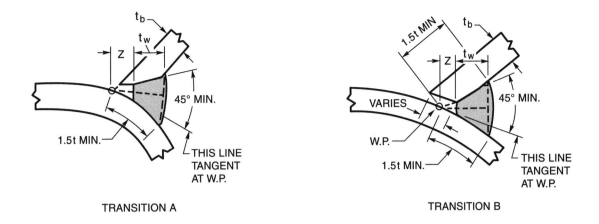

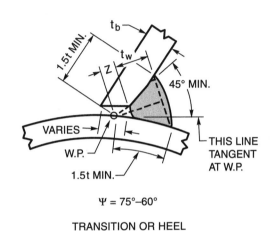

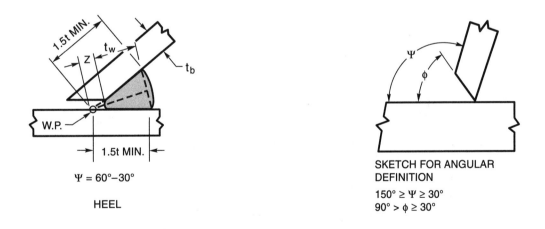

Figure 10.6 (Continued)—Prequalified Joint Details for PJP T-, Y-, and K-Tubular Connections (see 10.9.1)

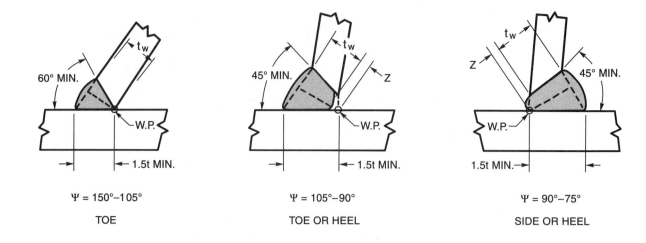

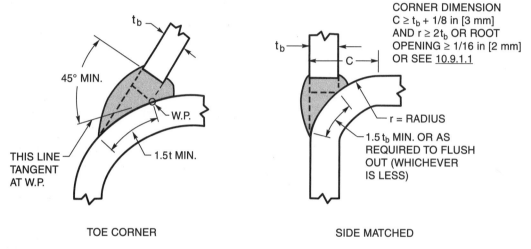

Notes:
1. t = thickness of thinner section.
2. Bevel to feather edge except in transition and heel zones.
3. Root opening: 0 in to 3/16 in [0 mm to 5 mm].
4. Not prequalified for under 30°.
5. Weld size (effective throat) $t_w \geq t$; Z Loss Dimensions shown in Table 10.5.
6. Calculations per 10.2.1 shall be done for leg length less than 1.5t, as shown.
7. For Box Section, joint preparation for corner transitions shall provide a smooth transition from one detail to another. Welding shall be carried continuously around corners, with corners fully built up and all weld starts and stops within flat faces.
8. See Clause 3 for definition of local dihedral angle, Ψ.
9. W.P. = work point.

Figure 10.6 (Continued)—Prequalified Joint Details for PJP T-, Y-, and K-Tubular Connections (see 10.9.1)

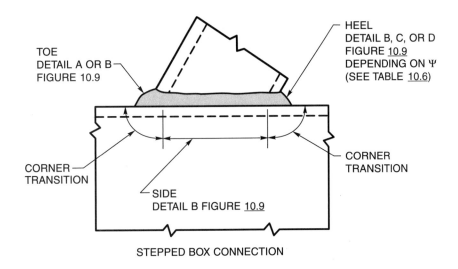
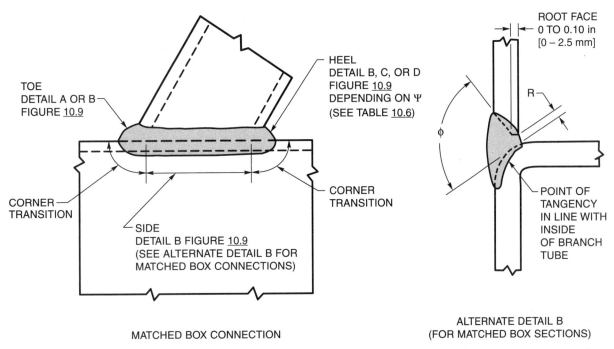

Notes:
1. Details A, B, C, D as shown in Figure 10.9 and all notes from Table 10.7 apply.
2. Joint preparation for corner welds shall provide a smooth transition from one detail to another. Welding shall be carried continuously around corners, with corners fully built up and all arc starts and stops within flat faces.
3. References to Figure 10.9 include Figures 10.10 and 10.11 as appropriate to thickness (see 10.2.3.7).

Figure 10.7—Prequalified Joint Details for CJP T-, Y-, and K-Tubular Connections (see 10.10.2)

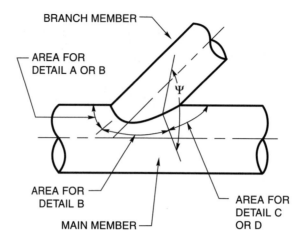

Figure 10.8—Definitions and Detailed Selections for Prequalified CJP T-, Y-, and K-Tubular Connections (see 10.10.2 and Table 10.6)

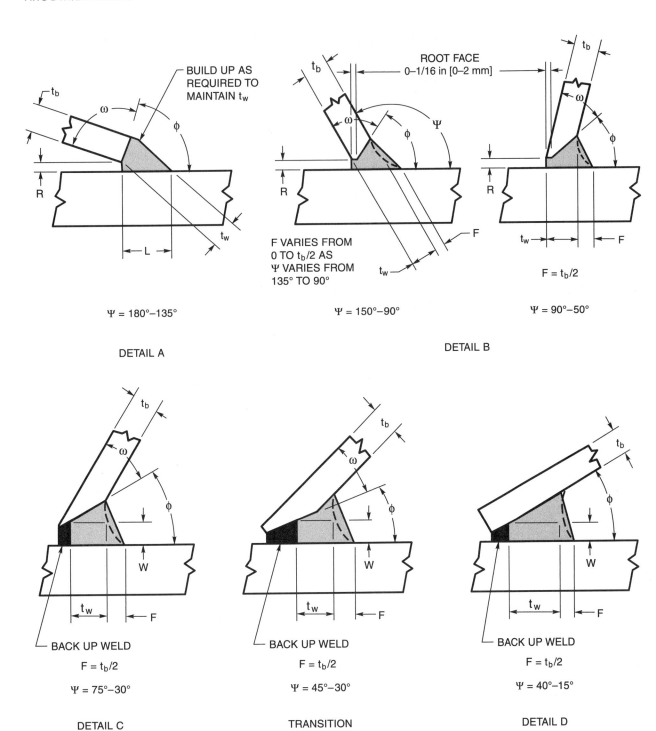

Notes:
1. See Table 10.7 for dimensions t_w, L, R, W, ω, ϕ.
2. Minimum standard flat weld profile shall be as shown by solid line.
3. A concave profile, as shown by dashed lines, shall also be applicable.
4. Convexity, overlap, etc. shall be subject to the limitations of 7.23.
5. Branch member thickness, t_b, shall be subject to limitations of 10.2.3.7.

Figure 10.9—Prequalified Joint Details for CJP Groove Welds in Tubular T-, Y-, and K-Connections—Standard Flat Profiles for Limited Thickness (see 10.10.2)

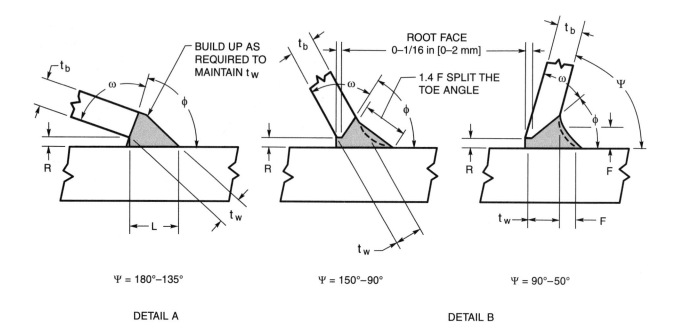

Ψ = 180°–135°

DETAIL A

Ψ = 150°–90°

Ψ = 90°–50°

DETAIL B

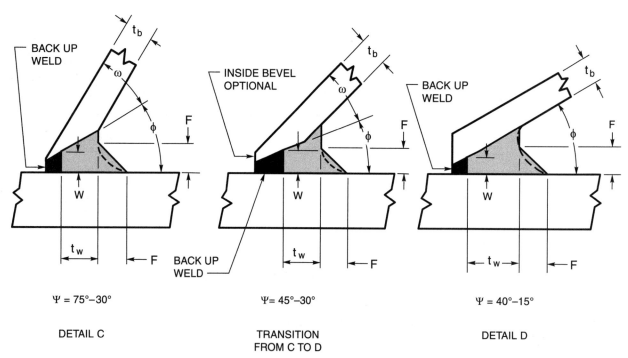

Ψ = 75°–30°

DETAIL C

Ψ = 45°–30°

TRANSITION FROM C TO D

Ψ = 40°–15°

DETAIL D

Notes:
1. Sketches illustrate alternate standard profiles with toe fillet.
2. See 10.2.3.7 for applicable range of thickness t_b.
3. Minimum fillet weld size, $F = t_b/2$, shall also be subject to limits of Table 7.7.
4. See Table 10.7 for dimensions t_w, L, R, W, ω, φ.
5. Convexity and overlap shall be subject to the limitations of 7.23.
6. Concave profiles, as shown by dashed lines shall also be acceptable.

Figure 10.10—Prequalified Joint Details for CJP Groove Welds in Tubular T-, Y-, and K-Connections—Profile with Toe Fillet for Intermediate Thickness (see 10.10.2)

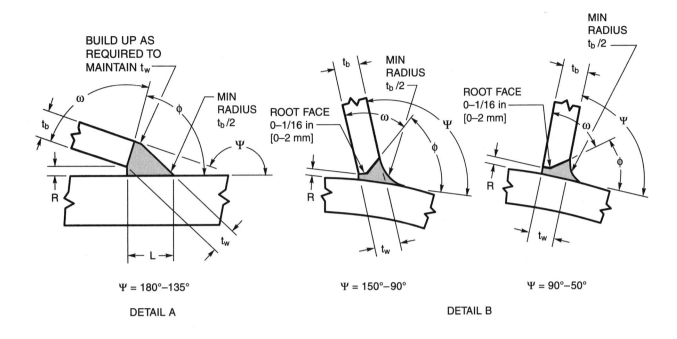

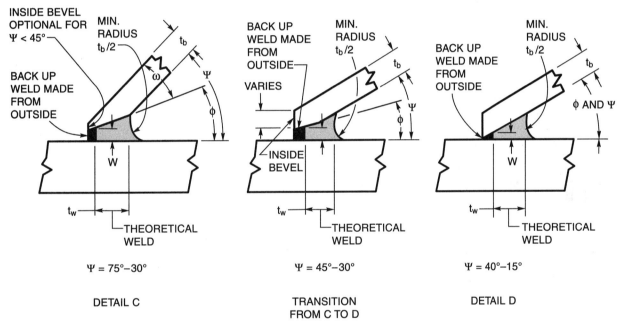

Notes:
1. Illustrating improved weld profiles for 10.2.3.6(1) as welded and 10.2.3.6(2) fully ground.
2. For heavy sections or fatigue critical applications as indicated in 10.2.3.7.
3. See Table 10.7 for dimensions t_b, L, R, W, ω, φ.

Figure 10.11—Prequalified Joint Details for CJP Groove Welds in Tubular T-, Y-, and K-Connections—Concave Improved Profile for Heavy Sections or Fatigue (see 10.10.2)

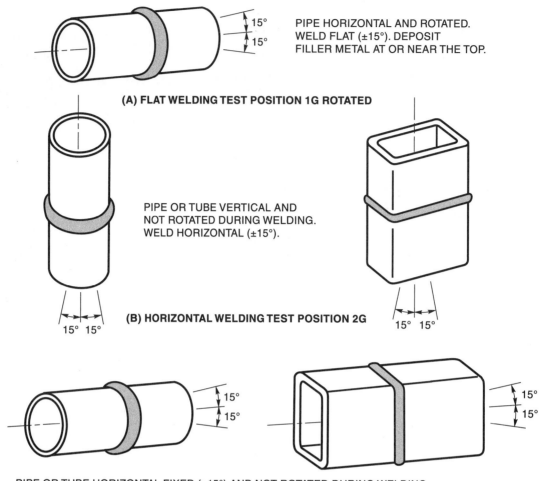
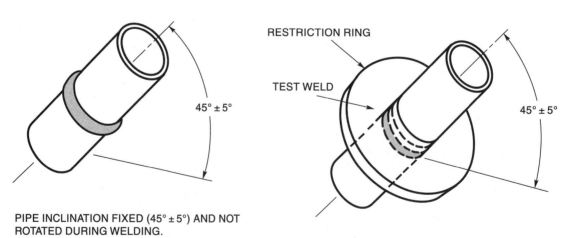

Figure 10.12—Positions of Test Pipe or Tubing for Groove Welds (see 10.11.1)

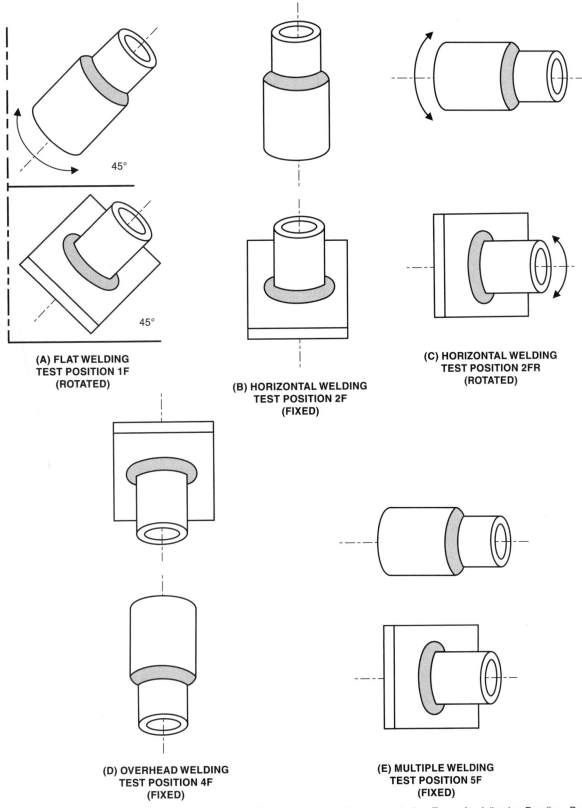

Reproduced from AWS A3.0M/A3.0:2010, *Standard Welding Terms and Definitions, Including Terms for Adhesive Bonding, Brazing, Soldering, Thermal Cutting, and Thermal Spraying*, Figure B.20, Miami: American Welding Society.

Figure 10.13—Positions of Test Pipes or Tubing for Fillet Welds (see 10.11.1)

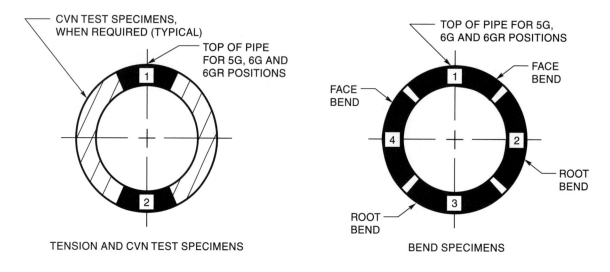

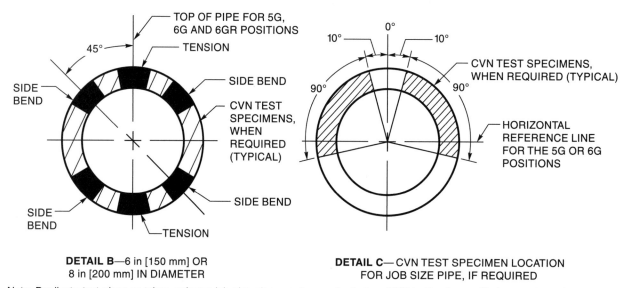

Figure 10.14—Location of Test Specimens on Welded Test Pipe—WPS Qualification (see 10.13)

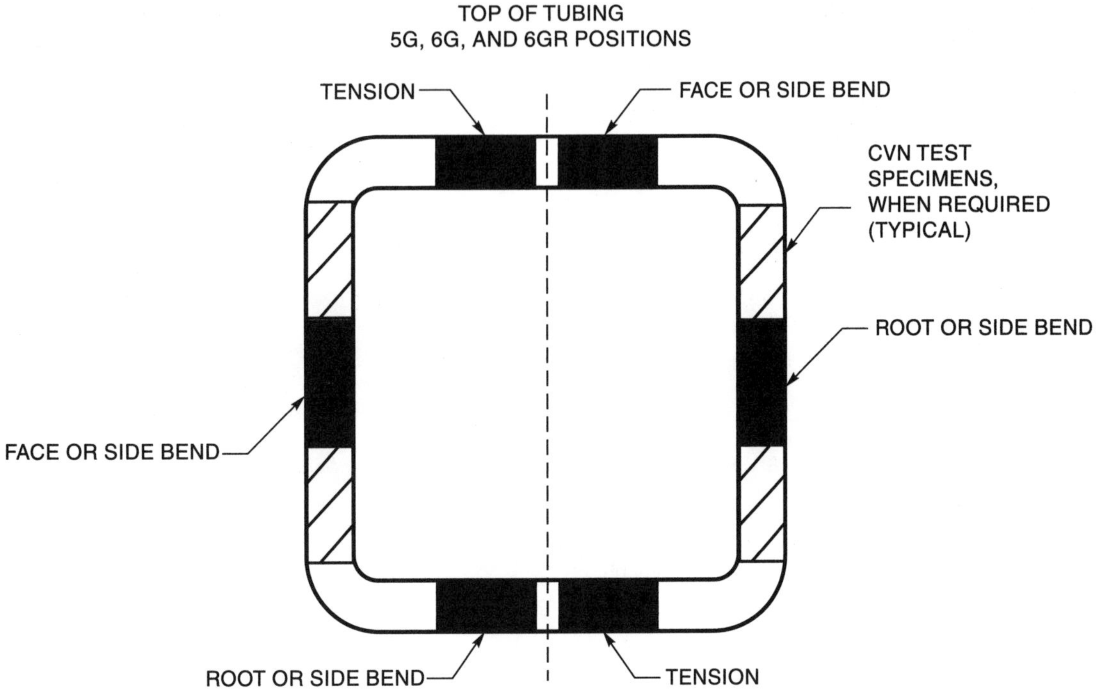

Figure 10.15—Location of Test Specimens for Welded Box Tubing—WPS Qualification (see 10.13)

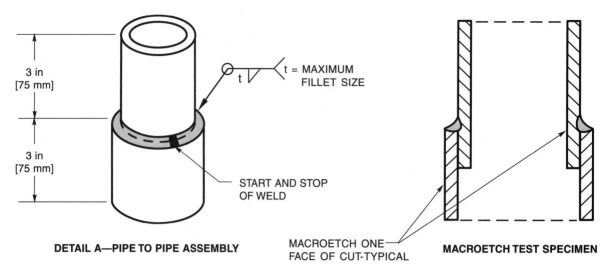

DETAIL A—PIPE TO PIPE ASSEMBLY

MACROETCH TEST SPECIMEN

Notes:
1. See Table 10.8 for position requirements.
2. Pipe shall be of sufficient thickness to prevent melt-through.

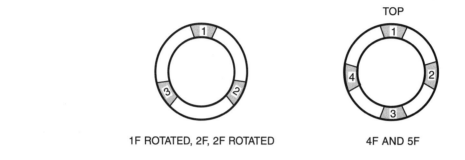

1F ROTATED, 2F, 2F ROTATED   4F AND 5F

LOCATION OF TEST SPECIMENS ON WELDED PIPE – WPS QUALIFICATION

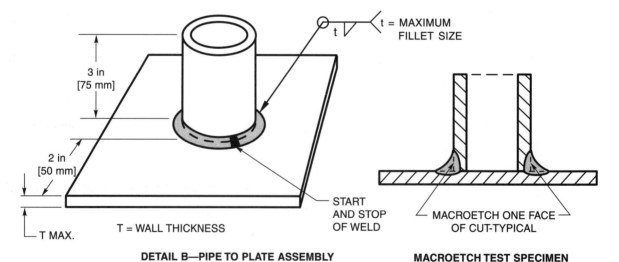

DETAIL B—PIPE TO PLATE ASSEMBLY   **MACROETCH TEST SPECIMEN**

Notes:
1. See Table 10.8 for position requirements.
2. Pipe shall be of sufficient thickness to prevent melt-through.
3. All dimensions are minimums.

Figure 10.16—Pipe Fillet Weld Soundness Test—WPS Qualification (see 6.13.2 and 10.15)

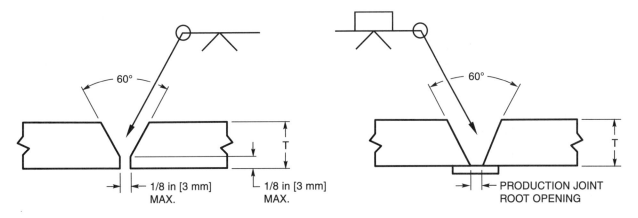

Note: T = qualification pipe or box tube wall thickness

Figure 10.17—Tubular Butt Joint—Welder Qualification with and without Backing (see 10.18)

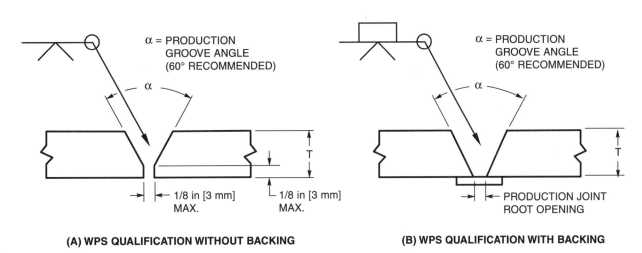

Note: T = qualification pipe or box tube wall thickness

Figure 10.18—Tubular Butt Joint—WPS Qualification with and without Backing (see 10.14.1 and 10.14.2)

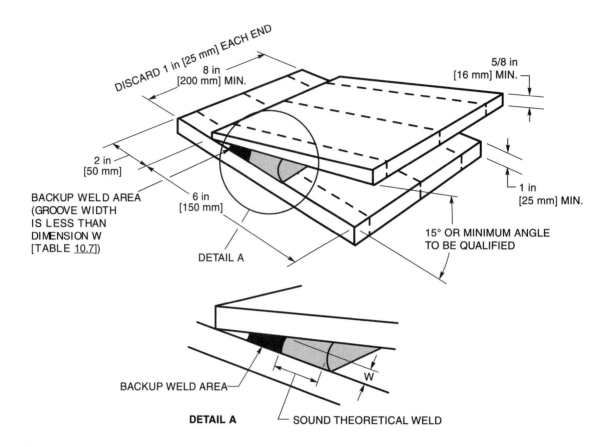

Figure 10.19—Acute Angle Heel Test (Restraints not Shown) (see 10.14.4.2)

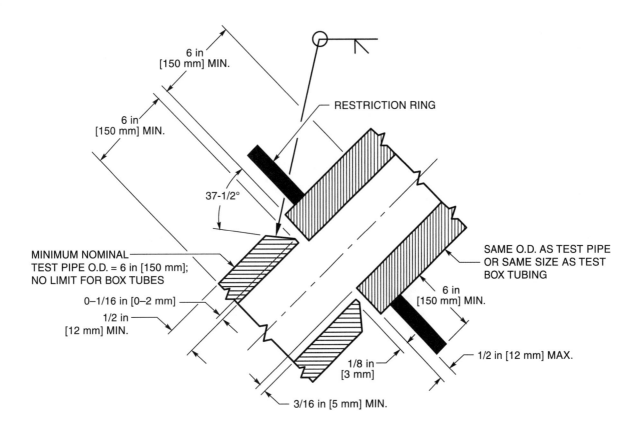

Figure 10.20—Test Joint for T-, Y-, and K-Connections without Backing on Pipe or Box Tubing (≥ 6 in [150 mm] O.D.)—Welder and WPS Qualification (see 10.14.4.1 and 10.18)

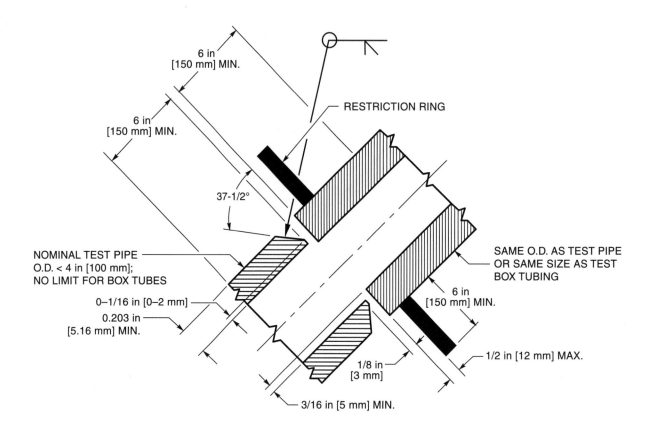

Figure 10.21—Test Joint for T-, Y-, and K-Connections without Backing on Pipe or Box Tubing (< 4 in [100 mm] O.D.)—Welder and WPS Qualification (see 10.14.4.1 and 10.18)

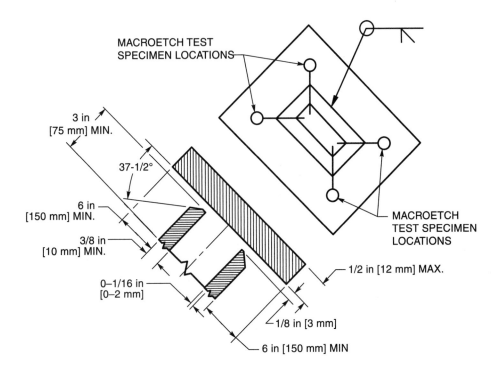

Figure 10.22—Corner Macroetch Test Joint for T-, Y-, and K-Connections without Backing on Box Tubing for CJP Groove Welds—Welder and WPS Qualification (see 10.14.4.1 and 10.18)

CLAUSE 10. TUBULAR STRUCTURES

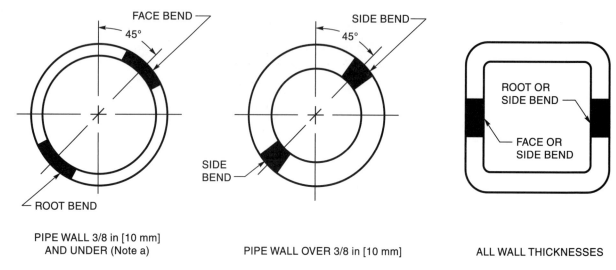

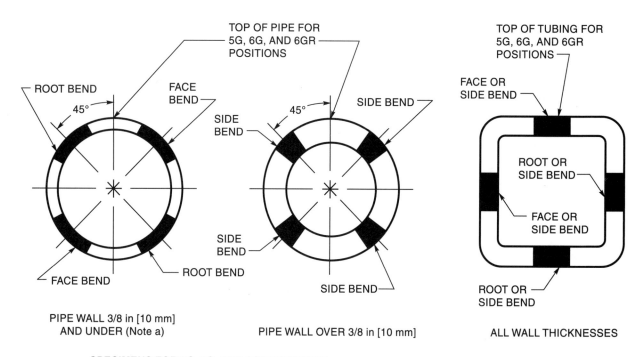

^a For 3/8 in [10 mm] wall thickness, a side-bend test may be substituted for each of the required face- and root-bend tests.

Figure 10.23—Location of Test Specimens on Welded Test Pipe and Box Tubing—Welder Qualification (see 10.16)

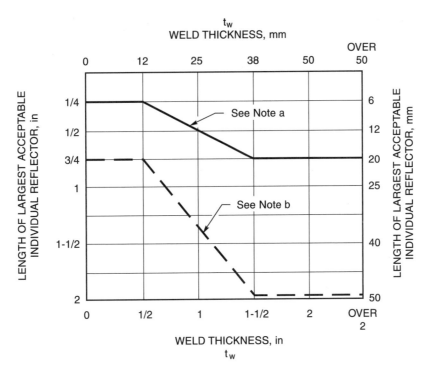

a Internal linear or planar reflectors above standard sensitivity.
b Minor reflectors (above disregard level up to and including standard sensitivity). Adjacent reflectors separated by less than their average length shall be treated as continuous.

Figure 10.24—Class R Indications (see 10.26.1.1)

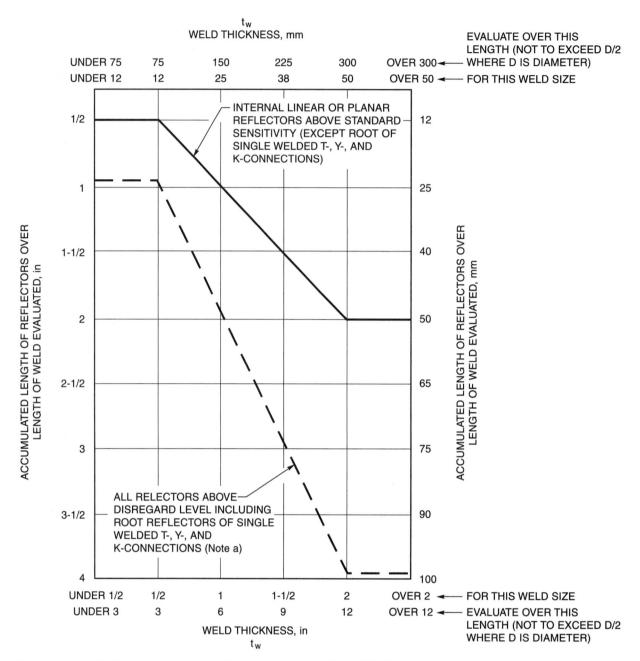

[a] Root area discontinuities falling outside theoretical weld (dimensions "t_w" or "L" in Figures 10.9, 10.10, and 10.11) are to be disregarded.

Figure 10.24 (Continued)—Class R Indications (see 10.26.1.1)

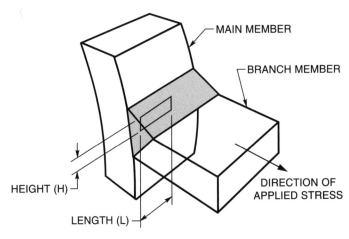

1. Aligned discontinuities separated by less than (L1 + L2)/2 and parallel discontinuities separated by less than (H1 + H2)/2 shall be evaluated as continuous.
2. Accumulative discontinuities shall be evaluated over 6 in [150 mm] or D/2 length of weld (whichever is less), where tube diameter = D.

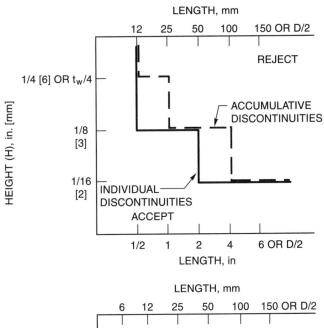

T-, Y-, AND K-ROOT DISCONTINUITIES

1. For CJP weld in single welded T-, Y-, and K-tubular connections made without backing.
2. Discontinuities in the backup weld in the root, Details C and D of Figures 10.9, 10.10, and 10.11 shall be disregarded.

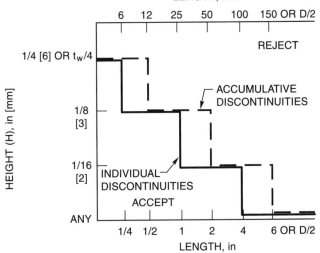

INTERNAL REFLECTORS AND ALL OTHER WELDS

Discontinuities that are within H or $t_w/6$ of the outside surface shall be sized as if extending to the surface of the weld.

Figure 10.25—Class X Indications (see 10.26.1.2)

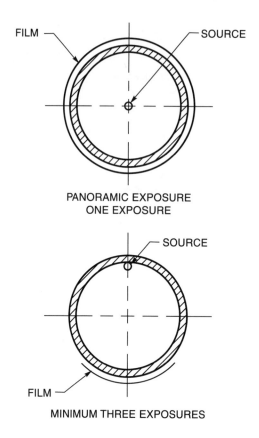

Figure 10.26—Single-Wall Exposure—Single-Wall View (see 10.28.1.1)

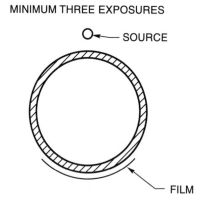

Figure 10.27—Double-Wall Exposure—Single-Wall View (see 10.28.1.2)

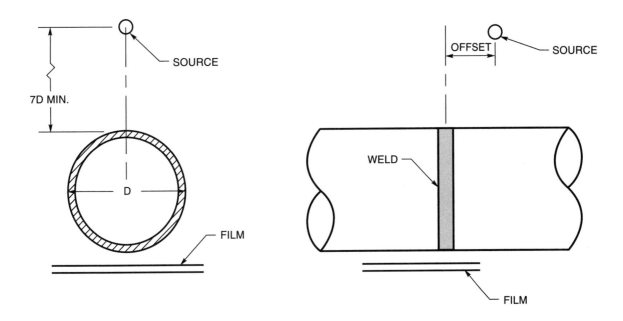

Figure 10.28—Double-Wall Exposure—Double-Wall (Elliptical) View, Minimum Two Exposures (see 10.28.1.3)

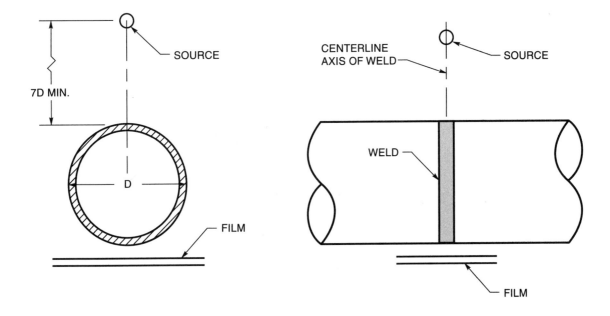

Figure 10.29—Double-Wall Exposure—Double-Wall View, Minimum Three Exposures (see 10.28.1.3)

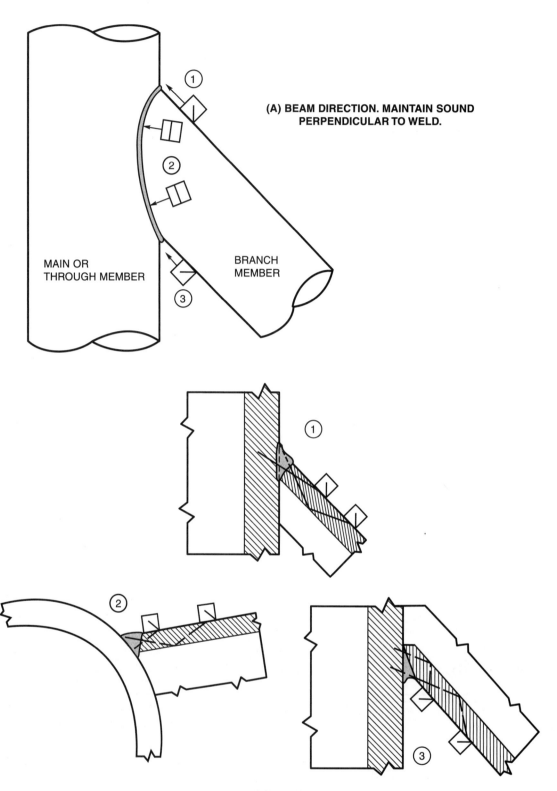

(A) BEAM DIRECTION. MAINTAIN SOUND PERPENDICULAR TO WELD.

(B) V-PATHS. USE SINGLE AND MULTIPLE LEGS AND VARIOUS ANGLES AS REQUIRED TO COVER THE COMPLETE WELD INCLUDING THE ROOT AREA.

Figure 10.30—Scanning Techniques (see 10.29.5)

11. Strengthening and Repair of Existing Structures

11.1 Scope

Strengthening or repairing an existing structure shall consist of modifications to meet design requirements specified by the Engineer. This clause includes requirements for a comprehensive plan for the work, including design, workmanship, inspection, and documentation. The use of fatigue life enhancement methods is also described.

11.2 General

The Engineer shall prepare a comprehensive plan for the work. Such plans shall include, but are not limited to, design, workmanship, inspection, and documentation. Except as modified in this section, all provisions of this code shall apply equally to the strengthening and repairing of existing structures, including heat straightening of distorted members.

11.3 Base Metal

11.3.1 Investigation. Before preparing drawings and specifications for strengthening or repairing existing structures, the types of base metal used in the original structure shall be determined either from existing drawings, specifications or from representative base metal tests.

11.3.2 Suitability for Welding. The suitability of the base metal for welding shall be established (see Table C-11.1 for guidance).

11.3.3 Other Base Metals. Where base metals other than those listed in Table 5.3 are to be joined, special consideration by the Engineer shall be given to the selection of filler metal and WPSs.

11.4 Design for Strengthening and Repair

11.4.1 Design Process. The design process shall consider applicable governing code provisions and other parts of the general specifications. The Engineer shall specify the type and extent of survey necessary to identify existing conditions that require strengthening or repair in order to satisfy applicable criteria.

11.4.2 Stress Analysis. An analysis of stresses in the area affected by the strengthening or repair shall be made. Stress levels shall be established for all in-situ dead and live load cases. Consideration shall be made for accumulated damage that members may have sustained in past service.

11.4.3 Fatigue History. Members subject to cyclic loading shall be designed according to the requirements for fatigue stresses. The previous loading history shall be considered in the design. When the loading history is not available, it shall be estimated.

11.4.4 Restoration or Replacement. Determination shall be made whether the repairs should consist of restoring corroded or otherwise damaged parts or of replacing entire members.

11.4.5 Loading During Operations. The Engineer shall determine the extent to which a member will be allowed to carry loads while heating, welding or thermal cutting is performed. When necessary, the loads shall be reduced. The local and general stability of the member shall be investigated, considering the effect of elevated temperature extending over parts of the cross-sectional area.

11.4.6 Existing Connections. Existing connections in structures requiring strengthening or repair shall be evaluated for design adequacy and reinforced as necessary.

11.4.7 Use of Existing Fasteners. When design calculations show rivets or bolts will be overstressed by the new total load, only existing dead load shall be assigned to them. If rivets or bolts are overstressed by dead load alone or are subject to cyclic loading, then sufficient base metal and welding shall be added to support the total load.

11.5 Fatigue Life Enhancement

11.5.1 Methods. The following methods of reconditioning critical weld details may be used when written procedures have been approved by the Engineer:

(1) Profile Improvement. Reshaping the weld face by grinding with a carbide burr to obtain a concave profile with a smooth transition from base material to weld.

(2) Toe Grinding. Reshaping only the weld toes by grinding with a burr or pencil grinder.

(3) Peening. Shot peening of weld surface, or hammer peening of weld toes.

(4) TIG Dressing. Reshaping of weld toe by the remelting of existing weld metal with heat from a GTAW arc (no filler metal used).

(5) Toe Grinding plus Hammer Peening. When used together, the benefits are cumulative.

11.5.2 Stress Range Increase. The Engineer shall establish the appropriate increase in the allowable stress range.

11.6 Workmanship and Technique

11.6.1 Base Metal Condition. Base metal to be repaired and surfaces of existing base metal in contact with new base metal shall be cleaned of dirt, rust and other foreign matter except adherent paint film as per SSPC SP2 (Surface Preparation Specification #2—Hand Tool Cleaning). The portions of such surfaces which will be welded shall be thoroughly cleaned of all foreign matter including paint for at least 2 in [50 mm] from the root of the weld.

11.6.2 Member Discontinuities. When required by the Engineer, unacceptable discontinuities in the member being repaired or strengthened shall be corrected prior to heat straightening, heat curving, or welding.

11.6.3 Weld Repairs. If weld repairs are required, they shall be made in conformance with 7.25, as applicable.

11.6.4 Base Metal of Insufficient Thickness. Base metal having insufficient thickness to develop the required weld size or required capacity shall be, as determined by the Engineer:

(1) built up with weld metal to the required thickness,

(2) cut back until adequate thickness is available,

(3) reinforced with additional base metal, or

(4) removed and replaced with base metal of adequate thickness or strength.

11.6.5 Heat Straightening. When heat straightening or heat curving methods are used, the maximum temperature of heated areas as measured using temperature sensitive crayons or other positive means shall not exceed 1100°F [600°C] for quenched and tempered steel, nor 1200°F [650°C] for other steels. Accelerated cooling of steel above 600°F [315°C] shall be prohibited.

11.6.6 Welding Sequence. In strengthening or repairing members by the addition of base metal or weld metal, or both, welding and weld sequencing shall, as far as practicable, result in a balanced heat input about the neutral axis to minimize distortion and residual stresses.

11.7 Quality

11.7.1 Visual Inspection. All members and welds affected by the work shall be visually inspected in conformance with the Engineer's comprehensive plan.

11.7.2 NDT. The method, extent, and acceptance criteria of NDT shall be specified in the contract documents.

Annexes
Normative Information

These annexes contain information and requirements that are considered a part of the standard.

Annex A	Effective Throats of Fillet Welds in Skewed T-Joints
Annex B	Guideline on Alternative Methods for Determining Preheat
Annex D	Temperature-Moisture Content Charts
Annex E	Flatness of Girder Webs—Statically Loaded Structures
Annex F	Flatness of Girder Webs—Cyclically Loaded Structures
Annex G	Qualification and Calibration of UT Units with Other Approved Reference Blocks
Annex H	Phased Array Ultrasonic Testing (PAUT)
Annex I	Symbols for Tubular Connection Weld Design

Informative Information

These annexes are not considered a part of the standard and are provided for informational purposes only.

Annex J	Sample Welding Forms
Annex K	Contents of Prequalified WPS
Annex L	Filler Metal Strength Properties
Annex M	AWS A5.36 Filler Metal Classifications and Properties
Annex N	Guide for Specification Writers
Annex O	UT Examination of Welds by Alternative Techniques
Annex P	UT Equipment Qualification and Inspection Forms
Annex Q	Local Dihedral Angle
Annex R	Preliminary Design of Circular Tube Connections
Annex S	List of Reference Documents
Annex T	Guidelines for the Preparation of Technical Inquiries for the Structural Welding Committee

This page is intentionally blank.

Annex A (Normative)
Effective Throats of Fillet Welds in Skewed T-Joints

This annex is part of this standard and includes mandatory elements for use with this standard.

Table A.1 is a tabulation showing equivalent leg size factors for the range of dihedral angles between 60° and 135°, assuming no root opening. Root opening(s) 1/16 in [2 mm] or greater, but not exceeding 3/16 in [5 mm], shall be added directly to the leg size. The required leg size for fillet welds in skewed joints shall be calculated using the equivalent leg size factor for correct dihedral angle, as shown in the example.

EXAMPLE
(U.S. Customary Units)

Given: Skewed T-joint, angle: 75°; root opening: 1/16 (0.063) in

Required: Strength equivalent to 90° fillet weld of size: 5/16 (0.313) in

Procedure:
(1) Factor for 75° from Table A.1: 0.86
(2) Equivalent leg size, w, of skewed joint, without root opening:
  w = 0.86 × 0.313   = 0.269 in
(3) With root opening of: + 0.063 in
(4) Required leg size, w   = 0.332 in
  of skewed fillet weld: [(2) + (3)]
(5) Rounding up to a practical dimension: w = 3/8 in

EXAMPLE
(SI Units)

Given: Skewed T-joint, angle 75°; root opening: 2 mm

Required: Strength equivalent to 90° fillet weld of size: 8 mm

Procedure:
(1) Factor for 75° from Table A.1: 0.86
(2) Equivalent leg size, w, of skewed joint, without root opening:
  w = 0.86 × 8   = 6.9 mm
(3) With root opening of: + 2.0 mm
(4) Required leg size, w   = 8.9 mm
  of skewed fillet weld: [(2) + (3)]
(5) Rounding up to a practical dimension: w = 9.0 mm

For fillet welds having equal measured legs (w_n), the distance from the root of the joint to the face of the diagrammatic weld (t_n) may be calculated as follows:

For root openings > 1/16 in [2 mm] and ≤ 3/16 in [5 mm], use

$$t_n = \frac{W_n - R_n}{2 \sin \frac{\psi}{2}}$$

For root openings <1/16 in [2 mm], use

$R_n = 0$ and $t'_n = t_n$

where the measured leg of such fillet weld (w_n) is the perpendicular distance from the surface of the joint to the opposite toe, and (R) is the root opening, if any, between parts (see Figure 5.4). Acceptable root openings are defined in 7.21.1.

Table A.1
Equivalent Fillet Weld Leg Size Factors for Skewed T-Joints

Dihedral angle, Ψ	60°	65°	70°	75°	80°	85°	90°	95°
Comparable fillet weld size for same strength	0.71	0.76	0.81	0.86	0.91	0.96	1.00	1.03
Dihedral angle, Ψ	100°	105°	110°	115°	120°	125°	130°	135°
Comparable fillet weld size for same strength	1.08	1.12	1.16	1.19	1.23	1.25	1.28	1.31

Annex B (Normative)
Guideline on Alternative Methods for Determining Preheat

This annex is part of this standard and includes mandatory elements for use with this standard.

B1. Introduction

The purpose of this guide is to provide some optional alternative methods for determining welding conditions (principally preheat) to avoid cold cracking. The methods are based primarily on research on small scale tests carried out over many years in several laboratories world-wide. No method is available for predicting optimum conditions in all cases, but the guide does consider several important factors such as hydrogen level and steel composition not explicitly included in the requirements of Table 5.8. The guide may therefore be of value in indicating whether the requirements of Table 5.8 are overly conservative or in some cases not sufficiently demanding.

The user is referred to the Commentary for more detailed presentation of the background scientific and research information leading to the two methods proposed.

In using this guide as an alternative to Table 5.8, careful consideration shall be given to the assumptions made, the values selected, and past experience.

B2. Methods

Two methods are used as the basis for estimating welding conditions to avoid cold cracking:

(1) HAZ hardness control

(2) Hydrogen control

B3. HAZ Hardness Control

B3.1 The provisions included in this guide for use of this method are restricted to fillet welds.

B3.2 This method is based on the assumption that cracking will not occur if the hardness of the HAZ is kept below some critical value. This is achieved by controlling the cooling rate below a critical value dependent on the hardenability of the steel. Hardenability of steel in welding relates to its propensity towards formation of a hard HAZ and can be characterized by the cooling rate necessary to produce a given level of hardness. Steels with high hardenability can, therefore, produce hard HAZ at slower cooling rates than a steel with lower hardenability.

Equations and graphs are available in the technical literature that relate the weld cooling rate to the thickness of the steel members, type of joint, welding conditions and variables.

B3.3 The selection of the critical hardness will depend on a number of factors such as steel type, hydrogen level, restraint, and service conditions. Laboratory tests with fillet welds show that HAZ cracking does not occur if the HAZ Vickers Hardness No. (HV) is less than 350 HV, even with high-hydrogen electrodes. With low-hydrogen electrodes, hardnesses of 400 HV could be tolerated without cracking. Such hardness, however, may not be tolerable in service where there is an increased risk of stress corrosion cracking, brittle fracture initiation, or other risks for the safety or serviceability of the structure.

The critical cooling rate for a given hardness can be approximately related to the carbon equivalent (CE) of the steel (see Figure B.2). Since the relationship is only approximate, the curve shown in Figure B.2 may be conservative for plain carbon and plain carbon-manganese steels and thus allow the use of the high hardness curve with less risk.

Some low-alloy steels, particularly those containing columbium (niobium), may be more hardenable than Figure B.2 indicates, and the use of the lower hardness curve is recommended.

B3.4 Although the method can be used to determine a preheat level, its main value is in determining the minimum heat input (and hence minimum weld size) that prevents excessive hardening. It is particularly useful for determining the minimum size of single-pass fillet welds that can be deposited without preheat.

B3.5 The hardness approach does not consider the possibility of weld metal cracking. However, from experience it is found that the heat input determined by this method is usually adequate to prevent weld metal cracking, in most cases, in fillet welds if the electrode is not a high-strength filler metal and is generally of a low-hydrogen type [e.g., low-hydrogen (SMAW) electrode, GMAW, FCAW, SAW].

B3.6 Because the method depends solely on controlling the HAZ hardness, the hydrogen level and restraint are not explicitly considered.

B3.7 This method is not applicable to quenched and tempered steels [see B5.2(3) for limitations].

B4. Hydrogen Control

B4.1 The hydrogen control method is based on the assumption that cracking will not occur if the average quantity of hydrogen remaining in the joint after it has cooled down to about 120°F [50°C] does not exceed a critical value dependent on the composition of the steel and the restraint. The preheat necessary to allow enough hydrogen to diffuse out of the joint can be estimated using this method.

B4.2 This method is based mainly on results of restrained PJP groove weld tests; the weld metal used in the tests matched the parent metal. There has not been extensive testing of this method on fillet welds; however, by allowing for restraint, the method has been suitably adapted for those welds.

B4.3 A determination of the restraint level and the original hydrogen level in the weld pool is required for the hydrogen method.

In this guide, restraint is classified as high, medium, and low, and the category must be established from experience.

B4.4 The hydrogen control method is based on a single low-heat input weld bead representing a root pass and assumes that the HAZ hardens. The method is, therefore, particularly useful for high strength, low-alloy steels having quite high hardenability where hardness control is not always feasible. Consequently, because it assumes that the HAZ fully hardens, the predicted preheat may be too conservative for carbon steels.

B5. Selection of Method

B5.1 The following procedure is recommended as a guide for selection of either the hardness control or hydrogen control method.

Determine carbon and carbon equivalent:

$$CE = C + \frac{(Mn+Si)}{6} + \frac{(Cr+Mo+V)}{5} + \frac{(Ni+Cu)}{15}$$

to locate the zone position of the steel in Figure B.1 (see B6.1.1 for the different ways to obtain chemical analysis).

B5.2 The performance characteristics of each zone and the recommended action are as follows:

(1) **Zone I.** Cracking is unlikely, but may occur with high hydrogen or high restraint. Use hydrogen control method to determine preheat for steels in this zone.

(2) **Zone II.** The hardness control method and selected hardness shall be used to determine minimum energy input for single-pass fillet welds without preheat.

If the energy input is not practical, use hydrogen method to determine preheat.

For groove welds, the hydrogen control method shall be used to determine preheat.

For steels with high carbon, a minimum energy to control hardness and preheat to control hydrogen may be required for both types of welds, i.e., fillet and groove welds.

(3) **Zone III.** The hydrogen control method shall be used. Where heat input is restricted to preserve the HAZ properties (e.g., some quenched and tempered steels), the hydrogen control method should be used to determine preheat.

B6. Detailed Guide

B6.1 Hardness Method

B6.1.1 The carbon equivalent shall be calculated as follows:

$$CE = C + \frac{(Mn+Si)}{6} + \frac{(Cr+Mo+V)}{5} + \frac{(Ni+Cu)}{15}$$

The chemical analysis may be obtained from:

(1) Mill test certificates

(2) Typical production chemistry (from the mill)

(3) Specification chemistry (using maximum values)

(4) User tests (chemical analysis)

B6.1.2 The critical cooling rate shall be determined for a selected maximum HAZ hardness of either 400 HV or 350 HV from Figure B.2.

B6.1.3 Using applicable thicknesses for "flange" and "web" plates, the appropriate diagram shall be selected from Figure B.3 and the minimum energy input for single-pass fillet welds shall be determined. This energy input applies to SAW welds.

B6.1.4 For other processes, minimum energy input for single-pass fillet welds can be estimated by applying the following multiplication factors to the energy estimated for the SAW process in B6.1.3:

Welding Process	**Multiplication Factor**
SAW	1
SMAW	1.50
GMAW, FCAW	1.25

B6.1.5 Figure B.4 may be used to determine fillet sizes as a function of energy input.

B6.2 Hydrogen Control Method

B6.2.1 The value of the composition parameter, P_{cm}, shall be calculated as follows:

$$P_{cm} = C + \frac{Si}{30} + \frac{Mn}{20} + \frac{Cu}{20} + \frac{Ni}{60} + \frac{Cr}{20} + \frac{Mo}{15} + \frac{V}{10} + 5B$$

B6.2.2 The hydrogen level shall be determined and shall be defined as follows:

(1) **H1 Extra-Low Hydrogen.** These consumables give a diffusible hydrogen content of less than 5 ml/100g deposited metal when measured using ISO 3690–1976 or, a moisture content of electrode covering of 0.2% maximum in conformance with AWS A5.1 or A5.5. This may be established by testing each type, brand, or wire/flux combination used after removal from the package or container and exposure for the intended duration, with due consideration of actual storage conditions prior to immediate use. The following may be assumed to meet this requirement:

(a) low-hydrogen electrodes taken from hermetically sealed containers, dried at 700°F–800°F [370°–430°C] for one hour and used within two hours after removal,

(b) GMAW with clean solid wires.

(2) **H2 Low Hydrogen.** These consumables give a diffusible hydrogen content of less than 10 ml/100g deposited metal when measured using ISO 3690–1976, or a moisture content of electrode covering of 0.4% maximum in conformance with AWS A5.1. This may be established by a test on each type, brand of consumable, or wire/flux combination used. The following may be assumed to meet this requirement:

(a) Low-hydrogen electrodes taken from hermetically sealed containers conditioned in conformance with 7.3.2.1 of the code and used within four hours after removal,

(b) SAW with dry flux.

(3) **H3 Hydrogen Not Controlled.** All other consumables not meeting the requirements of H1 or H2.

B6.2.3 The susceptibility index grouping from Table B1 shall be determined.

B6.2.4 Minimum Preheat Levels and Interpass. Table B.2 gives the minimum preheat and interpass temperatures that shall be used. Table B.2 gives three levels of restraint. The restraint level to be used shall be determined in conformance with B6.2.5.

B6.2.5 Restraint. The classification of types of welds at various restraint levels should be determined on the basis of experience, engineering judgment, research, or calculation.

Three levels of restraint have been provided:

(1) **Low Restraint.** This level describes common fillet and groove welded joints in which a reasonable freedom of movement of members exists.

(2) **Medium Restraint.** This level describes fillet and groove welded joints in which, because of members being already attached to structural work, a reduced freedom of movement exists.

(3) **High Restraint.** This level describes welds in which there is almost no freedom of movement for members joined (such as repair welds, especially in thick material).

Table B.1
Susceptibility Index Grouping as Function of Hydrogen Level "H" and Composition Parameter P_{cm} (see B6.2.3)

Hydrogen Level, H	Susceptibility Index[a] Grouping[b]					
	Carbon Equivalent = P^c_{cm}				
	<0.18	<0.23	<0.28	<0.33	<0.38
H1	A	B	C	D	E
H2	B	C	D	E	F
H3	C	D	E	F	G

[a] Susceptibility index—12 P_{cm} + log10 H.
[b] Susceptibility Index Groupings, A through G, encompass the combined effect of the composition parameter, P_{cm}, and hydrogen level, H, in conformance with the formula shown in footnote a.

The exact numerical quantities are obtained from the footnote a formula using the stated values of P_{cm} and the following values of H, given in ml/100g of weld metal [see B6.2.2, (1), (2), (3)]:

H1—5; H2—10; H3—30.

For greater convenience, Susceptibility Index Groupings have been expressed in the table by means of letters, A through G, to cover the following narrow ranges:

A = 3.0; B = 3.1–3.5; C = 3.6–4.0; D = 4.1–4.5; E = 4.6–5.0; F = 5.1–5.5; G = 5.6–7.0

These groupings are used in Table B.2 in conjunction with restraint and thickness to determine the minimum preheat and interpass temperature.

[c] $P_{cm} = C + \dfrac{Si}{30} + \dfrac{Mn}{20} + \dfrac{Cu}{20} + \dfrac{Ni}{60} + \dfrac{Cr}{20} + \dfrac{Mo}{15} + \dfrac{V}{10} + 5B$

Table B.2
Minimum Preheat and Interpass Temperatures for Three Levels of Restraint (see B6.2.4)

Level	Thickness[b] in	Minimum Preheat an Interpass Temperature (°F)[a]						
		Susceptibility Index Grouping						
		A	B	C	D	E	F	G
Low	< 3/8	< 65	< 65	< 65	< 65	140	280	300
	3/8– 3/4 incl.	< 65	< 65	65	140	210	280	300
	> 3/4– 1–1/2 incl.	< 65	< 65	65	175	230	280	300
	> 1–1/2– 3 incl.	65	65	100	200	250	280	300
	> 3	65	65	100	200	250	280	300
Medium	< 3/8	< 65	< 65	< 65	< 65	160	280	320
	3/8– 3/4 incl.	< 65	< 65	65	175	240	290	320
	> 3/4– 1–1/2 incl.	< 65	65	165	230	280	300	320
	> 1–1/2– 3 incl.	65	175	230	265	300	300	320
	> 3	200	250	280	300	320	320	320
High	< 3/8	< 65	< 65	< 65	100	230	300	320
	3/8– 3/4 incl.	< 65	65	150	220	280	320	320
	> 3/4– 1–1/2 incl.	65	185	240	280	300	320	320
	> 1–1/2– 3 incl.	240	265	300	300	320	320	320
	> 3	240	265	300	300	320	320	320

(Continued)

[a] "<" indicates that preheat and interpass temperatures lower than the temperature shown may be suitable to avoid hydrogen cracking. Preheat and interpass temperatures that are both lower than the listed temperature and lower than Table 5.8 shall be qualified by test.
[b] Thickness is that of the thicker part welded.

Table B.2 (Continued)
Minimum Preheat and Interpass Temperatures for Three Levels of Restraint (see B6.2.4)

Level	Thickness[b] mm	Minimum Preheat an Interpass Temperature (°C)[a]						
		Susceptibility Index Grouping						
		A	B	C	D	E	F	G
Low	< 10	< 20	< 20	< 20	< 20	60	140	150
	10–20 incl.	< 20	< 20	20	60	100	140	150
	> 20–38 incl.	< 20	< 20	20	80	110	140	150
	> 38–75 incl.	20	20	40	95	120	140	150
	> 75	20	20	40	95	120	140	150
Medium	< 10	< 20	< 20	< 20	< 20	70	140	160
	10–20 incl.	< 20	< 20	20	80	115	145	160
	> 20–38 incl.	20	20	75	110	140	150	160
	> 38–75 incl.	20	80	110	130	150	150	160
	> 75	95	120	140	150	160	160	160
High	< 10	< 20	< 20	20	40	110	150	160
	10–20 incl.	<20	20	65	105	140	160	160
	> 20–38 incl.	20	85	115	140	150	160	160
	> 38–75 incl.	115	130	150	150	160	160	160
	> 75	115	130	150	150	160	160	160

[a] "<" indicates that preheat and interpass temperatures lower than the temperature shown may be suitable to avoid hydrogen cracking. Preheat and interpass temperatures that are both lower than the listed temperature and lower than Table 5.8 shall be qualified by test.
[b] Thickness is that of the thicker part welded.

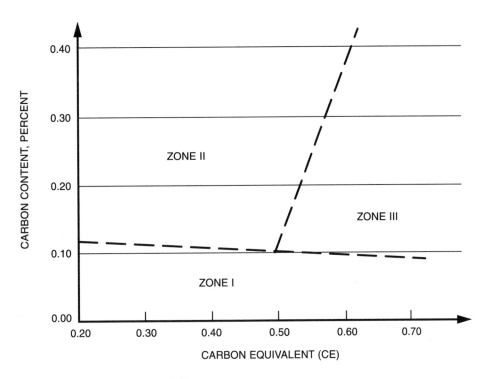

Notes:
1. CE = C + (Mn + Si)/6 + (Cr + Mo + V)/5 + (Ni + Cu)/15.
2. See B5.2(1), (2), or (3) for applicable zone characteristics.

Figure B.1—Zone Classification of Steels (see B5.1)

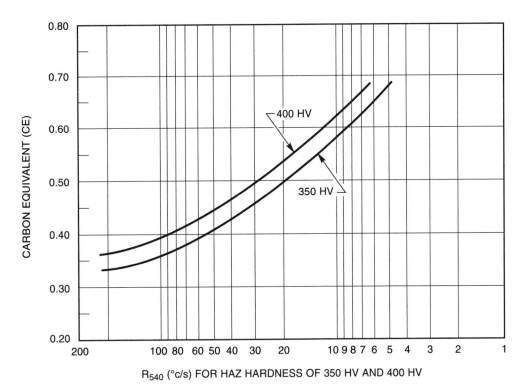

Note: CE = C + (Mn + Si)/6 + (Cr + Mo + V)/5 + (Ni + Cu)/15.

Figure B.2—Critical Cooling Rate for 350 HV and 400 HV (see B3.3)

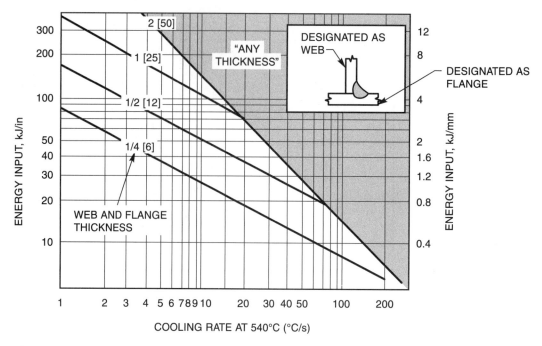

(A) SINGLE-PASS SAW FILLET WELDS WITH WEB AND FLANGE OF SAME THICKNESS

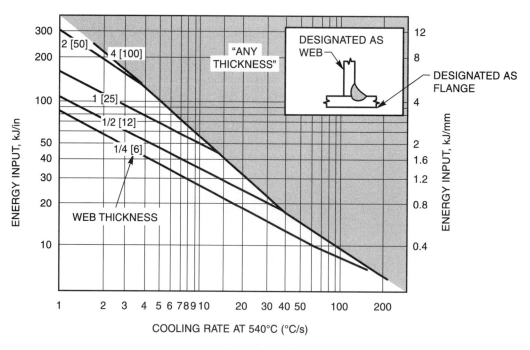

(B) SINGLE-PASS SAW FILLET WELDS WITH 1/4 in [6 mm] FLANGES AND VARYING WEB THICKNESSES

Figure B.3—Graphs to Determine Cooling Rates for Single-Pass SAW Fillet Welds (see B6.1.3)

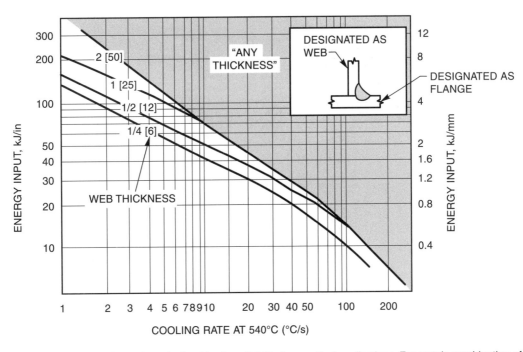

Note: Energy input determined from chart shall not imply suitability for practical applications. For certain combination of thicknesses melting may occur through the thickness.

(C) SINGLE-PASS SAW FILLET WELDS WITH 1/2 in [12 mm] FLANGES AND VARYING WEB THICKNESSES

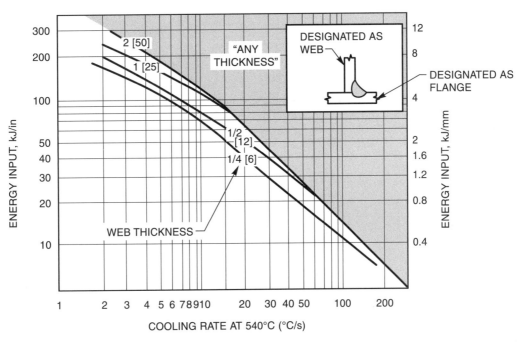

Note: Energy input determined from chart shall not imply suitability for practical applications. For certain combination of thicknesses melting may occur through the thickness.

(D) SINGLE-PASS SAW FILLET WELDS WITH 1 in [25 mm] FLANGES AND VARYING WEB THICKNESSES

Figure B.3 (Continued)—Graphs to Determine Cooling Rates for Single-Pass SAW Fillet Welds (B6.1.3)

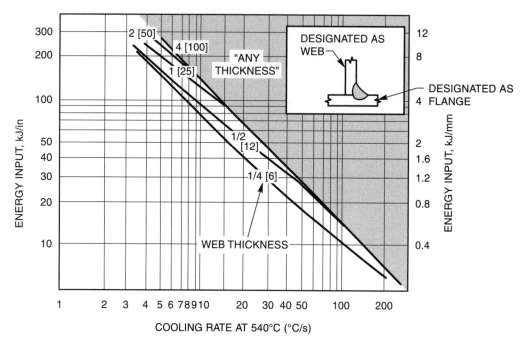

(E) SINGLE-PASS SAW FILLET WELDS WITH 2 in [50 mm] FLANGES AND VARYING WEB THICKNESSES

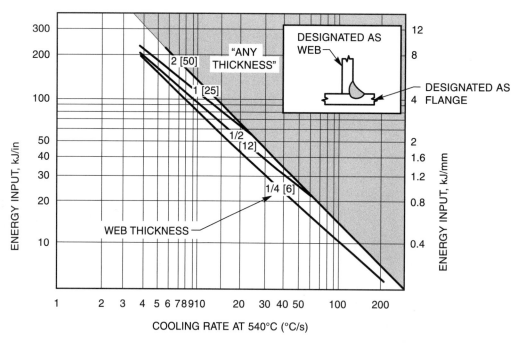

Note: Energy input determined from chart shall not imply suitability for practical applications. For certain combination of thicknesses melting may occur through the thickness.

(F) SINGLE-PASS SAW FILLET WELDS WITH 4 in [100 mm] FLANGES AND VARYING WEB THICKNESSES

Figure B.3 (Continued)—Graphs to Determine Cooling Rates for Single-Pass SAW Fillet Welds (see B6.1.3)

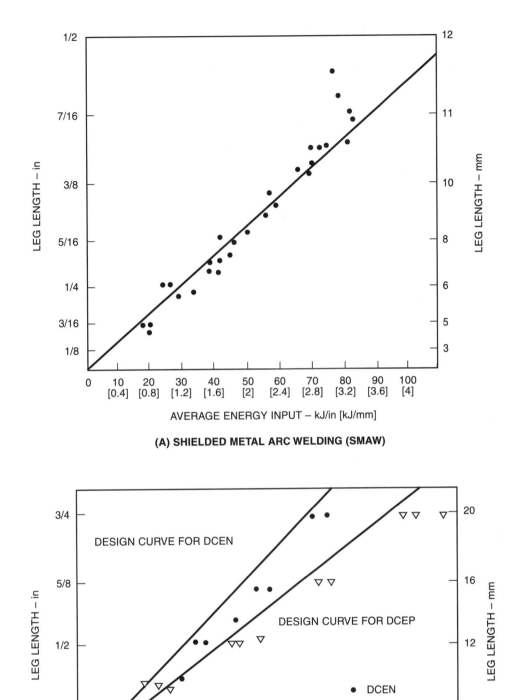

Figure B.4—Relation Between Fillet Weld Size and Energy Input (see B6.1.5)

Annex D (Normative)
Temperature-Moisture Content Charts

This annex is part of this standard and includes mandatory elements for use with this standard.

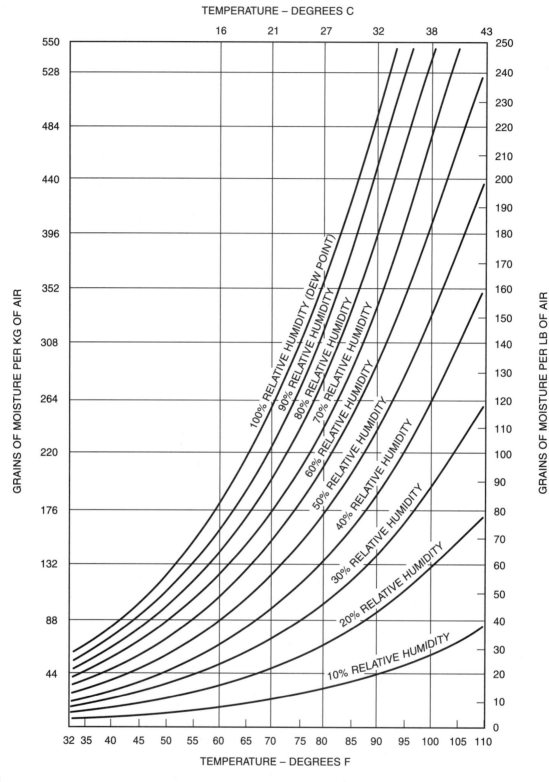

Notes:
1. Any standard psychrometric chart may be used in lieu of this chart.
2. See Figure D.2 for an example of the application of this chart in establishing electrode exposure conditions.

Figure D.1—Temperature-Moisture Content Chart to be Used in Conjunction with Testing Program to Determine Extended Atmospheric Exposure Time of Low-Hydrogen SMAW Electrodes (see 7.3.2.3)

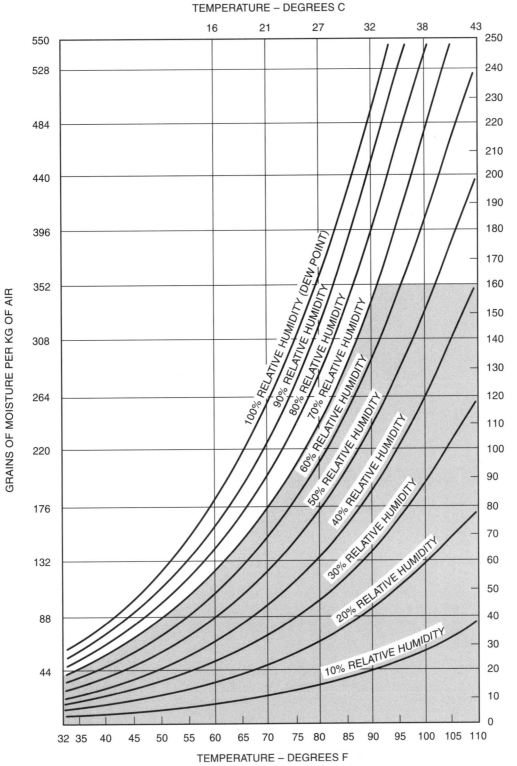

EXAMPLE: AN ELECTRODE TESTED AT 90°F [32°C] AND 70% RELATIVE HUMIDITY (RH) MAY BE USED UNDER THE CONDITIONS SHOWN BY THE SHADED AREAS. USE UNDER OTHER CONDITIONS REQUIRES ADDITIONAL TESTING.

Figure D.2—Application of Temperature-Moisture Content Chart in Determining Atmospheric Exposure Time of Low-Hydrogen SMAW Electrodes (see 7.3.2.3)

This page is intentionally blank.

Annex E (Normative)
Flatness of Girder Webs—Statically Loaded Structures

This annex is part of this standard and includes mandatory elements for use with this standard.

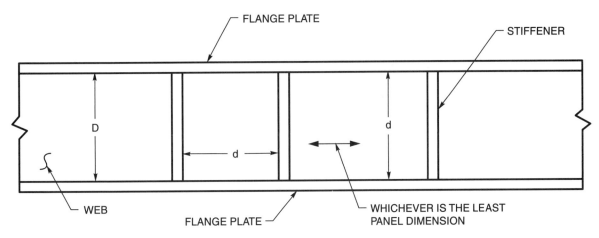

Notes:
1. D = Depth of web.
2. d = Least panel dimension.

Table E.1
Intermediate Stiffeners on Both Sides of Web

Thickness of Web, in	Depth of Web, in	Least Panel Dimension, in													
5/16	Less than 47	25	31	38	44	50								
	47 and over	20	25	30	35	40	45	50	55	60	65	70	75	80	85
3/8	Less than 56	25	31	38	44	50	56	63						
	56 and over	20	25	30	35	40	45	50	55	60	65	70	75	80	85
7/16	Less than 66	25	31	38	44	50	56	63	69					
	66 and over	20	25	30	35	40	45	50	55	60	65	70	75	80	85
1/2	Less than 75	25	31	38	44	50	56	63	69	75	81			
	75 and over	20	25	30	35	40	45	50	55	60	65	70	75	80	85
9/16	Less than 84	25	31	38	44	50	56	63	69	75	81	88		
	84 and over	20	25	30	35	40	45	50	55	60	65	70	75	80	85
5/8	Less than 94	25	31	38	44	50	56	63	69	75	81	88	94	
	94 and over	20	25	30	35	40	45	50	55	60	65	70	75	80	85
	Maximum Allowable Variation, in													
		1/4	5/16	3/8	7/16	1/2	9/16	5/8	11/16	3/4	13/16	7/8	15/16	1	1–1/16

Thickness of Web, mm	Depth of Web, m	Least Panel Dimension, meters													
8.0	Less than 1.19	0.63	0.79	0.97	1.12	1.27								
	1.19 and over	0.51	0.63	0.76	0.89	1.02	1.14	1.27	1.40	1.52	1.65	1.78	1.90	2.03	2.16
9.5	Less than 1.42	0.63	0.79	0.97	1.12	1.27	1.42	1.60						
	1.42 and over	0.51	0.63	0.76	0.89	1.02	1.14	1.27	1.40	1.52	1.65	1.78	1.90	2.03	2.16
11.1	Less than 1.68	0.63	0.79	0.97	1.12	1.27	1.42	1.60	1.75					
	1.68 and over	0.51	0.63	0.76	0.89	1.02	1.14	1.27	1.40	1.52	1.65	1.78	1.90	2.03	2.16
12.7	Less than 1.90	0.63	0.79	0.97	1.12	1.27	1.42	1.60	1.75	1.90	2.05			
	1.90 and over	0.51	0.63	0.76	0.89	1.02	1.14	1.27	1.40	1.52	1.65	1.78	1.90	2.03	2.16
14.3	Less than 2.13	0.63	0.79	0.97	1.12	1.27	1.42	1.60	1.75	1.90	2.05	2.24		
	2.13 and over	0.51	0.63	0.76	0.89	1.02	1.14	1.27	1.40	1.52	1.65	1.78	1.90	2.03	2.16
15.9	Less than 2.39	0.63	0.79	0.97	1.12	1.27	1.42	1.60	1.75	1.90	2.05	2.24	2.39	
	2.39 and over	0.51	0.63	0.76	0.89	1.02	1.14	1.27	1.40	1.52	1.65	1.78	1.90	2.03	2.16
	Maximum Allowable Variation, millimeters													
		6	8	10	11	12	14	16	18	20	21	22	24	25	27

Note: For actual dimensions not shown, use the next higher figure.

Table E.2
No Intermediate Stiffeners

Thickness of Web, in	Depth of Web, in																
Any	38	47	56	66	75	84	94	103	113	122	131	141	150	159	169	178	188
	Maximum Allowable Variation, in																
	1/4	5/16	3/8	7/16	1/2	9/16	5/8	11/16	3/4	13/16	7/8	15/16	1	1–1/16	1–1/18	1–3/16	1–1/4

Thickness of Web, mm	Depth of Web, meters																
Any	0.97	1.19	1.42	1.68	1.90	2.13	2.39	2.62	2.87	3.10	3.33	3.58	3.81	4.04	4.29	4.52	4.77
	Maximum Allowable Variation, millimeters																
	6	8	10	11	12	14	16	18	20	21	22	24	25	27	29	30	32

Note: For actual dimensions not shown, use the next higher figure.

Table E.3
Intermediate Stiffeners on One Side Only of Web

Thickness of Web, in	Depth of Web, in	Least Panel Dimension, in													
5/16	Less than 31	25	31												
	31 and over	17	21	25	29	34	38	42	46	50	54	59	63	67	71
3/8	Less than 38	25	31	38											
	38 and over	17	21	25	29	34	38	42	46	50	54	59	63	67	71
7/16	Less than 44	25	31	38	44										
	44 and over	17	21	25	29	34	38	42	46	50	54	59	63	67	71
1/2	Less than 50	25	31	38	44	50									
	50 and over	17	21	25	29	34	38	42	46	50	54	59	63	67	71
9/16	Less than 56	25	31	38	44	50	56								
	56 and over	17	21	25	29	34	38	42	46	50	54	59	63	67	71
5/8	Less than 63	25	31	38	44	50	56	63							
	63 and over	17	210	25	29	34	38	42	46	50	54	59	63	67	71
		\multicolumn{15}{c}{Maximum Allowable Variation, in}													
		1/4	5/16	3/8	7/16	1/2	9/16	5/8	11/16	3/4	13/16	7/8	15/16	1	1–1/16

Thickness of Web, mm	Depth of Web, m	Least Panel Dimension, meters													
8.0	Less than 0.78	0.63	0.79												
	0.78 and over	0.43	0.53	0.63	0.74	0.86	0.97	1.07	1.17	1.27	1.37	1.50	1.60	1.70	1.80
9.5	Less than 0.97	0.63	0.79	0.97											
	0.97 and over	0.43	0.53	0.63	0.74	0.86	0.97	1.07	1.17	1.27	1.37	1.50	1.60	1.70	1.80
11.1	Less than 1.12	0.63	0.79	0.97	1.12										
	1.12 and over	0.43	0.53	0.63	0.74	0.86	0.97	1.07	1.17	1.27	1.37	1.50	1.60	1.70	1.80
12.7	Less than 1.27	0.63	0.79	0.97	1.12	1.27									
	1.27 and over	0.43	0.53	0.63	0.74	0.86	0.97	1.07	1.17	1.27	1.37	1.50	1.60	1.70	1.80
14.3	Less than 1.42	0.63	0.79	0.97	1.12	1.27	1.42								
	1.42 and over	0.43	0.53	0.63	0.74	0.86	0.97	1.07	1.17	1.27	1.37	1.50	1.60	1.70	1.80
15.9	Less than 1.60	0.63	0.79	0.97	1.12	1.27	1.42	1.60							
	1.60 and over	0.43	0.53	0.63	0.74	0.86	0.97	1.07	1.17	1.27	1.37	1.50	1.60	1.70	1.80
		\multicolumn{15}{c}{Maximum Allowable Variation, millimeters}													
		6	8	10	11	12	14	16	18	20	21	22	24	25	27

Note: For actual dimensions not shown, use the next higher figure.

This page is intentionally blank.

Annex F (Normative)
Flatness of Girder Webs—Cyclically Loaded Structures

This annex is part of this standard and includes mandatory elements for use with this standard.

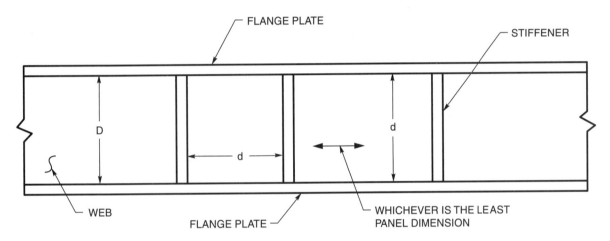

Notes:
1. D = Depth of web.
2. d = Least panel dimension.

Table F.1
Intermediate Stiffeners on Both Sides of Web, Interior Girders

Thickness of Web, in	Depth of Web, in	Least Panel Dimension, in													
5/16	Less than 47	29	36	43	50										
	47 and over	23	29	35	40	46	52	58	63	69	75	81	86	92	98
3/8	Less than 56	29	36	43	50	58									
	56 and over	23	29	35	40	46	52	58	63	69	75	81	86	92	98
7/16	Less than 66	29	36	43	50	58	65								
	66 and over	23	29	35	40	46	52	58	63	69	75	81	86	92	98
1/2	Less than 75	29	36	43	50	58	65	72	79						
	75 and over	23	29	35	40	46	52	58	63	69	75	81	86	92	98
9/16	Less than 84	29	36	43	50	58	65	72	79	86					
	84 and over	23	29	35	40	46	52	58	63	69	75	81	86	92	98
5/8	Less than 94	29	36	43	50	58	65	72	79	86	93				
	94 and over	23	29	35	40	46	52	58	63	69	75	81	86	92	98
Maximum Allowable Variation, in		1/4	5/16	3/8	7/16	1/2	9/16	5/8	11/16	3/4	13/16	7/8	15/16	1	1–1/16

Thickness of Web, mm	Depth of Web, m	Least Panel Dimension, meters													
8.0	Less than 1.19	0.74	0.91	1.09	1.27										
	1.19 and over	0.58	0.74	0.89	1.02	1.17	1.32	1.47	1.60	1.75	1.90	2.06	2.18	2.34	2.49
9.5	Less than 1.42	0.74	0.91	1.09	1.27	1.47									
	1.42 and over	0.58	0.74	0.89	1.02	1.17	1.32	1.47	1.60	1.75	1.90	2.06	2.18	2.34	2.49
11.1	Less than 1.68	0.74	0.91	1.09	1.27	1.47	1.65								
	1.68 and over	0.58	0.74	0.89	1.02	1.17	1.32	1.47	1.60	1.75	1.90	2.06	2.18	2.34	2.49
12.7	Less than 1.90	0.74	0.91	1.09	1.27	1.47	1.65	1.83	2.00						
	1.90 and over	0.58	0.74	0.89	1.02	1.17	1.32	1.47	1.60	1.75	1.90	2.06	2.18	2.34	2.49
14.3	Less than 2.13	0.74	0.91	1.09	1.27	1.47	1.65	1.83	2.00	2.18					
	2.13 and over	0.58	0.74	0.89	1.02	1.17	1.32	1.47	1.60	1.75	1.90	2.06	2.18	2.34	2.49
15.9	Less than 2.39	0.74	0.91	1.09	1.27	1.47	1.65	1.83	2.00	2.18	2.36				
	2.39 and over	0.58	0.74	0.89	1.02	1.17	1.32	1.47	1.60	1.75	1.90	2.06	2.18	2.34	2.49
Maximum Allowable Variation, millimeters		6	8	10	11	12	14	16	18	20	21	22	24	25	27

Note: For actual dimensions not shown, use the next higher figure.

Table F.2
Intermediate Stiffeners on One Side Only of Web, Fascia Girders

Thickness of Web, in	Depth of Web, in	Least Panel Dimension, in													
5/16	Less than 31	30	38												
	31 and over	20	25	30	35	40	45	50	55	60	65	70	75	80	85
3/8	Less than 38	30	38												
	38 and over	20	25	30	35	40	45	50	55	60	65	70	75	80	85
7/16	Less than 44	30	38	45											
	44 and over	20	25	30	35	40	45	50	55	60	65	70	75	80	85
1/2	Less than 50	30	38	45	53										
	50 and over	20	25	30	35	40	45	50	55	60	65	70	75	80	85
9/16	Less than 56	30	38	45	53	60									
	56 and over	20	25	30	35	40	45	50	55	60	65	70	75	80	85
5/8	Less than 63	30	38	45	53	60	68								
	63 and over	20	25	30	35	40	45	50	55	60	65	70	75	80	85
	Maximum Allowable Variation, in														
		1/4	5/16	3/8	7/16	1/2	9/16	5/8	11/16	3/4	13/16	7/8	15/16	1	1–1/16

Thickness of Web, mm	Depth of Web, m	Least Panel Dimension, meters													
8.0	Less than 0.78	0.76	0.97												
	0.78 and over	0.51	0.63	0.76	0.89	1.02	1.14	1.27	1.40	1.52	1.65	1.78	1.90	2.03	2.16
9.5	Less than 0.97	0.76	0.97												
	0.97 and over	0.51	0.63	0.76	0.89	1.02	1.14	1.27	1.40	1.52	1.65	1.78	1.90	2.03	2.16
11.1	Less than 1.12	0.76	0.97	1.14											
	1.12 and over	0.51	0.63	0.76	0.89	1.02	1.14	1.27	1.40	1.52	1.65	1.78	1.90	2.03	2.16
12.7	Less than 1.27	0.76	0.97	1.14	1.35										
	1.27 and over	0.51	0.63	0.76	0.89	1.02	1.14	1.27	1.40	1.52	1.65	1.78	1.90	2.03	2.16
14.3	Less than 1.42	0.76	0.97	1.14	1.35	1.52									
	1.42 and over	0.51	0.63	0.76	0.89	1.02	1.14	1.27	1.40	1.52	1.65	1.78	1.90	2.03	2.16
15.9	Less than 1.60	0.76	0.97	1.14	1.35	1.52	1.73								
	1.60 and over	0.51	0.63	0.76	0.89	1.02	1.14	1.27	1.40	1.52	1.65	1.78	1.90	2.03	2.16
	Maximum Allowable Variation, millimeters														
		6	8	10	11	12	14	16	18	20	21	22	24	25	27

Note: For actual dimensions not shown, use the next higher figure.

Table F.3
Intermediate Stiffeners on One Side Only of Web, Interior Girders

Thickness of Web, in	Depth of Web, in	Least Panel Dimension, in													
5/16	Less than 31	25	31												
	31 and over	17	21	25	29	34	38	42	46	50	54	59	63	67	71
3/8	Less than 38	25	31	38											
	38 and over	17	21	25	29	34	38	42	46	50	54	59	63	67	71
7/16	Less than 44	25	31	38	44										
	44 and over	17	21	25	29	34	38	42	46	50	54	59	63	67	71
1/2	Less than 50	25	31	38	44	50									
	50 and over	17	21	25	29	34	38	42	46	50	54	59	63	67	71
9/16	Less than 56	25	31	38	44	50	56								
	56 and over	17	21	25	29	34	38	42	46	50	54	59	63	67	71
5/8	Less than 63	25	31	38	44	50	56	63							
	63 and over	17	21	25	29	34	38	42	46	50	54	59	63	67	71
Maximum Allowable Variation, in		1/4	5/16	3/8	7/16	1/2	9/16	5/8	11/16	3/4	13/16	7/8	15/16	1	1–1/16

Thickness of Web, mm	Depth of Web, m	Least Panel Dimension, meters													
8.0	Less than 0.78	0.63	0.79												
	0.78 and over	0.43	0.53	0.63	0.74	0.86	0.97	1.07	1.17	1.27	1.37	1.50	1.60	1.70	1.80
9.5	Less than 0.97	0.63	0.79	0.97											
	0.97 and over	0.43	0.53	0.63	0.74	0.86	0.97	1.07	1.17	1.27	1.37	1.50	1.60	1.70	1.80
11.1	Less than 1.12	0.63	0.79	0.97	1.12										
	1.12 and over	0.43	0.53	0.63	0.74	0.86	0.97	1.07	1.17	1.27	1.37	1.50	1.60	1.70	1.80
12.7	Less than 1.27	0.63	0.79	0.97	1.12	1.27									
	1.27 and over	0.43	0.53	0.63	0.74	0.86	0.97	1.07	1.17	1.27	1.37	1.50	1.60	1.70	1.80
14.3	Less than 1.42	0.63	0.79	0.97	1.12	1.27	1.42								
	1.42 and over	0.43	0.53	0.63	0.74	0.86	0.97	1.07	1.17	1.27	1.37	1.50	1.60	1.70	1.80
15.9	Less than 1.60	0.63	0.79	0.97	1.12	1.27	1.42	1.60							
	1.60 and over	0.43	0.53	0.63	0.74	0.86	0.97	1.07	1.17	1.27	1.37	1.50	1.60	1.70	1.80
Maximum Allowable Variation, millimeters		6	8	10	11	12	14	16	18	20	21	22	24	25	27

Note: For actual dimensions not shown, use the next higher figure.

Table F.4
Intermediate Stiffeners on Both Sides of Web, Fascia Girders

Thickness of Web, in	Depth of Web, in	Least Panel Dimension, in													
5/16	Less than 47	33	41	49											
	47 and over	26	33	39	47	53	59	66	71	79	85	92	98	105	112
3/8	Less than 56	33	41	49	57										
	56 and over	26	33	39	47	53	59	66	71	79	85	92	98	105	112
7/16	Less than 66	33	41	49	57	65	73								
	66 and over	26	33	39	47	53	59	66	71	79	85	92	98	105	112
1/2	Less than 75	33	41	49	57	65	73	81							
	75 and over	26	33	39	47	53	59	66	71	79	85	92	98	105	112
9/16	Less than 84	33	41	49	57	65	73	81	89						
	84 and over	26	33	39	47	53	59	66	71	79	85	92	98	105	112
5/8	Less than 94	33	41	49	57	65	73	81	89	98					
	94 and over	26	33	39	47	53	59	66	71	79	85	92	98	105	112
		Maximum Allowable Variation, in													
		1/4	5/16	3/8	7/16	1/2	9/16	5/8	11/16	3/4	13/16	7/8	15/16	1	1–1/16

Thickness of Web, mm	Depth of Web, m	Least Panel Dimension, meters													
8.0	Less than 1.19	0.84	1.04	1.24											
	1.19 and over	0.66	0.84	0.99	1.19	1.35	1.50	1.68	1.83	2.01	2.16	2.34	2.49	2.67	2.84
9.5	Less than 1.42	0.84	1.04	1.24	1.45										
	1.42 and over	0.66	0.84	0.99	1.19	1.35	1.50	1.68	1.83	2.01	2.16	2.34	2.49	2.67	2.84
11.1	Less than 1.68	0.84	1.04	1.24	1.45	1.65	1.85								
	1.68 and over	0.66	0.84	0.99	1.19	1.35	1.50	1.68	1.83	2.01	2.16	2.34	2.49	2.67	2.84
12.7	Less than 1.90	0.84	1.04	1.24	1.45	1.65	1.85	2.06							
	1.90 and over	0.66	0.84	0.99	1.19	1.35	1.50	1.68	1.83	2.01	2.16	2.34	2.49	2.67	2.84
14.3	Less than 2.13	0.84	1.04	1.24	1.45	1.65	1.85	2.06	2.26						
	2.13 and over	0.66	0.84	0.99	1.19	1.35	1.50	1.68	1.83	2.01	2.16	2.34	2.49	2.67	2.84
15.9	Less than 2.39	0.84	1.04	1.24	1.45	1.65	1.85	2.06	2.26	2.49					
	2.39 and over	0.66	0.84	0.99	1.19	1.35	1.50	1.68	1.83	2.01	2.16	2.34	2.49	2.67	2.84
		Maximum Allowable Variation, millimeters													
		6	8	10	11	12	14	16	18	20	21	22	24	25	27

Note: For actual dimensions not shown, use the next higher figure.

Table F.5
No Intermediate Stiffeners, Interior or Fascia Girders

Thickness of Web, in	Depth of Web, in															
Any	38	47	56	66	75	84	94	103	113	122	131	141	150	159	169	178	188
	Maximum Allowable variation, in															
	1/4	5/16	3/8	7/16	1/2	9/16	5/8	11/16	3/4	13/16	7/8	15/16	1	1–1/16	1–1/18	1–3/16	1–1/4
Thickness of Web, mm	Depth of Web, meters															
Any	0.97	1.19	1.42	1.68	1.90	2.13	2.39	2.62	2.87	3.10	3.33	3.58	3.81	4.04	4.29	4.52	4.77
	Maximum Allowable Variation, millimeters															
	6	8	10	11	12	14	16	18	20	21	22	24	25	27	29	30	32

Note: For actual dimensions not shown, use the next higher figure.

Annex G (Normative)
Qualification and Calibration of UT Units with Other Approved Reference Blocks
(See Figure G.1)

This annex is part of this standard and includes mandatory elements for use with this standard.

G1. Longitudinal Mode

G1.1 Distance Calibration

G1.1.1 The transducer shall be set in position H on the DC block, or M on the DSC block.

G1.1.2 The instrument shall be adjusted to produce indications at 1 in [25 mm], 2 in [50 mm], 3 in [75 mm], 4 in [100 mm], etc., on the display.

NOTE: This procedure establishes a 10 in [250 mm] screen calibration and may be modified to establish other distances as allowed by 8.24.4.1.

G1.2 Amplitude. With the transducer in position described in G1.1, the gain shall be adjusted until the maximized indication from the first back reflection attains 50% to 75% screen height.

G2. Shear Wave Mode (Transverse)

G2.1 Sound Entry (Index) Point Check

G2.1.1 The search unit shall be set in position J or L on the DSC block; or I on the DC block.

G2.1.2 The search unit shall be moved until the signal from the radius is maximized.

G2.1.3 The point on the Search Unit that is in line with the line on the calibration block is indicative of the point of sound entry.

NOTE: This sound entry point shall be used for all further distance and angle checks.

G2.2 Sound Path Angle Check

G2.2.1 The transducer shall be set in position:

K on the DSC block for 45° through 70°

N on the SC block for 70°

O on the SC block for 45°

P on the SC block for 60°

G2.2.2 The transducer shall be moved back and forth over the line indicative of the transducer angle until the signal from the radius is maximized.

G2.2.3 The sound entry point on the transducer shall be compared with the angle mark on the calibration block (tolerance 2°).

G2.3 Distance Calibration

G2.3.1 The transducer shall be in position (Figure G.1) L on the DSC block. The instrument shall be adjusted to attain indications at 3 in [75 mm] and 7 in [180 mm] on the display.

G2.3.2 The transducer shall be set in position J on the DSC block (any angle). The instrument shall be adjusted to attain indications at 1 in [25 mm], 5 in [125 mm], 9 in [230 mm] on the display.

G2.3.3 The transducer shall be set in position I on the DC block (any angle). The instrument shall be adjusted to attain indication at 1 in [25 mm], 2 in [50 mm], 3 in [75 mm], 4 in [100 mm], etc., on the display.

NOTE: This procedure establishes a 10 in [250 mm] screen calibration and may be modified to establish other distances as allowed by 8.24.5.1.

G2.4 Amplitude or Sensitivity Calibration

G2.4.1 The transducer shall be set in position L on the DSC block (any angle). The maximized signal shall be adjusted from the 1/32 in [0.8 mm] slot to attain a horizontal reference line height indication.

G2.4.2 The transducer shall be set on the SC block in position:

N for 70° angle

O for 45° angle

P for 60° angle

The maximized signal from the 1/16 in [1.6 mm] hole shall be adjusted to attain a horizontal reference line height indication.

G2.4.3 The decibel reading obtained in G2.4.1 or G2.4.2 shall be used as the "reference level" "b" on the Test Report sheet (Annex P, Form P–11) in conformance with 8.22.1.

G3. Horizontal Linearity Procedure

NOTE: Since this qualification procedure is performed with a straight beam search unit which produces longitudinal waves with a sound velocity of almost double that of shear waves, it is necessary to double the shear wave distance ranges to be used in applying this procedure.

G3.1 A straight beam search unit, meeting the requirements of 8.21.6, shall be coupled in position:

G on the IIW type block (Figure 8.16)

H on the DC block (Figure G.1)

M on the DSC block (Figure G.1)

T or U on the DS block (Figure 8.16)

G3.2 A minimum of five back reflections in the qualification range being certified shall be attained.

G3.3 The first and fifth back reflections shall be adjusted to their proper locations with use of the distance calibration and zero delay adjustments.

G3.4 Each indication shall be adjusted to reference level with the gain or attenuation control for horizontal location examination.

G3.5 Each intermediate trace deflection location shall be correct within ± 2% of the screen width.

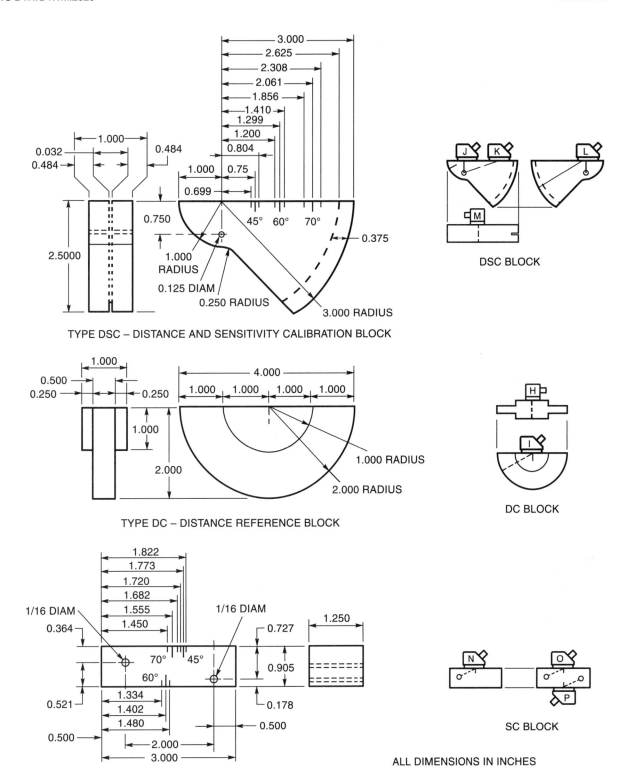

Notes:
1. The dimensional tolerance between all surfaces involved in referencing or calibrating shall be within ±0.005 in of detailed dimension.
2. The surface finish of all surfaces to which sound is applied or reflected from shall have a maximum of 125 µin r.m.s.
3. All material shall be ASTM A36 or acoustically equivalent.
4. All holes shall have a smooth internal finish and shall be drilled 90° to the material surface.
5. Degree lines and identification markings shall be indented into the material surface so that permanent orientation can be maintained.

Figure G.1—Other Approved Blocks and Typical Transducer Position (see G2.3.1)

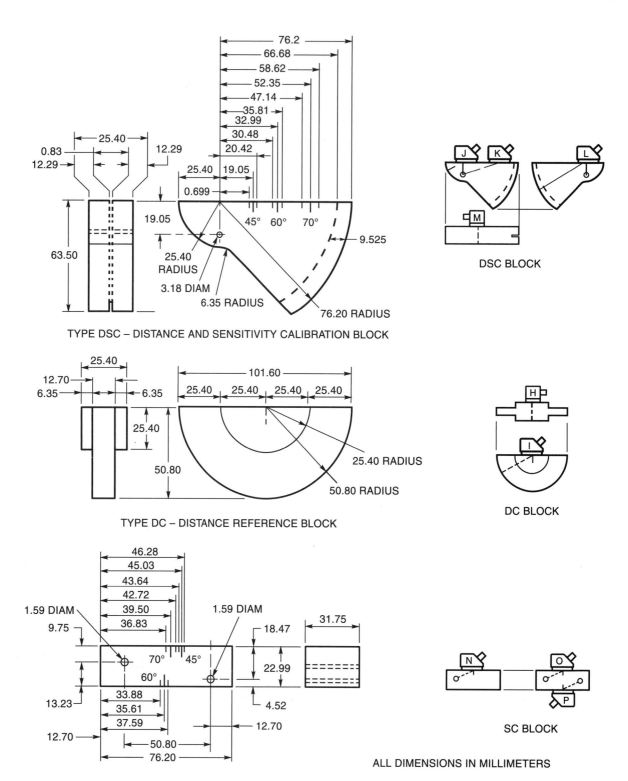

Notes:
1. The dimensional tolerance between all surfaces involved in referencing or calibrating shall be within ±0.13 mm of detailed dimension.
2. The surface finish of all surfaces to which sound is applied or reflected from shall have a maximum of 3.17 μm r.m.s.
3. All material shall be ASTM A36 or acoustically equivalent.
4. All holes shall have a smooth internal finish and shall be drilled 90° to the material surface.
5. Degree lines and identification markings shall be indented into the material surface so that permanent orientation can be maintained.

Figure G.1 (Continued)—Other Approved Blocks and Typical Transducer Position (see G2.3.1) (Metric)

Annex H (Normative)
Phased Array Ultrasonic Testing (PAUT)

This annex is part of this standard and includes mandatory elements for use with this standard

H1. Introduction

This annex provides mandatory requirements that shall apply when phased array ultrasonic testing (PAUT) is used. The alternative techniques presented in this annex require written procedures, advanced personnel training and qualification, and calibration methods specific to PAUT.

H2. Scope

The procedures and standards set forth in this annex govern PAUT examinations of groove welds, including heat-affected zones (HAZ), for thicknesses between 3/16 in and 8 in [5 mm and 200 mm] using encoded linear scanning.

These procedures and standards exclude the PAUT examination of tubular T, Y, and K connection welds.

H3. Definitions

H3.1 Azimuthal Scan. A focal law sequence that produces a fan-like series of beams through a defined range of angles using the same set of elements.

H3.2 Bandpass Filtering. A function of the receiver circuit in most modern UT and PAUT equipment designed to filter out unwanted returned sound frequencies outside of that used for sound wave generation. The frequencies of sound on return are of much broader range than the range of frequencies put into the test piece.

H3.3 Channel. A send/receive circuit in the phased array unit. The number of channels dictates the maximum number of elements that the phased array unit can support as a whole.

H3.4 Dead Elements. Individual elements that are no longer functional due to broken cables, connectors or element failure. This can also include elements with substandard performance.

H3.5 Electronic Scan. A focal law sequence that produces multiple beams of sound at a fixed angle by utilizing multiple sets of elements in a PAUT probe.

H3.6 Element. An individual crystal (piezo-composite material) within a phased array probe.

H3.7 Encoded. PAUT inspection data collected by use of an encoder.

H3.8 Encoder. A device that tracks the position of the search unit(s) as it is manually or automatically moved along the weld or component.

H3.9 Encoding. The use of an encoder during data acquisition.

H3.10 Focal law. A phased array operational file that defines search unit elements to be used, and time delays for transmitted/received signals which manipulates the sound beam direction. Focal laws are combined into sequences to form either multi-angle (azimuthal scans) or fixed angle (electronic scans).

H3.11 Full screen height (FSH). refers to the vertical axis of the instrument display and is a percentage measure of the sound amplitude return signal compared to the full height of the display screen.

H3.12 Imaging Views. Images defined by different plane views between the ultrasonic path (ultrasonic axis), beam movement (index axis), and probe movement (scan axis). See Figure H.1. Also called "scans" (see H3.12.1 to H3.12.5).

H3.12.1 A-Scan. A representation (view) of the received ultrasonic pulse amplitude versus time of flight in the ultrasonic path, also called a waveform.

H3.12.2 B-Scan. A 2-D view of the recorded A-Scan data showing the ultrasonic path (ultrasonic axis) along the probe movement path (scan axis) and is color coded to represent amplitude. The B-Scan is available in multiple forms including single angle/VPA (see H3.25) or merged angle/VPA presentations.

H3.12.3 C-Scan. A 2-D plan or top view of the recorded A-Scan data showing the beam movement (index axis) versus the probe movement (scan axis) path, using the maximum amplitude of the A-Scans at each transverse location. The C-Scan may be presented in the volume-corrected or uncorrected form.

H3.12.4 D-Scan. Is similar to the S-Scan imaging view as detailed below but less commonly used for weld inspection. Like the S-Scan, the D-Scan shows the ultrasonic path (ultrasonic axis) versus the beam movement axis (index axis). Unlike the S-Scan, the D-Scan is configured to show all the A-Scans collected through the entire or portions of the entire scan.

H3.12.5 S-Scan. A 2-D cross-sectional view of all A-Scans which are corrected for delay and refracted angle. The S-Scan shows the ultrasonic path (ultrasonic axis) versus the beam movement axis (index axis) and is presented as a single slice of focal laws (one focal law sequence) along the probe movement path (scan axis).

H3.13 Line Scan. The phased array scanning technique in which an electronic scan, azimuthal scan or combination thereof is performed with the beams directed perpendicular to the weld, at a fixed distance from the welds, in a manner demonstrated to provide full weld coverage. Also called a linear scan.

H3.14 PAUT. Phased array ultrasonic testing.

H3.15 Phased Array Instrument. A multichannel test instrument used with multiple element probes which enable the application of delay/focal laws when transmitting, and receiving, before summing.

H3.16 Phased Array Technique. A technique wherein UT data is generated by constructive phase interference formed by multiple elements controlled by accurate time-delayed pulses. This technique can perform beam sweeping through an angular range (azimuthal scans), beam scanning at fixed angle (electronic scans), beam focusing, lateral scanning and a variety of other scans depending on the array and programming.

H3.17 Phased Array Probe. A probe made up of several piezoelectric elements individually connected so that the signals they transmit or receive may be treated separately or combined as desired. The elements can be pulsed individually, simultaneously, or in a certain pattern relative to each other to create the desired beam angles or scan pattern.

H3.18 Pitch. The center to center distance between two successive phased array probe elements.

H3.19 Pulser. The instrument component which generates the electrical pulse. The number of pulsers dictates how many elements within a phased array probe may be applied within a given focal law.

H3.20 Saturated Signal. A signal in which the true peak amplitude cannot be measured in the stored data file due to bit depth of the phased array system.

H3.21 Scan Plan. A document specifying key process elements such as equipment detail, focal law settings and probe positions as necessary to complete an examination; also depicts weld and HAZ coverage.

H3.22 Scanner. A device used for holding the phased array probes in place while collecting data by means of an encoder. Scanners contain an encoder and may be automated or semi-automated types as described below.

H3.22.1 Automated Scanner. A mechanized device in which the PAUT probe movement is computerized or driven by remote control.

H3.22.2 Semi-Automated Scanner. A scanner which is manually driven along welds.

H3.23 Sound-Path or Depth Calibration (Horizontal Linearity). A specific action used to compensate and adjust instrument time delay over all focal laws for specific wedge geometry for a depth or sound-path calibration.

H3.24 Time Corrected Gain (TCG). Also called Time Varied Gain (TVG). A calibration technique in which the search unit computes the dB gain difference needed to balance standard calibration reflectors (i.e., side drilled holes) at various material depths to one set screen amplitude. When completed, all equivalently sized reflectors from all ultrasonic beams will equal the same approximate amplitude regardless of their varying sound path distances.

H3.25 Virtual Probe Aperture (VPA). The number of elements in a phased array probe used for the examination.

H3.26 Volume Correction. An image correction feature, in which corrections can be made to the A-Scan point locations displayed and based off of true positional information relative to the beam angle or angles used during the inspection.

H4 Personnel Requirements

H4.1 Personnel Qualification Requirements.

PAUT Level II and III personnel who collect and/or analyze PAUT data shall be qualified per 8.14.6.1 and 8.20. In addition, the PAUT inspector shall have documented a minimum of 320 hours of work time experience in PAUT applications. The practical exam as required in 8.20 shall consist of at least 2 flawed specimens representing joint types to be examined, each containing a minimum of 2 flaws.

Individuals not meeting these requirements may assist in PAUT data collection under the direct supervision of qualified PAUT personnel.

H4.2 Certification Requirements. Certification of Level II and III PAUT personnel shall be performed by a NDT UT Level III who meets the requirements of H4.1.

H5. Equipment

H5.1 Phased Array Instruments. Inspections shall be performed using phased array pulse-echo equipment which meets the requirements of 8.21, qualified in accordance with H8. Phased array instruments shall also meet the following requirements:

  H5.1.1 Number of Pulsers. The instrument shall be equipped with a minimum of 16 pulsers and channels (16:16 minimum). A minimum of 16:64 is required if electronic scans are to be used.

  H5.1.2 Imaging Views. The phased array instrument shall be equipped with sufficient display options to include A-Scan, B-Scan, C-Scan, and S-Scan views, and encoded scans to provide thorough data analysis through the entire scan length and through all beams.

H5.2 Straight-Beam (Longitudinal Wave) Probes. The straight-beam (longitudinal wave) phased array probe shall produce frequencies in the range of 1 to 6 MHz. Probe dimensions shall be small enough that standing wave signals do not appear on the display. The phased array probe shall be a linear array probe capable of meeting the requirements of 8.27.1.3. Alternatively, a conventional UT search unit meeting the requirements of 8.21.6 may be used.

H5.3 Angle-Beam Search Units. Angle-beam search units shall consist of a phased array probe and an angle wedge to produce the desired refracted angles.

  H5.3.1 Phased Array Probe. The probe shall be a linear array type with a minimum of 16 elements and shall produce frequencies between 1 and 6 MHz. Probe pitch dimensions shall be small enough that standing wave signals do not appear on the display.

  H5.3.2 Angled Phased Array Wedge The wedge shall be of a sufficient incident angle to produce shear waves in the material between 40° and 70°. Wedges shall be used within the angular range specified by the manufacturer.

H5.4 Encoder. The encoder shall be digital and capable of line scanning.

H5.5 Scanner. Encoding shall be performed by using a semi-automated or automated scanner as defined in H3.22.

H5.6 Couplant. A couplant material shall be used between the search unit and the test material. The same couplant used for calibration shall also be used in examination. Any commercial couplant, water, or oil may be used when performing calibrations and examinations.

H5.7 Basic Calibration Standard for Determining SSL. The standard reflector used to establish the Standard Sensitivity Level (SSL) shall be the 0.060 in [1.5 mm] diameter side drilled hole in an IIW type block in conformance with ASTM E164. The temperature of the calibration standard utilized shall be within ± 25 degrees F [± 14 degrees C] of the temperature of the part or component to be examined.

H5.7.1 Supplemental Reference Block. A supplemental reference block shall be used which allows a minimum of a 3 point TCG establishment throughout the range specified in the scan plan for all configured angles. The block shall be made of carbon steel and of sufficient thickness and length to allow calibration of reflectors throughout the entire examination volume to be tested. Each reference block shall have at least three side drilled holes at a range of depths to cover the entire material range to be tested. Each hole shall be detected by the PAUT system as configured per the scan plan and sensitivities shall be adjusted as necessary to provide sensitivity equal to or greater than SSL. An example of this block is shown in Figure H.2 but any block meeting the requirements of this clause may be used including a NAVSHIP or custom machined block.

H5.7.2 Mockup Verification Block. For materials over 2 in [50mm] in thickness, when required by the Engineer, or additionally at the option of PAUT personnel, detectability of the standard reflector (0.060 in [1.5 mm] side drilled hole) shall be verified in a mockup or the production part. When weld mockups and sections of production weldments are used, the reflector shall be in locations where it is difficult to direct sound beams, thereby ensuring detection of discontinuities in all areas of interest. Mock reflectors shall be placed a minimum of 0.060 in [1.5 mm] clear distance from edges of the fusion face. Example placements of the standard sensitivity reflector in a mockup or production part are shown in Figure H.3.

When this verification block is utilized, the standard sensitivty reflector shall be detectable above DRL established in H8.2.4.2. If the standard reflector is not detectable over DRL, the scan plan shall be adjusted until adequate detectability is obtained.

H6. Equipment Qualification

H6.1 System Linearity. System linearity verifications shall be validated at a maximum of 12 month intervals. Validation shall be performed as detailed in H14.

H6.2 Internal Reflections. Maximum internal reflections from each search unit shall be verified by PAUT personnel at a maximum time interval of 40 hours of instrument use, and checked in accordance with 8.28.3.

H6.3 Resolution Requirements. Testing of the resolution with the combination of search unit and instrument shall be performed and documented per 8.22.3.

H6.4 Probe Operability Checks. An element operability check shall be performed by PAUT personnel prior to initial calibration and use and weekly on each phased array probe to determine if dead (inactive) or defective elements are present. The probe operability check shall be performed by using the instrument auto-verification feature or manually by scanning through each element with the probe on the side of an IIW type block or any reference block and observing the back wall signal. No more than 10% of the elements may be dead in a given aperture, and no more than two adjacent elements may be dead within a given aperture. This check shall also be performed upon each 8-hour period of use. In addition, each element within a phased array probe shall be evaluated to check for comparable amplitude responses throughout the aperture. Each element shall be verified to be within 6 dB of the average response across the aperture. If the amplitude of any of the elements within the probe yields responses outside the 6 dB requirement, the element shall be declared dead.

H7. Scan Plans

H7.1 Scan Plans. A scan plan, as defined in H3.21, shall be developed for the welds to be examined. The scan plan shall provide the specific attributes necessary to achieve examination coverage, including those variables subject to material and geometric variation that are not addressed in a general procedure. Scan plan contents shall consider all essential variables listed in Table H.1.

H7.1.1 Scan Plan Qualification. For thicknesses greater than 2 in [50 mm], or when required by the Engineer, a mockup verification block shall be used to demonstrate effective PAUT sensitivity and weld volume coverage as referenced in paragraph H5.7.2.

H7.1.2 The scan plan shall demonstrate by plotting or computer simulation the appropriate refracted angles to be used during the examination for the groove weld geometry and areas of concern. The scan plan shall demonstrate and document coverage of the required examination volume. Performance shall be verified through the calibration (i.e., beam index point and beam angle verifications).

H7.1.3 When developing a scan plan, the essential variables of Table H.1 shall be established and documented with the initial calibration. The initial calibration shall be performed by a certified PAUT Level II or III to confirm adequate sound coverage throughout the configured ultrasonic range.

H7.2 Focal Law Configuration. Focal laws shall be configured to provide the necessary weld volume coverage requirements stipulated in H7.4. Azimuthal scans shall be used as the primary scan to optimize coverage and shall be configured in angular sweep increments of no greater than 1°. Electronic scans may be used as described below to supplement azimuthal scans but shall not be used as a sole inspection technique.

H7.2.1 Index Positions. A sufficient number of index positions shall be configured to accomplish the coverage requirements of H7.4. These may be multiple physical index positions, multiple electronic index positions (grouping), or a combination of both. Scans shall contain sufficient overlap to demonstrate full coverage in the scan plan.

H7.2.2 Focusing. When examining material 5/16 in [8 mm] thick, a true depth focus shall be applied at the depth corresponding to the depth where sound enters the weld volume. For examination of materials greater than 5/16 inches [8 mm] focusing may be performed provided a minimum of 4 to 1 signal to noise ratio is demonstrated on all TCG calibration reflectors.

H7.2.3 Supplemental Electronic Scans. Electronic scans may be used to supplement the azimuthal scans. When electronic scans are used, a minimum overlap of 50% of each VPA shall be configured as specified in the scan plan.

H7.2.4 Grouping. Combinations of multiple azimuthal or of azimuthal and electronic scans may be used through grouping features to assist in weld joint coverage. When combined, the minimum overlap between each scan shall be 10% of coverage.

H7.3 Adjustments to Scan Plans. Any changes to an essential variable in Table H.1 after the scan plan is established and approved shall require the development of a new scan plan, a new calibration and, if applicable, re-demonstration on the mock-up verification block.

H7.4 Phased Array Testing of Welds. The base metal through which ultrasound must travel to test the weld shall be tested for laminar reflectors using a straight-beam search unit conforming to H5.2. If any area of base metal exhibits total loss of back reflection or an indication equal to or greater than the original back reflection, refer to H9.2.2.

H7.4.1 The scan plan, utilizing array configurations specified in H7.2, shall demonstrate full ultrasonic coverage in two crossing directions to cover the heat affected zone (HAZ) with beam angles between 40 and 60 degrees and full weld volume including weld fusion face coverage within ±10° of perpendicular (90° to the weld fusion face) for azimuthal scans or ±5° of perpendicular for supplemental electronic scans where possible. The weld and the HAZ area shall be tested using a PAUT probe conforming to the requirements of H5.3.

H7.4.2 All welds in butt joints examined by PAUT shall be tested from the same face of the base metal and from both sides of the weld axis where access is possible. Welds in corner and T-joints shall be primarily tested from one side of the weld axis only. All welds shall be tested using applicable line scans or scanning patterns necessary to detect both longitudinal and transverse discontinuities.

H7.4.3 Scanning Near Edges. If edges and corners prevent access or result in other limitations for encoded PAUT, these areas may be scanned by running the scan in the opposite direction toward the edge, or by non-encoded PAUT using scanning patterns described in Clause 8. Use of non-encoded PAUT shall be noted in the examination report.

H7.4.4 Restricted Access Welds. Groove welds in butt joints that cannot be examined from both sides of the weld axis using the angle beam technique shall be scanned from alternate faces where possible to assure full coverage of the weld and HAZ is obtained. These situations shall be addressed with a modified scan plan and noted in the examination report.

H7.4.5 Backing. For joint configurations will contain backing that is left in place, the scan plan shall consider the effects of the backing (see C-8.25.12 for additional guidance on inspecting welds with steel backing).

H7.4.6 Inspection for Transverse Indications. Welds that are ground flush shall be inspected for transverse indications using scanning pattern D as shown in Figure 8.15. Scanning pattern E shall be used on welds with reinforcement. Encoding is not required for the transverse indication inspection.

H7.5 Setup File Storage. Scan plan parameters shall be configured on the phased array system and stored in a manner that will allow repeatability for subsequent examinations.

H8. Calibration for Testing

The phased array configuration as prepared in the scan plan shall be verified at intervals in accordance with 8.24.3 and as described below.

H8.1 Straight Beam Calibration. The ultrasonic range of the search unit shall be adjusted, using an electronic scan at 0° set-up (or conventional straight beam probe), such that it will produce the equivalent of at least two plate thicknesses on the display. The sensitivity shall be adjusted at a location free of indications so that the first back reflection from the far side of the plate will be 80% ± 5% of FSH. Minor sensitivity adjustments may be made to accommodate for surface roughness.

H8.2 Shear Wave Calibration

H8.2.1 Beam Angle Verification. PAUT personnel shall verify the sound beam angles to be within 2° of the minimum and maximum angles configured in azimuthal scans or within 2° of the first and last VPAs configured for electronic scans using the procedure stipulated in 8.27.2.2.

H8.2.2 Horizontal Sweep. The horizontal sweep shall be adjusted to represent the actual material path distance throughout all the configured angles using an IIW type block or other alternate block as detailed in 8.22.1. The screen range shall be set such that the entire area of interest under examination is visible in the A- and S-Scan displays.

H8.2.3 Time Corrected Gain (TCG). By use of the supplemental calibration block as specified in H5.7.1, a TCG shall be established throughout all configured angles at a minimum of three points throughout the material range to be tested. The TCG shall balance all calibration points within ±5% amplitude of each other.

H8.2.4 Standard Sensitivity Level (SSL). The standard sensitivity level shall be established at 50% ±5% FSH off of the 0.060 in [1.5 mm] reflector as specified in H5.7. When dynamic line scanning, an additional 4 dB shall be added to account for sound losses during data acquisition. This dB level shall be noted as the primary reference level sensitivity (SSL) dB. Additional dB is not required for manual supplemental scans.

H8.2.4.1 Automatic Reject Level (ARL). The ARL shall be defined as 5 dB over SSL, which equals 89% FSH (see Figure H.4).

H8.2.4.2 Disregard Level (DRL). The DRL shall be defined as 6 dB under SSL, which equals 25% FSH (see Figure H.4).

H8.3 Encoder Calibration. The encoder shall be verified by PAUT personnel, through daily in-process checks, to be within 1% of measured length for a minimum of half the total scan length. Encoder resolution shall be configured so that data is taken at 0.04 in [1 mm] increments or smaller.

H9. Examination Procedure

H9.1 X and Y. Coordinates shall be identified prior to scanning as required in 8.25.1 and 8.25.2.

H9.2 Straight Beam Scanning. Straight beam examination using a PAUT electronic scan or conventional UT scan at 0° shall be performed over the entire base metal area through which sound must pass in accordance with 8.24.4.

H9.2.1 If part of the weld is inaccessible to testing in conformance with the requirements of the scan plan due to laminar content recorded in conformance with 8.24.4, the testing shall be conducted using one or more of the following alternative procedures as necessary to attain full coverage. These alternative procedures shall be accounted for in the scan plan and noted in the examination report.

(1) Weld surface(s) shall be ground flush in conformance with 7.23.3.1

(2) Testing performed from alternative faces

(3) Additional scan types (electronic scans) or alternative index points

H9.2.2 Any indication evaluated as laminar reflectors in base material, which interfere with the scanning of examination volumes, shall require modification of the angle beam examination technique such that the maximum feasible volume is examined, and the modification shall be noted in the record of the examination. If any area of base metal exhibits total loss of back reflection or an indication equal to or greater than the original back reflection height is located in a position that will interfere with the normal weld scanning procedure, its size and location shall be determined and reported on the UT report, and an alternate weld scanning procedure shall be used.

H9.3 Angle Beam Scanning. Automatic computer recording of essential ultrasonic data in the manner of line scans shall be performed down the axial length of each weld. Scanning shall be performed in accordance with the documented and approved scan plan as detailed in H7.

H9.3.1 Scanning Gain. Scanning may be performed at primary reference level (SSL) sensitivity as configured in H8.2.4, provided soft gain or color palette alterations are made during evaluation to aid in detection. If scanning is performed at Standard Sensitivity Level (SSL), soft gain shall be increased by 6 dB or the color palette adjusted to end at 50% screen height during the evaluation of the weld data. If color palette adjustment or soft gain increase is not used, 6 dB of additional gain over primary reference level shall be applied during scanning. For manual supplemental examinations, e.g., transverse flaw detection, scanning shall be a minimum of 6 dB over Standard Sensitivity Level (SSL).

H9.3.2 Encoded Scanning. Except as permitted in H7.4, scanning shall be done with an encoder. Encoded line scanning shall be performed by using a mechanical fixture or apparatus which limits probe movement along the index axis while scanning.

H9.3.3 Scanning Speed. The indicated speed of acquisition which is established for the instrument for the given setup shall not be exceeded. If data dropout is noted, it shall not exceed 1% of the recorded data and no two consecutive lines of data shall be missed.

H9.3.4 Data Collection. PAUT personnel shall ensure that ultrasonic examination data is recorded in an unprocessed form. A full and complete data recording set of the original A-Scan data with no exclusionary gating or filtering other than receiver bandpass shall be included in the data record.

H9.3.5 Delayed Inspection Acceptance for ASTM A514, A517, and A709 Grade HPS 100W [HPS 690W] shall be based on PAUT performed not less than 48 hours after the completion of welding.

H10. Evaluation

H10.1 Evaluation Level Data shall be analyzed with consideration to scanning gain as noted in H9.3.1 above. Evaluation of indications for accept/reject disposition shall be performed with respect to the SSL level.

H10.2 Length Measurements Indication length shall be determined by using the 6 dB drop method described in 8.29.2. For saturated indications, in which the true peak amplitude measurement cannot be obtained, additional scans at lower gain levels shall be performed for Class B and C indications with near rejectable lengths or proximities to adjacent indications or weld intersections where applicable. The length may be determined from the stored data file.

H10.3 Acceptance Criteria

Welds shall be acceptable provided they have no planar indications, nor any indications whose amplitude or length exceeds that specified in Table H.2 for the applicable type of loading. Discontinuities shall be classified based on their maximum amplitude in accordance with Table H.3 (also see Figure H.4):

H10.3.1 Manual supplemental PAUT and/or alternative NDT techniques shall be used when necessary to verify questionable indications found in the collected data.

H10.3.2 Indications characterized as cracks shall be considered unacceptable regardless of length or amplitude.

H10.3.3 Class B and C indications shall be separated by at least 2L, L being the length of the longer indication, except that when two or more such indications are not separated by at least 2L, but the combined length of indications and their separation distance is equal to or less than the maximum allowable length under the provisions of Class B or C, the flaw shall be considered a single acceptable indication.

H10.3.4 Class B and C indications shall not begin at a distance less than 2L from the weld ends carrying primary tensile stresses, L being the indication length.

H10.3.5 For Class C indications, the depth of the discontinuity shall be determined by the location of the peak amplitude, at the angle producing the maximum signal amplitude.

H10.4 Identification of Rejected Areas. Each unacceptable discontinuity shall be indicated on the weld by a mark directly over the discontinuity for its entire length. The depth from the surface and the indication rating shall be noted on the nearby base material.

H10.5 Repair. Repairs to welds found unacceptable by PAUT shall be made in accordance with 7.25. Repaired areas shall be retested using the same scan plan and techniques as used for the original inspection, unless the scan plan does not provide coverage of the repaired area. In this case, a new scan plan shall be developed for the repair area. The minimum length of the repair to be re-inspected shall be the length of the repaired area plus 2 in [50 mm] on each end.

H11. Data Analysis

H11.1 Validation of Coverage. Recorded data shall be assessed to ensure full execution of the scan plan over 100% of the required examination length.

H11.2 Data Analysis and Recording Requirements. The following are requirements for evaluation of data:

(1) The entire examination volume as dictated by the scan plan shall be analyzed using gates and available cursors to locate and identify the source, location and nature for all indications. Alternately, manual plotting may be used to augment on-board analysis, e.g., nonparallel or inconsistent geometries.

(2) Responses resulting from weld root and weld cap geometries shall be investigated and the basis for this classification should be noted on the report form.

(3) Any indication warranting evaluation shall be recorded to support the resultant disposition. The extent of recording shall be sufficient for reviewers and subsequent examiners to repeat the result and should stand alone as a written record.

(4) Rejectable indications shall be reported. The report shall include peak amplitude, indication rating, length of indication, depth below the surface and relative position to provide adequate information for the repair. Cursor placement, measurement features, and annotations and comments shall clearly support disposition.

H12. Data Management

H12.1 Data Management System. There shall be a data management scheme established consistent with job requirements and size.

H12.2 File Nomenclature. A systematic file naming scheme shall be used to control data management of calibration and set-up files, phased array data files, and digitally generated data report forms.

H12.3 Raw Data. All phased array data shall be saved in the original raw A-Scan format as collected during the examination.

H12.4 Data Reviewing. Any review and evaluation of the phased array data shall not change or affect the original A-Scan data.

H13. Documentation and Reporting

H13.1 Reporting. Examination reports shall meet the requirements of 8.26 and may be output from the on-board reporting feature of the phased array unit provided all necessary information is included. Reports may also be produced in the written manual UT conventional format or by external computer-generation.

H13.2 If PAUT is being substituted for RT, the written report shall include, at a minimum, encoded C-Scans covering the entire length inspected and A-Scan, B-Scan, C-Scan and S-Scan views (see H3.12) of all reportable indications. All raw data for PAUT substituted for RT shall be retained for the same duration required for radiographic film.

H13.3 Repairs. Results from PAUT inspections of repaired welds shall be tabulated on the original form (if available) or additional report forms, and shall be indicated by the appropriate repair number (R1, R2, R3, etc.).

H13.4 Scan Plan Reporting. The scan plan used during inspection shall accompany the report form.

H14. System Linearity Verification

H14.1 General Requirements. Linearity verifications shall be conducted at a minimum of every 12 months and recorded on a form similar to that shown in Figure H.6. The verifications shall be conducted by a PAUT Level II or III, or, at the Contractor's option, the equipment may be sent to the manufacturer for verification.

H14.1.1 The phased array instrument shall be configured to display an A-Scan presentation.

H14.1.2 The time-base of the A-Scan shall be adjusted to a suitable range to display the pulse-echo signals selected for the particular linearity verification to be performed. A standard IIW type block (see Figure 8.13) or other linearity block similar to that described in ASTM E317 shall be selected to provide signals to assess linearity aspects of the instrument.

H14.1.3 Pulser parameters shall be selected for the frequency and bandpass filter to optimize the response from the probe used for the linearity verifications.

H14.1.4 The receiver gain shall be set to display non-saturating signals of interest for display height and amplitude control linearity assessments.

H14.2 Display Height Linearity Verification Procedure

(1) With the phased array instrument connected to a probe (shear or longitudinal) and coupled to any block that will produce two signals, the probe shall be adjusted such that the amplitude of the two signals are at 80% and 40% of the display screen height.

(2) The gain shall be increased using the receiver gain adjustment to obtain 100% FSH of the larger response. The height of the lower response is recorded at this gain setting as a percentage of FSH.

(3) The height of the higher response shall be reduced in 10% steps to 10% FSH and record the height of the second response for each step.

(4) The larger signal shall be returned to 80% to ensure that the smaller signal has not drifted from its original 40% level due to coupling variation. Repeat the test if variation of the second signal is greater than 41% or less than 39% FSH.

(5) For an acceptable tolerance, the responses from the two reflectors shall bear a 2 to 1 relationship to within ±3% of FSH throughout the range 10% to 100 % (99% if 100% is saturation) of FSH.

(6) The results shall be recorded on an instrument linearity form as shown in Figure H.6.

H14.3 Amplitude Control Linearity Verification Procedure

Each of the pulser-receiver components shall be checked to determine the linearity of the instrument amplification capabilities. For instruments configured to read amplitudes greater than can be seen on the display, a larger range of checkpoints may be used. For these instruments the gated output instead of the A-Scan display shall be verified for linearity.

(1) A flat (normal incidence) linear array phased array probe shall be selected having at least as many elements as the phased array ultrasonic instrument has pulsers.

(2) The phased-array ultrasonic instrument shall be configured using this probe to have an electronic scan at 0°. Each focal law will consist of one element. The scan will start at element number 1 and end at the element number that corresponds to the number of pulsers in the phased-array instrument.

(3) The probe shall be coupled to a suitable surface to obtain a pulse-echo response from each focal law. The backwall echo from the 1 in [25 mm] thickness of the IIW type block or similar block provides a suitable target option. Alternatively, immersion testing may be used.

(4) Channel 1 of the pulser-receivers of the phased-array instrument shall be selected. Using the A-Scan display, monitor the response from the selected target. Adjust the gain to bring the signal to 40% FSH.

(5) The gain shall be added to the receiver in the increments of 1 dB, then 2 dB, then 4 dB and then 6 dB. Remove the gain added after each increment to ensure that the signal has returned to 40% FSH. Record the actual height of the signal as a percentage of the display height.

(6) The signal shall be adjusted to 100% FSH, remove 6 dB gain and record the actual height of the signal as a percentage of the display height.

(7) Signal amplitudes shall fall within a range of +/- 3% of the display height required in the allowed height range of Figure H.6.

(8) The sequence from (4) to (7) shall be repeated for all other pulser-receiver channels and results shall be recorded on a linearity report form as shown in Figure H.6.

H14.4 Time-Base Linearity (Horizontal Linearity) Verification Procedure

(1) The phased array instrument shall be configured to display an A-Scan presentation using a 10 in [250 mm] range.

(2) Any longitudinal (compression) wave probe shall be selected and the phased-array instrument shall be configured to display a range obtaining at least ten multiple back reflections from the 1 in [25 mm] wall thickness of the calibration block.

(3) The phased-array instrument shall be set to an analog-to-digital conversion rate of at least 80 MHz.

(4) With the probe coupled to the block and the A-Scan displaying 10 clearly defined multiples as illustrated in Figure H.5, the display software shall be used to assess the interval between adjacent backwall signals.

(5) The acoustic velocity of the test block shall be set by calibration (ASTM E494 may be used as a guide), the acoustic velocity in the display software shall be entered, and the display shall be configured to read out in distance (thickness).

(6) Using the reference and measurement cursors, the interval between each multiple shall be determined and the interval of the first 10 multiples shall be recorded.

(7) Each intermediate trace deflection shall be correct within 2% of the screen width.

References for Annex H

The following references provide additional supplemental information relative to this Annex.

1. ASTM E494, *Standard Practice for Measuring Ultrasonic Velocity in Materials*

2. ASTM E2491, *Standard Guide for Evaluating Performance Characteristics of Phased-Array Ultrasonic Examination Instruments and Systems*

3. ASTM E2700, *Standard Practice for Contact Ultrasonic Testing of Welds Using Phased Arrays*

4. ISO 2400, *Non-destructive testing – Ultrasonic examination specification for calibration block No. 1*

Table H.1
Essential Variables for PAUT (see H7.1 and H7.3)

(1)	Element numbers used for focal laws
(2)	Angular range of azimuthal scan
(3)	Manufacturer's documented permitted wedge angular range
(4)	Weld configurations to be examined, including thickness dimensions and base material product form (pipe, plate, etc.)
(5)	Surface curvature along index axis (e.g., for longitudinal weld in tubular member)
(6)	The surfaces from which the examination is performed (top, bottom, etc.)
(7)	Techniques (straight beam, angle beam, contact)
(8)	Search unit types, frequencies, element sizes and shapes
(9)	Phased array unit type – manufacturer and model including acquisition software
(10)	Manual vs. automated/semi-automated scanning
(11)	Method for discriminating geometric from weld defect indications
(12)	Decrease in scan overlap
(13)	Method for determining focal/delay laws if other than on-board equipment algorithms included in the software revision specified
(14)	Acquisition or analysis software type
(15)	Probe manufacturer and model
(16)	Any increase in scanning speed
(17)	Couplant, if not listed in 8.25.4
(18)	Computer enhanced data analysis

Table H.2
PAUT Acceptance Criteria (see H10.3)

Maximum Discontinuity Amplitude Level Obtained	Maximum Discontinuity Lengths by Type of Loading	
	Static and Cyclic Compression	Cyclic Tension
Class A	None allowed	None allowed
Class B	3/4 in [20 mm]	1/2 in [12 mm]
Class C	2 in [50 mm]	Middle half of weld: 2 in [50 mm]
		Top or bottom quarter of weld: 3/4 in [20 mm]
Class D	Disregard	Disregard

Table H.3 (see H8.2.4)
Discontinuity Classification

Discontinuity Classification	Description
A	> ARL
B	> SSL, ≤ ARL
C	> DRL, ≤ SSL
D	≤ DRL

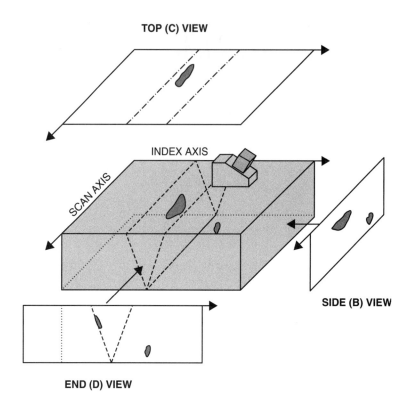

Figure H.1—Phased Array Imaging Views (see H3.12)

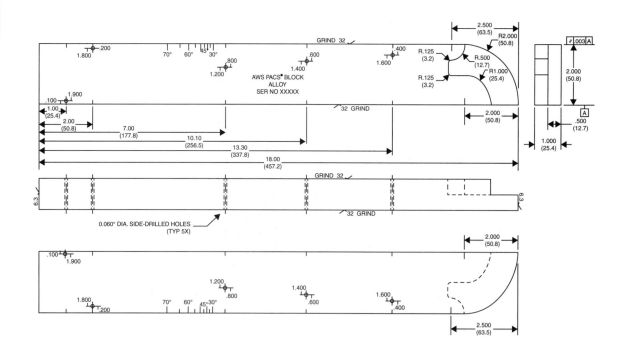

Figure H.2—Example of a Supplemental Reference Block (see H5.7.1)

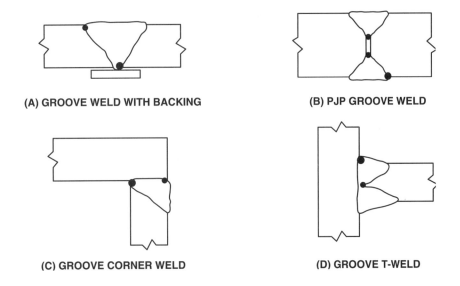

Figure H.3—Example Standard Reflector Locations in Weld Mockup (see H5.7.2)

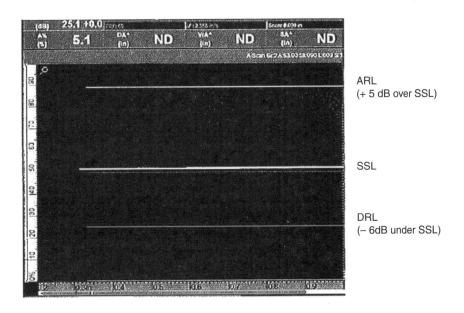

Figure H.4—Sensitivity Levels (see H8.2.4 and H10.3)

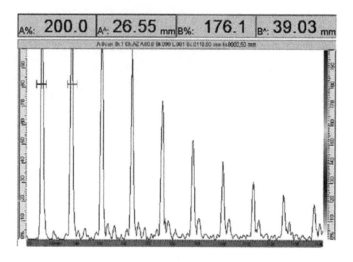

Figure H.5—Example of Time Based Linearity Verification (see H14.4)

AWS D1.1/D1.1M:2020 ANNEX H

Figure H.6
Linearity Verification Report Form (see H14)

Location:			Date:		
Operator Personnel:			Signature:		
Instrument:			Couplant:		
Pulser Voltage (V):		Pulse Duration (ns):	Receiver (band):		Receiver Smoothing:
Digitization Frequency (MHz):			Averaging		

Display Height Linearity			Amplitude Control Linearity		
Large %	Small Allowed Range %	Small Actual %	Ind. Height %	dB	Allowed Range %
100	47–53		40	+1	42–47
90	42–48		40	+2	48–52
80	40	40	40	+4	60–66
70	32–38		40	+6	77–83
60	27–33		40	–6	17–23
50	22–28		100	–6	47–53
40	17–23				
30	12–18				
20	7–13				
10	2–8				

Amplitude Control Linearity Channel Results: (Note any channels that do not fall in the allowed range)

Channel (Add more if required for 32 or 64 pulser-receiver units)

1	2	3	4	5	6	7	8	9	10	11	12	13	14	15	16

Time Based (Horizontal) Linearity (for 1 in [25 mm] IIW Blocks)										
Multiple	1	2	3	4	5	6	7	8	9	10
Thickness	1 in [25 mm]	2 in [50 mm]	3 in [75 mm]	4 in [100 mm]	5 in [125 mm]	6 in [150 mm]	7 in [175 mm]	8 in [200 mm]	9 in [225 mm]	10 in [250 mm]
Measured Interval										
Allowed deviation (Yes/No)										

This page is intentionally blank.

Annex I (Normative)
Symbols for Tubular Connection Weld Design

This annex is part of this standard and includes mandatory elements for use with this standard.

Symbols used in Clause 10, Part A, are as follows:

Symbol	Meaning
α	(alpha) chord ovalizing parameter
a	width of rectangular hollow section product
a_x	ratio of a to sin θ
b	transverse width of rectangular tubes
b_{et} ($b_{e(ov)}$)	branch effective width at through member
b_{eo} (b_e)	branch effective width at chord
b_{eoi} (b_{ep})	branch effective width for outside punching
b_{gap}	effective width at gap of K-connections
β	(beta) diameter ratio of d_b to D
	ratio of r_b to R (circular sections)
	ratio of b to D (box sections)
β_{gap}	dimensionless effective width at gap of K-connections
β_{eoi}	dimensionless ratio of b_{eoi} to the width of the main member
β_{eop}	dimensionless effective width for outside punching
β_{eff}	effective β for K-connection chord face plastification
c	corner dimension
D	outside diameter OD (circular tubes) or outside width of main member (box sections)
D	cumulative fatigue damage ratio, $\sum \frac{n}{N}$
d_b	diameter of branch member
η	(eta) ratio of a_x to D
ε_{TR}	(epsilon) total strain range
F	toe fillet weld size
F_{EXX}	classified minimum tensile strength of weld deposit
F_y	yield strength of base metal
F_{yo}	yield strength of main member
f_a	axial stress in branch member
f_a	axial stress in main member
f_b	bending stress in branch member

f_b	bending stress in main member
f_{by}	nominal stress, in-plane bending
f_{bz}	nominal stress, out-of-plane bending
f_n	nominal stress in branch member
g	gap in K-connections
H	web depth (box chord) in plane of truss
γ	(gamma) main member flexibility parameter; ratio R to t_c (circular sections); ratio of D to $2t_c$ (box sections)
$γ_b$	radius to thickness ratio of tube at transition
$γ_t$	thru member γ (for overlap conn.)
ID	inside diameter
K–	connection configuration
K_a	relative length factor
K_b	relative section factor
λ	(lambda) interaction sensitivity parameter
L	size of fillet weld dimension as shown in Figure 10.3
L	length of joint can
LF	load factor (partial safety factor for load in LRFD)
l_1	actual weld length where branch contacts main member
l_2	projected length (one side) of overlapping weld, measured perpendicular to the main member
M	applied moment
M_c	moment in chord
M_u	ultimate moment
n	cycle of load applied
N	number of cycles allowed at given stress range
OD	outside diameter
P	axial load in branch member
P_c	axial load in chord
P_u	ultimate load
$P_\perp$	individual member load component perpendicular to main member axis
p	projected footprint length of overlapping member
q	amount of overlap
ϕ	(phi) joint included angle
π	(pi) ratio of circumference to diameter of circle
Ψ	(psi) local dihedral angle. See definition Clause 3
$\bar{Ψ}$	(psi bar) supplementary angle to the local dihedral angle change at transition
Q_b	geometry modifier
Q_f	stress interaction term
Q_q	branch member geometry and load pattern modifier
R	outside radius, main member
R	root opening (joint fit-up)
r	corner radius of rectangular hollow sections as measured by radius gauge
r	effective radius of intersection

r_b	radius of branch
r_m	radius to the midpoint of the branch wall thickness
r_w	radius to midpoint of the effective throat
SCF	stress concentration factor
Σl_1	(sigma) summation of actual weld lengths
T–	connection configuration
TCBR	tension/compression or bending, or both, total range of nominal stress
t	wall thickness of tube
t_b	wall thickness of branch member branch member for dimensioning of CJP groove welds thinner member for dimensioning PJP groove welds and fillet welds
t_c	wall thickness of main member joint can thickness
t_w	weld size (effective throat)
t_w^1	the lesser of the weld size (effective throat) or the thickness t_b of the inner branch member
τ	(tau) branch-to-main relative thickness geometry parameter; ratio of t_b to t_c
τ_t	$t_{overlap}/t_{thru}$
θ	(theta) acute angle between two member axes angle between member center lines brace intersection angle
U	utilization ratio of axial and bending stress to allowable stress, at point under consideration in main member
V_p	punching shear stress
V_w	allowable stress for weld between branch members
χ	algebraic variable $\dfrac{1}{2\pi \sin \theta}$
Y–	connection configuration
y	algebraic variable $\dfrac{1}{3\pi} \cdot \dfrac{3-\beta^2}{2-\beta^2}$
Z	Z loss dimension
ζ	(zeta) ratio of gap to D

This page is intentionally blank.

Annex J (Informative)
Sample Welding Forms

This annex is not part of this standard but is included for informational purposes only.

This annex contains example forms for the recording of WPS information procedure qualification test data, qualification test results, welder qualification, welding operator qualification, and tack welder qualification data required by this code. Also included are laboratory report forms for recording the results of NDT of welds.

It is recommended that the qualification and NDT information required by this code be recorded on these forms or similar forms which have been prepared by the user. Variations of these forms to suit the user's needs are permitted.

J1. Use of Forms

For joint details on the WPS, a sketch or a reference to the applicable prequalified joint detail may be used (e.g., B-U4a).

The WPS forms contain lines for inclusion of information that may not be required by the code; these lines need not be completed in order to comply with code requirements.

J2. Prequalified WPSs

J2 is sufficient for the purpose of documenting prequalified WPSs.

J3. Example Forms

Examples of completed WPSs and a PQR have been included for informational purposes. Separate forms for WPSs and PQRs are provided for welding process categories GTAW/SMAW, GMAW/FCAW, and SAW. The names are fictional and the test data given is not from any actual test and should not be used. The Committee trusts that these examples will assist code users in producing acceptable documentation.

J4. WPSs Qualified by Testing

The WPS may be qualified by testing in conformance with the provisions of Clause 6. In this case, a supporting PQR is required in addition to the WPS. For the WPS, state the allowed ranges qualified by testing or state the appropriate tolerances on essential variable (e.g., 250 amps ± 10%).

For the PQR, the actual joint details and the values of essential variables used in the testing should be recorded. An example of a completed PQR form is provided for guidance in filling out the form. A copy of the Mill Test Report for the material tested should be attached. Also, Testing Laboratory Data Reports may also be included as backup information or a PQR Test Result Form similar to the example in this annex may be used. Cross references to the required mechanical tests as applicable to the WPS being qualified are provided on the form for ready reference. Note that not all tests referenced are required.

Index of Forms

Form No.	Title	Page No.
J–1	Procedure Qualification Record (PQR) forms	405
J–2	Welding Procedure Specification (WPS) forms	411
J–3	WPS Qualification Test Record for Electroslag and Electrogas Welding	416
J–4	Welder, Welding Operator, or Tack Welder Performance Qualification Test Record (Single-Process) forms	417
J–5	Welder, Welding Operator, or Tack Welder Performance Qualification Test Record (Multi-Process) forms	419
J–6	Report of Radiographic Examination of Welds	421
J–7	Report of Magnetic-Particle Examination of Welds	422
J–8	Stud Welding Application Qualification Test, Pre-Production Test, PQR, or WQR Form	423

AWS D1.1/D1.1M:2020 ANNEX J

Blank Sample PQR Form (GTAW & SMAW – page 1)
PROCEDURE QUALIFICATION RECORD (PQR)

Company Name _____ PQR No. _____ Rev. No. _____ Date _____

BASE METALS	Specification	Type or Grade	AWS Group No.	Thickness	Size (NPS)	Schedule	Diameter
Base Material							
Welded To							
Backing Material							
Other							

JOINT DETAILS	
Groove Type	
Groove Angle	
Root Opening	
Root Face	
Backgouging	
Method	

JOINT DETAILS (Sketch)

POSTWELD HEAT TREATMENT	
Temperature	
Time at Temperature	
Other	

PROCEDURE								
Weld Layer(s)								
Weld Pass(es)								
Process								
Type (Manual, Mechanized, etc.)								
Position								
Vertical Progression								
Filler Metal (AWS Spec.)								
AWS Classification								
Diameter								
Manufacturer/Trade Name								
Shielding Gas Compos. (GTAW)								
Flow Rate (GTAW)								
Nozzle Size (GTAW)								
Preheat Temperature								
Interpass Temperature								
Electrical Characteristics	—	—	—	—	—	—	—	—
Electrode Diameter (GTAW)								
Current Type & Polarity								
Amps								
Volts								
Cold or Hot Wire Feed (GTAW)								
Travel Speed								
Maximum Heat Input								
Technique	—	—	—	—	—	—	—	—
Stringer or Weave								
Multi or Single Pass (per side)								
Oscillation (GTAW Mech./Auto.)								
Traverse Length								
Traverse Speed								
Dwell Time								
Peening								
Interpass Cleaning								
Other								

Form J-1 (Front)   (See http://go.aws.org/D1forms)

ANNEX J   AWS D1.1/D1.1M:2020

Blank Sample PQR Form (GMAW & FCAW – page 1)
PROCEDURE QUALIFICATION RECORD (PQR)

Company Name _____   PQR No. _____   Rev. No. _____   Date _____

BASE METALS	Specification	Type or Grade	AWS Group No.	Thickness	Size (NPS)	Schedule	Diameter
Base Material							
Welded To							
Backing Material							
Other							

JOINT DETAILS	
Groove Type	
Groove Angle	
Root Opening	
Root Face	
Backgouging	
Method	

JOINT DETAILS (Sketch)

POSTWELD HEAT TREATMENT	
Temperature	
Time at Temperature	
Other	

PROCEDURE								
Weld Layer(s)								
Weld Pass(es)								
Process								
Type *(Semiautomatic, Mechanized, etc.)*								
Position								
Vertical Progression								
Filler Metal (AWS Spec.)								
AWS Classification								
Diameter								
Manufacturer/Trade Name								
Shielding Gas Composition								
Flow Rate								
Nozzle Size								
Preheat Temperature								
Interpass Temperature								
Electrical Characteristics	—	—	—	—	—	—	—	—
Current Type & Polarity								
Transfer Mode (GMAW)								
Power Source Type *(cc, cv, etc.)*								
Amps								
Volts								
Wire Feed Speed								
Travel Speed								
Maximum Heat Input								
Technique	—	—	—	—	—	—	—	—
Stringer or Weave								
Multi or Single Pass (per side)								
Oscillation *(Mechanized/Automatic)*								
Traverse Length								
Traverse Speed								
Dwell Time								
Number of Electrodes								
Contact Tube to Work Dist.								
Peening								
Interpass Cleaning								
Other								

Form J-1 (Front)   (See http://go.aws.org/D1forms)

AWS D1.1/D1.1M:2020 ANNEX J

Blank Sample PQR Form (SAW – page 1)
PROCEDURE QUALIFICATION RECORD (PQR)

Company Name _____ PQR No. _____ Rev. No. _____ Date _____

BASE METALS	Specification	Type or Grade	AWS Group No.	Thickness	Size (NPS)	Schedule	Diameter
Base Material							
Welded To							
Backing Material							
Other							

JOINT DETAILS	
Groove Type	
Groove Angle	
Root Opening	
Root Face	
Backgouging	
Method	

JOINT DETAILS (Sketch)

POSTWELD HEAT TREATMENT	
Temperature	
Time at Temperature	
Other	

PROCEDURE								
Weld Layer(s)								
Weld Pass(es)								
Process	SAW							
Type (Semiautomatic, Mechanized, etc.)								
Position								
Filler Metal (AWS Spec.)								
AWS Classification								
Electrode Diameter								
Electrode/Flux Classification								
Manufacturer/Trade Name								
Supplemental Filler Metal								
Preheat Temperature								
Interpass Temperature								
Electrical Characteristics	—	—	—	—	—	—	—	—
Current Type & Polarity								
Amps								
Volts								
Wire Feed Speed								
Travel Speed								
Maximum Heat Input								
Technique	—	—	—	—	—	—	—	—
Stringer or Weave								
Multi or Single Pass (per side)								
Number of Electrodes								
Longitudinal Spacing of Arcs								
Lateral Spacing of Arcs								
Angle of Parallel Electrodes								
Angle of Electrode (Mech./Auto.)								
Normal To Direction of Travel								
Oscillation (Mechanized/Automatic)								
Traverse Length								
Traverse Speed								
Dwell Time								
Peening								
Interpass Cleaning								
Other								

Form J-1 (Front) (See http://go.aws.org/D1forms)

ANNEX J   AWS D1.1/D1.1M:2020

Blank Sample PQR Form (Test Results – page 2)
PROCEDURE QUALIFICATION RECORD (PQR) TEST RESULTS

PQR No. _____   Rev. No. _____

TESTS

✓	Type of Tests	Clause/Figure(s) Reference	Acceptance Criteria	Result	Remarks
✓	Visual Inspection	6.10.1	6.10.1	Acceptable	
	Radiographic Examination	6.10.2.1	6.10.2.2	Acceptable	
	Ultrasonic Testing	6.10.2.1	6.10.2.2		
	2 Transverse Root Bends	6.10.3.1/Fig. 6.8	6.10.3.3		
	2 Transverse Face Bends	6.10.3.1/Fig. 6.8	6.10.3.3		
	2 Longitudinal Root Bends	6.10.3.1/Fig. 6.8	6.10.3.3		
	2 Longitudinal Face Bends	6.10.3.1/Fig. 6.8	6.10.3.3		
	2 Side Bends	6.10.3.1/Fig. 6.9	6.10.3.3		
✓	4 Side Bends	6.10.3.1/Fig. 6.9	6.10.3.3	Acceptable	< 1/16 in Opening
✓	2 Tensile Tests	6.10.3.1/Fig. 6.10	6.10.3.5	Acceptable	
✓	All-Weld-Metal Tensions	6.10.3.1/Figs. 6.14 and 6.18	6.15.1.3(2)	Acceptable	*See Note
	3 Macroetch	6.10.4	6.10.4.1		
	4 Macroetch	6.10.4	6.10.4.1		
✓	CVN Tests	6 Part D/Fig. 6.28	6.30 and Table 6.15		

TENSILE TEST DETAILS

Specimen Number	Width	Thickness	Area	Ultimate Tensile Load	Ultimate Unit Stress	Type of Failure and Location

TOUGHNESS TEST DETAILS

Specimen Number	Notch Location	Specimen Size	Test Temperature	Absorbed Energy	Percent Shear	Lateral Expansion	Average

CERTIFICATION

Welder's Name	ID Number	Stamp Number

Tests Conducted by	
Laboratory	
Test Number	
File Number	

We, the undersigned, certify that the statements in this record are correct and that the test welds were prepared, welded, and tested in accordance with the requirements of Clause 6 of AWS D1.1/D1.1M, (_____) *Structural Welding Code—Steel*.
(year)

Title	
Name	Signature
Date	

Form J-1 (Back)

(See http://go.aws.org/D1forms)

Example PQR (GMAW & FCAW – page 1)
PROCEDURE QUALIFICATION RECORD (PQR)

Red Inc.	231	0	01/18/2020
Company Name	PQR No.	Rev. No.	Date

BASE METALS	Specification	Type or Grade	AWS Group No.	Thickness	Size (NPS)	Schedule	Diameter
Base Material	ASTM A131	A	I	1 in	—	—	—
Welded To	ASTM A131	A	I	1 in	—	—	—
Backing Material	ASTM A131	A	I	1/4 in			
Other							

JOINT DETAILS	
Groove Type	Single V Groove Butt Joint
Groove Angle	35° included
Root Opening	1/4 in
Root Face	—
Backgouging	None
Method	—

POSTWELD HEAT TREATMENT	
Temperature	—
Time at Temperature	—
Other	—

JOINT DETAILS (Sketch)

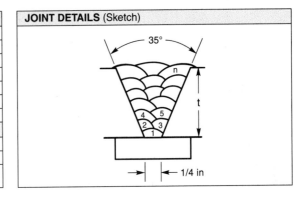

PROCEDURE								
Weld Layer(s)	—	—	—	—	—			
Weld Pass(es)	1	2–8	9–11	12–15	16			
Process	FCAW	FCAW	FCAW	FCAW	FCAW			
Type (Semiautomatic, Mechanized, etc.)	Semi-automatic	Semi-automatic	Semi-automatic	Semi-automatic	Semi-automatic			
Position	4G	4G	4G	4G	4G			
Vertical Progression	—	—	—	—	—			
Filler Metal (AWS Spec.)	A5.20	A5.20	A5.20	A5.20	A5.20			
AWS Classification	E71T-1C	E71T-1C	E71T-1C	E71T-1C	E71T-1C			
Diameter	0.045 in	0.045 in	0.045 in	0.045 in	0.045 in			
Manufacturer/Trade Name	—	—	—	—	—			
Shielding Gas Composition	100% CO_2	100% CO_2	100% CO_2	100% CO_2	100% CO_2			
Flow Rate	45–55 cfh	45–55 cfh	45–55 cfh	45–55 cfh	45–55 cfh			
Nozzle Size	#4	#4	#4	#4	#4			
Preheat Temperature	75° min.	75° min.	75° min.	75° min.	75° min.			
Interpass Temperature	75°–350°	75°–350°	75°–350°	75°–350°	75°–350°			
Electrical Characteristics	—	—	—	—	—	—	—	—
Current Type & Polarity	DCEP	DCEP	DCEP	DCEP	DCEP			
Transfer Mode (GMAW)	—	—	—	—	—			
Power Source Type (cc, cv, etc.)	—	—	—	—	—			
Amps	180	200	200	200	200			
Volts	26	27	27	27	27			
Wire Feed Speed	(Amps)	(Amps)	(Amps)	(Amps)	(Amps)			
Travel Speed	8 ipm	10 ipm	11 ipm	9 ipm	11 ipm			
Maximum Heat Input	—	—	—	—	—			
Technique	—	—	—	—	—	—	—	—
Stringer or Weave	Stringer	Stringer	Stringer	Stringer	Stringer			
Multi or Single Pass (per side)	Multipass	Multipass	Multipass	Multipass	Multipass			
Oscillation (Mechanized/Automatic)	—	—	—	—	—			
Number of Electrodes	1	1	1	1	1			
Contact Tube to Work Dist.	3/4–1 in	3/4–1 in	3/4–1 in	3/4–1 in	3/4–1 in			
Peening	None	None	None	None	None			
Interpass Cleaning	Wire Brush	Wire Brush	Wire Brush	Wire Brush	Wire Brush			
Other								

Form J-1 (Front)   (See http://go.aws.org/D1forms)

ANNEX J AWS D1.1/D1.1M:2020

Example PQR (GMAW & FCAW – page 2)
PROCEDURE QUALIFICATION RECORD (PQR) TEST RESULTS

PQR No. __231__   Rev. No. __0__

TESTS

✓	Type of Tests	Clause/Figure(s) Reference	Acceptance Criteria	Result	Remarks
✓	Visual Inspection	6.10.1	6.10.1	Acceptable	
✓	Radiographic Examination	6.10.2.1	6.10.2.2	Acceptable	
	Ultrasonic Testing	6.10.2.1	6.10.2.2		
	2 Transverse Root Bends	6.10.3.1/Fig. 6.8	6.10.3.3		
	2 Transverse Face Bends	6.10.3.1/Fig. 6.8	6.10.3.3		
	2 Longitudinal Root Bends	6.10.3.1/Fig. 6.8	6.10.3.3		
	2 Longitudinal Face Bends	6.10.3.1/Fig. 6.8	6.10.3.3		
	2 Side Bends	6.10.3.1/Fig. 6.9	6.10.3.3		
✓	4 Side Bends	6.10.3.1/Fig. 6.9	6.10.3.3	Acceptable	< 1/16 in Opening
✓	2 Tensile Tests	6.10.3.1/Fig. 6.10	6.10.3.5	Acceptable	
✓	All-Weld-Metal Tensions	6.10.3.1/Figs. 6.14 and 6.18	6.15.1.3(2)	Acceptable	*See Note
	3 Macroetch	6.10.4	6.10.4.1		
	4 Macroetch	6.10.4	6.10.4.1		
✓	CVN Tests	6 Part D/Fig 6.28	6.30 and Table 6.14		

TENSILE TEST DETAILS

Specimen Number	Width	Thickness	Area	Ultimate Tensile Load	Ultimate Unit Stress	Type of Failure and Location
231-1	0.75 in	100 in	0.75 in^2	52 500 lb	70 000 psi	Ductile/Weld Metal
231-2	0.75 in	100 in	0.75 in^2	52 275 lb	69 700 psi	Ductile/Weld Metal

TOUGHNESS TEST DETAILS

Specimen Number	Notch Location	Specimen Size	Test Temperature	Absorbed Energy	Percent Shear	Lateral Expansion	Average
231-7	BM	10 × 10 mm	−20°F	126 ft·lbf	50%	45 mils	
231-8	BM	10 × 10 mm	−20°F	124 ft·lbf	50%	45 mils	125/50/45
231-9	BM	10 × 10 mm	−20°F	125 ft·lbf	50%	45 mils	
231-10	HAZ	10 × 10 mm	−20°F	86 ft·lbf	50%	45 mils	
231-11	HAZ	10 × 10 mm	−20°F	84 ft·lbf	50%	45 mils	85 ft·lbf 50% 45 mils
231-12	HAZ	10 × 10 mm	−20°F	85 ft·lbf	50%	45 mils	
231-13	WM	10 × 10 mm	−20°F	27 ft·lbf	50%	45 mils	
231-14	WM	10 × 10 mm	−20°F	29 ft·lbf	50%	45 mils	28 ft·lbf 50% 45 mils
231-14	WM	10 × 10 mm	−20°F	28 ft·lbf	50%	45 mils	

CERTIFICATION

Welder's Name	ID Number	Stamp Number
W. T. Williams	261	–

Tests Conducted by	
Laboratory	Red Inc. & ABC Testing
Test Number	PQR 231 (per D. Miller)
File Number	WeldingForms/PQR231.pdf

We, the undersigned, certify that the statements in this record are correct and that the test welds were prepared, welded, and tested in accordance with the requirements of Clause 6 of AWS D1.1/D1.1M, (_____) *Structural Welding Code—Steel.*
(year)

Title	
Q.C. Mgr.	
Name	Signature
R. M. Boncrack	
Date	
12/15/2020	

Form J-1 (Back)

(See http://go.aws.org/D1forms)

AWS D1.1/D1.1M:2020 ANNEX J

Blank Sample WPS Form (GTAW & SMAW)
WELDING PROCEDURE SPECIFICATION (WPS)

Company Name _____ WPS No. _____ Rev. No. _____ Date _____

Authorized by _____ Date _____ Supporting PQR(s) _____ CVN Report _____

BASE METALS	Specification	Type or Grade	AWS Group No.
Base Material			
Welded To			
Backing Material			
Other			

BASE METAL THICKNESS	As-Welded	With PWHT
CJP Groove Welds		
CJP Groove w/CVN		
PJP Groove Welds		
Fillet Welds		
DIAMETER		

JOINT DETAILS	
Groove Type	
Groove Angle	
Root Opening	
Root Face	
Backgouging	
Method	

JOINT DETAILS (Sketch)

POSTWELD HEAT TREATMENT	
Temperature	
Time at Temperature	
Other	

PROCEDURE							
Weld Layer(s)							
Weld Pass(es)							
Process							
Type (Manual, Mechanized, etc.)							
Position							
Vertical Progression							
Filler Metal (AWS Spec.)							
AWS Classification							
Diameter							
Manufacturer/Trade Name							
Shielding Gas Compos. (GTAW)							
Flow Rate (GTAW)							
Nozzle Size (GTAW)							
Preheat Temperature							
Interpass Temperature							
Electrical Characteristics	—	—	—	—	—	—	—
Electrode Diameter (GTAW)							
Current Type & Polarity							
Amps							
Volts							
Cold or Hot Wire Feed (GTAW)							
Travel Speed							
Maximum Heat Input							
Technique	—	—	—	—	—	—	—
Stringer or Weave							
Multi or Single Pass (per side)							
Oscillation (GTAW Mech./Auto.)							
Traverse Length							
Traverse Speed							
Dwell Time							
Peening							
Interpass Cleaning							
Other							

Form J-2 (See http://go.aws.org/D1forms)

ANNEX J　　　　　　　　　　　　　　　　　　　　　　　　　　　　　　　　　　　　AWS D1.1/D1.1M:2020

Blank Sample WPS Form (GMAW & FCAW)
WELDING PROCEDURE SPECIFICATION (WPS)

Company Name _____　　WPS No. _____　Rev. No. _____　Date _____

Authorized by _____ Date _____　Supporting PQR(s) _____　CVN Report _____

BASE METALS	Specification	Type or Grade	AWS Group No.
Base Material			
Welded To			
Backing Material			
Other			

BASE METAL THICKNESS	As-Welded	With PWHT
CJP Groove Welds		
CJP Groove w/CVN		
PJP Groove Welds		
Fillet Welds		
DIAMETER		

JOINT DETAILS	
Groove Type	
Groove Angle	
Root Opening	
Root Face	
Backgouging	
Method	

JOINT DETAILS (Sketch)

POSTWELD HEAT TREATMENT	
Temperature	
Time at Temperature	
Other	

PROCEDURE								
Weld Layer(s)								
Weld Pass(es)								
Process								
Type (Semiautomatic, Mechanized, etc.)								
Position								
Vertical Progression								
Filler Metal (AWS Spec.)								
AWS Classification								
Diameter								
Manufacturer/Trade Name								
Shielding Gas (Composition)								
Flow Rate								
Nozzle Size								
Preheat Temperature								
Interpass Temperature								
Electrical Characteristics	—	—	—	—	—	—	—	—
Current Type & Polarity								
Transfer Mode								
Power Source Type (cc, cv, etc.)								
Amps								
Volts								
Wire Feed Speed								
Travel Speed								
Maximum Heat Input								
Technique	—	—	—	—	—	—	—	—
Stringer or Weave								
Multi or Single Pass (per side)								
Oscillation (Mechanized/Automatic)								
Traverse Length								
Traverse Speed								
Dwell Time								
Number of Electrodes								
Contact Tube to Work Distance								
Peening								
Interpass Cleaning								
Other								

Form J-2　　　　　　　　　　　　(See http://go.aws.org/D1forms)

AWS D1.1/D1.1M:2020 ANNEX J

Blank Sample WPS Form (SAW)
WELDING PROCEDURE SPECIFICATION (WPS)

Company Name _____ WPS No. _____ Rev. No. _____ Date _____

Authorized by _____ Date _____ Supporting PQR(s) _____ CVN Report _____

BASE METALS	Specification	Type or Grade	AWS Group No.
Base Material			
Welded To			
Backing Material			
Other			

BASE METAL THICKNESS	As-Welded	With PWHT
CJP Groove Welds		
CJP Groove w/CVN		
PJP Groove Welds		
Fillet Welds		
DIAMETER		

JOINT DETAILS	
Groove Type	
Groove Angle	
Root Opening	
Root Face	
Backgouging	
Method	

JOINT DETAILS (Sketch)

POSTWELD HEAT TREATMENT	
Temperature	
Time at Temperature	
Other	

PROCEDURE								
Weld Layer(s)								
Weld Pass(es)								
Process	SAW							
Type (Semiautomatic, Mechanized, etc.)								
Position								
Filler Metal (AWS Spec.)								
AWS Classification								
Electrode Diameter								
Electrode/Flux Classification								
Manufacturer/Trade Name								
Supplemental Filler Metal								
Preheat Temperature								
Interpass Temperature								
Electrical Characteristics	—	—	—	—	—	—	—	—
Current Type & Polarity								
Amps								
Volts								
Wire Feed Speed								
Travel Speed								
Maximum Heat Input								
Technique	—	—	—	—	—	—	—	—
Stringer or Weave								
Multi or Single Pass (per side)								
Number of Electrodes								
Longitudinal Spacing of Arcs								
Lateral Spacing of Arcs								
Angle of Parallel Electrodes								
Angle of Electrode (Mech./Auto.)								
Normal To Direction of Travel								
Oscillation (Mechanized/Automatic)								
Traverse Length								
Traverse Speed								
Dwell Time								
Peening								
Interpass Cleaning								
Other								

Form J-2 (See http://go.aws.org/D1forms)

ANNEX J AWS D1.1/D1.1M:2020

Example WPS (Prequalified)
WELDING PROCEDURE SPECIFICATION (WPS)

LECO		W2081		2	01/03/2020
Company Name		WPS No.		Rev. No.	Date
C. W. Hayes	01/03/2015	**None (Prequalified)**		No	
Authorized by	Date	Supporting PQR(s)		CVN Report	

BASE METALS	Specification	Type or Grade	AWS Group No.
Base Material	ASTM A36	–	II
Welded To	ASTM A36	–	II
Backing Material	ASTM A36	–	–
Other			

BASE METAL THICKNESS	As-Welded	With PWHT
CJP Groove Welds	> 3/4–2.5 in	–
CJP Groove w/CVN	–	–
PJP Groove Welds	–	–
Fillet Welds	–	–
DIAMETER	–	–

JOINT DETAILS	
Groove Type	Single V Groove Butt Joint
Groove Angle	20°
Root Opening	5/8 in
Root Face	–
Backgouging	None
Method	–

POSTWELD HEAT TREATMENT	
Temperature	N.A.
Time at Temperature	–
Other	–

JOINT DETAILS (Sketch)

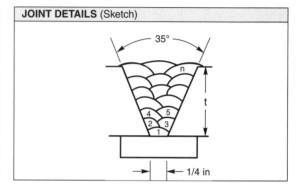

PROCEDURE									
Weld Layer(s)	All								
Weld Pass(es)	All								
Process	SAW								
Type (Semiautomatic, Mechanized, etc.)	Mechanized								
Position	F								
Filler Metal (AWS Spec.)	A5.17								
AWS Classification	EM12K								
Electrode Diameter	5/32 in								
Electrode/Flux Classification	F7A2-EM12K								
Manufacturer/Trade Name	(Flux XYZ)								
Supplemental Filler Metal	–								
Preheat Temperature	150°F min.								
Interpass Temperature	500°F max.								
Electrical Characteristics	—	—	—	—	—	—	—	—	
Current Type & Polarity	DCEP								
Amps	500–600								
Volts	26–30								
Wire Feed Speed	–								
Travel Speed	20–25 ipm								
Maximum Heat Input	–								
Technique	—	—	—	—	—	—	—	—	
Stringer or Weave	Stringer								
Multi or Single Pass (per side)	Multipass								
Number of Electrodes	1								
Longitudinal Spacing of Arcs	–								
Lateral Spacing of Arcs	–								
Angle of Parallel Electrodes	–								
Angle of Electrode (Mech./Auto.)	5° ± 2°								
Normal To Direction of Travel	90° ± 2°								
Oscillation (Mechanized/Automatic)	None								
Traverse Length	–								
Traverse Speed	–								
Dwell Time	–								
Peening	None								
Interpass Cleaning	Slag Removed								
Other	–								

Form J-2 (See http://go.aws.org/D1forms)

Example WPS (Single-Process)
WELDING PROCEDURE SPECIFICATION (WPS)

RED Inc.		2010	0	12/01/2020
Company Name		WPS No.	Rev. No.	Date
J. Jones	12/01/2015	231		No
Authorized by	Date	Supporting PQR(s)		CVN Report

BASE METALS	Specification	Type or Grade	AWS Group No.
Base Material	ASTM A131	A	I
Welded To	ASTM A131	A	I
Backing Material	ASTM A131	A	I
Other			

BASE METAL THICKNESS	As-Welded	With PWHT
CJP Groove Welds	3/4–1-1/2 in	–
CJP Groove w/CVN	–	–
PJP Groove Welds	–	–
Fillet Welds	–	–
DIAMETER	–	–

JOINT DETAILS	
Groove Type	Single V Groove Butt Joint
Groove Angle	35° included
Root Opening	1/4 in
Root Face	–
Backgouging	None
Method	–

JOINT DETAILS (Sketch)

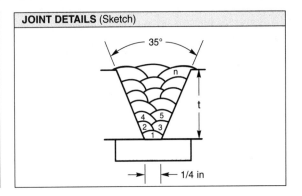

POSTWELD HEAT TREATMENT	
Temperature	None
Time at Temperature	–
Other	–

PROCEDURE								
Weld Layer(s)	All							
Weld Pass(es)	All							
Process	**FCAW**							
Type (Semiautomatic, Mechanized, etc.)	Semiauto							
Position	OH							
Vertical Progression	–							
Filler Metal (AWS Spec.)	A5.20							
AWS Classification	E71T-1C							
Diameter	0.045 in							
Manufacturer/Trade Name	–							
Shielding Gas (Composition)	100% CO_2							
Flow Rate	45–55 cfh							
Nozzle Size	#4							
Preheat Temperature	60° min.							
Interpass Temperature	60°–350°							
Electrical Characteristics	—							
Current Type & Polarity	DCEP							
Transfer Mode	–							
Power Source Type (cc, cv, etc.)	CV							
Amps	180–220							
Volts	25–26							
Wire Feed Speed	(Amps)							
Travel Speed	8–12 ipm							
Maximum Heat Input	–							
Technique	—							
Stringer or Weave	Stringer							
Multi or Single Pass (per side)	Multipass							
Oscillation (Mechanized, Automatic)	–							
Number of Electrodes	1							
Contact Tube to Work Dist.	1/2–1 in							
Peening	None							
Interpass Cleaning	Wire Brush							
Other	–							

Form J-2   (See http://go.aws.org/D1forms)

WPS QUALIFICATION TEST RECORD FOR ELECTROSLAG AND ELECTROGAS WELDING

PROCEDURE SPECIFICATION

Material specification _____
Welding process _____
Position of welding _____
Filler metal specification _____
Filler metal classification _____
Filler metal _____
Flux _____
Shielding gas _____ Flow rate _____
Gas dew point _____
Thickness range this test qualifies _____
Single or multiple pass _____
Single or multiple arc _____
Welding current _____
Preheat temperature _____
Postheat temperature _____
Welder's name _____
Guide tube flex _____
Guide tube composition _____
Guide tube diameter _____
Vertical rise speed _____
Traverse length _____
Traverse speed _____
Dwell _____
Type of molding shoe _____

VISUAL INSPECTION (Table 8.1, Cyclically loaded limitations)

Appearance _____
Undercut _____
Piping porosity _____
Test date _____
Witnessed by _____

TEST RESULTS

Reduced-section tensile test

Tensile strength, psi
1. _____
2. _____

All-weld-metal tension test

Tensile strength, psi _____
Yield point/strength, psi _____
Elongation in 2 in, % _____

Side-bend tests

1. _____ 3. _____
2. _____ 4. _____

Radiographic-ultrasonic examination

RT report no. _____
UT report no. _____

Impact tests

Size of specimen _____ Test temp _____
ft·lbf: 1. _____ 2. _____ 3. _____ 4. _____
5. _____ 6. _____ Avg. _____
High _____ Low _____
Laboratory test no. _____

WELDING PROCEDURE

Pass No.	Electrode Size	Welding Current		Joint Detail
		Amperes	Volts	

We, the undersigned, certify that the statements in this record are correct and that the test welds were prepared, welded, and tested in conformance with the requirements of Clause 6 of AWS D1.1/D1.1M, (_____) *Structural Welding Code—Steel*.
(year)

Procedure no. _____
Revision no. _____
Form J-3

Manufacturer or Contractor _____
Authorized by _____
Date _____

(See http://go.aws.org/D1forms)

AWS D1.1/D1.1M:2020   ANNEX J

Sample Welder Qualification Blank Form (Single-Process)
WELDER, WELDING OPERATOR, OR TACK WELDER
PERFORMANCE QUALIFICATION TEST RECORD

Name				Test Date		Rev.	
ID Number			OPTIONAL PHOTO ID	Record No.			
Stamp No.				Std. Test No.			
Company				WPS No.			
Division				Qualified To			

BASE METALS	Specification	Type or Grade	AWS Group No.	Size (NPS)	Schedule	Thickness	Diameter
Base Material							
Welded To							

VARIABLES	Actual Values		RANGE QUALIFIED	
Type of Weld Joint				
Base Metal				

	Groove	Fillet	Groove	Fillet
Plate Thickness				
Pipe/Tube Thickness				
Pipe Diameter				

Welding Process		
Type *(Manual, Semiautomatic, Mechanized, Automatic)*		
Backing		
Filler Metal (AWS Spec.)		
AWS Classification		
F-Number		
Position		
Groove – Plate & Pipe ≥ 24 in		
Groove – Pipe < 24 in		
Fillet – Plate & Pipe ≥ 24 in		
Fillet – Pipe < 24 in		
Progression		
GMAW Transfer Mode		
Single or Multiple Electrodes		
Gas/Flux Type		

TEST RESULTS

Type of Test	Acceptance Criteria	Results	Remarks

CERTIFICATION

Test Conducted by	
Laboratory	
Test Number	
File Number	

We, the undersigned, certify that the statements in this record are correct and that the test welds were prepared, welded, and tested in accordance with the requirements of Clause 6 of AWS D1.1/D1.1M (_____) *Structural Welding Code—Steel*.
(year)

Manufacturer or Contractor _____   Authorized by _____

Date _____

Form J-4   (See http://go.aws.org/D1forms)

ANNEX J   AWS D1.1/D1.1M:2020

Example Welder Qualification (Single-Process)
WELDER, WELDING OPERATOR, OR TACK WELDER
PERFORMANCE QUALIFICATION TEST RECORD

Name	Z. W. Elder		Test Date	12/12/2020	Rev.
ID Number	00-001-ZWE	OPTIONAL PHOTO ID	Record No.	WPQ-001	0
Stamp No.	ZWE-1		Std. Test No.	ST-001	0
Company	RED Inc.		WPS No.	WPS-001	0
Division	–		Qualified To	AWS D1.1	

BASE METALS	Specification	Type or Grade	AWS Group No.	Size (NPS)	Schedule	Thickness	Diameter
Base Material	ASTM A36	UNS K02600	I	–	–	3/8 in	–
Welded To	ASTM A36	UNS K02600	I	–	–	3/8 in	–

VARIABLES	Actual Values	RANGE QUALIFIED
Type of Weld Joint	Plate – Groove (Fig. 6.20) with Backing	Groove, Fillet, Plug, and Slot Welds (T-, Y-, K-Groove PJP only)
Base Metal	Group I to Group I	Any AWS D1.1 Qualified Base Metal

	Groove	Fillet		Groove	Fillet
Plate Thickness	3/8 in	–		1/8 in – 3/4 in	1/8 in min.
Pipe/Tube Thickness	–	–		1/8 in – 3/4 in	Unlimited
Pipe Diameter	–	–		24 in min.	Unlimited

Welding Process	GMAW	GMAW
Type (Manual, Semiautomatic, Mechanized, Automatic)	Semiautomatic	Semiautomatic, Mechanized, Automatic
Backing	With	With (incl. Backgouging and Backwelding)
Filler Metal (AWS Spec.)	A5.18	A5.xx
AWS Classification	ER70S-6	All
F-Number	–	–
Position	2G, 3G, and 4G	
Groove – Plate and Pipe ≥ 24 in		All
Groove – Pipe < 24 in		–
Fillet – Plate and Pipe ≥ 24 in		All
Fillet – Pipe < 24 in		All
Progression	Vertical Up	Vertical Up
GMAW Transfer Mode	Globular	Spray, Pulsed, Globular
Single or Multiple Electrodes	Single	Single
Gas/Flux Type	A5.32 SG-C	A5.xx Approved

TEST RESULTS

Type of Test	Acceptance Criteria	Results	Remarks
Visual Examination per 6.10.1	6.10.1	Acceptable	–
Each Position: 1 Root Bend per 6.10.3.1 and Fig. 6.8	6.10.3.3	Acceptable	–
Each Position: 1 Face Bend per 6.10.3.1 and Fig. 6.8	6.10.3.3	Acceptable	3G: Small (<1/16 in) Opening

CERTIFICATION

Test Conducted by	
Laboratory	Welding Forms Lab
Test Number	Fictitious Test XYZ
File Number	Welding Forms/Sample-WPQ-for-GMAW.pdf

We, the undersigned, certify that the statements in this record are correct and that the test welds were prepared, welded, and tested in accordance with the requirements of Clause 6 of AWS D1.1/D1.1M (_____) Structural Welding Code—Steel.
(year)

Manufacturer or Contractor   **Red Inc.**   Authorized by   **E. M. Ployee (Q.C. Mgr.)**

Date   **12/12/2020**

Form J-4   (See http://go.aws.org/D1forms)

AWS D1.1/D1.1M:2020   ANNEX J

Sample Welder Qualification Blank Form (Multi-Process)
WELDER, WELDING OPERATOR, OR TACK WELDER
PERFORMANCE QUALIFICATION TEST RECORD

Name				Test Date		Rev.
ID Number			OPTIONAL PHOTO ID	Record No.		
Stamp No.				Std. Test No.		
Company				WPS No.		
Division				Qualified To		

BASE METALS	Specification	Type or Grade	AWS Group No.	Size (NPS)	Schedule	Thickness	Diameter
Base Material							
Welded To							

VARIABLES	Actual Values		RANGE QUALIFIED	
Type of Weld Joint				
Base Metal				

	Groove	Fillet	Groove	Fillet
Plate Thickness				
Pipe/Tube Thickness				
Pipe Diameter				

Welding Process						
Type *(Manual, Semiautomatic, Mechanized, Automatic)*						
Backing						
Filler Metal (AWS Spec.)						
AWS Classification						
F-Number						
Position						
Groove – Plate and Pipe ≥ 24 in						
Groove – Pipe < 24 in						
Fillet – Plate and Pipe ≥ 24 in						
Fillet – Pipe < 24 in						
Progression						
GMAW Transfer Mode						
Single or Multiple Electrodes						
Gas/Flux Type						

TEST RESULTS

Type of Test	Acceptance Criteria	Results	Remarks

CERTIFICATION

Test Conducted by	
Laboratory	
Test Number	
File Number	

We, the undersigned, certify that the statements in this record are correct and that the test welds were prepared, welded, and tested in accordance with the requirements of Clause 6 of AWS D1.1/D1.1M (_____) *Structural Welding Code—Steel*.
(year)

Manufacturer or Contractor _____   Authorized by _____

                                                                                      Date _____

Form J-5   (See http://go.aws.org/D1forms)

ANNEX J   AWS D1.1/D1.1M:2020

Example Welder Qualification (Multi-Process)
WELDER, WELDING OPERATOR, OR TACK WELDER
PERFORMANCE QUALIFICATION TEST RECORD

Name	Z. W. Elder		Test Date	12/12/2020	Rev.
ID Number	00-001-ZWE	OPTIONAL PHOTO ID	Record No.	WPQ-003	0
Stamp No.	ZWE-01		Std. Test No.	ST-003	0
Company	RED Inc.		WPS No.	WPS-003	0
Division	–		Qualified To	AWS D1.1	

BASE METALS	Specification	Type or Grade	AWS Group No.	Size (NPS)	Schedule	Thickness	Diameter
Base Material	ASTM A36	UNS K02600	II	–	–	1 in	–
Welded To	ASTM A36	UNS K02600	II	–	–	1 in	–

VARIABLES	Actual Values		RANGE QUALIFIED	
Type of Weld Joint	Plate – Groove (Fig. 6.20) with Backing		Groove, Fillet, Plug, and Slot Welds (T-, Y-, K-Groove PJP only)	
Base Metal	Group II to Group II		Any AWS D1.1 Qualified Base Metal	

	Groove	Fillet		Groove	Fillet
Plate Thickness	1 in	–		1/8 in min.	1/8 in min.
Pipe/Tube Thickness	–	–		1/8 in min.	Unlimited
Pipe Diameter	–	–		24 in min.	Unlimited

Welding Process	GTAW	SMAW	FCAW	GTAW	SMAW	FCAW
Type (Manual, Semiautomatic, Mechanized, Automatic)	Manual	Manual	Semiauto	Man/Mech/Auto	Manual	Semi/Mech/Auto
Backing	With	With	With	With	With	With
Filler Metal (AWS Spec.)	A5.18	A5.1	A5.20	A5.xx	A5.xx	A5.xx
AWS Classification	ER70S-2	E7018	E70T-6	All	All	All
F-Number	–	4	–	–	1 thru 4	–
Position	1G	1G	1G			
Groove – Plate and Pipe ≥ 24 in				F	F	F
Groove – Pipe < 24 in				–	–	–
Fillet – Plate and Pipe ≥ 24 in				F, H	F, H	F, H
Fillet – Pipe < 24 in				F, H	F, H	F, H
Progression	–	–	–	–	–	–
GMAW Transfer Mode	–	–	–	–	–	–
Single or Multiple Electrodes	Single	–	Single	Single	–	Single
Gas/Flux Type	A5.32 SG-A	–	None	A5.xx Approved	–	A5.xx Approved

TEST RESULTS

Type of Test	Acceptance Criteria	Results	Remarks
Visual Examination per 6.10.1	6.10.1	Acceptable	–
2 Transverse Side Bends per 6.10.3.1 and Fig. 6.9	6.10.3.3	Acceptable	–

CERTIFICATION

Test Conducted by	
Laboratory	Welding Forms Lab
Test Number	Fictitious Test XYZ
File Number	Welding Forms/Sample-WPQ-for-GTAW-SMAW-FCAW.pdf

We, the undersigned, certify that the statements in this record are correct and that the test welds were prepared, welded, and tested in conformance with the requirements of Clause 6 of AWS D1.1/D1.1M (_____) Structural Welding Code—Steel.
(year)

Manufacturer or Contractor _____**Red Inc.**_____   Authorized by _____**E. M. Ployee (Q.C. Mgr.)**_____

Date _____**12/12/2020**_____

Form J-5   (See http://go.aws.org/D1forms)

REPORT OF RADIOGRAPHIC EXAMINATION OF WELDS

Project _____
Quality requirements—section no. _____
Reported to _____

WELD LOCATION AND IDENTIFICATION SKETCH

Technique
Source _____
Film to source _____
Exposure time _____
Screens _____
Film type _____

(Describe length, width, and thickness of all joints radiographed)

Date	Weld identification	Area	Interpretation		Repairs		Remarks
			Accept.	Reject	Accept.	Reject	

We, the undersigned, certify that the statements in this record are correct and that the test welds were prepared and tested in conformance with the requirements of AWS D1.1/D1.1M, (_____) *Structural Welding Code—Steel*.
(year)

Radiographer(s) _____   Manufacturer or Contractor _____

Interpreter _____   Authorized by _____

Test date _____   Date _____

Form J-6   (See http://go.aws.org/D1forms)

ANNEX J　　　　　　　　　　　　　　　　　　　　　　　　　　　　　　　　　AWS D1.1/D1.1M:2020

REPORT OF MAGNETIC-PARTICLE EXAMINATION OF WELDS

Project _____

Quality requirements—Section No. _____

Reported to _____

WELD LOCATION AND IDENTIFICATION SKETCH

Quantity: _____　Total Accepted: _____　Total Rejected: _____

Date	Weld identification	Area Examined		Interpretation		Repairs		Remarks
		Entire	Specific	Accept.	Reject	Accept.	Reject	

PRE-EXAMINATION

Surface Preparation: _____

EQUIPMENT

Instrument Make: _____　Model: _____　S. No.: _____

METHOD OF INSPECTION

☐ Dry　　☐ Wet　　☐ Visible　　☐ Fluorescent

How Media Applied: _____

☐ Residual　☐ Continuous　☐ True-Continuous

☐ AC　　　☐ DC　　　　☐ Half-Wave

☐ Prods　　☐ Yoke　　　☐ Cable Wrap　☐ Other _____

Direction for Field:　☐ Circular　☐ Longitudinal

Strength of Field: _____

(Ampere turns, field density, magnetizing force, number, and duration of force application.)

POST EXAMINATION

Demagnetizing Technique (if required): _____

Cleaning (if required): _____　Marking Method: _____

We, the undersigned, certify that the statements in this record are correct and that the test welds were prepared and tested in conformance with the requirements of AWS D1.1/D1.1M, (_____) *Structural Welding Code—Steel*.
　　　　　　　　　　　　　　　　　　　　　　　　　　　　(year)

Inspector _____　　Manufacturer or Contractor _____

Level _____　　Authorized By _____

Test Date _____　　Date _____

Form J-7　　　　　　　　　(See http://go.aws.org/D1forms)

AWS D1.1/D1.1M:2020   ANNEX J

STUD WELDING APPLICATION QUALIFICATION TEST DATA FORM PER SUBCLAUSE 9.6 Yes ☐
PRE-PRODUCTION TEST PER SUBCLAUSE 9.7.1 (WPS) Yes ☐
OR PROCEDURE QUALIFICATION RECORD (PQR) Yes ☐
OR WELDER QUALIFICATION RECORD (WQR) Yes ☐

Company name _____
Operator name _____
Test number _____
Weld stud material _____
Weld stud size and PN#/Manufacturer _____

Base Material
Specification _____
Alloy and temper _____
Surface condition   HR ☐   CR ☐
Coating _____
Cleaning method _____
Decking gage _____

Shape of Base Material
Flat ☐   Round ☐   Tube ☐
Angle ☐   Inside ☐   Outside ☐   Inside radius ☐
Thickness _____

Ferrule
Part No./Manufacturer _____
Ferrule description _____

Stud Base Sketch/Application Detail

Equipment Data
Application Settings, Current, and Time Settings
Make _____   Model _____
Stud gun: Make _____   Model _____
Weld time (seconds) _____
Current (amperage) _____
Polarity: DCEN _____   DCEP _____
Lift _____
Plunge (protrusion) _____
Weld cable size _____   Length _____
Number of grounds (workpiece leads) _____

Welding Position
Flat (Down hand) ☐   Horizontal (Side hand) ☐   Angular—degrees from normal ☐   Overhead ☐

Shielding Gas
Shielding gas(es)/Composition _____
Flow rate _____

WELD TEST RESULTS

Stud No.	Visual Acceptance	Option #1 Bend Test	Option #2 Tension Test	Option #3 Torque Test*
1				
2				
3				
4				
5				
6				
7				
8				
9				
10				

*Note: Torque test optional for threaded fasteners only.

Mechanical tests conducted by _____   Date _____
                                                  (Company)

We, the undersigned, certify that the statements in this record are correct and that the test welds were prepared, welded, and tested in conformance with the requirements of Clause 9 of AWS D1.1/D1.1M, (_____) *Structural Welding Code—Steel*.
                                                                                                    (year)

Signed by _____   Title _____   Date _____
           (Contractor/Applicator/Other)
Form J-8   Company _____

(See http://go.aws.org/D1forms)

Annex K (Informative)
Contents of Prequalified WPS

This annex is not part of this standard but is included for informational purposes only.

Prequalified welding requires a written WPS addressing the following code subclauses as applicable to weldments of concern. In addition to the requirements for a written WPS, this code imposes many other requirements and limitations on prequalified welding. The organization using prequalified welding is required by the code to conform with all the relevant requirements.

The specification of the WPS may meet the users needs. Items such as assembly tolerances may be referenced.

1.4	Limitations
4.4.1.4	Effective Size of Flare-Groove Welds
4.4.2.9	Maximum Weld Size in Lap Joints – Fillet Welds
4.4.4.2	Slot Ends
4.4.3	Skewed T-Joints (all subclauses)
5.2	General WPS Requirements (all subclauses)
5.2.1–5.8	Limitation of WPS Variables (all subclauses)
5.3	Base Metal
5.4.3.2	Fillet Weld Details—Skewed T-Joints
5.4.4	Plug and Slot Weld Requirements
5.5.1	Prequalified Processes
5.6.1	Filler Metal Matching
5.7	Preheat and Interpass Temperature Requirements (all subclauses)
5.4.1	CJP Groove Weld Requirements (all subclauses)
5.4.2	PJP Groove Weld Requirements (all subclauses)
Table 5.2	Essential Variables for Prequalified WPS
7.2.2	Base Metal for Weld Tabs, Backing, and Spacers
7.3.1.2	Suitability of Classification – Welding Consumables and Electrode Requirements
7.3.2	SMAW Electrodes (all subclauses)
7.3.3	SAW Electrodes and Fluxes (all subclauses)
7.3.4	GMAW/FCAW Electrodes
7.5	WPS Variables
7.7	Heat Input Control for Quenched and Tempered Steels
7.9	Backing (all subclauses)
7.13	Minimum Fillet Weld Sizes
7.14	Preparation of Base Metal (all subclauses)
7.21.1.1	Faying Surface
7.24	Technique for Plug and Slot Welds (all subclauses)
7.26	Peening (all subclauses)
7.29.1	In-Process Cleaning
9.5.5	FCAW, GMAW, SMAW Fillet Weld Option (all subclauses)
9.7.5	Removal Area Repair
10.9	PJP Groove Weld Requirements (all subclauses)
10.10	CJP Groove Weld Requirements (all subclauses)
10.22	Backing (all subclauses)

This page is intentionally blank.

Annex L (Informative)
Filler Metal Strength Properties

This annex is not part of this standard but is included for informational purposes only.

The data contained in this annex are copied from the appropriate A5 specification. Values shown are for reference purposes only and other process variables may have to be controlled in order to achieve the Nominal Tensile Strength and Nominal Yield Strength. (See the applicable AWS A5 Filler Metal Specification for a more specific description.)

AWS A5.1/A5.1M, Specification for Carbon Steel Electrodes for Shielded Metal Arc Welding[a, b]

AWS Classification		Tensile Strength		Yield Strength at 0.2% Offset		Elongation Percentage in 4x Diameter Length
A5.1	A5.1M	A5.1 (ksi)	A5.1M (MPa)	A5.1 (ksi)	A5.1M (MPa)	
E6010	E4310	60	430	48	330	22
E6011	E4311	60	430	48	330	22
E6012	E4312	60	430	48	330	17
E6013	E4313	60	430	48	330	17
E6018	E4318	60	430	48	330	22
E6019	E4319	60	430	48	330	22
E6020	E4320	60	430	48	330	22
E6022	E4322	60	430	Not Specified		Not Specified
E6027	E4327	60	430	48	330	22
E7014	E4914	70	490	58	400	17
E7015	E4915	70	490	58	400	22
E7016	E4916	70	490	58	400	22
E7018	E4918	70	490	58	400	22
E7024	E4924	70	490	58	400	17[c]
E7027	E4927	70	490	58	400	22
E7028	E4928	70	490	58	400	22
E7048	E4948	70	490	58	400	22
E7018M	E4918M	Footnote[d]	Footnote[d]	53–72[e]	370–500[e]	24

[a] Requirements are in the as-welded condition with aging.
[b] Single values are minimum.
[c] Weld metal from electrodes identified as E7024–1 [E4924–1] shall have elongation of 22% minimum.
[d] Tensile strength of this weld metal is a nominal 70 ksi [490 MPa].
[e] For 3/32 in [2.4 mm] electrodes, the maximum yield strength is 77 ksi [530 MPa].

AWS A5.5/A5.5M, Specification for Low-Alloy Steel Electrodes for Shielded Metal Arc Welding[a]

AWS Classification[b]		Tensile Strength		Yield Strength, at 0.2% Offset		Elongation	Postweld Condition[c]
A5.5	A5.5M	ksi	MPa	ksi	MPa	Percent	
E7010-P1	E4910-P1	70	490	60	415	22	AW
E7010-A1	E4910-A1	70	490	57	390	22	PWHT
E7010-G	E4910-G	70	490	57	390	22	AW or PWHT
E7011-A1	E4911-A1	70	490	57	390	22	PWHT
E7011-G	E4911-G	70	490	57	390	22	AW or PWHT
E7015-X	E4915-X	70	490	57	390	22	PWHT
E7015-B2L	E4915-B2L	75	520	57	390	19	PWHT
E7015-G	E4915-G	70	490	57	390	22	AW or PWHT
E7016-X	E4916-X	70	490	57	390	22	PWHT
E7016-B2L	E4916-B2L	75	520	57	390	19	PWHT
E7016-G	E4916-G	70	490	57	390	22	AW or PWHT
E7018-X	E4918-X	70	490	57	390	22	PWHT
E7018-B2L	E4918-B2L	75	520	57	390	19	PWHT
E7018-C3L	E4918-C3L	70	490	57	390	22	AW
E7018-W1	E4918-W1	70	490	60	415	22	AW
E7018-G	E4918-G	70	490	57	390	22	AW or PWHT
E7020-A1	E4920-A1	70	490	57	390	22	PWHT
E7020-G	E4920-G	70	490	57	390	22	AW or PWHT
E7027-A1	E4927-A1	70	490	57	390	22	PWHT
E7027-G	E4927-G	70	490	57	390	22	AW or PWHT
E8010-P1	E5510-P1	80	550	67	460	19	AW
E8010-G	E5510-G	80	550	67	460	19	AW or PWHT
E8011-G	E5511-G	80	550	67	460	19	AW or PWHT
E8013-G	E5513-G	80	550	67	460	16	AW or PWHT
E8015-X	E5515-X	80	550	67	460	19	PWHT
E8015-B3L	E5515-B3L	80	550	67	460	17	PWHT
E8015-G	E5515-G	80	550	67	460	19	AW or PWHT
E8016-X	E5516-X	80	550	67	460	19	PWHT
E8016-C3	E5516-C3	80	550	68 to 80[d]	470 to 550[d]	24	AW
E8016-C4	E5516-C4	80	550	67	460	19	AW
E8016-G	E5516-G	80	550	67	460	19	AW or PWHT
E8018-X	E5518-X	80	550	67	460	19	PWHT
E8018-B3L	E5518-B3L	80	550	67	460	17	PWHT
E8018-C3	E5518-C3	80	550	68 to 80[d]	470 to 550[d]	24	AW
E8018-C4	E5518-C4	80	550	67	460	19	AW
E8018-NM1	E5518-NM1	80	550	67	460	19	AW
E8018-P2	E5518-P2	80	550	67	460	19	AW
E8018-W2	E5518-W2	80	550	67	460	19	AW
E8018-G	E5518-G	80	550	67	460	19	AW or PWHT
E8045-P2	E5545-P2	80	550	67	460	19	AW
E9010-P1	E6210-P1	90	620	77	530	17	AW
E9010-G	E6210-G	90	620	77	530	17	AW or PWHT
E9011-G	E6211-G	90	620	77	530	17	AW or PWHT

(Continued)

AWS A5.5/A5.5M, Specification for Low-Alloy Steel Electrodes for Shielded Metal Arc Welding[a] (Continued)

AWS Classification[b]		Tensile Strength		Yield Strength, at 0.2% Offset		Elongation	Postweld Condition[c]
A5.5	A5.5M	ksi	MPa	ksi	MPa	Percent	
E9013-G	E6213-G	90	620	77	530	14	AW or PWHT
E9015-X	E6215-X	90	620	77	530	17	PWHT
E9015-G	E6215-G	90	620	77	530	17	AW or PWHT
E9016-X	E6216-X	90	620	77	530	17	PWHT
E9016-G	E6216-G	90	620	77	530	17	AW or PWHT
E9018M	E6218M	90	620	78 to 90[d]	540 to 620[d]	24	AW
E9018-P2	E6218-P2	90	620	77	530	17	AW
E9018-X	E6218-X	90	620	77	530	17	PWHT
E9018-G	E6218-G	90	620	77	530	17	AW or PWHT
E9045-P2	E6245-P2	90	620	77	530	17	AW
E10010-G	E6910-G	100	690	87	600	16	AW or PWHT
E10011-G	E6911-G	100	690	87	600	16	AW or PWHT
E10013-G	E6913-G	100	690	87	600	13	AW or PWHT
E10015-X	E6915-X	100	690	87	600	16	PWHT
E10015-G	E6915-G	100	690	87	600	16	AW or PWHT
E10016-X	E6916-X	100	690	87	600	16	PWHT
E10016-G	E6916-G	100	690	87	600	16	AW or PWHT
E10018M	E6918M	100	690	88 to 100[d]	610 to 690[d]	20	AW
E10018-X	E6918-X	100	690	87	600	16	PWHT
E10018-G	E6918-G	100	690	87	600	16	AW or PWHT
E10045-P2	E6945-P2	100	690	87	600	16	AW
E11010-G	E7610-G	110	760	97	670	15	AW or PWHT
E11011-G	E7611-G	110	760	97	670	15	AW or PWHT
E11013-G	E7613-G	110	760	97	670	13	AW or PWHT
E11015-G	E7615-G	110	760	97	670	15	AW or PWHT
E11016-G	E7616-G	110	760	97	670	15	AW or PWHT
E11018-G	E7618-G	110	760	97	670	15	AW or PWHT
E11018M	E7618M	110	760	98 to 110[d]	680 to 760[d]	20	AW
E12010-G	E8310-G	120	830	107	740	14	AW or PWHT
E12011-G	E8311-G	120	830	107	740	14	AW or PWHT
E12013-G	E8313-G	120	830	107	740	11	AW or PWHT
E12015-G	E8315-G	120	830	107	740	14	AW or PWHT
E12016-G	E8316-G	120	830	107	740	14	AW or PWHT
E12018-G	E8318-G	120	830	107	740	14	AW or PWHT
E12018M	E8318M	120	830	108 to 120[d]	745 to 830[d]	18	AW
E12018M1	E8318M1	120	830	108 to 120[d]	745 to 830[d]	18	AW

[a] Single values are minimum, except as otherwise specified.
[b] The letter suffix "X" as used in this table represents the suffixes (A1, B1, B2, etc.) which are tested in the PWHT condition only.
[c] "AW" signifies as-welded, which may or may not be aged, at the manufacturer's option. "PWHT" signifies postweld heat treated.
[d] For 3/32 in [2.5 mm] electrodes, the upper value for the yield strength may be 5 ksi [35 MPa] higher than the indicated value.

AWS A5.17/A5.17M, Specification for Carbon Steel Electrodes and Fluxes for Submerged Arc Welding

Flux-Electrode Classification[a]		Tensile Strength psi [MPa]	Yield Strength[b] psi [MPa]	Elongation[b] %
A5.17	A5.17M			
F6XX-EXXX	F43XX-EXXX	60 000–80 000 [430–560]	48 000 [330]	22
F7XX-EXXX	F48XX-EXXX	70 000–95 000 [480–660]	58 000 [400]	22

[a] The letter "S" will appear after the "F" as part of the classification designation when the flux being classified is a crushed slag or a blend of crushed slag with unused (virgin) flux. The letter "C" will appear after the "E" as part of the classification designation when the electrode being classified is a composite electrode. The letter "X" used in various places in this table stands for, respectively, the condition of heat treatment, the toughness of the weld metal, and the classification of the electrode.

[b] Minimum requirements. Yield strength at 0.2% offset and elongation in 2 in [51 mm] gauge length.

AWS A5.18/A5.18M, Specification for Carbon Steel Electrodes and Rods for Gas Shielded Arc Welding

AWS Classification[a]		Shielding Gas	Tensile Strength (minimum)		Yield Strength[b] (minimum)		Elongation[b] minimum)
A5.18	A5.18M		psi	MPa	psi	MPa	
ER70S–2 ER70S–3 ER70S–4 ER70S–6 ER70S–7	ER48S–2 ER48S–3 ER48S–4 ER48S–6 ER48S–7	CO_2[c]	70 000	480	58 000	400	22
ER70S-G	ER48S-G	Footnote[d]	70 000	480	58 000	400	22
ER70C–3X E70C–6X	E48C–3X E48C–6X	75–80% Ar/balance CO_2 or CO_2	70 000	480	58 000	400	22
E70C-G(X)	E48C-G(X)	Footnote[d]	70 000	480	58 000	400	22
E70C-GS(X)	E48C-GS(X)	Footnote[d]	70 000	480	Not Specified		Not Specified

[a] The final X shown in the classification represents a "C" or "M" which corresponds to the shielding gas with which the electrode is classified. The use of "C" designates 100% CO_2 shielding (AWS A5.32 Class SG-C); "M" designates 75–80% Ar/balance CO_2 (AWS A5.32 Class SG-AC-Y, where Y is 20 to 25). For E70C-G [E48C-G] and E70C-GS [E48C-GS], the final "C" or "M" may be omitted.

[b] Yield strength at 0.2% offset and elongation in 2 in [50 mm] gage length.

[c] CO_2 = carbon dioxide shielding gas (AWS A5.32 Class SG-C). The use of CO_2 for classification purposes is not to be construed to preclude the use of Ar/CO_2 (AWS A5.32 Class SG-AC-Y) or Ar/O_2 (AWS A5.32 Class SG-AO-X) shielding gas mixtures. A filler metal tested with gas blends, such as Ar/O_2, or Ar/CO_2, may result in weld metal having higher strength and lower elongation.

[d] Shielding gas is as agreed between purchaser and supplier, unless designated by the "C" or "M" suffix.

AWS A5.20/A5.20M, Specification for Carbon Steel Electrodes for Flux Cored Arc Welding

AWS Classification		Tensile Strength ksi [MPa]	Minimum Yield Strength[a] ksi [MPa]	Minimum % Elongation[b]	Minimum Charpy V-Notch Impact Energy
A5.20	A5.20M				
E7XT–1C, –1M	E49XT–1C, –1M	70–95 [490–670]	58 [390]	22	20 ft·lbf @ 0°F [27 J @ –20°C]
E7XT–2C[c], –2M[c]	E49XT–2C[c], –2M[c]	70 [490] min.	Not Specified	Not Specified	Not Specified
E7XT–3[c]	E49XT–3[c]	70 [490] min.	Not Specified	Not Specified	Not Specified
E7XT–4	E49XT–4	70–95 [490–670]	58 [390]	22	Not Specified
E7XT–5C, –5M	E49XT–5C, –5M	70–95 [490–670]	58 [390]	22	20 ft·lbf @ –20°F [27 J @ –30°C]
E7XT–6	E49XT–6	70–95 [490–670]	58 [390]	22	20 ft·lbf @ –20°F [27 J @ –30°C]
E7XT–7	E49XT–7	70–95 [490–670]	58 [390]	22	Not Specified
E7XT–8	E49XT–8	70–95 [490–670]	58 [390]	22	20 ft·lbf @ –20°F [27 J @ –30°C]
E7XT–9C, –9M	E49XT–9C, –9M	70–95 [490–670]	58 [390]	22	20 ft·lbf @ –20°F [27 J @ –30°C]
E7XT–10[c]	E49XT–10[c]	70 [490] min.	Not Specified	Not Specified	Not Specified
E7XT–11	E49XT–11	70–95 [490–670]	58 [390]	20[d]	Not Specified
E7XT–12C, –12M	E49XT–12C, –12M	70–90 [490–620]	58 [390]	22	20 ft·lbf @ –20°F [27 J @ –30°C]
E6XT–13[c]	E43XT–13[c]	60 [430] min.	Not Specified	Not Specified	Not Specified
E7XT–13[c]	E49XT–13[c]	70 [490] min.	Not Specified	Not Specified	Not Specified
E7XT–14[c]	E49XT–14[c]	70 [490] min.	Not Specified	Not Specified	Not Specified
E6XT-G	E43XT-G	60–80 [430–600]	48 [330]	22	Not Specified
E7XT-G	E49XT-G	70–95 [490–670]	58 [390]	22	Not Specified
E6XT-GS[c]	E43XT-GS[c]	60 [430] min.	Not Specified	Not Specified	Not Specified
E7XT-GS[c]	E49XT-GS[c]	70 [490] min.	Not Specified	Not Specified	Not Specified

[a] Yield strength at 0.2% offset.
[b] In 2 in [50 mm] gage length when a 0.500 in [12.5 mm] nominal diameter tensile specimen and nominal gage length to diameter ratio of 4:1 is used.
[c] These classifications are intended for single pass welding. They are not for multiple pass welding. Only tensile strength is specified.
[d] In 1 in [25 mm] gage length when a 0.250 in [6.5 mm] nominal diameter tensile specimen is used as permitted for 0.045 in [1.2 mm] and smaller sizes of the E7XT–11[E49XT–11] classification.

AWS A5.23/A5.23M, Specification for Low-Alloy Steel Electrodes and Fluxes for Submerged Arc Welding

Flux-Electrode Classification[a]		Tensile Strength[b] psi [MPa]	Yield Strength[b] (0.2% Offset) psi [MPa]	Elongation[b] (%)
A5.23	A5.23M			
Multiple Pass Classifications				
F7XX-EXX-XX	F49XX-EXX-XX	70 000–95 000 [490–660]	58 000 [400]	22
F8XX-EXX-XX	F55XX-EXX-XX	80 000–100 000 [550–700]	68 000 [470]	20
F9XX-EXX-XX	F62XX-EXX-XX	90 000–110 000 [620–760]	78 000 [540]	17
F10XX-EXX-XX	F69XX-EXX-XX	100 000–120 000 [690–830]	88 000 [610]	16
F11XX-EXX-XX	F76XX-EXX-XX	110 000–130 000 [760–900]	98 000 [680]	15[c]
F12XX-EXX-XX	F83XX-EXX-XX	120 000–140 000 [830–970]	108 000 [740]	14[c]
F13XX-EXX-XX	F90XX-EXX-XX	130 000–150 000 [900–1040]	118 000 [810]	14[c]
Two-Run Classifications				
F6TXX-EXX	F43TXX-EXX	60 000 [430]	50 000 [350]	22
F7TXX-EXX	F49TXX-EXX	70 000 [490]	60 000 [415]	22
F8TXX-EXX	F55TXX-EXX	80 000 [550]	70 000 [490]	20
F9TXX-EXX	F62TXX-EXX	90 000 [620]	80 000 [555]	17
F10TXX-EXX	F69TXX-EXX	100 000 [690]	90 000 [625]	16
F11TXX-EXX	F76TXX-EXX	110 000 [760]	100 000 [690]	15
F12TXX-EXX	F83TXX-EXX	120 000 [830]	110 000 [760]	14
F13TXX-EXX	F90TXX-EXX	130 000 [900]	120 000 [830]	14

[a] The letter "S" will appear after the "F" as part of the classification designation when the flux being classified is a crushed slag or a blend of crushed slag with unused (virgin) flux. The letter "C" will appear after the "E" as part of the classification designation when the electrode used is a composite electrode. *For two-run classifications, the letter "G" will appear after the impact designator (immediately before the hyphen) to indicate that the base steel used for classification is not one of the base steels prescribed in AWS A5.23/A5.23M but is a different steel, as agreed between purchaser and supplier.* The letter "X" used in various places in this table stands for, respectively, the condition of heat treatment, the toughness of the weld metal, and the classification of the weld metal.

[b] For multiple pass classifications, the requirements listed in the table for yield strength and % elongation (in 2 in [50 mm] gage length) are minimum requirements. *For two-run classifications, the requirements listed for tensile strength, yield strength and % elongation (in 1 in [25 mm] gage length) are all minimum requirements.*

[c] Elongation may be reduced by one percentage point for F11XX-EXX-XX, F11XX-ECXX-XX, F12XX-EXX-XX, F12XX-ECXX-XX, F13XX-EXX-XX, and F13XX-ECXX-XX [F76-EXX-XX, F76-ECXX-XX, F83XX-EXX-XX, F83XX-ECXX-XX, F90XX-EXX-XX, and F90XX-ECXX-XX] weld metals in the upper 25% of their tensile strength range.

AWS A5.28/A5.28M, Specification for Low-Alloy Steel Electrodes and Rods for Gas Shielded Arc Welding

AWS Classification		Shielding Gas[a]	Tensile Strength (minimum)		Yield Strength[b] (minimum)		Elongation[b] Percent (minimum)	Testing Condition
A5.28	A5.28M		psi	MPa	psi	MPa		
ER70S-B2L E70C-B2L ER70S-A1	ER49S-B2L E49C-B2L ER49S-A1	Argon/1–5% O_2 (Classes SG-AO–1 thru SG-AO–5)	75 000	515	58 000	400	19	PWHT
ER80S-B2 E80C-B2	ER55S-B2 E55C-B2		80 000	550	68 000	470	19	
ER80S-B3L E80C-B3L	ER55S-B3L E55C-B3L		80 000	550	68 000	470	17	
ER90S-B3 E90C-B3	ER62S-B3 E62C-B3		90 000	620	78 000	540	17	
ER80S-B6	ER55S-B6		80 000	550	68 000	470	17	
E80C-B6	E55C-B6		80 000	550	68 000	470	17	
ER80S-B8	ER55S-B8		80 000	550	68 000	470	17	
E80C-B8	E55C-B8		80 000	550	68 000	470	17	
ER90S-B9	ER62S-B9	Argon/5% O_2 (Class SG-AC–5)	90 000	620	60 000	410	16	
E90C-B9	E62C-B9	Argon/5–25% CO_2 (Classes SG-AC–5 thru SG-AC–25)						
E70C-Ni2	E49C-Ni2	Argon/1–5% O_2 (Classes SG-AO–1 thru SG-AO–5)	70 000	490	58 000	400	24	PWHT
ER80S-Ni1 E80C-Ni1	ER55S-Ni1 E55C-Ni1		80 000	550	68 000	470	24	As-Welded
ER80S-Ni2 E80C-Ni2 ER80S-Ni3 E80C-Ni3	ER55S-Ni2 E55C-Ni2 ER55S-Ni3 E55C-Ni3		80 000	550	68 000	470	24	PWHT
ER80S-D2	ER55S-D2	CO_2 (Class SG-C)	80 000	550	68 000	470	17	As-Welded
ER90S-D2 E90C-D2	ER62S-D2 E62C-D2	Argon/1–5% O_2 (Classes SG-AO–1 thru SG-AO–5)	90 000	620	78 000	540	17	
ER100S–1	ER69S–1	Argon/2% O_2 (Class SG-AO–2)	100 000	690	88 000	610	16	
ER110S–1	ER76S–1		110 000	760	95 000	660	15	
ER120S–1	ER83S–1		120 000	830	105 000	730	14	

(Continued)

AWS A5.28/A5.28M, Specification for Low-Alloy Steel Electrodes and Rods for Gas Shielded Arc Welding (Continued)

AWS Classification		Shielding Gas[a]	Tensile Strength (minimum)		Yield Strength[b] (minimum)		Elongation[b] Percent (minimum)	Testing Condition
A5.28	A5.28M		psi	MPa	psi	MPa		
E90C-K3	E62C-K3	Argon/5–25% CO_2 (Classes SG-AC–5 thru SG-AC–25)	90 000	620	78 000	540	18	As-Welded
E100C-K3	E69C-K3		100 000	690	88 000	610	16	
E110C-K3 E110C-K4	E76C-K3 E76C-K4		110 000	760	98 000	680	15	
E120C-K4	E83C-K4		120 000	830	108 000	750	15	
E80C-W2	E55C-W2		80 000	550	68 000	470	22	
ER70S-G E70C-G	ER49S-G E49C-G	Footnote c	70 000	490	Footnote d	Footnote d	Footnote d	Footnote d
ER80S-G E80C-G	ER55S-G E55C-G	Footnote c	80 000	550	Footnote d	Footnote d	Footnote d	Footnote d
ER90S-G E90C-G	ER62S-G E62C-G	Footnote c	90 000	620	Footnote d	Footnote d	Footnote d	Footnote d
ER100S-G E100C-G	ER69S-G E69C-G	Footnote c	100 000	690	Footnote d	Footnote d	Footnote d	Footnote d
ER110S-G E110C-G	ER76S-G E76C-G	Footnote c	110 000	760	Footnote d	Footnote d	Footnote d	Footnote d
ER120S-G E120C-G	ER83S-G E83C-G	Footnote c	120 000	830	Footnote d	Footnote d	Footnote d	Footnote d

[a] The use of a particular shielding gas for classification purposes is not to be construed to restrict the use of other gas mixtures. A filler metal tested with other gas blends, such as argon/O_2 or argon/CO_2 may result in weld metal having different strength and elongation.
[b] Yield strength at 0.2% offset and elongation in 2 in [51 mm] gage length.
[c] Shielding gas shall be as agreed to between purchaser and supplier.
[d] Not specified (as agreed to between purchaser and supplier).

AWS A5.29/A5.29M, Specification for Low-Alloy Steel Electrodes for Flux Cored Arc Welding[a, b]

AWS Classification[a, b]		Condition[c]	Tensile Strength ksi [MPa]	Yield Strength ksi [MPa]	% Elongation Minimum	Charpy V-Notch Impact Energy Minimum
A5.29	A5.29M					
E7XT5-A1C, -A1M	E49XT5-A1C, -A1M	PWHT	70–90 [490–620]	58 [400] min.	20	20 ft·lbf @ –20°F [27 J @ –30°C]
E8XT1-A1C, -A1M	E55XT1-A1C, -A1M	PWHT	80–100 [550–690]	68 [470] min.	19	Not Specified
E8XT1-B1C, -B1M, -B1LC, -B1LM	E55XT1-B1C, -B1M, -B1LC, -B1LM	PWHT	80–100 [550–690]	68 [470] min.	19	Not Specified
E8XT1-B2C, -B2M, -B2HC, -B2HM, -B2LC, -B2LM	E55XT1-B2C, -B2M, -B2HC, -B2HM, -B2LC, -B2LM	PWHT	80–100 [550–690]	68 [470] min.	19	Not Specified
E8XT5-B2C, -B2M, -B2LC, -B2LM	E55XT5-B2C, -B2M, -B2LC, -B2LM					
E9XT1-B3C, -B3M, -B3LC, -B3LM, -B3HC, -B3HM	E62XT1-B3C, -B3M, -B3LC, -B3LM, -B3HC, -B3HM	PWHT	90–110 [620–760]	78 [540] min.	17	Not Specified
E9XT5-B3C, -B3M	E62XT5-B3C, -B3M					
E10XT1-B3C, -B3M	E69XT1-B3C, -B3M	PWHT	100–120 [690–830]	88 [610] min.	16	Not Specified
E8XT1-B6C,[d] -B6M,[d] -B6LC,[d] -B6LM,[d]	E55XT1-B6C, -B6M, -B6LC, -B6LM	PWHT	80–100 [550–690]	68 [470] min.	19	Not Specified
E8XT5-B6C,[d] -B6M,[d] -B6LC,[d] -B6LM,[d]	E55XT5-B6C, -B6M, -B6LC, -B6LM					
E8XT1-B8C,[d] -B8M,[d] -B8LC,[d] -B8LM,[d]	E55XT1-B8C, -B8M, -B8LC, -B8LM	PWHT	80–100 [550–690]	68 [470] min.	19	Not Specified
E8XT5-B8C,[d] -B8M,[d] -B8LC,[d] -B8LM,[d]	E55XT5-B8C, -B8M, -B8LC, -B8LM					
E9XT1-B9C, -B9M	E62XT1-B9C, -B9M	PWHT	90–120 [620–830]	78 [540] min.	16	Not Specified
E6XT1-Ni1C, -Ni1M	E43XT1-Ni1C, -Ni1M	AW	60–80 [430–550]	50 [340] min.	22	20 ft·lbf @ –20°F [27 J @ –30°C]
E7XT6-Ni1	E49XT6-Ni1	AW	70–90 [490–620]	58 [400] min.	20	20 ft·lbf @ –20°F [27 J @ –30°C]
E7XT8-Ni1	E49XT8-Ni1	AW	70–90 [490–620]	58 [400] min.	20	20 ft·lbf @ –20°F [27 J @ –30°C]
E8XT1-Ni1C, -Ni1M	E55XT1-Ni1C, -Ni1M	AW	80–100 [550–690]	68 [470] min.	19	20 ft·lbf @ –20°F [27 J @ –30°C]
E8XT5-Ni1C, -Ni1M	E55XT5-Ni1C, -Ni1M	PWHT	80–100 [550–690]	68 [470] min.	19	20 ft·lbf @ –60°F [27 J @ –50°C]
E7XT8-Ni2	E49XT8-Ni2	AW	70–90 [490–620]	58 [400] min.	20	20 ft·lbf @ –20°F [27 J @ –30°C]
E8XT8-Ni2	E55XT8-Ni2	AW	80–100 [550–690]	68 [470] min.	19	20 ft·lbf @ –20°F [27 J @ –30°C]
E8XT1-Ni2C, -Ni2M	E55XT1-Ni2C, -Ni2M	AW	80–100 [550–690]	68 [470] min.	19	20 ft·lbf @ –40°F [27 J @ –40°C]
E8XT5-Ni2C,[e] -Ni2M[e]	E55XT5-Ni2C,[e] -Ni2M[e]	PWHT	80–100 [550–690]	68 [470] min.	19	20 ft·lbf @ –75°F [27 J @ –60°C]
E9XT1-Ni2C, -Ni2M	E62XT1-Ni2C, -Ni2M	AW	90–110 [620–760]	78 [540] min.	17	20 ft·lbf @ –40°F [27 J @ –40°C]
E8XT5-Ni3C,[e] -Ni3M[e]	E55XT5-Ni3C,[e] -Ni3M[e]	PWHT	80–100 [550–690]	68 [470] min.	19	20 ft·lbf @ –100°F [27 J @ –70°C]
E9XT5-Ni3C,[e] -Ni3M[e]	E62XT5-Ni3C,[e] -Ni3M[e]	PWHT	90–110 [620–760]	78 [540] min.	17	20 ft·lbf @ –100°F [27 J @ –70°C]

(Continued)

AWS A5.29/A5.29M, Specification for Low-Alloy Steel Electrodes for Flux Cored Arc Welding (Continued)

AWS Classification[a,b]		Condition[c]	Tensile Strength ksi [MPa]	Yield Strength ksi [MPa]	% Elongation Minimum	Charpy V-Notch Impact Energy Minimum
A5.29	A5.29M					
E8XT1-Ni3	E55XT11-Ni3	AW	80–100 [550–690]	68 [470] min.	19	20 ft·lbf @ 0°F [27 J @ –20°C]
E9XT1-D1C, -D1M	E62XT1-D1C, -D1M	AW	90–110 [620–760]	78 [540] min.	17	20 ft·lbf @ –40°F [27 J @ –40°C]
E9XT5-D2C, -D2M	E62XT5-D2C, -D2M	PWHT	90–110 [620–760]	78 [540] min.	17	20 ft·lbf @ –60°F [27 J @ –50°C]
E10XT5-D2C, -D2M	E69XT5-D2C, -D2M	PWHT	100–120 [690–830]	88 [610] min.	16	20 ft·lbf @ –40°F [27 J @ –40°C]
E9XT1-D3C, -D3M	E62XT1-D3C, -D3M	AW	90–110 [620–760]	78 [540] min.	17	20 ft·lbf @ –20°F [27 J @ –30°C]
E8XT5-K1C, -K1M	E55XT5-K1C, -K1M	AW	80–100 [550–690]	68 [470] min.	19	20 ft·lbf @ –40°F [27 J @ –40°C]
E7XT7-K2	E49XT7-K2	AW	70–90 [490–620]	58 [400] min.	20	20 ft·lbf @ –20°F [27 J @ –30°C]
E7XT4-K2	E49XT4-K2	AW	70–90 [490–620]	58 [400] min.	20	20 ft·lbf @ 0°F [27 J @ –20°C]
E7XT8-K2	E49XT8-K2	AW	70–90 [490–620]	58 [400] min.	20	20 ft·lbf @ –20°F [27 J @ –30°C]
E7XT11-K2	E49XT11-K2	AW	70–90 [490–620]	58 [400] min.	20	20 ft·lbf @ +32°F [27 J @ 0°C]
E8XT1-K2C, -K2M	E55XT1-K2C, -K2M	AW	80–100 [550–690]	68 [470] min.	19	20 ft·lbf @ –20°F [27 J @ –30°C]
E8XT5-K2C, -K2M	E55XT5-K2C, -K2M	AW				
E9XT1-K2C, -K2M	E62XT1-K2C, -K2M	AW	90–110 [620–760]	78 [540] min.	17	20 ft·lbf @ 0°F [27 J @ –20°C]
E9XT5-K2C, -K2M	E62XT5-K2C, -K2M	AW	90–110 [620–760]	78 [540] min.	17	20 ft·lbf @ –60°F [27 J @ –50°C]
E10XT1-K3C, -K3M	E69XT1-K3C, -K3M	AW	100–120 [690–830]	88 [610] min.	16	20 ft·lbf @ 0°F [27 J @ –20°C]
E10XT5-K3C, -K3M	E69XT5-K3C, -K3M	AW	100–120 [690–830]	88 [610] min.	16	20 ft·lbf @ –60°F [27 J @ –50°C]
E11XT1-K3C, -K3M	E76XT1-K3C, -K3M	AW	110–130 [760–900]	98 [680] min.	15	20 ft·lbf @ 0°F [27 J @ –20°C]
E11XT5-K3C, -K3M	E76XT5-K3C, -K3M	AW	110–130 [760–900]	98 [680] min.	15	20 ft·lbf @ –60°F [27 J @ –50°C]
E11XT1-K4C, -K4M	E76XT1-K4C, -K4M	AW	110–130 [760–900]	98 [680] min.	15	20 ft·lbf @ 0°F [27 J @ –20°C]
E11XT5-K4C, -K4M	E76XT5-K4C, -K4M	AW	110–130 [760–900]	98 [680] min.	15	20 ft·lbf @ –60°F [27 J @ –50°C]
E12XT1-K4C, -K4M	E83XT5-K4C, -K4M	AW	120–140 [830–970]	108 [745] min.	14	20 ft·lbf @ –60°F [27 J @ –50°C]
E12XT1-K5C, -K5M	E83XT1-K5C, -K5M	AW	120–140 [830–970]	108 [745] min.	14	Not Specified
E7XT5-K6C, -K6M	E49XT5-K6C, -K6M	AW	70–90 [490–620]	58 [400] min.	20	20 ft·lbf @ –75°F [27 J @ –60°C]
E6XT8-K6	E43XT8-K6	AW	60–80 [430–550]	50 [340] min.	22	20 ft·lbf @ –20°F [27 J @ –30°C]
E7XT8-K6	E49XT8-K6	AW	70–90 [490–620]	58 [400] min.	20	20 ft·lbf @ –20°F [27 J @ –30°C]

(Continued)

AWS A5.29/A5.29M, Specification for Low-Alloy Steel Electrodes for Flux Cored Arc Welding (Continued)

AWS Classification[a,b]		Condition[c]	Tensile Strength ksi [MPa]	Yield Strength ksi [MPa]	Elongation Minimum %	Charpy V-Notch Impact Energy Minimum
A5.29	A5.29M					
E10XT1-K7C, -K7M	E69XT1-K7C, -K7M	AW	100–120 [690–830]	88 [610] min.	16	20 ft·lbf @ −60°F [27 J @ −50°C]
E9XT8-K8	E62XT8-K8	AW	90–110 [620–760]	78 [540] min.	17	20 ft·lbf @ −20°F [27 J @ −30°C]
E10XT1-K9C, -K9M	E69XT1-K9C, -K9M	AW	100–120 [690–830][f]	82–97 [560–670]	18	35 ft·lbf @ −60°F [27 J @ −50°C]
E8XT1-W2C, -W2M	E55XT1-W2C, -W2M	AW	80–100 [550–690]	68 [470] min.	19	20 ft·lbf @ −20°F [27 J @ −30°C]
EXXTX-G,[g] -GC,[g] -GM[g]	EXXTX-G,[g] -GC,[g] -GM[g]	The weld deposit composition, condition of test (AW or PWHT) and Charpy V-Notch impact properties are as agreed upon between the supplier and purchaser. Requirements for the tension test, positionality, slag system and shielding gas, if any, conform to those indicated by the digits used.				
EXXTG-X[g]	EXXTG-X[g]	The slag system, shielding gas, if any, condition of test (AW or PWHT) and Charpy V-Notch impact properties are as agreed upon between the supplier and purchaser. Requirements for the tension test, positionality and weld deposit composition conform to those indicated by the digits used.				
EXXTG-G[g]	EXXTG-G[g]	The slag system, shielding gas, if any, condition of test (AW or PWHT), Charpy V-Notch impact properties and weld deposit composition are as agreed upon between the supplier and purchaser. Requirements for the tension test and positionality conform to those indicated by the digits used.				

[a] The "Xs" in actual classification will be replaced with the appropriate designators.
[b] The placement of a "G" in a designator position indicates that those properties have been agreed upon between the supplier and purchaser.
[c] AW = As Welded. PWHT = Postweld heat treated.
[d] These electrodes are presently classified E502TX-X or E505TX-X in AWS A5.22–95. With the next revision of A5.22 they will be removed and exclusively listed in this specification.
[e] PWHT temperatures in excess of 1150°F [620°C] will decrease the Charpy V-Notch impact properties.
[f] For this classification (E10XT1-K9C, -K9M [E69XT1-K9C, -K9M]) the tensile strength range shown is not a requirement. It is an approximation.
[g] The tensile strength, yield strength, and % elongation requirements for EXXTX-G, -GC, -GM [EXXTX-G, -GC, -GM]; EXXTG-X and EXXTG-G [EXXTG-X and EXXTG-G] electrodes are as shown in this table for other electrode classifications (not including the E10XT1-K9C, -K9M [E69XT1-K9C, -K9M] classifications) having the same tensile strength designator.

AWS D1.1/D1.1M:2020

Annex M (Informative)
AWS A5.36 Filler Metal Classifications and Properties

This annex is not part of this standard but is included for informational purposes only.

Annex M provides information and applicability to D1.1 code users of this new specification. Additionally, filler metals of the previous filler metal specifications and classifications in existing WPSs which have been reclassified with fixed or open classifications of AWS A5.36 are exempt from PQR requalification due to the re-classification of the electrode(s) if the electrode(s) meet all of the previous classification requirements. A comparison list of multiple-pass Annex L electrode classifications for the AWS A5.36 fixed and open classifications are at the end of this annex. With all of the variations of the AWS A5.36 open classification system, the listing was limited to the D1.1:2010 Annex V classifications. Other AWS A5.36 classifications may be used in this code.

AWS filler specification A5.36/A5.36M:2012, S*pecification for Carbon and Low-Alloy Steel Flux Cored Electrodes for Flux Cored Arc Welding and Metal Cored Electrodes for Gas Metal Arc Welding*, has two methods of classifying carbon steel FCAW and GMAW-metal cored electrodes, fixed classification and open classification. AWS A5.36 uses several different tables for the classification of electrodes (6 tables and 1 figure), which are included in this annex. Note that additional footnotes with restrictions may have been added to tables to facilitate compliance with prequalified and qualified WPS requirements for filler metals in the previous filler metal specifications for existing WPSs.

The fixed classification is for carbon steel electrodes only and was to be identical to the requirements of AWS A5.18 GMAW and AWS A5.20 FCAW specifications, however, certain classifications in both the GMAW-metal cored and the FCAW electrodes were not included, E70C–6C, E70C–3C, E70C–3M, and E7XT–11, but can be classified under the open classification system. Also, the optional designator "J" was eliminated and the "D" designator does not support the previous "D" designator of AWS A5.20 and therefore cannot be used interchangeably. While the T–12 FCAW electrodes have the same requirements as AWS A5.20 in the fixed classification, the T–12 in the open classification does not.

The open classification system is used for classification of carbon and low-alloy steel electrodes for both FCAW and GMAW-metal cored. Caution is necessary when comparing previous AWS A5.18, A5.20, A5.28, and A5.29 classifications and requirements directly with either the fixed or open classifications, as there are possible differences in mechanical and chemical requirements. Under the open classification system, there are few defined mandatory requirements for an electrode classification, but any shielding gas or any strength classification may be applied as well as an electrode may be classified either as As-Welded (AW) or Postweld Heat Treated (PWHT) or both. Previously carbon steel electrodes were always classified as AW, but under AWS A5.36 may be classified as PWHT. Likewise, low-alloy steel electrodes classified as PWHT now also may be classified in the AW condition.

Shielding gas requirements are a major change to previous specifications and classification requirements. For the purposes of the D1.1 code for prequalified WPSs, restrictions have been placed on AWS A5.36 classification requirements. The argon mixed shielding gas requirements have been restricted to the previous classification mixed shielding gas requirements of AWS A5.18, A5.20, A5.28, and A5.29 for prequalified WPSs or to other specific shielding gases of AWS A5.36 Table 5 used for classification, not to the entire range of a shielding gas designator (see 5.6.3 of the code).

Under the open classification system, other argon based shielding gases may be used for classification. However, changes to the argon content and minor gas(es) in the blend may affect the mechanical and chemical weld deposit, i.e., a 70 ksi [490 MPa] electrode classification under the previous specifications may now increase possibly to 90 ksi [620 MPa] or decrease to 60 ksi [430 MPa] in mechanical properties.

For those code users welding Demand Critical connections with the Optional "D" designator of AWS A5.20 and A5.29, the A5.36 "D" designator requirements for heat inputs have been changed. AWS A5.36 electrodes classified with optional "D" designator are not interchangeable with the AWS A5.20 or AWS A5.29 electrodes for seismic Demand Critical connections due to this difference in heat input requirements. AWS D1.8/D1.8M, *Structural Welding Code—Seismic Supplement*, will have similar restrictions on interchanging the two different "D" designators.

Table M.1
AWS A5.36/A5.36M Carbon Steel Electrode Classifications with Fixed Requirements[a, b]

Source Specification for Electrode Classification and Requirements	Classification Designation[b, c]	Electrode Type	Shielding Gas[d]	Weld Deposit Requirements	
				Mechanical Properties[e]	Weld Deposit[f]
AWS A5.20/A5.20M	E7XT–1C[g]	Flux Cored	C1	Tensile Strength: 70 ksi–95 ksi Minimum Yield Strength: 58 ksi[i] Min. Charpy Impact: 20 ft·lbf @ 0°F Minimum % Elongation: 22%[j]	CS1
	E7XT–1M[g]		M21[i]		
	E7XT–5C[g]		C1	Tensile Strength: 70 ksi–95 ksi Minimum Yield Strength: 58 ksi[i] Min. Charpy Impact: 20 ft·lbf @ –20°F Minimum % Elongation: 22%[j]	CS1
	E7XT–5M[g]		M21		
	E7XT–6[g]		None		CS3
	E7XT–8[g]				
	E7XT–9C[g]		C1		CS1
	E7XT–9M[g]		M21[i]		
	E7XT–12C[g]		C1	Tensile Strength: 70 ksi–90 ksi Minimum Yield Strength: 58 ksi[i] Min. Charpy Impact: 20 ft·lbf @ –20°F Minimum % Elongation: 22%[j]	CS2
	E7XT–12M[g]		M21[i]		
	E70T–4[g]		None	Tensile Strength: 70 ksi–95 ksi Minimum Yield Strength: 58 ksi[i] Min. Charpy Impact: Not Specified Minimum % Elongation: 22%[j]	CS3
	E7XT–7[g]				
AWS A5.18/A5.18M	E70C–6M[h]	Metal Cored	M21[i]	Tensile Strength: 70 ksi minimum Minimum Yield Strength: 58 ksi[i] Min. Charpy Impact: 20 ft·lbf @ –20°F Minimum % Elongation: 22%[j]	CS1

[a] These multiple pass electrodes are classified according to the fixed classification system utilized in AWS A5.20/A5.20M or AWS A5.18/A5.18M, as applicable, which has been carried over for these specific electrodes as a part of AWS A5.36/A5.36M. The mechanical property and weld deposit requirements are as defined in this table. These same electrodes may also be classified to the same requirements or to different requirements using the open classification system introduced in this specification. In this case, the classification designations are as prescribed in Figure M.1. See Table A.1 or Table A.3, as applicable, in Annex A of AWS A5.36/A5.36M for comparisons of the "fixed classification" designations and equivalent "open classification" designations for the above electrodes when both are classified to the requirements listed in this table.

[b] AWS A5.36 excluded the E7XT-11 multiple-pass electrodes in AWS A5.20 from the above table, but they can be classified under the open classification. Metal Cored classifications, E70C–6C, E70C–3C, E70C–3M, under AWS A5.18/A5.18M were also excluded, but can be classified under the open classification.

[c] The electrodes shown in the shaded panels are self-shielded.

[d] See Table M.5.

[e] Mechanical properties are obtained by testing weld metal from the groove weld shown in Figure 2 of AWS A5.36/A5.36M. Welding and testing shall be done as prescribed in AWS A5.36/A5.36M. The requirements for welding and testing are the same as those given in AWS A5.20/A5.20M. All mechanical property testing for the classifications listed in this table shall be done in the as-welded condition.

[f] See Table M.6.

[g] The "D," "Q," and "H" optional designators, which are not part of the electrode classification designation, may be added to the end of the designation as established in AWS A5.20/A5.20M, i.e., E7XT-XXD, E7XT-XXQ, E7XT-XXHX, E7XT-XXDHX, or E7XT-XXQHX, as applicable. The "J" optional, supplemental designator listed in AWS A5.20/A5.20M is no longer required. The open classification system introduced in this AWS A5.36/A5.36M specification eliminates the need for this designator.

[h] The "H" optional, supplemental designator, which is not part of the electrode classification designation, may be added to the end of the designation as established in AWS A5.18/A5.18M, i.e., E70C–6 MHZ. Provisions for the "J," "D," and "Q" optional, supplemental designators have not been established in AWS A5.18/A5.18M and, as a result, may not be used with the E70C–6M designation. However, that does not preclude their use with metal cored electrodes classified utilizing the open classification system under the AWS A5.36/A5.36M specification.

[i] For purposes of the D1.1 code, the prequalified argon/CO_2 mixed shielding gas shall be limited to SG-AC–20/25 as with the previous AWS A5.18 and AWS A5.20 classification shielding gas requirements, and not the shielding gas range of M21.

Source: Adapted from AWS A5.36/A5.36M:2012, *Specification for Carbon and Low-Alloy Steel Flux Cored Electrodes for Flux Cored Arc Welding and Metal Cored Electrodes for Gas Metal Arc Welding*, Table 1, American Welding Society.

Table M.2
AWS A5.36/A5.36M Tension Test Requirements[a]

Tensile Strength Designator		Single Pass Electrodes	For A5.36 Multiple Pass Electrodes U.S. Customary Units			For A5.36M Multiple Pass Electrodes International System of Units (SI)		
U.S. Customary Units	International System of Units (SI)	Minimum Tensile Strength ksi [MPa]	Tensile Strength (ksi)	Minimum Yield Strength[b] (ksi)	Minimum Percent Elongation[c]	Tensile Strength [MPa]	Minimum Yield Strength[b] [MPa]	Minimum Percent Elongation[c]
6	43	60 [430]	60–80	48	22	430–550	330	22
7	49	70 [490]	70–95	58	22	490–660	400	22
8	55	80 [550]	80–100	68	19	550–690	470	19
9	62	90 [620]	90–110	78	17	620–760	540	17
10	69	100 [690]	100–120	88	16	690–830	610	16
11	76	110 [760]	110–130	98	15[d]	760–900	680	15[d]
12	83	120 [830]	120–140	108	14[d]	830–970	740	14[d]
13	90	130 [900]	130–150	118	14[d]	900–1040	810	14[d]

[a] The fixed E7XT–12X classification tensile strength range is exceeded by open classification of E7XT12-C1A2-CS2/E7XT12-M21A2-CS2, 70 ksi to 90 ksi vs. 70 ksi to 95 ksi.
[b] Yield strength at 0.2% offset.
[c] In 2 in [50 mm] gage length when a 0.500 in [12.5 mm] nominal diameter tensile specimen and nominal gage length to diameter ratio of 4:1 (as specified in the Tension Test section of AWS B4.0 [AWS B4.0M]) is used. In 1 in [25 mm] gage length when a 0.250 in [6.5 mm] nominal tensile specimen is used as permitted for 0.045 in [1.2 mm] and smaller sizes of the E7XT11-AZ-CS3 [E49XT11-AZ-CS3].
[d] Elongation requirement may be reduced by one percentage point if the tensile strength of the weld metal is in the upper 25% of the tensile strength range.

Source: Adapted from AWS A5.36/A5.36M:2012, *Specification for Carbon and Low-Alloy Steel Flux Cored Electrodes for Flux Cored Arc Welding and Metal Cored Electrodes for Gas Metal Arc Welding*, Table 2, American Welding Society.

Table M.3
AWS A5.36/A5.36M Charpy Impact Test Requirements

A5.36 Requirements U.S. Customary Units			A5.36M Requirements International System of Units (SI)		
Impact Designator[a, b]	Maximum Test Temperature[c, d] (°F)	Minimum Average Energy Level	Impact Designator[a, b]	Maximum Test Temperature[c, d] (°C)	Minimum Average Energy Level
Y	+68		Y	+20	
0	0		0	0	
2	–20		2	–20	
4	–40		3	–30	
5	–50	20 ft·lbf	4	–40	27 J
6	–60		5	–50	
8	–80		6	–60	
10	–100		7	–70	
15	–150		10	–100	
Z	No Impact Requirements		Z	No Impact Requirements	
G	As agreed between supplier and purchaser				

[a] Based on the results of the impact tests of the weld metal, the manufacturer shall insert in the classification the appropriate designator from Table M.3 above, as indicated in Figure M.1.
[b] When classifying an electrode to AWS A5.36 using U.S. Customary Units, the Impact Designator indicates the maximum impact test temperature in degrees Fahrenheit. When classifying to AWS A5.36M using the International System of Units (SI), the Impact Designator indicates the maximum impact test temperature in degrees Celsius. With the exception of the Impact Designator "4," a given Impact Designator will indicate different temperatures depending upon whether classification is according to AWS A5.36 in U.S. Customary Units or according to AWS A5.36M in the International System of Units (SI). For example, a "2" Impact Designator when classifying to AWS A5.36M indicates a test temperature of –20°F, which is –29°C.
[c] Weld metal from an electrode that meets the impact requirements at a given temperature also meets the requirements at all higher temperatures in this table. For example, weld metal meeting the AWS A5.36 requirements for designator "5" also meets the requirements for designators 4, 2, 0, and Y. [Weld metal meeting the A5.36M requirements for designator "5" also meets the requirements for designators 4, 3, 2, 0, and Y.]
[d] Filler metal classification testing to demonstrate conformance to a specified minimum acceptable level for impact testing, i.e., minimum energy at specified temperature, can be met by testing and meeting the minimum energy requirement at any lower temperature. In these cases, the actual temperature used for testing shall be listed on the certification documentation when issued.

Source: Adapted from AWS A5.36/A5.36M:2012, *Specification for Carbon and Low-Alloy Steel Flux Cored Electrodes for Flux Cored Arc Welding and Metal Cored Electrodes for Gas Metal Arc Welding*, Table 3, American Welding Society.

Table M.4
Electrode Usability Characteristics

Electrode Usability Designator[a]	Process	General Description of Electrode Type[b,c]	Typical Positions of Welding[d,e]	Polarity[f]
T1	FCAW-G	Flux cored electrodes of this type are gas shielded and have a rutile base slag. They are characterized by a spray transfer, low spatter loss, and a moderate volume of slag which completely covers the weld bead.	H, F, VU, & OH	DCEP
T1S	FCAW-G	Flux cored electrodes of this type are similar to the "T1" type electrodes but with higher manganese or silicon, or both. They are designed primarily for single pass welding in the flat and horizontal positions. The higher levels of deoxidizers in this electrode type allow single pass welding of heavily oxidized or rimmed steel.	H, F, VU, & OH	DCEP
T3S	FCAW-S	Flux cored electrodes of this type are self shielded and are intended for single pass welding and are characterized by a spray type transfer. The titanium-based slag system is designed to make very high welding speeds possible.	H, F	DCEP
T4	FCAW-S	Flux cored electrodes of this type are self shielded and are characterized by a globular type transfer. Its fluoride-based basic slag system is designed to make very high deposition rates possible and to produce very low sulfur welds for improved resistance to hot cracking.	H, F	DCEP
T5	FCAW-G	Flux cored electrodes of this type are gas shielded and are characterized by a globular transfer, slightly convex bead contour, and a thin slag that may not completely cover the weld bead. They have a lime-fluoride slag system and develop improved impact properties and better cold cracking resistance than typically exhibited by the "T1" type electrodes.	H, F, VU, & OH	DCEP or DCEN[g]
T6	FCAW-S	Flux cored electrodes of this type are self shielded and are characterized by a spray transfer. Its oxide-based slag system is designed to produce good low temperature impacts, good penetration into the root of the weld, and excellent slag removal.	H & F	DCEP
T7	FCAW-S	Flux cored electrodes of this type are self shielded and are characterized by a small droplet to spray type transfer. The fluoride-based slag system is designed to provide high deposition rates in the downhand positions with the larger diameters and out of position capabilities with the smaller diameters.	H, F, VU, & OH	DCEN
T8	FCAW-S	Flux cored electrodes of this type are self shielded and are characterized by a small droplet to spray type transfer. The fluoride-based slag system is designed to provide improved out-of-position control. The weld metal produced typically exhibits very good low temperature notch toughness and crack resistance.	H, F, VD, VU, & OH	DCEN
T9	FCAW-G	Flux cored electrodes of this type are similar in design and application to the T1 types but with improved weld metal notch toughness capabilities.	H, F, VU, & OH	DCEP

(Continued)

Table M.4 (Continued)
Electrode Usability Characteristics

Electrode Usability Designator[a]	Process	General Description of Electrode Type[b,c]	Typical Positions of Welding[d,e]	Polarity[f]
T10S	FCAW-S	Flux cored electrodes of this type are self shielded and are characterized by a small droplet transfer. The fluoride-based slag system is designed to make single pass welds at high travel speeds on steel of any thickness.	H, F	DCEN
T11	FCAW-S	Flux cored electrodes of this type are self shielded and are characterized by a smooth spray type transfer, limited slag coverage, and are generally not recommended for the welding of materials over 3/4 in [20 mm] thick.	H, F, VD, & OH	DCEN
T12	FCAW-G	Flux cored electrodes of this type are similar in design and application to the T1 types. However, they have been modified for improved impact toughness and to meet the lower manganese requirements of the A-No. 1 Analysis Group in the ASME *Boiler and Pressure Vessel Code*, Section IX.	H, F, VU, & OH	DCEP
T14S	FCAW-S	Flux cored electrodes of this type are self shielded and are characterized by a smooth spray-type transfer. The slag system is designed for single pass welds in all positions and at high travel speeds.	H, F, VD, & OH	DCEN
T15	GMAW-C	Electrodes of this type are gas shielded composite stranded or metal cored electrodes. The core ingredients are primarily metallic. The nonmetallic components in the core typically total less than 1% of the total electrode weight. These electrodes are characterized by a spray arc and excellent bead wash capabilities. Applications are similar in many ways to solid GMAW electrodes.	H, F, OH, VD, & VU	DCEP or DCEN
T16	GMAW-C	This electrode type is a gas shielded metal cored electrode specifically designed for use with AC power sources with or without modified waveforms.	H, F, VD, VU, & OH	AC[h]
T17	FCAW-S	This flux cored electrode type is a self-shielded electrode specifically designed for use with AC power sources with or without modified waveforms.	H, F, VD, VU, & OH	AC[h]
G		As agreed between supplier and purchaser	Not specified	Not specified

[a] An "S" is added to the end of the Usability Designator when the electrode being classified is recommended for single pass applications only.
[b] For more information refer to A7, Description and Intended Use, in AWS A5.36/A5.36M Annex A.
[c] Properties of weld metal from electrodes that are used with external shielding gas will vary according to the shielding gas used. Electrodes classified with a specific shielding gas should not be used with other shielding gases without first consulting the manufacturer of the electrode.
[d] H = horizontal position, F = flat position, OH = overhead position, VU = vertical position with upward progression, VD = vertical position with downward progression.
[e] Electrode sizes suitable for out-of-position welding, i.e., welding positions other that flat and horizontal, are usually those sizes that are smaller than the 3/32 in [2.4 mm] size or the nearest size called for in Clause 10 for the groove weld. For that reason, electrodes meeting the requirements for the groove weld tests may be classified as EX1TX-XXX-X (where X represents the tensile strength, usability, shielding gas, if any, condition of heat treatment, impact test temperature, and weld metal composition designators) regardless of their size.
[f] The term "DCEP" refers to direct current electrode positive (dc, reverse polarity). The term "DCEN" refers to direct current electrode negative (dc, straight polarity).
[g] Some EX1T5-XXX-X electrodes may be recommended for use on DCEN for improved out-of-position welding. Consult the manufacturer for the recommended polarity.
[h] For this electrode type the welding current can be conventional sinusoidal alternating current, a modified AC waveform alternating between positive and negative, an alternating DCEP waveform, or an alternating DCEN waveform.

Source: Adapted from AWS A5.36/A5.36M:2012, *Specification for Carbon and Low-Alloy Steel Flux Cored Electrodes for Flux Cored Arc Welding and Metal Cored Electrodes for Gas Metal Arc Welding*, Table 4, American Welding Society.

Table M.5
AWS A5.36/A5.36M Composition Requirements for Shielding Gases

AWS A5.36/ A5.36M Shielding Gas Designator[a]	AWS A5.32M/A5.32 Composition Ranges for Indicated Main/Sub Group[b]		Nominal Composition of Shielding Gases to be Used for Classification of Gas Shielded Electrodes to AWS A5.36/A5.36M[c, d]		
	Oxidizing Components[e]			Oxidizing Components[c, g]	
	% CO_2	% O_2	ISO 14175 Designation[f]	% CO_2	% O_2
C1	100	—	C1	100	—
M12	$0.5 \leq CO_2 \leq 5$	—	M12 – ArC – 3	3	—
M13	—	$0.5 \leq O_2 \leq 3$	M13 – ArO – 2	—	2
M14	$0.5 \leq CO_2 \leq 5$	$0.5 \leq O_2 \leq 3$	M14 – ArCO – 3/2	3	2
M20	$5 < CO_2 \leq 15$	—	M20 – ArC – 10	10	—
M21	$15 < CO_2 \leq 25$	—	M21 – ArC – 20	20	—
M22	—	$3 < O_2 \leq 10$	M22 – ArO – 7	—	7
M23	$0.5 \leq CO_2 \leq 5$	$3 < O_2 \leq 10$	M23 – ArOC – 7/3	3	7
M24	$5 < CO_2 \leq 15$	$0.5 \leq O_2 \leq 3$	M24 – ArCO – 10/2	10	2
M25	$5 < CO_2 \leq 15$	$3 < O_2 \leq 10$	M25 – ArCO – 10/7	10	7
M26	$15 < CO_2 \leq 25$	$0.5 \leq O_2 \leq 3$	M26 – ArCO – 20/2	20	2
M27	$15 < CO_2 \leq 25$	$3 < O_2 \leq 10$	M27 – ArCO – 20/7	20	7
M31	$25 < CO_2 \leq 50$	—	M31 – ArC – 38	38	—
M32	—	$10 < O_2 \leq 15$	M32 – ArO – 12.5	—	12.5
M33	$25 < CO_2 \leq 50$	$2 < O_2 \leq 10$	M33 – ArCO – 38/6	38	6
M34	$5 < CO_2 \leq 25$	$10 < O_2 \leq 15$	M34 – ArCO – 15/12.5	15	12.5
M35	$25 < CO_2 \leq 50$	$10 < O_2 \leq 15$	M35 – ArCO – 38/12.5	38	12.5
Z	The designator "Z" indicates that the shielding gas used for electrode classification is not one of the shielding gases specified in this table but is a different composition as agreed upon between the supplier and purchaser.				

[a] The Shielding Gas Designators are identical to the Main group/Sub-group designators used in AWS A5.32M/A5.32:2011 [ISO 14175:2008 MOD], *Welding Consumables-Gases and Gas Mixtures for Fusion Welding and Allied Processes*, for these same shielding gases.

[b] Under AWS A5.32M/A5.32:2011, the inert gas used for the balance of the gas mixture may be either argon, helium, or some mixture thereof.

[c] For purposes of the D1.1 code for fixed classifications, the prequalified argon/CO2 mixed shielding gas for carbon steel FCAW and GMAW-metal cored electrodes shall be limited to SG-AC–20/25 as with the previous AWS A5.18 and AWS A5.20 classification shielding gas requirements, and not the shielding gas range of M21.

[d] For open classifications of carbon and low-alloy steel FCAW and GMAW-metal cored electrodes, the electrodes classified with the previous shielding gas requirements of AWS A5.18, A5.20, A5.28, and A5.29 shall be considered as prequalified. Other AWS A5.36, Table 5 shielding gas classifications used for classifying an electrode are limited to the specific shielding gas in the shielding gas classification range.

[e] The mixture tolerances are as follows:
For a component gas with a nominal concentration of >5%, ±10% of nominal. For a component gas with a nominal concentration of 1%–5%, ±0.5% absolute.
For a component gas with a nominal composition of <1%, not specified in this standard.

[f] AWS A5.32M/A5.32:2011 shielding gas designators begin with "AWS A5.32 (ISO 14175)." That part of the designation has been omitted from the Shielding Gas Designator for brevity.

[g] The inert gas to be used for the balance of the gas mixtures specified for the classification of gas shielded flux cored and metal cored electrodes shall be argon.

Source: Adapted from AWS A5.36/A5.36M:2012, *Specification for Carbon and Low-Alloy Steel Flux Cored Electrodes for Flux Cored Arc Welding and Metal Cored Electrodes for Gas Metal Arc Welding*, Table 5, American Welding Society.

ANNEX M AWS D1.1/D1.1M:2020

Table M.6
Weld Metal Chemical Composition Requirements[a]

Weld Metal Designation	UNS Number[b]	C	Mn	Si	S	P	Ni	Cr	Mo	V	Al	Cu	Other[d]
Weight Percent[c]													
Carbon Steel Electrodes												
CS1[e]	—	0.12	1.75	0.90	0.030	0.030	0.50[f]	0.20[f]	0.30[f]	0.08[f]	0.50[f]	0.35[f]	—
CS2[e,g]	—	0.12	1.60	0.90	0.030	0.030	0.50[f]	0.20[f]	0.30[f]	0.08[f]	0.50[f]	0.35[f]	—
CS3[e]	—	0.30	1.75	0.60	0.030	0.030	0.50[f]	0.20[f]	0.30[f]	0.08[f]	1.8[f,h]	0.35[f]	—
Molybdenum Steel Electrodes												
A1	W1703X	0.12	1.25	0.80	0.030	0.030	—	—	0.40–0.65	—	—	—	—
Chromium-Molybdenum Steel Electrodes												
B1	W5103X	0.05–0.12	1.25	0.80	0.030	0.030	—	0.40–0.65	0.40–0.65	—	—	—	—
B1L	W5113X	0.05	1.25	0.80	0.030	0.030	—	0.40–0.65	0.40–0.65	—	—	—	—
B2	W5203X	0.05–0.12	1.25	0.80	0.030	0.030	—	1.00–1.50	0.40–0.65	—	—	—	—
B2L	W5213X	0.05	1.25	0.80	0.030	0.030	—	1.00–1.50	0.40–0.65	—	—	—	—
B2H	W5223X	0.10–0.15	1.25	0.80	0.030	0.030	—	1.00–1.50	0.40–0.65	—	—	—	—
B3	W5303X	0.05–0.12	1.25	0.80	0.030	0.030	—	2.00–2.50	0.90–1.20	—	—	—	—
B3L	W5313X	0.05	1.25	0.80	0.030	0.030	—	2.00–2.50	0.90–1.20	—	—	—	—
B3H	W5323X	0.10–0.15	1.25	0.80	0.030	0.030	—	2.00–2.50	0.90–1.20	—	—	—	—
B6	W50231	0.05–0.12	1.20	1.00	0.030	0.025	0.40	4.0–6.0	0.45–0.65	—	—	0.35	—
B6L	W50230	0.05	1.20	1.00	0.030	0.025	0.40	4.0–6.0	0.45–0.65	—	—	0.35	—
B8	W50431	0.05–0.12	1.20	1.00	0.030	0.040	0.40	8.0–10.5	0.85–1.20	—	—	0.50	—
B8L	W50430	0.05	1.20	1.00	0.030	0.040	0.40	8.0–10.5	0.85–1.20	—	—	0.50	—
B91[i]	W50531	0.08–0.13	1.20[j]	0.50	0.015	0.020	0.80[j]	8.0–10.5	0.85–1.20	0.15–0.30	0.04	0.25	Nb: 0.02–0.10 N: 0.02–0.07
B92	—	0.08–0.15	1.20[j]	0.50	0.015	0.020	0.80[j]	8.0–10.5	0.30–0.70	0.15–0.30	0.04	0.25	Nb: 0.02–0.10 W: 1.5–2.0 B: 0.006 N: 0.02–0.08 Co[k]
Nickel Steel Electrodes												
Ni1	W2103X	0.12	1.75	0.80	0.030	0.030	0.80–1.10	0.15	0.35	0.05	1.8[h]	—	—
Ni2	W2203X	0.12	1.50	0.80	0.030	0.030	1.75–2.75	—	—	—	1.8[h]	—	—
Ni3	W2303X	0.12	1.50	0.80	0.030	0.030	2.75–3.75	—	—	—	1.8[h]	—	—

(Continued)

446

Table M.6 (Continued)
Weld Metal Chemical Composition Requirements[a]

Weight Percent[c]

Weld Metal Designation	UNS Number[b]	C	Mn	Si	S	P	Ni	Cr	Mo	V	Al	Cu	Other[d]
Manganese-Molybdenum Steel Electrodes													
D1	W1913X	0.12	1.25–2.00	0.80	0.030	0.030	—	—	0.25–0.55	—	—	—	—
D2	W1923X	0.15	1.65–2.25	0.80	0.030	0.030	—	—	0.25–0.55	—	—	—	—
D3	W1933X	0.12	1.00–1.75	0.80	0.030	0.030	—	—	0.40–0.65	—	—	—	—
Other Low-Alloy Steel Electrodes													
K1	W2113X	0.15	0.80–1.40	0.80	0.030	0.030	0.80–1.10	0.15	0.20–0.65	0.05	—	—	—
K2	W2123X	0.15	0.50–1.75	0.80	0.030	0.030	1.00–2.00	0.15	0.35	0.05	1.8[h]	—	—
K3	W2133X	0.15	0.75–2.25	0.80	0.030	0.030	1.25–2.60	0.15	0.25–0.65	0.05	—	—	—
K4	W2223X	0.15	1.20–2.25	0.80	0.030	0.030	1.75–2.60	0.20–0.60	0.20–0.65	0.03	—	—	—
K5	W2162X	0.10–0.25	0.60–1.60	0.80	0.030	0.030	0.75–2.00	0.20–0.70	0.15–0.55	0.05	—	—	—
K6	W2104X	0.15	0.50–1.50	0.80	0.030	0.030	0.40–1.00	0.20	0.15	0.05	1.8[h]	—	—
K7	W2205X	0.15	1.00–1.75	0.80	0.030	0.030	2.00–2.75	—	—	—	—	—	—
K8	W2143X	0.15	1.00–2.00	0.40	0.030	0.030	0.50–1.50	0.20	0.20	0.05	1.8[h]	—	—
K9	W23230	0.07	0.50–1.50	0.60	0.015	0.015	1.30–3.75	0.20	0.50	0.05	—	0.06	—
K10	—	0.12	1.25–2.25	0.80	0.030	0.030	1.75–2.75	0.20	0.50	—	—	0.50	—
K11	—	0.15	1.00–2.00	0.80	0.030	0.030	0.40–1.00	0.20	0.50	0.05	1.8[h]	—	—
W2	W2013X	0.12	0.50–1.30	0.35–0.80	0.030	0.030	0.40–0.80	0.45–0.70	—	—	—	0.30–0.75	—
G	—						As agreed upon between supplier and purchaser						
GS[n]	—	(m)					As agreed upon between supplier and purchaser						

[a] The weld metal shall be analyzed for the specific elements for which values are shown in this table.
[b] Refer to ASTM DS-56/SAE HS-1086, *Metals & Alloys in the Unified Numbering System.* An "X," when present in the last position, represents the usability designator for the electrode type used to deposit the weld metal. An exception to this applies to the "11" electrode type where a "9" is used instead of an "11."
[c] Single values are maximums.
[d] An analysis of the weld deposit for boron is required and shall be reported if this element is intentionally added or if it is known to be present at levels in excess of 0.0010%.
[e] The total of all the elements listed in this table for this classification shall not exceed 5%.
[f] The analysis of these elements shall be reported only if intentionally added.
[g] Meets the lower Mn requirements of the A-No. 1 Analysis Group in the ASME *Boiler and Pressure Vessel Code*, Section IX Welding and Brazing Qualifications, QW-422.
[h] Applicable to self-shielded electrodes only. Electrodes intended for use with gas shielding normally do not have significant additions of aluminum.
[i] The "B91" designation is a new designation, replacing the "B9" designation previously used for this alloy type.
[j] Mn + Ni = 1.40% maximum. See A7.16.2 in Annex A of AWS A5.36/A5.36M.
[k] Analysis for Co is required to be reported if intentionally added, or if it is known to be present at levels greater than 0.20%.
[l] The limit for gas shielded electrodes is 0.18% maximum. The limit for self-shielded electrodes is 0.30% maximum.
[m] The composition of weld metal is not particularly meaningful since electrodes in this category are intended only for single pass welds. Dilution from the base metal in such welds is usually quite high. See A7.2 in Annex A of AWS A5.36/A5.36M.

Source: Reproduced from AWS A5.36/A5.36M:2012, *Specification for Carbon and Low-Alloy Steel Flux Cored Electrodes for Flux Cored Arc Welding and Metal Cored Electrodes for Gas Metal Arc Welding*, Table 6, American Welding Society.

Table M.7
AWS A5.20/A5.20M Procedure Requirements for "D" Optional Supplemental Designator

Optional Supplemental Designator	Procedure Heat Input (Fast or Slow Cooling Rate)	Preheat Temperature °F [°C]	Interpass Temperature °F [°C]	Heat Input Requirement for Any Single Pass[a]	Required Average Heat Input for All Passes[a]
D	low (fast cooling rate)	70°F ± 25°F [20°C ± 15°C]	200°F ± 25°F [90°C ± 15°C]	For electrode diameters < 3/32 in [2.4 mm]	
				33 kJ/in [1.3 kJ/mm] maximum	30 +2, –5 kJ/in [1.2 +0.1, –0.2 kJ/mm]
				For electrode diameters ≥ 3/32 in [2.4 mm]	
				44 kJ/in [1.7 kJ/mm] maximum	40 +2, –5 kJ/in [1.6 +0.1, –0.2 kJ/mm]
	high (slow cooling rate)	300°F ± 25°F [150°C ± 15°C]	500°F ± 50°F [260°C ± 25°C]	75 kJ/in [3.0 kJ/mm] minimum	80 +5, –2 kJ/in [3.1 +0.2, –0.1 kJ/mm]

[a] Does not apply to first layer. The first layer may have one or two passes.

Source: Adapted from AWS A5.20/A5.20M:2005, *Specification for Carbon Steel Electrodes for Flux Cored Arc Welding*, Table 9, American Welding Society.

Table M.8
AWS A5.36/A5.36M Procedure Requirements for "D" Optional Supplemental Designator

Optional Supplemental Designator	Procedure Heat Input (fast or slow cooling rate)	Preheat Temperature °F [°C]	Interpass Temperature °F [°C]	Heat Input Requirement for Any Single Pass[a]	Required Average Heat Input for All Passes[a]
D	Low (fast cooling rate)	**120°F [50°C] max.**	**250°F [120°C] max.**	For electrode diameters < 3/32 in [2.4 mm]	
				38 kJ/in [1.5 kJ/mm] max.	**24 kJ/in–36 kJ/in [0.9 kJ/mm–1.4 kJ/mm]**
				For electrode diameters ≥ 3/32 in [2.4 mm]	
				44 kJ/in [1.7 kJ/mm] max.	35 kJ/in–42 kJ/in [1.4 kJ/mm–1.6 kJ/mm]
	High (slow cooling rate)	**250°F [120°C] min.**	**450°F [240°C] min.**	**65 kJ/in [2.6 kJ/mm] min.**	**65 kJ/in–85 kJ/in [2.6 kJ/mm–3.3 kJ/mm]**

[a] Does not apply to first layer. The first layer may have one or two passes.

Note: The changes in "D" designator requirements for preheat and interpass temperatures, and heat inputs are **bolded** and underlined.

Source: Adapted from AWS A5.36/A5.36M:2012, *Specification for Carbon and Low-Alloy Steel Flux Cored Electrodes for Flux Cored Arc Welding and Metal Cored Electrodes for Gas Metal Arc Welding*, Table 10, American Welding Society.

Table M.9
Comparison of Classifications of AWS A5.18, A5.20, A5.28, and A5.29 Specifications to AWS A5.36 Fixed and Open Classifications for Multiple-Pass FCAW and GMAW—Metal Cored Electrodes

Annex L		A5.36		
Specification	Classification	Fixed Classification	Open Classification	Annex V Shielding Gas Classification Requirements
A5.18	E70C–3C	Not Applicable	E7XT15-C1A0-CS1	SG-C
	E70C–3M	Not Applicable	E7XT15-M21A0-CS1	SG-AC–20–25
	E70C–6C	Not Applicable	E7XT15-C1A2-CS1	SG-C
	E70C–6M	E70C–6M	E7XT15-M21A2-CS1	SG-AC–20–25
	E70C-G	Not Applicable	E7XT15-XAX-G	As Agreed
A5.20	E7XT–1C	E7XT–1C	E7XT1-C1A0-CS1	SG-C
	E7XT–1M	E7XT–1M	E7XT1-M21A0-CS1	SG-AC–20–25
	E7XT–4	E70T–4	E7XT4-AZ-CS3	
	E7XT–5C	E7XT–5C	E7XT5-C1A2-CS1	SG-C
	E7XT–5M	E7XT–5M	E7XT5-M21A2-CS1	SG-AC–20–25
	E7XT–6	E7XT–6	E7XT6-A2-CS3	
	E7XT–7	E7XT–7	E7XT7-AZ-CS3	
	E7XT–8	E7XT–8	E7XT8-A2-CS3	
	E7XT–9C	E7XT–9C	E7XT9-C1A2-CS1	SG-C
	E7XT–9M	E7XT–9M	E7XT9-M21A2-CS1	SG-AC–20–25
	E7XT–11	Not Applicable	E7XT11-AZ-CS3	
	E7XT–12C	E7XT–12C	Not Applicable	SG-C
	E7XT–12M	E7XT–12M	Not Applicable	SG-AC–20–25
	E6XT-G	Not Applicable	E6XTG-XXX-G	As Agreed
	E7XT-G	Not Applicable	E7XTG-XXX-G	As Agreed
A5.28	E70C-B2L	Not Applicable	E7XT15-M13PZ-B2L E7XT15-M22PZ-B2L	SG-AO–1–5
	E80C-B2	Not Applicable	E8XT15-M13PZ-B2 E8XT15-M22PZ-B2	SG-AO–1–5
	E80C-B3L	Not Applicable	E8XT15-M13PZ-B3L E8XT15-M22PZ-B3L	SG-AO–1–5
	E90C-B3	Not Applicable	E9XT15-M13PZ-B3 E9XT15-M22PZ-B3	SG-AO–1–5
	E80C-B6	Not Applicable	E8XT15-M13PZ-B6 E8XT15-M22PZ-B6	SG-AO–1–5
	E80C-B8	Not Applicable	E8XT15-M13PZ-B8 E8XT15-M22PZ-B8	SG-AO–1–5
	E90C-B9	Not Applicable	E9XT15-M20PZ-B9 E9XT15-M21PZ-B9	SG-AC–5–25
	E70C-Ni2	Not Applicable	E7XT15-M13P8-Ni2 E7XT15-M22P8-Ni2	SG-AO–1–5

(Continued)

Table M.9 (Continued)
Comparison of Classifications of AWS A5.18, A5.20, A5.28, and A5.29 Specifications to AWS A5.36 Fixed and Open Classifications for Multiple-Pass FCAW and GMAW—Metal Cored Electrodes

Annex L		A5.36		
Specification	Classification	Fixed Classification	Open Classification	Annex V Shielding Gas Classification Requirements
A5.28 (Cont'd)	E80C-Ni1	Not Applicable	E8XT15-M13A5-Ni1 E8XT15-M22A5-Ni1	SG-AO–1–5
	E80C-Ni2	Not Applicable	E8XT15-M13P8-Ni2 E8XT15-M22P8-Ni2	SG-AO–1–5
	E80C-Ni3	Not Applicable	E8XT15-M13P10-Ni3 E8XT15-M22P10-Ni3	SG-AO–1–5
	E90C-D2	Not Applicable	E9XT15-M13A2-D2 E9XT15-M22A2-D2	SG-AO–1–5
	E90C-K3	Not Applicable	E9XT15-M12A6-K3 E9XT15-M20A6-K3 E9XT15-M21A6-K3	SG-AC–5–25
	E100C-K3	Not Applicable	E10XT15-M12A6-K3 E10XT15-M20A6-K3 E10XT15-M21A6-K3	SG-AC–5–25
	E110C-K3	Not Applicable	E11XT15-M12A6-K3 E11XT15-M20A6-K3 E11XT15-M21A6-K3	SG-AC–5–25
	E110C-K4	Not Applicable	E11XT15-M12A6-K4 E11XT15-M20A6-K4 E11XT15-M21A6-K4	SG-AC–5–25
	E120C-K4	Not Applicable	E12XT15-M12A6-K4 E12XT15-M20A6-K4 E12XT15-M21A6-K4	SG-AC–5–25
	E80C-W2	Not Applicable	E8XT15-M12A2-W2 E8XT15-M20A2-W2 E8XT15-M21A2-W2	SG-AC–5–25
	E70C-G	Not Applicable	E7XTG-XXX-X	As Agreed
	E80C-G	Not Applicable	E8XTG-XXX-X	As Agreed
	E90C-G	Not Applicable	E9XTG-XXX-X	As Agreed
	E100C-G	Not Applicable	E10XTG-XXX-X	As Agreed
	E110C-G	Not Applicable	E11XTG-XXX-X	As Agreed
	E120C-G	Not Applicable	E12XTG-XXX-X	As Agreed
A5.29	E7XT5-A1C, -A1M	Not Applicable	E7XT5-C1P2-A1 E7XT5-M21P2-A1	SG-C SG-AC–20–25
	E8XT1-A1C, -A1M	Not Applicable	E8XT1-C1PZ-A1 E8XT1-M21PZ-A1	SG-C SG-AC–20–25
	E8XT1-B1C, -B1M, -B1LC, -B1LM	Not Applicable	E8XT1-C1PZ-B1 E8XT1-M21PZ-B1 E8XT1-C1PZ-B1L E8XT1-M21PZ-B1L	SG-C SG-AC–20–25 SG-C SG-AC–20–25

(Continued)

Table M.9 (Continued)
Comparison of Classifications of AWS A5.18, A5.20, A5.28, and A5.29 Specifications to AWS A5.36 Fixed and Open Classifications for Multiple-Pass FCAW and GMAW—Metal Cored Electrodes

Annex L		A5.36		
Specification	Classification	Fixed Classification	Open Classification	Annex V Shielding Gas Classification Requirements
A5.29 (Cont'd)	E8XT1-B2C, -B2M, -B2HC, -B2HM, -B2LC, -B2LM	Not Applicable	E8XT1-C1PZ-B2 E8XT1-M21PZ-B2 E8XT1-C1PZ-B2H E8XT1-M21PZ-B2H E8XT1-C1PZ-B2L E8XT1-M21PZ-B2L	SG-C SG-AC–20–25 SG-C SG-AC–20–25 SG-C SG-AC–20–25
	E8XT5-B2C, -B2M, -B2LC, -B2LM	Not Applicable	E8XT5-C1PZ-B2 E8XT5-M21PZ-B2 E8XT5-C1PZ-B2L E8XT5-M21PZ-B2L	SG-C SG-AC–20–25 SG-C SG-AC–20–25
	E9XT1-B3C, -B3M, -B3LC, -B3LM, -B3HC, -B3HM	Not Applicable	E9XT1-C1PZ-B3 E9XT1-M21PZ-B3 E9XT1-C1PZ-B3L E9XT1-M21PZ-B3L E9XT1-C1PZ-B3H E9XT1-M21PZ-B3H	SG-C SG-AC–20–25 SG-C SG-AC–20–25 SG-C SG-AC–20–25
	E9XT5-B3C, -B3M	Not Applicable	E9XT5-C1PZ-B3 E9XT5-M21PZ-B3	SG-C SG-AC–20–25
	E10XT1-B3C, -B3M	Not Applicable	E10XT1-C1PZ-B3 E10XT1-M21PZ-B3	SG-C SG-AC–20–25
	E8XT1-B6C, -B6M, -B6LC, -B6LM	Not Applicable	E8XT1-C1PZ-B6 E8XT1-M21PZ-B6 E8XT1-C1PZ-B6L E8XT1-M21PZ-B6L	SG-C SG-AC–20–25 SG-C SG-AC–20–25
	E8XT5-B6C, -B6M, -B6LC, -B6LM	Not Applicable	E8XT5-C1PZ-B6 E8XT5-M21PZ-B6 E8XT5-C1PZ-B6L E8XT5-M21PZ-B6L	SG-C SG-AC–20–25 SG-C SG-AC–20–25
	E8XT1-B8C, -B8M, -B8LC, -B8LM	Not Applicable	E8XT1-C1PZ-B8 E8XT1-M21PZ-B8 E8XT1-C1PZ-B8L E8XT1-M21PZ-B8L	SG-C SG-AC–20–25 SG-C SG-AC–20–25
	E8XT5-B8C, -B8M, -B8LC, -B8LM	Not Applicable	E8XT5-C1PZ-B8 E8XT5-M21PZ-B8 E8XT5-C1PZ-B8L E8XT5-M21PZ-B8L	SG-C SG-AC–20–25 SG-C SG-AC–20–25
	E9XT1-B9C, -B9M	Not Applicable	E9XT1-C1PZ-B9 E9XT1-M21PZ-B9	SG-C SG-AC–20–25
	E6XT1-Ni1C, -Ni1M	Not Applicable	E6XT1-C1A2-Ni1 E6XT1-M21A2-Ni1	SG-C SG-AC–20–25
	E7XT6-Ni1	Not Applicable	E7XT6-A2-Ni1	
	E7XT8-Ni1	Not Applicable	E7XT8-A2-Ni1	

(Continued)

Table M.9 (Continued)
Comparison of Classifications of AWS A5.18, A5.20, A5.28, and A5.29 Specifications to AWS A5.36 Fixed and Open Classifications for Multiple-Pass FCAW and GMAW—Metal Cored Electrodes

Annex L		A5.36		
Specification	Classification	Fixed Classification	Open Classification	Annex V Shielding Gas Classification Requirements
A5.29 (Cont'd)	E8XT1-Ni1C, -Ni1M	Not Applicable	E8XT1-C1A2-Ni1 E8XT1-M21A2-Ni1	SG-C SG-AC–20–25
	E8XT5-Ni1C, -Ni1M	Not Applicable	E8XT5-C1P6-Ni1 E8XT5-M21P6-Ni1	SG-C SG-AC–20–25
	E7XT8-Ni2	Not Applicable	E7XT8-A2-Ni2	
	E8XT8-Ni2	Not Applicable	E8XT8-A2-Ni2	
	E8XT1-Ni2C, -Ni2M	Not Applicable	E8XT1-C1A4-Ni2 E8XT1-M21A4-Ni2	SG-C SG-AC–20–25
	E8XT5-Ni2C, -Ni2M	Not Applicable	E8XT5-C1P8-Ni2[a] E8XT5-M21P8-Ni2[a]	SG-C SG-AC–20–25
	E9XT1-Ni2C, -Ni2M	Not Applicable	E9XT1-C1A4-Ni2 E9XT1-M21A4-Ni2	SG-C SG-AC–20–25
	E8XT5-Ni3C, -Ni3M	Not Applicable	E8XT5-C1P10-Ni3 E8XT5-M21P10-Ni3	SG-C SG-AC–20–25
	E9XT5-Ni3C, -Ni3M	Not Applicable	E9XT5-C1P10-Ni3 E9XT5-M21P10-Ni3	SG-C SG-AC–20–25
	E8XT11-Ni3	Not Applicable	E8XT11-A0-Ni3	
	4E9XT1-D1C, -D1M	Not Applicable	E9XT1-C1A4-D1 E9XT1-M21A4-D1	SG-C SG-AC–20–25
	E9XT5-D2C, -D2M	Not Applicable	E9XT5-C1P6-D2 E9XT5-M21P6-D2	SG-C SG-AC–20–25
	E10XT5-D2C, -D2M	Not Applicable	E10XT5-C1P4-D2 E10XT5-M21P4-D2	SG-C SG-AC–20–25
	E9XT1-D3C, -D3M	Not Applicable	E9XT1-C1A2-D3 E9XT1-M21A2-D3	SG-C SG-AC–20–25
	E8XT5-K1C, -K1M	Not Applicable	E8XT5-C1A4-K1 E8XT5-M21A4-K1	SG-C SG-AC–20–25
	E7XT7-K2	Not Applicable	E7XT7-A2-K2	
	E7XT4-K2	Not Applicable	E7XT4-A0-K2	
	E7XT8-K2	Not Applicable	E7XT8-A2-K2	
	E7XT11-K2	Not Applicable	E7XT11-A-K2[b]	
	E8XT1-K2C, -K2M	Not Applicable	E8XT1-C1A2-K2 E8XT1-M21A2-K2	SG-C SG-AC–20–25
	E8XT5-K2C, -K2M	Not Applicable	E8XT5-C1A2-K2 E8XT5-M21A2-K2	SG-C SG-AC–20–25
	E9XT1-K2C, -K2M	Not Applicable	E9XT1-C1A0-K2 E9XT1-M21A0-K2	SG-C SG-AC–20–25
	E9XT5-K2C, -K2M	Not Applicable	E9XT5-C1A6-K2 E9XT5-M21A6-K2	SG-C SG-AC–20–25

(Continued)

Table M.9 (Continued)
Comparison of Classifications of AWS A5.18, A5.20, A5.28, and A5.29 Specifications to AWS A5.36 Fixed and Open Classifications for Multiple-Pass FCAW and GMAW—Metal Cored Electrodes

Annex L		A5.36		
Specification	Classification	Fixed Classification	Open Classification	Annex V Shielding Gas Classification Requirements
A5.29 (Cont'd)	E10XT1-K3C, -K3M	Not Applicable	E10XT1-C1A0-K3 E10XT1-M21A0-K3	SG-C SG-AC–20–25
	E10XT5-K3C, -K3M	Not Applicable	E10XT5-C1A6-K3 E10XT5-M21A6-K3	SG-C SG-AC–20–25
	E11XT1-K3C, -K3M	Not Applicable	E11XT1-C1A0-K3 E11XT1-M21A0-K3	SG-C SG-AC–20–25
	E11XT5-K3C, -K3M	Not Applicable	E11XT5-C1A6-K3 E11XT5-M21A6-K3	SG-C SG-AC–20–25
	E11XT1-K4C, -K4M	Not Applicable	E11XT1-C1A0-K4 E11XT1-M21A0-K4	SG-C SG-AC–20–25
	E11XT5-K4C, -K4M	Not Applicable	E11XT5-C1A6-K4 E11XT5-M21A6-K4	SG-C SG-AC–20–25
	E12XT5-K4C, -K4M	Not Applicable	E12XT5-C1A6-K4 E12XT5-M21A6-K4	SG-C SG-AC–20–25
	E12XT1-K5C, -K5M	Not Applicable	E12XT1-C1AZ-K5 E12XT1-M21AZ-K5	SG-C SG-AC–20–25
	E7XT5-K6C, -K6M	Not Applicable	E7XT5-C1A8-K6[a] E7XT5-M21A8-K6[a]	SG-C SG-AC–20–25
	E6XT8-K6	Not Applicable	E6XT6-A2-K6	
	E7XT8-K6	Not Applicable	E7XT8-A2-K6	
	E10XT1-K7C, -K7M	Not Applicable	E10XT1-C1A6-K7 E10XT1-M21A6-K7	SG-C SG-AC–20–25
	E9XT8-K8	Not Applicable	E9XT8-A2-K8	
	E10XT1-K9C, -K9M	Not Applicable	E10XT1-C1A6-K9[c] E10XT1-M21A6-K9[c]	SG-C SG-AC–20–25
	E8XT1-W2C, -W2M	Not Applicable	E8XT1-C1A2-W2 E8XT1-M21A2-W2	SG-C SG-AC–20–25
	EXXTX-G	Not Applicable	EXXTX-XX-G	
	EXXTX-GC, -GM	Not Applicable	EXXTX-C1XX-G EXXTX-M21XX-G	SG-C SG-AC–20–25
	EXXTG-X	Not Applicable	EXXTG-XXX-X	As Agreed
	EXXTG-G	Not Applicable	EXXTG-XXX-G	As Agreed

[a] The existing AWS A5.29 CVN temperature is –75°F [–60°C], for which there is not a designator in A5.36, therefore, the open classification comparison is an approximation.

[b] The existing AWS A5.29 CVN temperature is +32°F [0°C], for which there is not a designator in A5.36, therefore, the open classification comparison is not complete due to lack of CVN test temperature.

[c] AWS A5.36 does not address the specific requirements for a K9 electrode, designed for military applications, in which the yield strength is controlled and the tensile strength is reported as a reference. AWS A5.29 reported the tensile as an approximation (100 ksi–120 ksi [690 MPa–830 MPa]), not a requirement and controlled the yield at 82 ksi–97 ksi [570 MPa–670 MPa].

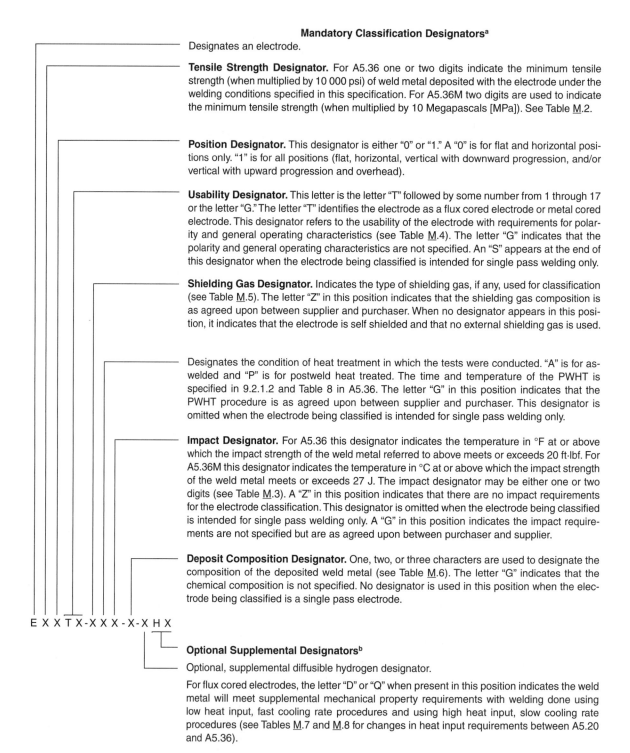

Figure M.1—A5.36/A5.36M Open Classification System

Annex N (Informative)
Guide for Specification Writers

This annex is not part of this standard but is included for informational purposes only.

A statement in a contract document that all welding be done in conformance with AWS D1.1, *Structural Welding Code—Steel*, covers only the mandatory welding requirements. Other provisions in the code are optional. They apply only when they are specified. The following are some of the more commonly used optional provisions and examples of how they may be specified.

Optional Provision	Typical Specification
Fabrication/Erection Inspection [When not the responsibility of the Contractor (8.1.2)]	Fabrication/Erection inspection will be performed by the Owner. *or* Fabrication/Erection inspection will be performed by testing agency retained by the Owner. *NOTE: When fabrication/erection inspection is performed by the Owner or the Owner's testing agency, complete details on the extent of such testing shall be given.*
Verification Inspection (8.1.2.2)	Verification inspection (8.1.2.2) shall be performed by the Contractor. *or* Verification inspection shall be performed by the Owner. *or* Verification inspection shall be performed by a testing agency retained by the Owner. *or* Verification inspection shall be waived.
Nondestructive Testing	**NDT General:** For each type of joint and (other than visual [8.14] and type of stress [tension, compression and shear]) indicate type of NDT to be used, extent of inspection, any special techniques to be used, and acceptance criteria. Specific examples (to be interpreted as examples and not recommendations) follow. The Engineer shall determine the specific requirements for each condition.
	Statically Loaded Structure Fabrication: Moment Connection Tension
	Groove Welds in Butt Joints—25% UT inspection of each of the first four joints, dropping to 10% of each of the remaining joints. Acceptance criteria— Table 8.2.
	Fillet welds—MT—Inspection of 10% of the length of each weld. Acceptance criteria—Table 8.1 for nontubular welds and Table 10.15 for tubular welds.
	Cyclically Loaded Structure Fabrication: Tension Butt Splices—100% UT, or 100% RT—Acceptance criteria—UT: 8.13.2; RT: 8.12.2.
	Full Penetration Corner Welds in Axially Loaded Members: Tension Stresses—100% UT, Scanning Patterns D or E—Acceptance criteria—Table 8.3.
	Compression Stresses—25%, UT, Scanning Movements A, B, or C. Acceptance criteria—Table 8.1.
(8.15.3)	Fillet Welds—MT—Inspection of 10% of the length of each weld—Acceptance criteria—8.12.2. *or* Rejection of any portion of a weld inspected on a less than 100% basis shall require inspection of 100% of that weld.
(8.15.3)	*or* Rejection of any portion of a weld inspected on a partial length basis shall require inspection of the stated length on each side of the discontinuity.

Annex O (Informative)
UT Examination of Welds by Alternative Techniques

This annex is not part of this standard but is included for informational purposes only.

O1. General

The purpose of this annex is to describe alternative techniques for UT of welds. The techniques described are proven methods currently being used for other applications but not presently detailed in the code. The alternative techniques presented require qualified, written procedures, special UT operator qualifications, and special calibration methods needed to obtain the required accuracy in discontinuity sizing. The use of this annex and the resulting procedures developed, including the applicable acceptance criteria, are subject to approval by the Engineer.

This annex is nonmandatory unless specified in the contract documents. When so specified, however, the entire requirements contained herein (as applicable) shall be considered mandatory unless specifically modified in the contract documents.

Applicable requirements of the code regarding instrumentation and operator qualifications, except as amended herein, may be used to supplement this annex. However, it is not intended that these techniques be used to supplement the existing requirements of Clause 8 of the code since the procedures and techniques specified therein are complete and represent a different approach for the UT of welds.

Part A
Basic UT Procedures

O.2 Introduction

The basic UT procedure, instrumentation and operator requirements contained in this part A are necessary to ensure maximum accuracy in discontinuity evaluation and sizing. The methods described herein are not new. They have been used by other industries, including the shipbuilding and offshore structures, for the past 25 years. Although they have not been prohibited, they have not been organized and specifically made available for use in AWS documents. Some of the methods included in this section are also contained in the American Petroleum Institute's API RP 2X, *Recommended Practices for Ultrasonic Examination of Offshore Structural Fabrication and Guidelines for Qualification of Ultrasonic Technicians*. Additional, useful information can be obtained by reference. For maximum control of discontinuity sizing, emphasis has been placed upon: the UT procedure which shall be written and qualified; UT technician special requirements; and UT instrumentation and calibration requirements. AWS recognizes the inherent limitations and inconsistencies of UT examination for discontinuity characterization and sizing. The accuracies obtainable are required to be proven by the UT technician using the applicable procedures and equipment.

Procedure qualification results should be furnished to the Engineer. AWS makes no claim for accuracies possible for using the methods contained herein.

O3. UT Procedure

All UT shall be performed in conformance with a written procedure which shall contain a minimum of the following information regarding the UT method and examination techniques:

(1) The types of weld joint configurations to be examined

(2) Acceptance criteria for the types of weld joints to be examined (additional criteria when the acceptance criteria of Clause 8, Part C are not invoked by the Engineer)

(3) Type of UT equipment (manufacturer, model number, serial number)

(4) Type of transducer, including frequency, size, shape, angle and type of wedge if it is different than that in 8.21.6 or 8.21.7

(5) Scanning surface preparation and couplant requirements

(6) Type of calibration test block(s) with the appropriate reference reflectors

(7) Method of calibration and calibration interval

(8) Method for examining for laminations prior to weld evaluation if the method is different from 8.25.5

(9) Weld root index marking and other preliminary weld marking methods

(10) Scanning pattern and sensitivity requirements

(11) Methods for determining discontinuity location height, length and amplitude level

(12) Transfer correction methods for surface rough- ness, surface coatings and part curvature, if applicable

(13) Method of verifying the accuracy of the completed examination. This verification may be by re-UT by others (audits), other NDE methods, macroetch specimen, gouging or other visual techniques as may be approved by the Engineer

(14) Documentation requirements for examinations, including any verifications performed

(15) Documentation retention requirements.

The written procedure shall be qualified by testing mock-up welds which represent the production welds to be examined. The mock-up welds shall be sectioned, properly examined, and documented to prove satisfactory performance of the procedure. The procedure and all qualifying data shall be approved by an individual who has been certified Level III in UT by testing in conformance with ASNT SNT-TC–1A and who is further qualified by experience in examination of the specific types of weld joints to be examined.

O4. UT Operator and Equipment

In addition to the requirements of 8.14.6, 8.20, and 10.29.2, the UT operator shall demonstrate ability to use the written procedure, including all special techniques required and, when discontinuity height and length are required, shall establish ability and accuracy for determining these dimensions.

UT equipment shall meet the requirements of 8.21 and as required in this annex. Alternate equipment which utilizes computerization, imaging systems, mechanized scanning, and recording devices, may be used, when approved by the Engineer. Transducers with frequencies up to 6 MHz, with sizes down to 1/4 in [6 mm] and of any shape may be used, provided they are included in the procedure and properly qualified.

O5. Reference Standard

The standard reflector shall be a 1.5 mm diameter side drilled hole or equivalent. The reflector may be placed in any design of calibration block, weld mock-up or actual production part at the option of the user. Orientation and tolerances for placement of the reflector are shown in Figure O.1. A recommended calibration block is shown in Figure O.2. Alternate possible uses of the reflector are shown in Figure O.3. When placed in weld mock-ups and sections of production weldments, the reflector should be in locations where it is difficult to direct sound beams, thereby ensuring detection of discontinuities in all areas of interest.

O6. Calibration Methods

Calibration methods described herein are considered acceptable and are to be used for accomplishing these alternate UT procedures. The code recognizes that other calibration methods may be preferred by the individual user. If other methods are used, they should produce results which can be shown to be at least equal to the methods recommended herein. The standard reflector described in O5 should be considered the standard reflector for these and for all other methods which might be used.

O6.1 Standard Sensitivity. Standard sensitivity should consist of the sum of the following:

(1) **Basic Sensitivity.** The maximized indication from the standard reflector, plus

(2) **Distance Amplitude Correction.** Determined from indications from multiple standard reflectors at depths representing the minimum, middle and maximum to be examined, plus

(3) **Transfer Correction.** Adjustment for material type, shape and scanning surface conditions as described below:

For precise sensitivity standardization, transfer correction should be performed. This will ensure that the differences in acoustical properties, scanning surfaces and part shape between the calibration standard and the calibration block are utilized when performing the standard sensitivity calibration. Transfer correction values should be determined initially before examination and when material type, shape, thickness and scanning surfaces vary such that different values exceeding ± 25% of the original values are expected. The transfer correction values should be determined as shown in Figure O.4.

O6.1.1 Scanning Sensitivity. Scanning sensitivity should be standard sensitivity + approximately 6 dB – 12 dB or as required to verify sound penetration from indications of surface reflections. Indication evaluation should be performed with reference to the standard sensitivity except that standard sensitivity is not required if higher or lower sensitivity is more appropriate for determining the maximum discontinuity size (height and length).

O6.2 Compression Wave

O6.2.1 Depth (Horizontal Sweep). Indications from multiple reflections obtained from the thickness of the calibration standard or from a gaged area of a mock-up or production weldment should be used, as shown in Figure O.5. Accuracy of calibration should be within ± 5% of actual thickness for examination of base metal for laminations and ± 2% for determining discontinuity size (height) and location.

O6.2.2 Sensitivity Calibration (Standard). The search unit should be placed over the standard reflectors at a minimum of 3 depths to ensure coverage throughout the thickness to be examined in conformance with Figure O.6. The dB values obtained from the maximized indications from each reflector should be recorded. A distance amplitude curve (DAC) should be established or electronic methods used to know the display indication locations which represent the standard reflector at the various thicknesses to be examined.

O6.3 Shear Wave

O6.3.1 Depth (Horizontal Sweep). Indications from the selected standard reflectors should be used to cover the maximum depth to be used during examination in conformance with Figure O.7. Accuracy should be within ± 1% to facilitate the most accurate discontinuity height measurement. The delay technique should be used for discontinuities with depth greater than approximately 1.5 in to maximize the most accurate discontinuity depth reading (and discontinuity height) accuracy.

O6.3.2 Sensitivity (Standard). Standard reflectors located at the minimum, middle, and maximum depths below the surface to be used for examination should be used in conformance with Figure O.7. Indications should be maximized and a DAC established or electronic methods used to locate the display indications which represent the standard reflector at the various depths selected. The DAC should be adjusted, based upon the results of the transfer correction. The sensitivity calibration methods described herein are not essential when actual discontinuity size (height and length) is required. In this case, it is only necessary to maintain sufficient sensitivity throughout the part being examined so that all discontinuities are found and properly evaluated.

O7. Scanning

Scanning shall be as described in 8.30 and 10.29.5. In addition, for special applications not covered in the above code references, the scanning methods of Figure O.8 should be used, as applicable.

O8. Weld Discontinuity Characterization Methods

O8.1 Discontinuities should be characterized as follows:

(1) Spherical (individual pores and widely spaced porosity, nonelongated slag)

(2) Cylindrical (elongated slag, aligned pores of porosity, hollow beads)

(3) Planar (incomplete fusion, inadequate joint penetration, cracks)

O8.2 The following methods should be used for determining basic discontinuity characteristics:

O8.2.1 Spherical. Sound is reflected equally in all directions. Indication remains basically unchanged as the search unit is moved around the spherical discontinuity as shown in Figure O.9.

O8.2.2 Cylindrical. Sound is reflected equally in one direction but is changed in other directions. Indication remains basically unchanged when the search unit is moved in one direction but is drastically changed when moved in other directions as shown in Figure O.10.

O8.2.3 Planar. Sound is reflected at its maximum from only one angle of incidence with one plane. Indication is changed with any angular movement of the search unit as shown in Figure O.11. Indications from cracks typically have multiple peaks as a result of the many discontinuity facets usually present.

O9. Weld Discontinuity Sizing and Location Methods

O9.1 Calibration. Calibration should be based upon depth from the surface in conformance with O6. Discontinuities may be sized with the highest achievable level of accuracy using the methods described in this section; however, the user is reminded that UT, like all other NDT methods, provides relative discontinuity dimensions. Discontinuity orientation and shape, coupled with the limitations of the NDT method, may result in significant variations between relative and actual dimensions.

O9.2 Height. The discontinuity height (depth dimension) should be determined using the following methods:

O9.2.1 The indication height should be maximized by moving the search unit to and from the discontinuity in conformance with A of Figure O.12. The indication height should be adjusted to a known value (e.g., 80% of full screen height [FSH]).

O9.2.2 The search unit should be moved towards the discontinuity until the indication height begins to drop rapidly and continuously towards the base line. The location of the leading (left) edge of the indication at location B in Figure O.12 in relation to the display horizontal base line scale should be noted. A 0.10 in [2.5 mm] division scale or metric scale should be used.

O9.2.3 The search unit should be moved away from the discontinuity until the indication height begins to drop rapidly and continuously towards the base line. The location of the leading edge of the indication at location C in Figure O.12 in relation to the display horizontal base-line scale should be noted.

O9.2.4 The mathematical difference between B and C should be obtained to determine the height dimension of the discontinuity.

O9.3 Length. The discontinuity length should be determined using the following methods:

O9.3.1 The orientation of the discontinuity should be determined by manipulation of the search unit to determine the plane and direction of the strongest indication in conformance with A of Figure O.13.

O9.3.2 The search unit should be moved to one end of the discontinuity while keeping part of the indication visible on the display at all times until the indication drops completely to the base line. The search unit should be moved back towards the discontinuity until the indication height reaches 50% of the maximum height originally obtained near the end in conformance with B of Figure O.13. The location should be marked on the end of the discontinuity on the scanning surface or welded in line with the search unit maximum indication mark. This marking should be performed carefully using a fine-line marking method.

O9.3.3 The steps above should be repeated for locating the opposite end of the discontinuity in conformance with C of Figure O.13 and should be marked carefully.

O9.3.4 The length of the discontinuity should be obtained by measuring the distance between the two marks in conformance with Figure O.13.

O9.4 Location—Depth Below the Scanning Surface. The depth location of discontinuities can be read directly from the display horizontal base-line scale when using the methods described above for determining discontinuity height. The reported location should be the deepest point determined, unless otherwise specified, to assist in removal operations.

O9.5 Location—Along the Length of the Weld. The location of the discontinuity from a known reference point can be determined by measuring the distance from the reference point to the discontinuity length marks established for the length. Measurement should be made to the beginning of the discontinuity unless otherwise specified.

O10. Problems with Discontinuities

Users of UT for examinations of welds should be aware of the following potential interpretation problems associated with weld discontinuity characteristics:

O10.1 Type of Discontinuity. Ultrasonic sound has variable sensitivity to weld discontinuities depending upon their type. Relative sensitivity is shown in the following tables and should be considered during evaluation of discontinuities. The UT technician can change sensitivity to all discontinuity types by changing UT instrument settings, search unit frequency, and size and scanning methods, including scanning patterns and coupling.

Discontinuity Type	Relative UT Sensitivity
(1) Incomplete fusion	Highest
(2) Cracks (surface)	.
(3) Inadequate penetration	.
(4) Cracks (sub-surface)	.
(5) Slag (continuous)	.
(6) Slag (scattered)	.
(7) Porosity (piping)	.
(8) Porosity (cluster)	.
(9) Porosity (scattered)	Lowest

O10.2 General classification of discontinuities may be compared as follows:

General Classification of Discontinuity	Relative UT Sensitivity
(a) Planar	Highest
(b) Linear	.
(c) Spherical	Lowest

NOTE: The above tabulation assumes best orientation for detection and evaluation.

O10.3 Size. Discontinuity size affects accurate interpretation. Planar-type discontinuities with large height or very little height may give less accurate interpretation than those of medium height. Small, spherical pores are difficult to size because of the rapid reflecting surface changes which occur as the sound beam is moved across the part.

O10.4 Orientation. Discontinuity orientation affects UT sensitivity because the highest sensitivity is one that reflects sound more directly back to the search unit. Relative sensitivities in regards to orientation and discontinuity types are opposite those shown in the previous tables. The UT technician can increase sensitivity to discontinuity orientation by selecting a sound beam angle which is more normal to the discontinuity plane and reflecting surface. The selection of angles which match the groove angle will increase sensitivity for planar and linear-type discontinuities which are most likely to occur along that plane.

O10.5 Location. Discontinuity location within the weld and adjacent base metal can influence the capability of detection and proper evaluation. Discontinuities near the surface are often more easily detected but may be less easily sized.

O10.6 Weld Joint Type and Groove Design. The weld joint type and groove design are important factors affecting the capabilities of UT for detecting discontinuities.

The following are design factors which can cause problems and should be considered for their possible affects:

(1) Backings

(2) Bevel angles

(3) Joint member angles of intercept

(4) PJP welds

(5) Tee welds

(6) Tubular members

(7) Weld surface roughness and contour

O11. Discontinuity Amplitude Levels and Weld Classes Discontinuity Amplitude Levels

The following discontinuity amplitude level categories should be applied in evaluation of acceptability:

Level	Description
1	Equal to or greater than SSL (see Figure O.14)
2	Between the SSL and the DRL (see Figure O.14)
3	Equal to or less than the DRL (see Figure O.14)

SSL = Standard Sensitivity Level—per Clause 8.
DRL= Disregard Level = 6 dB less than the SSL.

Weld Classes. The following weld classes should be used for evaluation of discontinuity acceptability:

Weld Class	Description
S	Statically loaded structures
D	Cyclically loaded structures
R	Tubular structures (substitute for RT)
X	Tubular T-, Y-, K-connections

O12. Acceptance-Rejection Criteria

O12.1 Amplitude. The acceptance-rejection criteria of Table O.1 should apply when amplitude and length are the major factors and maximum discontinuity height is not known or specified.

O12.2 Size. When maximum allowable discontinuity size (height and length) are known and are specified by the Engineer, the actual size (both height and length) along with location (depth and along the weld) should be determined and reported. Final evaluation and acceptance/ rejection should be by the Engineer.

O13. Preparation and Disposition of Reports

A report shall be made which clearly identifies the work and the area of examination by the UT operator at the time of examination. The report, as a minimum, shall contain the information shown on the sample report form, Figure O.15. UT discontinuity characterization and subsequent categorization and reporting should be limited to spherical, cylindrical, and planar only.

When specified, discontinuities approaching unacceptable size, particularly those about which there is some doubt in their evaluation, should also be reported.

Before a weld subject to UT by the Contractor for the Owner is accepted, all report forms pertaining to the weld, including any that show unacceptable quality prior to repair, should be submitted to the Owner upon completion of the work. The Contractor's obligation to retain UT reports should cease (1) upon delivery of a full set to the Owner, or (2) one full year after completion of the Contractor's work, provided the Owner is given prior written notice.

Table O.1
Acceptance-Rejection Criteria (see O12)

Maximum Discontinuity Amplitude Level Obtained	Maximum Discontinuity Lengths by Weld Classes			
	Statically Loaded	Cyclically Loaded	Tubular Class R	Tubular Class X
Level 1—Equal to or greater than SSL (see O6.1 and Figure O.14)	> 5 dB above SSL = none allowed 0 thru 5 dB above SSL = 3/4 in [20 mm]	> 5 dB above SSL = none allowed 0 thru 5 dB above SSL = 1/2 in [12 mm]	See Figure 10.24	See Figure 10.25 (Utilizes height)
Level 2—Between the SSL and the DRL (see Figure O.14)	2 in [50 mm]	Middle 1/2 of weld = 2 in [50 mm] Top & bottom 1/4 of weld = 3/4 in [20 mm]	See Figure 10.24	See Figure 10.25 (Utilizes height)
Level 3—Equal to or less than the DRL (see Figure O.14)	Disregard (when specified by the Engineer, record for information)			

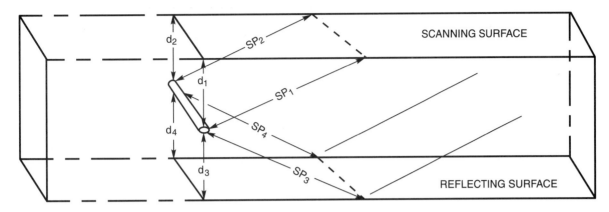

Notes:
1. $d_1 = d_2 \pm 0.5$ mm $d_3 = d_4 \pm 0.5$ mm
   $SP_1 = SP_2 \pm 1$ mm $SP_3 = SP_4 \pm 1$ mm
2. The above tolerances should be considered as appropriate. The reflector should, in all cases, be placed in a manner to allow maximizing the reflection and UT indication. (This is a general comment for all notes in Annex O.)

Figure O.1—Standard Reference Reflector (see Q5)

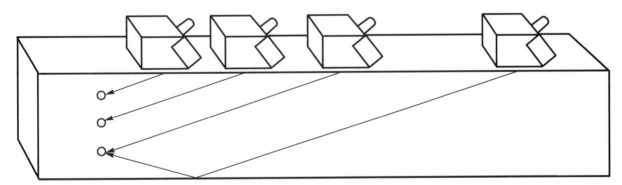

Note: Dimensions should be required to accommodate search units for the sound path distances required.

Figure O.2—Recommended Calibration Block (see Q5)

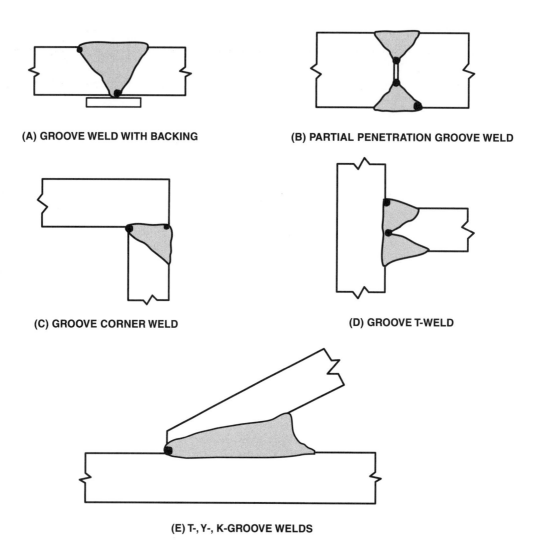

Figure O.3—Typical Standard Reflector (Located in Weld Mock-Ups and Production Welds) (see O5)

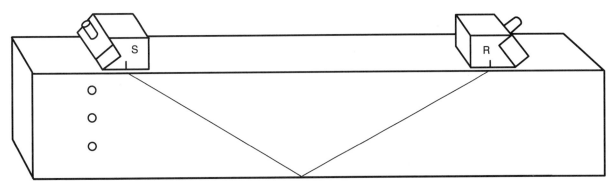

Procedure:
1. Place two similar angle beam search units on the calibration block or mock-up to be used in the position shown above.
2. Using through transmission methods, maximize the indication obtained and obtain a dB value of the indication.
3. Transfer the same two search units to the part to be examined, orient in the same direction in which scanning will be performed, and obtain a dB value of indications as explained above from the least three locations.
4. The difference in dB between the calibration block or mock-up and the average of that obtained from the part to be examined should be recorded and used to adjust the standard sensitivity.

Figure O.4—Transfer Correction (see O6.1)

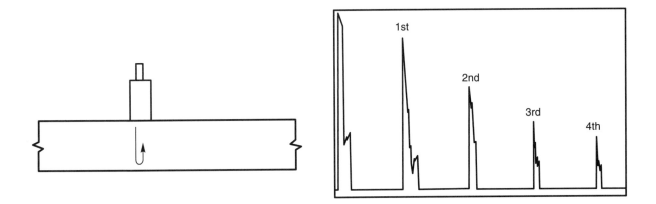

Figure O.5—Compression Wave Depth (Horizontal Sweep Calibration) (see O6.2.1)

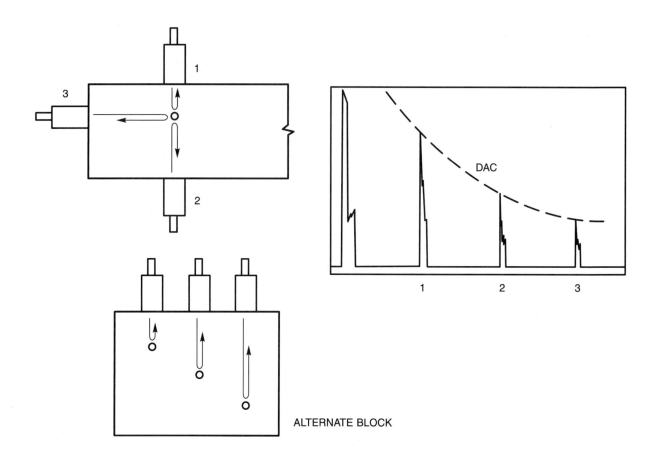

Figure O.6—Compression Wave Sensitivity Calibration (see O6.2.2)

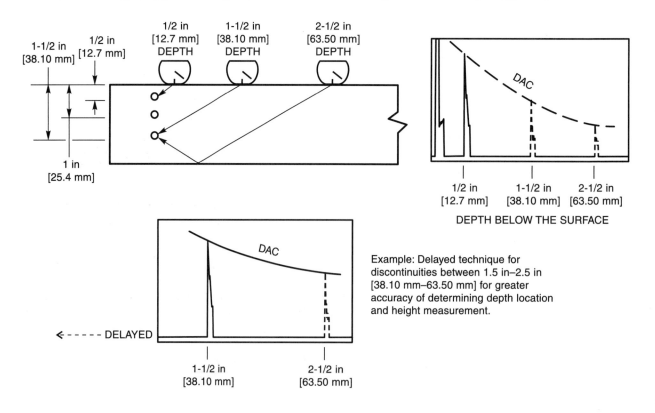

Figure O.7—Shear Wave Distance and Sensitivity Calibration (see O6.3.1)

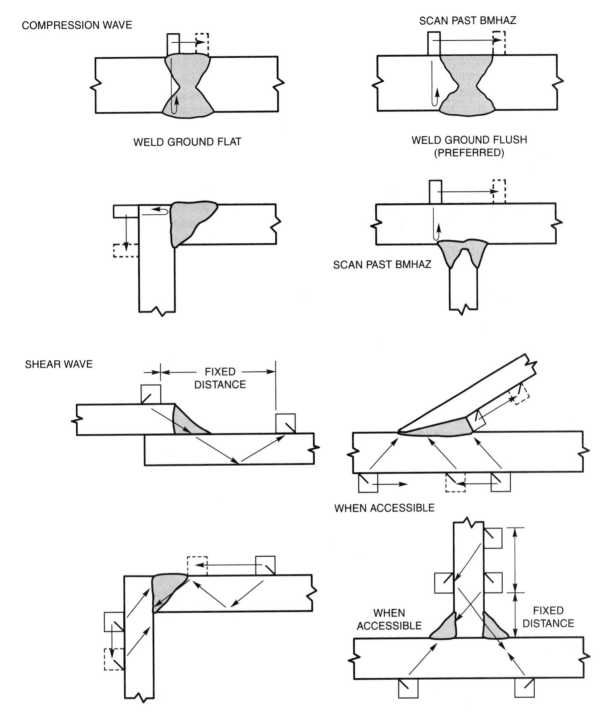

Notes:
1. ⊟→⊡ ⬚→⬚ Denote scanning, otherwise search unit should be at a fixed distance from the weld while scanning down the weld.
2. Cross section scanning is shown. It is assumed that scanning will also be performed completely down the length of the weld with a minimum of 25% overlap to ensure 100% coverage. All scanning positions shown may not be required for full coverage. Optional positions are given in case that inaccessibility prevents use of some positions.

Figure O.8—Scanning Methods (see Q7)

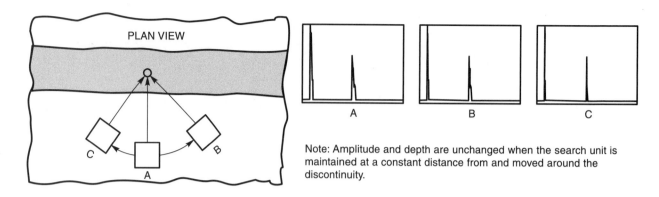

Figure O.9—Spherical Discontinuity Characteristics (see O8.2.1)

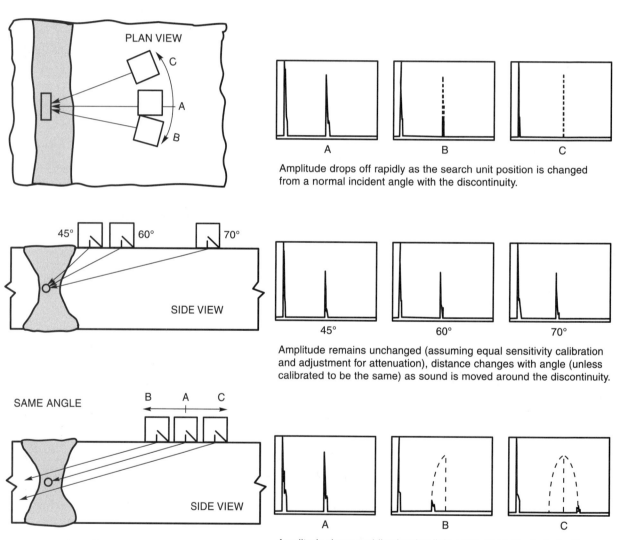

Figure O.10—Cylindrical Discontinuity Characteristics (see O8.2.2)

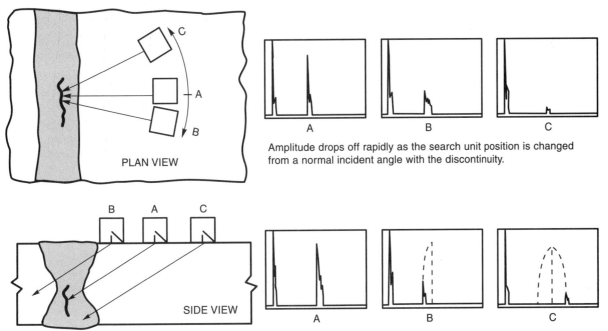

Figure O.11—Planar Discontinuity Characteristics (see O8.2.3)

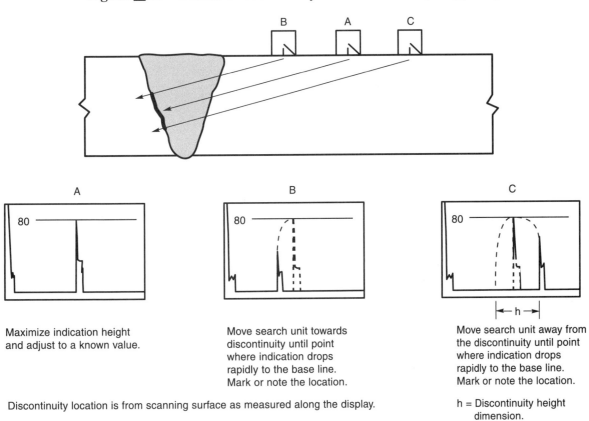

Figure O.12—Discontinuity Height Dimension (see O9.2)

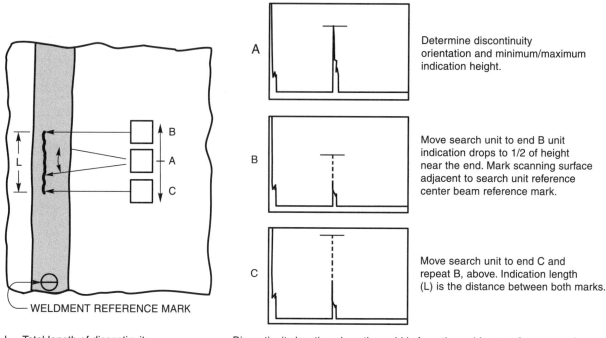

Figure O.13—Discontinuity Length Dimension (see O9.3)

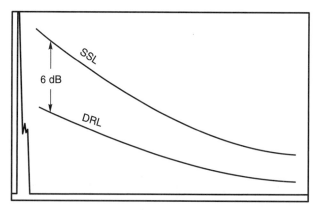

Note: The display screen may be marked to show SSl established during sensitivity calibration with the DRL located 6 dB below.

Figure O.14—Display Screen Marking (see O11)

Page _____ of _____

Project _____ Report No. _____

Weld I.D. _____ Thickness _____ Class _____

UT Procedure No. _____ Technique _____

UT Instrument _____

Search Unit: No. _____ Angle _____ Freq. _____ Size _____

RESULT (identify and describe each discontinuity)

No.	Location from	Ampl. Level	Length	Height	Comments

Sketch (identify each discontinuity listed above)

NDT Tech. _____ Contractor _____

Date Examined _____ Approved _____

                                           Date Approved _____

Figure O.15—Report of UT (Alternative Procedure) (see O13)

Annex P (Informative)
UT Equipment Qualification and Inspection Forms

This annex is not part of this standard but is included for informational purposes only.

This annex contains examples for use of three forms, P–8, P–9, and P–10, for recording of UT test data. Each example of forms P–8, P–9, and P–10 shows how the forms may be used in the UT inspection of welds. Form P–11 is for reporting results of UT inspection of welds.

Ultrasonic Unit Calibration Report—AWS

Ultrasonic Unit Model _____ Serial No. _____

Search Unit—Size _____ Type _____ Frequency _____ MHz

Calibration—Date _____ Interval _____ Method _____

Block Serial No. _____ Data _____ As Found _____ As Adjusted

SUPPLEMENTAL INSTRUCTIONS

- Start with the lowest dB level that you can obtain a 40 percent display height indication from directly over the two in section of the DS block. Add 6 dBs and record this dB reading "a" and display height "b" as the starting point on the tabulation chart.

- After recording these values in Rows "a" and "b," slide the transducer to obtain a new 40 percent display height. Without moving the transducer add 6 dBs and record the new dB reading and the new display height in the appropriate row. Repeat this step as many times as the unit allows.

- Find the average % screen values from Row "b" by disregarding the first 3 and the last 3 tabulations. Use this as $\%_2$ in calculating the corrected reading.

- The following equation is used to calculate Row "c":
  - $\%_1$ is Row "b"
  - $\%_2$ is the average of Row "b" disregarding the first and last three tabulations.
  - dB_1 is Row "a"
  - dB_2 is Row "c"

$$dB_2 = 20 \times \log\left(\frac{\%_2}{\%_1}\right) + dB_1$$

- The dB Error "d" is established by subtracting Row "c" from Row "a": (a − c = d).

- The Collective dB Error "e" is established by starting with the dB Error "d" nearest to 0.0, collectively add the dB Error "d" values horizontally, placing the subtotals in Row "e."

- Moving horizontally, left and right from the Average % line, find the span in which the largest and smallest Collective dB Error figures remain at or below 2 dB. Count the number of horizontal spaces of movement, subtract one, and multiply the remainder by six. This dB value is the acceptable range of the unit.

- In order to establish the acceptable range graphically, Form P-8 should be used in conjunction with Form P-9 as follows:
  1. Apply the collective dB Error "e" values vertically on the horizontal offset coinciding with the dB reading values "a."
  2. Establish a curve line passing through this series of points.
  3. Apply a 2 dB high horizontal window over this curve positioned vertically so that the longest section is completely encompassed within the 2 dB Error height.
  4. This window length represents the acceptable dB range of the unit.

Row	Number	1	2	3	4	5	6	7	8	9	10	11	12	13
a	dB Reading													
b	Display Height													
c	Corrected Reading													
d	dB Error													
e	Collective dB Error													

Accuracy Required: Minimum allowable range is _____ . $\%_2$ (Average) _____ %

Equipment is: Acceptable for Use _____ Not Acceptable for Use _____ Recalibration Due Date _____

Total qualified range _____ dB to _____ dB = _____ dB Total error _____ dB (From the Chart above)

Total qualified range _____ dB to _____ dB = _____ dB Total error _____ dB (From Form L-9)

Calibrated by _____ Level _____ Location _____

Form P-8

AWS D1.1/D1.1M:2020   ANNEX P

Ultrasonic Unit Calibration Report—AWS

Ultrasonic Unit Model __USN-50__   Serial No. __47859-5014__

Search Unit—Size __1" ROUND__   Type __SAB__   Frequency __2.25__ MHz

Calibration—Date __June 17, 1996__   Interval __2 Months__   Method __AWS D1.1__

Block Serial No. __1234-5678__   Data __XX__ As Found _____ As Adjusted

SUPPLEMENTAL INSTRUCTIONS

- Start with the lowest dB level that you can obtain a 40 percent display height indication from directly over the two in section of the DS block. Add 6 dBs and record this dB reading "a" and display height "b" as the starting point on the tabulation chart.

- After recording these values in Rows "a" and "b," slide the transducer to obtain a new 40 percent display height. Without moving the transducer add 6 dBs and record the new dB reading and the new display height in the appropriate row. Repeat this step as many times as the unit allows.

- Find the average % screen values from Row "b" by disregarding the first 3 and the last 3 tabulations. Use this as %$_2$ in calculating the corrected reading.

- The following equation is used to calculate Row "c":
  %$_1$ is Row "b"
  %$_2$ is the average of Row "b" disregarding
         the first and last three tabulations. $dB_2 = 20 \times \log\left(\frac{\%_2}{\%_1}\right) + dB_1$
  dB_1 is Row "a"
  dB_2 is Row "c"

- The dB Error "d" is established by subtracting Row "c" from Row "a": (a − c = d).

- The Collective dB Error "e" is established by starting with the dB Error "d" nearest to 0.0, collectively add the dB Error "d" values horizontally, placing the subtotals in Row "e."

- Moving horizontally, left and right from the Average % line, find the span in which the largest and smallest Collective dB Error figures remain at or below 2 dB. Count the number of horizontal spaces of movement, subtract one, and multiply the remainder by six. This dB value is the acceptable range of the unit.

- In order to establish the acceptable range graphically, Form P-8 should be used in conjunction with Form P-9 as follows:
  (1) Apply the collective dB Error "e" values vertically on the horizontal offset coinciding with the dB reading values "a."
  (2) Establish a curve line passing through this series of points.
  (3) Apply a 2 dB high horizontal window over this curve positioned vertically so that the longest section is completely encompassed within the 2 dB Error height.
  (4) This window length represents the acceptable dB range of the unit.

ROW	NUMBER	1	2	3	4	5	6	7	8	9	10	11	12	13
a	dB Reading	6	12	18	24	30	36	42	48	54	60	66	72	78
b	Display Height	69	75	75	77	77	77	77	78	77	78	79	80	81
c	Corrected Reading	7.1	12.3	18.3	24.1	30.1	36.1	42.1	48.0	54.1	60.0	65.9	71.8	77.7
d	dB Error	−1.1	−0.3	−0.3	−0.1	−0.1	−0.1	−0.1	0.0	−0.1	0.0	+0.1	+0.2	+0.3
e	Collective dB Error	−2.2	−1.1	−0.8	−0.5	−0.4	−0.3	−0.2	−0.1	−0.1	0.0	+0.1	+0.3	+0.6

Accuracy Required: Minimum allowable range is 60 dB.   %$_2$ (Average) __78__ %

Equipment is: Acceptable for Use __XX__   Not Acceptable for Use _____   Recalibration Due Date __August 17, 1996__

Total qualified range __12__ dB to __78__ dB = __66__ dB   Total error __1.7__ dB (From the Chart above)

Total qualified range __11__ dB to __80__ dB = __69__ dB   Total error __2.0__ dB (From Form L-9)

Calibrated by _____   Level _____   Location _____

Form P-8

ANNEX P   AWS D1.1/D1.1M:2020

dB Accuracy Evaluation

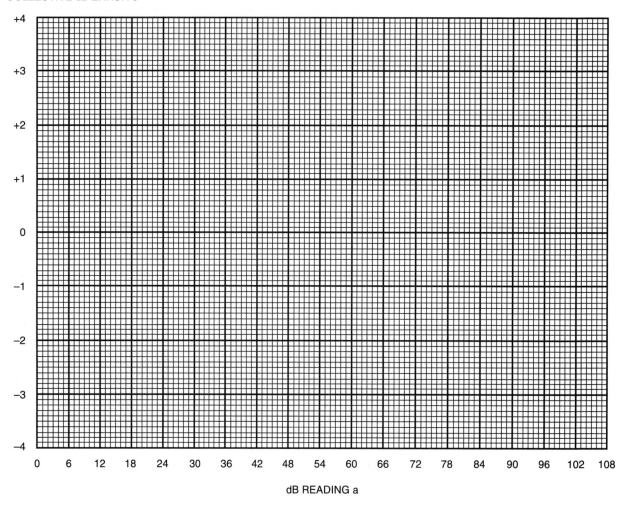

Form P-9

dB Accuracy Evaluation—AWS

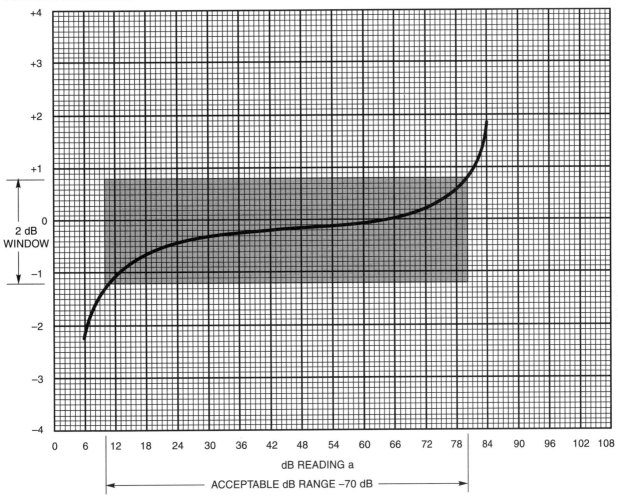

THE CURVE ON FORM P-9 EXAMPLE IS DERIVED FROM CALCULATIONS FROM FORM P-8.
THE SHADED AREA ON THE GRAPH ABOVE SHOWS THE AREA OVER WHICH THE EXAMPLE UNIT QUALIFIES TO THIS CODE.

Note: The first line of example of the use of Form P–8 is shown in this example.

Form P-9

Decibel (Attenuation or Gain) Values Nomograph

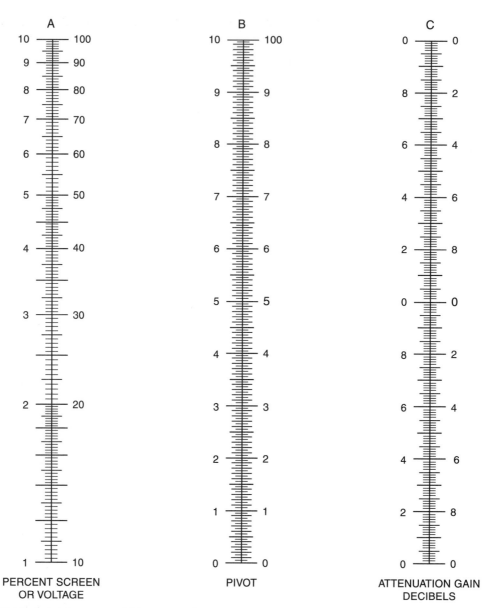

Note: See 8.28.3 for instruction on use of this nomograph.

Form P-10

Decibel (Attenuation or Gain) Values Nomograph—AWS

THE USE OF THE NOMOGRAPH IN RESOLVING NOTE 3 IS AS SHOWN ON THE FOLLOWING EXAMPLE.

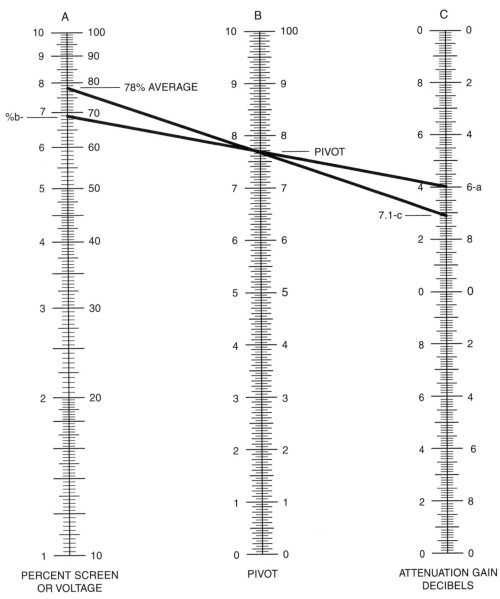

Procedure for using the nomograph:
- Extend a straight line between the decibel reading from Row "a" applied to the C scale and the corresponding percentage from Row "b" applied to the A scale.
- Use the point where the straight line above crosses the pivot line B as a pivot line for a second straight line.
- Extend a straight line from the average sign point on scale A, through the pivot point developed above, and onto the dB scale C.
- This point on the C scale is indicative of the corrected dB for use in Row "c."

Notes:
1. The 6 dB reading and 69% scale are derived from the instrument reading and become dB_1 "b" and $\%_1$ "c," respectively.
2. $\%_2$ is 78-constant.
3. dB_2 (which is corrected dB "d") is e3qual to 20 times X log (78/69) + 6 or 7.1.

Form P-10

ANNEX P   AWS D1.1/D1.1M:2020

Report of UT of Welds

Project _____   Report no. _____

Weld identification _____
Material thickness _____
Weld joint AWS _____
Welding process _____
Quality requirements—section no. _____
Remarks _____

Line number	Indication number	Transducer angle	From Face	Leg[a]	Decibels				Discontinuity					Discontinuity evaluation	Remarks
					Indication level	Reference level	Attenuation factor	Indication rating	Length	Angular distance (sound path)	Depth from "A" surface	Distance			
					a	b	c	d				From X	From Y		
1															
2															
3															
4															
5															
6															
7															
8															
9															
10															
11															
12															
13															
14															
15															
16															
17															
18															
19															
20															
21															
22															
23															
24															
25															
26															

We, the undersigned, certify that the statements in this record are correct and that the welds were prepared and tested in conformance with the requirements of Clause 8, Part F of AWS D1.1/D1.1M, (_____) *Structural Welding Code—Steel*.
(year)

Test date _____   Manufacturer or Contractor _____

Inspected by _____   Authorized by _____

Note: This form is applicable to Clause 4, Parts B or C (Statically   Date _____
and Cyclically Loaded Nontubular Structures). Do **NOT** use this
form for Tubular Structures (Clause 10, Part A).

Form P-11

a use Leg I, II, or III (see Clause 3 for definition of Leg).

Notes:
1. In order to attain Rating "d"
    a. With instruments with gain control, use formula $a - b - c = d$.
    b. With instruments with attenuation control, use formula $b - a - c = d$.
    c. A plus or minus sign shall accompany the "d" figure unless "d" is equal to zero.
2. Distance from X is used in describing the location of a weld discontinuity in a direction perpendicular to the weld reference line. Unless this figure is zero, a plus or minus sign shall accompany it.
3. Distance from Y is used in describing the location of a weld discontinuity in a direction parallel to the weld reference line. This figure is attained by measuring the distance from the "Y" end of the weld to the beginning of said discontinuity.
4. Evaluation of Retested Repaired Weld Areas shall be tabulated on a new line on the report form. If the original report form is used, R_n shall prefix the indication number. If additional forms are used, the R number shall prefix the report number.

This page is intentionally blank.

Annex Q (Informative)
Local Dihedral Angle

This annex is not part of this standard but is included for informational purposes only.

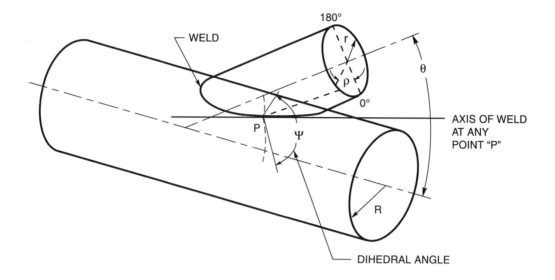

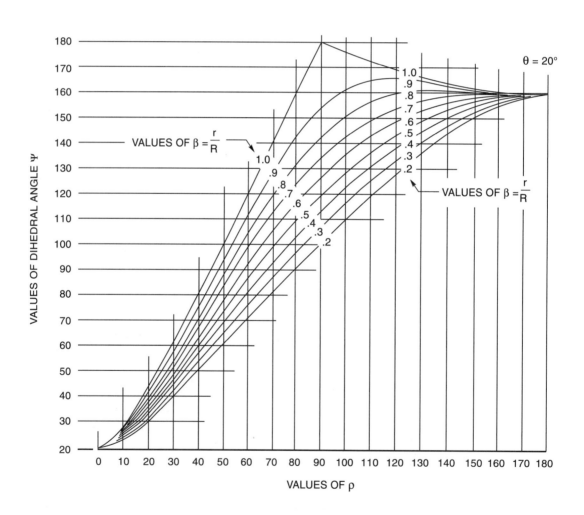

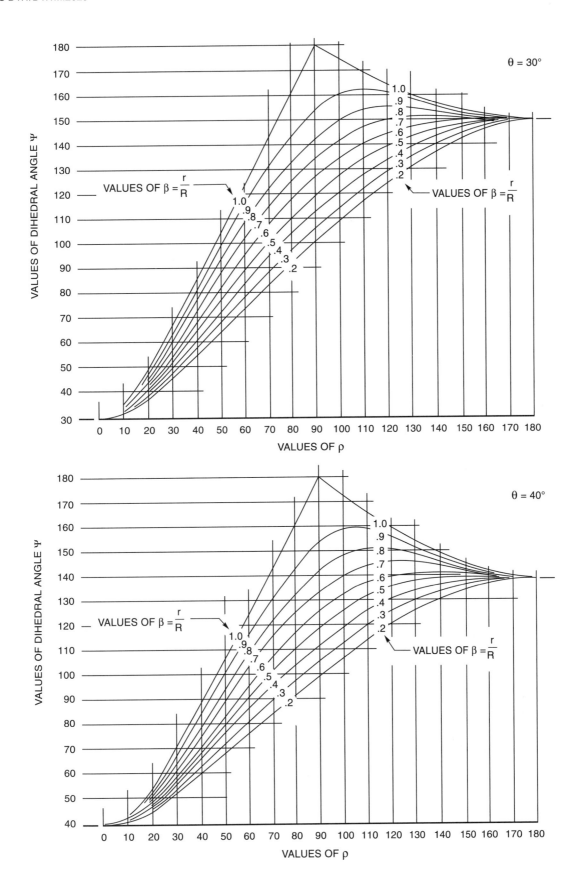

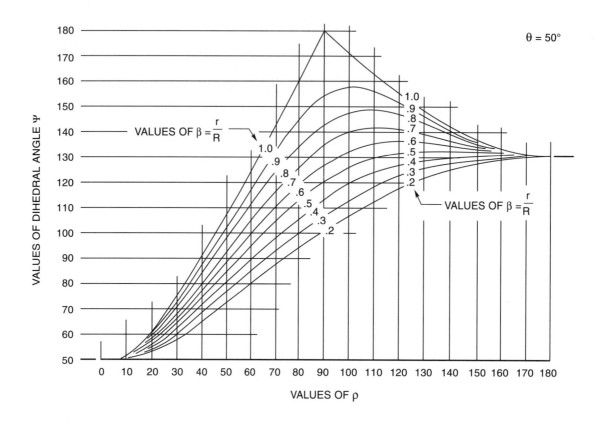

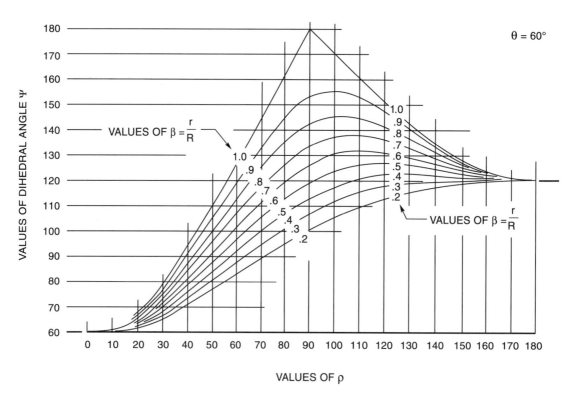

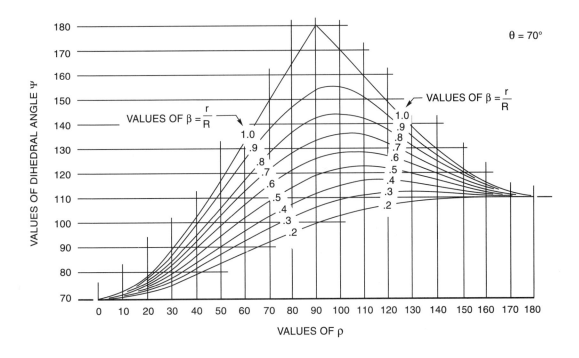

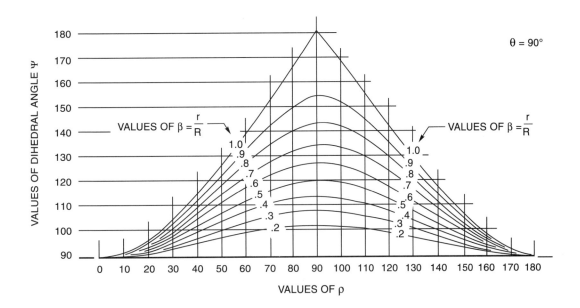

This page is intentionally blank.

Annex R (Informative)
Preliminary Design Circular Tube Connections

This annex is not part of this standard but is included for informational purposes only.

R1 General

T-, Y-, K-connections subjected primarily to axial load are subjected to local base metal stresses around the circumference of the connection through the wall thickness of the main or chord member. These stresses, often referred to as punching shear stresses, may limit the usable strength of the welded connection. These stresses depend not only on the strength of the main member but also on the geometry of the connection.

The optional efficiency check described herein is intended for preliminary static design of tubular trusses and space frame structures, in which the branch members are primarily subject to axial loads, and which may or may not be cyclically loaded.

The Engineer should confirm that all tubular connections have either been designed for required strength according to *AISC 360*, or to another applicable standard. All load combinations and interactions at all branch-chord intersections should be considered.

"Connection efficiency" is defined as the available strength of a connection divided by the available strength of a branch member. Preliminary design may be to 60% to 100% connection efficiency using the simplified criteria herein for circular tube connections. The need for subsequent strengthening during later design stages is minimized by designing all the branch and chord members with a utilization less than 0.8 times the connection efficiency, to accommodate interaction.

"Connection efficiency is different from the connection utilization ratio. The utilization ratio is the factored load for a particular load combination divided by the connection available strength."

Considering rational proportions for axially loaded branch members, the Engineer should consider the following:

    (a) Prohibiting the welding of thick onto thin (i.e. using the thinner member as the chord or thru member.

    (b) Prohibiting fabrication of T-, Y-, or K-connections having less than 60% efficiency. For connection efficiency less than 60%, the connection may fail although the member strength checks were adequate.

R2 Punching Shear Efficiency Check

Punching shear, Figure R.1, is a useful concept for visualizing the local failure mode of a circular tubular connection. Other failure modes i.e. ovalizing or general collapse, can also be represented in this format. Reliability of this formerly normative AWS criteria is demonstrated in Figure R.2. Axial capacities are similar to AISC and CIDECT, even though the algebraic format looks quite different (see Reference R1).

Connection efficiency = V_{pc} (punching shear available for connection) / V_{pb} (punching shear required to fully develop yield in the branch), where

$V_{pc} = 3 Q_q F_{yo} / \gamma$ ... but not more than $F_{yo}/\sqrt{3}$

$V_{pb} = \tau F_{yb} \sin \theta$

Terms used in the foregoing equations are defined as follows:

τ, θ, γ, β and other parameters of connection geometry are defined in Figure 10.2.

F_{yb} is the specified minimum yield strength of the branch member. For long slender secondary bracings, whose design would be governed by compression, F_{yb} may be taken as the ultimate column buckling strength.

F_{yo} is the specified minimum yield strength of the main member chord, but not more than 80% of UTS.

Qq, is geometry modifier term given in Table R.1 or calculated per Figure R.3.

R3 Ovalizing Parameter Alpha

R3.1 Multiplaner Connections

Figure R.3 gives a formula and defines the terms used for composing a value of the chord ovalizing parameter alpha α when designing multiplanar tubular joints. The values of alpha obtained are compatible with both static strength design and fatigue using the punching shear format.

Alpha is evaluated separately for each branch for which punching shear is checked (the "reference brace"), and for each load case, with summation being carried out for all braces present at the node, each time alpha is evaluated. In the summation, the cosine term expresses the influence of braces as a function of position around the circumference, and the exponential decay term expresses the lessening influence of braces as distance L_1 increases; this cosine term is unity for the reference brace, which also appears in the denominator. In complex space frames, the repetitive calculation may be incorporated into a connection design postprocessor to perform the computer analysis.

For hand calculations, the Engineer might prefer the simpler forms of alpha given in Table R.1. However, these do not cover multiplanar cases where higher values of alpha may apply (i.e, 3.8 for a hubstyle cross joint with four branches), and may require a somewhat arbitrary classification of joint types. For joints whose load pattern falls in between two cases (e.g., part of the load is carried as in a K-connection and part as a T-connection) interpolated values of alpha should be determined. Computing alpha using the equation in Figure R.3 is also acceptable.

However, for similarly loaded connections in adjacent planes no increase in capacity over that of the corresponding uniplanar connections should be taken.

R3.2 General Collapse of Cross Connections. Strength and stability for loads transferred across the main member in a tubular connection, with any reinforcement, should be investigated using available technology in conformance with the applicable design code. General collapse may be particularly severe at truss end bearing supports and connections subjected to crushing loads [see Figure 10.2 (G) and (J)]. Such connections may be reinforced by locally increasing the main member thickness, or by use of diaphragms, rings, or collars. For the simplified alpha punching shear check, α of 2.4 should be applied.

R4 Weld Design

Weld stress calculations per 10.5 are not necessary where any one of the following provisions are met:

(1) Due to differences in the relative flexibilities of the main member loaded normal to its surface, versus the branch member carrying membrane stresses parallel to its surface, transfer of load across the weld is highly non-uniform, and local yielding can be expected before the connection reaches its design load. To prevent "unzipping" or progressive failure of the weld and ensure ductile behavior of the joint, the minimum welds provided in simple T-, Y-, or K-connections should be capable of developing, at their ultimate breaking strength, the lesser of the brace member yield strength or local strength V_p (capacity) / sin θ of the main member.

(2) The foregoing requirement may be presumed to be met by the prequalified joint details of Figures 10.6 (PJP) and 10.9–10.11 (CJP) when matching materials (Table 5.4) are used.

(3) Compatible strength of welds may also be presumed with the prequalified fillet weld details of Figure 10.5, when the following effective throat requirements are met:

(a) $E = 0.7 t_b$ for elastic design of mild steel circular steel tubes (F_{yb} < 40 ksi [280 MPa]) joined with overmatched welds

(b) $E = 1.0 \, t_b$ for connections of mild steel tubes or Group I steels, with welds made using the matching strength requirements of Tables 5.3 and 5.4.

(c) $E =$ lesser of t_c or $1.07 \, t_b$ for all other cases

References for Annex R

1. P. W. Marshall, Review of tubular joint criteria. Proceedings of the 5th International Workshop on Connections in Steel Structures, Amsterdam, The Netherlands, June 2004.

2. Rodabaugh, E.C. "Review of data relevant to the design of tubular joints for use in fixed offshore platforms." *WRC Bulletin 256*, January 1980.

Table R.1
Terms for Strength of Connections (Circular Sections) (See R.2)

Branch member Geometry and load modifier Q_q	$Q_q = \left(\dfrac{1.7}{\alpha} + \dfrac{0.18}{\beta}\right) Q_\beta^{0.7(\alpha-1)}$	For axial loads in branch members of trusses and space frames[a]
Q_β (needed for Q_q)	$Q_\beta = 1.0$	For $\beta \leq 0.6$
	$Q_\beta = \dfrac{0.3}{\beta(1 - 0.833\beta)}$	For $\beta > 0.6$
Chord ovalizing parameter α (needed for Q_q)	$\alpha = 1.0 + 0.7\, g/d_b$ $1.0 \leq \alpha < 1.7$	For axial load in gap K-connections having all members in the same plane and loads transverse to the main member essentially balanced (within 20%)[b]
	$\alpha = 1.7$	For axial load in T-and Y-connections
	$\alpha = 2.4$	For axial load in cross connections

[a] For general collapse (transverse compression) see also R4.1.
[b] Gap g is defined in Figures 10.2(E), (F) and (H); d_b is branch diameter.

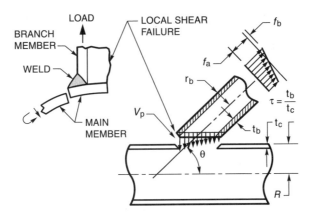

Notes:
1. Acting $V_p = Tf_n \sin \theta$
2. f_n is the nominal axial stress (f_a) + primary bending stress (f_b), however, within the context of this Annex, secondary bending stress may be ignored. Secondary bending moments may be ignored in trusses and space frames which have all pin-connected branch members. Primary moments (i.e. cantilever arms and sway frames) should be considered in final design when present, but are outside the scope of this Annex.

Figure R.1—Simplified Concept of Punching Shear (See R2)

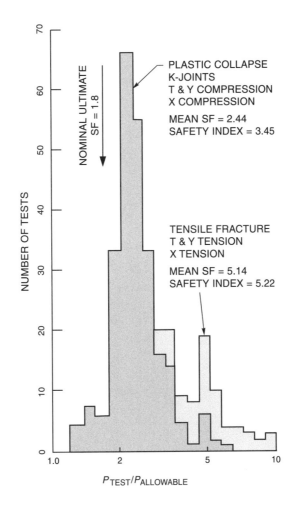

Notes:
1. SF = Safety Factor
2. Data base: 306 (nonoverlap) joints (see Reference 2)

Figure R.2—Reliability of Punching Shear Criteria Using Computed Alpha (See R2)

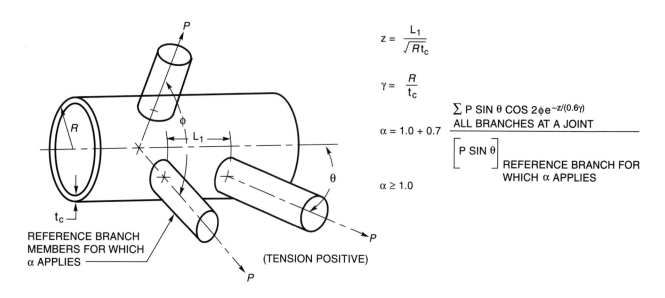

Figure R.3—Definition of Terms for Computed Alpha (See R3)

Annex S (Informative)
List of Reference Documents

This annex is not part of this standard but is included for informational purposes only.

American Welding Society (AWS) standards:

AWS B1.10M/B1.10:2009, *Guide for Nondestructive Examination of Welds*

AWS B2.1/B2.1M, *Specification for Welding Procedure and Performance Qualification*

AWS B4.0M, *Standard Methods for Mechanical Testing of Welds (Metic customary units)*

AWS D1.3/D1.3M:2018, *Structural Welding Code—Sheet Steel*

AWS D1.6/D1.6M:2017, *Structural Welding Code—Stainless Steel*

AWS *Welding Handbook*, Volume 1, 9th Edition, Chapter 13

American Institute of Steel Construction (AISC) standard:

Specification for Structural Steel Buildings

American Society of Mechanical Engineers (ASME) standard:

ASME *Boiler and Pressure Vessel Code*, Section IX.

American Society for Testing and Materials (ASTM) standards:

ASTM E1025, *Standard Practice for Design, Manufacture, and Material Grouping Classification of Hole-Type Image Quality Indicators (IQI) Used for Radiography*

ASTM E1936, *Standard Reference Radiograph for Evaluating the Performance of Radiographic Digitization Systems*

Annex T (Informative)
Guidelines for the Preparation of Technical Inquiries for the Structural Welding Committee

This annex is not part of this standard but is included for informational purposes only.

T1. Introduction

The American Welding Society (AWS) Board of Directors has adopted a policy whereby all official interpretations of AWS standards are handled in a formal manner. Under this policy, all interpretations are made by the committee that is responsible for the standard. Official communication concerning an interpretation is directed through the AWS staff member who works with that committee. The policy requires that all requests for an interpretation be submitted in writing. Such requests will be handled as expeditiously as possible, but due to the complexity of the work and the procedures that must be followed, some interpretations may require considerable time.

T2. Procedure

All inquiries shall be directed to:

Director
Standards Development
American Welding Society 8669 NW 36 Street, # 130
Miami, FL 33166

All inquiries shall contain the name, address, and affiliation of the inquirer, and they shall provide enough information for the committee to understand the point of concern in the inquiry. When the point is not clearly defined, the inquiry will be returned for clarification. For efficient handling, all inquiries should be typewritten and in the format specified below.

T2.1 Scope. Each inquiry shall address one single provision of the code, unless the point of the inquiry involves two or more interrelated provisions. The provision(s) shall be identified in the scope of the inquiry along with the edition of the code that contains the provision(s) the inquirer is addressing.

T2.2 Purpose of the Inquiry. The purpose of the inquiry shall be stated in this portion of the inquiry. The purpose can be either to obtain an interpretation of a code's requirement, or to request the revision of a particular provision in the code.

T2.3 Content of the Inquiry. The inquiry should be concise, yet complete, to enable the committee to quickly and fully understand the point of the inquiry. Sketches should be used when appropriate and all paragraphs, figures, and tables (or the Annex), which bear on the inquiry shall be cited. If the point of the inquiry is to obtain a revision of the code, the inquiry must provide technical justification for that revision.

T2.4 Proposed Reply. The inquirer should, as a proposed reply, state an interpretation of the provision that is the point of the inquiry, or the wording for a proposed revision, if that is what inquirer seeks.

T3. Interpretation of Code Provisions

Interpretations of code provisions are made by the Structural Welding Committee. The secretary of the committee refers all inquiries to the chair of the particular subcommittee that has jurisdiction over the portion of the code addressed by the inquiry. The subcommittee reviews the inquiry and the proposed reply to determine what the response to the inquiry should be. Following the subcommittee's development of the response, the inquiry and the response are presented to the entire Structural Welding Committee for review and approval. Upon approval by the committee, the interpretation is an official interpretation of the Society, and the secretary transmits the response to the inquirer and to the *Welding Journal* for publication.

T4. Publication of Interpretations

All official interpretations shall appear in the *Welding Journal* and will be posted on the AWS website.

T5. Telephone Inquiries

Telephone inquiries to AWS Headquarters concerning the *Structural Welding Code* should be limited to questions of a general nature or to matters directly related to the use of the code. The AWS Board of Directors' policy requires that all AWS staff members respond to a telephone request for an official interpretation of any AWS standard with the information that such an interpretation can be obtained only through a written request. Headquarters staff cannot provide consulting services. However, the staff can refer a caller to any of those consultants whose names are on file at AWS Headquarters.

T6. The Structural Welding Committee

The activities of the Structural Welding Committee regarding interpretations are limited strictly to the interpretation of code provisions or to consideration of revisions to existing provisions on the basis of new data or technology. Neither AWS staff nor the committees are in a position to offer interpretive or consulting services on: (1) specific engineering problems, or (2) code requirements applied to fabrications outside the scope of the code or points not specifically covered by the code. In such cases, the inquirer should seek assistance from a competent engineer experienced in the particular field of interest.

AWS D1.1/D1.1M:2020

Commentary on Structural Welding Code—Steel

19th Edition

Prepared by the
AWS D1 Committee on Structural Welding

Under the Direction of the
AWS Technical Activities Committee

Approved by the
AWS Board of Directors

This page is intentionally blank.

Foreword

This foreword is not part of the Commentary of AWS D1.1/D1.1M:2020, Structural Welding Code—Steel, but is included for informational purposes only.

This commentary on AWS D1.1/D1.1M:2020 has been prepared to generate better understanding in the application of the code to welding in steel construction.

Since the code is written in the form of a specification, it cannot present background material or discuss the Structural Welding Committee's intent; it is the function of this commentary to fill this need.

Suggestions for application as well as clarification of code requirements are offered with specific emphasis on new or revised sections that may be less familiar to the user.

Since publication of the first edition of the code, the nature of inquiries directed to the American Welding Society and the Structural Welding Committee has indicated that there are some requirements in the code that are either difficult to understand or not sufficiently specific, and others that appear to be overly conservative.

It should be recognized that the fundamental premise of the code is to provide general stipulations applicable to any situation and to leave sufficient latitude for the exercise of engineering judgment.

Another point to be recognized is that the code represents the collective experience of the committee and while some provisions may seem overly conservative, they have been based on sound engineering practice.

The committee, therefore, believes that a commentary is the most suitable means to provide clarification as well as proper interpretation of many of the code requirements. Obviously, the size of the commentary had to impose some limitations with respect to the extent of coverage.

This commentary is not intended to provide a historical background of the development of the code, nor is it intended to provide a detailed resume of the studies and research data reviewed by the committee in formulating the provisions of the code.

Generally, the code does not treat such design considerations as loading and the computation of stresses for the purpose of proportioning the load-carrying members of the structure and their connections. Such considerations are assumed to be covered elsewhere, in a general building code, bridge specification, or similar document.

As an exception, the code does provide allowable stresses in welds, fatigue provisions for welds in cyclically loaded structures and tubular structures, and strength limitations for tubular connections. These provisions are related to particular properties of welded connections.

The Committee has endeavored to produce a useful document suitable in language, form, and coverage for welding in steel construction. The code provides a means for establishing welding standards for use in design and construction by the Owner or the Owner's designated representative. The code incorporates provisions for regulation of welding that are considered necessary for public safety.

The committee recommends that the Owner or Owner's representative be guided by this commentary in application of the code to the welded structure. The commentary is not intended to supplement code requirements, but only to provide a useful document for interpretation and application of the code; none of its provisions are binding.

It is the intention of the Structural Welding Committee to revise the commentary on a regular basis so that commentary on changes to the code can be promptly supplied to the user. In this manner, the commentary will always be current with the edition of the *Structural Welding Code—Steel* with which it is bound.

Changes in the commentary have been indicated by underlining. Changes to illustrations are indicated by vertical lines in the margin.

This page is intentionally blank.

Commentary on Structural Welding Code—Steel

C-1. General Requirements

Users of this code may notice that there is no Annex C. This was done intentionally to avoid any confusion with commentary references which use "C-" to identify text, tables, and figures that are part of the commentary. Identifying commentary on code text is relatively straightforward. If there exists commentary on code text, for example 4.3.2 of the code, you can identify it in Clause C-4 of the commentary labeled C-4.3.2. The "C-" indicates it is commentary, the "4.3.2" identifies the portion of the code on which the commentary refers. Likewise, C-Table 5.2 indicates commentary on Table 5.2, and C-Figure 5.6 indicates commentary on Figure 5.6. If a code subclause, table, or figure does not have commentary associated with it, then there will be no labeled section in the commentary for that code component. This is why when reading through the commentary the numbering can seem erratic, and it seems some sections are missing. For example, you might read the following three sections of the commentary in order: C-7.22.6.1, C-7.22.6.2, C-7.22.6.4. You may notice in this sequence that C-7.22.6.3 is missing. This is not a typo but rather an indication that no commentary exists on 7.22.6.3 of the code. The commentary has tables and figures of its own. These are identified in a slightly different way, be careful not to confuse these with commentary on code table and figures. For example, the first table in Clause C-11 supporting commentary is labeled "Table C-11.1." Notice that the "C-" does not come before "Table" but instead after it. "C-Table 5.2" is commentary on Table 5.2 in the code; "Table C-11.1" is the first commentary table in Clause C-11. The same is true for the labeling of figures, e.g., "C-Figure 5.1" and "Figure C-5.1."

C-1.1 Scope

The *Structural Welding Code—Steel,* hereinafter referred to as the code, provides welding requirements for the construction of steel structures. It is intended to be complimentary with any general code or specification for design and construction of steel structures.

When using the code for other structures, Owners, architects, and Engineers should recognize that not all of its provisions may be applicable or suitable to their particular structure. However, any modifications of the code deemed necessary by these authorities should be clearly referenced in the contractual agreement between the Owner and the Contractor.

C-1.2 Standard Units of Measurement

D1.1 has a two unit system: U.S. Customary and SI (metric) Units. Throughout the code, the user will find dimensions in U.S. Customary Units followed by SI (metric) in brackets [ ]. The SI Units are "soft" conversions of the U.S. Customary Units; that is each "soft' conversion value has been rounded off from the SI value using a conversion that is close as opposed to a rational value based on a conversion factor. For example, the soft conversion of 1/2 in is 12 mm, and the hard conversion is 12.7 mm. Similarly, the soft conversion for each inch is 25 mm and the hard conversion is 25.4 mm. It is inappropriate to pick and choose between U.S. Customary and SI tolerances; each system of units should be used as a whole, and the system used should be the same as that used in the shop drawings. In terms of WPSs, fabricators should not be required to rerun PQRs for a change in units. However, WPSs should be drafted in the appropriate units.

C-1.5 Responsibilities

C-1.5.1 Engineer's Responsibilities. The Engineer is responsible at the time of contract preparation for providing recommendations to the Owner or contracting authority with respect to the suitability of the code to meet the particular requirements of a specific structure. The Engineer may change any code requirements, but the basis for such changes should be well documented and take into consideration the suitability for service using past experience, experimental evidence or engineering analysis, considering material type, load effects and environmental factors.

The Engineer may recommend, from time to time during the course of the project, additional changes to code provisions for the good of the project. Such changes should be documented. The effect upon the contractual relationship should be resolved between the parties involved.

Common examples of contract-letting code modifications include resolution of unforeseen project difficulties, handling of minor nonconformances, and dealing with specific code violation issues. For example, acceptance of a minor nonconformance with due consideration of service requirements may be more desirable for the overall project than mandating a repair that would result in full code conformance, but a less desirable final product.

The fundamental premise of the code is to provide general stipulations applicable to most situations. Acceptance criteria for production welds different from those specified in the code may be used, but there should be a basis of such alternate acceptance criteria, such as past experience, experimental evidence, or engineering analysis.

After the contract is awarded, the Engineer can change requirements in the code, but the changes should be documented and agreed upon between the parties involved. The Engineer cannot unilaterally modify or change any provision of the code after the contracts are awarded without potentially creating conflict with the terms of the contract. These types of modifications should be mutually agreed upon between the parties involved in order to satisfactorily address unexpected circumstances.

The Engineer is required to determine the suitability of the particular joint detail for the specific welded assembly. Prequalified joint details, as well as the particular joint details that may be qualified by testing, may not be suitable for all loading conditions or restraint conditions. Consideration should be made to through-thickness properties of steels, likelihood of lamellar tearing, sizes and proportions of members being joined, and other factors.

C-1.5.1(1) Certain provisions of the code are mandatory only when specified by the Engineer. This is required by the code to be done in contract documents.

C-1.5.1(2) The Engineer has the authority and the responsibility to determine what NDT (if any) will be specified for a specific project. The Engineer should take into consideration the consequences of failure, the applicability of the inspection process to the specific weld involved, and recognize limitations of the NDT methods specified and the extent of that NDT.

C-1.5.1(3) Verification inspection is not required by the code and, if used, is required by the code to be specified by the Engineer (see 8.1.2.2). The Engineer may elect to have no verification inspection, verification inspection of only a portion of the fabrication, or verification inspection that totally replaces the Contractor's inspection. However, when the Engineer elects to eliminate the Contractor's inspection, the Engineer should be aware that there are a great number of responsibilities assigned to the Contractor's Inspector that include activities that may not be traditionally considered as part of verification inspection (see 8.1.2.1, 8.2, 8.3, 8.5, 8.9, and 10.24). These activities are important for controlling weld quality. It should not be assumed that NDT, no matter how extensive, will eliminate the need for control of these activities.

C-1.5.1(5) Notch toughness for weld metal, base metal, and/or HAZs is not mandated by this code. Such requirements, when necessary, are required by the code to be specified in the contract documents.

C-1.5.1(6) The code contains provisions for both statically and cyclically loaded, nontubular applications. The criteria for such fabrications differ, and as such, the applicable steel form and loading conditions are required by the code to be specified in contract documents.

C-1.5.1(8) The Engineer is responsible to specify additional fabrication and inspection requirements that are not necessarily addressed in the code. These additional requirements may be necessary because of conditions such as: extreme operating temperatures (either hot or cold) of the structure, unique material fabrication requirements, etc.

C-1.5.1(9) For OEM applications (see Clause 3), some of the responsibilities of the Engineer are performed by the Contractor. The code requires that contract documents define these responsibilities (see C-Clause 3).

C-1.5.2 Contractor's Responsibilities. The abbreviated list in 1.5.2 highlights major areas of the Contractor's responsibilities, and is not exhaustive. Responsibilities for Contractors are contained throughout the code.

C-1.5.3 Inspector's Responsibilities. Subclause 1.5.3 highlights major areas of responsibility for the various inspectors and is not exhaustive. Clause 8 highlights specific responsibilities.

C-1.7 Mandatory and Nonmandatory Provisions

C-1.7.1.2 Should. "Should" provisions are advisory (see 7.28 for example—arc strikes should be avoided, but they are not prohibited). However, if they are present, they "shall" (i.e., are required to) be removed.

Certain code provisions are options that are given to the Contractor (see 7.26 as an example where peening is allowed (may) but not required (shall) on intermediate weld layers).

C-1.7.1.3 May. Some code provisions are not mandatory unless the Engineer invokes them in the contract documents.

C-3. Terms and Definitions

C-Engineer. The code does not define the Engineer in terms of education, professional registration, professional license, area of specialization, or other such criterion. The code does not provide for a test of the Engineer's competence or ability. However, the assumption throughout the code as it relates to responsibilities and authorities assigned to the Engineer is that the individual is competent and capable of executing these responsibilities. Applicable building codes may have requirements to be met by the Engineer. These requirements may include, but not be limited to, compliance with local jurisdictional laws and regulations governing the Practice of Engineering.

C-Contractor's Inspector. In past editions of this code, the term "fabrication-erection inspector" was used to designate the individual who oversees the Contractor's work. Specific responsibilities of the Contractor's Inspector are defined in 8.1. In some industries, this may be referred to as "quality control" or "QC inspection."

C-Verification Inspector. The Verification Inspector's duties are identified by the Engineer. The Engineer has the responsibility to determine whether or not verification inspection will be required for a specific project, and when required, to define the responsibilities of that Inspector. In some industries, this type of inspection is called "quality assurance" or "QA inspection." Building codes may specify verification inspection requirements. The Engineer should then identify such requirements in contract documents.

C-Inspector(s) (unmodified). When the word "Inspector" is used without the modifying term "Contractor's" or "Verification," the provision is equally applicable to both types of Inspectors (see 8.1.4 as an example).

C-OEM (Original Equipment Manufacturer). The primary industries and applications governed by this code typically involve separate entities, fitting into the broad categories of Contractor and Engineer. For some applications of this code, one entity functions as both the Engineer and Contractor. In this code, this is referred to as an Original Equipment Manufacturer (OEM). Examples would include metal building systems, equipment skids and platforms, material storage systems, transmission towers, light poles, and sign structures. For these situations, contract documents should define how the various responsibilities are handled. By definition, this code separates the functions of the Engineer from those of the Contractor, and yet these are merged for OEM applications. Many possible arrangements exist, but the following general categories capture many examples of OEM applications:

- OEM 1—The OEM assumes responsibility for "turn-key products," and the Owner has no involvement in engineering or inspection issues.

- OEM 2—A turnkey product is delivered, but the Owner supplies his own Verification Inspector who reports findings to the Owner.

- OEM 3—The duties of the code-defined Engineer are addressed by both the OEM and the Owner's Engineer.

To address each of the preceding situations, examples of possible contractual language are included below. These should be reviewed to be certain they are applicable to the specific situation.

Sample Language for OEM 1:

"D1.1 shall be used. The Contractor's Engineer shall assume the responsibilities of the Engineer as defined in Clause 3. Deviations from the code requirements as described in 1.5.1, shall not be allowed."

Sample Language for OEM 2:

"D1.1 shall be used. The Contractor's Engineer shall assume the responsibilities of the Engineer as defined in Clause 3, except all references to the "Engineer" in Clause 8 shall mean the "Owner." Verification inspection shall be as determined by the Owner, and the Verification Inspector shall report the results to the Owner. In addition, decisions made by the Contractor's Engineer that require changes to the code as described by 1.5.1 shall be submitted to the Owner for approval."

Sample Language for OEM 3:

No specific suggested language is supplied here because the number of permutations is too great. The user is encouraged to look at each reference to the Engineer and resolve how each situation is to be handled. As an example, the contents of Clauses 1, 4, and 8 may be assigned to the Owner's Engineer, and the responsibilities of Clauses 5, 6, 7, and 9 assigned to the Contractor's Engineer.

C-4. Design of Welded Connections

C-4.3 Contract Plans and Specifications

C-4.3.2 Notch Toughness Requirements. Notch toughness is a material property which provides a measure of its sensitivity to brittle fracture. The CVN test is the most common method of measuring notch toughness. Other tests are available and may be more reliable, but they are also more complex and expensive. More precise measures of toughness are not justified unless fracture mechanics methods are used in design.

The demand for toughness depends upon load type, rate of application of the load, temperature and other factors. Redundancy and the consequences of fracture may also be considered in determining CVN test requirements for a welded joint. Many applications do not require a measure of notch toughness. In applications where a minimum CVN test value is required, the specification of a filler metal classification that includes CVN test values may suffice. Many filler metal classifications are available that provide CVN test criteria. Most filler metals used in structural field applications are not tested for CVN test values. Of the filler metals that are tested for CVN test values and used in structural applications, the most common meet 20 ft·lbf at –20° or 0°F [27 J at –29° or –18°C]. In more severe cases, WPSs can be qualified to meet CVN test values. It should be recognized that a CVN test criteria in filler metal or in a WPS qualification relates to a material's susceptibility to brittle fracture but is not a precise measure of the material property in a production joint. The goal of most CVN test requirements is to provide some assurance that the material is not on its lower shelf of notch toughness at the structure's service temperature.

Structural shapes and plates have been surveyed and CVN test resulted in values of 15 ft·lbf [20 J] or higher at 40°F [4°C]. These surveys were conducted at the request of mill producers in order to show that CVN testing of base metal was unnecessary for most building applications (see Reference 6). Subclauses 6.8.1, C-10.14.4.4, and Clause 6, Part D contain information on CVN test values (see also *Fracture and Fatigue Control in Structures*, Barsom and Rolfe [Reference 1]).

C-4.3.4 Weld Size and Length. The Engineer preparing contract design drawings cannot specify the depth of groove "D" without knowing the welding process and the position of welding. The code is explicit in stipulating that only the weld size "(S)" is to be specified on the design drawings for PJP groove welds This allows the Contractor to produce the weld size assigning a depth of preparation to the groove shown on the shop drawings as related to the Contractor's choice of welding process and position of welding.

The root penetration will generally depend on the angle at the root of the groove in combination with the root opening, the welding position, and the welding process. For joints using bevel and V-groove welds, these factors determine the relationship between the depth of preparation and the weld size for prequalified PJP groove welds.

The strength of fillet welds is dependent upon the throat size; however, leg size of fillet welds is more useful and measurable dimension for execution of the work. On both the contract documents and shop drawings, when the parts joined meet at an angle between 80° and 100°, the effective size is taken to be the throat dimension of a 90° fillet weld, and is designated on both contract documents and on shop drawings by leg size.

In the acute angle side of significantly skewed T-joints [see Figure 5.4(A), (B), and (D)], the relationship between leg size and effective throat is complex. When the parts meet at angles less than 80° or greater than 100°, the contract documents show the required effective throat to provide for design conditions, and shop drawings show leg size required to provide specified effective throat.

When the acute angle is between 30 and 60°, the effective weld size is dependent upon the Z-loss reduction [see Figure 5.4(D)] which is dependent upon the welding process and position. Specifying only the effective throat size required to satisfy design conditions on the contract documents enables the fabricator, using welding processes suitable to his equipment and practice, to indicate his intent and instructions by appropriate WPSs, and symbols on the shop drawings.

C-4.3.5.4 Prequalified Detail Dimensions. The background and basis for prequalification of joints is explained in C-5.5.1 Designers and detailers should note that the prequalification of joint geometries is based upon proven satisfactory conditions of shape, clearances, welding position and access to a joint between plate elements for a qualified welder to deposit sound weld metal well fused to the base metal. Other design considerations important to the suitability of a particular joint for a particular application are not part of the prequalified status. Such considerations include, but are not necessarily limited to:

(1) the effect of restraint imposed by rigidity of connected base metal on weld metal contraction,

(2) the potential for causing lamellar tearing by large weld deposits under restrained conditions on base metal stressed in the through-thickness direction,

(3) limitations on welder access to the joint for proper positioning and manipulation of the electrode imposed by base metal nearby but not part of the joint,

(4) the potential for biaxial or triaxial state of stress at intersecting welds,

(5) limitations on access to allow reliable UT or RT inspection,

(6) effect of tensile residual stresses from weld shrinkage,

(7) effect of larger than necessary welds on distortion.

C-4.4 Effective Areas

C-4.4.1.4 Effective Size of Flare-Groove Welds. Rectangular hollow structural sections are formed in a manner that may not result in a 90° angle. The research supporting Table 4.1 accounts for this practice and 2t was found to be acceptable.

C-4.4.2.5 Maximum Effective Length. When longitudinal fillet welds parallel to the stress are used to transmit the load to the end of an axially loaded member, the welds are termed "end loaded." Typical examples of such welds would include, but are not necessarily limited to, longitudinally welded lap joints at the end of axially loaded members, welds attaching bearing stiffeners, welds attaching transverse stiffeners to the webs of girders designed on basis of tension field action, and similar cases. Typical examples of longitudinally loaded fillet welds which are not considered end loaded include, but are not limited to, welds that connect plates or shapes to form a built-up cross sections in which the shear force is applied to each increment of length of weld stress depending upon the distribution of shear load along the length of the member, welds attaching beam web connection angles and shear plates because the flow of shear force from the beam or girder web to the weld is essentially uniform throughout the weld length, that is, the weld is not end-loaded despite the fact that it is loaded parallel to the weld axis. Neither does the reduction factor apply to welds attaching stiffeners to webs designed on basis of conventional beam shear because the stiffeners and welds are not subject to calculated axial stress but merely serve to keep the web flat. The distribution of stress along the length of end loaded fillet welds is far from uniform and is dependent upon complex relationships between the stiffness of longitudinal fillet weld relative to the stiffness of the connected base metals. Beyond some length, it is nonconservative to assume that the average stress over the total length of the weld may be taken as equal the full allowable stress. Experience has shown that when the length of the weld is equal to approximately 100 times the weld size or less, it is reasonable to assume the effective length is equal to the actual length. For weld lengths greater than 100 times the weld size, the effective length should be taken less than the actual length. The reduction coefficient, β, provided in 4.4.2.5 is the equivalent (in U.S. Units and terminology) of Eurocode 3, which is a simplified approximation to exponential formulas developed by finite element studies and tests performed in Europe over many years. The criteria is based upon combined consideration of ultimate strength for fillet welds with leg size less than 1/4 in [6 mm] and upon judgment based serviceability limit of slightly less than 1/32 in [1 mm] displacement at the end of the weld for welds with leg size 1/4 in [6 mm] and larger. Mathematically, multiplication of the actual length by the β factor leads to an expression which implies that the effective length reaches a maximum when the actual length is approximately 300 times the leg size; therefore the maximum effective length of an end loaded fillet weld is taken as 180 times the weld leg size.

C-4.6 Stresses

C-4.6.1 Calculated Stresses. It is intended that the calculated stresses to be compared with the allowable stresses be nominal stresses determined by appropriate analysis methods and not "hot spot" stresses which might be determined by finite element analysis using a mesh finer than approximately one foot. Some applicable invoking design specifications require that certain joints be designed to provide, not only for the calculated forces due to applied loads, but also for a certain minimum percentage of the strength of the member, regardless of the magnitude of the forces applied to the joint. Examples of such requirements are to be found in the AISC Specifications.

C-4.6.2 Calculated Stresses due to Eccentricity. Tests have shown that balancing welds about the neutral axis of a single angle or double-angle member or similar members does not increase the load carrying capacity of the connection. Therefore, unbalanced welds are allowed. It should be noted that end returns are not necessary, as tearing is not a problem (see Figure C-4.1).

C-4.6.4 Allowable Weld Metal Stresses. The philosophy underlying the code provisions for stresses in welds can be described by the following principles:

(1) The weld metal in CJP groove welds subject to tension stresses normal to the effective area should have mechanical properties closely comparable to those of the base metal. This, in effect, provides nearly homogenous weldment of unreduced cross section so that stresses used in proportioning the component parts may be used in and adjacent to the deposited weld metal. For stresses resulting from other directions of loading, lower strength weld metal may be used, provided that the strength requirements are met.

(2) For fillet welds and PJP groove welds, the designer has a greater flexibility in the choice of mechanical properties of weld metal as compared with those components that are being joined. In most cases, the force to be transferred by these welds is less than the capacity of the components. Such welds are proportioned for the force to be transferred. This can be achieved with weld metal of lower strength than the base metal, provided the throat area is adequate to support the given force. Because of the greater ductility of the lower strength weld metal, this choice may be preferable.

A working stress equal to 0.3 times the tensile strength of the filler metal, as designated by the electrode classification, applied to the throat of a fillet weld has been shown by tests (Reference 7) to provide a factor of safety ranging from 2.2 for shear forces parallel to the longitudinal axis of the weld, to 4.6 for forces normal to the axis under service loading. This is the basis for the values given in Table 4.3.

(3) The stresses on the effective throat of fillet welds is always considered to be shear. Although a resistance to failure of fillet welds loaded perpendicular to their longitudinal axis is greater than that of fillet welds loaded parallel to this axis, higher load capacities have not been assigned in Table 4.3 for fillet welds loaded normal to their longitudinal axis.

Alternative criteria allowing higher allowable stresses for fillet welds loaded obliquely to the longitudinal axis of the weld is provided in 4.6.4.2.

(4) The load-carrying capacity of any weld is determined by the lowest of the capacities calculated in each plane of stress transfer. These planes for shear in fillet and groove welds are illustrated in Figure C-4.2.

    (a) Plane 1-1, in which the capacity may be governed by the allowable shear stress for material "A."

    (b) Plane 2-2, in which the capacity is governed by the allowable shear stress of the weld metal.

    (c) Plane 3-3 in which the capacity may be governed by the allowable shear stress for material "B."

C-4.6.4.2 Alternative Allowable Fillet Weld Stress. It has long been recognized that the strength and deformation performance of fillet weld elements is dependent on the angle Θ that the force makes with the axis of the weld element. Transversely loaded fillet welds have approximately 50% greater strength than longitudinally loaded welds. Conversely, it has been known that transversely loaded fillet welds have less distortion capacity prior to fracture than longitudinally loaded fillet welds. Following the tests by Higgins and Preece, Welding Journal Research Supplement, October 1968 (Reference 7), in the interests of simplicity and because the methods for handling interaction between longitudinal and transverse loading cases were not available, the allowable stress on fillet welds in the code has been limited to $0.3 F_{EXX}$.

This value is based upon the lower strength test results for welds loaded longitudinally with a factor safety against rupture of approximately 2.2 to 2.7. The same basic criteria still applies; however, the code now provides the option, of a higher allowable stress for fillet welds based upon calculation of a value specific to the angle of loading.

The ultimate shear strength of a single fillet weld element at various angles of application of load was originally obtained from the load-deformation relationships by Butler (1972) for E60 electrodes (Reference 3). Curves for E70 electrodes were obtained by Lesik (1990) (Reference 4). The strength and deformation performance of welds is dependent on the angle Θ that the resultant elemental force makes with the axis of the weld element (see Figure C-4.3). The actual load deformation relationship for fillet welds taken from Lesik is shown in Figure C-4.4.

The following is the formula for weld ultimate stress, F_v.

$$F_v = 0.852\ (1.0 + 0.50\ \sin^{1.5}\Theta)\ F_{EXX}$$

Because the allowable stress is limited to $0.3\ F_{EXX}$ for longitudinally loaded welds ($\Theta = 0°$), the test results indicate that formulas in 4.6.4.2 and 4.6.4.3 provide a safety factor greater than the commonly accepted value of 2.

C-4.6.4.3 Instantaneous Center of Rotation. When weld groups are loaded in shear by an external load that does not act through the center of gravity of the group, the load is eccentric and will tend to cause a relative rotation and translation between the parts connected by the welds. The point about which rotation tends to take place is called the instantaneous center of rotation. Its location is dependent upon the load eccentricity, geometry of the weld group and the deformation of the weld at different angles of the resultant elemental force relative to the weld axis. The individual resistance force of each unit weld element can be assumed to act on a line perpendicular to a ray passing through the instantaneous center of rotation and the elements location (see Figure C-4.3).

The total resistance of all weld elements combine to resist the eccentric load, and when the correct location of the instantaneous center of rotation has been selected, the in plane equations of statics (Σ_x, Σ_y, ΣM) will be satisfied. A complete explanation of the procedure, including sample problems is given in Tide (Reference 5). Numerical techniques, such as those given in Brandt (Reference 2), have been developed to locate the instantaneous center of rotation to convergence tolerance. To eliminate possible computational difficulties, the maximum deformation in the weld elements is limited to the lower bound value of 0.17W. For design convenience, a simple elliptical formula is used for $F(\rho)$ to closely approximate the empirically derived polynomial in Lesik (Reference 4).

C-4.6.4.4 Concentrically Loaded Weld Groups. A weld loaded transverse to its longitudinal axis has a greater capacity but a lower reduced deformation capability as compared to a weld loaded along its longitudinal axis. Therefore, when welds loaded at varying angles are combined in a single weld group the designer must account for their relative deformation capabilities to ensure that the entire weld group has strain compatibility. The procedure in 4.6.4.3 accounts for strain compatibility.

Since the introduction of the equation in 4.6.4.2, which gives the strength of a linear fillet weld loaded in-plane through its center of gravity as: $Fv = 0.3\ F_{EXX}\ [1 + 0.5\ \sin^{1.5}(\Theta)]$ many engineers have simply added the capacity of the individual line welds in a weld group while neglecting load-deformation compatibility. This is an unconservative practice.

The design of a concentrically but obliquely loaded weld group, in addition to being performed using 4.6.4.4 as demonstrated in Figure C-4.5, can be accomplished graphically using Figure C-4.6, the load-deformation curves. For example, to find the strength of the concentrically loaded weld group shown in Figure C-4.5, first the weld element with the most limited deformation capacity is determined. In this case it is the transversely loaded weld. By drawing a vertical line from the point of fracture, the strength increase or decrease for the remaining elements can be determined.

C-4.7 Joint Configuration and Details

C-4.7.1 General Considerations. In general, details should minimize constraint which would inhibit ductile behavior, avoid undue concentration of welding, as well as afford ample access for depositing the weld metal.

C-4.7.3 Base Metal Through-Thickness Loading. The rolling of steel to produce shapes and plates for use in steel structures causes the base metal to have different mechanical properties in the different orthogonal directions. This makes it necessary for the designers, detailers and the fabricators to recognize the potential for laminations and/or lamellar tearing to affect the integrity of the completed joints, especially when thick base metal is involved.

Laminations do not result from the welding. They are a result of the steel manufacturing processes. They generally do not affect the strength of base metal when the plane of the lamination is parallel to the stress field, that is, stressed in the longitudinal or transverse direction. They do have direct effect upon the ability of base metal at T- and corner joints to transmit through-thickness forces.

Lamellar tears, if and when they occur, generally are the result of the contraction of large weld metal deposits under conditions of high restraint. Lamellar tears rarely occur when the weld size is less than about 3/4 in to 1 in [20 mm to 25 mm]. Lamellar tears rarely occur under fillet welds. Lamellar tears do not occur in the absence of restraint to contraction of hot solidified weld metal; however, in large welds, the solidified initial weld passes deposited in the root area of the weld, can provide an internal rigid abutment to tensile contraction strains of the subsequently deposited weld passes.

Because lamellar tears are caused by solidified weld metal contraction that is forced to be accommodated within a short gage length by localized balancing compressive restraint, the unit through-thickness direction strains in the base metal can be many times larger than yield point strain. Lamellar tears may result. The localized strains that can produce lamellar tears occur upon cooling during fabrication and constitute the most severe condition that will be imposed upon the base metal in the vicinity of the joint in the life of the structure. Because the compressive and tensile stresses within or, in close proximity to, the joint are self equilibrating, and because the strains associated with applied design stresses are a small fraction of those associated with weld shrinkage, externally applied loads do not initiate lamellar tears; however, if tears have been initiated by the welding, existing lamellar tears may be extended.

The design and detailing of T- and corner joints establish the conditions which may increase or decrease the potential for lamellar tearing, and make the fabrication of a weldment a straight forward operation or a difficult or virtually impossible one. Therefore, attention on the part of all members of the team, designer, detailer, fabricator and welder is necessary to minimize the potential for lamellar tearing.

Definitive rules cannot be provided in the code to assure lamellar tearing will not occur; hence, this commentary is intended to provide understanding of the causes and to provide guidance on means to minimize the probability of occurrence. The following precautions have been demonstrated in tests and experience to minimize the risk of tearing:

(1) Base metal thickness and weld size should be adequate to satisfy design requirements; however, designing joints on basis of stresses lower than the code allowable stresses, rather than providing a conservative design, results in increased restraint and increased weld size and shrinkage strain that will need to be accommodated. Therefore such a practice increases rather than diminishes the potential for lamellar tearing.

(2) Use low-hydrogen electrodes when welding large T- and corner joints. Absorbed hydrogen is not deemed to be a principal cause for lamellar tearing initiation, but the use of low-hydrogen electrodes on large joints (longitudinal, transverse or through thickness) to minimize the tendency for hydrogen-induced cold cracking is good practice in any case. Use of nonlow-hydrogen electrodes may invite trouble.

(3) Application of a layer of "buttering" weld passes approximately 1/8 in to 3/16 in [3 mm to 5 mm] thick to the face of the base metal to be stressed in the through thickness direction prior to assembly of the joint has been demonstrated in tests and experience to reduce the likelihood of lamellar tearing. Such a "buttered" layer provides tough weld metal with a cast grain structure in lieu of the fibrous anisotropic rolled steel grain structure at the location of the most intense weld shrinkage strains.

(4) In large joints, sequence weld passes in a manner that builds out the surface of base metal stressed in the longitudinal direction prior to depositing weld beads against the face of the base metal stressed in the through-thickness direction. This procedure allows a significant part of the weld shrinkage to take place in the absence of restraint.

(5) On corner joints, where feasible, the beveled joint preparation should be on the base metal stressed in the through-thickness direction so that the weld metal fuses to the base metal on a plane into the thickness of the base metal to the maximum degree practical.

(6) Double-V and double-bevel joints require deposition of much less weld metal than single-V and single bevel joints, and therefore, reduce the amount of weld shrinkage to be accommodated by approximately one-half. Where practical, use of such joints may be helpful.

(7) In weldments involving several joints of different thickness base metal, the larger joints should be welded first so that the weld deposits which may involve the greatest amount of weld shrinkage may be completed under conditions of lowest restraint possible. The smaller joints, although welded under conditions of higher restraint, will involve a smaller amount of weld shrinkage to be accommodated.

(8) The area of members to which large welds will transfer stresses in the through-thickness direction should be inspected during layout to assure that joint weld shrinkage does not apply through-thickness strains on base metal with preexisting laminations or large inclusions (see ASTM A578).

(9) Properly executed peening of intermediate weld passes has been demonstrated to reduce the potential for lamellar tearing. Root passes should not be peened in order to avoid the possibility of introducing cracks in the initial thin weld passes which may go undetected and subsequently propagate through the joint. Intermediate passes should be peened with a round nosed tool with sufficient vigor to plastically deform the surface of the pass and change the tensile residuals to compressive residual stresses, but not so vigorously to cause a chopped up surface or overlaps. Finish passes should not be peened.

(10) Avoid the use of over-strength filler metal.

(11) When practical, use base metal with low (< 0.006%) sulfur or base metal with improved through-thickness properties.

(12) Critical joints should be examined by RT or UT after the joint has cooled to ambient temperature.

(13) If minor discontinuities are detected the Engineer should carefully evaluate whether the discontinuities can be left unrepaired without jeopardizing the suitability for service or structural integrity. Gouging and repair welding will add additional cycles of heating and cooling and weld contraction under restraint conditions that are likely to be more severe than the conditions under which the joint was initially welded. Repair operations may cause a more detrimental condition.

(14) When lamellar tears are identified and repair is deemed advisable, the work should not be undertaken without first reviewing the WPS and an effort made to identify the cause of the unsatisfactory result. A special WPS or a change in the joint detail may be required.

C-4.7.4 Combinations of Welds. Fillet welds deposited over groove welds do not directly increase the effective throat of the joint; therefore the strength of the joint may not be taken as the algebraic sum of the strength of the groove weld and the strength of the fillet.

C-4.7.5 Butt, Corner, and T-Joint Surface Contouring. Reinforcing and contouring fillet welds serve a useful purpose in T- and corner joints, and in butt joints joining parts of unequal width or thickness. They provide a fillet which reduces the severity of the stress concentration that would exist at the ninety degree geometric change in section.

C-4.7.6 Weld Access Holes. Weld access holes are not required nor even desirable for every application. However it is important to recognize that any transverse joint in the flange of wide flange, H and similar cross sections made without the use of a weld access hole cannot be considered as a prequalified CJP groove welded joint. This is true, because prequalified CJP groove welded joints are limited to the cases of plain plate elements to plain plate elements shown in Figure 5.1. The decision to use prequalified CJP joints or to use non-prequalified joints without access holes depends upon consideration of several factors which include but may not be limited to the following:

(1) The size of the members being joined.

(2) Whether the joint is a shop or field weld, that is, whether the parts may be positioned for welding so that overhead welding may be avoided and reinforcing fillets may be readily deposited at the location of peak stress concentrations.

(3) The variation in the restraint to weld shrinkage and the distribution of applied stress along the length of transverse joint due to geometry of parts being joined.

For example, the high restraint due to the column web in the region of the column flange centerline as compared to lower restraint away from the centerline causes welding residual stresses and applied stresses to peak sharply in the difficult-to-weld region at the middle of beam flange.

(5) Whether, in the case of geometry that affords more uniform restraint without a "hard spot" along the length of the joint, the probability of increased number of small internal weld metal discontinuities, but without the large discontinuity of the weld access hole, might provide a higher strength joint. For example, tests of end plate moment connections (Murray 1996) have shown that joints between beam ends and end plates made without weld access holes, but with unrepaired discontinuities in the region of the web to flange junction provide higher strength than similar connections made using access holes but with fewer internal discontinuities.

Research, thought and ingenuity are being directed toward improved details for welding of beam-to-column moment connections. Alternative joint design and details to provide the strength and suitability for service should be considered where they are applicable. Engineering judgment is required.

When weld access holes are required, the minimum requirements of 7.16 apply. The minimum required size to provide clearances for good workmanship and sound welds may have a significant effect upon net section properties of the connected members.

C-4.7.7 Welds with Rivets or Bolts. In previous editions of the code, it was permitted to share loading between welds and high-strength bolts when the joint was designed and the bolts were installed as a slip-critical connection. Limitations on such use were provided in the Specifications for Structural Joints Using ASTM A325 or A490 Bolts of the Research Council on Structural Connections (RCSC Specification). Based upon research by Kulak and Grondin, (Strength of Joints that Combine Bolts and Welds; Geoffrey L. Kulak and Gilbert Y. Grondin; Engineering Journal, Second Quarter, 2003, page 89), permission was withdrawn in the 2000 version of the RCSC Specification, and the RCSC Specification provided no guidance. The AISC Specification for Structural Steel Buildings (July 2016), section J1.8, provides new guidance for the sharing of loads between bolts and welds.

C-4.8 Joint Configuration and Details—Groove Welds

C-4.8.1 Transitions in Thicknesses and Widths. In tension applications, stress concentrations that occur at changes in material thickness or width of stressed elements, or both, are dependent upon the abruptness of transition, with stress concentration factors varying between 1 and 3. In statically loaded applications, such geometric changes are not of structural significance when a notch-free transition is provided. The Engineer should consider the use of added contouring fillet welds or other details to improve the stress flow continuity in static applications at locations such as butt splices in the axially aligned tension chords of long-span trusses, joints subjected to cold temperature applications where brittle fracture is a concern, or other high-stress or severe service locations. Fatigue provisions provide for the effects of geometrical discontinuities in cyclic load applications.

C-4.9 Joint Configuration and Details—Fillet Welded Joints

C-4.9.1.1 Transverse Fillet Welds. Because transversely loaded fillet welded lap joints involve eccentricity, the applied force tends to open the joint and cause prying action on the root of the weld as shown in Detail B of Figure C-4.7 unless restrained by a force, R, shown in Detail A. The code requires that this mode of action be prevented by double fillet welds or other means.

C-4.9.2 Longitudinal Fillet Welds. The transfer of force by longitudinal fillet welds alone at the ends of members causes an effect known as shear lag in the transition region between the joint where shear stress is concentrated along the edges of the member to the location where stress in the member may be considered uniform across the cross section. The disposition of the longitudinal welds relative to the shape of the cross section affects the design of the member as well as the strength of the connection. For the simple case of flat bar and plate type cross sections, experience as well as theory have shown the requirements of 4.9.2 assures the adequacy of the connection as well as the connected parts. For other cross sections, the effective area of the connected member is dependent upon the disposition of the connecting welds at the end; therefore, reference should be made to the applicable specification for member and structure design.

C-4.9.3 Fillet Weld Terminations.

C-4.9.3.1 General. In most cases, whether fillet welds terminate at the ends or sides of a member has no effect upon the suitability for service of a joint, thus, this is the default case; however, in several situations the manner of termination is important. Separate rational rules are provided for individual cases.

C-4.9.3.2 Lap Joints Subject to Tension. When a joint is made between members in which one connected part extends beyond the edge or end of the other part, it is important that notches are avoided in the edge of a part subject to calculated tension stress. A good practice for avoiding such notches in critical locations is to strike the arc for welding slightly back from the edge and then proceed with the deposition of the weld bead in the direction away from the edge to be protected against notches.

C-4.9.3.3 Maximum End Return Length. For framing angles and simple end-plate connections in which the flexibility of the connection assumed in the design of the member is important, tests have shown that the static strength of the connection is not dependent upon the presence or absence of an end return. Therefore, a weld made along the outstanding leg of the connection (generally the vertical weld) may be stopped short of the end, or carried to the extreme top and bottom ends of the angle or returned slightly along the horizontal ends. If returns are used however, it is important to ensure that the length be limited in order that flexibility of the connection is not impaired.

C-4.9.3.4 Transverse Stiffener Welds. Experience has shown that, when stiffeners are not welded to the flanges, it is important to stop the stiffener-to-web welds a short distance away from the toe of the web-to-flange weld. If this is not done, slight twisting of the flange during normal handling and shipment will induce extremely high bending stresses in the extremely short gage length between the termination of the stiffener weld and the toe of the web-to-flange weld. A few cycles of these noncalculated stresses into the inelastic range initiate cracking which ultimately may propagate through the web or flange in service. The nonwelded length has a maximum limit to avoid column buckling in the nonstiffened portion of the web.

C-4.9.3.5 Opposite Sides of a Common Plane. An attempt to tie two fillet welds deposited on opposite sides of a common plane of contact between two parts could result in notches or masking of poor fit-up.

Certain design or application requirements may necessitate that welds be continuous. Welds may be required to be continuous for corrosion resistance, for sanitary wash down conditions, or for hot dipped galvanized applications, and others. Subclause 4.9.3.5 requires interrupted welds unless an exception is noted in the contract documents. As a general practice, fillet welds should be interrupted on the opposite sides of a common plane in order to eliminate the potential of workmanship defects, such as undercut. Due to the inherent challenges associated with making quality continuous welds, special inspection may be justified, particularly for cyclically loaded structures. MT or PT of the welds in the corners may be justified; when this is the case, this clause requires the inspection method and frequency to be specified in contract documents.

C-4.12 Built-Up Members

C-4.12.2.1 General. Irrespective of the requirement for sufficient welding to assure that parts act in unison, a maximum spacing for intermittent fillet welds is specified to assure tightness of joints that will allow painting to seal nonwelded portions of the joint and to prevent unsightly "quilting" of base metal between welds for joints that are protected from corrosion by enclosure within a building.

C-4.12.2.2 Compression Members. The criteria for spacing of intermittent fillet welds attaching outside plates of members subject to compression are derived from classical elastic plate buckling theory and are consistent with the criteria provided in the AISC *Specification for Structural Steel Buildings*.

C-4.12.2.3 Unpainted Weathering Steel. For unpainted weathering steel which will be exposed to atmospheric corrosion, based upon experience and testing, closer spacing is required to provide for resistance to bulking of corrosion products between parts causing unsightly "quilting" between welds and the potential for initiating cracking at weld terminations.

C-4.13 General

C-4.13.1 Applicability. The provisions of Part C of the code apply to structures and weldments subject to many cycles of application, removal and re-application of the live load within the elastic range of stress. This type of loading is generally termed high cycle fatigue. The maximum calculated design stresses allowed under the code are in the range of $0.60\ F_y$ or as similarly allowed by other standard invoking codes and specifications. Thus, the maximum range of stress due to application and removal of live load is generally some fraction of this stress level. Although these conditions were not adopted for the purpose of limiting the scope of applicability of the code provisions, they do establish natural boundaries which should be recognized.

Design for fatigue resistance is not normally required for building structures; however, cases involving cyclic loading which might cause crack initiation and fatigue crack propagation include but are not necessarily limited to the following:

(1) Members supporting lifting equipment.

(2) Members supporting rolling traffic loads.

(3) Members subject to wind induced harmonic vibration.

(4) Support for reciprocating machinery.

To place some perspective upon the scope of applicability, for example, if the stress range on the connected base metal of a relatively sensitive welded detail (the end of a partial length cover plate) due to cycles of application and removal of full live load is 30 ksi [210 MPa], the predicted life to fatigue failure is 36,000 cycles (4 applications per day for 25 years). For the same detail, if the range of stress is less than 4.5 ksi [32 MPa], infinite life would be expected. Therefore, if the cycles of application of full live load is less than a few thousand cycles or the range of resulting stress is less than the threshold stress range, fatigue need not be a concern.

C-4.14 Limitations

C-4.14.2 Low Cycle Fatigue. Because earthquake loading involves a relatively low number of cycles of high stress into the inelastic range, reliance upon the provisions of this Part C for design for earthquake loading is not appropriate.

C-4.14.4 Redundant–Nonredundant Members. The concept of recognizing a distinction between redundant and nonredundant members and details is not based upon consideration of any difference in the fatigue performance of any given member or detail, but rather upon the consequences of failure. Prior to the adoption of AASHTO/AWS D1.5, AASHTO Specifications provided criteria for fracture critical members which included special base metal and inspection requirements, but which incorporated by reference, reduced allowable stress range curves as in D1.1 as it related to dynamically loaded structures (discontinued 1996). The reduced allowable stress range curves, designated for nonredundant structures, were derived by arbitrarily limiting the fatigue stress ranges to approximately 80% of the stress range curves for redundant members and details. With the adoption of AASHTO/AWS D1.5, reference to AWS D1.1Code as it related to dynamically loaded structures (discontinued 1996) for allowable stresses and stresses and stress ranges has been eliminated and shifted to the AASHTO design specifications. Subsequently, within AASHTO it was decided that specifying allowable stress ranges that were only 80% of the mean minus 2 standard deviation curves for fatigue detail test data, in addition to special base metal and inspection requirements, constituted a double conservatism. Hence, in the current AASHTO LRFD Specification for the design of bridges, the lowered allowable stress ranges for design of nonredundant members and details has been eliminated while special base metal and inspection requirements have been retained.

C-4.15 Calculation of Stresses

C-4.15.1 Elastic Analysis. The criteria contained in Tables 4.5 and 10.1 are based upon fatigue testing of full size specimens typical of the cases presented. The effects of local geometrical stress concentrations are accounted for by the stress categories.

C-4.16 Allowable Stresses and Stress Ranges

C-4.16.2 Allowable Stress Range. The stress range cycle life curves criteria provided by Formulas (2) through (4) and plotted graphically in Figure 4.16 were developed through research sponsored by National Cooperative Highway Research Program on actual details which incorporated realistic geometrical discontinuities making it inappropriate to amplify calculated stresses to account for notch effect. This research is published as research reports 102 and 147 "Effects of Weldments on Strength of Beams" and "Strength of Steel Beams With Welded Stiffeners and Attachments." Subsequent research in the United States and abroad on other real details support cases contained in Tables 4.5 and 10.1 but not contained in the NCHRP test program.

When a plate element, which is connected by a transverse CJP or PJP weld or by a pair of transverse fillet welds on opposite sides of the plate, is subjected to cyclically applied load, the toes of the transverse welds are generally the critical location for crack initiation into the connected base metal. The critical stress range for crack initiation at this location is the same for each of the joint types and may be determined by Formula (2), and Category C criteria.

If the transverse weld is a PJP weld or a pair of fillet welds, consideration is given to the potential for crack initiation from the root of the welds, as well as crack initiation from the opposite weld toe when the base metal to which the load is delivered is subject to bending tensile stress. The maximum stress range for the joint to account for cracking from the root is determined by multiplying the allowable stress range criteria for cracking from the toe by a reduction factor. The relative size of the nonwelded thickness dimension of the joint to the plate thickness is the essential parameter in the reduction factor incorporated in Formula (4). For the case of a pair of fillet welds on opposite sides of the plate, $2a/t_p$ becomes unity and the reduction factor formula reduces to Formula (5) (see Reference 8).

C-4.17 Detailing, Fabrication, and Erection

C-4.17.6 Fillet Weld Terminations. In fillet welded angle brackets, angle beam seats, framing angles and similar connections, in which the applied load tends to separate the connected parts and apply a prying tensile stress at the root of the weld, the weld is required to be returned to protect the root at the start of the weld against crack initiation.

References for Clause C-4

1. Rolfe, S. T. and Barsom, J. M. *Fracture and Fatigue Control in Structures*. Prentice Hall, 1977.

2. Brandt, G. D. "A General Solution for Eccentric Loads on Weld Groups." AISC Engineering Journal, 3rd Qtr., 1982.

3. Butler, L., Pal, J. S., and Kulak, G. L. "Eccentrically Loaded Welded Connections." Journal of The Structural Division, ASCE 98 (ST5) 1972.

4. Lesik, D. F. and Kennedy, D. J. L. "Ultimate Strength of Fillet Welded Connections Loaded In-Plane." Canadian Journal of Civil Engineering, 17(1) 1990.

5. Tide, R. H. R. "Eccentrically Loaded Weld Groups—AISC Design Tables." AISC Engineering Journal 17(4) 1980.

6. R. J. Dexter, et al. University of Minnesota. 7/20/2000 MTR Survey of Plate Material Used in Structural Fabrication; R. L. Brockenbrough; AISC; 3/1/2001 Statistical Analysis of Charpy V-Notch Toughness for Steel Wide Flange Structural Shapes; J. Cattan; AISC; 7/95.

7. Higgins, T. R. and Preece, F. R. "Proposed Working Stresses of Fillet Welds in Building Construction." *Welding Journal* Supplement, October 1968.

8. Frank, K. H. and Fisher, J. W. "Fatigue Strength of Fillet Welded Cruciform Joints." Journal of The Structural Division, ASCE 105 (ST9), September 1979, pp. 1727–1740.

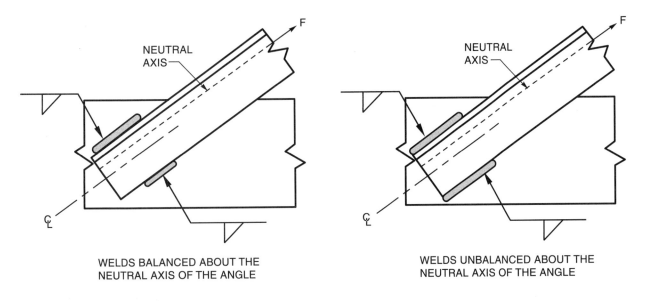

Figure C-4.1—Balancing of Fillet Welds About a Neutral Axis (see C-4.6.2)

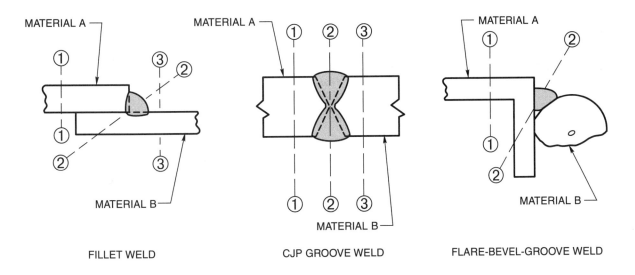

Figure C-4.2—Shear Planes for Fillet and Groove Welds (see C-4.6.4)

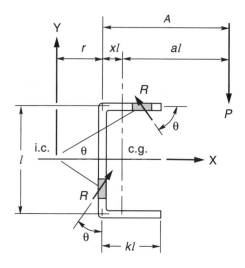

Figure C-4.3—Eccentric Loading (see C-4.6.4.2 and C-4.6.4.3)

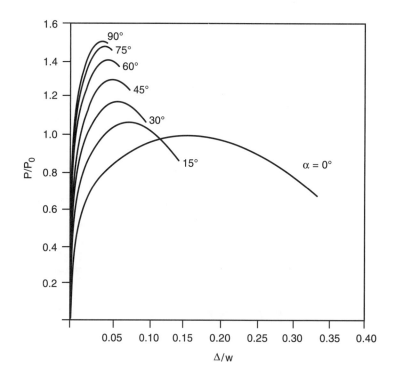

Figure C-4.4—Load Deformation Relationship for Welds (see C-4.6.4.2 and C-4.6.4.3)

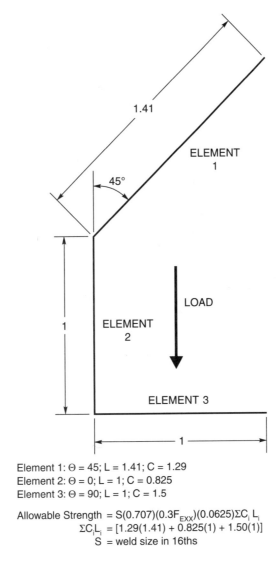

Element 1: Θ = 45; L = 1.41; C = 1.29
Element 2: Θ = 0; L = 1; C = 0.825
Element 3: Θ = 90; L = 1; C = 1.5

Allowable Strength = $S(0.707)(0.3F_{EXX})(0.0625)\Sigma C_i L_i$
$\Sigma C_i L_i = [1.29(1.41) + 0.825(1) + 1.50(1)]$
S = weld size in 16ths

Figure C-4.5—Example of an Obliquely Loaded Weld Group (see C-4.6.4.4)

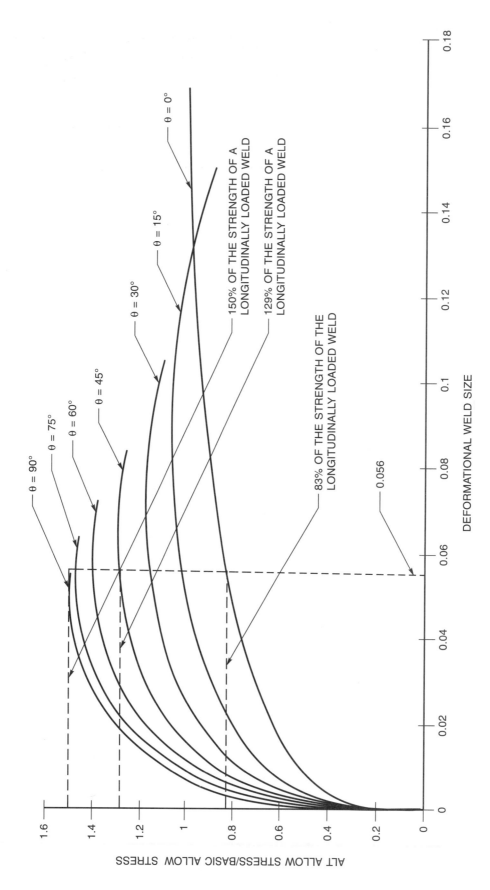

Figure C-4.6—Graphical Solution of the Capacity of an Obliquely Loaded Weld Group (see C-4.6.4.4)

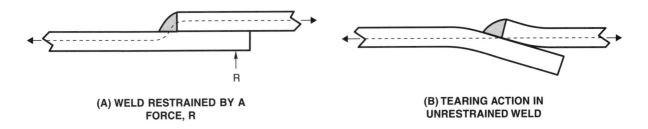

Figure C-4.7—Single Fillet Welded Lap Joints (see C-4.9.1.1

C-5. Prequalification of WPSs

Part A
WPS Development

C-5.1 Scope

Prequalified provisions are given in Clause 5, which includes WPSs, with specific reference to preheat, filler metals, electrode size, and other pertinent requirements. Additional requirements for prequalified joints in tubular construction are given in Clause 10.

The use of prequalified joints and WPSs does not necessarily guarantee sound welds. Fabrication capability is still required, together with effective and knowledgeable welding supervision to consistently produce sound welds.

C-Table 5.1 Maximum Root Pass Thickness (Variable). The thickness of a root pass can be measured by subtracting the unfilled depth from the nominal plate thickness or depth of the groove for double-sided joint preparation.

C-Table 5.1 Maximum Fill Pass Thickness (Variable). The thickness of a fill pass can be measured by subtracting the unfilled depth from the plate surface of a new fill pass/layer from the depth of the previously deposited pass/layer. Penetration into the preceding pass/layer is not counted as part of thickness.

C-Table 5.1 Electrical Limitations. Tests have demonstrated that an empirical relation appears to exist between the angle at the root of the groove and the maximum current that can be used without producing weld profiles prone to cracking, as shown in Figure C-5.1. Under these circumstances, only prequalified bevel and V-grooves without backing are effective.

J- and U-grooves have a greater angle at the root than the groove angle and, in their case, the probability of an undesirable crack-prone weld nugget is very small. However, the code makes no distinction between V-grooves and J- and U-grooves in this regard. It makes the requirements of Table 5.1 applicable to all grooves. Since the use of J-and U-grooves is less frequent, this requirement does not appear to be unreasonable.

The empirical relation defines the acceptable amount of current, in amperes, as approximately ten times the included groove angle. This applies primarily to prequalified joints welded without backing using bevel and V grooves. Since the included angle for such prequalified joints is 60°, the maximum amperage allowed by the code is 600 A; for a 90° fillet weld, the maximum current allowed is 1000 A. This limitation applies only to passes fusing both faces of the joint, except for the cover pass.

C-Table 5.1 Requirement for Multiple Electrode SAW. When using GMAW plus SAW in tandem (see Table 5.1), the maximum 15 in [380 mm] spacing between the gas metal arc and the leading submerged arc is required to preserve the preheating effects of the first arc for the subsequent main weld deposited by the remaining two high deposition rate submerged arcs. The short spacing also provides a better condition for remelting the first pass.

C-Table 5.1 Requirements for GMAW/FCAW. This section provides the requirements for GMAW and FCAW WPSs when prequalified WPSs are used.

The gas shielding at the point of welding is to be protected from the wind to prevent interruption in shielding and resulting contamination of the weld by the atmosphere.

The prequalified provisions apply only to GMAW using spray and globular transfer modes of metal deposition. GMAW-S is not prequalified and therefore must be qualified in conformance with Clause 6. Experience has shown frequent cases of lack of penetration and fusion with this mode of metal transfer. A common reason for this unreliability is the low-heat input per unit of deposited weld metal resulting in a tendency toward little or no melting of the base metal. Therefore, each user is required to demonstrate the ability of the selected WPS to produce sound welds when using GMAW-S.

C-Table 5.2 Prequalified WPS Variables

(1) **Purpose of Table 5.2.** This commentary provides explanation and examples for the application of Table 5.2 for a prequalified WPS. Current Table 5.2 was included in D1.1 in 2010 as Table 3.8, eliminating the need for the user to utilize current Table 6.5 (Table 4.5 in D1.1:2010) for prequalified WPSs. Table 5.2 identifies what must be included in a prequalified WPS. Some of the variables listed in Table 6.5 were not applicable to prequalified WPSs. An "X" in Table 5.2 indicates that a change in that variable, or a change outside the range indicated for that variable, requires a new prequalified WPS.

(2) **Where to get values.** The amperage (or wire feed speed if amperage is not controlled), voltage, and travel speed identified on the prequalified WPS is required to meet the conditions of Table 5.1, and be suitable for the material thickness, welding position, and be capable of meeting the quality requirements of this code. Specific values or ranges of values that meet these provisions may be listed on the prequalified WPS.

(3) **What combination of variables can be listed on a single WPS.** It is permissible for one WPS to list multiple combinations of variables on a single WPS. For example, a separate WPS is not required for each and every grade of steel for which the WPS is applicable. A single WPS can list, for example, A36, A53, and A106 in that all are Group I steels (see Table 5.3). Furthermore, A131 and A572 Grade 50 could also be listed on the aforementioned WPS, even though they are in Group II, so long as the other welding variables (electrode type, for example) are applicable to all the steels listed.

(4) **Changes that can be made to one prequalified WPS.** Once a prequalified WPS has been written, changes can be made to that WPS, within the limits prescribed by Table 5.2. For example, if the initial WPS listed only E7016, and a Contractor wishes to use E7018, then the initial WPS would need to be modified to reflect the acceptability of using E7018, per item 5 of Table 5.2. Until E7018 is added to the WPS, the use of that electrode would not be acceptable, even though it is a prequalified filler metal, per Table 5.4. However, a single WPS could list both E7016 and E7018, providing both are equally acceptable for the application. Though such changes are a matter of paperwork, rather than a retest as would be the case for qualified WPS, it is important to note such changes to the WPS will likely require resubmittal of the revised WPS. It is generally advantageous for the Contractor to include in a single WPS a range of conditions wherein the same prequalified WPS can be used (multiple steels, or reference to the Table 5.3 Group number, for example) so as to minimize the amount of paperwork involved. The range of permitted conditions, however, must be within the bounds of Table 5.2 and other code requirements. Overly inclusive WPSs may preclude clear communication to the welder and the inspector.

(5) **Changes that require a new or modified prequalified WPS.** If an existing prequalified WPS lists an amperage range of 300 amps, plus or minus 30 amps, and the Contractor wishes to use 400 amps, the WPS would need to be modified to reflect the acceptability of the use of this higher amperage since this is outside the range permitted by Table 5.2, item 9. Also, with this higher amperage, the travel speed value or range may need to be adjusted as well so as to comply with the requirements of Table 5.2, for example. The existing WPS can be modified, adding a new combination of variables that will result in a weld that meets the requirements of this code but with the higher amperage range.

(6) **Changes that require a qualified WPS.** Some changes to a WPS may push the WPS out of the range of prequalification. For example, if the welding process on the prequalified WPS is GMAW with variables that result in spray transfer, it would not be acceptable to add to the prequalified WPS variables that would result in short circuit transfer. However, such parameters could be qualified by test in accordance with Clause 6.

Part C
Weld Joints

C-5.4 Complete Joint Penetration (CJP) Groove Weld Details

C-5.4.1.1 Joint Dimensions. After preparation, the second side of double welded joints may not exactly match the sketches shown for prequalified welded joints in Figure 5.1 due to inherent limitations of the backgouging process. U- and J-shapes may appear to be combined with V- and bevel shapes. This is an acceptable condition.

C-5.4.1.6 and C-5.4.2.5 Corner Joint Preparation. The code allows an alternative option for preparation of the groove in one or both members for all bevel- and J-groove welds in corner joints as shown in Figure C-5.2. This provision was prompted by lamellar tearing considerations allowing all or part of the preparation in the vertical member of the joint. Such groove preparation reduces the residual tensile stresses, arising from shrinkage of welds on cooling, that act in the through-thickness direction in a single vertical plane, as shown in prequalified corner joints diagrammed in Figures 5.1, 5.2, and 5.4. Therefore, the probability of lamellar tearing can be reduced for these joints by the groove preparation now allowed by the code. However, some unprepared thickness, "a," as shown in Figure C-5.2, should be maintained to prevent melting of the top part of the vertical plate. This may easily be done by preparing the groove in both members (angle β).

C-Figures 5.1 and 5.2. Notes for Figures 5.1 and 5.2. Note "o" allows prequalified weld status for steel backed weld joints where the orientation of the members exceeds the allowances in Note j (see Figure 5.5). Grinding, machining, or forming may be utilized to modify the backing bar.

C-Figure 5.2 Prequalified PJP Groove Welded Joint Details

Effective Weld Size of Flare-Bevel-Groove Welded Joints. Tests have been performed on cold formed ASTM A500 material exhibiting a "c" dimension as small as T_1 with a nominal radius of 2t. As the radius increases, the "c" dimension also increases.

The corner curvature may not be a quadrant of a circle tangent to the sides. The corner dimension, "c," may be less than the radius of the corner.

C-5.4.4 Plug and Slot Weld Requirements

Plug and slot welds conforming to the dimensional requirements of 4.5, welded by techniques described in 7.24 and using materials listed in Table 5.3 or Table 6.9 are considered prequalified and may be used without performing joint WPS qualification tests.

Part D
Welding Processes

C-5.5 Welding Processes

C-5.5.1 Prequalified Processes. Certain SMAW, SAW, GMAW (excluding GMAW-S), and FCAW WPSs in conjunction with certain related types of joints have been thoroughly tested and have a long record of proven satisfactory performance. These WPSs and joints are designated as prequalified and may be used without tests or qualification (see Clause 6).

The code does not prohibit the use of any welding process. It also imposes no limitation on the use of any other type of joint; nor does it impose any procedural restrictions on any of the welding processes. It provides for the acceptance of such joints, welding processes, and WPSs on the basis of a successful qualification by the Contractor conducted in conformance with the requirements of the code (see Clause 6).

Short circuiting transfer is a type of metal transfer in GMAW-S in which melted material from a consumable electrode is deposited during repeated short circuits.

Short circuiting arc welding uses the lowest range of welding currents and electrode diameters associated with GMAW. Typical current ranges for steel electrodes are shown in Table C-5.1. This type of transfer produces a small, fast freezing weld pool that is generally suited for the joining of thin sections, for out-of-position welding, and for the filling of large root openings. When weld heat input is extremely low, plate distortion is small. Metal is transferred from the electrode to the work only during a period when the electrode is in contact with the weld pool. There is no metal transfer across the arc gap.

The electrode contacts the molten weld pool at a steady rate in a range of 20 to over 200 times each second. The sequence of events in the transfer of metal and the corresponding current and voltage is shown in Figure C-5.3. As the wire touches the weld metal, the current increases. It would continue to increase if an arc did not form, as shown at E in Figure C-5.3. The rate of current increase must be high enough to maintain a molten electrode tip until filler metal is transferred. Yet, it should not occur so fast that it causes spatter by disintegration of the transferring drop of filler metal. The rate of current increase is controlled by adjustment of the inductance in the power source. The value of inductance required depends on both the electrical resistance of the welding circuit and the temperature range of electrode melting. The open circuit voltage of the power source must be low enough so that an arc cannot continue under the existing welding conditions. A portion of the energy for arc maintenance is provided by the inductive storage of energy during the period of short circuiting.

As metal transfer only occurs during short circuiting, shielding gas has very little effect on this type of transfer. Spatter can occur. It is usually caused by either gas evolution or electromagnetic forces on the molten tip of the electrode.

C-5.5.4 GMAW and FCAW Power Sources. CV power sources have volt-ampere relationships that yield large current changes from small arc voltage changes. Constant current (CC) power sources have a volt-ampere relationship that yield small current changes from large arc voltage changes. Best welding results for GMAW and FCAW are normally achieved using CV output. CC may be used for GMAW or FCAW, providing WPSs are qualified by test in accordance with Clause 6.

Part E
Filler Metals and Shielding Gases

C-5.6 Filler Metal and Shielding Gas

Filler metals with designators listed in Note c of Table 5.4 obtain their classification tensile strength by PWHT at 1275°F or 1350°F [690°C or 730°C]. In the as-welded condition their tensile strengths may exceed 100 ksi [690 MPa].

The electrodes and electrode-flux combinations matching the approved base metals for use in prequalified joints are listed in Table 5.4. In this table, groups of steel specifications of Table 5.3 specifications are matched with filler metal classifications having similar tensile strengths in Table 5.4. In joints involving base metals that differ in tensile strengths, electrodes applicable to the lower strength material may be used provided they are of the low-hydrogen type if the higher strength base metal requires the use of such electrodes.

AWS A5.36/A5.36M, *Specification for Carbon and Low-Alloy Steel Flux Cored Electrodes for Flux Cored Arc Welding and Metal Cored Electrodes for Gas Metal Arc Welding*, is the first specification combining carbon and low-alloy steel electrodes into a single specification with fixed and open classification systems. A5.36 combines all tubular electrodes, both FCAW and GMAW, for welding carbon steel and low-alloy steel. The specification has two classification systems for carbon steel electrodes, fixed and open classifications, for multipass electrodes, except that T11 FCAW and E70C-6C, E70C-3C, E70C-3M, and E70C-G GMAW are classified under open classification system. Low-Alloy steel FCAW and GMAW-metal cored can only be classified by the open classification system. Annex M has more detailed data for mechanical and chemical classifications for electrodes under the A5.36 specification.

Because of the two classification systems, the D1.1:2010 Table 3.1 is not able to accommodate the new A5.36 classifications along with the existing specifications. In the 2020 edition of D1.1, Table 5.3 is split into Table 5.3 for base materials and Table 5.4 for filler metals, still retaining the material groups as the previous table. December, 2015 is the date set for A5.36 to be the governing specification for these FCAW and GMAW electrodes; however, existing contracts with fabricators may still exist for many years beyond this date with the current electrode specifications A5.18, A5.20, A5.28, and A5.29 in approved WPSs, and must be maintained within the Code to allow fulfillment of contract requirements.

In Table 5.4, the AWS A5.36 fixed and open classification of electrodes are intended to be identical or as similar as possible to the existing classifications of A5.18, A5.20, A5.28, and A5.29 specifications in Table 5.4 for consistency with existing prequalified and qualified WPSs. In this way these prequalified and qualified WPS may be "Grandfathered" with editing changes only. AWS A5.36/A5.36M:2012, Specification for Carbon and Low-Alloy Steel Flux Cored Electrodes for Flux Cored Arc Welding and Metal Cored Electrodes for Gas Metal Arc Welding, filler metal with fixed or open classifications that are identical to the mechanical and chemical properties of the previous specifications A5.18, A5.20, A5.28, and A5.29, and having the same shielding gas requirements for classification of these specifications are considered "grandfathered" for existing valid WPSs. Editing of the prequalified and qualified WPSs with the new A5 specification and classification is the only change required for the WPSs.

AWS A5.36 does not identify any particular shielding gas for qualification of electrodes except for the carbon steel fixed classifications. The A5.36-2012 classification systems use shielding gas ranges for classification; however, the nominal shielding gas percentage of the range is the actual classification gas per Table 5 of A5.36.

To provide continuity, the shielding gases identified in the classifications shown in D1.1:2015 Table 3.2, note b and Table 6.9 are identical to the previous A5.18, A5.20, A5.28, and A5.29 in order to maintain existing prequalified and qualified WPSs. Other shielding gases may be used for classification but, for the purposes of the code provisions, are to be limited to the specific shielding gas or mixture used for classification, and not the range of the shielding gas designator of Table 5 of A5.36.

Carbon steel FCAW T-12 classification under the A5.20 specification has a controlled chemical composition and mechanical properties. Under the new specification A5.36, these requirements remain identical, however, under the open classification the chemical composition requirement remains as in A5.20, but the mechanical requirement range is expanded to an upper tensile strength of 95 ksi, rather than the previous 90 ksi of A5.20 or the fixed classification of A5.36.

The "J" optional designator does not exist in A5.36, and for those FCAW electrodes that were previously classified with the "J" designator, the open classification is required to classify an electrode with the lower CVN temperature requirement of a "J" designator.

The optional "D" designator requirements for Demand Critical applications can now be applied to the GMAW metal cored electrodes. However, the heat input level tested in filler metals complying with AWS A5.36 is not the same as those tested in AWS A5.20. Therefore, WPSs with electrodes using the "D" designator of AWS A5.36 will need to be evaluated and potentially modified to use heat inputs within the heat input range tested in AWS A5.36. D1.8, Seismic Supplement will also reflect the differences in interchangeability of the two "D" designators as it pertains to heat input classification requirements.

In the open classification of electrodes, both carbon and low-alloy steel may be classified in the as-welded or postweld heat treated condition, or both. The letter "A" for as-welded or the letter "P" for postweld heat treated conditions is located in between the shielding gas designator, if applicable, and the CVN test temperature (e.g., E8XT1-C1A8-CS1; E7XT15-M21P6-CS2).

There are both subtle and major changes in A5.36, too numerous to elaborate on in this commentary. Obtaining a copy of A5.36 by Engineers and manufacturers, fabricators, or Contractors for their reference is highly advisable.

C-5.6.2 Weathering Steel Requirements. The requirements in this subclause are for exposed, bare, unpainted applications of weathering steel where atmospheric corrosion resistance and coloring characteristics similar to those of the base metal are required. The filler metals specified in Table 5.6 meet this requirement. When welding these steels for other applications, the electrode, the electrode-flux combination, or grade of weld metal specified in Table 5.4 is satisfactory.

The use of filler metals other than those listed in Table 5.6 for welding weathering steel (used in bare, exposed applications) is allowed for certain size single-pass fillets (related to welding process), as shown in 5.6.2. Here, the amount of weld metal-base metal admixture results in weld metal coloring and atmospheric corrosion characteristics similar to the base metal. In multiple-pass welds, a filler metal from Table 5.4 may be used to fill the joint except for the last two layers. Filler metal as specified in Table 5.6 should be used for the last two surface layers and ends of welds.

Surface capping option allows use of noncorrosion matching filler material for interior weld passes on multipass groove welds. The last fill pass and cap pass are welded with Table 5.6 filler to minimize dilution of the underlying filler metal on the surface of the weld.

Part F
Preheat and Interpass Temperature Requirements

C-5.7 Preheat and Interpass Temperature Requirements

The principle of applying heat until a certain temperature is reached and then maintaining that temperature as a minimum is used to control the cooling rate of weld metal and adjacent base metal. The higher temperature allows more rapid hydrogen diffusion and reduces the tendency for cold cracking. The entire part or only the metal in the vicinity of the joint to be welded may be preheated (see Table 5.8). For a given set of welding conditions, cooling rates will be faster for a weld made without preheat than for a weld made with preheat. The higher preheat temperatures result in slower cooling rates. When cooling is sufficiently slow, it will effectively reduce hardening and cracking.

For quenched and tempered steels, slow cooling is not desirable and is not recommended by the steel producer.

The preheat temperatures in Table 5.8 are minimum temperatures. Preheat and interpass temperatures are to be sufficiently high to ensure sound welds. The amount of preheat needed to slow down cooling rates so as to produce crack-free welds will depend on many factors including:

(1) The ambient temperature

(2) Heat from the arc

(3) Heat dissipation of the joint

(4) Chemistry of the steel (weldability)

(5) Hydrogen content of deposited weld metal

(6) Degree of restraint in the joint

Point 1 is considered above.

Point 2 is not presently considered in the code.

Point 3 is partly expressed in the thickness of material.

Point 4 is expressed indirectly in grouping of steel designations.

Point 5 is presently expressed either as nonlow-hydrogen welding process or a low-hydrogen welding process.

Point 6 is least tangible and only the general condition is recognized in the provisions of Table 5.8.

Based on these factors, the requirements of Table 5.8 should not be considered all-encompassing, and the emphasis on preheat and interpass temperatures as being minimum temperatures assumes added validity.

Caution is to be used when preheating and welding quenched and tempered steel since these steels may have minimum and maximum interpass temperatures, as well as heat input restrictions (see Table 5.8 note b).

Part G
WPS Requirements

C-5.8 WPS Requirements

C-5.8.2.1 Width/Depth Pass Limitation. The weld nugget or bead shape is an important factor affecting weld cracking. Solidification of molten weld metal due to the quenching effect of the base metal starts along the sides of the weld metal and progresses inward until completed. The last liquid metal to solidify lies in a plane through the centerline of the weld. If the weld depth is greater than the width of the face, the weld surface may solidify prior to center solidification. When this occurs, the shrinkage forces acting on the still hot, semi-liquid center or core of the weld may cause a centerline crack to develop, as shown in Figures C-5.1(A) and (B). This crack may extend throughout the longitudinal length of the weld and may or may not be visible at the weld surface. This condition may also be obtained when fillet welds are made simultaneously on both sides of a joint with the arcs directly opposite each other, as shown in Figure C-5.1(C).

The purpose of limiting the ratio of bead width to bead depth is to prevent shrinkage cracking of weld passes which are too deep for their width. Weld bead dimensions may best be measured by sectioning and etching a sample weld. WPSs which produce narrow beads that do not comply with this provision may be qualified per Clause 6.

Part H
Postweld Heat Treatment

C-5.9 Postweld Heat Treatment

Historically, the D1.1 requirements for PWHT have been largely based on experience with ASME code fabrication of plain carbon-manganese steels. The structural steel industry is increasingly moving away from carbon manganese steels to newer steels which are metallurgically more complex such as low alloy and microalloyed [e.g., Cb(Nb) and V additions] steels. The newer steels can be furnished in the as-rolled condition or with heat treatments such as quenching and tempering (Q&T), quenching and self-tempering (QST), or thermo-mechanically controlled processing (TMCP) to obtain higher yield strength. In general, Cb(Nb) and V additions are not used in pressure vessel steels, and when included, they are usually restricted to low values. The exception in pressure vessel steels is SA-737 which has additions of Cb(Nb) or V depending upon the grade. There have been a least seven Welding Research Council (WRC) Bulletins on topics relevant to the subjects of PWHT and microalloyed steels. A summary of the overall conclusions indicates that:

(1) PWHT (at 1150°F [620°C] for a few hours) of as-rolled or normalized carbon-manganese and low alloy steels (having a 50 ksi [345 MPa] or lower yield strength) does not adversely affect strength. PWHT, regardless of temperature or duration, degrades the notch toughness of Cb(Nb) or V microalloyed base metals and HAZs. Degradation varies in severity and may or may not affect the suitability for service.

(2) Steels manufactured by Q&T, QST, or TMCP processing need to have the development of their PWHT based on the specific material and processing. PWHT may reduce strength and notch toughness properties. The response to PWHT is very dependent on composition. Some Japanese data indicate that 1025°F [550°C] may be a more appropriate PWHT temperature for certain TMCP steels. The optimum PWHT temperature is dependent on specific composition, strength, and notch toughness requirements.

(3) ASTM A710 Grade A, age hardening Ni, Cu, Cr, Mo, Cb(Nb) steel is susceptible to cracking in the HAZ during PWHT. Grades B and C have not been studied. Some grades of ASTM A514/A517 are marginal for PWHT due to low ductility and possible HAZ cracking during PWHT as well as loss of strength and toughness. Some specifications place specific limits on PWHT such as ASTM A913 or "High-Strength Low-Alloy Steel Shapes of Structural Quality, Produced by the Quenching and Self-Tempering Process (QST)" which requires that "shapes shall not be formed and postweld heat treated at temperatures exceeding 1100°F [600°F]." The API offshore structures specifications 2W for TMCP steels and 2Y for Q& T steels have similar warnings regarding "Post Manufacturing Heating" which need to be considered when PWHT is contemplated.

References for Clause C-5

1. Stout, R. D. Postweld Heat Treatment of Pressure Vessels." WRC Bulletin 302(1), February 1985.
2. Shinohe, N., Sekizawa, M., and Pense, A. W. "Long Time Stress Relief Effects in ASTM A737 Grade B and Grade C Microalloyed Steels." WRC Bulletin 322(4), April 1987.
3. Konkol, P. J. "Effect of Long-Time Postweld Heat Treatment of Properties of Constructional-Steel Weldments." WRC Bulletin 330(2), January 1988.
4. Pense, A. W. "Mechanical Property Characterization of A588 Steel Plates and Weldments." WRC Bulletin 332(2), April 1988.
5. Japanese Pressure Vessel Research Council, "Characterization of the PWHT Behavior of 500 N/mm2 Class TMCP Steels." WRC Bulletin 371, April 1992.
6. Xu, P., Somers, B. R. and Pense, A. W. "Vanadium and Columbium Additions to Pressure Vessel Steels." WRC Bulletin 395, September 1994.
7. Spaeder Jr., C. E. and Doty, W. D. "ASME Post-Weld Heat Treating Practices: An Interpretive Report." WRC Bulletin 407(2), December 1995.

Table C-5.1
Typical Current Ranges for GMAW-S on Steel (see C-5.5.1)

Electrode Diameter		Welding Current, Amperes (Electrode Positive)			
		Flat and Horizontal Positions		Vertical and Overhead Positions	
in	mm	min.	max.	min.	max.
0.030	0.8	50	150	50	125
0.035	0.9	75	175	75	150
0.045	1.2	100	225	100	175

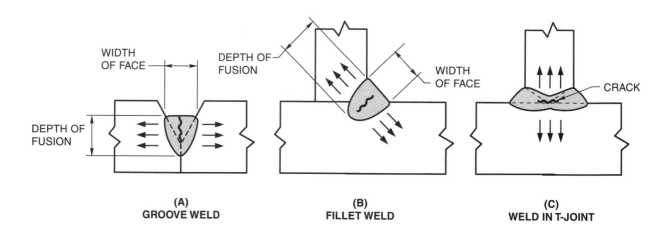

Figure C-5.1—Examples of Centerline Cracking (see C-5.8.2.1)

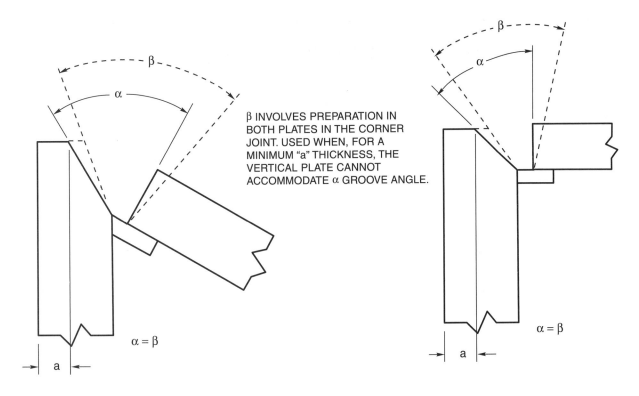

Figure C-5.2—Details of Alternative Groove Preparations for Prequalified Corner Joints (see C-5.4.1.6 and C-5.4.2.5)

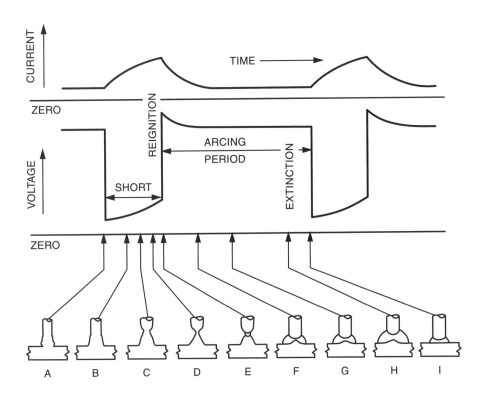

Figure C-5.3—Oscillograms and Sketches of GMAW-S Metal Transfer (see C-5.5.1)

C-6. Qualification

Part A
General Requirements

C-6.2 General

C-6.2.1.1 Qualification Responsibility. All Contractors are responsible for their final product. Therefore, it is their responsibility to comply with the qualification requirements of the code relative to WPSs. For properly documented WPSs conducted by the Contractor in conformance with this code, it is recommended by the code that they be accepted by the Engineer for the contract.

C-6.2.2 Performance Qualification of Welding Personnel. The qualification tests are especially designed to determine the ability of the welders, welding operators, and tack welders to produce sound welds by following a WPS. The code does not imply that anyone who satisfactorily completes qualification tests can do the welding for which they are qualified for all conditions that might be encountered during production welding. It is essential that welders, welding operators, and tack welders have some degree of training for these differences.

Ideally, welders, welding operators and tack welders welding quenched and tempered high-strength steels should have experience welding such base metals. In lieu of such experience, the Contractor should ensure that the Contractor's personnel receive instruction and training in the welding of such steels. It is further recommended that other personnel, such as fitters and thermal cutters (burners) involved in fabrication utilizing quenched and tempered high-strength steel be experienced or receive instruction and training prior to the start of thermal cutting operations.

C-6.2.2.1 Previous Performance Qualification. Standards other than D1.1 have welder and welding operator qualification tests that are similar or even identical to those required by this standard. This provision allows the Engineer to accept other qualification tests if in his/her judgment, the skills measured by these alternate tests are essentially the same as the tests prescribed by this code. For example, AWS B2.1, *Specification for Welding Procedure and Performance Qualification*, as well as the ASME code, may be used in this manner for qualification of welding personnel. The performance qualification tests of other D1 standards, such as D1.2, D1.3, D1.4, D1.5, and D1.6, may be used in a similar manner. The Engineer is assigned the responsibility of determining the acceptability of qualification to other standards as there are differences between these standards, and such differences may be significant, depending on the specific structure, or service conditions, or both.

C-6.3 Common Requirements for WPS and Welding Personnel Performance Qualification

C-6.3.4 Positions of Test Welds. This subclause defines welding positions for qualification test welds and production welds. Position is an essential variable for all of the WPSs, except for the EGW and ESW processes which are made in only one position. Relationships between the position and configuration of the qualification test weld and the type of weld and positions qualified are shown in Table 6.1. It is essential to perform testing and evaluation of the welds to be encountered in construction prior to their actual use on the job. This will assure that all the necessary positions are tested as part of the qualification process.

Part B
Welding Procedure Specification (WPS) Qualification

C-6.5 Type of Qualification Tests

Table 6.2 summarizes the requirements for the number and type of test specimens and the range of thicknesses qualified. A test plate thickness of 1 in [25 mm] or over qualifies a WPS for unlimited thickness. The 1 in [25 mm] thickness has been shown to generally reflect the influence of weld metal chemistry, heat input, and preheat temperature on the weld metal and HAZ. Although the term *direction of rolling* is deemed recommended, the mechanical properties of steel plate may vary significantly with the direction of rolling and may affect the test results. For example, tensile strength and impact toughness are often greater in the longitudinal direction than in the transverse direction unless cross rolling is used. Similarly, the rolling direction shown in the sketches often gives better results in the bend tests. For some applications, toughness results are required and the direction of rolling should be referenced on the test results.

C-6.7 Preparation of WPS

The written WPS and PQR may follow any convenient format (see Annex J for examples).

C-6.8 Essential Variables

This code allows some degree of departure from the variables used to qualify a WPS. However, departure from variables that affect the mechanical or chemical composition of material properties, or soundness of the weldment is allowed without requalification. These latter variables are referred to as essential variables. The base metal essential variables are listed in 6.8.3. The welding process essential variables are listed in 6.8.1. The positions of test welds are listed in 6.3.4 and 10.11.1. Changes in these variables beyond the variation allowed by the subject subclauses require requalification of the WPS. Similarly, changes beyond those shown in 6.8.2 require requalification using RT or UT only.

These essential variables are to be specific in the WPS document and followed in welding fabrication.

Table 6.7 lists the supplementary essential variable requirements for PQR qualification where CVN testing is required by contract documents. The requirements of Table 6.7 are in addition to the essential variable requirements of Table 6.5. The essential variable requirements of Table 6.5 should be fully met for procedure qualification without CVN testing and the essential variables of Table 6.7 should be fully met to further qualify the PQR to meet CVN testing requirements. The requirements of Table 6.7 do not change any of the requirements in Table 6.5. In Table 6.7, the Essential Variable on line 6 eliminates the need for CVN testing in all positions but does not change the requirements of Table 6.5 regarding welding positions. All WPS qualification requirements for strength and soundness according to Table 6.5 are still required.

C-6.8.1 SMAW, SAW, GMAW, GTAW, and FCAW. Travel speed affects heat input, weld cooling rates, and weld metallurgy, which are important for the HAZ, for fracture toughness control, and for welding quenched and tempered steels. Proper selection of travel speed is also necessary to avoid incomplete fusion and slag entrapment.

Electrode extension or contact tube to work distance is an important welding variable that affects the amperage as well as the transfer mode. At a set wire feed speed, using a constant voltage power source, longer electrode extensions cause the welding current to decrease. This may reduce weld penetration and heat input and cause fusion discontinuities. Shorter extension causes an increase in welding current. A variation in electrode extension may cause a spray transfer to change to globular or short circuiting modes. It is important to control electrode extension as well as other welding variables.

Semiautomatic welding processes may be controlled by using wire feed speed, electrode extension and arc length, or voltage. For machine operation, electrode extension may be premeasured; for manual welding, it is visually estimated. Welding on pipe (or tubing) material product forms does not necessarily mean that pipe welding is being performed. There is obviously a difference between welding around a pipe as opposed to welding along a pipe parallel to the pipe axis (centerline). A girth weld in a butt joint is completely different from a longitudinal groove weld that joins rolled plate to make a pipe; a socket joint with a fillet weld is completely different from a fillet weld along the pipe length attaching a plate plug. Obviously, the skills for straight line progression parallel to the pipe axis are no different from the skills for welding plate wrought shapes using a straight line progression; therefore, the pipe product form limitation does not apply in these straight line cases. Refer to Figure C-6.1.

C-6.8.5 Heat Input Calculations in Waveform-Controlled Welding. Advances in microprocessor controls and welding power source technology have resulted in the ability to develop waveforms for welding that improve the control of droplet shape, penetration, bead shape and wetting. Some welding characteristics that were previously controlled by the welder or welding operator are controlled by software or firmware internal to the power source. It is recognized that the use of controlled waveforms in welding can result in improvements in productivity and quality. The intention of this Code is to enable their use with both new and existing procedure qualifications.

The AWS D1.1 heat input measurement method described in 6.8.5.1 equation (1) was developed at a time when welding power source output was relatively constant. The heat input of welds made using waveform-controlled welding is not accurately represented by this equation due to the rapidly-changing outputs, phase shifts, and synergic changes, but is correctly represented by 6.8.5.1 equations (2) and (3). During waveform-controlled welding, current and voltage and values observed on conventional ammeters and voltmeters are no longer valid for heat input determination, and must be replaced by total instantaneous energy (joules) or power (joules/second or watts) to correctly calculate heat input. The calculations in 6.8.5.1 equations (2) and (3) more accurately reflect heat input changes when performing waveform-controlled welding. 6.8.5.1 equations (2) and (3) are also suitable for heat input calculations using non-waveform-controlled (conventional) welding.

Power sources that support rapidly pulsing processes (e.g., GMAW-P) are the most common waveform-controlled power sources. Power sources that are marketed as synergic, programmable, or microprocessor-controlled are generally capable of waveform-controlled welding. In these cases, heat input is calculated by the methods outlined in 6.8.5.1 equations 2 and 3 when performing WPS qualification or to determine compliance with a qualified procedure. If any doubt exists on whether waveform-controlled welding is being performed, the welding equipment manufacturer should be consulted. It is recognized that waveform controls may not be active for all of the welding processes or equipment settings for a particular power source. When the waveform control features of the equipment are not used, any of the heat input determination methods of 6.8.5.1 may be used.

When a welding power source does not display total instantaneous energy or average instantaneous power, an external device with high frequency sampling capable of determining total instantaneous energy or average instantaneous power is typically used, or the welding power source may be upgraded or modified to display these values.

Welding power sources or external meters are capable of displaying total instantaneous energy as cumulative measurements of total instantaneous energy, i.e., the sum of total instantaneous energy measurements made during a welding period such as trigger-on to trigger-off. The units of measurement may be Joules (J). Other conveniently obtained units of energy such as calories or British thermal units (Btu) may be used with the appropriate conversion factors. The other measurement needed to perform the calculations given in 6.8.5.1 equation (2) is weld length.

Welding power sources or external meters typically display average instantaneous power as the average measurements, i.e., the average value of instantaneous power measurements made during a welding period such as trigger-on to trigger-off. The typical unit of measurement is watts (W). Because power must be multiplied by time to obtain energy, the arc time needs to be recorded and the distance traveled during that time needs to be measured; with these data, the calculations shown in 6.8.5.1 can be made. Equation (2) or (3) may be used interchangeably depending on whether total instantaneous energy (TIE) or average instantaneous power (AIP) is displayed.

C-6.10 Methods of Testing and Acceptance Criteria for WPS Qualification

C-6.10.3.2 Longitudinal Bend Specimens. Provision has been made in this subclause for longitudinal bend tests when material combinations differ markedly in mechanical bending properties.

C-6.10.3.3 Acceptance Criteria for Bend Tests. The new, more definitive wording for bend test acceptance was added to aid the interpretation of the test results. The purpose of the bend test is to prove the soundness of the weld. The statement regarding the total quantity of indications was added to restrict the accumulative amount of discontinuities.

A maximum limit on tears originating at the corners was added to prevent the case where the corner cracks might extend halfway across the specimen, and under the previous criteria, would be judged acceptable.

C-6.12 PJP Groove Welds for Nontubular Connections

C-6.12.1 Type and Number of Specimens to be Tested. This subclause addresses the requirements for qualification of PJP groove welds that require qualification by the Contractor because the joint design and WPS to be used in construction do not meet prequalified status as described in 5.1, or a WPS qualified to produce CJP welds utilizing a specific joint design is proposed for use as a PJP weld. The intent is to establish the weld size that will be produced using the joint design and WPS proposed for construction. Certain joint designs in combination with a specific welding process and position may show that the groove preparation planned will not give the desired weld size (S).

Macroetch test specimens are only required for WPS qualifications that meet the requirements of 6.12.2 or 6.12.3. Additional testing is required for those WPS that fall under the criteria of 6.12.1. These test requirements are shown in Table 6.3.

C-6.13 Fillet Welds

C-6.13.1 Type and Number of Specimens. When single-pass fillet welds are to be used, one test weld is required as shown in Figures 6.15 and 10.16 using the maximum size single-pass fillet weld. If multiple-pass fillet welds only are used, then one test weld is required, as shown in Figures 6.15 and 10.16, using the minimum size multiple-pass fillet weld to be used. Each of these tests is presumed to evaluate the most critical situation.

C-6.15 Welding Processes Requiring Qualification

The code does not restrict welding to the prequalified WPSs described in 5.1. As other WPSs and new ideas become available, their use is allowed, provided they are qualified by the requirements described in Clause 6, Part B. Where a Contractor has previously qualified a WPS meeting all the requirements described in Part B of this section, the Structural Welding Committee recommends that the Engineer accept properly documented evidence of a previous test and not require the test be performed again. Proper documentation means that the Contractor has complied with the requirements of Clause 6, Part B, and the results of the qualification tests are recorded on appropriate forms such as those found in Annex J. When used, the form in Annex J should provide appropriate information listing all essential variables and the results of qualification tests performed.

There are general stipulations applicable to any situation. The acceptability of qualification to other standards is the Engineer's responsibility to be exercised based on the specific structures and service conditions. The Structural Welding Committee does not address qualification to any other welding standard.

C-6.15.1.3 WPS Requirements (ESW/EGW). The welding processes, procedures, and joint details for ESW and EGW are not accorded prequalified status in the code. The WPSs are to be established in conformance with Clause 6. Welding of quenched and tempered steels with either of these processes is prohibited since the high-heat input associated with them has been known to cause serious deterioration of the mechanical properties of the HAZ.

Part C
Performance Qualification

C-6.16 General

The welder qualification test is specifically designed to determine a welder's ability to produce sound welds in any given test joint. After successfully completing the welder qualification tests, the welder should be considered to have minimum acceptable qualifications.

Knowledge of the material to be welded is beneficial to the welder in producing a sound weldment; therefore, it is recommended that before welding quenched and tempered steels, welders should be given instructions relative to the properties of this material or have had prior experience in welding the particular steel.

From time to time, the Contractor may upgrade or add new control equipment. The previously qualified welding operator may need training to become familiar with this new equipment. The emphasis is placed on the word "training" rather than "requalification" since several beads on a plate or a tube, as appropriate, may be sufficient. The intention is that the Contractor would train the welding operator to weld using the new equipment.

C-Table 6.10. See C-Table 10.12 and Figure C-6.1.

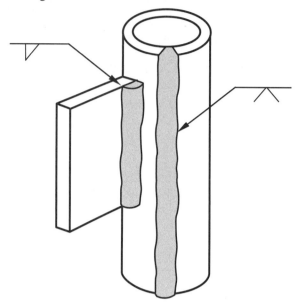

Note: Pipe qualification is not required and plate qualification is acceptable for groove and fillet welds in the flat, horizontal, vertical, and overhead positions.

Figure C-6.1—Type of Welding on Pipe That Does Not Require Pipe Qualification (see Tables 6.10 and 10.12)

C-6.20 Essential Variables

The ability of a welder to produce a sound weld is considered by the code to be dependent upon certain essential variables, and these are listed in Table 6.12.

C-Table 6.13. Electrodes for SMAW are grouped relative to the skill required of the welder. The F Group designation allows a welder qualified with an electrode of one group designation to use other electrodes listed in a numerically lower designation. For example, a welder qualified with an E6010 electrode will also be qualified to weld an E6011 electrode, group designation F3 and is allowed to weld with electrodes having group designation F2 and F1; the welder is not qualified to weld with electrodes having a group designation F4.

Part D
Requirements for CVN Toughness Testing.

C-6.26 General: CVN Testing

A number of test methods exist for determination of toughness. The test method utilized in this Code is the Charpy V-notch (CVN) test. References to this method are designated as "CVN testing."

The Engineer may specify in the contract documents alternative fracture toughness testing methods, in which case full details of how the testing is to be performed should be included.

When CVN testing is specified in contract documents, testing in accordance with Part D is required. The Engineer is responsible for specifying the toughness requirements (see 4.3.2).

C-6.26.1.2 FCAW-S. When weld metal first deposited with the FCAW-S process is covered with weld metal deposited using another welding process there is the potential for the formation of an area of reduced toughness. FCAW-G is considered a separate process from FCAW-S.

C-6.26.3 Test Requirements. When CVN testing is required, it is possible that the minimum test coupon dimensions shown in Figures 6.5, 6.6, 6.7, 10.14 and 10.15) will not provide sufficient material to produce the number of CVN specimens required. Multiple processes in a single joint, anticipated retests and other factors will determine the number of specimens and hence affect the length of the test coupon.

Rolling direction may also affect the outcome of CVN testing in the HAZ. Orienting the plate rolling direction perpendicular (transverse) to the weld axis may result in higher HAZ CVN energy values. See Figures 6.6 and 6.7.

C-6.27 CVN Tests.

Where testing is performed by an outside testing laboratory, it is recommended that they be certified to ISO 17025 or an equivalent certification. However, the contractor may choose to do CVN testing in-house.

C-6.27.2 Number of Specimens. CVN acceptance criteria are based on a 3-specimen test. Due to the tendency to get a number of outlier CVN values, the code allows a 5-specimen test, discarding high and low values to minimize some of the scatter normally associated with CVN testing of welds and HAZs.

C-6.27.6.1 Test Coupon Thickness 7/16 in [11 mm] or greater. The absorbed energy and test temperature requirements for Charpy V-notch impact tests typically assume a standard bar size, i.e., 10 mm × 10 mm × 55 mm. When full sized bars cannot be used such as due to geometry effects and the thickness across the notch is less than 80% of the material thickness, a reduction is made in the required test temperature. The temperature reduction is intended to make the smaller, less restrained fracture area behave more like the thicker 10 mm bar.

The absorbed energy requirement is reduced proportionally to the actual CVN bar size relative to a full size 10 mm bar regardless of the test coupon material thickness (Table 6.15, Charpy V-Notch Test Acceptance Criteria for Various Sub-Size Specimens).

Example (1): Design drawings or specifications require that CVN tests achieve an absorbed energy of 20 ft-lbf [27 J] at 32°F [0°C]. The WPS coupon thickness is 3/4 in [18 mm]. Due to physical constraints the largest CVN specimen that can be extracted is a 10 mm × 5 mm sub-sized CVN specimen. The ratio of the CVN face width [5mm] to the WPS coupon thickness (3/4 in [18 mm]) is less than 80%. The actual CVN test temperature based on Table 6.14 would be 12°F [–11°C]: (32°F – 20°F) = 12°F [(0°C – 11°C) = –11°C]. The absorbed energy requirement for the sub-size CVN test specimen would be, from Table 6.15, 10 ft-lbf [14 J] (50% of 20 ft-lbf [27 J]).

C-6.27.6.2 Test Coupon Thickness less than 7/16 in [11 mm]. When the test coupon material thickness is less than 7/16 in [11 mm] extrapolating to a similar 10 mm x 10 mm CVN bar is not permitted and it is only necessary to adjust the test temperature relative to the CVN specimen face width for the actual test coupon thickness. The reason for this is that the impact properties of thinner materials are equivalent to those provided for the reduced face width of the sub-size specimens. Thus a sub-size CVN specimen whose face is nearly the same thickness as the test coupon, even if the material is thinner than 10 mm, does not require a test temperature reduction. If the CVN bar is less than 80% of the test coupon thickness, the temperature reduction is the difference between the temperature reduction value listed in Table 6.14 for the bar size and the temperature reduction listed for the thickness of the (thinner) test coupon.

The absorbed energy requirement is reduced proportionally to the actual CVN bar size relative to a full size 10 mm bar regardless of the material thickness (Table 6.15, Charpy V-Notch Test Acceptance Criteria for Various Sub-Size Specimens).

Example (1): Design drawings or specifications require that CVN tests achieve an absorbed energy of 20 ft-lbf [27 J] at 32°F [0°C]. The qualification test is conducted using 1/4 in [6 mm] thick test plate material. The largest CVN specimen that can be extracted is a 10 mm × 5 mm sub-sized CVN specimen. The ratio of the CVN face width [5 mm] and the test coupon thickness (1/4 in [6 mm]) is greater than 80%.

No Table 6.14 temperature differential is permitted and the specified test temperature of 32°F [0°C] is required. The absorbed energy requirement for the sub-size test specimen would be, from Table 6.15, 10 ft-lbf [14 J] (50% of 20 ft-lbf [27 J]).

Example (2): Design drawings or specifications require that CVN tests achieve an absorbed energy of 20 ft-lbf [27 J] at 32°F [0°C]. A 6 in, schedule 40 pipe (168.3 mm outside diameter × 7.11 mm wall thickness) is qualified to the requirements of Clause 6. Due to the curvature of the pipe, the largest CVN specimen that can be extracted is a 10 mm × 5 mm sub-sized CVN specimen. The ratio of the CVN face width [5 mm] to the WPS coupon thickness [7.11 mm] is 70%. The actual CVN test temperature based on Table 6.14 would be 20°F [–7°C]: (32°F – (20°F – 8°F)) = (32°F – 12°F) = 20°F [0°C – (11°C – 4°C) = (0°C – 7°C) = –7°C] which is the difference between the reductions listed in Table 6.14 for a 7 mm test coupon and 5 mm CVN bars. The absorbed energy requirement for the sub-size test specimen would be, from Table 6.15, 10 ft-lbf [14 J] (50% of 20 ft-lbf [27 J]).

C-6.28 Purpose

Many intermix tests are performed by filler metal manufacturers.

C-7. Fabrication

C-7.1 Scope

The criteria contained in Clause 7 are intended to provide definition to the producer, supervisor, engineer and welder of what constitutes good workmanship during fabrication and erection. Compliance with the criteria is achievable and expected. If the workmanship criteria are not generally met, it constitutes a signal for corrective action.

C-7.2 Base Metal

The ASTM A6 and A20 specifications govern the delivery requirements for steels, provide for dimensional tolerances, delineate the quality requirements, and outline the type of mill conditioning.

Material used for structural applications is usually furnished in the as-rolled condition. The Engineer should recognize that surface imperfections (seams, scabs, etc.) acceptable under A6 and A20 may be present on the material received at the fabricating shop. Special surface finish quality, when needed in as-rolled products, should be specified in the information furnished to the bidders.

The steels listed as approved in Table 5.3 or Table 6.9 of the code include those considered suitable for welded cyclically loaded structures and statically loaded structures as well as tubular structures. Also listed are other ASTM specifications, American Bureau of Shipping (ABS) specifications, and American Petroleum Institute (API) specifications that cover types of materials that have been used in tubular structures. All of the steels approved are considered weldable by the WPSs that may be qualified or prequalified to this code. Every code-approved steel is listed in Table 5.3 or Table 6.9.

The ASTM specifications for grades of structural steel used in building construction for which WPSs are well established are listed in Tables 5.3 and 6.9 together with other ASTM specifications covering other types of material having infrequent application but which are suitable for use in statically loaded structures. The ASTM A588, A514, and A517 specifications contain grades with chemistries that are considered suitable for use in the unpainted or weathered condition. ASTM A618 is available with enhanced corrosion resistance.

Structural steels that are generally considered applicable for use in welded steel cyclically loaded structures are listed in Table 5.3 and Table 6.9 as approved steels. Other ASTM specifications for other types of steel having infrequent applications, but suitable for use in cyclically loaded structures, are also listed as approved steels. Steels conforming to these additional ASTM specifications, A500, A501, and A618, covering structural tubing, and A516 and A517 pressure vessel plates are considered weldable and are included in the list of approved steels for cyclically loaded structures.

The complete listing of approved steels in Table 5.3 and Table 6.9 provides the designer with a group of weldable steels having a minimum specified yield strength range from 30 ksi to 100 ksi [210 MPa to 690 MPa], and in the case of some of the materials, notch toughness characteristics which make them suitable for low-temperature application.

Other steels may be used when their weldability has been established according to the qualification procedure required by Clause 6.

The code restricts the use of steels to those whose specified minimum yield strength does not exceed 100 ksi [690 MPa].

C-7.3 Welding Consumables and Electrode Requirements

C-7.3.1.3 Shielding Gas. From information supplied by the manufacturers of shielding gas, it has been determined that a dew point of –40°F [–40°C] is a practical upper limit providing adequate moisture protection. A dew point of –40°F

[–40°C] converts to approximately 128 parts per million (ppm) by volume of water vapor or about 0.01% available moisture. This moisture content appears very low but, when dissociated at welding temperatures, could contribute hydrogen to that already associated with the electrode. Therefore, it is mandatory to have –40°F [–40°C] or lower dew point in shielding gas. Shielding gas purchased as certified to AWS A5.32M/A5.32 meets this requirement.

C-7.3.2 SMAW Electrodes. The ability of low-hydrogen electrodes to prevent underbead cracking is dependent on the moisture content in the coating. During welding, the moisture dissociates into hydrogen and oxygen; hydrogen is absorbed in the molten metal, and porosity and cracks may appear in the weld after the weld metal solidifies. The provisions of the code for handling, storage, drying, and use of low-hydrogen electrodes should be strictly adhered to in order to prevent moisture absorption by the coating material.

C-7.3.2.1 Low-Hydrogen Electrode Storage Conditions. The term "low-hydrogen" is widely used in the welding industry, and yet, this term is not officially defined in industry documents, such as AWS A3.0M/A3.0. As a result, a great deal of confusion and misinterpretation has become associated with these words. It may be useful to provide a brief summary of its evolution.

The phenomenon of Hydrogen Induced Cracking (HIC) is usually associated with the simultaneous interaction of three factors in a weldment: (1) High applied or residual stress, (2) the presence of a steel microstructure that is susceptible to HIC, and (3) the presence of a high concentration of diffusible (atomic) hydrogen. In the case of welded structures, the first factor is usually fixed by the strength levels of the filler metals and base metals being used. The remaining two items are typically controlled through preheat (to promote less HIC-sensitive microstructures) and the use of adequate welding processes and techniques to minimize hydrogen. In order to avoid HIC, there are many combinations of preheat and hydrogen control that may be used. For example, higher levels of diffusible hydrogen can be tolerated with higher levels of preheat and interpass temperature. Conversely, with lower levels of diffusible hydrogen, lower preheat and interpass temperatures may be used.

For years the only real option for arc welding was SMAW. Its use of cellulosic or rutile electrodes (e.g., E6010) were a vast improvement over the previously used bare electrodes, greatly reducing the nitrogen content of weld deposits. However, the same coatings that provided for this improvement in weld quality did so by allowing a large amount of hydrogen to enter the weldment as a result of the electrode coverings' decomposition in the arc. The amount varied, but 40 ml–80 ml/ 100 grams of diffusible hydrogen in the weld metal was not unusual. This high level of diffusible hydrogen was acceptable, and crack-free welds were possible, when welding was restricted to lower strength steels that were less susceptible to HIC. For more sensitive steels, and thicker steels, the HIC tendencies were overcome with additional preheat and interpass temperature. However, HIC was a not uncommon occurrence, especially in thicker weldments where inadequate preheating and very high residual stresses made cracking almost inevitable.

The welding industry met this challenge by developing SMAW electrodes with low-hydrogen coatings. Such coatings had, and still have today, specific limits on the maximum permitted moisture content in the electrode coating. Such electrodes included the AWS A5.1 and A5.5 E7015, E7016, E7018, and E7028 classifications. These products became known as "low-hydrogen electrodes," even though the term was not defined by an AWS standard. Nevertheless, the contrast between these "low-hydrogen electrodes" and the alternative cellulosic or rutile electrodes was distinct: the weld deposits made with electrodes having the low-hydrogen coatings deposited weld metal with substantially less hydrogen. This allowed "hard to weld steels" to be welded with less or no preheat, and greatly reduced HIC. The subsequent development of semiautomatic processes, such as GMAW and FCAW, enabled weldments to be made with similar hydrogen levels as the low-hydrogen SMAW electrodes. However, the term "low-hydrogen" was never used for these processes and their electrodes, since it never went through the evolutionary process of hydrogen minimization. To capture these processes, terms such as "low- hydrogen processes" began to be used, again without formal definition.

There continues to be much debate about what numbers actually represent adequate "low-hydrogen." Today, the diffusible hydrogen content of actual weld metal can be measured, and is typically expressed in units of ml/100 g of deposited weld metal. Rather than debating the definition of "low-hydrogen," it is more common today to discuss hydrogen contents in terms of this unit of measurement. For the vast majority of typical industrial applications governed by this code, the use of these semiautomatic processes will provide adequately low levels of hydrogen introduction to prevent HIC. The stipulation of "low-hydrogen" electrodes in contract documents has sometimes been strictly interpreted to exclude the semiautomatic processes and instead require the use of the SMAW electrodes with low-hydrogen coatings. Indeed, such contract language is typically rooted in the thought pattern of the 1950s, to distinguish between SMAW electrodes that did not deposit weld metal of a low-hydrogen content. Typically, there is no reason to restrict the use of semiautomatic processes. Further, the AWS D1.1 Code provides the necessary restrictions on acceptable filler metals, steels and preheat limits, and there is generally no need for contract documents to add additional restrictions with respect to acceptable filler metals. For preheat and other purposes, D1.1 treats GMAW, FCAW, SAW, GTAW, ESW, and EGW in the same way as "low-hydrogen" SMAW.

Tests have shown there can be a wide variation in the moisture absorption rate of various brands of electrodes representing a given AWS classification. Some electrodes absorb very little moisture during standard exposure times while others absorb moisture very rapidly. The moisture control requirements of 7.3.2.1 are necessarily conservative to cover this condition and ensure that sound welds can be produced.

The time restrictions on the use of electrodes after removal from a storage oven may seem overly restrictive to some users. The rate of moisture absorption in areas of low humidity is lower than that encountered in areas of high humidity. The code covers the most restrictive situations.

C-7.3.2.5 Low-Hydrogen Electrode Restrictions for ASTM A514 or A517 Steels. For carbon steel low-hydrogen electrodes, AWS A5.1/A5.1M, *Specification for Carbon Steel Electrodes for Shielded Metal Arc Welding*, specifies a moisture limit of 0.6% by weight in the low-hydrogen electrode covering either "as-received" or "conditioned," except for E7018M, which has a specified moisture limit of 0.1%. Moisture-resistant low-hydrogen electrodes, which use the optional designator R, have a specific moisture limit of 0.3%. Alloy steel low-hydrogen electrodes covered in AWS A5.5/A5.5M, *Specification for Low-Alloy Steel Electrodes for Shielded Metal Arc Welding*, have a specified maximum moisture content in the as manufactured condition. For the E70XX-X class electrodes, it is 0.4%; for E80XX-X electrodes, it is 0.2%; for the E90XX-X, E100XX-X, E110XX-X, and E120XX-X class electrodes, it is 0.15%. Electrodes with the optional H4 designator provide sufficient limits of coating moisture to assure the actual diffusible hydrogen level is limited to a maximum value of 4 ml/100 g of deposited weld metal, regardless of the electrode tensile classification.

Experience has shown that the limits specified above for moisture contents in electrode coverings are not always sufficiently restrictive for some applications using the E80XX-X and lower classes. Electrodes of classifications lower than E90XX-X are subject to more stringent moisture level requirements when used for welding the high-strength quenched and tempered steels, ASTM A514 and A517. All such electrodes are required to be dried between 700°F and 800°F [370°C and 430°C] before use. Electrodes of classifications below E90XX-X are not required by AWS A5.5/A5.5M to have a moisture content less than 0.15%, and the required drying will achieve at least this moisture level. This precaution was necessary because of the sensitivity of high-strength steels and weld metal to hydrogen cracking.

C-7.3.3.1 Electrode-Flux Combination Requirements. Electrodes and fluxes conforming to the classification designation of these specifications may be used as prequalified, provided the provisions of 5.6, Table 5.3 and Table 5.4 are observed. The Contractor should follow the supplier's recommendations for the proper use of fluxes.

C-7.3.3.2 Condition of Flux. The requirements of this section are necessary to assure that the flux is not a medium for introduction of hydrogen into the weld because of absorbed moisture in the flux. Whenever there is a question about the suitability of the flux due to improper storage or package damage, the flux should be discarded or dried in conformance with the manufacturer's recommendations.

C-7.3.3.3 Flux Reclamation. For recovery of the unfused flux through the vacuum recovery system, a distinction has to be made between fused and bonded fluxes. Fused fluxes, in general, tend to become more coarse as they are recycled (especially where particles are less than 200 mesh). In this case, the vacuum system generally filters out some of the fines—and hence at least 25% virgin material should be added to replenish the fines before it is reused. Bonded fluxes on the other hand, because of their method of manufacture, tend to break up in the flux recovery system giving rise to a greater proportion of smaller particles. In order to compensate for the flux break-up, at least 25% virgin material (although 50% is more common among users) needs to be added to the recycled flux before it is reused. For both categories of fluxes, it is essential to separate out any possible metallics (from plate rust or mill scale) before recycling the flux.

The quality of recovered flux from manual collection systems is dependent on the consistency of that collection technique. The Contractor should follow a procedure that assures that a consistent ratio of virgin flux is added and mixed with the recovered flux.

C-7.3.3.4 Crushed Slag. The slag formed during SAW may not have the same chemical composition as unused (virgin) flux. Its composition is affected by the composition of the original flux, the base metal plate and electrode composition, and the welding parameters.

Although it may be possible to crush and reuse some SAW slag as a welding flux, the crushed slag, regardless of any addition of virgin flux to it, may be a new chemically different flux. It can be classified under the AWS A5.17/A5.17M or A5.23/A5.23M specification, but should not be considered to be the same as virgin flux.

C-7.4 ESW and EGW Processes

The procedures to be used for ESW and EGW are detailed in 7.4, and the essential variables for these procedures are given in 6.8.2.

The code requires the qualification of WPSs since welding variables influence the operation of the process with respect to adequate penetration, complete fusion of the joint area, and ability to produce a sound weld.

These are relatively new processes, and insufficient experience is the justification for not according a prequalified status to them.

C-7.4.1 Process Limitations. The ESW and EGW processes can be advantageous because both deposit weld metal in very thick plates in one pass at very high deposition rates and with minimum distortion. The relatively high heat input and slow cooling rate of ESW and EGW processes can produce welds with softened heat-affected zones (HAZs) in quench and tempered (Q&T), thermomechanical control process (TMCP), and precipitation hardened (PH) steels. Reduced fracture toughness, i.e., local brittle zones (LBZ), may occur in the HAZ of these welds.

When fatigue cracks occur, the initiation is often at the toe of welds where there is a stress concentration. Fatigue crack growth progresses perpendicular to the principal stress field and is a function of the live load stress range. The larger size of the HAZ and any LBZ in ESW and EGW process welds will mean larger fatigue cracks can remain in the region of lower fracture toughness. Furthermore, high loads may concentrate any plastic straining in softened HAZ regions. These two factors, either individually or in combination, will increase the risk of brittle fracture. The square groove joint design used with the ESW and EGW processes compound this problem because it orients the HAZ and LBZ perpendicular to the tensile stresses. Standard tensile and bend test procedure qualifications will not reveal this condition. If these issues apply for the intended application, the designer or Engineer should consider additional testing such as CTOD (crack tip opening displacement) fracture toughness and fatigue tests.

The Engineer and Contractor who desire to use these processes are encouraged to investigate the potential consequences with respect to material and process selection. ESW-NG, a specific variation of the process, has relatively lower heat input and therefore reduces the size of the HAZ and impairment in the HAZ notch toughness.

C-7.5 WPS Variables

It is the intent of the code that welders, welding operators, and tack welders be able to properly use the WPS. This may be accomplished through experience, training, or instruction, as necessary.

C-7.7 Heat Input Control for Quenched and Tempered Steels

The strength and toughness of the HAZ of welds in quenched and tempered steels are related to the cooling rate. Contrary to principles applicable to other steels, the fairly rapid dissipation of welding heat is needed to retain adequate strength and toughness. The austenitized HAZ must be cooled rapidly to ensure the formation of the hardening constituents in the steel microstructure. Overheating of quenched and tempered steel followed by slow cooling prevents the formation of a hardened microstructure.

The deposition of many small weld beads improves the notch toughness of the weld by grain refining and the tempering action of ensuing passes. A weave bead, with its slower travel speed, increases heat input and is therefore not recommended. Because the maximum heat input for various quenched and tempered steels varies over a wide range, heat input as developed and recommended by the steel producers should be strictly observed.

C-7.8 Stress Relief Heat Treatment

This subclause provides for two PWHT methods for stress relief of a welded assembly. The first method requires the assembly to be heated to 1100°F [600°C] maximum. for quenched and tempered steels, and between 1100°F and 1200°F

[600°C to 650°C] for other steels. The assembly is held at this temperature for the time specified in Table 7.2. In 7.8.2, an alternative method allows a decrease in temperature below the minimum specified in the first method, when the holding time is increased. The alternative method is used when it is impractical to PWHT the welded assembly at higher temperatures. These temperatures are sufficiently below the critical temperature to preclude any change in properties.

If the purpose of the PWHT is to stress relieve the weld, the holding time is based on the weld metal thickness even though some material in the weldment is thicker than the weld. If the purpose of the PWHT is to maintain dimensional stability during subsequent machining, the holding time is based on the thickest component in the weldment. Certain quenched and tempered steels, if stress relieved as a carbon or low-alloy steel, may undergo undesirable changes in microstructure, causing a deterioration of mechanical properties or cracking, or both. Such steels should only be stress relieved after consultation with the steel producer and in strict conformance with the producer's recommendations.

PRECAUTIONARY NOTE: Consideration should be given to possible distortion due to stress relief.

C-7.9 Backing

All prequalified CJP groove welds made from one side only, except as allowed for tubular structures, are required to have complete fusion of the weld metal with a steel backing. Other backing, such as listed in 7.9.3, may be used, if qualified in conformance with Clause 6. When steel backing is used, it is required to be continuous for the entire length of the weld except as provided in 7.9.1.2 or 10.22.1. When not continuous, the unwelded butt joint of the backing will act as a stress raiser that may initiate cracking.

C-7.9.1.2 Full-Length Backing. Experience has shown that a tightly fitted but unwelded square butt joint in steel backing constitutes a severe notch and crack-like condition that potentially leads to transverse cracks in the weld when the unfused area is perpendicular to the stress field. Such cracks will, in many cases, propagate into the base metal.

Discontinuities in backing are not avoidable in some details where piece geometry requires a discontinuity and the lack of access prevents welding the backing. Backing welded parallel to the axis of the member will strain with the member. The unwelded discontinuity will provide a stress concentration and defect when it is perpendicular to the stress field. The stress concentration may initiate cracking. Residual stresses, high restraint and cyclic loads increase the risk of such cracking. Compressive stresses decrease the risk of cracking unless there is stress reversal in the applied stress cycle. The code permits discontinuous backing in selected specific applications or where the Engineer approves.

C-7.9.1.3 Backing Thickness. Recommended minimum nominal thickness of steel backing is shown in the following table:

Process	Thickness, min.	
	in	mm
GTAW	1/8	3
SMAW	3/16	5
GMAW	1/4	6
FCAW-S	1/4	6
FCAW-G	3/8	10
SAW	3/8	10

Commercially available steel backing for pipe and tubing is acceptable, provided there is no evidence of melting on exposed interior surfaces.

C-7.9.1.4 Cyclically Loaded Nontubular Connections. Steel backing transverse to applied stress forms a point of stress concentration and may be a source of fatigue crack initiation in cyclically loaded structures. Therefore, the provisions of 7.9.1.4 require the removal of backing that is transverse to the direction of computed stress in cyclically loaded structures.

C-7.11 Welding Environment

C-7.11.2 Minimum Ambient Temperature. Experience has shown that welding personnel cannot produce optimum results when working in an environment where the temperature is lower than 0°F [–20°C]. Reference is made in 7.11.2

relative to the use of a heated structure or shelter to protect the welder, and the area being welded, from inclement weather conditions. If the temperature in this structure or shelter provides an environment at 0°F [–20°C], or above, the prohibition of 7.11.2 is not applicable. The environmental conditions inside the structure or shelter do not alter the preheat or interpass temperature requirements for base metals stated elsewhere in the code.

C-7.12 Conformance with Design

Either or both legs of fillet welds may be oversized without correction, provided the excess does not interfere with satisfactory end use of a member. Attempts to remove excess material from oversized welds serve no purpose. Adequacy of throat dimension and conformance to the weld profiles of Figure 7.4 should be the only acceptance criteria.

C-7.13 Minimum Fillet Weld Sizes

The code specifies minimum fillet weld sizes based upon two independent considerations.

(1) For nonlow-hydrogen processes, the minimum size specified is intended to ensure sufficient heat input to reduce the possibility of cracking in either the HAZ or weld metal.

(2) When possibility of cracking is reduced by use of low-hydrogen processes or by nonlow-hydrogen processes using a procedure established in conformance with 6.8.4, the specified minimum is intended to maintain reasonable proportionality with the thinner connected parts.

In both cases, the minimum size applies if it is larger than the size required to satisfy design requirements.

C-7.14 Preparation of Base Metal

C-7.14.1 General. For quality welds, base metal cleanliness is important. However, it is neither required nor necessary for base metal to be perfectly clean before welds are made. It is difficult both to establish quantifiable limits of cleanliness and to measure to those limits; therefore this provision uses the practical standard of the resultant weld quality. If the base metal is sufficiently clean so as to facilitate a weld being made that meets the requirements of this code, it is clean enough. If the resultant welds do not meet the quality requirements of this code, cleaner base metal may be required.

C-7.14.2 Mill-Induced Surface Defects. The base metal to which welds are attached must be sufficiently sound so as to not affect the strength of the connection. Base metal defects may be repaired prior to the deposition of the prescribed weld. This subclause does not limit base metal repairs by welding. Defects that may be exposed on cut edges are governed by 7.14.5.

C-7.14.3 Scale and Rust. Excessive rust or scale can negatively affect weld quality. The code permits welding on surfaces that contain thin mill scale, provided:

(1) the mill scale remains intact after wire brushing, and

(2) the resultant weld quality is not adversely affected (see C-7.14.1).

C-7.14.4 Foreign Materials. This subclause prohibits volumetric (three-dimensional) quantities of contaminants from being left in place on the surface to be welded and adjacent areas. Surfaces contaminated by the materials listed in 7.14.4.1 must be cleaned, such as by wiping prior to welding. Special consideration should be given to the removal of surface contaminants containing hydrocarbons or condensed moisture, as the hydrogen released into the molten weld pool can cause serious weld imperfections, e.g., cracking. The cleaning operations, which may involve just wiping, need not remove all foreign contaminants nor do they require the use of solvents; welding through thin layers of remaining contaminants is acceptable, unless they degrade the quality requirements of this code resulting in unacceptable welds.

C-7.14.5.2 Repair. Mill-induced defects observed on cut surfaces are caused by entrapped slag or refractory inclusions, deoxidation products, or blow holes. The repair procedures for discontinuities of cut surfaces may not be adequate where tension is applied in the through-thickness direction of the material. For other directions of loading, this article allows some lamination-type discontinuities in the material. Experience and tests have shown that laminations parallel to the direction of tensile stresses do not generally adversely affect the load carrying capacity of a structural member. The user should note that the repair procedures of 7.14.5.2 are only intended for correction of material with sheared or thermal cut edges.

C-7.14.6 Joint Preparation. Oxygen gouging on steels that are quenched and tempered or thermomechanically controlled processed (TMCP) is prohibited because of the high/heat input of the process (see C-7.7).

C-7.14.8.3 Roughness Requirements. Corrections are allowed for thermal cut surfaces that exceed the maximum allowable surface roughness values. Occasional notches or gouges of limited depth may be corrected, the deeper ones only with approval. Depth limitations represent the collective judgment of the Committee and reflect the structural requirements and typical workmanship capability of the Contractor.

By referring to "occasional notches and gouges," the Committee refrained from assigning any numerical values on the assumption that the Engineer–being the one most familiar with the specific conditions of the structure–will be a better judge of what is acceptable. The Engineer may choose to establish the acceptance criteria for occasional notches and gouges.

C-7.15 Reentrant Corners

Minimum radii are specified for reentrant corners to avoid effects of stress concentrations. The code does not specify a minimum radius for corners of beam copes of hot-rolled beams or welded built-up cross sections because any arbitrarily selected minimum radius would extend up into the beam fillet or the bottom of the flange, in some cases, making the radius extremely difficult or impossible to provide. Further, the peak stress can be accommodated only by localized yielding, and the magnitude of the elastic stress concentration factors is not significantly affected by the differences in radii of any practical size.

Statically loaded and tubular structures allow, and generally require, a smaller reentrant corner radius than is allowed for cyclically loaded structures. The smaller radius is necessary for some standard bolted or riveted connections.

See Figure C-7.1 for examples of unacceptable reentrant corners.

C-7.16 Weld Access Holes, Beam Copes, and Connection Material

C-7.16.1 Weld Access Hole Dimensions. Solidified but still hot weld metal contracts significantly as it cools to ambient temperature. Shrinkage of large welds between elements which are not free to move to accommodate the shrinkage-induced strains in the material adjacent to the weld can exceed the yield point strain. In thick material, the weld shrinkage is restrained in the thickness directions as well as in the width and length directions, causing triaxial stresses to develop that may inhibit the ability of ductile steel to deform in a ductile manner. (Figure C-7.2 shows examples of good practice for forming copes.)

Under these conditions, the possibility of brittle fracture increases. Generously sized weld access holes, Figure 7.2, are required to provide increased relief from concentrated weld shrinkage strains, to avoid close juncture of welds in orthogonal directions, and to provide adequate clearance for the exercise of high-quality workmanship in hole preparation, welding, and ease of inspection. Welded closure of weld access holes is not recommended. When weld access holes are required to be closed for cosmetic or corrosion protection reasons, sealing by use of mastic materials may be preferable to welding.

C-7.16.2 Shapes to be Galvanized. The grinding of sections to be galvanized is intended to reduce the formation of a hard surface layer and the tendency to initiate cracks.

C-7.16.3 Copes and Access Holes in Heavy Shapes. The preheat prior to cutting copes and access holes in heavy sections is intended to reduce the formation of a hard surface layer and the tendency to initiate cracks.

Code editions through 2015 required MT or PT on thermally cut surfaces of beam copes and weld access holes because of the susceptibility of these areas to cracking. Other changes to the Code were made to increase resistance to cracking in these areas by controlling access hole shape, profile, and fabrication methods. AISC removed the requirement in 2016.

C-7.18 Camber in Built-Up Members

Heat upsetting (also referred to as *flame shrinking*) is deformation of a member by application of localized heat. It is allowed for the correction of moderate variations from specified dimensions. The upsetting is accomplished by careful application of heat with the resulting temperature not exceeding the maximum temperature specified in 7.25.2.

C-7.20 Control of Distortion and Shrinkage

C-7.20.2 Sequencing. When making splices between sections, whether in the shop or field, the welding sequence should be reasonably balanced between the web and flange welds as well as about the major and minor axes of the member.

C-7.21 Tolerance of Joint Dimensions

C-7.21.1 Fillet Weld Assembly. Except for the separation of faying surfaces in lap joints and backing bars, a gap of 3/16 in [5 mm] maximum is allowed for fillet welding material not exceeding 3 in [75 mm] in thickness. For material over 3 in [75 mm], the maximum allowable gap is 5/16 in [8 mm].

These gaps are necessitated by the allowable mill tolerances and inability to bring thick parts into closer alignment. The code presupposes straightening of material prior to assembly or an application of external load mechanism to force and keep the material in alignment during assembly.

These gaps may require sealing either with a weld or other material capable of supporting molten weld metal. It should be realized that upon release of any external jacking loads, additional stresses may act upon the welds. Any gap 1/16 in [2 mm] or greater in size requires an increase in size of fillet by the amount of separation.

C-7.21.2 PJP Groove Weld Assembly. See C-7.21.1.

C-7.21.3 Butt Joint Alignment. Typical sketches of the application of the alignment requirements for abutting parts to be joined in welds in butt joints are shown in Figures C-7.3 and C-7.4.

C-7.21.4.2 Correction. Root openings wider than those allowed by Figure 7.3 and Table 10.14 may be corrected by building up one or both sides of the groove faces by welding. In correcting root openings, the user is cautioned to obtain the necessary approvals from the Engineer where required. The final weld is to be made only after the joint has been corrected to conform to the specified root opening tolerance, thus keeping shrinkage to a minimum.

C-7.22 Dimensional Tolerances of Welded Structural Members

C-7.22.2 Beam and Girder Straightness (No Camber Specified) and C-7.22.3 Beam and Girder Camber (Typical Girder). Allowable variation in straightness of welded built-up members are the same as those specified in ASTM A6 for hot-rolled shapes.

C-7.22.4 Beam and Girder Camber (without Designed Concrete Haunch). The cambering of welded beams or girders is used to eliminate the appearance of sagging or to match elevation of adjacent building components when the member is fully loaded.

Although the tolerance on camber is of less importance than camber per se, for consistency, allowable variation in camber is based upon the typical loading case of distributed load which causes a parabolic deflected shape.

The tolerances shown are to be measured when members are assembled to drill holes for field splices or to prepare field welded splices (see Figure C-7.6).

When the deck is designed with a concrete haunch, the 1–1/2 in [40 mm] tolerance at mid-span is based upon an assumed 2 in [50 mm] design haunch. The 1/2 in [12 mm] difference is for field deviations and other contingencies.

When the Contractor checks individual members, care should be exercised to assure that the tolerances of the assembly will be met.

There are two sets of tolerances for allowable variation from specified camber. The first set of tolerances applies to all welded beams and girders, except members whose top flange is embedded in concrete without a designed concrete haunch. Here the camber tolerance is positive with no minus tolerance allowed.

The second set of tolerances applies to welded members where the top flange is embedded in concrete without a designed haunch; the variation allowed has both a plus and minus tolerance.

C-7.22.6.1 Measurements. Allowable tolerances for variations from flatness of dynamically loaded girder webs are given in the code separately for interior and fascia girders. The stricter tolerance for fascia girders is based only on appearance as there are no structural requirements for the difference. Even fascia girder distortion allowed will be somewhat noticeable, particularly when members are painted with a glossy finish. The fascia tolerances are considered satisfactory for most requirements. If more stringent tolerances are needed for appearance, they should be included in contract documents as stated in 7.22.6.5, but some degree of distortion is unavoidable. Variations from flatness in girder webs are determined by measuring offset from the nominal web centerline to a straight edge whose length is greater than the least panel dimension and placed on a plane parallel to the nominal web plane (see Figure C-7.5 for typical measurement method).

C-7.22.6.2 Statically Loaded Nontubular Structures. The flatness tolerances for webs with intermediate stiffeners on both sides and subject to dynamic loading is the same as that for interior bridge girders (see 7.22.6.3). When subject to static loading only, the tolerance is somewhat more liberal. The tolerance given for intermediate stiffeners placed only on one side of the web is the same for either cyclic or static loading and is the same as that for interior bridge girders.

NOTE: The AISC Specification for Design, Fabrication, and Erection of Structural Steel for Buildings states that the tolerances for flatness of girder webs given in 7.22.6.2 need not apply for statically loaded girders.

C-7.22.6.4 Excessive Distortion. Web distortions of twice the amount allowed for interior or fascia girder panels are allowed in end panels of girders if the installation of field bolted splice plates will reduce the distortion to the level otherwise allowed. To avoid the possibility of costly field correction, the Contractor should determine by a shop assembly that the bolted splice plate will reduce the distortion to acceptable limits.

C-7.22.8 Flange Warpage and Tilt. The combined warpage and tilt Δ of the flange of welded beams and girders is measured as shown in Figure C-7.7. In the Committee's judgment, this tolerance is easier to use than the ASTM A6 specification criteria, although both sets of tolerances are in reasonable agreement.

Tolerance on twist is not specified because the torsional stiffness of open (nonbox) shapes is very low, such that twist is readily eliminated by interconnection with other members during erection. Members of box cross sections are approximately 1000 times as stiff in torsion as an open I or W shape with equivalent bending and area section properties. Once a closed box section has been welded, it is extremely difficult to correct any twist that may have been built in without cutting one corner apart and rewelding. Because twist resulting from welding is not entirely predictable and extremely difficult to correct in closed box members, the following apply.

(1) Appropriate provisions should be incorporated in design to ensure reliable service performance of such members with some arbitrary measure of twist.

(2) Due cognizance should be taken of the size of the element, of the effect of the twist when placing concrete on the structure, and the use of such connection details that will satisfactorily accommodate the twist.

C-7.22.10 Bearing at Points of Loading. Figure C-7.8 illustrates application of the code requirement.

C-7.22.12 Other Dimensional Tolerances. Tolerances specified in 7.22 are limited to routinely encountered cases. Dimensional tolerances not covered in 7.22 should be established to reflect construction or suitability for service requirements.

C-7.23 Weld Profiles

The 1982 edition changed the fillet weld convexity requirements in such a way that the maximum convexity formula applies not only to the total face width of the weld, but also to the width of an individual bead on the face of a multiple-pass weld. This was done to eliminate the possibility of accepting a narrow "ropey" bead on the face of an otherwise acceptable weld. The new formula, which is based on the "width of face," provides the same convexity requirement as the previous formula which was based on "leg size."

When a fillet weld is started, the weld metal, due to its surface tension, is rounded at the end. Sometimes this is such that there is a slight curve inward. Also, at both the start and finishing ends, this curve prevents the weld from being full size to the very end. Therefore, these portions are not included as part of the effective weld length. If the designer has any concern relative to the notch effects of the ends, a continuous fillet weld should be specified which would generally reduce the required weld size.

C-7.25 Repairs

C-7.25.1 Contractor Options. The code allows the Contractors, at their option, to either repair or remove and replace an unacceptable weld. It is not the intent of the code to give the Inspector authority to specify the mode of correction.

C-7.25.2 Localized Heat Repair Temperature Limitations. Application of localized heat is allowed for straightening members; however, this should be done carefully so as not to exceed temperature limitations that would adversely affect the properties of the steel. Quenched and tempered steels that are heated above 1100°F [600°C] may experience deterioration of mechanical properties as a result of the formation of an undesirable microstructure when cooled to room temperature. Other steels that are heated above 1200°F [650°C] may experience the possibility of undesirable transformation products or grain coarsening, or both. However, these maximums are sufficiently below the metal lower transformation temperature to allow some tolerance in temperature measurement method.

C-7.25.5 Welded Restoration of Base Metal with Mislocated Holes. Welded restoration of holes may be required for structural function, for corrosion prevention, or for aesthetics. The technique for making plug welds set forth in 7.24.1 of this code is not satisfactory for restoring the entire cross section of the base metal at mislocated holes. Restoration of base metal by filling mislocated holes requires different techniques than those required for making plug welds. Plug welds are intended to transmit shear from one plane surface to another and not to develop the full cross section of the hole.

The following may be found useful when addressing mislocated holes:

(1) A procedure similar to the following may be used to restore base metal not in tension zones of cyclically loaded members:

(a) Provide a chamfer bevel to the bottom of the affected hole if the base metal thickness exceeds 5/16 in [8 mm], or if the hole diameter is not at least 5/16 in [8 mm] more than the base metal thickness. Install steel backing conforming to 7.2.2.2 and 7.9.1.3 at the base of the affected hole. Then, maintaining the preheat and interpass temperatures in accordance with Table 5.8, deposit filler metal, using the technique provisions of 7.24.1.1 through 7.24.1.3 as applicable per position.

(b) Completely remove the backing and excavate the second side to sound metal as necessary prior to back welding.

(2) A procedure similar to the following may be used to restore base metal under any loading condition: Insert a tightly fitted cylindrical steel filler meeting the requirements of 7.2.2.2 into the hole to one-half the hole depth. Then, beginning on the unfilled side, prepare an elongated U-groove cavity, preferably longitudinal to the direction of primary stress, with a depth slightly into the steel filler. The sides and ends of the cavity should slope appropriately from the surface of the plate. The cavity may be made by grinding or air carbon arc gouging followed by grinding. Maintaining the preheat and interpass temperatures in accordance with Table 5.8, completely fill the cavity with weld metal in accordance with an approved welding procedure using the stringer bead technique. Excavate the second side, completely removing all remnants of the steel filler and forming a cavity of the same profile as the first side, extending into sound weld and base metal. Fill this cavity with weld metal using the procedure described for the initial side.

C-7.26 Peening

Except as provided in 10.2.3.6(3), peening of the surface layer of the weld is prohibited because mechanical working of the surface may mask otherwise unacceptable surface discontinuities. For similar reasons, the use of lightweight vibrating tools for slag removal should be used with discretion.

C-7.27 Caulking

The code has historically prohibited any plastic deformation of the weld or base metal surfaces for the purpose of obscuring or sealing discontinuities. However, since some minor discontinuities may interfere with the integrity of the coating system, limited caulking may now be used for the softer welds and base metals when approved by the Engineer.

There are no prohibitions against the use of mastic or nonmetallic fillers for cosmetic reasons provided that all required inspections of the weld and base metal have been completed and accepted prior to application.

C-7.28 Arc Strikes

Arc strikes result in heating and very rapid cooling. When located outside the intended weld area, they may result in hardening or localized cracking, and may serve as potential sites for initiating fracture.

C-7.29 Weld Cleaning

The removal of slag from a deposited weld bead is mandatory to prevent the inclusion of the slag in any following bead and to allow for visual inspection.

C-7.30 Weld Tabs

The termination, start or stop, of a groove weld tends to have more discontinuities than are generally found elsewhere in the weld. This is due to the mechanism of starting and stopping the arc. Hence, weld tabs should be used to place these zones outside the finished, functional weld where they can be removed as required by 7.30.2 or 7.30.3. Weld tabs will also help maintain the full cross section of the weld throughout its specified length. It is important that they be installed in a manner that will prevent cracks from forming in the area where the weld tab is joined to the member.

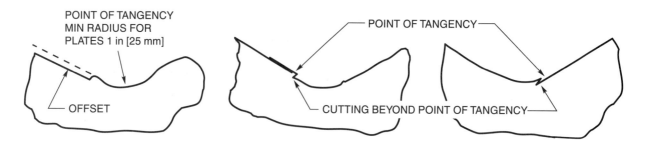

Figure C-7.1—Examples of Unacceptable Reentrant Corners (see C-7.15)

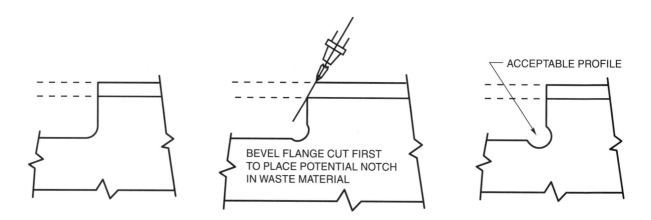

Figure C-7.2—Examples of Good Practice for Cutting Copes (see C-7.16)

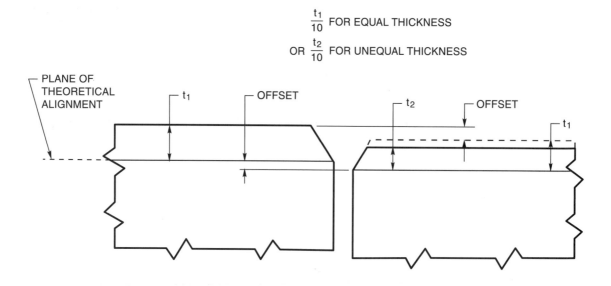

Figure C-7.3—Permissible Offset in Abutting Members (see C-7.21.3)

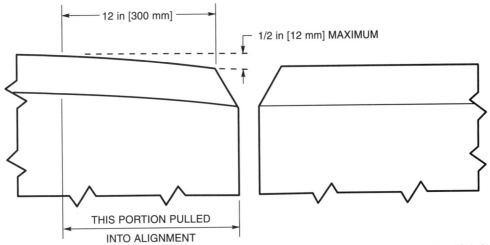

Note: In correcting misalignment that exceeds the allowance, the parts are not to be drawn to a slope greater than 1/2 in [12 mm] in 12 in [300 mm].

Figure C-7.4—Correction of Misaligned Members (see C-7.21.3)

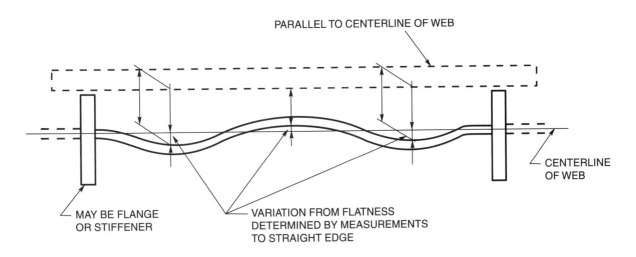

Figure C-7.5—Typical Method to Determine Variations in Girder Web Flatness (see C-7.22.6.1)

COMMENTARY

AWS D1.1/D1.1M:2020

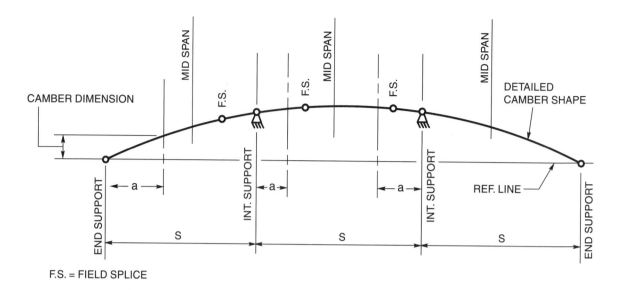

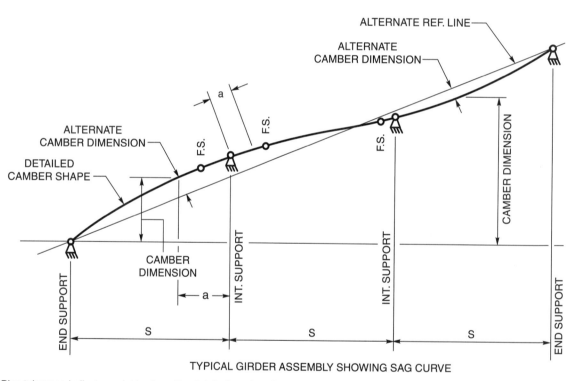

Note: Plus tolerance indicates point is above the detailed camber shape. Minus tolerance indicates point is below the detailed camber surface.

Figure C-7.6—Illustration Showing Camber Measurement Methods (see C-7.22.4)

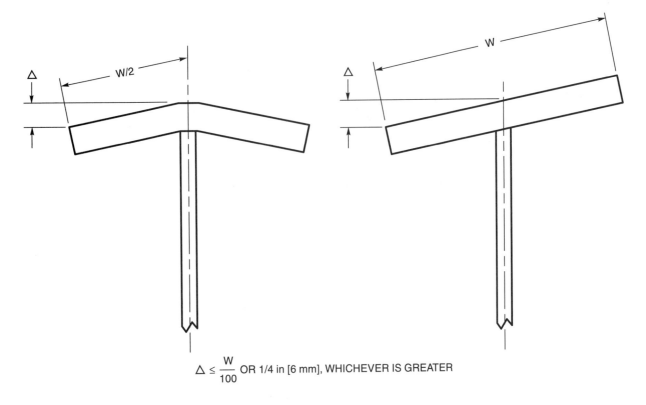

$$\Delta \leq \frac{W}{100} \text{ OR 1/4 in [6 mm], WHICHEVER IS GREATER}$$

Figure C-7.7—Measurement of Flange Warpage and Tilt (see C-7.22.8)

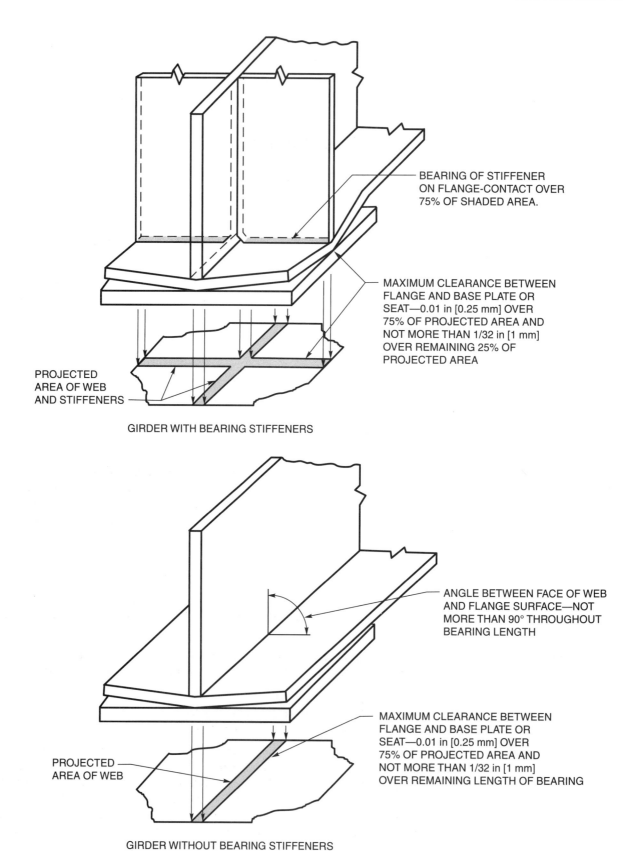

Figure C-7.8—Tolerances at Bearing Points (see C-7.22.10)

C-8. Inspection

C-8.1 Scope

This clause of the code is designed to clarify the separate responsibilities of the Contractor/fabricator/erector, as opposed to the Owner/Engineer, etc.

The provisions clarify the basic premise of contractual obligations when providing product and services. Those who submit competitive bids or otherwise enter into a contract to provide materials and workmanship for structural weldments in conformance with the provisions of the code assume an obligation to furnish the products as specified in the contract documents and are fully responsible for product quality.

In this clause, the term *fabrication/erection inspection* is separated from verification inspection. In the original draft of this section, these separate functions were designated as quality control and quality assurance, respectively. These terms were replaced with the broader terms now contained in the code to avoid confusion with the usage in some industries (e.g., nuclear). Quality assurance means specific tasks and documentation procedures to some users of the code. It was advantageous to use more general terms that place greater emphasis on timely inspection. Inspection by the Owner should be planned and timely if it is to improve the quality of the construction.

C-8.1.1 Information Furnished to Bidders. It is essential that the Contractor know in advance which welds are subject to NDTs and which testing procedures will be used. Unless otherwise provided in the contract documents, the quality criteria for acceptance of welds are stated in Clause 8, Part C and Clause 10, Part F for tubulars. It is not necessary to write in the contract documents exactly which weld or what portions of specific welds will be examined by a specific test method. A general description of weld test requirements may be specified (e.g., "10% of the length of all fillet welds shall be inspected by MT," or "All CJP butt joint welds in tension flanges of girders shall be radiographed").

If the location of tension flange butt joint welds is not obvious, their location should be designated on the plans.

When spot checking is specified (e.g., 10% of all fillet welds), it should not be taken to imply that the Contractor be notified prior to welding which specific welds or portion of welds will be tested. It is a basic premise of the specifications that if random tests or spot tests are made, there should be a sufficient number of random tests to give a reliable indication of weld quality.

There are different acceptance criteria for statically loaded structures, cyclically loaded structures, and tubular structures. The basic difference in acceptance criteria for each of these structures is based upon the difference between static, and fatigue loading.

When fatigue crack growth is anticipated, initial weld discontinuity sizes are smaller. All criteria are established in an attempt to preclude weld failure during the anticipated service life of the weldment.

C-8.1.2 Inspection and Contract Stipulations. This subclause describes the responsibility of the Contractor for fabrication/erection inspection and testing, which is basically the quality control responsibility described in other contract documents. The Owner has the right, but generally not the responsibility, to provide independent inspection to verify that the product meets specified requirements. This quality assurance function may be done independently by the Owner or their representative or, when provided in the contract, verification inspection may be waived or it may be stipulated that the Contractor perform both the inspection and the verification. When this is done, quality control and quality assurance remain separate functions. Verification inspection should be performed independently by personnel whose primary responsibility is quality assurance and not production.

C-8.1.3 Definition of Inspector Categories. This subclause describes the difference between the Inspector representing the Owner and the Inspector representing the Contractor.

C–8.1.4.1 Engineer's Responsibilities. The Engineer may wish to specify in the contract documents a requirement for the qualifications of the Contractor's Inspectors or Verification Inspectors, or both, and their Assistant Inspectors if they are other than as provided in 8.1.4.2.

In some building codes, the authority having jurisdiction (Building Official) may also have requirements for inspector certification and/or approval.

C–8.1.4.3 Alternative Inspector Qualifications. Documentation of the specified alternative qualification of inspection personnel should include records of compliance with all of the alternative requirements specified. This evidence may include records of training, education, eye examination, experience and testing of code knowledge.

Documentation should include training, education, eye examination, experience and testing of code knowledge.

C-8.1.5 Inspector Responsibility. This subclause requires that the Inspector verify that all fabrication and erection by welding is performed in conformance with the requirements of the contract documents. This includes not only welding but also materials, assembly, preheating, NDT, and all other requirements of the code and provisions of the contract documents.

C-8.1.6 Items to be Furnished to the Inspector. Inspectors need a complete set of approved drawings to enable them to properly do their work. They need be furnished only the portion of the contract documents describing the requirements of products that they will inspect. Much of the contract documents deal with matters that are not the responsibility of the Inspector; these portions need not be furnished.

C-8.1.7 Inspector Notification. If the Inspectors are not notified in advance of the start of operations, they cannot properly perform the functions required of them by the code.

C-8.2 Inspection of Materials and Equipment

This code provision is all-encompassing. It requires inspection of materials and review of materials certification and mill test reports. It is important that this work be done in a timely manner so that unacceptable materials are not incorporated in the work.

C-8.3 Inspection of WPS Qualification

The requirements of this section, including any qualification testing required by Clause 6, should be completed before any welding is begun on any weldments required by the contract documents. Qualification should always be done before work is started, but all qualification does not have to be completed before any work can be started.

C-8.4 Inspection of Welder, Welding Operator, and Tack Welder Qualifications

C-8.4.1 Determination of Qualification. It is important that the Inspector determine that all welders are qualified before work is begun on the project. If discovered after welding has begun, lack of welder qualification documentation may cause serious delays in the acceptance of weldments.

C-8.4.2 Retesting Based on Quality of Work. Examples of methods the inspector may choose to demonstrate the ability to produce sound welds include the following:

(1) an interview of the welder,

(2) increased visual inspection for a limited time period,

(3) observation of the welding, or a simplified weld test developed to evaluate the issue of concern,

(4) requalification in compliance with Clause 6 or Clause 10 for tubulars.

C-8.4.3 Retesting Based on Qualification Expiration. If evidence cannot be supplied that shows a welder, welding operator, or tack welder has used the welding process within the last six months, he or she is not considered qualified to weld using that process without new qualification testing. Because active welders can maintain their certification as long as they continue to do good work, it is important that Inspectors regularly evaluate the quality of the welds produced by each welder, welding operator, and tack welder.

C-8.5 Inspection of Work and Records

Except for final visual inspection, which is required for every weld, the Inspector inspects the work at suitable intervals to ensure that the requirements of the applicable sections of the code are met. Such inspections should be conducted, on a sampling basis, prior to assembly, during assembly, and during welding. The inspector identifies final acceptance or rejection of the work either by marking on the work or with other recording methods. The method of identification should not be destructive to the weldment.

C-8.5.4 Inspector Identification of Inspections Performed. Die stamp marks may form sites for crack initiation.

C-8.6 Obligations of the Contractor

C-8.6.1 Contractor Responsibilities. The Contractor conducts inspection to the extent necessary to ensure conformance with the code, except as provided in 8.6.5.

Part C
Acceptance Criteria

C-8.8 Engineer's Approval for Alternate Acceptance Criteria

The criteria provided in Clause 7, Fabrication, are based upon knowledgeable judgment of what is achievable by a qualified welder. The criteria in Clause 7 should not be considered as a boundary of suitability for service. Suitability for service analysis would lead to widely varying workmanship criteria unsuitable for a standard code. Furthermore, in some cases, the criteria would be more liberal than what is desirable and producible by a qualified welder. In general, the appropriate quality acceptance criteria and whether a deviation produces a harmful product should be the Engineer's decision. When modifications are approved, evaluation of suitability for service using modern fracture mechanics techniques, a history of satisfactory service in similar structures, or experimental evidence is recognized as a suitable basis for alternate acceptance criteria for welds.

C-Tables 8.1 and 10.15, Item 8—Piping Porosity. Tables 8.1 and 10.15 contain visual acceptance criteria for surface-breaking piping porosity, since this is visually detectable, and may significantly reduce the cross-sectional areas of the weld available to resist applied loads. Other forms of surface-breaking porosity do not reduce the cross-sectional area as significantly (see Clause 3 for a definition of piping porosity).

C-8.9 Visual Inspection

This subclause makes visual inspection of welds mandatory and contains the acceptance criteria for it. The fabrication requirements of Clause 7, are also subject to visual inspection. Allowable depth of undercut was revised in the 1980 edition of the code to more accurately reflect an acceptable percentage reduction of cross-sectional area for three categories of stress. The undercut values are for structures and individual members that are essentially statically loaded.

Undercut values for cyclically loaded structures or tubular structures (8.9 and 10.24) have not been changed and should be specified for structures and individual members subject to cyclic loading.

C-8.11 NDT

Except as required in 10.25.1, the weld quality requirements for NDT are not a part of the contract unless NDT is specified in information furnished to the bidders or is subsequently made a part of the contract. Both the Owner and Contractor should give careful attention to the provisions of 8.6.5 and 8.14.1. When, in addition to the requirement for visual inspection, NDT is specified, the acceptance criteria of 8.11, 8.12, or 8.13 apply. The acceptance criteria for ASTM A514, A517, and A709, Grade HPS 100W high-strength quenched and tempered steels are based on inspection, visual or NDT, conducted at least 48 hours after completion of the weld. Since high-strength steels, when welded, and weld metals are susceptible to delayed cracking caused by hydrogen embrittlement, stress rupture, etc., it has been necessary to impose this time restriction to assure that any delayed cracking has a reasonable chance of being discovered during inspection.

C-8.12 RT

C-8.12.1 Discontinuity Acceptance Criteria for Statically Loaded Nontubular Connections. See Figures C-8.7 and C-8.8.

C-8.12.2.1 Cyclically Loaded Nontubular Connections in Tension. See Figure C-8.9.

C-8.13 UT

C-8.13.1 Acceptance Criteria for Statically Loaded Nontubular Connections. In the notes in Table 8.2, the key words that are most often misinterpreted are "... from weld ends carrying primary tensile stress." This phrase generally refers to the ends of groove welds that are subject to applied tensile design loads that are normal to the weld throat. When box columns are used with moment connection members welded to the outside surface and diaphragm plates welded on the inside to transfer the primary stress through the box column member, the ends of the moment plate-to-box column plate welds are subject to the 2L distance from the weld end, but the welds on diaphragm plates on the inside of the box are not subject to this restriction. The weld ends of the diaphragm plates do not carry primary tensile stress because this stress is carried through the width of the adjacent box member plates.

The note of Table 8.2 on scanning levels was added because experience with UT acceptance level provisions previously required by the code resulted in acceptance of some rather large gas pockets and piping porosity that can occur in ESW and EGW welds. The shape of these gas defects, which are peculiar to ESW and EGW welds, is such that they reflect less ultrasound than the usual weld discontinuities. Testing at 6 dB more sensitive than standard testing amplitudes will not guarantee accurate evaluation of gas defects in ESW or EGW welds. This type of discontinuity is easily evaluated by RT, which is recommended if indications of pipe or other gas discontinuities are seen at scanning levels.

C-8.13.2 Acceptance Criteria for Cyclically Loaded Nontubular Connections. See Clause C-8, Part F. The code provides acceptance criteria for welds subject to tensile stresses that differ from those subject only to compressive stresses. Groove welds subject to compressive stresses only and which are indicated on design or shop drawings are required to conform to the acceptance criteria of Table 8.2. Groove welds subject to tensile stresses under any condition of loading and welds subject only to compressive stresses but not specifically designated as such on design or shop drawings are required to conform to the acceptance criteria of Table 8.3, which are up to 6 dB higher than those in Table 8.2.

Part D
NDT Procedures

C-8.14 Procedures

In addition to visual inspection, which is always necessary to achieve compliance with code requirements, four NDT methods are provided for in the code: (1) RT, (2) UT, (3) MT, and (4) PT.

RT and UT are used to detect both surface and internal discontinuities. MT is used to detect surface and near surface discontinuities. PT is used to detect discontinuities open to the surface. Other NDT methods may be used upon agreement between Owner and Contractor.

C-8.14.6 Personnel Qualification. In addition to Level II technicians, Level III individuals may also perform NDT tests provided they meet the requirements of NDT Level II. NDT Level III engineers and technicians are generally supervisors and may not be actively engaged in the actual work of testing. Since there is no performance qualification test for individuals qualified to NDT Level III, all individuals providing testing services under the code are required to be qualified to Level II, which has specific performance qualification requirements.

C-8.15 Extent of Testing

It is important that joints to be nondestructively examined be clearly described in the information furnished to bidders as explained in Part A of this Commentary.

Part E
Radiographic Testing

C-8.16 RT of Groove Welds in Butt Joints

C-8.16.1 Procedures and Standards (RT). The procedures and standards set forth in this section are primarily designed for the RT of CJP groove welds in cyclically loaded structures and statically loaded structures. Typical geometries for structural connections and design requirements for these structures were taken into account in the preparation of the specification. An effort was made to incorporate the methodology of ASTM and to utilize procedures described in the ASME *Boiler and Pressure Vessel Code* whenever possible.

C-8.16.1.1 Digital Radiography. There is an important distinction between direct radiography (DR) using a DDA and computed radiography (CR). The ASTM standards for these methods are not interchangeable.

C-8.16.2 Variations. Phased array ultrasonic testing (PAUT) in accordance with Annex H may be substituted for RT when approved by the Engineer. Just as conventional UT and RT have different detection capabilities (each may find indications the other will not), the RT and PAUT methods in Part B and Annex H have different detection capabilities; however, both are expected to yield similar reliability in inspected product. PAUT should not be substituted for RT of EGW or ESW because of the concerns raised in C–8.13.1 and C–8.19.3.

C-8.17 RT Procedure

The single source of inspecting radiation is specified to avoid confusion or blurring of the radiographic image. Elsewhere in the code, limits are placed on the size of the source to limit geometric unsharpness. RT sensitivity is judged solely on the quality of the IQI image(s), as in both ASTM and ASME.

C-8.17.3 Removal of Reinforcement. When the Owner wishes weld surfaces to be ground flush or otherwise smoothed in preparation for RT, it should be stated in the contract documents. The Owner and the Contractor should attempt to agree in advance on which weld surface irregularities will not be ground unless surface irregularities interfere with interpretation of the radiograph. It is extremely difficult and often impossible to separate internal discontinuities from surface discontinuities when reviewing radiographs in the absence of information describing the weld surface. When agreement can be reached on weld surface preparation prior to RT, rejections and delays will generally be reduced.

C-8.17.3.1 Tabs. Weld tabs are generally removed prior to RT so that the radiograph will represent the weld as finished and placed in service. Contraction cracks are commonly found in the weld at the interface between weld tabs and the edge of the plate or shape joined by the weld. These cracks are hard to identify in the radiograph under the best conditions. It is considered necessary to remove the weld tabs before attempting to radiograph the boundaries of the welded joint (see also C-8.17.13).

C-8.17.3.3 Reinforcement. When weld reinforcement, or backing, or both, is not removed, shims placed under the IQIs are required so that the IQI image may be evaluated on the average total thickness of steel (weld metal, backing, reinforcement) exposed to the inspecting radiation.

C-8.17.4 Radiographic Film. Provisions of this section are to provide fine-grain film and to avoid coarseness in the image that may result from the use of fluorescent screens.

C-8.17.5 Technique. The source of radiation is centered with respect to the portion of the weld being examined to avoid as much geometric distortion as possible.

C-8.17.5.1 Geometric Unsharpness. This subclause is provided to limit geometric unsharpness, which causes distortion and blurring of the radiographic image.

C-8.17.5.2, C-8.17.5.3 Source-to-Subject Distance and Limitations. These sections are intended to limit geometric distortion of the object as shown in the radiograph. An exception is made for panoramic exposures in tubular structures, which are covered by 10.28 of the code.

C-8.17.6 Sources. This subclause intends that X-ray units, 600 kVp maximum, and iridium 192 sources may be used for all RT, provided they have adequate penetrating ability and can produce acceptable radiographic sensitivity based upon IQI image as provided in 8.17.7 and 10.27.2. Since cobalt 60 produces poor radiographic contrast in materials of limited thickness, it is not approved as a radiographic source when the thickness of steel being radiographed is equal to, or less than, 2–1/2 in [65 mm]. When the thickness of steel being radiographed exceeds 2–1/2 in [65 mm], cobalt 60 is often preferred for its penetrating ability. Care should be taken to ensure that the effective size of the radiograph source is small enough to preclude excessive geometric unsharpness.

C-8.17.7 IQI Selection and Placement. Since radiographic sensitivity and the acceptability of radiographs are based upon the image of the required IQIs, care is taken in describing the manufacture and use of the required IQIs. IQIs are placed at the extremities of weld joints where geometric distortion is anticipated to contribute to lack of sensitivity in the radiograph as shown in Figures 8.6–8.9.

IQIs may only be placed on the source side unless otherwise approved by the Engineer. Failure to place the IQIs on the source side during the radiographic exposure, without prior approval of the Engineer, should be cause for rejection of the radiographs.

C–8.17.10 Quality of Radiographic Film and Digital Images. Standard imaging plates, as in any radiography, may be masked using lead shielding. Masking may be necessary to eliminate internal and backscatter radiation. In digital radiography, backscatter and unnecessary exposure may also contribute to the deterioration of the panel. Particularly, scattered radiation that may strike the digital detector array crystals horizontally may cause more rapid deterioration.

Histogram normalization "smooths" the small integral steps of the digital image. This is similar to taking the noise out of a digital image. Gamma correction is normalization of the various shades of gray. All normalized images are saved as separate files from the original raw image. If there is any concern regarding the integrity of the image after normalization enhancements, the viewer can refer to the raw image, which cannot be overwritten.

Sub-images are useful for facilitating inspection, such as when non deleterious aberrations occur. For example, an inspector may verify that indications that appear to be scratches are the result of grinding, the inspector can note this on a sub-image, thereby documenting this evaluation without modifying the original image.

Digital artifacts may include artifacts resulting from dead pixels, or blooming pixels (where the charge spreads to adjacent pixels).

C-8.17.8.3 Backscatter. Backscattered radiation can cause general fogging and produce artifacts in the radiograph. The method described in this section will identify backscattered radiation so that corrective steps can be taken.

C-8.17.11.1.1 H & D Density. It is the intent of the specification to use radiographic films within the full limits of the useful film density. An effort is made in this code to avoid the necessity of making multiple exposures or using films of more than one exposure speed when examining welded joints routinely expected to be encountered in cyclically loaded structures and statically loaded structures.

C–8.17.11.1.2 Transitions. The weld transitions in thickness provided for in this subclause are expected to be gradual with a maximum slope of 1 on 2–1/2 as shown in Figure 4.2.

C–8.17.11.2 Digital Image Sensitivity Range. The SPIP/DDA must be calibrated for sensitivity. SPIP/DDA calibrations may be performed by the operator using the "standard" built into the software and an ASTM IQI per the standardized written procedure.

ASTM E1936 describes the relationship between monitor calibration and the SPIP/DDA.

ASTM E2698 also describes SPIP/DDA calibration.

Radiographic film transmitted density measures the amount of light that passes through the film; however, this is not applicable to digital radiography. In digital images, light balance in the image is controlled through the brightness and contrast settings.

C-**8**.17.12 **Identification Marks.** This subclause describes all information required to identify the radiograph and also provides methods for matching the radiograph to the weld joint, so that weld repairs, when necessary, may be made without repetitive or unnecessarily large excavations.

C-**8**.17.13 **Edge Blocks.** Flange-to-flange welded butt joints that join segments of thick flanges in beams and girders are particularly difficult to radiograph due to geometric distortion and undercut from scattered radiation at the ends of the weld that represent the flange edges. Weld defects at these critical locations are limited under the provisions of **8**.12.

On weldments over 1/2 in [12 mm] in thickness, it was demonstrated by using drilled holes and lead indicators near the top edge of a weldment that a substantial portion of this edge was over exposed and could not be shown, which left the possibility of not showing defects. By using edge blocks and a standard source alignment, lead indicators and drilled holes could be shown on a radiograph at the plate edge.

C–**8**.17.14 **Linear Reference Comparators.** Hole-type IQIs may be used as a linear reference comparator in digital radiography, using the cursor measuring tool built into the software. Wire-type IQIs are typically not used as comparators because the wires are relatively thin and the ends may be hard to locate, making measurement difficult.

C-**8**.18 Examination, Report, and Disposition of Radiographs

C-**8**.18.1 **Equipment Provided by Contractor.** A suitable, variable intensity illuminator with spot review or masked spot review capability is required since more accurate film viewing is possible when the viewers' eyes are not subjected to light from portions of the radiograph not under examination. The ability to adjust the light intensity reduces eye discomfort and enhances visibility of film discontinuities. Subdued light in the viewing area allows the reviewer's eyes to adjust so that small discontinuities in the radiographic image can be seen. Film review in complete darkness is not advisable since the contrast between darkness and the intense light from portions of the radiograph with low density cause discomfort and loss of accuracy. In general, within the limits of density approved by the code, the greater the film density, the greater the radiographic sensitivity.

C-**8**.18.3 **Reports and Retention.** The term *a full set of radiographs* as used in **8**.18.3 means one radiograph of acceptable quality from each radiographic exposure required for complete RT. If Contractors elect to load more than one film in each cassette to produce an extra radiograph for their own use or to avoid possible delays, extra exposures due to film artifacts, or both, the extra radiographs, unless otherwise specified, are the property of the Contractor.

Part F
UT of Groove Welds

C-**8**.19 General

C-**8**.19.1 **UT Procedures and Standards.** The UT provisions are written as a precise, direct method of testing weldments. These provisions were designed to ensure reproducibility of test results when examining specific reflectors. Most groove welds may be satisfactorily tested using the provisions of Clause **8**, Part F.

Provisions for UT of welds in T-, Y-, and K-tubular connections can be found in 10.26.1 and 10.29. Detailed procedures have not been included in this section of the code because of the complex geometry associated with these welds. UT procedures for these welded joints should be approved by both the Engineer and Contractor.

The UT of fillet welds was not included in the code because of the inability to formulate a simple procedure giving satisfactory results. Considerable information can be obtained about the location of a discontinuity in a fillet weld, as well as its size and orientation, when using special techniques. The complexity and limitations of UT increase as the size of the fillet weld decreases. Fillet weld sizes less than 3/4 in [20 mm] usually require the use of miniature search units for complete evaluation. The frequency for miniature search units should be higher than the 2.25 MHz nominal frequency normally required, in order to control the sound beam divergence. This frequency change would also affect the 2 decibels per inch [25 mm] attenuation factor used for indication evaluation. Variations from the code provisions for UT may be acceptable upon agreement from the Engineer. It is recommended that details of such agreements be in writing so that all parties know how the welds are to be inspected.

C-8.19.2 Variations. When used in accordance with Annex H, PAUT is an effective test method with noted advantages, without many of the limitations of conventional UT. In addition, encoding the scan allows for informative imaging, post analysis and a permanent electronic record.

UT through paint layers on painted surfaces has been changed to an essential variable requiring approval by the Engineer. Although the code prohibits routine UT through paint layers, it does not necessarily mean that a good, tight, uniform coat of paint will interfere with the application of UT procedure. When paint is present, it should be measured and reported.

During routine fabrication of structural steel, all welds should be inspected and accepted prior to being painted. Most testing where painted surfaces are involved is on members that have been in service, and the condition of that test surface should be considered before routine testing is done.

C-8.19.3 Piping Porosity. The general nature of piping porosity in ESW and EGW welds is usually such that holes in the central portion of the weld may be masked by other surrounding holes. The branches or tunnels of piping porosity have a tendency to tail out toward the edges of the weld nugget. UT can only effectively evaluate the first major reflector intercepted by the sound path. Some discontinuities may be masked in this manner; this is true for all UT.

RT should be used to evaluate suspected piping porosity in ESW and EGW welds used in building construction (see Note 4 of Table 8.2). No mention of additional RT is presently made with reference to testing ESW and EGW welds in Table 8.3 since these processes are not presently accepted for tension welds. UT of ESW and EGW welds at higher scanning levels will give intermittent responses from piping porosity. This indicates RT should be used as described above.

The pitch-and-catch technique for evaluating incomplete fusion by UT in ESW and EGW welds is intended to be used only as a secondary test to be conducted in an area along the original groove face in the middle half of the plate thickness. This test is specified to further evaluate an UT indication in this area which appears on the display at scanning level but is acceptable by indication rating. The expected pitch-catch amplitude response from such a reflector is very high, making it unnecessary to use the applicable amplitude acceptance levels. However, since no alternative is provided, these decibel ratings are to be used. Since only a specific location is being evaluated, predetermined positioning of the probe can be made. Probe-holding fixtures are most helpful in this operation.

The use of the 70° probe in the primary application is adequate in testing ESW and EGW weld fusion surfaces of material 2–1/2 in [65 mm] and less in thickness because acceptance levels are such that proper evaluation can be expected.

C-8.21 UT Equipment

Standards are established for UT discontinuity detectors to ensure adequate mechanical and electrical performance when used in conformance with the requirements of the code.

Subclauses 8.21.1 through 8.21.5 cover the specific equipment features that are to be considered for equipment qualification; 8.22.1 covers the reference standards;

8.23.1 through 8.23.4 cover the time interval requirements and references to the applicable 8.27 reference block usage; and 8.28 presents detailed qualification procedures. Examples of these applications are included in Annex P, Form P–8.

C-8.21.6 Straight Beam (Longitudinal Wave) Search Unit. The size limitations of the active areas of straight beam transducers have not been changed; however, the sizes being given as ½ in² [323 mm²] and 1 in² [645 mm²] have been misinterpreted as being ½ in [12.7 mm] square and 1 in [25.4 mm] square, instead of the intended 1/2 square in and 1 square in, respectively.

These active area requirements are now written out to eliminate the confusion.

C-8.21.7.2 Transducer Dimensions. Tables 8.2 and 8.3 of Clause 8 were developed within a limited range of parameters. Modifications to the equipment parameters or procedures of Clause 8 may result in test results significantly different than those intended. Users are cautioned that any change in testing parameters (e.g., transducer sizes, angles, shapes, frequencies, equipment calibrations, etc.) may invalidate the applicability of Tables 8.2 and 8.3. Approval of the Engineer for modifications to the testing equipment or procedures should be made only when the effects of such changes upon acceptance criteria have been established.

C-8.22 Reference Standards

C-8.22.1 IIW Standard. The code no longer requires the Type 1 IIW UT Reference Block. Any of the IIW "type" blocks may be used.

The standard shown in Figure 8.13 is commonly known in the US as an IIW type reference block. IIW is an acronym for the International Institute of Welding. It is referred to as an IIW "type" reference block because it was patterned after the "true" IIW block but does not conform to IIW requirements in IIS/IIW–23–59. "True" IIW blocks are only made out of steel (to be precise, killed, open hearth or electric furnace, low-carbon steel in the normalized condition with a grain size of McQuaid-Ehn #8) where IIW type" blocks can be commercially obtained in a selection of materials. The dimensions of "true" IIW blocks are in metric units while IIW "type" blocks usually have US Customary units. IIW "type" blocks may also include additional calibration and references features such as notches, circular grooves, and scales that are not specified by IIW.

C-Figure 8.13 All of the notes shown herewith pertain to all of the reference blocks in both Figure 8.13 and Annex G.

C-8.22.2 Prohibited Reflectors. The code prohibits the use of square corners for calibration purposes because of the inability of acquiring amplitude standardization from various corners that are called "square." Factors that can affect amplitude standardization are the size of the fillet or chamfer on the corner, if any; the amount the corner is out of square (variation from 90°); and surface finish of the material. When a 60° probe is used, it is very difficult to identify the indication from the corner due to high amplitude wave mode conversions occurring at the corner.

C-8.23 Equipment Qualification

C-8.23.1 Horizontal Linearity. The use of ASTM E317 for horizontal linearity qualification has been eliminated, and a step-by-step procedure outlined in 8.28.1 is used for certification.

C-8.23.2 Gain Control. Caution is needed when applying alternate methods for vertical linearity certification. Normal ways of translating voltage ratios to dB graduations generally cannot be used due to potentiometer loading and capacitance problems created by the high frequency current transfer. A high degree of shielding is also needed in all wiring.

C-8.23.4 Calibration of Angle Beam Search Units. Since the contact surfaces of search units wear and cause loss of indication location accuracy, the code requires accuracy checks of the search unit after a maximum of eight hours use. The responsibility for checking the accuracy of the search unit after this time interval is placed on the individual performing the work.

C-8.24 Calibration for Testing

C-8.24.4.1 Sweep. Indications of at least two plate thicknesses are to be displayed in order to ensure proper distance calibration because the initial pulse location may be incorrect due to a time delay between the transducer crystal face and the search unit face.

C-8.24.5.1 Horizontal Sweep. At least two indications other than the initial pulse are also to be used for this distance calibration due to the built-in time delay between the transducer face and the face of the search unit.

NOTE 1: The initial pulse location will always be off to the left of the zero point on the display.

NOTE 2: Caution is to be exercised to ensure that the pulse at the left side of the screen is the initial pulse and not one from a reference reflector. (Verify by removing search unit from workpiece.)

The note has been added to the end of this subclause to ensure duplication of location data.

C-8.25 Testing Procedures

C-8.25.4 Couplants. It is recognized that couplants, other than those specifically required in the code, may work equally well or better for some applications. It is beyond the scope of the code to list all fluids and greases that could be acceptable couplant materials. Any couplant material, other than those listed in the code, that has demonstrated its capability of performing to code requirements, may be used in inspection with the approval of the Engineer and the UT operator.

Tests should be conducted to determine if there is a difference in responses from the reference reflector, due to differences between the couplant used for calibration compared to the couplant used in actual testing. Any measurable difference should be taken into account in discontinuity evaluation.

See Annex, P Form P–11 for a sample UT report form.

C-8.25.5 Extent of Testing. The provision to search the base metal for laminar reflectors is not intended as a check of the acceptability of the base metal, but rather to determine the ability of the base metal to accept specified UT procedures.

C-8.25.5.1 Reflector Size. A procedure for lamellar size evaluation is now included in 8.29.1.

C-8.25.5.2 Inaccessibility. The requirement in this subclause to grind the weld surface or surfaces flush is necessary only to obtain geometric accessibility for an alternate UT procedure when laminar discontinuities in the base metal prohibit testing using standard procedure. Contract documents may require flush grinding of tension groove welds to improve fatigue performance and facilitate more accurate RT and UT.

C-Table 8.7 The procedure chart was established on the basis that a search unit angle of 70° will best detect and more accurately evaluate discontinuities having a major dimension oriented normal or near normal to the combined residual and applied tensile stresses (most detrimental to weld integrity). It should be assumed that all discontinuities could be oriented in this direction, and the 70° probe should be used whenever possible. For optimum results, a 10 in [250 mm] sound path has been established as a routine maximum. There are, however, some joint sizes and configurations that require longer sound paths to inspect the weld completely.

Testing procedures 6, 8, 9, 12, 14, and 15 in the procedure legend of Table 8.7, identified by the top quarter designation GA or the bottom quarter designation GB, require evaluation of discontinuities directly beneath the search unit. More accurate results may be obtained by testing these large welds from both face A and face B, as also provided for in this table.

The procedure chart was developed taking into account the above factors. The reason for the very exacting requirements of the code with respect to the application of the search unit (frequency, size, angle) is to maintain the best condition for reproducibility of results. It is the intent of the code that welds be examined using search unit angles and weld faces specified in Table 8.7. Use of other angles or weld faces may result in a more critical examination than established by the code.

Legend "P" 60° probes are prohibited for evaluation when using the pitch-and-catch method of testing because of the high energy loss that is possible due to wave mode conversion.

C-8.25.6 Testing of Welds. When required by Tables 8.2 and 8.3 as applicable, the sensitivity for scanning is increased by at least four decibels above the maximum reject level at the maximum testing sound path. This increased sensitivity assures that unacceptable discontinuities are not missed during scanning. Scanning at the upper dB levels required by Tables 8.2 and 8.3 may require specific scanning techniques that differ from the techniques used when testing with a lesser dB level. It is important to note that the scanning level is determined by sound path distance and not material thickness. For example, assume ultrasonic testing on a 2 in [50 mm] thick weld joint in a cyclically loaded nontubular connection (Table 8.3). Using Table 8.7 procedure 4, the weld is divided into three separate testing areas. They are the top quarter, the middle half, and the bottom quarter of the weld thickness. The sound path distance for the top quarter would be 7 in to 8 in [180 mm to 200 mm] (testing in leg 2 using a 60 transducer), the sound path distance for the middle half would be 1–1/2 in to 4–3/8 in [38 mm to 111 mm], and the sound path distance for the bottom quarter would be 4–3/8 in to 5–7/8 in [111 mm to 150 mm]. The scanning levels, as determined by Table 8.3, are 35 dB for the top quarter, 25 dB for the middle half, and 35 dB for the bottom quarter. In order to minimize the problems associated with scanning at high dB levels, it is often necessary to perform the testing in multiple steps, scanning each of the three areas (top quarter, middle half, and bottom quarter) as separate activities using the scanning level applicable to the sound path required for each. Failure to use the required scanning level may result in improper acceptance of serious flaws.

C-8.25.6.4 Attenuation Factor. The attenuation rate of 2 decibels per inch [2 dB per 25 mm] of sound travel, excluding the first inch [25 mm], is established to provide for the combination of two factors: the distance square law and the attenuation (absorption) of sound energy in the test material. The sound path used is the dimension shown on the display. The rounding off of numbers to the nearest decibel is accomplished by maintaining the fractional or decimal values throughout the calculation, and at the final step, advancing to the nearest whole decibel value when values of one-half decibel or more are calculated or by dropping the part of the decibel less than one-half. The result of this calculation is converted to decibels.

For units with digital displays, the sound path is recorded in thousandths of an inch (0.001 in.) or three decimal place accuracy.

For units with digital displays in SI units, the sound path can be recorded in tenths (0.1) or hundredths (0.01) of a mm or two decimal place accuracy.

The formula allows the result of the attenuation factor to be rounded to one significant decimal place (0.1) and may result in a value less than 1. This provides consistency when using either US Customary or SI Units.

C–8.25.6.5 Indication Rating. Rounding by as much as ½ dB occurs at the last step for the equation a-b-c=d, or b-a-c=d. Rounding each component of the equation to a whole number may increase the resulting indication rating enough to unnecessarily alter the rating.

C-8.25.7 Length of Discontinuities. The required six decibel drop in sound energy may be determined by adding six decibels of gain to the indication level with the calibrated gain control and then rescanning the weld area until the amplitude of the discontinuity indication drops back to the reference line.

When evaluating the length of a discontinuity that does not have equal reflectivity over its full length, its length evaluation could be misinterpreted. When a six decibel variation in amplitude is obtained by probe movement and the indication rating is greater than that of a minor reflector, the operator should record each portion of the discontinuity that varies by ± 6 dB as a separate discontinuity to determine whether it is acceptable under the code based on length, location, and spacing.

C-8.25.8 Basis for Acceptance or Rejection. In procedures specified for UT, the zero reference level for discontinuity evaluation is the maximum indication reflected from a 0.06 in [1.5 mm] diameter hole in the IIW ultrasonic reference block. When actual testing of welds is performed, the minimum acceptable levels are given in decibels for various weld thicknesses. In general, the higher the indication rating or acceptance level, the smaller the cross-sectional area of the discontinuity normal to the applied stress in the weld.

Indication ratings up to 6 dB more sensitive than unacceptable are to be recorded on the test report for welds designated as being "Fracture Critical" so that future testing, if performed, may determine if there is discontinuity growth.

C-8.25.12 Steel Backing. Steel backing is considered by many UT operators to be a deterrent to effective UT of groove welds due to the spurious indications that result on the screen. However, the reflection from the steel backing may be used by the UT operator as a confirmation that the ultrasound waves are penetrating through the entire cross section of the weld root area. The presence of the steel backing indication and absence of any other trace on the UT screen is evidence of a weld free of major discontinuities. It also proves that the weld zone, in which the sound wave is passing through, is free of those discontinuities that could disrupt the sound waves normal path. The sound wave can be disrupted by attenuation, reflection, or refraction so as to prevent return of the sound wave to the transducer, which can result in acceptance of a weld containing a critical size discontinuity.

UT of complex welded joints can be performed reliably and economically. Weld joint mock-ups, UT operator training, and knowledge of the weld joint and applicable UT equipment will ensure testing reliability and economy.

Spurious indications from steel backing will result from a variety of configurations. The following examples include combined inspection procedures and techniques.

(1) **T- or Corner Joints**

(a) **90° Dihedral Angle.** The end of the steel backing in Figure C-8.1 will act as a reflector ("R_B"), provided the root gap and penetration depth is as large as shown. "R_B" will result in a horizontal trace at approximately an equal sound path distance to a welding discontinuity at point "D."

Resolution Technique:

(1) Use straight beam UT from point "C" to determine if discontinuity "D" exists (if "C" is accessible).

(2) Determine if the indication is relatively continuous for the length of the weld joint. *NOTE: Most welding discontinuities are not relatively uniform.*

(3) Evaluate weld from point "B" to determine if "D" exists. *NOTE: Point "F" may require modification by flush grinding to effect ultrasound access to point "D."*

(4) Increase transducer angle to provide better access to "D."

(5) Remove a small section of the backing so that "R_B" is not accessible to the sound wave to confirm that "D" actually exists or "R_B" is the source of the indication.

(6) Select an area of largest discontinuity rating for exploratory grinding or gouging to determine if "D" exists.

(b) Skewed T- or Corner Joints. The interpretation of a T-joint becomes more complex as the dihedral angle changes. The increase in complexity is due to an increase in the steel backing reflection and the position of the end of the steel backing in relation to the upper toe of the weld. As shown in Figure C-8.2(A), the reflection of "R_B" can also be interpreted as an under-bead crack ("C_U").

With the dihedral angle greater than 90° as shown in Figure C-8.2(B), "R_B" is now at an equal sound path distance to a slag inclusion ("D"). Resolution of these conditions is the same as for the 90° T- or corner joints [see C-8.25.12(1)(a)].

(2) **Butt Joints**

(a) Separation Between Backing and Joint. The most common spurious indication ("I_S") is caused by an offset of the joined parts (fit-up problem) or joining two plates of different thickness resulting in a separation of the faying surface between the steel backing and the plate. Based on the sound path distance and depth, the indication in Figure C-8.3 appears to be a root discontinuity such as a crack or lack of fusion, when tested from point "A."

Resolution Technique:

(1) Accurately mark the location ("L") of the indication.

(2) Repeat the UT from point "A1."

(3) An "L" indication from point "A1" is verification that a discontinuity does exist at the root.

(4) The lack of the "L" indication from point "A1" is evidence that "I_S" is the source of the reflection.

(b) **Surface Geometry and Backing with Similar Sound Paths.** Another source of confusion is the surface weld profile and the steel backing resulting in a reflection at the same sound path distance. The root opening in Figure C-8.4(A) is large enough in this weld joint to allow the sound wave to transmit to the steel backing resulting in reflection and a large indication from point "R_B."

In Figure C-8.4(B) the root opening is tighter and the sound wave entrance is slightly farther from the "A" side of the weld joint, which results in sound wave reflection and a large indication from the surface of the weld reinforcement ("WR").

At this stage in the UT process the UT operator is faced with a complex interpretation of the indications; the sound path distance is the same for both (A) and (B). Is the indication a surface discontinuity, the weld reinforcement, or the edge of the steel backing?

Resolution Technique:

(1) UT weld (B) from point "A.1" to determine if there is a discontinuity in the area of "WR."

(2) Any indication at "WR" is justification for examination by grinding to specifically identify the discontinuity and judge criticality.

(3) If no indication results from the test at "A.1," then repeat the test from "A."

- Confirmation that the indication from "WR" is the weld reinforcement is done by first manipulating the transducer until maximum screen trace height is obtained, then wet "WR" with couplant and rub it with a finger while introducing sound waves from "A."

- If "WR" is the reflector, the trace on the screen will become unstable corresponding to the movement of the finger. *NOTE: This technique works best on thicker plate. Sound waves tend to flood very thin plate, which can influence the UT operator to accept indications that result from a discontinuity.*

(4) If "WR" is not the reflector, the steel backing can be verified as the source of reflection as follows:

- Position the transducer at "A.1" on (A) to obtain maximum screen trace height;
- Calculate the projected surface distance from the transducer exit point to the reflector;
- Mark that dimension on the opposite side of the weld from the transducer, which is now "L."
- Measure the dimension from "L" to "WR"—if the UT unit is calibrated correctly, this dimension should be the width of the steel backing. *NOTE: It is important the UT operator be knowledgeable regarding the size of the steel backing used and the basic root opening dimension to remove some of the questions regarding the source of the reflection.*

(5) As a general rule, it is advisable to divide the weld into two parts, as shown in (B), by the centerline ("CL") mark:

- Reflectors should be evaluated from the same side of the weld the transducer is on to minimize spurious indications.

(3) **Seal Welded Steel Backing.** The contract may require seal welds on all steel backing. The seal weld can result in the inability to transmit ultrasound through the entire cross section of the groove weld. The NDT Level III should determine the most practical width of the steel backing and the complementary shear wave transducer angle for testing, prior to fabrication.

In Figure C-8.6(A) the location of the ends of the steel backing is critical because it interferes with the reflection of the sound wave to the upper portion of the weld joint. Location of the end of the steel backing in the general region of "B" to "B.1" results in the sound wave entering the steel backing and either returning as "R_B" indication, or, not returning at all if the conditions are right as for "A.1."

In Figure C-8.6(B) the same condition exists when the sound wave enters the steel backing at "B" and continues to propagate through the bar and into the perpendicular plate. If an indication is noted on the screen, it is very likely to be spurious.

Resolution Techniques:

(1) Changing the specified dimensions of the steel backing to be seal welded by increasing the width will minimize this problem.

(2) Or, decrease the angle of the transducer if (1) above is not practical.

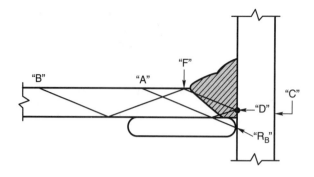

Figure C-8.1—90° T- or Corner Joints with Steel Backing

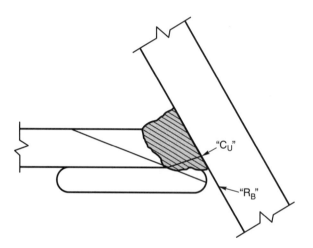

(A) LESS THAN 90° DIHEDRAL ANGLE

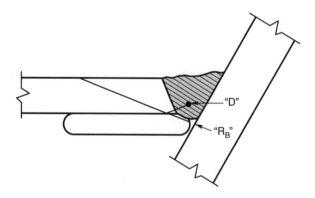

(B) GREATER THAN 90° DIHEDRAL ANGLE

Figure C-8.2—Skewed T- or Corner Joints

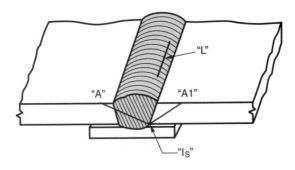

Figure C-8.3—Butt Joints with Separation Between Backing and Joint

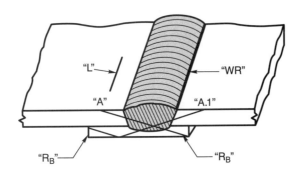

(A) WIDE ROOT OPENINGS

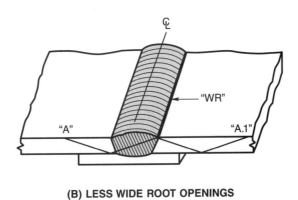

(B) LESS WIDE ROOT OPENINGS

Figure C-8.4—Effect of Root Opening on Butt Joints with Steel Backing

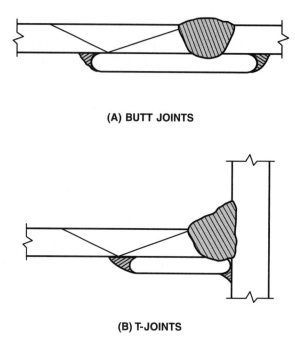

(A) BUTT JOINTS

(B) T-JOINTS

Figure C-8.5—Resolutions for Scanning with Seal Welded Steel Backing

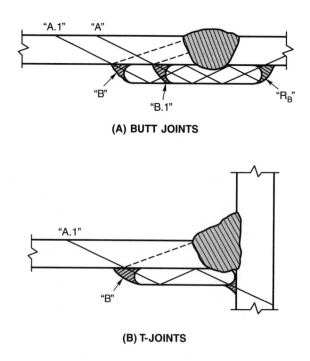

(A) BUTT JOINTS

(B) T-JOINTS

Figure C-8.6—Scanning with Seal Welded Steel Backing

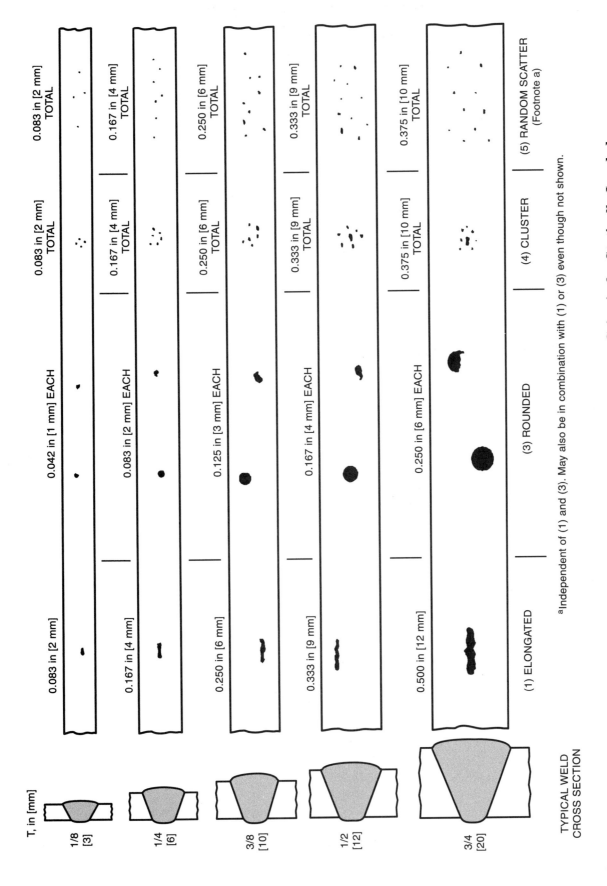

Figure C-8.7—Illustration of Discontinuity Acceptance Criteria for Statically Loaded Nontubular and Statically or Cyclically Loaded Tubular Connections (see 8.12.1)

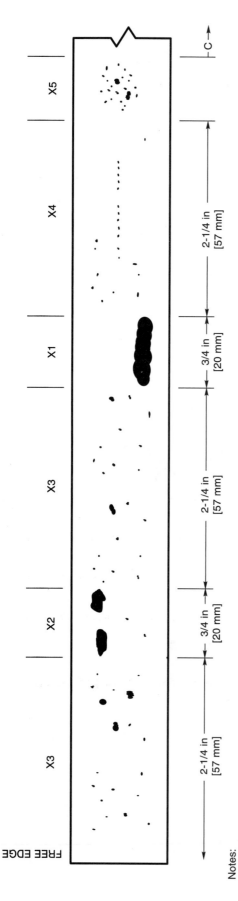

Notes:
1. C—Minimum clearance allowed between edges of discontinuities 3/32 in [2.5 mm] or larger (per Figure 8.1). Larger of adjacent discontinuities governs.
2. X1—Largest permissible elongated discontinuity for 1-1/8 in [30 mm] joint thickness (see Figure 8.1).
3. X2—Multiple discontinuities within a length allowed by Figure 8.1 may be handled as a single discontinuity.
4. X3–X4—Rounded-type discontinuity less than 3/32 in [2.5 mm].
5. X5—Rounded-type discontinuities in a cluster. Such a cluster having a maximum of 3/4 in [20 mm] for all pores in the cluster is treated as requiring the same clearance as a 3/4 in [20 mm] long discontinuity of Figure 8.1.
6. Interpretation: Rounded and elongated discontinuities are acceptable as shown. All are within the size limits and the minimum clearance allowed between discontinuities or the end of a weld joint.

Figure C-8.8—Illustration of Discontinuity Acceptance Criteria for Statically Loaded Nontubular and Statically or Cyclically Loaded Tubular Connections 1-1/8 in [30 mm] and Greater, Typical of Random Acceptable Discontinuities (see 8.12.1)

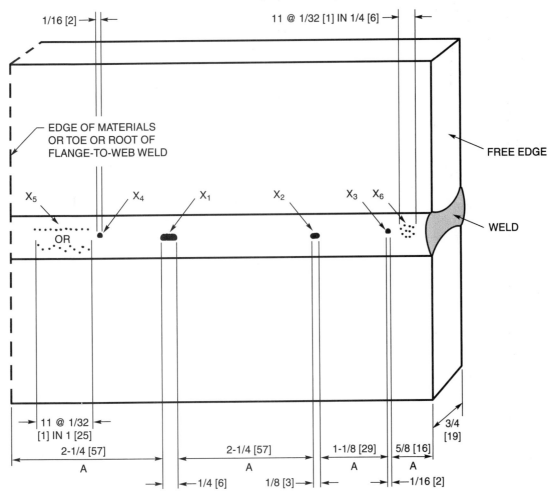

Notes:
1. A—Minimum clearance allowed between edges of porosity or fusion-type discontinuities 1/16 in [2 mm] or larger. Larger of adjacent discontinuities governs.
2. X_1—Largest allowable porosity or fusion-type discontinuity for 3/4 in [20 mm] joint thickness (see Figure 8.2).
3. X_2, X_3, X_4—Porosity or fusion-type discontinuity 1/16 in [2 mm] or larger, but less than maximum allowable for 3/4 in [20 mm] joint thickness.
4. X_5, X_6—Porosity or fusion-type discontinuity less than 1/16 in [2 mm].
5. Porosity or fusion-type discontinuity X_4 is not acceptable because it is within the minimum clearance allowed between edges of such discontinuities (see 8.12.2.1 and Figure 8.2). Remainder of weld is acceptable.
6. Discontinuity size indicated is assumed to be its greatest dimension.

Figure C-8.9—Illustration of Discontinuity Acceptance Criteria for Cyclically Loaded Nontubular Connections in Tension (see 8.12.2.1)

C-9. Stud Welding

C-9.1 Scope

Stud welding is unique among the approved welding processes in this code in that not only are the arc length and the weld time automatically controlled, but it also lends itself to a significant production proof test. Once the equipment is properly set, the process is capable of a large number of identical sound welds when attention is given to proper workmanship and techniques. Many millions of studs have been successfully applied. For other reasons outlined above, formal procedure qualifications are not required when studs are welded in the flat (down-hand) position to materials listed in Table 5.3, Group I and II. Procedures developed under the application qualification requirements of 9.6 are an exception to the foregoing. Since this constitutes the basic change from other approved welding processes in this code, stud welding has been moved to Clause 9.

There are provisions for the following:

(1) Tests to establish mechanical properties and the qualification of stud bases by the stud manufacturer

(2) Tests to establish or verify the welding setup (essential variables) and to qualify the operator and applications

(3) Tests for inspection requirement

C-9.2 General Requirements

C-9.2.5 Stud Finish. Heads of shear connectors or anchor studs are subject to cracks or bursts, which are names for the same thing. Cracks or bursts designate an abrupt interruption of the periphery of the stud head by radial separation of the metal. Such interruptions do not adversely affect the structural strength, corrosion resistance, or other functional requirements of headed studs.

A typical example of calculating the crack length or burst is as follows:

0.5 in [13 mm] headed anchor
H-Head Diameter = 1.0 in [25 mm] C-Shank Diameter = 0.5 in [13 mm] CL-Crack Length

$CL \leq (H - C)/4$ $CL \leq (1.0 - 0.5)/4$
$CL \leq 0.125$ in [3.2 mm]

C-9.3 Mechanical Requirements

The subclause on mechanical requirements has been expanded to show three strength levels of studs. The lower strength level, Type A, is used for general purpose studs and the higher strength level, Type B, is used as an essential component of composite beam design and construction. Type B studs are the most used in composite construction for highway bridges.

C-9.4 Workmanship/Fabrication

Several items of cleanliness are needed to produce sound quality studs. There is new emphasis on keeping the studs. Type B studs are used as an essential component in composite beam construction for highway bridges and buildings. Type C studs are commonly used as embedded connections in concrete/steel construction.

C-9.4.6 Arc Shield Removal and C-9.4.7 Acceptance Criteria. These subclauses clearly call for used arc shields to be removed and a visual inspection to be made by the applicator. Good judgment would call for this check to be performed as soon as practical after the stud is welded to avoid a large number of defective studs in the case of equipment malfunction.

The expelled metal around the base of the stud is designated as flash in conformance with the definition of flash in Clause 3 of this code. It is not a fillet weld such as those formed by conventional arc welding. The expelled metal, which is excess to the weld required for strength, is not detrimental but, on the contrary, is essential to provide a good weld. The containment of this excess molten metal around a welded stud by the ferrule (arc shield) assists in securing sound fusion of the entire cross section of the stud base. The stud weld flash may have nonfusion in its vertical leg and overlap on its horizontal leg; and it may contain occasional small shrink fissures or other discontinuities that usually form at the top of the weld flash with essentially radial or longitudinal orientation, or both, to the axis of the stud. Such nonfusion on the vertical leg of the flash and small shrink fissures are acceptable.

C-9.5 Technique

C-9.5.1 Automatic Mechanized Welding. Technique is a subclause that covers the requirements for equipment and initial settings.

C-9.5.5 Fillet Weld Option. The code also allows studs to be fillet welded, at the option of the Contractor, although the use of automatically-timed stud welding equipment is generally preferred.

C-9.6 Stud Application Qualification Requirements

Studs applied to a vertical surface may require modified arc shields and modified arc shields may also be required when welding to other than flat surfaces. Since this and other special cases are not covered by the manufacture's stud base qualification, the Contractor should be responsible for the performance of these tests. Test data serve the same purpose as WPS qualification for other processes. Inspectors should accept evidence of previous special application tests based on satisfactory preproduction tests with the specific stud welding set up in use.

C-9.6.2 WPSs Qualified by Testing. The testing under this section is required to assure weld quality for welding conditions that are not simulated by the manufacturer's stud base qualification.

Studs that are applied through decking have been added because of problems associated only at the production site that are not related to the Manufacturer's Stud Base Qualification. These include, but are not limited to, deck fit-up, weather conditions, the flange thickness of beams, coating thickness on beam flanges, deck coating thickness and number of deck plies. It is recommended that application qualification be performed at the job site, using the same welding equipment, operators, materials, coating thickness, and beam conditions as those at the site. It is recommended that the heaviest metal deck thickness, whether one or two plies, be tested along with the thickest coating (galvanized if used) to qualify work for the project. While the welding variables developed for this worst case would not necessarily apply to every stud to be used on the project, the equipment to be used would have been proven for the worst case. The operators would be qualified, the PQR/WQR records would be established for the site conditions, and the preproduction testing of 9.7.1 could then be used for each other set up and production period.

It is not necessary to requalify each field condition by this qualification test. Records of qualification of previous WPSs for specific production site conditions will serve to support future through-deck welding applications on the same and other production sites with the same welding variables. All of these qualified welding procedures remain subject to the start of shift preproduction testing of 9.7.1.

It is recommended that the Engineer accept properly documented evidence of weld through decking application tests where new work would fall within previous qualification limits.

The application test for other than Group I or II steels has been added to serve as a reminder that the Engineer should evaluate each such application.

Most steels in Group III of Table 5.3 and the steels in Table 6.9 are heat-treated steels, and the heat from stud welding can lead to reduced base plate static or dynamic physical properties. For example, thin quenched-and-tempered steels may have reduced tensile properties, and thicker quenched-and-tempered steels are more likely to have reduced notch toughness in the stud weld HAZ. The Engineer should particularly evaluate the application where studs will be welded in members subject to cyclic tensile stress or to stress reversal. The application test will serve to prove only that the stud itself is acceptable with the metal used.

C-9.7 Production Control

Applicator testing is required for the first two studs in each day's production or any change in the set up such as changing of any one of the following: stud gun, timer, power source, stud diameter, gun lift and plunge, total welding lead length, or changes greater than 5% in current (amperage) and time. This subclause does not apply to fillet welded studs in 9.5.5. Users who are unfamiliar with any of these terms are encouraged to refer to the latest edition of AWS Welding Handbook, Part 1, Vol. 2, Chapter 9, *Arc Stud Welding*. At the very high currents used in stud welding, it is very important to have adequate lead size and good lead connections.

C-9.7.1.4 Bend Test. Bending some stud and base materials at temperatures below 50°F [10°C], creates inadequate toughness to pass a hammer test.

C-9.8 Fabrication and Verification Inspection Requirements

In addition to visual and bend tests by the applicator, studs are to be visually inspected and bend tested by the inspector.

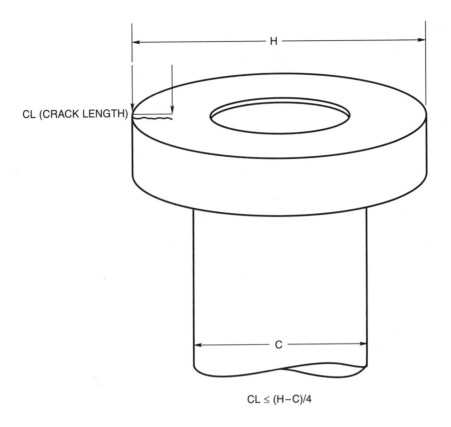

Figure C-9.1—Allowable Defects in the Heads of Headed Studs

C-10. Tubular Structures

Part A
Design of Tubular Connections

C-10.1 Scope

The tubular provisions of this code originally evolved from a background of practices and experience with fixed offshore platforms of welded tubular construction. Like bridges, these are subject to a moderate amount of cyclic loading. Like conventional building structures, they are redundant to a degree which keeps isolated joint failures from being catastrophic. The requirements of Clause 10, Part A, are intended to be generally applicable to a wide variety of tubular structures.

C-10.2.1 Statically Loaded Tubular Connections. The strength of tubular connections may be controlled by the properties of the main member or chord, as shown in Figure 10.8.

C-10.2.2 Weld Stresses. The available unit stresses in welds are presented in Table 10.2. This table is a consolidated and condensed version which lists for each type of weld the allowable unit stress for tubular application and the kind of stress the weld will experience. The required weld metal strength level is also specified. This table is presented in the same format as Table 4.3.

C-10.2.3 Cyclically Loaded Tubular Connections. While this Clause is sufficient for the majority of designs, the user should be aware of other important references endorsed for the fatigue design of tubular structures, including but not limited to, ISO 14347, Fatigue—Design Procedure for Welded Hollow-Section Joints-Recommendations; CIDECT Design Guide No. 8, Design Code for Circular and Rectangular Hollow Section Welded Joints under Fatigue Loading. For offshore structures refer to API RP 2A, Planning, Design, and Constructing Fixed Offshore Platforms and or ISO 19902 Fixed Platforms —Steel.

C-10.2.3.2 Fatigue Stress Categories. The basis for the fatigue stress categories can be found in Reference 1. These were derived from the data on circular sections and provide only approximate guidance for box sections.

The stress categories and fatigue curves have been revised in order to be consistent with current cyclically loaded structure provisions 4.16.2 and the latest revision of API RP 2A (Reference 3).

The sloping portion of most of the early curves has been retained. Following API, curves X and K have been split into two curves each. The upper curve represents the small-scale laboratory quality specimens in the historical (pre-1972) data base, while the lower curve represents recent large scale tests having welds without profile control. In interpreting the latter, earlier editions of the American codes emphasized weld profile while proposed British rules (Reference 6) emphasize thickness effects. The current hypothesis is that both weld profile and size effects are important to understanding fatigue performance, and that they are interrelated. This is also an area where design and welding cannot be separated, and 10.2.3.7 makes reference to a consistent set of "standard" weld profile control practices and fatigue category selections, as a function of thickness. Improved profiles and grinding are discussed 10.2.3.6 along with peening as an alternative method of fatigue improvement.

The endurance limits on most of the curves have been delayed beyond the traditional two million cycles. The historical data base did not provide much guidance in this area, while more recent data from larger welded specimens clearly shows that the sloping portion should be continued. The cutoffs are consistent with those adopted for cyclically loaded structures and atmospheric service. For random loading in a sea environment, API adopted a cutoff of 200 million cycles; however, this need not apply to AWS applications.

With the revised cutoffs, a single set of curves can be used when the provisions of 10.2.3.5 are taken into account.

For Category K (punching shear for K-connections), the empirical design curve was derived from tests involving axial loads in branch members. The punching shear formula based on gross static considerations and geometry does not always produce results consistent with what is known about the influence of various modes of loading on localized hot spot stress, particularly where bending is involved. Since some of the relevant parameters (e.g., the gap between braces) are not included, the following simplified approximations appear to be more appropriate for typical connections with $0.3 \leq \beta \leq 0.7$.

In these formulae, the nominal branch member stresses f_a, f_{by}, f_{bz} correspond to the modes of loading shown in Figure C-10.1. The α factor on the f_a has been introduced to combine the former curves K and T into a single curve. Other terms are illustrated in Figure C-10.3.

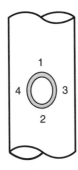

AT LOCATIONS 1 AND 2
Cyclic $V_p = \tau \sin \theta \,[\alpha \, f_a \pm 0.67 \, f_{by}]$

AT LOCATIONS 3 AND 4
Cyclic $V_p = \tau \sin \theta \,[\alpha \, f_a \pm 1.5 \, f_{bz}]$

AT THE POINT OF HIGHEST STRESS
Cyclic $V_p = \tau \sin \theta \,[\alpha \, f_a + \sqrt{(0.67 \, f_{by})^2 + (1.5 \, f_{bz})^2}]$

C-10.2.3.3 Basic Allowable Stress Limitation. Fatigue data characteristically show a large amount of scatter. The design curves have been drawn to fall on the safe side of 95% of the data points. The AWS design criteria are appropriate for fail-safe structures in which localized fatigue failure of a single connection does not lead immediately to collapse. For critical members whose sole failure would be catastrophic, the cumulative fatigue damage ratio, D, as defined in 10.2.3.4, is limited to a fractional value (i.e., 1/3) to provide an added safety factor. This statement presumes there is no conservative bias or hidden safety factor in the spectrum of applied loads used for fatigue analysis (many codes include such bias). References 2 and 3 discuss application of these criteria to offshore structures, including modifications that may be appropriate for high-cycle fatigue under random loading and corrosive environments.

C-10.2.3.6 Fatigue Behavior Improvement. The fatigue behavior of as-welded joints can be improved by reducing the notch effect at the toe of the weld, or by reducing the tensile residual stresses, neither of which is included in the measured hot spot strain range which designers use. Various methods for improving the fatigue behavior of welded joints, as discussed in Reference 5, are as follows:

   (1) improving the as-welded profile (including the use of special electrodes designed to give a smooth transition at the weld toe),

   (2) full profile grinding,

   (3) weld toe grinding,

   (4) weld toe remelting (GTAW dressing or plasma arc dressing),

   (5) hammer peening, and

   (6) shot peening.

A long established (but not universally used) offshore industry practice for improved weld profile is shown in Figure C-10.2. The desired profile is concave, with a minimum radius of one-half the branch member thickness, and merges smoothly with the adjoining base metal. Achieving the desired profile as-welded generally requires the selection of welding materials having good wetting and profile characteristics, along with the services of a capping specialist who has mastered the stringer bead wash pass technique for various positions and geometries to be encountered. Difficulties in achieving this are often experienced with high deposition rate processes in the overhead and vertical positions. Inspection of the finished weld profile is mostly visual, with the disk test being applied to resolve borderline cases. Notches relative to the desired weld profile are considered unacceptable if a 0.04 in [1 mm] wire can be inserted between the disk of the specified radius and the weld, either at the toe of the weld or between passes.

Earlier editions of AWS D1.1 contained a less stringent weld profile requirement. Surprisingly poor weld profiles could pass this test, with the relative notch effect becoming increasingly more severe as the thickness of the members increased. Recent European research has shown the earlier D1.1 to be inadequate in distinguishing between welded tubular connections which meet the performance of AWS Fatigue Classification X1, and those which fall short (References 5 and 6).

Notch stress analysis and fracture mechanics considerations, while confirming the inadequacy of the old profile requirements for heavy sections, also indicate that the tighter requirements of Figure C-10.2 are more effective in maintaining Class X1 fatigue performance over a wide range of thicknesses (Reference 7). Figure C-10.2 also suggests the use of light grinding to correct toe defects, such as excessive notch depth or undercut. Once grinding starts, note that the allowable notch depth is reduced to 0.01 in [0.25 mm]; merely flattening the tops of the individual weld passes, while leaving sharp canyons in between, does little to improve the fatigue performance, even though it would meet the letter of the disk test.

Since the toes of welds frequently contain microscopic cracks and other crack-like defects, MT is necessary to ensure these defects have been eliminated. Judicious use of grinding to resolve MT indication, often done routinely as part of the inspection, also enhances the weld profile.

Depending upon circumstances, it may be more cost effective to grind the entire weld profile smooth. This would avoid the use of special welding techniques, profile checking, corrective grinding, and MT, as described above, for controlling the as-welded profile. For tubular connections, with multiple concave pass caps, fatigue cracks may start in the notch between passes; here, weld toe grinding alone is not as effective as with flat-fillet-weld profiles that were used in much of the research.

Weld toe remelting techniques can improve the geometry of the notch at the weld toe, and have been shown in the laboratory to improve the fatigue performance of welded connections. However, unless carefully controlled, the rapid cycle of heating and cooling tends to produce unacceptably hard HAZs, with possible susceptibility to stress corrosion cracking in aggressive environments (e.g., seawater).

Hammer peening with a round-nose tool also improves the weld toe geometry; this additionally induces a compressive residual stress in the surface layers where fatigue cracks would otherwise be initiated. Excessive deformation of the base metal may render it susceptible to strain embrittlement from subsequent nearby welding. Also, surface layers may be so smeared as to obscure or obliterate pre-existing cracks; thus the requirement for MT. Shot peening is less radical in its deformation effects, but also less effective in improving geometry.

It should be emphasized that, for many tubular structure applications, the performance of fatigue Classifications X2, K2, and ET will suffice, and the foregoing measures taken to improve fatigue performance are not required. Furthermore, the "standard" weld profile practices described in 10.10.2 can achieve the performance of fatigue Classifications X1, K1, and DT for all but the heaviest sections.

C-10.2.3.7 Size and Profile Effects. The adverse size effect in the fatigue of welded connections is well documented (recent References 5, 6, and 7, as well as many earlier ones). For welded joints with a sharp notch at the weld toe, scaling up the size of the weld and the size of the notch results in a decrease in fatigue performance. When the application exceeds the scale of the data base, size effect should be accounted for in design. Reference 12 suggests decreasing the fatigue strength in proportion to

$$\left[\frac{(\text{size})}{(\text{size limit})}\right]^{-0.25}$$

Other authorities (Reference 8) indicate a milder size effect, approximating an exponent of –0.10.

The geometric notch effect largely responsible for the size effect in welds is not present in fully ground profiles and is relatively minor for those profiles which merge smoothly with the adjoining base metal (Fatigue Categories B and C1). The stated size limits (beyond which we are outside the historical data base) for most of the other categories are similar to those cited in Reference 6, except that the dimensions in inches have been rounded off. The larger size limits for Categories X2, K2, and DT reflect the fact that these S-N curves have already been drawn to fall below the recent large-scale test data.

Reference 7 discusses the role of size effect relative to weld profile, at various levels of fatigue performance. The "standard" weld profile practices for T-, Y-, and K-connections referred to in 10.2.3.7 vary with thickness so as to define two fatigue performance levels which are size-independent. However, where an inferior profile is extended beyond its standard range, the size effect (reduction in performance) would come into play. "Improved" weld profiles which meet the requirements of 10.2.3.6(1) keep the notch effect constant over a wide range of thicknesses, thereby mitigating the size effect. The smooth surface profile of fully ground welds also exhibit no size effects. Since peening only improves a relative limited volume of the welded joint, the size effect would be expected to show up fairly soon if peening is the only measure taken; however, peening should not incur a size effect penalty where it is done in addition to profile control.

The size effect may also exhibit itself in static ultimate strength behavior, since the design rules are based in part on tests to tensile fracture. For tubular T-, Y-, and K-connections involving high-strength steels of low or unknown notch toughness, the Level I profile selections are recommended in preference to larger notches allowed by Level II.

C-10.3 Identification and Parts of Tubular Connections

(1) The term "T-, Y-, and K-connections" is often used generically to describe tubular connections tubular connections in which branch members are welded to a main member, or chord, at a structural node. Specific criteria are also given for X connections. N-connections are a special case of K-connections in which one of the branches is perpendicular to the chord; but the same design criteria apply. For multiplanar connections see Annex R.

(2) Connection classifications as T-, Y-, K-, or X-should apply to individual branch members according to the load pattern for each load case. To be considered a K-connection, the punching load in a branch member should be essentially balanced by loads on other braces in the same plane on the same side of the connection. In T- and Y-connections the punching load is reacted as beam shear in the chord. In X connections the punching load is carried through the chord to branch members that carry part of their load as K-connections, and part as T-, Y-, or X connections, interpolate based on the portion of each in total, or use computed alpha (see Annex R).

C-10.5.1 Weld Effective Lengths. The flexibility of the connecting wall of the main (through) member in a tubular connection, as well as that of the connecting wall of the through branch in an overlapped tubular connection, creates an uneven distribution of load along the welds to branch members.

C-10.5.5 Weld Effective Lengths in Circular-to-Circular Tubular Connections. Load transfer across the weld can be highly non-uniform along its length in these connections.

C-10.7 Material Limitations

C-10.7.1 Material Limitations for Tubular T-, Y-, K-Connections.

A rational approach to the ultimate strength of stepped box connections can be taken, using the upper bound theorem of limit analysis (see Figure C-10.3) and yield line patterns similar to those shown in Figure C-10.4. Various yield line patterns should be assumed in order to find the minimum computed capacity, which may be equal to or greater than the true value. Fan corners (as shown for the T-joint) often produce lower capacity than plain corners shown for the other cases. For T- and Y-connections, the geometry modifier is found to be a function of η as well as β,. For K-connections, the gap parameter ξ also should be taken into account. The dimensionless geometry parameters, η, β, and ξ are defined in Figure 10.2.

For gaps approaching 0 and for very large β approaching unity, yield line analysis indicates extremely and un-realistically high joint capacity.

Tubular connections are subject to stress concentrations which may lead to local yielding and plastic strains. Sharp notches and discontinuities at the toe of the welds, and fatigue cracks which initiate under cyclic loading, place additional demands on the ductility and notch toughness of the steel, particularly under cyclic loads. These demands are particularly severe in the main member of tubular T-, Y-, and K-connections. Cold-formed box tubing (e.g., ASTM A500 and tubing fabricated from bent plates) is susceptible to degraded toughness due to strain aging in the corners, when these severely deformed regions are subjected to even moderate heat of nearby welding. Suitability of such tubing for the intended service should be evaluated using tests representing their final condition (i.e., strained and aged, if the tubing is not normalized after forming) (see C-10.7.2.2 for a discussion of CVN testing requirements).

C-10.7.2 Tubular Base Metal Notch Toughness. Some steels are listed by strength group (Groups I, II, III, IV, and V) and toughness class (Classes A, B, and C) in Tables C-10.1, C-10.2, and C-10.3. These listings are for guidance to designers, and follow long-established practice for offshore structures, as described in Reference 3 and the following:

Strength Groups. Steels may be grouped according to strength level and welding characteristics as follows (also see 5.6 and 5.7):

(1) Group I designates mild structural carbon steels with specified minimum yield strengths of 40 ksi [280 MPa] or less. Carbon equivalent (defined in Annex H, H6.1.1) is generally 0.40% or less, and these steels may be welded by any of the welding processes as described in the code.

(2) Group II designates intermediate strength low alloy steels with specified minimum yield strengths of over 40 ksi through 52 ksi [280 MPa through 360 MPa]. Carbon equivalent ranges up to 0.45% and higher, and these steels require the use of low-hydrogen welding processes.

(3) Group III designates high-strength low-alloy steels with specified minimum yield strengths in excess of 52 ksi through 75 ksi [360 MPa through 515 MPa]. Such steels may be used, provided that each application is investigated with regard to the following:

(a) Weldability and special WPSs which may be required. Low-hydrogen WPSs would generally be presumed.

(b) Fatigue problems which may result from the use of higher working stresses, and

(c) Notch toughness in relation to other elements of fracture control, such as fabrication, inspection procedures, service stress, and temperature environment.

(4) Groups IV and V include higher strength constructional steels in the range of over 75 ksi through 100 ksi yield [515 MPa through 690 MPa]. Extreme care should be exercised with regard to hydrogen control to avoid cracking and heat input to avoid loss of strength due to over-tempering.

Toughness Class. Toughness classifications A, B, and C may be used to cover various degrees of criticality shown in the matrix of Table C-10.4, and as described below:

Primary (or fracture critical) structure covers elements whose sole failure would be catastrophic.

Secondary structure covers elements whose failure would not lead to catastrophic collapse, under conditions for which the structure could be occupied or capable of major off-site damages (e.g., pollution), or both.

For tubular space-frame structures, fracture of a single brace or its end connection is not likely to lead to collapse under normal or even moderately severe loads. The strength is reduced somewhat, however, and the risk of collapse under extreme overload increases correspondingly.

(1) Class C steels are those which have a history of successful application in welded structures at service temperatures above freezing, but for which impact tests are not specified. Such steels are applicable to structural members involving limited thickness, moderate forming, low restraint, modest stress concentration, quasi-static loading (rise time 1 second or longer) and structural redundancy such that an isolated fracture would not be catastrophic. Examples of such applications are piling, braces in space frames, floor beams, and columns.

(2) Class B steels are suitable for use where thickness, cold work, restraint, stress concentration, and impact loading or lack of redundancy, or both, indicate the need for improved notch toughness. Where impact tests are specified, Class B steels should exhibit CVN energy of 15 ft·lbf [20 J] for Group I, 25 ft·lbf [34 J] for Group II, and 35 ft·lbf [48 J] for Group III, at the lowest anticipated service temperature. Steels listed herein as Class B can generally meet these CVN requirements at temperatures ranging from 50°F to 32°F [10°C to 0°C].

Examples of such applications are connections in secondary structure, and bracing in primary structure. When impact tests are specified for Class B steel, heat-lot testing in conformance with ASTM A673, Frequency H, is normally used. However, there is no positive assurance that Class B toughness will be present in pieces of steel that are not tested.

(3) Class A steels are suitable for use at subfreezing temperatures and for critical applications involving adverse combinations of the factors cited above. Critical applications may warrant CVN testing at 36°F–54°F [20°C–30°C] below the lowest anticipated service temperature. This extra margin of notch toughness prevents the propagation of brittle fractures from large discontinuities, and provides for crack arrest in thicknesses of several inches. Steels enumerated herein as

Class A can generally meet the CVN requirements stated above at temperatures ranging from –4°F to –40°F [–20°C to –40°C]. Impact testing frequency for Class A steels should be in conformance with the specification under which the steel is ordered; in the absence of other requirements, heat-lot testing may be used.

C-10.7.2.1 CVN Test Requirements. These minimal notch toughness requirements for heavy-section tension members follow the provisions recently proposed by AISC. They rely to a considerable extent on the temperature-shift phenomenon described by Barsom (Reference 9). The temperature-shift effect is that statically loaded materials exhibit similar levels of ductility as cyclically loaded CVN test specimens tested at a higher temperature. For higher strength steels, Groups III, IV, and V, the temperature-shift is less effective; also fracture mechanics strain energy release considerations would suggest higher required energy values. Testing as-rolled steels on a heat-lot basis leaves one exposed to considerable variation within the heat, with CVN tests showing more scatter than strength properties. However, it is better than no testing at all.

C-10.7.2.2 LAST Requirements. The main members in tubular connections are subject to local stress concentrations which may lead to local yielding and plastic strains at the design load. During the service life, cyclic loading may initiate fatigue cracks, making additional demands on the ductility of the steel. These demands are particularly severe in heavy-wall joint-cans designed for punching shear.

Part B
Prequalification of Welding Procedure Specification (WPSs)

C-10.10 CJP Groove Weld Requirements

C-10.10.1 Butt Joints. Where groove details in T-, Y-, and K-connections deviate from the prequalified details of 5.4.1, or there is some question as to the suitability of these joint details for a new WPS, then a mock-up or sample joint is called for, in order to validate the WPS.

Part C
Welding Procedure Specification (WPS) Qualification

C-10.11 Common Requirements for WPS and Welding Personnel Performance Qualification

C-10.11.1 Positions of Test Welds. This subclause defines welding positions for qualification test welds and production welds. Position is an essential variable for all of the WPSs, except for the EGW and ESW processes which are made in only one position. Each WPS shall be qualified for each position for which it will be used in fabrication. Relationships between the position and configuration of the qualification test weld and the type of weld and positions qualified are shown in Table 10.8. It is essential to perform testing and evaluation of the welds to be encountered in construction prior to their actual use on the job. This will assure that all the necessary positions are tested as part of the qualification process.

C-10.13 Type of Qualification Tests, Methods of Testing and Acceptance Criteria for WPS Qualification

C-Table 10.9 WPS Qualification—CJP Groove Welds; Number and Type of Test Specimens and Range of Thickness and Diameter Qualified. The WPS qualification for pipe includes conditions for large diameter job size pipe. This is intended for WPS qualification of large diameter pipe by automatic welding processes, such as SAW, and may be applied to any welding process that can be used on large diameter pipe, but not on 8 in [200 mm] Sch. 120 pipe.

C-10.14 CJP Groove Welds for Tubular Connections

Welding on tubular members differs from that in conventional plate and wide flange construction in several important aspects. Position often changes continuously in going around the joint; in T-, Y-, and K-connections, the joint geometry also changes. Often there is no access to the root side of the weld; and circumstances may preclude the use of backing (e.g., the use of tubes as a conduit, or the complicated geometry of T-, Y-, and K-connections). Yet, for many structures, the conditions of service demand that these welds meet the strength and fatigue performance qualities conventionally associated with CJP groove welds. To meet these needs, a specialized set of practices regarding WPS and welder qualifications, as well as prequalified joint details, has evolved for tubular structures. These provisions supplement those given elsewhere in the code.

Several specialized tubular applications are defined in which CJP groove welds are allowed to be welded from the outside only, without backing:

(1) **Pipe Butt Joints.** In butt joints, CJP groove welds made from one side are prohibited under the conventional provisions for cyclically loaded structures and statically loaded structures, yet they are widely used in pressure piping applications. They are now allowed for tubular structures, but only when all the special provisions of 10.14.2 are followed.

(2) **T-, Y-, and K-Connections.** Where groove details in T-, Y-, and K-connections differ from the prequalified details of 10.10.2, or there is some question as to the suitability of the joint details for WPS, then a mock-up or sample joint in conformance with 10.14.4.1 is required, in order to validate the WPS.

Additional WPS qualification tests may be required on account of some essential variable other than joint design. These circumstances include (but are not limited to) the following:

(a) The use of a process outside the prequalified range (e.g., GMAW-S).

(b) The use of base metal or welding materials outside the prequalified range (e.g., the use of proprietary steels or a nonlow-hydrogen root pass on thick material).

(c) The use of welding conditions outside the prequalified range (e.g., amps, volts, preheat, speed, and direction of travel).

(d) The need to satisfy special Owner testing requirements (e.g., impact tests).

Qualification for CJP welds using tubular box sections detailed with single-welded T, Y-, and K-connections requires additional tests as stated in Table 10.8 and shown in Figure 10.22. In this test, the welder demonstrates the skill and technique to deposit sound weld metal around the corners of a box tube member. This macroetch test is not required for fillet or PJP groove welds (see Commentary C-10.18 for further discussion).

For these tests, the joint configurations of Figures 10.20 and 10.22 are used in order to simulate the root condition and limited access of T-, Y-, and K-connections. Conventional specimens for mechanical testing are then prepared in conformance with Table 10.9.

PJP T-, Y-, and K-connections are also provided for. They can be executed by welders having the common pipe qualifications 2G plus 5G. This could be advantageous in areas where 6GR qualified welders are not readily available. Although lower fatigue allowables apply, the static strength of such joints is almost the same as for CJP, particularly where mild steel is used with E70 filler metal.

C-10.14.4 T-, Y-, and K-Connections without Backing Welded from One Side Only. Under carefully described conditions (see Figures 10.7 and Figures 10.9–10.11), the code allows CJP groove welds in tubular T-, Y-, and K-connections to be made from one side without backing. Lack of access and complex geometry preclude more conventional techniques. A very high level of welder skill (as demonstrated by the 6GR test) is required. When matching materials (see Table 5.3) are used, such joints may be presumed to equal the strength of the sections joined subject to the limitations of 10.5 and 10.2.3.

In making the weld in a T-, Y-, or K-connection, the geometry and position vary continuously as one progresses around the joint. The details shown in Figures 10.7 and Figures 10.9–10.11 were developed from experience with all-position SMAW and fast-freezing GMAW-S. These details are also applicable to FCAW processes with similar fast-freezing characteristics. The wider grooves (and wider root openings) shown for GMAW were found necessary to accommodate the shrouded tip of the welding gun. Although the later process is not prequalified for GMAW-S, the joint details are still applicable to such GMAW WPSs.

In many applications, particularly with small tubes, the PJP of 10.9.1 will be entirely adequate. Although requiring additional strength checks by the designer, the less stringent requirements for fit-up and welder's skill result in significant economies on the job. For very large tubes in which inside access is possible, the conventional CJP groove welds made from both sides are applicable.

For applications where increased fatigue performance associated with CJP groove welds is needed for T-, Y-, and K-connections, the code refers to a consistent set of "standard" weld profiles, as described earlier in C-10.2.3.7. Once learned, these should become a natural progression with thickness for the welders to follow. They have evolved from the following experience.

For very thin tubular connections, flat profiles (Figure 10.9) represent those commonly obtained on small tubular connections used for onshore applications. They also are similar to the profiles obtained on some of the scale models used to develop the historical fatigue data base. Here the entire weld cap is made in one pass, with weaving as required. Using E6010 electrodes, the more artistic capping specialist could make this a concave profile, merging smoothly with the adjoining base metal. With the advent of higher strength steels and heavier sections, requiring low-hydrogen electrodes, and with the introduction of high deposition rates, semiautomatic welding processes, this seems to have become a lost art.

For heavier thicknesses, a definite fillet is added at the weld toe as required to limit the weld toe notch effect to that of a 45° fillet weld (see Figure 10.10). These fillet welds are scaled to the branch member thickness so as to approximate a concave weld shape. However, we are also constrained by the need to maintain minimum fillet weld sizes to avoid creating dangerously high hardnesses in the HAZ at the weld toe (this is also the location of the "hot spot" which may experience localized yielding at the design load levels). This alternative "standard" profile is easier to communicate to the welders, and easier for them to achieve out of position than the idealized concave weld profile shown in earlier editions of the code. The resulting weld profile is much like that observed on early Gulf of Mexico offshore platforms, whose fatigue performance over several decades of service has been consistent with Categories X1, K1, and DT.

For branch member thicknesses in excess of 0.625 in [16mm] (typically associated with chord thicknesses in excess of 1.25 in [32 mm]) designers are going beyond the historical fatigue data base and the experience of early Gulf of Mexico platforms.

The size effect begins to manifest itself, and fatigue performance would begin to decline toward the lower level defined by fatigue Categories X2 and K2, unless the profile is further improved. Branch members of 1.5 in [38 mm] and chord thicknesses of 3 in [75 mm], represent the limits of the recent large-scale European tests, and further adverse size effects (performance below X2 and K2) would be expected if sharply notched weld profiles were to be scaled up even further. Figure 10.11 describes a concave weld profile which merges smoothly with the adjoining base metal, mitigating the notch effect and providing an improved level of fatigue performance for heavier sections.

The standardized pipe butt joint test specimens, specified in Part B of Clause 4 for WPS qualification, are satisfactory for establishing metallurgical soundness of WPSs and materials. They cannot cover the full range of continuously varying geometry and position encountered in structural T-, Y-, and K-connections.

The prequalified joint details given in 10.10.2 are based on experience with full scale mock-ups of such connections that often reveal practical problems that do not show up in the standard test specimen. The Code requires qualification of processes not prequalified and WPSs with essential variables outside prequalified ranges meeting the provisions of 10.14.4.1. This subclause provides for sample joint or tubular mock-up tests. WPS for box sections may be based on either plate or pipe tests for position and compatibility. When mock-up tests for box sections for T-, Y-, and K-connections are considered, box tubes should be used.

Additional tests are required for connections with groove angles less than 30° as outlined in 10.14.4.2.

C-10.14.4.4 Weldments Requiring CVN Toughness. Weld metal and HAZ toughness should be based on the same engineering considerations as used to establish the base metal toughness requirements. However, fracture avoidance, by increasing toughness alone, may not be cost effective. Consideration should be given to fatigue cracking, hydrogen-induced cold cracking, and solidification hot cracking. Other parts of the code address these other problems, via design, qualification, technique, and inspection requirements. Notch toughness just helps us live with imperfect solutions.

Weld Metal. Notch tough base metals should be joined with filler metals possessing compatible properties. The test temperatures and minimum energy values in Table C-10.6 are recommended for matching the performance of the various steel grades as listed in Tables C-10.1, C-10.2, and C-10.3. When WPS qualification by test is required (i.e., when the WPS is not prequalified, when comparable impact performance has not been previously demonstrated, or when the welding consumables are to be employed outside the range of essential variables covered by prior testing), qualification should include CVN testing of the as deposited weld metal. Specimens should be removed from the test weld, and CVN tested, in conformance with Clause 6, Part D, Requirements for CVN Testing. Single specimen energy values (one of three) may be 5 ft·lbf [7 J] lower without requiring retest.

Since AWS WPS requirements are concerned primarily with tensile strength and soundness (with minor emphasis on fracture toughness), it is appropriate to consider additional essential variables which have an influence on fracture toughness—e.g., specific brand wire/flux combinations, and the restriction of SAW consumables to the limits actually tested for AWS classification. Note that, for Class A steels, specified energy levels higher than the AWS classifications require that all WPSs be qualified by test, rather than having prequalified status.

CVN testing is a method for qualitative assessment of material toughness. Although lacking the fracture mechanics basis of crack tip opening displacement (CTOD) testing, the method has been and continues to be a reasonable measure of fracture safety, when employed with a definitive program of NDT to eliminate weld area defects. The recommendations contained herein are based on practices which have generally provided satisfactory fracture experience in structures located in moderate temperature environments (e.g., 40°F [4°C] sea water and 14°F [−10°C] air exposure). For environments which are either more or less hostile, impact testing temperatures should be reconsidered, based on local temperature exposures.

For critical welded connections, the more technical CTOD test is appropriate. CTOD tests are run at realistic temperatures and strain rates, representing those of the engineering application, using specimens having the full prototype thickness. This yields quantitative information useful for engineering fracture mechanics analysis and defect assessment, in which the required CTOD is related to anticipated stress levels (including residual stress) and discontinuity sizes.

Representative CTOD requirements range from 0.004 in at 40°F [0.10 mm at 4°C] to 0.015 in at 14°F [0.38 mm at −10°C]. Achieving the higher levels of toughness may require some difficult trade-offs against other desirable attributes of the welding process—for example, the deep penetration and relative freedom from trapped slag of up-hill passes, versus the lower heat input and highly refined weld layers of downhill passes.

HAZ. In addition to weld metal toughness, consideration should be given to controlling the properties of the HAZ. Although the heat cycle of welding sometimes improves as-rolled base metals of low toughness, this region will often have degraded toughness properties. The HAZ is often the site of hydrogen-induced underbead cracking. A number of early failures in welded tubular joints involved fractures which either initiated in or propagated through the HAZ, often before significant fatigue loading.

Clause 6, Part D gives requirements for sampling both weld metal and HAZ, with CVN test energy and temperature to be specified in contract documents. The average HAZ values in Table C-10.7 have been found by experience to be reasonably attainable, where single-specimen energy values (one of three) 5 ft·lbf [7 J] lower are allowed without requiring retest.

As criticality of the component's performance increases, lower testing temperatures (implying more restrictive WPSs) would provide HAZs which more closely match the performance of the adjoining weld metal and parent material, rather than being a potential weak link in the system. The Owner may also wish to consider more extensive sampling of the HAZ than the single set of CVN tests required by Clause 6, Part D, e.g., sampling at 0.4 mm, 2 mm, and 5 mm from the fusion line. (These dimensions may change with heat input.) More extensive sampling increases the likelihood of finding local brittle zones with low toughness values.

Since HAZ toughness is as much dependent on the steel as on the welding parameters, a preferable alternative for addressing this issue is through weldability prequalification of the steel. Reference 10 of Clause C-2 spells out such a prequalification procedure, using CTOD as well as CVN testing. This prequalification testing is presently being applied as a supplementary requirement for high-performance steels such as API Specs 2W and 2Y, and is accepted as a requirement by some producers.

Caution: Clause 6 *of this code allows testing one 50 ksi [345 MPa] steel to qualify all other grades of 50 ksi [345 MPa] and below. Consequently, selection of API-2H-50-Z (very low sulfur, 200 ft·lbf [270 J] upper shelf CVNs) for qualification test plates will virtually assure satisfying a HAZ CVN test requirement of 25 ft·lbf [34 J], even when welded with high-heat inputs and high interpass temperatures. There is no reasonable way to extrapolate this test to ordinary A572 Grade 50 with the expectation of reproducing either the HAZ impact energies or the 8:1 degradation of the test on API-2H-50-Z. Thus, separate CVN testing of different steel grades, thickness ranges, and processing routes should be considered, if HAZ toughness is being addressed via WPS testing.*

Local Brittle Zones (LBZ). Within the weld HAZs there may exist locally embrittled regions. Under certain conditions, those LBZs may be detrimental. The engineer should consider the risk of LBZs and determine if counter measures should be employed to limit the extent of LBZs and their influence on structural performance. Some counter measures and mitigating circumstances in offshore practice are listed below:

(1) The use of steels with moderate crack-arrest capabilities, as demonstrated by no-break in the NRL drop-weight test (small discontinuity)

(2) Overmatch and strain hardening in conventional normalized 42 ksi to 50 ksi [290 MPa to 345 MPa] carbon-manganese steels in which the weld metal and HAZ have higher yield strength than adjacent base metal, forcing plastic strains to go elsewhere

(3) The tendency for fatigue cracks in welded tubular joints to grow out of the HAZ before they reach appreciable size (assuming one avoids unfavorable tangency of joining can weld seam with the brace footprint)

(4) Prequalified limits on weld layer thickness in welding procedures, which along with observing limits on heat input, promote grain refinement in the HAZ and minimize the extent LBZ

(5) Composition changes, e.g., reduced limits on vanadium and nitrogen, and increased titanium

Part D
Performance Qualification

C-**10.16** Production Welding Positions, **Thicknesses, and Diameters** Qualified

C-Table 10.12. Welding on pipe (or tubing) material product forms does not necessarily mean that pipe welding is being performed. There is obviously a difference between welding around a pipe as opposed to welding along a pipe parallel to the pipe axis (centerline). A girth weld in a butt joint is completely different from a longitudinal groove weld that joins rolled plate to make a pipe; a socket joint with a fillet weld is completely different from a fillet weld along the pipe length attaching a plate plug. Obviously, the skills for straight line progression parallel to the pipe axis are no different from the skills for welding plate wrought shapes using a straight line progression; therefore, the pipe product form limitation does not apply in these straight line cases. Refer to Figure C-6.1.

Qualification of welders using job size pipe or tubing is allowed because pipe sizes specified in Table 10.13 for welder qualification are not always available to the Contractor.

C-**10.18** CJP Groove Welds for Tubular Connections

Because of the special skills required to successfully execute a CJP groove weld in tubular T-, Y-, and K-connections, the 6GR level of welder qualification for the process being used is always required. Also, where groove angles less than 30° are to be used, the acute angle sample joint test of 10.14.4.2 is also required for each welder. When box sections are used in performance qualification, bend tests taken from the faces do not evaluate the welder's ability to carry sound weld metal around the relatively abrupt corners. These bend tests do not fulfill the needs of CJP groove welds in T-, Y-, and K-connections because the corners in these connections may be highly stressed. Due to the concerns for welders to demonstrate their skill to weld the corners of box tubes when CJP is required, the corner macroetch test of Figure 10.22 was developed.

The corner macroetch test shown in Figure 10.22 is an additional performance test required for welders expected to make CJP groove welds in box tube T-, Y-, and K-connections.

For this case, qualified 6GR welders tested on round tubes or pipe per Figure 10.20 would only be required to pass the additional corner macroetch test per Figure 10.22, provided all the requirements of Table 10.12 and 10.14.4.2 are met.

If the Contractor wishes to qualify a welder without existing 6GR status for CJP groove welds in T-, Y-, and K-connections using box tubes, the welder is to weld the 6GR test assembly of Figure 10.20 and Figure 10.21 using either a round or box tube in conformance with the limitations of Table 10.13. In addition, the welder is to pass the corner macroetch test using Figure 10.22 or, if box sections were used, Figure 10.20 or Figure 10.21, by passing a macroetch test on the corner sections of the test weldment.

Qualification on 2G plus 5G or 6G pipe tests also qualifies for butt joints in box sections (with applicability based on thickness, neglecting diameter) but not vice versa. For these butt joints, the macroetch corner test of Figure 10.22 is not necessary because all production joints require NDT per 10.25.2.

Table 10.13 does not differentiate between pipe (circular tubing) and box sections. For this reason, the following interpretation is appropriate:

(1) Qualification on the 6GR pipe test also qualifies for T-, Y-, and K-connections and groove welds in box sections.

(2) Qualification on 5G and 2G pipe tests also qualifies for box sections (with applicability based on thickness, neglecting diameter), but not vice versa.

(3) Qualification for groove welds in box sections also qualifies for plate (and vice versa if within the limitation of Table 10.12 and 6.20 of the code).

(4) When box sections are used in qualification, bend tests taken from the faces do not evaluate the welder's ability to carry sound welding around corners. These bend tests do not fulfill the needs of T-, Y-, and K-connections, because the corners in these connections are highly stressed. Where a 6GR test utilizes box sections, RT is recommended to evaluate the corners.

PJP T-, Y-, and K-connections are also provided for. They can be executed by welders having the common pipe qualifications 2G plus 5G.

Part E
Fabrication

C-10.22 Backing

C-10.22.1 Full-Length Backing. Experience has shown that a tightly fitted, but unwelded square butt joint in steel backing constitutes a severe notch and crack-like condition that potentially leads to transverse cracks in the weld when the unfused area is perpendicular to the stress field. Such cracks will, in many cases, propagate into the base metal.

Discontinuities in backing are not avoidable in some details where piece geometry requires a discontinuity and the lack of access prevents welding the backing. Backing welded parallel to the axis of the member will strain with the member. The unwelded discontinuity will provide a stress concentration and defect when it is perpendicular to the stress field. The stress concentration may initiate cracking. Residual stresses, high restraint and cyclic loads increase the risk of such cracking. Compressive stresses decrease the risk of cracking unless there is stress reversal in the applied stress cycle. The code permits discontinuous backing in selected specific applications or where the Engineer approves.

C-10.23 Tolerance of Joint Dimensions

C-10.23.2.1 Correction. Root openings wider than those allowed by Figure 7.3 and Table 10.14 may be corrected by building up one or both sides of the groove faces by welding. In correcting root openings, the user is cautioned to obtain the necessary approvals from the Engineer where required. The final weld is to be made only after the joint has been corrected to conform to the specified root opening tolerance, thus keeping shrinkage to a minimum.

Part F
Inspection

C-10.25 NDT

C-10.25.1 Weld quality requirements for tubular structures are essentially the same as for statically loaded structures. RT can generally not be applied successfully to inspection of tubular T-, Y-, and K-connections.

C-10.25.2 Tubular Connection Requirements. The static RT requirements of 8.12.1 are applicable to longitudinal groove welds in butt joints with the reinforcement left on. For cyclic loading, this is consistent with the reduced allowable stress range of Category C details. If the weld reinforcement is removed to upgrade fatigue performance, upgrading the cyclic RT criteria to 8.12.2 would also be consistent.

When UT is used as an alternative to RT for longitudinal groove welds in butt joints, the nontubular criteria are typically used, following the logic of the previous paragraph. The tubular UT requirements of 10.26.1.1 Class R are similar to ASME *Boiler and Pressure Vessel Code*, Section VIII, and may also be considered for longitudinal groove welds. The relaxed provisions for the root condition of longitudinal groove welds are not appropriate for circumferential groove welds that are cyclically loaded in tension.

Because of the complex geometry, RT is not applicable to welds in tubular to tubular T-, Y-, and K-connections. The internal UT acceptance criteria of 10.26.1.2 Class X is appropriate since the fatigue performance is similar to that of basic welded profiles with the reinforcement left in place. When surface profile improvements in 10.2.3.7 are specified by the Engineer to improve the fatigue performance to Level I in Table 10.4, an upgrade of the internal inspection criteria to Class R may also be appropriate.

See P. W. Marshall, "Experience-Based, Fitness-for-Purpose Ultrasonic Reject Criteria for Tubular Structures," Fitness for Purpose in Welded Construction, Proc. AWS/WRC/WI Conference, Atlanta, May 1982 (also Proc. 2nd Int'l. Conference on Welding of Tubular Structures, IIW, Boston, 1984).

C-10.26 UT

C-10.26.1 UT Acceptance Criteria for Tubular Connections. The UT procedures and acceptance criteria set forth in Clause 8, i.e., 8.13.1 and 8.13.2, are not applicable to tubular T-, Y-, and K-connections. Acceptance criteria for the latter are set forth in 10.26.1. Contract documents should state the extent of testing, which of the acceptance criteria apply (Class R or Class X), and where applicable. Because of the complex geometry of tubular T-, Y-, and K-connections, standardized step-by-step UT procedures, such as those given in Clause 8, do not apply. Any variety of equipment and techniques may be satisfactory providing the following general principles are recognized.

The inspection technique should fully consider the geometry of the joint. This can be simplified by idealizing localized portions of welds as joining two flat plates, in which case the principal variables are local dihedral angles, material thickness, and bevel preparation; curvature effects may then be reintroduced as minor corrections. Plotting cards superimposing the sound beam on a cross-sectional view of the weld are helpful. Inspections should be referenced to the local weld axis rather than to the brace axis. Every effort should be made to orient the sound beams perpendicular to the weld fusion line; in some cases this will mean multiple inspection with a variety of transducer angles.

The use of amplitude calibrations to estimate discontinuity size should consider sound path attenuation, transfer mechanism (to correct differences in surface roughness and curvature), and discontinuity orientation (e.g., a surface discontinuity may produce a larger echo than an interior discontinuity of the same size). Transfer correction is described in 3.6.5 of Reference 4.

Amplitude calibration becomes increasingly difficult for small diameter (under 12 in [300 mm]) or with thin wall (under 1/2 in [12 mm]), or both. In the root area of tubular T-, Y-, and K-connections, prominent corner reflectors are often present which cannot be evaluated solely on the basis of amplitude; in this case, beam boundary techniques are useful for determining the size of the larger discontinuities of real concern. Beam boundary techniques are described in 3.8.3.2 of Reference 4 of C-4.

The UT acceptance criteria should be applied with the judgement of the Engineer, considering the following factors:

(1) For tubular T-, Y-, and K-connections having CJP groove welds made from the outside only (see Figures 10.7 and Figures 10.9–10.11), root discontinuities are less detrimental and more difficult to repair than those elsewhere in the weld.

(2) It should be recognized that both false alarms (UT discontinuities that are not subsequently verified during the repair) and occasionally missed discontinuities may occur. The former is a part of the cost of the inspection, while the latter emphasize the need for structural redundancy and notch-tough steel.

C-10.27 RT Procedures

C-10.27.2 IQI Selection and Placement. Since radiographic sensitivity and the acceptability of radiographs are based upon the image of the required IQIs, care is taken in describing the manufacture and use of the required IQIs. IQIs are placed at the extremities of weld joints where geometric distortion is anticipated to contribute to lack of sensitivity in the radiograph as shown in Figures 8.6–8.9.

C-10.29 UT of Tubular T-, Y-, and K-Connections

The UT requirements of this section represent the state of the art available for examination of tubular structures, especially T-, Y-, and K-connections. Height determination of elongated reflectors with a dimension (H) less than the beam height (see Figure 10.25) is considerably less accurate than length determination where the reflectors extend beyond the beam boundaries, and requires more attention in regards to procedure qualification and approval, and in the training and certification of UT operators.

This section sets forth requirements for procedures, personnel, and their qualifications. It is based largely on practices that have been developed for fixed offshore platforms of welded tubular construction. These are described in detail in Reference 4.

References for Clause C-10

1. Marshall, P. W. and Toprac, A. A. "Basis for tubular joint design." *Welding Journal*. Welding Research Supplement, May 1974. (Also available as American Society for Civil Engineers preprint 2008.)

2. Marshall, P. W. "Basic considerations for tubular joint design in offshore construction." *WRC Bulletin 193*, April 1974.

3. American Petroleum Institute. *Recommended Practice for Planning, Designing, and Constructing Fixed Offshore Platforms*. API RP 2A, 17th Ed. Dallas: American Petroleum Institute, 1987.

4. American Petroleum Institute. *Recommended Practice for Ultrasonic Examination of Offshore Structural Fabrication and Guidelines for Qualification of Ultrasonic Techniques*. API RP 2X, 1st Ed. Dallas: American Petroleum Institute, 1980.

5. Haagensen, P. J. "Improving the fatigue performance of welded joints." *Proceedings of International Conference on Offshore Welded Structures*, 36. London, November 1982.

6. Snedden, N. W. *Background to Proposed New Fatigue Design Rules for Welded Joints in Offshore Structures*." United Kingdom: United Kingdom Department of Energy, AERE Harwell, May 1981.

7. Marshall, P. W. "Size effect in tubular welded joints." ASCE Structures Congress 1983, Session ST6. Houston, October 1983.

8. Society of Automotive Engineers. *Society of Automotive Engineers Fatigue Design Handbook*, AE-4. Warrendale: Society of Automotive Engineers, 1968.

9. Rolfe, S. T. and Barsom, J. M. *Fracture and Fatigue Control in Structures*. Prentice Hall, 1977.

10. American Petroleum Institute. *Recommended Practice for Pre-Production Qualification of Steel Plates for Offshore Structures*, API RP2Z, 1st Edition. Dallas: American Petroleum Institute, 1987.

Table C-10.1
Structural Steel Plates (see C-10.7.2)

Strength Group	Toughness Class	Specification and Grade			Yield Strength ksi	Yield Strength MPa	Tensile Strength ksi	Tensile Strength MPa
I	C	ASTM A36	(to 2 in [50 mm] thick)		36	250	58–80	400–550
		ASTM A131	Grade A (to 1/2 in [12 mm] thick)		34	235	58–71	440–490
I	B	ASTM A131	Grades B, D		34	235	58–71	400–490
		ASTM A573	Grade 65		35	240	65–77	450–550
		ASTM A709	Grade 36T2		36	250	58–80	400–550
I	A	ASTM A131	Grades CS, E		34	235	58–71	400–490
II	C	ASTM A242	(to 1/2 in [12 mm] thick)		50	345	70	480
		ASTM A572	Grade 42 (to 2 in [50 mm] thick)		42	290	60	415
		ASTM A572	Grade 50 (to 1/2 in [12 mm] thick)[a]		50	345	65	450
		ASTM A588	(4 in [100 mm] and under)		50	345	70 min.	485 min.
II	B	ASTM A709	Grades 50T2, 50T3		50	345	65	450
		ASTM A131	Grade AH32		45.5	315	68–85	470–585
		ASTM A131	Grade AH36		51	350	71–90	490–620
		ASTM A808	(strength varies with thickness)		42–50	290–345	60–65	415–450
		ASTM A516	Grade 65		35	240	65–85	450–585
II	A	API Spec 2H	Grade 42		42	290	62–80	430–550
			Grade 50	(to 2–1/2 in [65 mm] thick)	50	345	70–90	483–620
				(over 2–1/2 in [65 mm] thick)	47	325	70–90	483–620
		API Spec 2W	Grade 42	(to 1 in [25 mm] thick)	42–67	290–462	62	427
				(over 1 in [25 mm] thick)	42–62	290–427	62	427
			Grade 50	(to 1 in [25 mm] thick)	50–75	345–517	65	448
				(over 1 in [25 mm] thick)	50–70	345–483	65	448
			Grade 50T	(to 1 in [25 mm] thick)	50–80	345–522	70	483
				(over 1 in [25 mm] thick)	50–75	345–517	70	483
		API Spec 2Y	Grade 42	(to 1 in [25 mm] thick)	42–67	290–462	62	427
				(over 1 in [25 mm] thick)	42–62	290–462	62	427
			Grade 50	(to 1 in [25 mm] thick)	50–75	345–517	65	448
				(over 1 in [25 mm] thick)	50–70	345–483	65	448
			Grade 50T	(to 1 in [25 mm] thick)	50–80	345–572	70	483
				(over 1 in [25 mm] thick)	50–75	345–517	70	483
		ASTM A131	Grades DH32, EH32		45.5	315	68–85	470–585
			Grades DH36, EH36		51	350	71–90	490–620
		ASTM A537	Class I (to 2–1/2 in [65 mm] thick)		50	345	70–90	485–620
		ASTM A633	Grade A		42	290	63–83	435–570
			Grades C, D		50	345	70–90	485–620
		ASTM A678	Grade A		50	345	70–90	485–620
III	C	ASTM A633	Grade E		60	415	80–100	550–690
III	A	ASTM A537	Class II	(to 2–1/2 in [65 mm] thick)	60	415	80–100	550–690
		ASTM A678	Grade B		60	415	80–100	550–690
		API Spec 2W	Grade 60	(to 1 in [25 mm] thick)	60–90	414–621	75	517
				(over 1 in [25 mm] thick)	60–85	414–586	75	517
		API Spec 2Y	Grade 60	(to 1 in [25 mm] thick)	60–90	414–621	75	517
				(over 1 in [25 mm] thick)	60–85	414–586	75	517
		ASTM A710	Grade A Class 3 (quenched and precipitation heat treated)					
			thru 2 in [50 mm]		75	515	85	585
			2 in [50 mm] to 4 in [100 mm]		65	450	75	515
			over 4 in [100 mm]		60	415	70	485
IV	C	ASTM A514	(over 2–1/2 in [65 mm] thick)		90	620	110–130	760–890
		ASTM A517	(over 2–1/2 in [65 mm] thick)		90	620	110–130	760–896
V	C	ASTM A514	(to 2–1/2 in [65 mm] thick)		100	690	110–130	760–895
		ASTM A517	(to 2–1/2 in [65 mm] thick)		100	690	110–130	760–895

[a] To 2 in [50 mm] Thick for Type 1 or 2 Killed, Fine Grain Practice.
Note: See list of Referenced Specifications for full titles of the above.

Table C-10.2
Structural Steel Pipe and Tubular Shapes (see C-10.7.2)

Group	Class	Specification and Grade			Yield Strength ksi	Yield Strength MPa	Tensile Strength ksi	Tensile Strength MPa
I	C	API Spec 5L	Grade B[a]		35	240	60	415
		ASTM A53	Grade B		35	240	60	415
		ASTM A139	Grade B		35	240	60	415
		ASTM A500	Grade A	(round)	33	230	45	310
				(shaped)	39	270	45	310
		ASTM A500	Grade B	(round)	42	290	58	400
				(shaped)	46	320	58	400
		ASTM A501	Grade A	(round and shaped)	36	250	58	400
		API Spec 5L	Grade X42	(2% max. cold expansion)	42	290	60	415
I	B	ASTM A106	Grade B	(normalized)	35	240	60	415
		ASTM A524	Grade I	(through 3/8 in [10 mm] w.t.)	35	240	60	415
			Grade II	(over 3/8 in [10 mm] w.t.)	30	205	55–80	380–550
I	A	ASTM A333	Grade 6		35	240	60	415
		ASTM A334	Grade 6		35	240	60	415
II	C	API Spec 5L	Grade X42	(2% max. cold expansion)	52	360	66	455
		ASTM A618			50	345	70	485
		ASTM A501	Grade B		50	345	70	485
II	B	API Spec 5L	Grade X52 with SR5, SR6, or SR8		52	360	66	455
III	C	ASTM A595	Grade A (tapered shapes)		55	380	65	450
		ASTM A595	Grades B and C (tapered shapes)		60	410	70	480

[a] Seamless or with longitudinal seam welds.

Notes:
1. See list of Referenced Specifications for full titles of the above.
2. Structural pipe may also be fabricated in accordance with API Spec 2B, ASTM A139+, ASTM A252+, or ASTM A671 using grades of structural plate listed in Table C-10.1 except that hydrostatic testing may be omitted.
3. With longitudinal welds and circumferential butt welds.

Table C-10.3
Structural Steel Shapes (see C-10.7.2)

Group	Class	Specification and Grade			Yield Strength ksi	Yield Strength MPa	Tensile Strength ksi	Tensile Strength MPa
I	C	ASTM A36	(to 2 in [50 mm] thick)		36	250	58–80	400–550
		ASTM A131	Grade A	(to 1/2 in [12 mm] thick)	34	235	58–80	400–550
I	B	ASTM A709	Grade 36T2		36	250	58–80	400–550
II	C	ASTM A572	Grade 42	(to 2 in [50 mm] thick)	42	290	60	415
		ASTM A572	Grade 50	(to 1/2 in [12 mm] thick)	50	345	65	480
		ASTM A588	(to 2 in [50 mm] thick)		50	345	70	485
II	B	ASTM A709	Grades 50T2, 50T3		50	345	65	450
		ASTM A131	Grade AH32		46	320	68–85	470–585
		ASTM A131	Grade AH36		51	360	71–90	490–620

Notes:
1. To 2 in [50 mm] Thick for Type 1 or 2 Killed, Fine Grain Practice
2. This table is part of the commentary on toughness considerations for tubular structures (or composites of tubulars and other shapes), e.g., used for offshore platforms. It is not intended to imply that unlisted shapes are unsuitable for other applications.

Table C-10.4
Classification Matrix for Applications (see C-10.7.2)

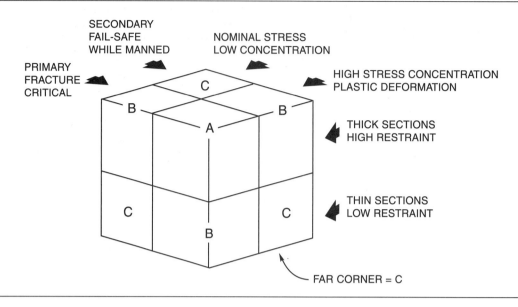

Table C-10.5
CVN Testing Conditions (see C-10.7.2.2)

Diameter/Thickness	Test Temperature	Test Condition
Over 30	36°F [20°C] below LAST[a]	Flat plate
20–30	54°F [30°C] below LAST	Flat plate
Under 20	18°F [10°C] below LAST	As fabricated

[a] LAST = Lowest Anticipated Service Temperature.

Table C-10.6
CVN Test Values (see C-10.14.4.4)

Steel Group	Steel Class	CVN Test Temperature	Weld Metal Avg. ft·lbf	(Joules)
I	C	0°F [−18°C]	20	[27]
I	B	0°F [−18°C]	20	[27]
I	A	−20°F [−29°C]	20	[27]
II	C	0°F [−18°C]	20	[27]
II	B	−20°F [−29°C]	20	[27]
II	A	−40°F [−40°C]	25	[34]
III	C	−20°F [−29°C]	20	[27]
III	B	−40°F [−40°C]	20	[27]
III	A	−40°F [−40°C]	30	[40]
IV and V		Special Investigation		

Note: Code requirements represent the lowest common denominator from the foregoing table.

Table C-10.7
HAZ CVN Test Values (see 10.14.1.1)

Steel Group	Steel Class	CVN Test Temperature	HAZ ft·lbf	(Joules)
I	C	50°F [10°C]	For information only	
I	B	40°F [4°C]	15	[20]
I	A	14°F [−10°C]	15	[20]
II	C	50°F [10°C]	For information only	
II	B	40°F [4°C]	15	[20]
II	A	14°F [−10°C]	25	[34]
III	A	14°F [−10°C]	30	[40]

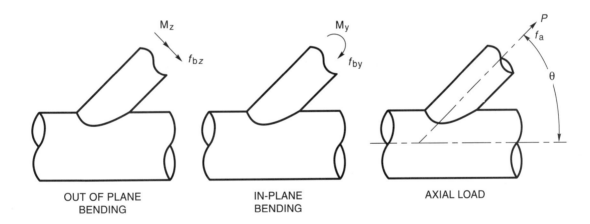

OUT OF PLANE BENDING

IN-PLANE BENDING

AXIAL LOAD

Figure C-10.1—Illustrations of Branch Member Stresses Corresponding to Mode of Loading (see C-10.2.3.2)

COMMENTARY AWS D1.1/D1.1M:2020

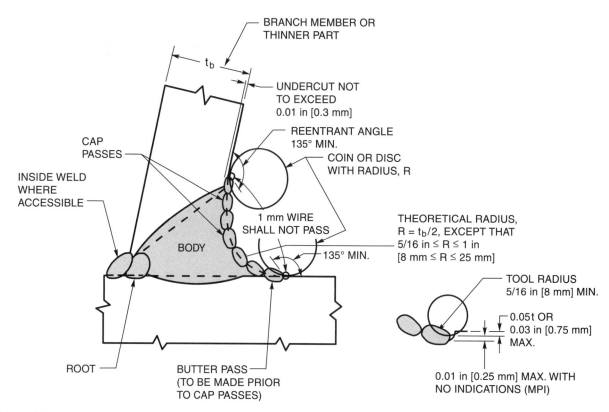

Figure C-10.2—Improved Weld Profile Requirements (see C-10.2.3.6)

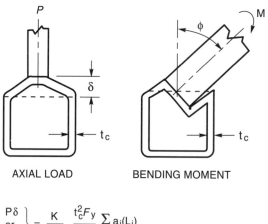

$$\left.\begin{array}{r}P\delta \\ \text{or} \\ M\phi\end{array}\right\} = \frac{K}{SF} \frac{t_c^2 F_y}{4} \sum_{\text{All Yield Lines}} a_i(L_i)$$

Key:
K = Reserve strength factor for strain hardening, triaxial stress, large deflection behavior, etc.
SF = Safety Factor
F_y = Specified yield strength of main member
a_i = Regular rotation of yield line i as determined by geometry of mechanism
L_i = Length of yield line segment
t_c = Wall thickness of chord

Figure C-10.3—Upper Bound Theorem (see C-10.2.3.2 and C-10.7)

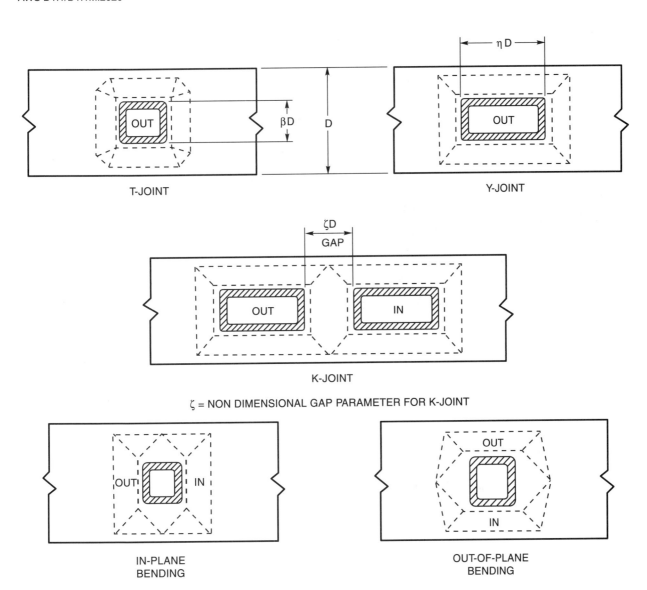

Figure C-10.4—Yield Line Patterns (see C-10.7)

C-11. Strengthening and Repair of Existing Structures

C-11.2 General

There are many technical and workmanship conditions that are common to strengthening, repairing, and heat straightening steel members, and as a result, Clause 11 has been expanded to include heat straightening, a form of repairing steel members.

Clause 11 of this document is not intended to replace ASTM A6 provisions for conditioning new steel, but to provide recommendations for repair and strengthening of members in existing structures.

C-11.3 Base Metal

C-11.3.1 Investigation. The first essential requirement in strengthening, repairing, and heat straightening existing structures is the identification of the material.

Weldability of the existing steel is of primary importance. Together with the mechanical properties of the material, it will provide information essential for the establishment of safe and sound WPSs. Only then will realistic data be available for reliable cost estimates. Should poor weldability make such cost economically prohibitive, other means of joining should be considered by the Engineer.

Mechanical properties may be subject to variability, determined by tests of representative samples taken from the existing structure. Hardness testing may also provide, by correlation, an estimation of the tensile properties of the material.

In cases of unknown weldability, References 1 and 2 provide examples of simple and inexpensive techniques to make a preliminary determination whether or not the base metal is suitable for welding.

Low melting temperature base metal elements such as sulfur, phosphorus, copper, tin, lead, and zinc can cause solidification or "hot" cracking. Utilizing low admixture WPSs and joint details that do not rely on penetration to gain strength may help minimize hot cracking tendencies. Higher levels of carbon, coupled with higher levels of alloys, whether intentionally or unintentionally added, increase steel hardenability and increase hydrogen-related or "cold" cracking tendencies. Low-hydrogen practice, higher preheat and interpass temperature, as well as postheat operations, reduce cold cracking tendencies. The material may range from easily weldable to unacceptable weldability. Investigation into the relative weldability is essential (see References 3, 5, and 6).

C-11.3.2 Suitability for Welding. Welding to stainless steel, wrought iron, and cast iron is not addressed in the general body of this code. However, these materials are sometimes encountered in older structures that are being renovated. As shown in Table C-11.1, a WPS and qualified welding supervision is needed in each case because of the inherent difficulty in welding. Guidance for welding stainless steel is given in References 3 and 4. Guidance for welding wrought iron is given in Reference 5. Guidance for welding cast iron is given in References 3, 4, 5, and 6.

C-11.4 Design for Strengthening and Repair

C-11.4.1 Design Process. It is strongly recommended that locations that are considered for welding or heating be inspected. Thermal expansion associated with either process can extend any existing crack further into the member.

C-11.4.3 Fatigue History. Generally, in the case of cyclically loaded structures, sufficient data regarding past service are not available to estimate the remaining fatigue life. A conservative estimate of remaining fatigue life should be made based on whatever loading history is available. Practical methods to extend the expected fatigue life of a member include: reducing the stress or stress range, providing connection geometry less susceptible to fatigue failure, and using fatigue life enhancement techniques.

C-11.4.5 Loading During Operations. Repair, strengthening and heat straightening of existing structures differ from new construction inasmuch as these operations may have to be executed with the structure or the structural element under some condition of working stress.

There is guidance in the literature (References 1, 2, 7, 8, 9, 10, 11, and 12) with respect to welding of structural members under load. Each situation should be evaluated on its own merits, and sound engineering judgment exercised.

Research (References 7, 11, and 19) indicates that residual stresses due to mill rolling practice and welding only have an effect on member capacity for some specific structural functions. Because of offsetting tensile and compressive residual force levels, residual stresses result in no quantifiable degradation of a member's flexural capacity. A similar condition exists for tensile members, provided that the heating or welding affects only a portion of the member's cross-sectional area. Compression members are more sensitive to residual stress distribution because of overall and local buckling possibilities. Balanced flame heating or welding about the neutral axis may be essential to avoid this type of problem. Regardless of the stress conditions, the heating or welding should not be performed over the entire cross section at the same time.

C-11.4.7 Use of Existing Fasteners. The retrofitting provisions for combining welds with overstressed rivets or bolts is more restrictive than 4.7.3 which deals with connecting elements not overstressed at the time of retrofitting (see Reference 22).

C-11.5 Fatigue Life Enhancement

C-11.5.1 Fatigue Life Enhancement. When properly administered, these reconditioning methods may be used for enhancing the fatigue life of existing structures, particularly when the applied stress is normal to the axis of the weldment. The following techniques affect fatigue life only from the point of view of failure from the weld toe. The possibility of fatigue crack initiation from other features of the weld, e.g., the root area, should not be overlooked. Typical uses include the repair of fatigue cracks and the extension of fatigue life of existing buildings and equipment.

Welded joints represent particularly severe stress concentrations. Research at The Welding Institute (TWI), Cambridge, England, identified an acute line of microscopic slag intrusions along the toes of all welds made by all arc processes except GTAW. All processes however, were found to produce some degree of undercut at the toe, notwithstanding ideal weld profiles (see Figure C-11.1). The practical implication was that all welds have a pre-existing discontinuity in the form of either microscopic undercut or slag intrusions, or both. Normal inspection methods cannot detect these discontinuities, which in any case are unavoidable when using existing welding technologies (see References 28, 32 [Chapters 1 and 2], and 33).

In plain material, fatigue life is spent in crack initiation and propagation. In weldments, however, it is assumed that crack-like discontinuities already exist. Therefore, the fatigue life of welds is spent solely in crack propagation. This, along with residual tensile stresses at or near the yield point, is the essential reason why weldments can endure fewer cycles to fatigue failure than a similarly loaded plain material (see Figure C-11.2).

Fatigue life enhancement can be obtained by reconditioning the weld toes. The small pre-existing discontinuities are either removed or the sharp openings dulled (see Figure C-11.5). Toe grinding and GTAW (TIG) dressing extend fatigue life by restoring a crack initiation phase. Peening, by the introduction of a compressive stress, retards the rate of crack propagation. The resulting weld profile also complements the overall joint resistance to fatigue cracking by reducing the geometric stress concentration. When these pre-existing toe discontinuities are perpendicular to the applied stress, fatigue life enhancement methods are most effective (Figure C-11.4) (see Reference 32 [Chapter 4]).

(1) Profile improvement for round tubular sections is to conform to 10.2.3.6, 10.2.3.7, and is to conform to the corresponding sections from the Commentary, C-10.2.3.6 and C-10.2.3.7. Acceptable profiles in conformance with Figure 10.11 may be attained by (1) adding a capping layer, (2) grinding the weld surface, and (3) peening the weld toe with a blunt instrument. Figure C-4.8 provides precise profile criteria. Fatigue category limitations on weld size or thickness and weld profile are required to meet the criteria of Table 10.4, Level I.

(2) Toe grinding should be done along the centerline of the weld toe for both tubular connections (T-, Y-, or K-) and nontubular joints. The recommended tools include a high-speed grinder for use with a tungsten carbide burr. The tip radius should be scaled to the plate thickness according to Table C-11.2. These radii are the minimum recommended, larger sizes may prove more beneficial.

Grinding should be carried out to a minimum depth of 0.03 in–0.04 in [0.8 mm–1.0 mm] below the plate surface or approximately 0.02 in–0.03 in [0.5 mm–0.8 mm] below the deepest undercut to a maximum total depth of 1/16 in [2 mm] or 5% of the plate thickness, whichever is greater. The axis of the burr should be at approximately 45° to the main plate (see Figure C-11.3). The angle of the burr axis should be a maximum 45° of the direction of travel to ensure that the grinding marks are nearly perpendicular to the weld toe line (parallel to the direction of stress). The ends of longitudinally stressed welds require special care to be effective (see Figure C-11.6). The finishing pass should be light to obtain a good surface finish. Check visually and with MT or PT for any remaining undercut or other discontinuities (see References 28, 29, 32 [Chapter 2], 34, and 37).

(3) Hammer peening applies to steels with yield strengths up to 115 ksi [800 MPa] and thicknesses not less than 3/8 in [10 mm]. Steel hammer bits should have approximately hemispherical tips with diameters between 1/4 in and 1/2 in [6 mm and 12 mm]. The indentation should be centered on the weld toe so that metal on each side (both weld metal and base metal) is deformed, resulting in a smooth surface free from obvious individual blows. The hammer should be held at 45° to the plate surface and approximately perpendicular to the direction of travel. The indentation in mild steel (yield strength up to 36 ksi [250 MPa]) should be approximately 0.02 in [0.5 mm]; in medium-strength steel (yield strength between 36 ksi and 65 ksi [250 MPa and 450 MPa]) 0.01 in [0.25 mm]; and in high-strength steel (yield strength between 65 ksi and 115 ksi [450 MPa and 800 MPa]) 0.004 in [0.1 mm] (see Figure C-11.7). These depths are roughly equivalent to four peening passes. The weld is to be checked visually and with MT or PT prior to peening (see References 28, 29, 32 [Chapter 2], 34, and 36).

The benefit of hammer peening is derived from the introduction of compressive residual stresses; thus, it is critical to ensure that nothing which will cause stress relief (e.g., PWHT) be performed after peening. Also, hammer peening should be applied when the joint is "in place" and carrying dead-load.

(4) GTAW (TIG) dressing consists of remelting the existing weld metal to a depth of approximately 1/16 in [2 mm] along the weld toe without the addition of filler metal. The weld surface should be free from rust, slag, and scale. The tip of the electrode should be kept sharp and clean. The tip should be located horizontally 0.02 in to 0.06 in [0.5 mm to 1.5 mm] from the weld toe (see Figure C-11.8). Where toughness of the HAZ may create problems, a modified technique using a second tempering pass may be used (see References 28, 32 [Chapters 2 and 4], and 35).

(5) Toe grinding followed by hammer peening inhibits fatigue crack initiation and the rate of crack propagation. Thus, for critical joints, this combined treatment offers superior resistance to fatigue failure. The weld surface should be checked visually and by MT for surface discontinuities prior to peening. During peening operations, visually check after each pass (see References 30, 31, and 34).

C-11.5.2 Stress Range Increase. The allowable stress range for cyclically loaded connections may be increased by a factor of 1.3 along the S-N design curve, which is equivalent to a factor of 4.2 on cycle life, for an S/N slope of approximately 1/3, when toe grinding, hammer peening, or GTAW (TIG) dressing is used. However, the effect of toe grinding and hammer peening is cumulative. A factor of 1.5 on the stress range may be allowed at high cycles ($N = 10^7$), but reduced to a factor of 1.0 (no benefit) at low cycles ($N = 10^4$). For nontubular joints, the improvement factor should not exceed the highest as-welded fatigue design category.

Tubular sections for T-, Y-, and K-joints are discussed in 10.2.3.6 and 10.2.3.7 (see References 28, 29, 30, 32, 34, 35, and 36).

NOTE: Current research at Lehigh University on beams with welded cover plates suggests that this joint does not respond to enhancement techniques (especially hammer peening) as well as previous testing had indicated.

C-11.6 Workmanship and Technique

C-11.6.2 Member Discontinuities. Welding or applying heat to steel in the presence of existing cracks can result in crack propagation because of the stress levels that may exist at the crack tip. As a result, it is strongly recommended that any crack-like discontinuities be removed prior to applying heat or welding. This can be accomplished by drilling a hole at the termination of the crack or by grinding. References 23 and 26 provide guidance for repairing many commonly observed conditions. Procedures for repairing existing welds are also required to conform to the provisions of Clause 7.

C-11.6.4 Base Metal of Insufficient Thickness. Corrosion or wear with resultant section loss may reduce the thickness of the parts below that required to provide adequate weld size. Building up the edge of the thin section may be performed by welding, provided the thickness of the overall section is adequate to carry the load.

Corrosion or wear may reduce the thickness of parts below that required to support the load. Similarly, increased loads may require additional member thickness. Increasing member thickness by facing with weld metal would generally be ineffective except for small, localized regions. Reinforcement of the member by the use of additional plates or similar attachments is preferred.

C-11.6.5 Heat Straightening. Heat straightening of steel members requires the sequencing of various heating patterns. Literature (References 1, 8, 9, 13, 14, 15, 16, 17, and 18) exists that provides guidance as to the mechanics of heat straightening. The actual process remains one of operator experience.

Limits (including significant safety factors) are placed on the temperature to which steel can be heated to avoid possible metallurgical changes in the steel when it subsequently cools down to an ambient temperature. The chemistry and prior tempering of the steel dictates the critical temperature.

Rapid cooling of steel from elevated temperatures down to about 600°F [315°C] is not recommended because undesirable metallurgical transformations can occur (see 7.8 for a more detailed discussion of this subject).

Reference 20 (AASHTO Div. II, Section 11.4.12.2.3) also considers 600°F [315°C] as the critical temperature. Furthermore, Reference 21 shows that at this temperature, the moduli of elasticity and yield point of steel are not significantly reduced from that of steel at ambient temperature.

Cooling rates suggested in other parts of the general specification for the appropriate grade and thickness of steel are recommended. A water mist, wet rags, or forced air is considered to be accelerated cooling and may only be used when the steel temperature is below 600°F [315°C].

C-11.6.6 Welding Sequence. Welding procedures should be adjusted so that the total heat input per unit length of the weld for a given thickness and geometry of the material will maintain the temperature isotherms relatively narrow and minor in relation to the cross section of the stress-carrying member.

C-11.7 Quality

The Engineer determines the level of inspection and NDT as appropriate for the job conditions. It is recommended that the contract document requirements be made compatible with Clause 8 and Clause 10 of this code.

Examination of rivets and bolts affected by the heat induced by welding or straightening should be considered.

References for Clause C-11

1. Blodgett, O. W. *Design of Welded Structures,* 12th Printing. Cleveland, OH: The James F. Lincoln Arc Welding Foundation, 1982.

2. Ricker, D. T. "Field welding to existing steel structures." *Engineering Journal* 25(1): Chicago, IL. American Institute of Steel Construction 1988.

3. American Welding Society. *Welding Handbook, Volume 4.* Miami, FL: American Welding Society, 1982.

4. Lincoln Arc Welding. *The Procedure Handbook of Arc Welding,* 12th Ed. Cleveland, OH: The James F. Lincoln Arc Welding Foundation, June 1973.

5. Jefferson, T. B. and Woods, G. *Metals and How to Weld Them,* 2nd Ed. Cleveland, OH: W. E. Publications, The James F. Lincoln Arc Welding Foundation, 1972.

6. American Welding Society, *Guide for Welding Iron Castings,* AWS D11.2. Miami, FL: American Welding Society, 1989.

7. Nagaraja Rao, N. R. and Tall, L. "Columns reinforced under load." *Welding Journal,* Vol. 42, Research Supplement 177-s, April 1963.

8. Shanafelt, G. O. and Horn, W. B. "Guidelines for evaluation and repair of damaged steel bridge members." *Report No. 271.* Washington, DC: National Cooperative Highway Research Program, Transportation Research Board, June 1984.

9. Tide, R. H. R. "Basic considerations when reinforcing existing steel structures." *Conference Proceedings, National Engineering Conference and Conference of Operating Personnel, New Orleans, LA.* Chicago, IL: American Institute of Steel Construction, April 1987.

10. Tide, R. H. R. "Effects of fabrication on local stress conditions." *Conference Proceedings, ASCE Seventh Structures and Pacific Rim Engineering Conference, San Francisco, CA.* New York, NY: American Society of Civil Engineers, May 1989.

11. Marzouk, H. and Mohan, S. "Strengthening of wide-flange columns under load." *Canadian Journal of Civil Engineering* 17(5), October 1990. Ottawa, Ontario, Canada: Canadian Society of Civil Engineers.

12. Tide, R. H. R. "Reinforcing steel members and the Effects of Welding." *Engineering Journal* 27(4): 1990.

13. Stitt, J. R. "Distortion control during welding of large structures." *Paper 844B, Air Transport and Space Meeting.* New York, NY: Society of Automotive Engineers, Inc. and American Society of Mechanical Engineers, April 1964.

14. Holt, Richard E. "Primary concepts of flame bending." *Welding Journal,* June 1971.

15. United States Steel Corporation. *Fabrication Aids for Continuously Heat-Curved Girders,* ADUSS 88-5538-01. Pittsburgh, PA: United States Steel Corporation, April 1972.

16. Stewart, J. P. *Flame Straightening Technology,* LaSalle, QUE, 1981.

17. Roeder, C. W. "Experimental study of heat induced deformation." *Journal of Structural Engineering* 112(ST10) 1986. New York, NY: American Society of Civil Engineers.

18. Avent, R. R. "Heat straightening of steel: from Art to Science." *Conference Proceedings, National Steel Construction Conference, Miami Beach, FL.* Chicago, IL: American Institute of Steel Construction, June 1988.

19. American Society of Civil Engineers. *Plastic Design in Steel, a Guide and Commentary.* New York, NY: ASCE, 1971.

20. American Association of State Highway and Transportation Officials. *Standard Specifications for Highway Bridges,* 15th Edition. Washington, DC: AASHTO, 1992.

21. United States Steel Corporation. *Steel Design Manual,* ADUSS 27-3400-04. Pittsburgh, PA: United States Steel Corporation, January 1981.

22. Jarosch, K. H. and Bowman, M.D. "Tension butt joints with bolts and welds in combination." *Engineering Journal* 23(1) 1986.

23. Fisher, J. W. *Fatigue and Fracture in Steel Bridges.* New York, NY: John Wiley and Sons, Inc., 1984.

24. Fisher, J. W. *Bridge Fatigue Guide—Design and Details.* Chicago, IL: American Institute of Steel Construction, 1977.

25. *Welding in the World.* "Recommendations for repairs and or strengthening of steel structures," 26(11/12), Great Britain.

26. Haagensen, P. J. "Improving the fatigue performance of welded joints," *Proceedings, International Conference on Offshore Welded Structures.* London, England, November 1982.

27. Maddox, S. J. *Aspects of the Improvement in Fatigue Strength of Fillet Welds by Peening.* Cambridge, England: The Welding Institute, 1985.

28. Maddox, S. J. *Fatigue Strength of Welded Structures*, Second Edition. Abington Publishing, 1992.

29. Haagensen, P. J. IIW DOC. XIII-WG2-22-93, *Repair Methods and Life Extension,* 1993.

30. Haagensen, P. J. *The Effect of Grinding and Peening on the Fatigue Strength of Welded T-Joints.* IIW Doc. XIII-1510-93. 1993.

31. Hagensen, P. J. *Life Extension and Repair by Grinding and Peening,* IIW Doc. XIII-1511-93, 1993.

32. The Welding Institute. *Improving the Fatigue Performance of Welded Joints.* Cambridge, England: The Welding Institute, 1983.

    Maddox, S. J., Chapter 1 "An introduction to the fatigue of welded joints"

    Booth, G. S., Chapter 2 "A review of fatigue strength improvement techniques"

    Woodley, C. C, Chapter 4 "Practical applications of weld toe grinding"

    Haagensen, P. J., Chapter 5 "Effect of TIG dressing on fatigue performance and hardness of steel weldments"

33. Maddox, S. J. "International efforts on fatigue of welded construction." *Welding & Metal Fabrication,* December 1992.

34. Commission IIW Working Group 2—Improvement Techniques, *Proposed IIW Specification for Weld Toe Improvement by Hammer Peening or Burr Grinding,* 1993.

35. Takenouchi et. al. *Fatigue Performances of Repairing Welds with TIG-Dressing for Fatigue Damaged Highway Bridges,* IIW Doc. XIII-1509-93. Japan, 1993.

36. Welsch, W. "Peening improves fatigue life." *Welding Design & Fabrication,* September 1990.

37. Connect, No. 12, The Welding Institute, Cambridge, England, (from AWS *Welding Journal,* February 12, 1991).

38. CETM/Centre Technique Des Industries Mecaniques. *Improving the Fatigue Strength of Welded Joints by TIG or Plasma Dressing.* IIW Doc. XIII-WG2-10-91, 1991.

Table C–11.1
Guide to Welding Suitability (see C-11.3.2)

Structure Category	Base Metal			
	ASTM, ABS, and API Steels per 5.3 and Table 5.3	Discontinued, Unknown Steels, Cast Steels, and Stainless Steels	Wrought Iron	Cast Iron
Static or Cyclic Nontubular Clause 4, Part B	Check for prequalified status per Clause 5. Prequalified WPSs may be used per Clause 5.	ASTM A7, A373, A441—Use Table 5.3 (Group II) and Clause 5. Others, see Footnote a.	Footnotes a and b apply.	Footnotes a and b apply.
Cyclic Nontubular Clause 4, Part C	Check for prequalified status per Clause 5. Prequalified WPSs may be used per Clause 5.	ASTM A7, A373, A441—Use Table 5.3 (Group II) and Clause 5. Others, see Footnote a.	Footnotes a and b apply.	Not recommended.
Tubular Clause 10, Part A	Prequalified WPSs may be used per Clause 10.			
Static Tubular	Check for prequalified status per Clause 10.	Footnote a applies.	Footnotes a and b apply.	Footnotes a and b apply. Not recommended.
Cyclic Tubular	Check for prequalified status per Clause 10.	Footnote a applies.	Footnotes a and b apply.	Not recommended.

[a] Established Welding Suitability: Existence of previous satisfactory welding may justify the use of Table 5.4 (Group II) filler metals. If not previously welded, obtain samples and prepare WPS qualification. Conduct in place weld test on safe area of structure if samples are not available.
[b] Persons qualified to establish welding suitability must provide a written WPS and should monitor welding operations.

Note: A written WPS is required and must have Engineer approval.

Table C-11.2
Relationship Between Plate Thickness and Burr Radius
[see C-11.5.1(2)]

Plate Thickness (in)	Plate Thickness (mm)	Burr Radius (mm)
<0.79	<20	5
0.79–1.14	20–29	6
1.18–1.54	30–39	8
1.57–1.93	40–49	10
1.97–2.52	50–64	12
2.56–3.11	65–79	16
3.15–3.90	80–99	18
3.94–4.69	100–119	20
4.72–5.87	120–149	25
5.91–7.09	150–180	30

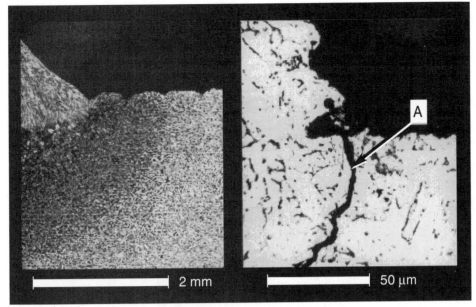

Courtesy of The Welding Institute UK, 1980.

Note: Microscopic intrusions at weld toe act as pre-existing discontinuities (see C-11.5.1).

Figure C-11.1—Microscopic Intrusions

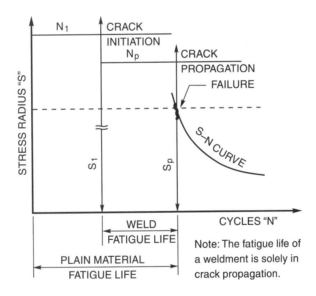

Figure C-11.2—Fatigue Life (see C-11.5.1)

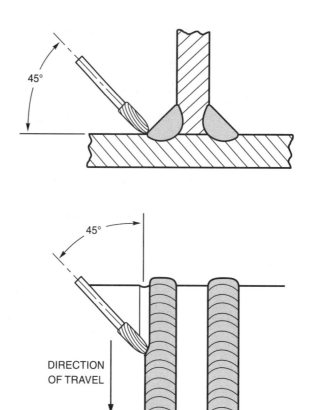

Figure C-11.3—Toe Dressing with Burr Grinder (see C-11.5.1)

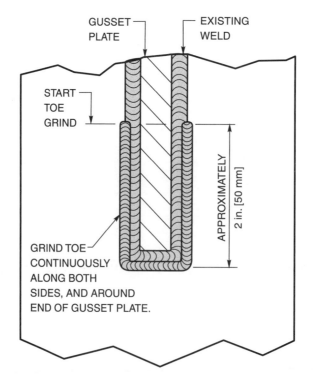

Figure C-11.4—Toe Dressing Normal to Stress (see C-11.5.1)

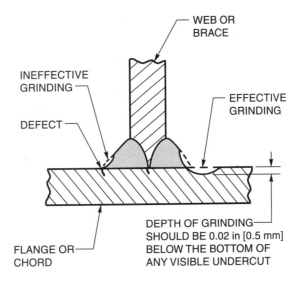

Figure C-11.5—Effective Toe Grinding (see C-11.5.1)

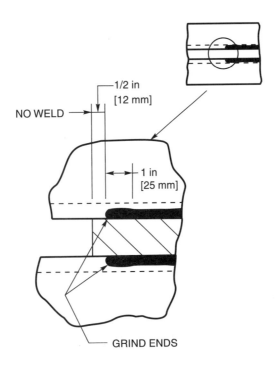

Figure C-11.6—End Grinding [see C-11.5.1(2)]

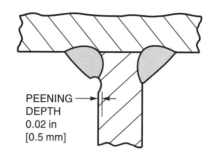
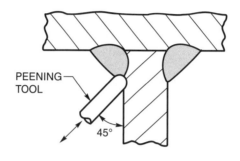
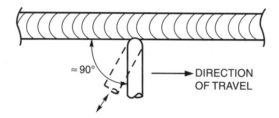

Figure C-11.7—Hammer Peening [see C-11.5.1(3)]
(Courtesy of S. Maddox, IIW, Com. XIII)

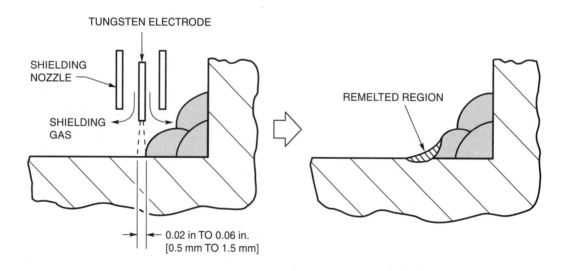

Figure C-11.8—Toe Remelting [see C-11.5.1(4)]
(Courtesy of P. Haagensen, IIW, Com. XIII)

… AWS D1.1/D1.1M:2020

C-Annex B
Guidelines on Alternative Methods for Determining Preheat

Preheat—Background Review and Discussion

General Observations. The probability of hydrogen cracking depends on a number of factors. Some of these can be classed as global (e.g., chemical composition and thickness) and can therefore be defined, while others which are local factors (e.g., the details of the weld root geometry, or local segregation of certain chemical elements) cannot be defined.

In some cases, these factors may dominate, and this makes it virtually impossible to predict in any rational manner the precise preheating conditions that are necessary to avoid hydrogen cracking. These situations should be recognized from experience and conservative procedures adopted. However, in the majority of cases, it is possible with present day knowledge of the hydrogen cracking phenomenon to predict a preheat and other welding procedure details to avoid hydrogen cracking that will be effective in the majority of cases without being overly conservative.

The preheat levels predicted from such a system should be compatible with experience. The requirements should allow fabricators to optimize preheating conditions for the particular set of circumstances with which they are concerned. Thus, rather than calling for a certain preheat for a given steel specification, the alternative guide allows preheats to be based on the chemistry of the plate being welded, as determined from mill reports or analysis. Fabricators may then, through knowledge of the particular set of circumstances they have, be able to use lower preheats and a more economical welding procedure. On the other hand, the requirements should provide better guidance for more critical joints; e.g., high restraint situations that will allow fabricators to undertake adequate precautions.

Basis of Predicting Preheat. Research has shown that there are the following four basic prerequisites for hydrogen cracking to occur:

  (1) susceptible microstructure (hardness may give a rough indication of susceptibility)

  (2) appropriate level of diffusible hydrogen

  (3) appropriate level of restraint

  (4) suitable temperature

One or more of these prerequisites may dominate, but the presence of all is necessary for hydrogen cracking to occur. Practical means to prevent this cracking, such as preheat, are designed to control one or more of these factors.

In the past, two different approaches have been taken for predicting preheat. On the basis of a large number of fillet weld controlled thermal severity (CTS) tests, a method based on critical HAZ hardness has been proposed (References 1 and 2). By controlling the weld cooling rate so that the hardness of HAZs does not exceed the critical level, the risk of hydrogen cracking could be removed.

The acceptable critical hardness can be a function of the hydrogen content. This approach does not recognize the effect of preheat on the removal of hydrogen from the weld during cooling; although recommended in the guide for predicting a minimum energy input for welding without preheat, it tends to be overly conservative when predicting preheat levels.

The second method for predicting preheat is based on the control of hydrogen. Recognizing the effect of the low temperature cooling rate, i.e., cooling rate between 572°F and 212°F [300°C and 100°C], empirical relationships between the critical cooling rate, the chemical composition, and hydrogen content have been determined using high restraint groove weld tests (Reference 3).

More generalized models have been proposed by other researchers (References 4, 5, and 6) using simple hydrogen diffusion models. Hydrogen content is usually included as a logarithmic term. The advantage of this approach is that the composition of the steel and the hydrogen content of the weld can be grouped together in one parameter, which may be considered to represent the susceptibility to hydrogen embrittlement. A relationship then exists between the critical cooling time and this parameter, for a given restraint level. It is possible to index the lines for various restraint levels by reference to large scale tests or experience, and for other types of fillet welds (Reference 7). In developing the method, relations between the specific preheat and the cooling time should be assumed.

It is important to recognize that the preheats predicted from these models depend upon the type of test used to provide the experimental data. The condition usually examined in these tests is that of a single root pass in a butt joint. This is considered the most critical and is used to determine the preheat; but there are situations where it is possible to weld the second pass before the first pass cools down (stove pipe welding for girth welds in pipes), and with these special procedures, the weld can be made with lower preheats that would be predicted. However, for general application, it is considered that the preheat is properly determined by that required to make the root pass. For this reason, energy input does not enter explicitly into this hydrogen control method.

Scope of Proposed Preheat Requirements. An important feature that is omitted in all of the proposed methods for predicting preheat is weld metal cracking. It is assumed that preheat is determined by HAZ cracking (and hence parent metal composition), but in some cases, particularly with modern high strength low alloy steels, the weld metal may be more susceptible. There has been insufficient research on this problem to include it in the present guidelines, and in such cases testing may be necessary.

Restraint

The major problem in determining preheats using the hydrogen control approach is in selecting a value for the restraint. In the guide three restraint levels are considered. The first represents a low restraint and is considered to be independent of thickness. The low restraint corresponds to an intensity of restraint, k, less than 1000 N/ mm/mm and this coincides with the fillet weld results. Many welds in practice would be in this category. The medium restraint is based on a value of k = 150 × plate thickness (in mm) and corresponds to a value covering most of the measured values of restraint that have been reported. The high restraint table is based on k = 400 × plate thickness (in mm) and represents a severe level of restraint. It is noticed that in the medium and higher restraint conditions, the restraint is considered to increase with plate thickness.

Restraint will directly affect the required level of preheat. The reference to it in Table 5.8 of the code is included in Footnote a under the table. There it may not fully convey the significance in preheat considerations given to it internationally.

Annex B draws the user's attention to the restraint aspect of welded joints by suggesting three generally described levels. With continuing alertness on the part of users within and outside an industry-conducted surveillance program, restraint will eventually be more precisely defined, in terms of actual detail or structural framing situations.

The fact it was impossible to define restraint more explicitly at this time was not taken as sufficiently valid ground not to address restraint, recognize its pronounced influence and provide the presently best available means to accommodate it.

NOTE: A concerted industry-sponsored surveillance program designed for an efficient and rapid exchange of experience so as to allow eventual classification and listing of specific structural details and situations under the three restraint levels, merits full consideration.

Restraint data collected from fabrication and engineering practice could provide grounds for more realistic evaluation of restraint and more reliable determination of preheats following the recommendations of these guidelines.

The present requirements for welding procedure qualification in structural work, except for some cases of tubular construction, rely on standard test assemblies to "prove" the adequacy of preheat for the same joints as parts of production assemblies. One should be aware that under these circumstances "restraint" is not being considered in the qualification. A shift towards qualification using "joint simulated test assemblies" would result in a much more reliable indication of performance under service conditions and additionally allow collection of reliable restraint data.

Relation Between Energy Input and Fillet Leg Size

Although the heat input to the plate is of prime consideration in regard to cooling rate and potential HAZ hardness, it is often more practical to specify weld size. The relation between energy input and fillet weld size (i.e., leg length) is not unique but depends on process, polarity, and other factors. Some workers have suggested that relationships exist between cooling rate and the total cross-sectional area of fused metal. The latter, however, is difficult to measure and would not be a suitable way of specifying weld sizes in practice.

The weld dimensions and welding conditions have been measured in fillet weld tests and these data used to make plots of leg lengths squared versus energy input. Another source is information derived from the deposition rate data where it has been assumed that all of the metal deposited went into forming an ideal fillet. Where a root opening was present, the leg length was smaller for the same energy input than for the condition of perfect fit up. The results of these plots are shown in Figure B.4.

For manual covered electrodes with large quantities of iron powder in the covering, a larger fillet size for the same energy is produced. For SAW, electrode polarity and electrode extensions have a marked effect, as would be expected. For the normal practical range of welding conditions, a single scatter band can be considered, and a lower bound curve selected as a basis for welding procedure design.

Application

It should be clear that the proposed methods presuppose a good engineering understanding of the concepts involved as well as sound appreciation of the influence of the basic factors and their interplay built into the preheat methodology.

Engineering judgment should be exercised in the selection of the applicable hardness curve and a realistic evaluation of the restraint level should be part of the judgment.

The method of measuring effective preheat remains an independent matter and requires separate and continuous attention.

The effectiveness of preheat in preventing cracking will depend significantly on the area preheated and the method used.

Since the objective is to retard the cooling rate to allow the escape of hydrogen, a larger preheated area will stay hot longer and be more effective.

There appears no need to change the reference in Note a under Table 5.8 to preheating within a 3 in [75 mm] radius from the point of welding, as other work has confirmed the validity of this requirement.

The methods of preheating (equipment, gases) should be the subject of another investigation with major input from fabricators with the objective to report on their economy and effectiveness.

References for C-Annex B

1. British Standards 5135–1974.

2. Coe, F. R. *Welding steels without hydrogen cracking.* Welding Institute of Canada, 1973.

3. Graville, B. A. "Determining requirements for preheat." *Technology Focus.* Welding Institute of Canada, October 1980

4. Ito, F. and Bessyo, K. *A prediction of welding procedures to avoid HAZ cracking.* IIW document No. IX–631–69.

5. McParlan, M. and Graville, B. A. *Welding Journal* 55(4): p. 92-s. Res. Suppl., April 1976.

6. Suzuki, M. *Cold cracking and its prevention in steel welding.* IIW document No. IXC–107478.

7. Tersaki et al., Trans. JWS 10(1), April 1979.

C-Annex H
Phased Array Ultrasonic Testing (PAUT)

C-H8.2.2 Horizontal Sweep. The screen range can be verified using the A- and S-Scan displays in combination with the generated scan plan to verify if adequate coverage is obtained. A general rule of thumb for range is setting at 3 times the part thickness at the minimum angle or 2 times the part thickness at the maximum angle but these general rules may produce undesirable results on thicker materials. PAUT personnel may choose to optimize the range setting as to not compress the area of interest and reduce signal resolution but in no case should the range setting exclude a portion of the area of interest as designed in the scan plan.

This page is intentionally blank.

List of AWS Documents on Structural Welding

Designation	Title
D1.1/D1.1M	*Structural Welding Code—Steel*
D1.2/D1.2M	*Structural Welding Code—Aluminum*
D1.3/D1.3M	*Structural Welding Code—Sheet Steel*
D1.4/D1.4M	*Structural Welding Code—Reinforcing Steel*
D1.5M/D1.5	*Bridge Welding Code*
D1.6/D1.6M	*Structural Welding Code—Stainless Steel*
D1.7/D1.7M	*Guide for Strengthening and Repairing Existing Structures*
D1.8/D1.8M	*Structural Welding Code—Seismic Supplement*
D1.9/D1.9M	*Structural Welding Code—Titanium*

Index

A

AASHTO/AWS D1.5 specifications, 1.4, C–14.4.4
acceptance criteria
  bend tests, 6.10.3.3, 9.8.3, C–6.10.3.3
  CVS toughness testing, 6.27.7, Table 6.15
  cyclically loaded nontubular connections, 8.12.2, 8.13.2, C–8.13.2, Figs. 8.2 and 8.3, Tables 8.2 and 8.3
  destructive testing, 6.24.2
  discontinuities, 7.14.5.1, Fig. 7.1, Table 7.4
  engineer's approval of alternate criteria, 8.8, C–8.8, Tables C–8.1 and C–10.15
  FCAW-S testing, 6.28.6
  fillet weld break test, 6.23.4.1, Figs. 6.25, 6.27
  inspections, 8.7
  macroetch test, 6.10.4.2, 6.23.2.2, 10.21.1.1, Fig. 6.25
  nondestructive testing, 8.11, C–8.11
  phased array ultrasonic testing, Table H.2
  radiographic testing (RT), 8.12, C–8.12
  reduced-section tension test, 6.10.3.5
  reinforcement testing (RT), 6.10.2.2, 6.23.2.2
  RT or UT, 6.10.2.2
  statically loaded nontubular connections, 8.13.1, C–8.13.1, Table 8.2
  stud welding, 9.4.7, 9.8.3, 9.9.9, C–9.4.7
  tack welders qualification, 6.24, Fig. 6.23
  tubular connections, ultrasonic testing, 10.26.1, C–10.26.1
  ultrasonic testing, 8.13, Annex O12, C–8.13
  visual acceptance criteria, 6.24.1
  visual inspection, 8.9, C–8.9, Table 8.1, Table 10.15
welders and welding operator qualification, 6.23, 10.21
WPS qualification, 6.7, 6.10, 10.13, C–10.13, Figs. 6.5 to 6.10, 6.14, 10.14 and 10.15, Tables 10.9 to 10.11, C–6.10, C–10.9
aging of test specimens, performance qualification, 6.3.2
AISC Design Guide, 10.2.1, 10.5.6, Annex R, C–4.7.7, C–4.12.2.2, C–10.7.2.1
alignment
  butt joints, 7.21.3, C–7.21.3, Fig. C–7.3 and C–7.4
  methods, 7.21.6
allowable stresses
  base metal, 4.6.3
  cyclically loaded structures, 4.16, C–4.16, Figure 4.16, Table 4.3
  fillet welds, 4.6.4.2, C–4.6.4.2, Fig. C–4.2
  increase, 4.6.5
  ranges, C–4.16.2, C–11.4.2
  weld metal, 4.6.4, C–4.6.4, Fig. C–4.2, Table 4.3
  Z-loss allowance, 4.4.2.6, 4.4.3.3, Table 4.2
all-weld-metal tension test, 6.10.3.6, 6.15.1.3, Fig. 6.14
American Society for Nondestructive Testing requirements, 8.14.6.1
amplitude calibration, ultrasonic testing, 8.25.6.4, 8.27.1.2, 8.27.2.4, 10.29.4, Annex O, C–8.25.6.4
angle-beam search units
  calibration, 8.23.4, 8.24.5, 8.27.2.1, 8.27.2.2, C–8.23.2
  discontinuity size evaluation, 8.29.2
  qualification, 8.23.4, 8.27.2.1, 8.27.2.2, C–8.23.4
  shear wave mode, 8.27.2.2, 8.29.2
  ultrasonic testing, 8.21.7, 10.29.6, Table 8.7
angle members, cyclically loaded structures, 4.15.4
ANSI/AISC 360, 10.2.1, 10.2.2
ANSI Z49.1, 1.3
architecture, web flatness and, 7.22.6.5
arc shields, stud welding, 9.2.2, 9.4.6, C–9.4.6
arc strikes, 7.28, C–7.28
artifacts, radiographic testing, 8.17.10.1, Figs. 8.4 to 8.9, Table 8.4
ASME B46.1, 7.23.3.2
ASME *Boiler and Pressure Vessel Code*, C–8.16.1, C–10.25
ASNT (American Society for Nondestructive Testing), 8.1.4.5, 8.14.6.1
  SNT-TC-1A, 8.35.1, O3
assistant inspector, 8.1.4.4
ASTM A6, C–7.2, C–7.22.8, C–11.1
ASTM A20, C–7.2
ASTM A29/A29M, 9.2.6, 9.9.3
ASTM A36, 6.28.2, 9.6.4.1, 9.9.5.1
ASTM A109, 7.2.2.2
ASTM A325, C–4.7.7
ASTM A370, 6.10.3.6, 6.26.2, 9.3.2
ASTM A435, 7.14.5.1
ASTM A490, C–4.7.7
ASTM A500, C–Fig. 5.2
ASTM A514, 7.3.2.5, 7.8.3, 8.11, C–7.2, C–7.3.2.5, C–8.11, H9.3.5
ASTM A517, 7.3.2.5, 7.8.3, 8.11, C–7.2, C–7.3.2.5, C–8.11, H9.3.5
ASTM A572, 6.28.2
ASTM A588/A588M, 7.4.7, C–7.2
ASTM A618, C–7.2
ASTM A653, 9.9.5.1
ASTM A673, 10.7.2.1, 10.7.2.2, C–10.7.2
ASTM A709, 7.8.3, 8.11, C–8.11, H9.3.5
ASTM A710, C–5.9
ASTM A913, 6.28.2, C–5.9
ASTM E23, 6.26.2
ASTM E92, 5.7.3
ASTM E94, 8.16.1, 8.17.4
ASTM E140, 5.7.3
ASTM E164, 8.22.1, H5.7
ASTM E165, 8.14.5
ASTM E317, C–8.23.1, H14.1.2
ASTM E494, H14.4
ASTM E709, 8.14.4
ASTM E1032, 8.16.1
ASTM E1254, 8.18.4.1

ASTM E1936, C–8.17.11.2
ASTM E2033, 8.16.1.1
ASTM E2339, 8.16.1
ASTM E2445, 8.16.1.1
ASTM E2698, 8.16.1.1, C–8.17.11.2
ASTM E2699, 8.18.4.2
atmospheric time periods, 7.3.2.2, 7.3.2.3, Annex D, Figs. D.1 and D.2, Table 7.1
attachments
 auxiliary, engineer's approval of, 5.3.1, Tables 5.3 and 5.4
 backing, 7.9.1
attenuation factor, ultrasonic testing, 8.25.6.4, 8.27.1.5, C–8.25.6.4
automatic mechanized welding, 9.5.1, C–9.5.1
AWS A2:4:2012, 1.8
AWS A5:32M/A5.32, 5.6.3, 7.3.1.3
AWS A5:36/A5.36M, 5.6.3
AWS A5.1/A5.1M, 7.3.2, 7.3.5, B6.2.2
AWS A5.5/A5.5M, 7.3.2, 7.3.5, B6.2.2
AWS A5.18/A5.18M, 5.6.3
AWS A5.25, 6.15.1.3
AWS A5.26, 6.15.1.3
axial stress and bending, 4.15.2
azimuthal scan, phased array ultrasonic testing, Annex H3.1

B

backgouging
 butt joints, 10.14.1, Fig.10.18
 T-, Y- or K-connections, 10.14.3, Fig.10.18
backing
 attachment, 7.9.1
 base metal, 7.2.2.2
 CJP butt joints, 4.17.2.3, 10.14.1, Fig.10.18
 CJP groove welds, 5.4.1.2, 5.4.1.3, Fig. 5.1
 CJP T and corner joints, 4.17.2.2
 contouring weld, corner and T-joints, 4.17.3
 fabrication, 7.9, 10.22, C–7.9, C–10.22
 full-length, 7.9.1.2, 10.22.1, C–7.9.1.2, C–10.22.1
 fusion, 7.9.1.1
 longitudinal groove welds and corner joints, 4.17.2.4
 non-steel, 7.9.3
 prohibited joints and welds, 4.18.4
 radiographic testing, 8.17.3.2
 steel attachment, 4.17.2.1, 8.25.12, C–8.25.12, Table 4.5
 T-, Y- or K-connections, 10.14.3, 10.14.4, C–10.14.4, Figs.10.17 and 10.18

thickness, 7.9.1.3, C–7.9.1.3
backscatter, radiographic testing, 8.17.8.3, C–8.17.8.3
baking electrodes, 7.3.2.4
bandpass filtering, phased array ultrasonic testing, Annex H3.2
base metal
 allowable stresses, 4.6.3
 cracks, 7.25.1.4
 fabrication, 7.2, C–7.2, Table 5.3, Table 6.9
 insufficient thickness, 11.5.4, C–11.5.4
 mislocated holes, welded restoration, 7.25.5, C–7.25.5
 notch toughness, tubular connections, 10.7.2, C–10.7.2, Table C–10.1 to C.10.4
 preparation, 7.14, C–7.14
 prequalified WPSs, 5.3, Tables 5.3 and 5.4
 qualification, 6.8.3, C–6.8, Table 5.3, Tables 6.8 and 6.9
 repair and strengthening, 11.2, C–11.2, Table C–11.1
 straight-beam testing, 8.24.4
 stud welding, 9.4.3, 9.5.4
 thickness combination, 5.7.2, Table 5.8
 through-thickness loading, 4.7.3, C–4.7.3
 ultrasonic testing, 8.25, 10.29.4, C–8.25
 workmanship, 11.5.1
base plates, connections and splices, 4.7.2
beams
 camber, 7.22.3, 7.22.4, C–7.22.4, Fig. C–7.6, Tables 7.5 and 7.6
 depth variation, 7.22.9
 straightness, 7.22.2, C–7.22.2
bearing
 loading points, 7.22.10, C–7.22.10, Fig. C–7.8
 stiffeners, 7.22.11.3, C–11.3
bend testing, 6.10.3.1, 6.10.3.2, 6.10.3.3, C–6.10.3.2, Figs. 6.6 to 6.13
 acceptance criteria, 6.10.3.3, 9.8.3, C–6.10.3.3
 RT as substitution for, 6.17.1.1
 stud welding, 9.6.7.1, 9.7.1.4, 9.8.3, 9.9.7.2, C–9.7.1.4, Figs. 9.4 and 9.5
bevel groove welds, 4.4.2.7, C–Fig. 5.2, Figs. 4.2 to 4.6
bidders, information furnished to, 8.1.1, C–8.1.1
bolts, 4.7.7, C–4.7.7
break test, fillet welds, 6.23.4, Figs. 6.25, 6.27
built-up members, 4.12, C–4.12
 camber in, 7.18.1, C–7.18

splices, 7.19.2
weld access holes, 7.16.1.2, Fig. 7.2
butt joints
 alignment, 7.21.3, C–7.21.3, Fig. C–7.3 and C.7.4
 circumferential groove welds, 10.28.1
 CJP welds, backing, 4.17.2.3, 10.14.1, C–10.14, Fig.10.18
 prequalified WPSs, 10.10.1, C–10.10.1
 radiographic testing (RT), 8.16, C–8.16
 single-side backing, 10.14.2, Fig.10.18
 steel backing, C–8.25.12, Fig. C–8.3 and C–8.4
 surface contouring, 4.7.5, C–4.7.5
 T-, Y- or K-connections, 10.15, Fig. 10.16 and 10.17, Tables 10.10 and 10.11
 testing, 8.25.6.2, Fig. 8.15
 thickness transitions, 4.17.1.1, Fig. 4.17
 transversely loaded, 4.17.5
 weld tabs, 7.30.4
 width transitions, 4.17.1.2, Fig. 4.18

C

calibration
 angle-beam testing, 8.24.5
 inspection testing, 8.24, 8.27
 phased array ultrasonic testing, Annex H8
 for testing, 8.24, C–8.24
 ultrasonic testing, 8.21.4, 8.24, 8.27, 10.29.3
camber
 beams and girders, 7.22.3, 7.22.4, C–7.22.4, Fig. C–7.6, Tables 7.5 and 7.6
 built-up sections, 7.18.1, C–7.18
caulking, 7.27, C–7.27
Centerline cracking, C–5.8.2.1, Fig. C–5.1
center of rotation, instantaneous, 4.6.4.3, C–4.6.4.3
certification, nondestructive testing personnel, 8.14.6.2
channels, phased array ultrasonic testing, Annex H3.3
Charpy impact testing, Annex M, Table M.3
circular-to-circular tubular connections
 effective lengths, 10.5.5, C–10.5.5
 preliminary design, Annex R
 stress categories, 10.2.3.2, C–10.2.3.2, Figs. C–10.1 and C–10.3, Table 10.3
circumferential groove welds, RT requirements, 10.28.1

cleaning
    completed welds, 7.29.2
    in-process, 7.29.1
cleanliness
    stud welding, 9.4.1
    ultrasonic testing, 8.25.3
coating, stud welding, 9.4.2
Code approved processes, prequalified WPSs, 5.5.2
code interpretations, Engineer's responsibilities, C–1.5.1(1)
columns, straightness, 7.22.1
Complete joint penetration (CJP) groove welds
    allowable stress ranges, 4.16.2, C–4.16.2
    butt joints, backing, 4.17.2.3
    effective size, 4.4.1.2, 10.5.3.2, Table 10.7
    full-length backing, 10.22.1, C–10.22.1
    partial length prohibition, 4.8.2
    prequalified WPSs, 5.4.1, 10.10, C–10.10, Figs. 5.1 and 5.5
    prohibited joints and welds, 4.18.4
    qualification requirements, 6.11, 6.21, Table 6.2(1)
    T-, Y- or K-connections, 10.10.2, 10.14.4.2, 10.14.4.3, Figs. 10.7 to 10.11, 10.19 and 10.20, Table 10.7
    T and corner joints, backing, 4.17.2.2
    tubular connections, 10.10, 10.14, 10.18, C–10.10, C–10.14, C–10.18, Figs. 10.17, 10.20 to 10.22, Table 10.13
    visual inspection, 6.10.1.1
    WPS qualification, 10.13, C–10.13, Tables 10.9, C–10.9
compression members
    connections and splices, 4.7.2
    cyclically loaded structures, 8.12.2.2, Fig. 8.3
    intermittent weld, maximum spacing, 4.12.2.2, C–4.12.2.2
concavity, repairs, 7.25.1.2
concentrically loaded weld groups, 4.6.4.4, C–4.6.4.4, Table 4.4
construction aid welding, 7.17
consumables
    fabrication, 7.3, C–7.3
    verification test, 6.13.3, Fig. 6.18
contract documents
    inspection stipulations, 8.1.2, C–8.1.2
    plans and specifications, 4.3
contractor
    inspection obligations, 1.5.3.1, 8.1.2.1, 8.6
    nondestructive testing responsibilities, 8.6.4, 8.6.5
    qualification responsibilities, 6.2.1.1, 6.2.2.2, C–6.2.1.1
    radiographic testing equipment provision, 8.18.1, C–8.18.1
    repairs, 7.25.1, C–7.25.1
    responsibilities of, 1.5.2, 8.6.1, C–1.5.2, C–8.6.1
    shrinkage or distortion responsibilities, 7.20.3
    stud welding requirements, 9.3.3
    testing responsibilities, 9.6.3
    WPS preparation, 6.7
convexity, 7.25.1.1
cooling rates, Annex B.3.3, B6.1.3, Figs. B.2, B.3
cope holes, 7.16.3, C–7.16.3, Fig C–7.2
corner joints
    backing, 4.17.2.2, 4.17.2.4, 4.17.3
    preparation, 5.4.1.6, 5.4.2.5, C–5.4.1.6, C–5.4.2.5, C–Figs. 5.1 and 5.2
    prequalified WPSs, Fig. 5.5
    prohibited joints and welds, 4.18.4
    qualification requirements, 6.11.1.1
    steel backing, C–8.25.12, Fig. C–8.1 and C–8.2
    surface contouring, 4.7.5, C–4.7.5
corner reflectors, prohibition of, 8.22.2, C–8.22.2
corrosion protection, cyclically loaded structures, 4.14.3
couplants
    phased array ultrasonic testing, Annex H5.6
    ultrasonic testing, 8.25.3, C–8.25.4
cracks, weld or base metal, 7.25.1.4
cratering, repairs, 7.25.1.2
crushed slag, fabrication, 7.3.3.4, C–7.3.3.4
cumulative damage, cyclically loaded tubular structures, 10.2.3.4
cutting equipment, 7.10
    thermal cutting process, 7.14.8
CVN toughness testing
    acceptance criteria, 6.27.7, Table 6.15
    FCAW-S testing, 6.28.5, Figs. 6.31–6.32
    general requirements, 6.26, C–6.26, Figs. 6.5 to 6.7, Tables 6.1 to 6.2, and 6.5 and 6.7
    locations, 6.27.1, Fig. 6.28
    notch location, 6.27.4, Fig. 6.28
    number of specimens, 6.27.2, C–6.27.2
    reporting requirements, 6.29
    retesting, 6.27.8
    specimen size, 6.27.3, Table 6.14
    sub-size specimens, 6.27.6, Table 6.15
    test temperature, 6.27.5, Table 6.14
    tubular connections, 10.7.2.1, 10.14.4.4, C–10.7.2.1, C–10.14.4.4
cyclically loaded structures
    allowable stresses and stress ranges, 4.16.2, C–4.16.2, C–11.4.2
    backing, 7.9.1.4, C–7.9.1.4, Table 4.5
    code provisions, nontubular structures, C–1.5.1(6)
    in compression, 8.12.2.2, Fig. 8.3
    detailing, fabrication and erection, 4.17
    discontinuities, acceptance criteria, 8.12.2, C–8.12.2.1, Fig. C–8.9, Figs. 8.2 and 8.3
    general requirements, 4.13, C–4.13
    girder web flatness, Annex F
    inspection, 4.19
    limitations, 4.14, C–14
    material trimming, 7.14.7
    prohibited joints and welds, 4.18
    splices, 7.19.1.2
    stress calculations, 4.15, 10.2.3.1, 10.2.3.2, 10.2.3.3, C–10.2.3.2, C–10.2.3.3, Figs. 10.1, C–10.1, C–10.3, Tables 10.1 and 10.3
    in tension, 8.12.2.1, Fig. 8.2
    tubular connections, 10.2.3, C–10.2.3
    ultrasonic testing, acceptance criteria, 8.13.2, C–8.13.2, Tables 8.2 and 8.3
    web flatness, 7.22.6.3
    weld tabs, 7.30.3

D

dB accuracy
    Annex P nomograph, 8.28.2.4
    decibel equation, 8.28.2.2
    inspection, 8.31
    qualification procedures, 8.28.2
dead elements, phased array ultrasonic testing, Annex H3.4
decibel equation, 8.28.2.2
definitions, 3
depth variation, beams and girders, 7.22.9
design requirements
    conformance, 7.12, C–7.12, Table 8.1, Table 10.15
    strengthening and repair, 11.3, C–11.3
    tubular connections, 10.2, 10.5, Table 10.1
    welded connections
        contract plans and specifications, 4.3, C–4.3
        scope, 4.1
destructive testing, acceptance criteria, 6.24.2, Fig. 6.23
detailing, cyclically loaded structures, 4.17, C–4.17

digital image sensitivity range, 8.17.11.2, C–8.17.11.2, C–11.2
digital radiography
  equipment, 8.18.1.2
  radiographic testing (RT), 8.16.1.1, C–8.16.1.1
dihedral angle
  prequalified WPSs limitations, 5.4.3.3, Annex Q, Fig. 5.4
  steel backing, C–8.25.12
dimensional tolerance
  groove dimensions, 10.23.2
  joint configuration, 7.21, C–5–4.1.1, C–7.21
  stiffeners, 7.22.12
  stud welding, 9.2.1, Fig. 9.1
  welded structural members, 7.22, C–7.22
discontinuities
  acceptance criteria, 7.14.5.1, 8.12.1, 8.12.2, 8.25.8, C–8.12.1, C–8.25.8, Fig. 7.1, Fig. C–8.7 and C–8.8, Figs. 8.1 to 8.3, Table 7.4, Tables 8.2 and 8.3
  cyclically loaded structures, 8.12.2, C–8.12.2.1, Fig. C–8.9, Figs. 8.2 and 8.3
  length measurement, 8.25.7, C–8.25.7
  longitudinal scanning patterns, 8.30.1
  members, 11.5.2, C–11.5.2
  mill-induced, 7.14.5, Table 7.4
  phased array ultrasonic testing, Table H.3
  repairs, 7.4.6, 7.14.5.2, C–7.14.5.2, Table 7.4
  size evaluation, 8.29
  statically loaded nontubular connections, 8.12.1, C–8.12.1, Fig. 8.1, Figs. C–8.7 and C–8.8
  tension cyclically loaded structures, 8.12.2.1, Fig. 8.2
  testing procedures, 8.25, C–8.25
  transverse discontinuities, 8.30.2
  ultrasonic testing, 10.29.7, Annex O8, O9, O10, O11
display range, ultrasonic testing, 8.21.5
distance calibration, ultrasonic testing, 8.21.7.6, 8.27.1.1, 8.27.2.3, 10.29.3.1
distortion
  control, 7.20, C–7.20
  excessive, 7.22.6.4, C–7.22.6.4
double-sided groove preparation, 5.4.1.4
drawings
  contract documents, 4.3.1
  filler plates, 4.11.3
  inspector's access to, 8.1.6, C–8.1.6
  shop drawings, 4.3.5

E

earthquake loading, C–14.4.2
eccentricity, calculated stresses, 4.6.2, C–4.6.2, Fig. C–4.3
edge blocks, radiographic testing, 8.17.13, C–8.17.13, Fig. 8.10
edge discontinuities, 7.14.5.1, Fig. 7.1, Table 7.4
edge distance, ultrasonic testing, 8.21.7.6
effective areas, 4.4. See also specific welds, e.g., Groove welds
  effective length, 4.4.2.2, 4.4.2.5, 10.5.1, 10.5.5, 10.5.6, C–4.4.2.5, C–10.5.1, C–10.5.5
  effective throat, 4.4.2.6, 4.4.3.7, Annex A, Fig. 4.1
  minimum length, 4.4.2.3
elastic analysis, 4.15.1, C–4.15.1
electrodes
  atmospheric time periods, 7.3.2.2, Table 7.1
  baking, 7.3.2.4
  carbon steel, Annex M, Table M.1
  certification, 7.3.1.1
  classification groups, 7.3.1.2, Table 6.13
  composition requirements, Annex M, Table M.6
  condition, 7.3.1.5
  crushed slag, 7.3.3.4, C–7.3.3.4
  EGW, 7.4.2
  ESW, 7.4.2
  FCAW, 7.3.4, Annex M, C–Table 5.1, Table M.9
  flux combinations, 7.3.1.1, 7.3.3.1, C–7.3.3.1
  flux conditions, 7.3.3.2, C–7.3.3.2
  flux reclamation, 7.3.3.3, C–7.3.3.3
  GMAW, 7.3.4, Annex M, C–Table 5.1, Table M.9
  GTAW, 7.3.5
  low-hydrogen storage conditions, 7.3.2.1, 7.3.2.5, Annex D, C–7.3.2.1, C–7.3.2.5, Figs. D.1 and D.2
  metal cored, Annex M, Table M.9
  SAW, 5.5.5, 7.3.3
  shielding gases, 5.6.3, 7.3.1.3, C–7.3.1.3
  SMAW, 7.3.2, C–7.3.2
  storage, 7.3.1.4, 7.3.2.1, C–7.3.2.1
  usability characteristic, Annex M, Table M.4
Electrogas welding (EGW)
  electrodes and guide tubes, 7.4.2
  processes, 7.4, C–7.4
  qualification, 6.8.2, 6.15.1, 6.15.1.3, C–6.15.1.3, Table 6.6
  qualification test plates, 6.10, 6.21.2.2, Figs. 6.5, 6.24
  scanning patterns, 8.30.3
Electroslag welding (ESW)
  electrodes and guide tubes, 7.4.2
  processes, 7.4, C–7.4
  qualification, 6.8.2, 6.15.1, 6.15.1.3, C–6.15.1.3, Table 6.6
  scanning patterns, 8.30.3
  test plates, 6.10, 6.21.2.2, Figs. 6.5, 6.24
end returns, maximum length, 4.9.3.3, C–4.9.3.3, Fig. 4.12
Engineer's responsibilities
  acceptance criteria, 8.8, C–8.8, Tables C–8.1 and C–10.15
  auxiliary attachments approval, 5.3.1, Tables 5.3 and 5.4
  cyclically loaded structures, 4.13.3
  groove dimensions, 7.21.4.3
  inspection personnel, 8.1.4.1, C–8.1.4.1
  inspection responsibilities, 8.6.3
  repairs, 7.25.3
  Structural Welding Code requirements, 1.5.1, C.1.5.1
  stud welding, 9.3.3
equipment
  inspection of, 8.2
  peening, 7.26.1, C–7.26
  phased array ultrasonic testing, Annex H5.6
  qualification, 8.23, 8.28, C–8.23
  radiographic testing, 8.18.1, C–8.18.1
  stud welding, 9.9.5.2
  ultrasonic testing, 8.21, Annex O, C–8.21, Table 8.8
  welding and cutting, 7.10
erection, cyclically loaded structures, 4.17
excessive convexity, 7.25.1.1
existing structures
  base metal repair, 11.2, C–11.2
  design for strengthening and repair, 11.3, C–11.3
  fatigue life enhancement, 11.4, C–11.4
  quality controls, 11.6, C–11.6
  strengthening and repair requirements, 11.1, C–11.1
  workmanship and technique, 11.5, C–11.5
exposures, tubular connections
  double-wall, 10.28.1.2, 10.28.1.3, Figs.10.27 to 10.29
  single-wall, 10.28.1.1, Fig. 10.26
eye examination, inspectors, 8.1.4.6

F

fabrication
    arc strikes, 7.28, C–7.28
    backing, 7.9, 10.22, C–7.9, C–10.22
    base metal, 7.2, 7.14, C–7.14
    camber, 7.18, C–7.18
    caulking, 7.27, C–7.27
    construction aid welds, 7.17
    consumables and electrode requirements, 7.3, C–7.3
    copes, heavy shapes, 7.16.3, C–7.16.3, Fig. C–7.2
    cyclically loaded structures, 4.17, C–4.17
    design conformance, 7.12, C–7.12
    dimensional tolerance, 7.22, C–7.22
    distortion and shrinkage control, 7.20, C–7.20
    environmental conditions, 7.11, C–7.11
    equipment, 7.10
    ESW/EGW processes, 7.4, C–7.4
    fillet weld sizes, 7.13, C–7.13, Table 7.7
    galvanized shapes, 7.16.2, C–7.16.2
    joint tolerance, 7.21, C–7.21
    peening, 7.26, C–7.26
    plug welds, 7.24
    preheat and interpass temperatures, 7.6
    quenched and tempered steels, 7.7
    reentrant corners, 7.15, C–7.15, Fig. C–7.1
    repairs, 7.25, C–7.25
    scope, 7.1, C–7.1
    slot welds, 7.24
    splices, 7.19
    stress-relief heat treatment, 7.8, C–7.8
    stud welding, 9.4, 9.8, C–9.4, C–9.8
    tack welds, 7.17
    weld access holes, 7.16.1, C–7.16.1
    weld cleaning, 7.29, C–7.29
    weld profiles, 7.23, C–7.23, Fig. 7.4, Tables 7.8 and 7.9
    weld tabs, 7.30, C–7.30
    WPS variables, 7.5, C–7.5
face bends, mechanical testing, 6.10.3.1, Fig. 6.8, 6.9, Figs. 6.11 to 6.13
failure
    requalification retest, 6.25.1.4
    stud weld testing, 9.7.1.5
fatigue
    analysis, for strengthening and repair, 11.3.3, C–11.3.3
    cyclically loaded structures, 4.14.2, 4.16.2, 10.2.3.2, 10.2.3.6, C–4.14.2, C–10.2.3.2, C–10.2.3.6, Figs. 4.16, C–10.1 to C–10.3, Tables 10.1, 10.3 and 10.4
    life enhancement methods, 11.4.1, C–11.4.1, Fig. C–11.2, Table C–11.2
    low cycle fatigue, C–14.4.2
    redundant-nonredundant members, C–14.4.4
    stress design parameters, 4.14.1, 4.17.2.1, Table 4.5
faying surface, 7.21.1.1
filler metals
    AWS A5.36 classifications and properties, Annex M
    prequalified WPSs, 5.6, 5.6.2, C–5.6, Tables 5.3, 5.4 and 5.6
    strength properties, Annex L
    variables testing, 6.28.1, Table 6.9, Table 6.16
filler plates
    joint configuration, 4.11
    thick plates, 4.11.2, Fig. 4.15
    thin plates, 4.11.1, Fig. 4.14
fillet welds
    allowable stress ranges, 4.16.2, C–4.16.2, C–11.4.2
    assembly, 7.21.1, C–7.21.1
    backing, 4.17.6
    break test, 6.23.4, Figs. 6.25, 6.27
    common plane, 4.9.3.5, C–4.9.3.5, Fig. 4.13
    drawing, 4.3.5.2
    effective area, 4.4.2, 10.5.2.1
    effective throat, 4.4.2.6, Annex A, Fig. 4.1
    equivalent strength coefficients, oblique loading, Table 4.4
    holes, 4.4.4, 4.9.4
    intermittent, 4.9.5, 7.23.2
    joint configuration, 4.9, 7.21.1, C–4.9, C–7.21.1
    longitudinal, 4.9.1.2, 4.9.2, C–4.9.1.2, Fig. 4.10
    macroetch test, 6.23.2.1, Figs. 6.25 to 6.26
    maximum weld size, 4.4.2.9, Fig. 4.7
    minimum size, 4.4.2.8, 7.13, C–7.13, Tables 7.7 and 9.2
    positions, test plates, 6.3.4, C–6–3.4, Fig. 6.4
    prequalified WPSs, 5.4.3, 10.8, Figs. 5.3, 10.3, 10.5, Table 7.7
    profile, 7.23.1, Fig. 7.4, Table 8.1, Table 10.15, Tables 7.8–7.9
    prohibited joints, 4.18.3
    qualification requirements, 6.13, 6.22.2, C–6.13, Figs. 6.15, 6.22, Tables 6.4, 6.11
    reinforcement of, 4.4.2.7, Figs. 4.2 to 4.6, 5.2
    repairs, 9.5.5.9
    SAW cooling rates, Annex B.3.3, B6.1.3, Figs. B.2, B.3
    single-pass, filler metals, 5.6.2.1, Table 5.4
    size and energy input, Annex B6.1.5, C–Annex B, Fig. B.4
    slots, 4.4.4, 4.9.4
    stresses, 4.6.4.1, 4.6.4.2, 10.5.4, Table 10.2
    stud welding, 9.5.5, C–9.5.5
    surface contouring, 4.7.5, C–4.7.5
    T-, Y- or K-connections, 10.15, Fig. 10.16, Tables 10.10 and 10.11
    terminations, 4.9.3, 4.9.3.2, 4.17.6, C–4.9.3, C–4.9.3.2, C–4.17.6, Fig. 4.11
    testing methods, 6.13.2, Fig. 6.15, 10.16, Table 6.4
    transverse, 4.9.1.1, C–4.9.1.1, Fig. 4.9, Fig. C–4.7
    tubular connections, 10.8, 10.15, 10.20, Fig.10.5 and 10.6, Table 10.13
    visual inspection, 6.10.1.2, Fig. 7.4
    WPS qualification, 6.13, 10.13, C–6.13, C–10.9, C–10.13, Tables 6.4 and 10.11
film equipment, radiographic testing, 8.18.1.1
film length and width, radiographic testing, 8.17.8.1, 8.17.9
film quality, radiographic testing, 8.17.10, 8.17.11.1, C–8.17
finish methods and values, 7.23.3.2
fit-up tolerances, 5.4.1.8, 5.4.2.7, Figs. 5.1 and 5.2
flame cut edges, 4.17.4
flange
    centerlines, 7.22.7
    warpage and tilt, 7.22.8, C–7.22.8, Fig. C–7.7
flare-groove welds
    effective size, 4.4.1.4, 4.4.2.7, C–4.4.1.4, C–5.2, Figs. 4.5 to 4.6, Table 4.1
    qualification of, 6.12.4, Fig. 6.29
flat position
    groove welds, 4.18.2
    plug welds, 7.24.1.1
flush surfaces, groove welds, 7.23.3.1
flux
    conditions, 7.3.3.2, 7.4.3, C–7.3.3.2
    electrode combinations, 7.3.1.1
    reclamation, 7.3.3.3, C–7.3.3.3
    stud welding, 9.2.3
Flux Cored Arc Welding (FCAW)
    CJP groove joints, 10.10.2.1
    electrodes, 7.3.4
    power sources, 5.5.4, C–5.5.4
    prequalified WPSs, 5.5.1, C–5.5.1

qualification, 6.8, C–6.8, Table 6.5, Table 6.7
shielding gases, 5.6.3, Table 5.7
SMAW joints, 5.4.1.5, 5.4.2.4
Flux Cored Arc Welding-G (FCAW-G), shielding gases, 5.6.3, Table 5.7
Flux Cored Arc Welding-S (FCAW-S)
CVN toughness testing, 6.26.1.2, C–6.26.1.2
testing procedures, 6.28, Table 6.16
focal law, phased array ultrasonic testing, Annex H3.10
foreign materials, 7.14.4, C–7.14.4
forms preparation
qualification procedures, 6.19
sample forms, Annex J
ultrasonic testing, 10.29.8.1
UT equipment qualification and inspection, Annex P
Fracture Critical discontinuities, 8.25.8, C–8.25.8, Tables 8.2 and 8.3
frequency, ultrasonic testing transducer, 8.21.7.1
full screen height (FSH), phased array ultrasonic testing, Annex H3.10
fusion
backing, 7.9.1.1
incomplete, 7.25.1.3

G

gain control, UT instrument qualification, 8.21.4, 8.23.2, 8.27.1.5, C–8.23.2
galvanized shapes, 7.16.2, C–7.16.2
Gas Metal Arc Welding (GMAW)
electrodes, 7.3.4, C–Table 5.1
power sources, 5.5.4, C–5.5.4
prequalified WPSs, 5.5.1, C–5.5.1, Table C–5.1
qualification, 6.8, 6.15.1, C–6.8, Table 6.5, Table 6.7
root pass, 5.5.5
shielding gases, 5.6.3, Table 5.7
SMAW joints, 5.4.1.5, 5.4.2.4
Gas Metal Arc Welding-S (GMAW-S)
CJP groove joints, 10.10.2.1
metal transfer, C–5.5.1, Fig. C–5.3
Gas Tungsten Arc Welding (GTAW), qualification, 6.8, 6.15.1, 6.15.1.2, C–6.8, Table 6.5, Table 6.7
geometric unsharpness, radiographic testing, 8.17.5.1, C–8.17.5.1
girders
camber, 7.22.3, 7.22.4, C–7.22.4, Fig. C–7.6, Tables 7.5 and 7.6
depth variation, 7.22.9
straightness, 7.22.2, C–7.22.2
web flatness, Annexes E, F

girth weld alignment, 10.23.1
gouges
grooves, 7.21.5, Figs. 5.1–5.2
limitations, 7.14.8.4
groove welds. *See also* CJP groove welds, flare-groove welds, PJP groove welds
dimensions, 7.21.4, 10.23.2, Table 10.14
double-sided groove preparation, 5.4.1.4
effective areas, 4.4, 10.5.3
flat position, 4.18.2
gouged grooves, 7.21.5
joint configuration, 4.8, 5.4.1, C–4.8, Fig. 4.8, Fig. 5.1
longitudinal, 4.17.2.4
one-sided, 4.18.1
profiles, 7.23.3, Fig. 7.4, Tables 7.8 and 7.9
radiographic testing (RT), 8.16, C–8.16
single-pass, filler metals, 5.6.2.1, Table 5.4
ultrasonic testing (UT), 8.19, C–8.19
unreinforced bevel groove weld, 4.4.2.7, Fig. 4.2
visual inspection, 6.10.1.1, Fig. 7.4
ground welds, discontinuities, 8.30.2.1
guided bend test jig, 6.10.3.1, Fig. 6.11 to 6.13

H

hardness requirements, preheat and interpass temperatures, 5.7.4, Annex B
H&D density, radiographic testing, 8.17.11.1.1, C–8.17.11.1.1
Heat-affected zone (HAZ)
CVN toughness, C–10.14.4.4
hardness requirements, 5.7.4, Annex B
test values, Fig. C–10.3, Table C–10.7
heat input
calculation methods, 6.8.5.1, C–6.8.5
fillet leg size, C–Annex B
maximum heat input, multi-position WPSs, 6.8.5.2
quenched and tempered steels, 7.7, C–7.7
heat treatment
CVN testing, 6.8.5, C–6.8.5
heat repair temperature limitations, 7.25.2, C–7.25.2
postweld, 5.9, 7.8.2, 7.8.3, C–5.9, Table 7.2
repair and strengthening, 11.5.5, C–11.5.5

stress-relief requirements, 7.8, Table 7.2
high current testing, stud welds, 9.9.6.1
holes
fillet welds, 4.4.4, 4.9.4
IQI requirements, 8.17.1, 10.27.1, Figs. 8.4 to 8.9, Tables 8.4 and 10.16
mislocated holes, welded restoration, 7.25.5, C–7.25.5
weld access holes, 4.7.6, C–4.7.6
horizontal linearity
instrument qualification, 8.23.1, 8.27.1.4, 8.28.1, C–8.23.1
ultrasonic testing, 8.21.1, Annex G3
horizontal sweep, 8.24.5.1, Annex H, C–8.24.5.1, C–Annex H
hydrogen control, preheat requirements, Annex B, Table B.1
Hydrogen Induced Cracking (HIC), C–7.3.2.1

I

identification mark, radiographic testing, 8.17.12, C–8.17.12
image enhancement, radiation imaging systems, 8.35.3
immediate retest, 6.25.1.1
inaccessibility, ultrasonic testing, 8.25.5.2, C–8.25.5.2, Table 8.7, Table C–8.7
indication rating, ultrasonic testing, 8.25.6.5, C–8.25.6.5
information aces
for bidders, 8.1.1, C–8.1.1
nondestructive testing personnel, 8.15.4
inspection
acceptance criteria, 8.7
advanced ultrasonic systems, 8.34
calibration for testing, 8.24, 8.27, C–8.24
contract documents, 4.3.5.6, 8.1.2
contractors' responsibilities, 8.6, C–8.6
cyclically loaded structures, 4.19
dB accuracy certification, 8.31
discontinuity size evaluation, 8.29
Engineer's approval, alternate acceptance criteria, 8.8, C–8.8, Tables C–8.1 and C–10.15
equipment qualification, 8.23, 8.28, C–8.23
extent of testing, 8.15, C–8.15
image enhancement, 8.35.3
incomplete, reporting requirements, 10.29.8.3
materials and equipment, 8.2, C–8.2

nondestructive testing, 8.11, 8.14, 8.15,
    8.32, C–8.11, C–8.14, C–8.15
penetrant/particle testing, 8.10
personnel qualification, 8.35.2
prior inspection reports, 8.26.2
procedure qualification, 8.35.1
qualification requirements, 8.20
radiation imaging systems, 8.33, 8.35.4
radiographic testing (RT), 8.12, 8.16,
    8.17, 8.18, C–8.12, C–8.17,
    C–8.18
records, radiation imaging, 8.35.4
reference standards, 8.22, C–8.22
report preparation and disposition, 8.26
scanning patterns, 8.30, Fig. 8.15
scope, 8.1, C–8.1
testing procedures, 8.25, C–8.25
ultrasonic (UT) testing, 8.13, 8.19,
    8.21, 8.27, C–8.13, C–8.19,
    C–8.21
verification, 8.1.2.2
visual weld inspection, 6.10.1, 6.23.1,
    8.9, 11.6.1, C–8.9, Table 8.1,
    Table 10.15
of welder, welding operator and tack
    welder qualifications, 8.4, C–8.4
work and records, 8.5, C–8.5
of WPSs, 8.3, C–8.3
inspector
    assistant, 8.1.4.4
    categories, 8.1.3, C–8.1.3
    contractor inspection by, 1.5.3.1
    contractor's inspector, 8.1.3.1
    identification of inspections performed,
        8.5.4, C–8.5.4
    items furnished to, 8.1.6, C–8.1.6
    qualifications, 8.1.4, C–8.1.4.3
    requests from, 8.6.2
    responsibilities of, 1.5.3, 8.1.4.8,
        C–1.5.3, C–8.1.5
    verification inspection, 1.5.3.2, 8.1.3.2
instantaneous center of rotation, 4.6.4.3,
    C–4.6.4.3
instrument requirements, ultrasonic
    testing, 8.21.3
intermediate stiffeners, 7.22.11.1,
    7.22.11.2
intermittent welds
    fillet welds, 4.9.5, 7.23.2
    maximum spacing, 4.12.2
    PJP groove welds, 4.8.3
internal reflections, ultrasonic testing,
    8.21.7.5, 8.28.3
International Institute of Welding (IIW)
    reference block
    calibration of UT unit, 8.27
    ultrasonic testing, 8.21.7.7, 8.22.1,
        Annex G, C–8.22.1, Fig. 8.12,
        Fig. 8.13, Fig. C–8.13

interpass temperature requirements
    fabrication, 7.6
    prequalified WPSs, 5.7, C–5.7
    qualification, 6.8.4, Table 5.8
    restraint levels, Annex B6.2.4,
        Table B.2
IQI requirements
    digital image sensitivity, 8.17.11.2,
        C–8.17.11.2
    hole-type requirements, 8.17.1,
        10.27.1, Figs. 8.4 to 8.9,
        Table 8.4, Table 10.16
    radiation imaging systems, 8.33.2
    radiographic testing, 8.17.7, 10.27.2,
        C–8.17.7, C–10.27.2 Table 10.18
    selection and placement, 8.17.7,
        C–8.17.7, Figs. 8.4 to 8.9,
        Table 8.6
    single- and double-wall exposures,
        10.28.1.2, 10.28.1.3, Figs.10.26 to
        10.29
    wire requirements, 8.17.1, 10.27.1,
        Figs. 8.4 to 8.9, Tables 8.5 and
        10.17

J

J-groove preparation, 5.4.1.4, 5.4.1.9,
    5.4.2.8
joint configuration, 4.7, C–4.7. *See also*
    specific joints
    fillet welds, 4.9, C–4.9
    groove welds, 4.8, C–4.8, Fig. 4.8
    plug and slot welds, 4.10
    preparation, 7.14.6, C–7.14.6
    prequalified WPSs, 5.4, C–5.4
    prohibited joints and welds, 4.18
    T-, Y- or K-tubular connections,
        10.5.2.2, 10.5.3.1, 10.5.3.2,
        Table 10.5
    tolerance of dimensions, 7.21, 10.23,
        C–7.21, C–10.23
    tubular connections, 10.18.1,
        Figs. 10.21 and 10.22

L

lap joints
    configuration, 4.9, Fig. 4.9
    maximum weld size, 4.4.2.9, Fig. 4.7
    prequalified WPSs, 10.8.1, Figs. 10.3,
        10.5
    tension, 4.9.3.2, C–4.9.3.2, Fig. 4.11
    tubular connections, 10.5.2.3, Fig. 10.3
linear reference comparators,
    radiographic testing, 8.17.14,
    C–8.17.14

loading points
    bearing, 7.22.10, C–7.22.10,
        Fig. C–7.8
    deformation, C–4.6.4.2, C–4.6.4.3,
        Figs. C–4.3 to C–4.6
    strengthening and repair, 11.3.5,
        C–11.3.5
Local Brittle Zones (LBZ), C–10.14.4.4
longitudinal ultrasonic testing
    discontinuity size evaluation, 8.29.1
    scanning patterns, 8.30.1
    straight-beam search units, 8.21.6,
        C–8.12.6
    unit calibration, 8.27.1
longitudinal welds
    bend specimen testing, 6.10.3.2,
        C–6.10.3.2
    fillet welds, 4.9.1.2, 4.9.2, C–4.9.1.2,
        Fig. 4.10
low current testing, stud welds, 9.9.6.6
low cycle fatigue, C–14.4.2
Lowest Anticipated Service Temperature
    (LAST), tubular connections,
    10.7.2.2, 10.14.4.4, C–10.7.2.2,
    C–10.14.4.4, Table C–10.5 and
    C–10.6
low-hydrogen electrodes
    ASTM A514 or A517 steels,
        restrictions, 7.3.2.5, C–7.3.2.5
    storage conditions, 7.3.2.1, C–7.3.2.1

M

macroetch test, 6.10.4, Fig. 6.25
    acceptance criteria, 6.10.4.2, 6.23.2.2,
        10.21.1.1, Fig. 6.25
    flare-groove welds, 6.12.4.2, Fig. 6.29
    PJP qualification, 6.12.2.2, Fig. 6.29
    plug and fillet welds, 6.23.2.1, Figs.
        6.25 to 6.26
    T-, Y- or K-connections, 10.21.1, Figs.
        10.21 to 10.22
    welders and welding operators
        qualification, 6.23.2
magnetic particle testing (MT), 8.10
manufacturers
    stud welding base qualification
        requirements, 9.9
    stud welding ID, 9.2.1.1
matched box connections, 10.9.1.1,
    Figs. 10.5 to 10.7
materials
    inspection of, 8.2, C–8.2
    stud welding, 9.5.5.2, 9.9.5.1,
        Tables 5.3 and 6.9
    trimming, 7.14.7
    tubular connections, 10.7, C–10.7,
        Figs. C–10.3 and C–10.4

maximum indication, ultrasonic testing, 8.25.6.3
maximum weld size
   cyclically loaded tubular structures, 10.2.3.4
   fillet welds, 4.4.2.9, Fig. 4.7
measurement units, standard units, 1.2, C–1.2
mechanical testing (MT), 6.10.3
   procedures and technique, 8.14.4
   RT as substitution for, 6.17.1.1
metal deck, stud welding, 9.9.6.3
microscopic intrusions, Fig. C–11.1
mill-induced discontinuities, 7.14.2, 7.14.5, C–7.14.2, Table 7.4
minimum holding time, 7.8.1, Table 7.2
minimum size, fillet weld, 4.4.2.8, 7.13, 9.5.5.6, C–7.13, Tables 7.7 and 9.2
moisture
   stud welding, 9.4.4
   temperature-moisture content charts, Annex D, Figs. D.1 and D.2
multiple electrodes, SAW, 5.5.5, C–Table 5.1

N

nomograph, dB accuracy, 8.28.2.5
nondestructive testing (NDT)
   acceptance criteria, 8.11, C–8.11
   contractors' responsibilities, 8.6.4, 8.6.5
   Engineer's responsibilities, C–1.5.1(2)
   extent of, 8.15, C–8.15
   full testing, 8.15.1
   general requirements, 8.32
   methods, 6.10.2
   partial testing, 8.15.2
   personnel qualification, 8.14.6, 8.35.2, C–8.14.6
   procedures, 8.14, 8.35.1, C–8.14
   qualifications, 8.35
   quality control, 11.6.2
   spot testing, 8.15.3
   tubular connections, 10.25, C–10.25
non-steel backing, 7.9.3
nontubular connections. *See also* tubular connections
   code provisions, C–1.5.1(6)
   cross-sectional variations, 7.21.4.1
   cyclically loaded structures, 4.13 to 4.19, 7.9.1.4, 7.22.6.3, C–4.13, C–7.9.1.4, Table 4.5
   qualification requirements, 6.21, 6.22
   statically loaded nontubular structures, 4.5 to 4.12, 7.22.6.2, C–7.22.6.2
   weld tabs, 7.30.2

Z loss dimension, 4.4.2.6, 4.4.3.3, Table 4.2
non-waveform-controlled power sources, WPSs qualification, 6.9.1
notch toughness
   base metals, tubular connections, 10.7.2, C–10.7.2, Table C–10.1 to C–10.4
   contract plans, 4.3.2, C–1.5.1(5)
   design requirements, welded connections, C–4.3.2
   limitations, 7.14.8.4
   ultrasonic Class X testing, 10.26.1.2, Figs. 10.9 and 10.25
notifications, to inspector, 8.1.7, C–8.1.7

O

OEM (Original Equipment Manufacturer) applications, C–1.5.1(9)
one-sided groove welds, 4.18.1
overhead position, plug welds, 7.24.1.3
overlap
   film, radiographic testing, 8.17.8.2
   minimum, fillet welded joints, 4.9.1.1, C–4.9.1.1, Fig. 4.9, Fig. C–4–7
   repair, 7.25.1.1

P

parallel electrodes, SAW, 5.5.5
Partial Joint Penetration (PJP) groove welds
   allowable stress ranges, 4.16.2, C–4.16.2, C–11.4.2
   assembly, 7.21.2, C–7.21.2
   drawing, 4.3.5.1
   effective areas, 4.4.2.7, 10.5.3.1, Figs. 4.2 to 4.6, 5.2, Table 10.5
   intermittent welds, 4.8.3
   minimum size, 4.4.1.3, 5.4.2.3, Table 5.5
   prequalified WPSs, 5.4.2, 10.6, 10.7, 10.9, Fig.10.6, Figs. 5.1 and 5.2
   qualification requirements, 6.12, 6.22.1, C–6.12, Fig 5.2
   stresses, 10.5.4, Table 10.2
   T-, Y- or K-connections, 10.15, Figs. 10.6 and 10.16, Tables 10.10 and 10.11
   tubular connections, 10.9, 10.15, 10.19
   WPS qualification methods, 6.12.1, 6.12.2, 6.12.3, 10.13, C–10.13, Fig. 5.2, Fig. 6.29, Tables 6.3, 6.5 and 6.6, Tables 10.10, C–6.12.1, C–10.9

peening, 7.26, 11.4.1, C–7.26, C–11.4.1, Fig. C–11.7
penetrant testing (PT), 8.10, 8.14.5
performance qualification
   essential variables, 6.20, C–6.20, Fig. C–6.1, Table 6.12
   general requirements, 6.16, C–6.16
   preparation of forms for, 6.19
   production welding, 10.16, C–10.16, Tables 10.13, C–10.12
   tack welders, 6.2.3.2, 6.16.1.2, 6.16.2.2, 10.16.2
   tubular structures, welding personnel qualifications for, 10.11
   welders, 6.2.2, 6.2.3.1, 6.16, 6.18, 6.19, 10.16.1, C–6.2.2, C–6.16, Fig. 10.23, Fig. C–6.1, Tables 6.10, 6.11 to 10.13
   welding operators, 6.2.2, 6.2.3.1, 6.16, 6.18, 6.19, 10.16.1, C–6.2.2, C–6.16, Fig. 10.23, Fig. C–6.1, Tables 6.10, 6.11 to 10.13
   weld types for, 6.18, 10.17
personnel requirements. *See also* welding personnel
   certification, nondestructive testing, 8.14.6.2
   information access for, 8.15.4
   inspections, 8.1.4.1, 8.35.2, C–8.1.4.1, C–8.1.4.3
   nondestructive testing, 8.14.6, 8.35.2
   phased array ultrasonic testing, Annex H4
   ultrasonic testing, 10.29.2
   UT operators, Annex O
Phased array ultrasonic testing (PAUT), Annex H, C–Annex H
   essential variables, C–8.19.2, Table H.1
plate length, radiographic testing, 8.17.8.1
plug welds
   effective area measurements, 4.4.5
   fabrication, 7.24.1
   joint configuration, 4.10
   macroetch tests, 6.23.2.1, Figs. 6.25 to 6.26
   minimum spacing, 4.10.1
   prequalified WPSs, 5.4.4, C–5.4.4
   qualification requirements, 6.22.3, Fig. 6.26, Fig. C–6.1, Table 6.10
porosity
   excessive, 7.25.1.3
   piping, C–8.19.3
positions of welds
   performance qualifications, 6.3.4, 6.4, 10.11.1, C–6.3.4, C–10.11.1, Fig. C–6.1, Figs. 6.1 to 6.4, Figs. 10.11 to 10.13, Table 6.1, Table 6.10, Table 10.8
   plug welds, 7.24.1.1, 7.24.1.2, 7.24.1.3

production positions, 10.12, 10.16, C–10.16, Tables 10.8 and 10.12, C–10.12
postweld heat treatment (PWHT)
  alternative, 7.8.2
  prequalified WPSs, 5.9, C–5.9
  steels not recommended for, 7.8.3
power
  non-waveform-controlled sources, 6.9.1
  total instantaneous power, 6.8.5.3
practice, retesting after, 6.25.1.2
preheat temperature requirements
  alternative methods, Annex B, C–Annex B
  ESW/ESG processes, 7.4.5
  fabrication, 7.6
  minimum preheat, Annex B6.2.4, Table B.2
  prequalified WPSs, 5.7, C–5.7
  qualification, 6.8.4, Table 5.8
  stud welding, 9.5.5.7
pre-production testing, 9.7.1
Prequalification of WPSs. *See also* Qualification
  base metals, 5.3, Tables 5.3
  butt joints, 10.10.1, C–10.10.1
  CJP groove welds, 5.4.1, 10.10, C–5.4, C–10.10, Figs. 5.1 and 5.5
  contents, Annex K
  essential variables for, 5.2.1, C–Table 5.2
  filler metal and shielding gas, 5.6, Table 5.4 Tables 5.6 and 5.7
  fillet weld requirements, 5.4.3, 10.8, Figs. 5.3, 10.3, 10.5, Table 7.7
  general WPS requirements, 5.2, 5.8, C–5.8, C–Table 5.2, Fig. 5.6, Table 5.8, Tables 5.1 and 5.2
  inspection of, 8.3.1
  limitation of variables, 5.8.2, Table 5.1
  postweld heat treatment, 5.9, C–5.9
  preheat and interpass temperature requirements, 5.7, C–5.7, Table 5.8
  sample forms, Annex J
  scope, 5.1, C–5.1, C–Table 5.1
  stud welds, 9.6.1, Table 5.3
  welding processes, 5.5
  weld joints, 5.4, Figs. 5.1 to 5.5, Table 5.5
prequalified joints
  detail dimensions, C–4.3.5.4
  drawing, 4.3.5.4
  plug and slot welds, 4.10.3
procedure qualification record (PWR), WPS preparation, 6.7, 6.8.2, Tables 6.6, 6.7
production control, stud welding, 9.5.5.1, 9.7, C–9.7
profile accuracy, 7.14.8.2

Q

QC1 requirements, exemption from, 8.14.6.3
Qualification
  aging of test specimens, 6.3.2
  assistant inspector, 8.1.4.5
  CJP groove welds, 6.21, Fig. C–6.1, Table 6.10
  common requirements, WPS and Welding Personnel Performance, 6.3
  CVN toughness testing, 6.2.1.3, 6.26, 6.27, C–6.26, C–6.27
  equipment, 8.23, 8.28, C–8.23
  essential variables, 6.8, 6.20, C–6.8, C–6.20, Fig. C–6.1, Table 6.12, Tables 6.7, 6.8
  existing non-waveform or waveform WPSs, 6.9
  FCAW-S testing, 6.28
  fillet weld specifications, 6.13, 6.22.2, C–6.13, Fig. 6.22, Table 6.4
  forms preparation, 6.19
  general requirements, 6.2
  groove weld specifications, 6.11
  inspection of, 8.4, C–8.4
  inspection personnel, 8.1.4, C–8.1.4.3
  lapse of, retesting after, 6.25.1.3
  manufacturers' stud welding base qualification requirements, 9.9
  nondestructive testing, 8.35
  nondestructive testing personnel, 8.14.6, C–8.14.6
  performance qualification, welding personnel, 6.2.2, 6.16, 6.18 to 6.19, C–6.2.2, C–6.16
  PJP groove weld specifications, 6.12, C–6.12
  plug weld specifications, 6.14
  previous performance qualification, 6.2.2.1, 6.3.1, 6.15.1.3, C–6.2.2.1, C–6.15.1.3
  production welding positions, 6.4, 6.9
  reporting, 6.29
  required tests, 6.17
  responsibility, 6.2.1.1, 6.2.2.2, C–6.2.1.1
  retesting, 6.25
  scope, 6.1
  slot weld specifications, 6.14
  stud application requirements, 9.6, 9.9.7.2, C–9.6, Fig. 9.5
  tack welders, 6.17.2, 6.24, Fig. 6.23, Fig 6.27
  test categories, 6.5, C–6.5
  testing methods and acceptance criteria, 6.10, 6.23, 6.24, 10.13, C–10.13, Table C–10.9
  ultrasonic testing requirements, 8.20
  welders and welding operators, 6.17.1, 6.18, 6.23, Fig. 6.24, Fig. 10.23, Figs. 6.8 to 6.10, Fig. 6.14, 6.16 to 6.17, 6.19 to 6.22, Table 6.11
  welding processes requiring, 6.15, C–6.15
  weld types, 6.6, 6.18
  WPS preparations, 6.2.1, 6.7, 6.10, 10.13, C–10.13, Figs. 6.5 to 6.10, 6.14, 10.14 and 10.15, Tables 10.9 to 10.11, C–10.9
quality control tests, 9.3.4, 11.6, C–11.6
  film quality, 8.17.10, 8.17.11.1, C–8.17
quenched and tempered steels
  heat input, 7.7, C–7.7
  plug and slot welds, 4.10.4

R

radiation imaging systems, 8.33
radiographic film, 8.17.4, C–8.17.4
Radiographic testing (RT)
  acceptance criteria, 6.10.2.2, 6.23.3.2, 8.12, C–8.12
  artifacts, 8.17.10.1
  density limitations, 8.17.11
  groove welds in butt joints, 8.16, C–8.16
  procedures and technique, 6.23.3.1, 8.14.2, 8.17, C–8.17, Figs. 8.4 to 8.9
  repairs, 7.25.1.1
  sources, C–8.17.6
  substitution for guided bend tests, 6.17.1.1
  tubular structures, 10.27, 10.28, C–10.27, Figs. 8.4 and 8.5, Tables 10.16 and 10.17
  variations in, 8.16.2, C–8.16.2
  welders and welding operators qualification, 6.23.3, 10.21.2
  WPS qualification, 6.10.2.1
range. *See* distance calibration
records
  of inspection, maintenance of, 8.5.5
  inspection of, 8.5, C–8.5
  performance test results, 6.3.3
  phased array ultrasonic testing, Annex H13
  radiation imaging systems, 8.35.4
  radiographic film/image archive, 8.18.4
  radiographic film/image retention, 8.18.3, C–8.18.3

reduced-section tension test, 6.10.3.4, 6.10.3.5, Fig. 6.10
redundant-nonredundant members, C–14.4.4
reentrant corners, 7.15, C–7.15, Fig. C–7.1
reference standards
  alternative ultrasonic testing, Annex O
  CVN toughness testing, 6.26.2, 6.26.3, C–6.26, C–6.26.3
  documents, Annex S
  inspection, 8.22, C–8.22
  measurement units, 1.2
  ultrasonic testing calibration, 8.27, Annex G
  WPSs qualification, 6.2.1.2
  zero reference level, 8.24.5.2
reflector size, ultrasonic testing, 8.25.5.1, 10.26.1.1, C–8.25.5.1, Fig. 10.24
reinforcement, 7.25.1.1, 8.17.3, C–8.17.3
reject control calibration, 8.24.1
rejected discontinuities, 8.25.8, 8.25.9, Annex O12, C–8.25.8, Tables 8.2 and 8.3
repairs, 7.25, C–7.25
  base metal, 11.2, C–11.2
  contractor options, 7.25.1, C–7.25.1
  design for, 11.3, C–11.3
  existing structures, 11.1 to 11.6, C–11.1 to C–11.6
  fillet welded studs, 9.5.5.9
  removal area, 9.7.5
  stud welds, 9.7.3, 9.7.5
  ultrasonic testing and, 8.25.10
  weld discontinuities, 7.4.6, 7.14.5.2, C–7.14.5.2, Table 7.4
replacement, for strengthening and repair, 11.3.4
reporting requirements
  completed reports, 8.26.3
  content, 8.26.1
  CVN toughness testing, 6.29
  inspections, 8.26
  phased array ultrasonic testing, Annex H13
  preparation and disposition of reports, 8.26
  prior inspection reports, 8.26.2
  radiographic testing, 8.18.2
  retest reports, 8.25.11
  specification writing guide, Annex N
  ultrasonic testing, 10.29.8, Annex O13
resolution requirements, ultrasonic testing, 8.22.3, 8.27.1.3, 8.27.2.5, C–8.25.12, Fig. 8.14
restoration, for strengthening and repair, 11.3.4
restraint, minimization, 7.20.5, C–Annex B
retesting, 6.10.5, 6.25
  CVN toughness testing, 6.27.8
  qualification expiration, 8.4.3, C–8.4.3
  reports, 8.25.11
  stud welds, 9.9.8
  tack welders, 6.25.2, C–8.4.2
  welder and welding operators, 6.25.1, C–8.4.2
rivets, 4.7.7, C–4.7.7
rolled sections, weld access holes, 7.16.1.1
rolling directions, qualification testing, C–6.5
root openings
  correction, 7.21.4.2, C–7.21.4.2
  mechanical testing, 6.10.3.1, Fig. 6.8, 6.9
  prequalified WPSs, 5.4.1.7, 5.4.2.6
root pass, 5.5.5, Table C–5.1
roughness requirements, 7.14.8.3, C–7.14.8.3
rust, 7.14.3, C–7.14.3

S

Safety and health, Structural Welding Code - Steel, safety precautions, 1.3
safety requirements, radiographic testing (RT), 8.17.2
scale, 7.14.3, C–7.14.3
scanning patterns, 8.25.6.1, 8.30, 10.29.5, Figs. 8.15, 10.30, Table 8.7
  phased array ultrasonic testing, Annex H
  ultrasonic testing, Annex O
seal welded steel backing, C–8.25.12, Fig. C–8.5 and C–8.6
sensitivity
  base metal straight-beam testing, 8.24.4.1, C–8.24.4.1
  calibration, 8.27.2.4, 10.29.3.2
sequencing of welds, 7.20.2, 11.5.6, C–7.20.2, C–11.5.6
shear wave mode, ultrasonic testing, 8.27.2
shelf bars
  base metal, 7.2.2.2
  profiles, 7.23.4
Shielded Metal Arc Welding (SMAW)
  CJP groove joints, 10.10.2.1
  electrodes, 7.3.2, C–7.3.2
  filler metals, 5.6, Table 5.4
  joint preparation, GMAW/FCAW, 5.4.1.5, 5.4.2.4
  prequalified WPSs, 5.5.1, C–5.5.1
  qualification, 6.8, C–6.8, Table 6.5, Table 6.7

temperature-moisture content charts, Annex D, Figs. D.1 and D.2
shielding gas
  composition requirements, Annex M, Table M.5
  electrodes, 5.6.3, 7.3.1.3, C–7.3.1.3
  prequalified WPSs, 5.6.3, C–5.6, Table 5.7
shop drawing requirements, 4.3.5
shop splices, 7.19.1.1
shrinkage, control, 7.20, C–7.20
side bends, mechanical testing, 6.10.3.1, Fig. 6.8, 6.9, Figs. 6.11 to 6.13
skewed T-joints. *See also* T-joints
  drawing, 4.3.5.2
  fillet welds, effective throats, Annex A
  measurements, 4.4.3, Fig. 5.4
  prequalified WPSs, 5.4.3.2, 5.4.3.3, 5.4.3.4, Fig. 5.4, Table 7.7
slag inclusion, 7.25.1.3
slot welds
  effective area measurements, 4.4.5
  fabrication, 7.24.2
  fillet welds as, 4.9.4
  joint configuration, 4.10
  minimum spacing, 4.10.2
  prequalified WPSs, 5.4.4
  qualification requirements, 6.22.3, Fig. 6.26, Fig. C–6.1, Table 6.10
source-to-subject distance, radiographic testing, 8.17.5.2, 8.17.5.3, C–8.17.5.2
spacers, 7.2.2.3
spacing requirements, stud welding, 9.4.5
specific welding requirements, contract plans, 4.3.3
specimen preparation
  manufacturers' stud welding base qualification requirements, 9.9.5
  stud weld testing, 9.6.4
splices
  built-up members, 7.19.2
  cyclically loaded, 7.19.1.2
  shop locations, 7.19.1.1
  subassembly, 7.19.1
  thick filler plates, 4.11.2, Fig. 4.15
  thin filler plates, 4.11.1, Fig. 4.14
statically loaded structures
  code provisions, nontubular structures, C–1.5.1(6)
  girder web flatness, Annex E, C–7.22.6.1, Fig. C–7.5
  nontubular structures
    backing, 7.9.1.5
    built-up members, 4.12
    discontinuities, 8.12.1, C–8.12.1, Fig. 8.1, Figs. C–8.7 and C–8.8
    filler plates, 4.11

fillet weld joint configuration, 4.9, C–4.9
general requirements, 4.5
groove weld joint configuration, 4.8, C–4.8
joint configuration, 4.7
plug and slot weld, joint configuration, 4.10
stresses, 4.6
thickness transition, 4.7.5, 4.8.1, C–4.8.1, Fig. 4.8
ultrasonic testing, 8.13.1, C–8.13.1, Table 8.2
web flatness, 7.22.6.2, C–7.22.6.2
tubular structures, 10.2.1, C–10.2.1
steels. *See also* quenched and tempered steels
ASTM A514 or A517 steels, low-hydrogen electrodes, 7.3.2.5, C–7.3.2.5
backing, 4.17.2.1, 8.25.12, C–8.25.12, Table 4.5
base metal qualification, 6.8.3, C–6.8, Table 5.3, Tables 6.8 and 6.9
postweld heat treatment, exclusion of, 7.8.3
zone classification, Annex B5.1, Fig. B1
stiffeners
girder web flatness, Annexes E and F, C–7.22.6.1, F, Fig. C–7.5
tolerance on, 7.22.11
Stiffeners, transverse welds, 4.9.3.4, C–4.9.3.4
storage, consumables, 7.3.1.4
straight-beam search units
base metal testing, 8.24.4
discontinuity size evaluation, 8.29.1
ultrasonic testing, 8.21.6, C–8.12.6
straightness
beams and girders, 7.22.2, C–7.22.2
columns and trusses, 7.22.1
strengthening procedures, existing structures, 11.1 to 11.6, C–11.1 to C–11.6
stresses. *See also* allowable stresses
analysis, for strengthening and repair, 11.3.2
axial stress and bending, 4.15.2
calculated stresses, C–4.6.1, C–4.6.2
calculation, 4.15, C–4.15
cyclically loaded nontubular structures, 4.15, 4.16
cyclically loaded tubular structures, 10.2.3.1, 10.2.3.2, 10.2.3.3, C–10.2.3.2, C–10.2.3.3, Figs. 10.1, C–10.1, C–10.3, Tables 10.1 and 10.3
eccentricity, 4.6.2, C–4.6.2
heat treatment relief, 7.8, C–7.8 Table 7.2
PJP and fillet welds, 10.5.4, Table 10.2
statically loaded nontubular structures, 4.6
statically loaded tubular structures, 10.2.2, C–10.2.2, Table 10.2
stress range threshold, 4.14.1, Table 4.5
allowable ranges, 4.16.2, C–4.16.2, C–11.5.2
cyclically loaded tubular connections, 10.2.3.1
Structural Welding Code - Steel
approval, 1.6
contractor's responsibilities, 1.5.2
engineer's responsibilities, 1.5.1
limitations, 1.4
mandatory and nonmandatory provisions, 1.7, C–1.7
safety precautions, 1.3
scope, 1.1, C–1.1
standard units of measurement, 1.2
terms and definitions, 3, C–3
welding symbols in, 1.8
Structural Welding Committee, inquiry preparation guidelines, Annex T
stud welding
acceptance criteria, 9.4.7, 9.8.3, C.9.4.7
allowable defects, Fig. C–9.1
application qualification requirements, 9.6, C–9.6
base metal thickness, 9.2.7
bases, 9.2.4
corrective action, 9.8.5
dimension and tolerances, 9.2.1, Fig. 9.1
fabrication, 9.4, 9.8, C–9.4
fillet welds, 4.4.2.8, 7.13, 9.5.5.6, C–7.13, Tables 7.7 and 9.2
finishes, 9.2.5
flash requirement, 9.7.1.3
general requirements, 9.2, C–9.2
manufacturers' stud base qualification requirements, 9.9
materials, 9.2.6
mechanical requirements, 9.3, C–9.3, Table 9.1
production control, 9.5.5.1, 9.7, C–9.7
repairs, 9.5.5.9
scope, 9.1, C–9.1
technique, 9.5, C–9.5
verification inspection, 9.8, C–9.8
visual inspection, 9.8.1
subassembly splices, 7.19.1
Submerged Arc Welding (SAW)
cooling rates, fillet welds, Annex B.3.3, B6.1.3, Figs. B.2, B.3
electrodes, 7.3.3
filler metals, 5.6, Table 5.4
fluxes, 7.3.3
multiple electrodes, 5.5.5, C–Table 5.1
parallel electrodes, 5.5.5
preheat and interpass temperatures, 5.7.3, Table 5.8
prequalified WPSs, 5.5.1, C–5.5.1
qualification, 6.8, C–6.8, Table 6.5, Table 6.7
tack weld incorporation, 7.17.5
surface contouring
backing, 4.17.3
fillet welds, 4.7.5, C–4.7.5
stud welding, 9.5.5.3
surface defects, mill-induced, 7.14.2, C–7.14.2
susceptibility index, hydrogen control, Annex B6.2.3, Table B.1
sweep
base metal straight-beam testing, 8.24.4.1, C–8.24.4.1
beams and girders, 7.22.5
horizontal, 8.24.5.1, Annex H, C–8.24.5.1, C–Annex H
symmetrical cross sections, cyclically loaded structures, 4.15.3

T

T-, Y- or K-connections
backing or backgouging, 10.14.3, Fig.10.18
CJP groove welds, 10.14.4.2, 10.14.4.3, 10.18, C–10.18, Figs. 10.7 to 10.11, 10.17, 10.19 to 10.22
fatigue improvement, 10.2.3.6, C–10.2.3.6, Fig. C–10.2
fillet welds, 10.15, Fig. 10.16, Tables 10.10 and 10.11
macrotech test, 10.21.1, Figs. 10.21 to 10.22
material limitations, 10.7.1, C–10.7.1
PJP welds, 10.15, Fig. 10.16, Tables 10.10 and 10.11
prequalified details, 10.5.2.2, 10.5.3.1, 10.5.3.2, Table 10.5
requirements, 10.10.2, Figs. 10.7 to 10.11, Tables 10.6 and 10.7
tolerance of joint dimensions, 10.23.2.1, Table 10.14
ultrasonic Class X testing, 10.26.1.2, Figs. 10.9 and 10.25
ultrasonic testing procedure, 10.29.1
weld effective lengths, 10.5.6
tack welders and welding
fabrication processes, 7.17
inspection of qualification, 8.4, C–8.4

performance qualification, 6.2.3.2, 6.16.1.2, 6.16.2.2, 10.16.2
required qualification tests, 6.17.2
retesting requirements, 6.25.2
SAW incorporation of, 7.17.5
stud welds, 9.5.5.5
testing methods and acceptance criteria, 6.24, Fig. 6.23
temperature requirements
  CVN toughness testing, 6.27.5, Table 6.14
  heat repair temperature limitations, 7.25.2, C–7.25.2
  Lowest Anticipated Service Temperature (LAST), 10.7.2.2, 10.14.4.4, C–10.7.2.2, C–10.14.4.4, Table C–10.5 and C–10.6
  minimum ambient temperature, 7.11.2, C–7.11.2, C–11.3
  prequalified WPSs, 5.7, C–5.7
  shrinkage or distortion, 7.20.6
  stud welding, 9.4.4
  temperature-moisture content charts, Annex D, Figs. D.1 and D.2
tensile strength, reduced-section tension specimens, 6.10.3.4, Fig. 6.10
tension testing
  AWS A5.36/A5.36M requirements, Annex M, Table M.2
  loaded plate elements, allowable stress ranges, 4.16.2, C–4.16.2, C–11.5.2
  stud welding, 9.3.2, 9.6.7.3, 9.9.7.1, Fig. 9.2
term of effectiveness, inspectors, 8.1.4.7
testing methods. See also specific testing methods
  atmospheric exposure time periods, 7.3.2.3, Table 7.1
  calibration for, 8.24, C–8.24
  consumables verification test, 6.13.3, 6.18
  CVN toughness testing standards, 6.26.2, 6.26.3, C–6.26.3
  electrode atmospheric time periods, 7.3.2.3, Table 7.1
  fillet welds, 6.13.2, Fig. 6.15, Fig. 10.16, Table 6.4
  Flux Cored Arc Welding-S (FCAW-S), 6.28
  macroetch test, 6.10.4, Fig. 6.25
  magnetic particle testing, 8.10
  manufacturers' stud welding base qualification requirements, 9.9
  mechanical testing, 6.10.3
  nondestructive testing, 6.10.2
  penetrant testing, 8.10
  pre-production testing, 9.7.1
  procedures, 8.25, C–8.25

qualification tests, 6.17, 10.13, C–10.13, Table C–10.9
quality control tests, 9.3.4
retesting, 6.10.5
specimen preparation, 9.6.4
stud welding, 9.3.2, 9.6.2, 9.6.3, 9.8.2, C–9.6.2, Fig. 9.2
tack welders, 6.24, Fig. 6.23
torque test, 9.6.6.2, 9.7.1.3, Fig. 9.3
welders and welding operator qualification, 6.23, 10.21
welds, 8.25.6, C–8.25.6, Tables 8.2, 8.3 and 8.7
weld-through-deck stud specimens, 9.9.7.3
WPS qualification, 6.2.1, 6.5, 6.7, 6.10, 8.3.2, 10.13, 10.14, 10.15, C–6.5, C–6.10, C–10.13, C–10.14, Figs. 6.5 to 6.10, 6.14, Table C–10.9, Tables 6.2 to 6.4, Tables 10.9 to 10.11
test plate weld
  consumables verification test, 6.13.3.2, 6.13.3.3, Figs. 6.16 to 6.18
  FCAW-S testing, 6.28.2, 6.28.3, 6.28.4, Fig. 6.30
  fillet welds, 6.3.4, C–6.3.4, Fig. 6.4
  groove welds, 6.3.4, C–6–3.4, Fig. 6.3
  qualification testing, 6.10, 6.21.1, Figs. 6.6, 6.7, 6.16, 6.19, 6.20
thermal cutting process, 7.14.8
through-thickness loading, base metal, 4.7.3, C–4.7.3
TIG dressing, 11.5.1, C–11.5.1, Fig. C–11.3 to Fig. C–11.6
tilt, flange, 7.22.8, C–7.22.8, Fig. C–7.7
time corrected gain (TCG), phased array ultrasonic testing, Annex H3.24
T-joints. See also skewed T-joints
  backing, 4.17.2.2, 4.17.3
  prequalified WPSs, Fig. 5.5
  prohibited joints and welds, 4.18.4
  qualification requirements, 6.11.1.1
  steel backing, C–8.2, C–8.25.12
toe grinding, 11.5.1, C–11.5.1, Figs. C–11.3 to C–11.6
tolerance
  dimensional tolerance, welded structures, 7.22, C–7.22
  fit-up tolerances, 5.4.1.8, Fig. 5.1
  groove welds, 7.21.4.1, Fig. 7.3
  joint dimensions, 7.21, 10.23, C–7.21, C–10.23
  on stiffeners, 7.22.11
  stud welding, 9.2.1, Fig. 9.1
tools. See equipment
torque test, stud welding, 9.6.6.2, 9.7.1.3, Fig. 9.3

total instantaneous energy or power, 6.8.5.3
training, retesting after, 6.25.1.2
transducers
  index point, 8.27.2.1
  qualification and calibration, Annex G
  ultrasonic testing, 8.21.7.1, 8.21.7.2, 8.22, 8.27, 8.28, C–8.21.7.2, C–8.22, Figs. 8.11, 8.16
transitions, radiographic testing, 8.17.11.1.2, C–8.17.11.1.2
transitions of thickness or widths
  backing, 7.9.1.3, C–7.9.1.2
  butt joints, 4.17.1.1, Fig. 4.17
  statically loaded nontubular structures, 4.7.5, 4.8.1, C–4.8.1, Fig. 4.8
  tubular connections, 10.6, Fig. 10.4
transverse welds
  butt joints, 4.17.5
  discontinuities, 8.30.2
  fillet welds, 4.9.1.1, C–4.9.1.1, Fig. 4.9, Fig. C–4.7
  stiffeners, 4.9.3.4, C–4.9.3.4
trusses, straightness, 7.22.1
tubular connections. See also nontubular connections
  acceptance criteria, 10.21
  backing, 10.22, C–10.22
  CJP groove weld requirements, 10.10, 10.14, 10.18, C–10.10, C–10.14, C–10.18, Figs. 10.17, 10.19 to 10.22, Figs. 10.19 and 10.20, Tables 10.7 and 10.13
  cross-sectional variations, 10.23.2.1, Table 10.14
  cyclically loaded structures, 10.2.3, C–10.2.3
  design criteria, 10.2
  fatigue limitations, 10.2.3.7, C–10.2.3.7, Table 10.4
  fillet welds, 10.8, 10.15, 10.20, Fig.10.5, Table 10.13
  identification and parts, Fig.10.2
  joint configuration, 10.18.1, Figs. 10.21 and 10.22
  materials, 10.7, C–10.7, Figs. C–10.3 and C–10.4
  nondestructive testing, 10.25, C–10.25
  parts, 10.3, C–10.3, Fig. 10.2
  PJP groove welds, 10.9, 10.15, 10.19
  production welding positions, 10.12, 10.16, C–10.16, Table 10.8, C–10.12
  radiographic testing, 10.27, 10.28, C–10.27, Figs. 8.4 and 8.5, Tables 10.16 and 10.17
  scope, 10.1, C–10.1
  stresses, 10.2.2, C–10.2.2, Table 10.2
  symbols, Annex I, Fig. 10.4

T-, Y-, and K-connections, 10.29, C–10.29
testing methods, 10.13, 10.21, C–10.13, Table C–10.9
thickness transition, 10.6, Fig. 10.4
tolerance of joint dimensions, 10.23, C–10.23
ultrasonic testing, 10.26, 10.29, C–10.26, C–10.29
visual inspection, 10.24, Table 10.15
weld design, 10.5
welding personnel performance qualification, 10.11, 10.17
WPS requirements, 10.11
tubular structures, statically loaded structures, 10.2.1, C–10.2.1

U

U-groove preparation, 5.4.1.4, 5.4.1.9, 5.4.5.8
ultrasonic testing (UT)
    acceptance criteria, 6.10.2.2, 8.13
    advanced systems, 8.34
    alternative techniques, Annex O
    angle criteria, 8.21.7, 8.25.5.2, C–8.25.5.2, Table 8.7, Table C–8.7
    approach distance, search unit, 8.27.2.6
    calibration, 8.24, 8.27, 10.29.3, Annex G, Annex O
    class R criteria, 10.26.1.1, Fig.10.24
    class X criteria, 10.26.1.2, Figs. 10.9 and 10.25
    equipment, 8.21, 8.23, C–8.21, C–8.23, Table 8.8
    equipment qualification and inspection forms, Annex P
    groove welds, 8.19, C–8.19
    indications, 8.13.2.1
    phased array, Annex H
    procedures, 8.14.3, 8.25, C–8.19.1, C–8.25
    qualification requirements, 8.20, 8.23, Annex G, C–8.23
    recalibration, 8.24.3
    report preparation and disposition, 8.26
    resolution, 8.22.3, 8.27.1.3, 8.27.2.5, Fig. 8.14
    retest reports, 8.25.11
    scanning patterns, 8.13.2.2, 8.25.6.1, 8.30, Fig. 8.15, Table 8.2 and 8.3, Table 8.7
    standards, C–8.19.1
    steel backing, 8.25.12, C–8.25.12
    T-, Y- or K-connections, 10.29, C–10.29
    transducer positions, 8.21.7.1, 8.21.7.2, 8.22, 8.27, 8.28, C–8.21.7.2, C–8.22, Figs. 8.11, 8.16
    tubular connections, 10.26, 10.29, C–10.26, C–10.29
    WPS qualification, 6.10.2.1
undercutting, 7.25.1.2
undersize welds, 7.25.1.2
unground welds, discontinuities, 8.30.2.2
Upper Bound Theorem, 10.7, C–10.7, Fig. C–10.3

V

verification inspection
    Engineer's responsibilities, C–1.5.1(3)
    stud welding, 9.8, C–9.8
vertical position, plug welds, 7.24.1.2
vertical-up welding, prequalified WPSs, 5.8.1.1
Vickers hardness number, preheat and interpass temperatures, 5.7.3, Table 5.8
visual acceptance criteria, tack welders, 6.24.1
visual inspection
    acceptance criteria, 8.9, C–8.9, Table 8.1, Table 10.15
    fillet welded studs, 9.5.5.10
    procedures, 6.10.1, 6.23.1
    quality control, 11.7.1
    stud welds, 9.8.1
    tubular connections, 10.24, Table 10.15

W

warpage, flange, 7.22.8, C–7.22.8, Fig. C–7.7
waveform equipment, production welds with non-waveform using, 6.9.2
weathering steel
    ESW/ESG processes, 7.4.7, Table 5.6
    prequalified WPSs, 5.6.2, C–5.6.2, Table 5.6
    unpainted, 4.12.2.3, C–4.12.2.3
web flatness, 7.22.6, C–7.22.6.1, Fig. C–7.5
    flange centerlines and, 7.22.7
weld access holes
    built-up sections, 7.16.1.2, Fig. 7.2
    dimension, 7.16.1, C–7.16.1
    heavy shapes, 7.16.3, C–7.16.3
    rolled sections, 7.16.1.1
Weld bead, prequalified WPSs, 5.8.2.1, Fig. 5.6
welders
    CJP groove weld qualification plates, 6.21.1, Figs. 6.16, 6.17 and 6.19 to 6.21
    inspection of qualifications, 8.4, C–8.4
    performance qualification, 6.2.2, 6.2.3.1, 6.16, 6.18, 6.19, 10.16.1, C–6.2.2, C–6.16, Fig. 10.23, C–6.1, Table 6.10 and 6.11, 10.13
    radiographic testing (RT), 6.23.3, 10.21.2
    required qualification tests, 6.17.1, Table 16.11
    retesting requirements, 6.25.1
    testing methods and acceptance criteria, 6.23, 10.21
    weld types for performance qualification, 6.18
    WPS qualification, 6.16.3
welding guns, 9.5.3
welding operators
    CJP qualification test plates, 6.21.1, Fig. 6.16, 6.17, 6.19 to 6.21
    inspection of qualifications, 8.4, C–8.4
    performance qualification, 6.2.2, 6.2.3.1, 6.16, 6.18, 6.19, 10.16.1, C–6.2.2, C–6.16, Fig. 10.23, Fig. C–6.1, Tables 6.10 and 10.13
    radiographic testing (RT), 6.23.3, 10.21.2
    required qualification tests, 6.17.1, Table 16.11
    retesting requirements, 6.25.1
    stud welding qualification, 9.7.4
    testing methods and acceptance criteria, 6.23, 10.21
    weld types for performance qualification, 6.18
    WPS qualification, 6.16.3
welding personnel
    common qualification requirements, 6.3
    performance qualification, 6.2.2, 6.16, 6.18, 6.19, C–6.2.2, C–6.16
    tubular structures, performance qualifications for, 10.11
Welding Procedure Specifications (WPSs)
    acceptance criteria and testing methods, 6.7, 6.10, 10.13, C–10.13, Figs. 6.5 to 6.10, 6.14, 10.14 and 10.15, C–6.1, Tables 10.9 to 10.11, C–6.10, C–10.9
    common qualification requirements, 6.3
    CVN toughness testing, 6.26.1.1
    development of, 5.2, Tables 5.1 and 5.2
    exemption from qualification requirements, 1.1

existing non-waveform or waveform WPSs, 6.9
fillet welds, 6.13, 10.13, C–10.13, Tables 6.4 and 10.11, C–6.13, C–10.9
GMAW-S, 6.15.1.1, Table 6.5
GTAW, 6.15.1.2
inspection of, 8.3, C–8.3
maximum heat input, multi-position WPSs, 6.8.5.2
PJP qualification methods, 6.12.1, 6.12.2, 6.12.3, 10.13, C–6.12.1, C–10.9, C–10.13, Fig. 5.2, Fig. 6.29, Tables 6.3, 6.5 and 6.6, Tables 10.10
preparation, 6.7, Annex L, C–6.7
prequalification (*See* Prequalification of WPSs)
qualification testing, 6.2.1, 6.7, 6.10, 10.13, C–6.5, C–10.9, C–10.13, Figs. 6.5 to 6.10, 6.14, 10.14 and 10.15, C–6.1, Tables 10.9 to 10.11
qualification to other standards, 6.2.1.2
stud welds, 9.6.2, C–9.6.2
tubular structures, 10.11
variables, 7.5, C–7.5
weld types, 6.6
without prequalification, 10.14.4.1, Figs.10.20 to 10.22
welding processes
environment, 7.11, C–7.11
equipment, 7.10

miscellaneous, 7.14.8.1
prequalified WPSs, 5.5, C–5.5
starts and stops, 7.4.4
Welding Suitability Guide, Table C–11.1
welding symbols, 1.8
contract documents, 4.3.5.3
tubular connections, Annex I, Fig. 10.4
welds
backing, 7.9.2
cleaning, 7.29, C–7.29
cracks, 7.25.1.4
inaccessibility, unacceptable welds, 7.25.4
inspection of, 8.5.1
metal, allowable stresses, 4.6.4, Table 4.3
performance qualification, types of, 6.18, 10.17
profiles, 7.23, 11.5.1, C–7.23, C–10.2.3.7, C–11.5.1, Fig. 7.4. 10.2.3.7, Fig. C–11.2, Table 8.1, Table 10.4, 10.15, Table C–11.2, Tables 7.8–7.9
progression, 7.20.4
prohibited joints and welds, 4.18
size and length, 4.3.4, 8.5.1, C–4.3.4
testing, 8.25.6, C–8.25.6, Tables 8.2, 8.3 and 8.7
weld tabs
base metal, 7.2.2.1, Table 5.3, Table 6.9
fabrication, 7.30, C–7.30
radiographic testing (RT), 8.17.3.1, C–8.17.3.1

removal, 4.8.4
weld-through-deck stud testing, 9.9.7.3
width/depth pass limitation, prequalified WPSs, 5.8.2.1, C–5.8.2.1
wind velocity, 7.11.1
wire requirements, IQI requirements, 8.17.1, 10.27.1, Figs. 8.4 to 8.9, Tables 8.5 and 10.17
work inspection, 8.5, C–8.5

X

"X" line testing, 8.25.1

Y

"Y" line testing, 8.25.2

Z

zero reference level, ultrasonic testing, 8.24.5.2
Z-loss allowance
effective throat calculation, 4.4.2.6
prequalified PJP T-, Y-, and K-tubular connections, 10.5.3.1, Table 10.5
skewed T-joints, 4.4.3.3, Table 4.2